INDUSTRIAL ORGANIC CHEMICALS

Starting Materials and Intermediates

VOLUME 3

Weinheim · New York · Chichester · Brisbane · Singapore · Toronto

INDUSTRIAL ORGANIC CHEMICALS

VOLUME 1
Acetaldehyde to **Aniline**

VOLUME 2
Anthracene to **Cellulose Ethers**

VOLUME 3
Chlorinated Hydrocarbons to **Dicarboxylic Acids, Aliphatic**

VOLUME 4
Dimethyl Ether to **Fatty Acids**

VOLUME 5
Fatty Alcohols to **Melamine and Guanamines**

VOLUME 6
Mercaptoacetic Acid and Derivatives to **Phosphorus Compounds, Organic**

VOLUME 7
Phthalic Acid and Derivatives to **Sulfones and Sulfoxides**

VOLUME 8
Sulfonic Acids, Aliphatic to **Xylidines**

Index

INDUSTRIAL ORGANIC CHEMICALS

Starting Materials and Intermediates

VOLUME 3

Chlorinated Hydrocarbons
to **Dicarboxylic Acids, Aliphatic**

Weinheim · New York · Chichester · Brisbane · Singapore · Toronto

> This book was carefully produced. Nevertheless, authors and publisher do not warrant the information contained therein to be free of errors. Readers are advised to keep in mind that statements, data, illustrations, procedural details or other items may inadvertently be inaccurate.

Library of Congress Card No.: Applied for.
British Library Cataloguing-in-Publication Data: A catalogue record for this book is available from the British Library.

Die Deutsche Bibliothek – CIP-Einheitsaufnahme
Industrial organic chemicals : starting materials and intermediates ;
an Ullmann's encyclopedia. – Weinheim ; New York ;
Chichester ; Brisbane ; Singapore ; Toronto : Wiley-VCH
  ISBN 3-527-29645-X
Vol. 3. Chlorinated Hydrocarbons to Dicarboxylic Acids, Aliphatic. – 1. Aufl. – 1999.

© WILEY-VCH Verlag GmbH, D-69469 Weinheim (Federal Republic of Germany), 1999
Printed on acid-free and chlorine-free paper.
All rights reserved (including those of translation in other languages). No part of this book may be reproduced in any form – by photoprinting, microfilm, or any other means – nor transmitted or translated into machine language without written permission from the publishers. Registered names, trademarks, etc. used in this book, even when not specifically marked as such, are not to be considered unprotected by law.

Composition and Printing: Rombach GmbH, Druck- und Verlagshaus, D-79115 Freiburg.
Bookbinding: Wilhelm Osswald & Co., D-67433 Neustadt (Weinstraße)
Cover design: mmad, Michel Meyer, D-69469 Weinheim.
Printed in the Federal Republic of Germany.

Contents

1 Chlorinated Hydrocarbons

1. Chloromethanes 1252
2. Chloroethanes 1286
3. Chloroethylenes 1330
4. Chloropropanes 1373
5. Chlorobutanes 1377
6. Chlorobutenes 1380
7. Chlorinated Paraffins 1396
8. Nucleus-Chlorinated Aromatic Hydrocarbons 1406
9. Side-Chain Chlorinated Aromatic Hydrocarbons 1447
10. Toxicology and Occupational Health 1467
11. References 1486

2 Chloroacetaldehydes

1. Monochloroacetaldehyde 1522
2. Dichloroacetaldehyde 1529
3. Trichloroacetaldehyde 1531
4. Toxicology 1534
5. References 1536

3 Chloroacetic Acids

1. Introduction 1539
2. Chloroacetic Acid 1540
3. Dichloroacetic Acid. 1549
4. Trichloroacetic Acid 1552
5. Environmental Protection 1556
6. Chemical Analysis 1557
7. Containment Materials, Storage, and Transportation 1558
8. Economic Aspects. 1559
9. Toxicology and Occupational Health 1560
10. References. 1561

4 Chloroamines

1. Introduction 1565
2. Chemical Properties 1565
3. Reactions 1566
4. N-Chloroisocyanuric Acids 1568
5. Other Organic N-Chloroamines. . . . 1569
6. Toxicology. 1571
7. References. 1572

5 Chloroformic Esters

1. Introduction 1575
2. Physical Properties 1575
3. Chemical Properties 1577
4. Production 1577
5. Safety. 1578
6. Quality Specifications and Analysis . 1579
7. Storage and Transportation 1579
8. Uses and Economic Aspects 1580
9. Toxicology. 1581
10. References. 1581

6 Chlorohydrins

1. Introduction 1585
2. Physical Properties 1585
3. Chemical Properties 1590
4. Production 1590
5. Environmental Protection 1595
6. Chemical Analysis 1596
7. Storage and Transportation 1596
8. Uses 1596
9. Toxicology and Occupational Health 1597
10. References 1600

7 Chlorophenols

1. Introduction 1605
2. Physical Properties 1606
3. Chemical Properties 1607
4. Production 1608
5. Analysis 1611
6. Quality Specifications 1612
7. Economic Aspects 1612
8. Uses 1613
9. Storage and Transportation 1614
10. Environmental Considerations 1614
11. Toxicology 1616
12. References 1617

8 Chlorophenoxyalkanoic Acids

1. Introduction 1619
2. Production 1620
3. Quality Specifications 1623
4. Uses 1624
5. Toxicology and Occupational Health 1625
6. Biochemical and Environmental Aspects 1628
7. References 1629

9 Choline

1. Properties 1633
2. Production 1634
3. Analysis 1634
4. Salts 1634
5. Uses 1635
6. References 1636

10 Cinnamic Acid

1. Introduction 1639
2. Physical Properties 1640
3. Chemical Properties 1640
4. Occurrence in Nature 1640
5. Production 1641
6. Uses 1642
7. Quality Specifications 1643
8. Economic Aspects 1643
9. Toxicology 1643
10. References 1643

11 Citric Acid

1. Introduction 1645
2. Physical Properties 1647
3. Chemical Properties 1648
4. Production 1650
5. Quality Specifications and Uses. . . . 1653
6. Economic Aspects. 1654
7. References. 1654

12 Cresols and Xylenols

1. Cresols 1656
2. Xylenols 1691
3. Environmental Protection 1699
4. Toxicology and Occupational Health 1700
5. References. 1702

13 Crotonaldehyde and Crotonic Acid

1. Crotonaldehyde 1711
2. Crotonic Acid 1715
3. Toxicology and Occupational Health 1719
4. References. 1720

14 Crown Ethers

1. Classification and Nomenclature . . . 1723
2. Physical Properties 1726
3. Chemical Properties 1726
4. Crown Ether Complexes 1726
5. Production 1728
6. Uses. 1730
7. Toxicology and Occupational Health 1731
8. References. 1732

15 Cyanamides

1. Introduction 1735
2. Calcium Cyanamide 1736
3. Cyanamide 1745
4. Dicyandiamide. 1754
5. Economic Aspects. 1761
6. Toxicology and Occupational Health 1762
7. References. 1763

16 Cyanuric Acid and Cyanuric Chloride

1. Introduction 1767
2. Cyanuric Acid 1768
3. Cyanuric Chloride. 1773
4. References. 1779

17 Cyclododecanol, Cyclodecanone and Laurolactam

1. Cyclododecanol and Cyclododecanone 1783
2. Laurolactam 1786
3. Toxicology 1787
4. References 1787

18 Cyclododecatriene and Cyclooctadiene

1. Introduction 1789
2. Physical Properties 1790
3. Chemical Properties 1790
4. Production 1791
5. Storage and Transportation 1791
6. Uses 1791
7. References 1792

19 Cyclohexane

1. Introduction 1795
2. Physical Properties 1795
3. Chemical Properties 1796
4. Production 1797
5. Quality Specifications and Tests 1802
6. Uses, Economic Aspects 1803
7. Homologues 1803
8. Toxicology and Occupational Health 1804
9. References 1805

20 Cyclohexanol and Cyclohexanone

1. Introduction 1807
2. Physical and Chemical Properties 1808
3. Production 1808
4. Plant Safety 1814
5. Quality Specifications and Chemical Analysis 1815
6. Storage and Transportation 1815
7. Uses and Trade Names 1816
8. Derivatives 1816
9. Economic Aspects 1820
10. Toxicology and Occupational Health 1820
11. References 1821

21 Cyclopentadiene and Cyclopentene

1. Cyclopentadiene and Dicyclopentadiene 1825
2. Cyclopentene 1832
3. Toxicology 1837
4. References 1839

22 Dextran

1. Introduction 1843
2. Structure, Chemical, and Physicochemical Properties 1844
3. Production 1846
4. Quality Specifications 1848
5. Storage 1849
6. Uses 1849
7. Economic Aspects 1850
8. Toxicology 1851
9. References 1851

23 Dialkyl Sulfates and Alkylsulfuric Acids

1. Introduction 1853
2. Physical Properties 1854
3. Chemical Properties 1855
4. Production 1857
5. Quality Specifications and Analysis . 1860
6. Storage, Transportation, and Handling 1862
7. Environmental Protection 1864
8. Uses 1864
9. Economic Aspects. 1870
10. Toxicology and Occupational Health 1870
11. References 1871

24 Diazo Compounds and Diazo Reactions

1. Introduction 1875
2. Aliphatic Diazo Compounds. 1876
3. Aromatic Diazo Compounds 1876
4. Diazo Reactions 1892
5. Environmental Protection 1899
6. Safety 1900
7. Toxicology 1900
8. References 1901

25 Dicarboxylic Acids, Aliphatic

1. Introduction 1903
2. Saturated Dicarboxylic Acids 1904
3. Unsaturated Dicarboxylic Acids ... 1919
4. Quality Specifications and Analysis . 1923
5. Storage, Transportation, and Handling 1924
6. References 1924

Chlorinated Hydrocarbons

MANFRED ROSSBERG, Hoechst Aktiengesellschaft, Frankfurt/Main, Federal Republic of Germany (Chaps. 1.1–1.3)

WILHELM LENDLE, Hoechst Aktiengesellschaft, Frankfurt/Main, Federal Republic of Germany (Chap. 1.4 and 1.5)

GERHARD PFLEIDERER, Hoechst Aktiengesellschaft, Frankfurt/Main, Federal Republic of Germany (Chap. 1.6)

ADOLF TÖGEL, Hoechst Aktiengesellschaft, Frankfurt/Main, Federal Republic of Germany (Chap. 1.7)

EBERHARD-LUDWIG DREHER, Dow Chemical GmbH, Stade, Federal Republic of Germany (Chaps. 2 and 3)

ERNST LANGER, BASF Aktiengesellschaft, Ludwigshafen, Federal Republic of Germany (Chap. 4)

HEINZ RASSAERTS, Chemische Werke Hüls AG, Marl, Federal Republic of Germany (Chaps. 5, 6.6, and 6.7)

PETER KLEINSCHMIDT, Bayer AG, Dormagen, Federal Republic of Germany (Chaps. 6.1–6.5)

HEINZ STRACK, formerly Dynamit Nobel AG (Chap. 7)

RICHARD COOK, ICI Chemicals and Polymers, Runcorn, United Kingdom (Chap. 7)

UWE BECK, Bayer AG, Leverkusen, Federal Republic of Germany (Chap. 8)

KARL-AUGUST LIPPER, Bayer AG, Krefeld, Federal Republic of Germany (Chap. 9)

THEODORE R. TORKELSON, Dow Chemical, Midland, Michigan 48674, United States (Chap. 10.1)

ECKHARD LÖSER, Bayer AG, Wuppertal, Federal Republic of Germany (Chap. 10.2)

KLAUS K. BEUTEL, Dow Chemical Europe, Horgen, Switzerland (Chap. 10.1.5)

1.	Chloromethanes	1252	1.5.	Storage, Transport, and Handling ... 1280
1.1.	Physical Properties	1254	1.6.	Behavior of Chloromethanes in the Environment ... 1282
1.2.	Chemical Properties	1257	1.6.1.	Presence in the Atmosphere ... 1282
1.3.	Production	1260	1.6.2.	Presence in Water Sources ... 1283
1.3.1.	Theoretical Bases	1260	1.7.	Applications of the Chloromethanes and Economic Data ... 1284
1.3.2.	Production of Monochloromethane	1265		
1.3.3.	Production of Dichloromethane and Trichloromethane	1268	2.	Chloroethanes ... 1286
1.3.4.	Production of Tetrachloromethane	1273	2.1.	Monochloroethane ... 1290
			2.1.1.	Physical Properties ... 1290
1.4.	Quality Specifications	1278	2.1.2.	Chemical Properties ... 1291
1.4.1.	Purity of the Commercial Products and Their Stabilization	1278	2.1.3.	Production ... 1291
			2.1.4.	Uses and Economic Aspects ... 1294
1.4.2.	Analysis	1279	2.2.	1,1-Dichloroethane ... 1295

2.2.1.	Physical Properties	1295	3.1.3.1. Vinyl Chloride from Acetylene	1334

- 2.2.1. Physical Properties 1295
- 2.2.2. Chemical Properties 1296
- 2.2.3. Production 1296
- 2.2.4. Uses and Economic Aspects 1297
- **2.3. 1,2-Dichloroethane** 1298
- 2.3.1. Physical Properties 1298
- 2.3.2. Chemical Properties 1299
- 2.3.3. Production 1299
- 2.3.4. Uses and Economic Aspects 1309
- **2.4. 1,1,1-Trichloroethane** 1309
- 2.4.1. Physical Properties 1310
- 2.4.2. Chemical Properties 1310
- 2.4.3. Production 1311
- 2.4.4. Uses and Economic Aspects 1316
- **2.5. 1,1,2-Trichloroethane** 1317
- 2.5.1. Physical Properties 1317
- 2.5.2. Chemical Properties 1317
- 2.5.3. Production 1318
- 2.5.4. Uses and Economic Aspects 1320
- **2.6. 1,1,1,2-Tetrachloroethane** ... 1320
- 2.6.1. Physical Properties 1321
- 2.6.2. Chemical Properties 1321
- 2.6.3. Production 1321
- **2.7. 1,1,2,2-Tetrachloroethane** ... 1321
- 2.7.1. Physical Properties 1322
- 2.7.2. Chemical Properties 1323
- 2.7.3. Production 1323
- 2.7.4. Uses and Economic Aspects 1325
- **2.8. Pentachloroethane** 1325
- 2.8.1. Physical Properties 1326
- 2.8.2. Chemical Properties 1326
- 2.8.3. Production 1327
- 2.8.4. Uses and Economic Aspects 1327
- **2.9. Hexachloroethane** 1328
- 2.9.1. Physical Properties 1328
- 2.9.2. Chemical Properties 1328
- 2.9.3. Production 1329
- 2.9.4. Uses and Economic Aspects 1329
- **3. Chloroethylenes** 1330
- **3.1. Vinyl Chloride (VCM)** 1330
- 3.1.1. Physical Properties 1331
- 3.1.2. Chemical Properties 1332
- 3.1.3. Production 1332
- 3.1.3.1. Vinyl Chloride from Acetylene ... 1334
- 3.1.3.2. Vinyl Chloride from 1,2-Dichloroethane 1336
- 3.1.3.3. Vinyl Chloride from Ethylene by Direct Routes 1343
- 3.1.3.4. Vinyl Chloride from Ethane 1345
- 3.1.3.5. Vinyl Chloride by Other Routes . 1346
- 3.1.4. Uses and Economic Aspects 1347
- **3.2. 1,1-Dichloroethylene (Vinylidene Chloride, VDC)** .. 1348
- 3.2.1. Physical Properties 1349
- 3.2.2. Chemical Properties 1349
- 3.2.3. Production 1350
- 3.2.4. Uses and Economic Aspects 1352
- **3.3. 1,2-Dichloroethylene** 1353
- 3.3.1. Physical Properties 1353
- 3.3.2. Chemical Properties 1354
- 3.3.3. Production 1355
- 3.3.4. Uses and Economic Aspects 1355
- **3.4. Trichloroethylene** 1356
- 3.4.1. Physical Properties 1356
- 3.4.2. Chemical Properties 1357
- 3.4.3. Production 1357
- 3.4.4. Uses and Economic Aspects 1361
- **3.5. Tetrachloroethylene** 1361
- 3.5.1. Physical Properties 1361
- 3.5.2. Chemical Properties 1362
- 3.5.3. Production 1363
- 3.5.4. Uses and Economic Aspects 1368
- **3.6. Analysis and Quality Control of Chloroethanes and Chloroethylenes** 1369
- **3.7. Storage and Transportation of Chloroethanes and Chloroethylenes** 1370
- **3.8. Environmental Aspects in the Production of Chloroethanes and Chloroethylenes** 1371
- **4. Chloropropanes** 1373
- **4.1. 2-Chloropropane** 1373
- **4.2. 1,2-Dichloropropane** 1374
- **4.3. 1,2,3-Trichloropropane** 1376

5.	**Chlorobutanes**	1377		8.1.	**Chlorinated Benzenes**	1407
5.1.	**1-Chlorobutane**	1378		8.1.1.	Physical Properties	1407
5.2.	**tert-Butyl Chloride**	1379		8.1.2.	Chemical Properties	1411
5.3.	**1,4-Dichlorobutane**	1380		8.1.3.	Production	1412
6.	**Chlorobutenes**	1380		8.1.3.1.	Monochlorobenzene	1417
6.1.	**1,4-Dichloro-2-butene**	1381		8.1.3.2.	Dichlorobenzenes	1417
6.2.	**3,4-Dichloro-1-butene**	1382		8.1.3.3.	Trichlorobenzenes	1418
6.3.	**2,3,4-Trichloro-1-butene**	1382		8.1.3.4.	Tetrachlorobenzenes	1420
6.4.	**2-Chloro-1,3-butadiene**	1383		8.1.3.5.	Pentachlorobenzene	1420
6.4.1.	Physical Properties	1383		8.1.3.6.	Hexachlorobenzene	1420
6.4.2.	Chemical Properties	1384		8.1.4.	Quality and Analysis	1421
6.4.3.	Production	1385		8.1.5.	Storage and Transportation	1421
6.4.3.1.	Chloroprene from Butadiene	1385		8.1.6.	Uses	1422
6.4.3.2.	Chloroprene from Acetylene	1387		8.2.	**Chlorinated Toluenes**	1423
6.4.3.3.	Other Processes	1388		8.2.1.	Physical Properties	1423
6.4.4.	Economic Importance	1388		8.2.2.	Chemical Properties	1426
6.5.	**Dichlorobutadiene**	1389		8.2.3.	Production	1427
6.5.1.	2,3-Dichloro-1,3-butadiene	1389		8.2.3.1.	Monochlorotoluenes	1429
6.5.2.	Other Dichlorobutadienes	1389		8.2.3.2.	Dichlorotoluenes	1430
6.6.	**3-Chloro-2-methyl-1-propene**	1390		8.2.3.3.	Trichlorotoluenes	1431
6.6.1.	Physical Properties	1390		8.2.3.4.	Tetrachlorotoluenes	1432
6.6.2.	Chemical Properties	1391		8.2.3.5.	Pentachlorotoluene	1432
6.6.3.	Production	1391		8.2.4.	Quality and Analysis	1433
6.6.4.	Quality Specifications and Chemical Analysis	1394		8.2.5.	Storage and Transportation	1433
6.6.5.	Storage and Shipment	1394		8.2.6.	Uses	1434
6.6.6.	Uses	1394		8.3.	**Chlorinated Biphenyls**	1435
6.7.	**Hexachlorobutadiene**	1395		8.3.1.	Physical and Chemical Properties	1436
7.	**Chlorinated Paraffins**	1396		8.3.2.	Disposal	1436
7.1.	**Physical Properties**	1397		8.3.3.	Analysis	1437
7.2.	**Chemical Properties and Structure**	1398		8.3.4.	Storage and Transportation	1438
7.3.	**Production**	1400		8.3.5.	Uses	1438
7.4.	**Analysis and Quality Control**	1403		8.4.	**Chlorinated Naphthalenes**	1439
7.5.	**Storage and Transportation**	1403		8.4.1.	Physical Properties	1440
7.6.	**Health and Safety Regulations**	1404		8.4.2.	Chemical Properties	1440
7.7.	**Uses**	1405		8.4.3.	Production	1441
8.	**Nucleus-Chlorinated Aromatic Hydrocarbons**	1406		8.4.4.	Quality and Analysis	1443
				8.4.5.	Storage and Transportation	1443
				8.4.6.	Use	1444
				8.5.	**Environmental Protection**	1445
				8.6.	**Economic Facts**	1446
				9.	**Side-Chain Chlorinated Aromatic Hydrocarbons**	1447
				9.1.	**Benzyl Chloride**	1448
				9.1.1.	Physical Properties	1448

9.1.2.	Chemical Properties	1449	9.4.4.	Uses	1465
9.1.3.	Production	1450	9.5.	**Ring-Chlorinated Derivatives**	1465
9.1.4.	Quality Specifications and Analysis	1454	9.6.	**Economic Aspects**	1467
9.1.5.	Storage and Transportation	1455	10.	**Toxicology and Occupational Health**	1467
9.1.6.	Uses	1455			
9.2.	**Benzal Chloride**	1456	10.1.	**Aliphatic Chlorinated Hydrocarbons**	1467
9.2.1.	Physical Properties	1456			
9.2.2.	Chemical Properties	1457	10.1.1.	Chloromethanes	1470
9.2.3.	Production	1457	10.1.2.	Chlorinated C_2 Hydrocarbons	1472
9.2.4.	Quality Specifications and Analysis	1457	10.1.3.	Chloropropanes and Chloropropenes	1477
9.2.5.	Storage and Transportation	1458	10.1.4.	Chlorobutadienes	1478
9.2.6.	Uses	1458	10.1.5.	Ecotoxicology and Environmental Degradation	1478
9.3.	**Benzotrichloride**	1458			
9.3.1.	Physical Properties	1459	10.2.	**Chlorinated Aromatic Hydrocarbons**	1480
9.3.2.	Chemical Properties	1460			
9.3.3.	Production	1460	10.2.1.	Chlorinated Benzenes	1480
9.3.4.	Quality Specifications and Analysis	1461	10.2.2.	Chlorotoluenes	1481
			10.2.3.	Polychlorinated Biphenyls	1481
9.3.5.	Storage and Transportation	1462	10.2.4.	Chlorinated Naphthalenes	1483
9.3.6.	Uses	1462	10.2.5.	Benzyl Chloride	1484
9.4.	**Side-Chain Chlorinated Xylenes**	1462	10.2.6.	Benzoyl Chloride	1485
			10.2.7.	Benzotrichloride	1485
9.4.1.	Physical and Chemical Properties	1463	10.2.8.	Side-Chain Chlorinated Xylenes	1485
9.4.2.	Production	1463			
9.4.3.	Storage and Transportation	1465	11.	**References**	1486

1. Chloromethanes

Among the halogenated hydrocarbons, the chlorine derivatives of methane monochloromethane (methyl chloride) [74-87-3], dichloromethane (methylene chloride) [75-09-2], trichloromethane (chloroform) [67-66-3], and tetrachloromethane (carbon tetrachloride) [56-23-5] play an important role from both industrial and economic standpoints. These products find broad application not only as important chemical intermediates, but also as solvents.

Historical Development. *Monochloromethane* was produced for the first time in 1835 by J. DUMAS and E. PELIGOT by the reaction of sodium chloride with methanol in the presence of sulfuric acid. M. BERTHELOT isolated it in 1858 from the chlorination of marsh gas (methane), as did C. GROVES in 1874 from the reaction of hydrogen chloride with methanol in the presence of zinc chloride. For a time, monochloromethane was produced commercially from betaine hydrochloride obtained in the course of beet sugar manufacture. The earliest attempts to produce methyl chloride by the chlorination of methane occurred before World War I, with the intent of hydrolyzing it to methanol. A commercial methane chlorination facility was first put into operation by the former Farbwerke Hoechst in 1923. In the meantime, however, a high-pressure methanol synthesis based on carbon monoxide and hydrogen had been developed, as a result of which the opposite process became practical — synthesis of methyl chloride from methanol.

Dichloromethane was prepared for the first time in 1840 by V. REGNAULT, who successfully chlorinated methyl chloride. It was for a time produced by the reduction of trichloromethane (chloroform) with zinc and hydrochloric acid in alcohol, but the compound first acquired significance as a solvent after it was successfully prepared commercially by chlorination of methane and monochloromethane (Hoechst AG, Dow Chemical Co., and Stauffer Chemical Co.).

Trichloromethane was synthesized independently by two groups in 1831: J. VON LIEBIG successfully carried out the alkaline cleavage of chloral, whereas M. E. SOUBEIRAIN obtained the compound by the action of chlorine bleach on both ethanol and acetone. In 1835, J. DUMAS showed that trichloromethane contained only a single hydrogen atom and prepared the substance by the alkaline cleavage of trichloroacetic acid and other compounds containing a terminal CCl_3 group, such as β-trichloroacetoacrylic acid. In analogy to the synthetic method of M. E. SOUBEIRAIN, the use of hypochlorites was extended to include other compounds containing acetyl groups, particularly acetaldehyde. V. REGNAULT prepared trichloromethane by chlorination of monochloromethane. Already by the middle of the last century, chloroform was being produced on a commercial basis by using the J. VON LIEBIG procedure, a method which retained its importance until ca. the 1960s in places where the preferred starting materials methane and monochloromethane were in short supply. Today, trichloromethane — along with dichloromethane — is prepared exclusively and on a massive scale by the chlorination of methane and/or monochloromethane. Trichloromethane was introduced into the field of medicine in 1847 by J. Y. SIMPSON, who employed it as an inhaled anaesthetic. As a result of its toxicologic properties, however, it has since been totally replaced by other compounds (e.g., Halothane).

Tetrachloromethane was first prepared in 1839 by V. REGNAULT by the chlorination of trichloromethane. Shortly thereafter, J. DUMAS succeeded in synthesizing it by the chlorination of marsh gas. H. KOLBE isolated tetrachloromethane in 1843 when he treated carbon disulfide with chlorine in the gas phase. The corresponding liquid phase reaction in the presence of a catalyst, giving CCl_4 and S_2Cl_2, was developed a short time later. The key to economical practicality of this approach was the discovery in 1893 by MÜLLER and DUBOIS of the reaction of S_2Cl_2 with CS_2 to give sulfur and tetrachloromethane, thereby avoiding the production of S_2Cl_2.

Currently, tetrachloromethane is produced on an industrial scale by one of two general approaches. The first is the methane chlorination process, using methane or mono-chloromethane as starting materials. The other involves either perchlorination or chlorinolysis. Starting materials in this case include C_1 to C_3 hydrocarbons and their chlorinated derivatives as well as Cl-containing residues obtained in other chlorination processes (vinyl chloride, propylene oxide, etc.).

Originally, tetrachloromethane played a role only in the dry cleaning industry and as a fire extinguishing agent. Its production increased dramatically, however, with the introduction of chlorofluoromethane compounds 50 years ago, these finding wide application as non-toxic refrigerants, as propellants for aerosols, as foam-blowing agents, and as specialty solvents.

Table 1. Physical properties of chloromethanes

	Unit	Monochloro-methane	Dichloro-methane	Trichloro-methane	Tetrachlor-omethane
Formula		CH_3Cl	CH_2Cl_2	$CHCl_3$	CCl_4
M_r		50.49	84.94	119.39	153.84
Melting point	°C	− 97.7	− 96.7	− 63.8	− 22.8
Boiling point at 0.1 MPa	°C	− 23.9	40.2	61.3	76.7
Vapor pressure at 20 °C	kPa	489	47.3	21.27	11.94
Density of liquid at 20 °C	kg/m^3	920 (0.5 MPa)	1328.3	1489	1594.7
Density of vapor at bp	kg/m^3	2.558	3.406	4.372	5.508
Enthalpy of formation $\Delta H°_{298}$	kJ/mol	− 86.0	− 124.7	− 132.0	− 138.1
Specific heat capacity of liquid at 20 °C	kJ kg^{-1}K^{-1}	1.595	1.156	0.980	0.867
Enthalpy of vaporization at bp	kJ/mol	21.65	28.06	29.7	30.0
Critical temperature	K	416.3	510.1	535.6	556.4
Critical pressure	MPa	6.68	6.17	5.45	4.55
Cubic expansion coeff. of liquid (0–40 °C)	K^{-1}	0.0022	0.00137	0.001399	0.00116
Thermal conductivity at 20 °C	W K^{-1}m^{-1}	0.1570	0.159	0.1454	0.1070
Surface tension at 20 °C	N/m	16.2×10^{-3}	28.76×10^{-3}	27.14×10^{-3}	26.7×10^{-3}
Viscosity of liquid at 20 °C	Pa · s	2.7×10^{-4} (0.5 MPa)	4.37×10^{-4}	5.7×10^{-4}	13.5×10^{-4}
Refractive index n_D^{20}			1.4244	1.4467	1.4604
Ignition temperature	°C	618	605	–	–
Limits of ignition in air, lower	vol%	8.1	12	–	–
Limits of ignition in air, upper	vol%	17.2	22	–	–
Partition coefficient air/water at 20 °C	$\frac{mg/L\ (air)}{mg/L\ (water)}$	0.3	0.12	0.12	0.91

1.1. Physical Properties

The most important physical properties of the four chloro derivatives of methane are presented in Table 1; Figure 1 illustrates the vapor pressure curves of the four chlorinated methanes.

The following sections summarize additional important physical properties of the individual compounds making up the chloromethane series.

Monochloromethane is a colorless, flammable gas with a faintly sweet odor. Its solubility in water follows Henry's law; the temperature dependence of the solubility at 0.1 MPa (1 bar) is:

t, °C	15	30	45	60
g of CH_3Cl/kg of H_2O	9.0	6.52	4.36	2.64

Monochloromethane at 20 °C and 0.1 MPa (1 bar) is soluble to the extent of 4.723 cm^3 in 100 cm^3 of benzene, 3.756 cm^3 in 100 cm^3 of tetrachloromethane, 3.679 cm^3 in 100 cm^3 of acetic acid, and 3.740 cm^3 in 100 cm^3 of ethanol. It forms

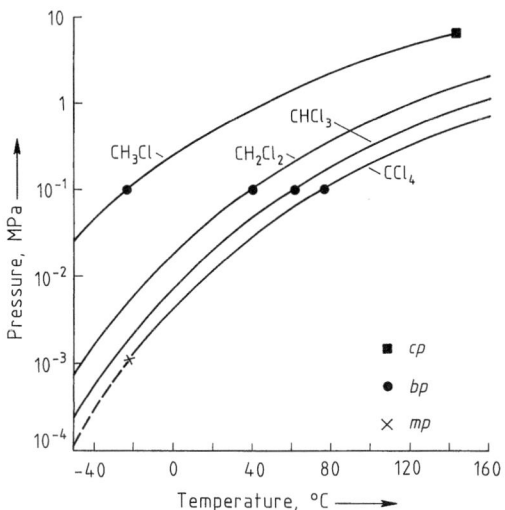

Figure 1. Vapor pressure curves of chloromethanes

azeotropic mixtures with dimethyl ether, 2-methylpropane, and dichlorodifluoromethane (CFC 12).

Dichloromethane is a colorless, highly volatile, neutral liquid with a slightly sweet smell, similar to that of trichloromethane. The solubility of water in dichloromethane is:

t, °C	− 30	0	+ 25
g of H_2O/kg of CH_2Cl_2	0.16	0.8	1.98

The solubility of dichloromethane in water and in aqueous hydrochloric acid is presented in Table 2.

Dichloromethane forms azeotropic mixtures with a number of substances (Table 3).

Dichloromethane is virtually nonflammable in air, as shown in Figure 2, which illustrates the range of flammable mixtures with oxygen–nitrogen combinations [11], [12]. Dichloromethane thereby constitutes the only nonflammable commercial solvent with a low boiling point. The substance possesses no flash point according to the definitions established in DIN 51 755 and ASTM 56-70 as well as DIN 51 758 and ASTM D 93-73. Thus, it is not subject to the regulations governing flammable liquids. As a result of the existing limits of flammability (CH_2Cl_2 vapor/air), it is assigned to explosion category G 1 (VDE 0165). The addition of small amounts of dichloromethane to flammable liquids (e.g., gasoline, esters, benzene, etc.) raises their flash points; addition of 10 – 30% dichloromethane can render such mixtures nonflammable.

Trichloromethane is a colorless, highly volatile, neutral liquid with a characteristic sweet odor. Trichloromethane vapors form no explosive mixtures with air [12]. Trichloromethane has excellent solvent properties for many organic materials, including alka-

Table 2. Solubility of dichloromethane in water and aqueous hydrochloric acid (in wt%)

Solvent	Temperature, °C			
	15	30	45	60
Water	2.50	1.56	0.88	0.53
10% HCl	2.94	1.85	1.25	0.60
20% HCl	–	2.45	1.20	0.65

Table 3. Azeotropic mixtures of dichloromethane

wt%	Compound	Azeotropic boiling point, in °C, at 101.3 kPa
30.0	acetone	57.6
11.5	ethanol	54.6
94.8	1,3-butadiene	–5.0
6.0	*tert*-butanol	57.1
30.0	cyclopentane	38.0
55.0	diethylamine	52.0
30.0	diethyl ether	40.80
8.0	2-propanol	56.6
7.3	methanol	37.8
51.0	pentane	35.5
23.0	propylene oxide	40.6
39.0	carbon disulfide	37.0
1.5	water	38.1

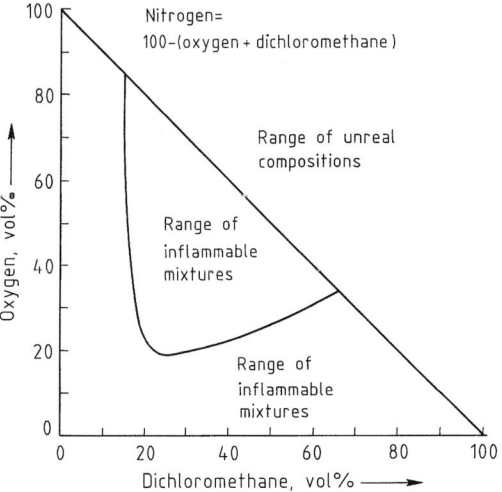

Figure 2. Range of flammability of mixtures of CH_2Cl_2 with O_2 and N_2 [11]

loids, fats, oils, resins, waxes, gums, rubber, paraffins, etc. As a result of its toxicity, it is increasingly being replaced as a solvent by dichloromethane, whose properties in this general context are otherwise similar. In addition, trichloromethane is a good solvent for iodine and sulfur, and it is completely miscible with many organic solvents. The solubility of trichloromethane in water at 25 °C is 3.81 g/kg of H_2O, whereas 0.8 g of H_2O is soluble in 1 kg of $CHCl_3$.

Table 4. Azeotropic mixtures of trichloromethane

wt%	Compound	Azeotropic boiling point, in °C, at 101.3 kPa
15.0	formic acid	59.2
20.5	acetone	64.5
6.8	ethanol	59.3
13.0	ethyl formate	62.7
96.0	2-butanone	79.7
2.8	n-hexane	60.0
4.5	2-propanol	60.8
12.5	methanol	53.4
23.0	methyl acetate	64.8
2.8	water	56.1

Important azeotropic mixtures of chloroform with other compounds are listed in Table 4.

Ternary azeotropes also exist between trichloromethane and ethanol – water (boiling point 55.5 °C, 4 mol% ethanol + 3.5 mol% H_2O), methanol – acetone, and methanol – hexane.

Tetrachloromethane is a colorless neutral liquid with a high refractive index and a strong, bitter odor. It possesses good solubility properties for many organic substances, but due to its high toxicity it is no longer employed (e.g., as a spot remover or in the dry cleaning of textiles). It should be noted that it does continue to find application as a solvent for chlorine in certain industrial processes.

Tetrachloromethane is soluble in water at 25 °C to the extent of 0.8 g of CCl_4/kg of H_2O, the solubility of water in tetrachloromethane being 0.13 g of H_2O/kg of CCl_4.

Tetrachloromethane forms constant-boiling azeotropic mixtures with a variety of substances; corresponding data are given in Table 5.

1.2. Chemical Properties

Monochloromethane, as compared to other aliphatic chlorine compounds, is thermally quite stable. Thermal decomposition is observed only at temperatures in excess of 400 °C, even in the presence of metals (excluding the alkali and alkaline-earth metals). The principal products of photooxidation of monochloromethane are carbon dioxide and phosgene.

Monochloromethane forms with water or water vapor a snowlike gas hydrate with the composition $CH_3Cl \cdot 6\ H_2O$, the latter decomposing into its components at + 7.5 °C and 0.1 MPa (1 bar). To the extent that monochloromethane still finds application in the refrigeration industry, its water content must be kept below 50 ppm. This specification is necessary to prevent potential failure of refrigeration equipment pressure release valves caused by hydrate formation.

Table 5. Azeotropic mixtures of tetrachloromethane

wt%	Compound	Azeotropic boiling point, in °C, at 101.3 kPa
88.5	acetone	56.4
17.0	acetonitrile	71.0
11.5	allyl alcohol	72.3
81.5	formic acid	66.65
43.0	ethyl acetate	74.8
15.85	ethanol	61.1
71.0	2-butanone	73.8
2.5	butanol	76.6
21.0	1,2-dichloroethane	75.6
12.0	2-propanol	69.0
20.56	methanol	55.7
11.5	propanol	73.1
4.1	water	66.0

Monochloromethane is hydrolyzed by water at an elevated temperature. The hydrolysis (to methanol and the corresponding chloride) is greatly accelerated by the presence of alkali. Mineral acids show no influence on the compound's hydrolytic tendencies.

Monochloromethane is converted in the presence of alkali or alkaline-earth metals, as well as by zinc and aluminum, into the corresponding organometallic compounds (e.g., CH_3MgCl, $Al(CH_3)_3 \cdot AlCl_3$). These have come to play a role both in preparative organic chemistry and as catalysts in the production of plastics.

Reaction of monochloromethane with a sodium–lead amalgam leads to tetramethyllead, an antiknocking additive to gasoline intended for use in internal combustion engines. The use of the compound is declining, however, as a result of ecological considerations.

A very significant reaction is that between monochloromethane and silicon to produce the corresponding methylchlorosilanes (the Rochow synthesis), e.g.:

$$2\ CH_3Cl + Si \longrightarrow SiCl_2(CH_3)_2$$

The latter, through their subsequent conversion to siloxanes, serve as important starting points for the production of silicones.

Monochloromethane is employed as a component in the Wurtz-Fittig reaction; it is also used in Friedel-Crafts reactions for the production of alkylbenzenes.

Monochloromethane has acquired particularly great significance as a methylating agent: examples include its reaction with hydroxyl groups to give the corresponding ethers (methylcellulose from cellulose, various methyl ethers from phenolates), and its use in the preparation of methyl-substituted amino compounds (quaternary methylammonium compounds for tensides). All of the various methylamines result from its reaction with ammonia. Treatment of CH_3Cl with sodium hydrogensulfide under pressure and at elevated temperature gives methyl mercaptan.

Dichloromethane is thermally stable to temperatures above 140 °C and stable in the presence of oxygen to 120 °C. Its photooxidation produces carbon dioxide, hydrogen chloride, and a small amount of phosgene [13]. Thermal reaction with nitrogen dioxide gives carbon monoxide, nitrogen monoxide, and hydrogen chloride [14]. In respect to most industrial metals (e.g., iron, copper, tin), dichloromethane is stable, exceptions being aluminum, magnesium, and their alloys; traces of phosgene first arise above 80 °C.

Dichloromethane forms a hydrate with water, $CH_2Cl_2 \cdot 17\ H_2O$, which decomposes at 1.6 °C and 21.3 kPa (213 mbar).

No detectable hydrolysis occurs during the evaporation of dichloromethane from extracts or extraction residues. Only on prolonged action of steam at 140–170 °C under pressure are formaldehyde and hydrogen chloride produced.

Dichloromethane can be further chlorinated either thermally or photochemically. Halogen exchange leading to chlorobromomethane or dibromomethane can be carried out by using bromine and aluminum or aluminum bromide. In the presence of aluminum at 220 °C and 90 MPa (900 bar), it reacts with carbon monoxide to give chloroacetyl chloride [15]. Warming to 125 °C with alcoholic ammonia solution produces hexamethylenetetramine. Reaction with phenolates leads to the same products as are obtained in the reaction of formaldehyde and phenols.

Trichloromethane is nonflammable, although it does decompose in a flame or in contact with hot surfaces to produce phosgene. In the presence of oxygen, it is cleaved photochemically by way of peroxides to phosgene and hydrogen chloride [16], [17]. The oxidation is catalyzed in the dark by iron [18]. The autoxidation and acid generation can be slowed or prevented by stabilizers such as methanol, ethanol, or amylene. Trichloromethane forms a hydrate, $CHCl_3 \cdot 17\ H_2O$, whose critical decomposition point is +1.6 °C and 8.0 kPa (80 mbar).

Upon heating with aqueous alkali, trichloromethane is hydrolyzed to formic acid, orthoformate esters being formed with alcoholates. With primary amines in an alkaline medium the isonitrile reaction occurs, a result which also finds use in analytical determinations. The interaction of trichloromethane with phenolates to give salicylaldehydes is well-known as the Reimer-Thiemann reaction. Treatment with benzene under Friedel-Crafts conditions results in triphenylmethane.

The most important reaction today of trichloromethane is that with hydrogen fluoride in the presence of antimony pentahalides to give monochlorodifluoromethane (CFC 22), a precursor in the production of polytetrafluoroethylene (Teflon, Hostaflon, PTFE).

When treated with salicylic anhydride, trichloromethane produces a crystalline addition compound containing 2 mol of trichloromethane. This result finds application in the preparation of trichloromethane of the highest purity. Under certain conditions, explosive and shocksensitive products can result from the combination of trichloromethane with alkali metals and certain other light metals [19].

Tetrachloromethane is nonflammable and relatively stable even in the presence of light and air at room temperature. When heated in air in the presence of metals (iron), phosgene is produced in large quantities, the reaction starting at ca. 300 °C [20]. Photochemical oxidation also leads to phosgene. Hydrolysis to carbon dioxide and hydrogen chloride is the principal result in a moist atmosphere [21]. Liquid tetrachloromethane has only a very minimal tendency to hydrolyze in water at room temperature (half-life ca. 70 000 years) [22].

Thermal decomposition of dry tetrachloromethane occurs relatively slowly at 400 °C even in the presence of the common industrial metals (with the exception of aluminum and other light metals). Above 500–600 °C an equilibrium reaction sets in

$$2\ CCl_4 \rightleftharpoons C_2Cl_4 + 2\ Cl_2 \tag{1}$$

which is shifted significantly to the right above 700 °C and 0.1 MPa (1 bar) pressure. At 900 °C and 0.1 MPa (1 bar), the equilibrium conversion of CCl_4 is > 70 % (see Chaps. 3.5, cf. Fig. 6).

Tetrachloromethane forms shock-sensitive, explosive mixtures with the alkali and alkaline-earth metals. With water it forms a hydratelike addition compound which decomposes at +1.45 °C.

The telomerization of ethylene and vinyl derivatives with tetrachloromethane under pressure and in the presence of peroxides has acquired a certain preparative significance [23]–[25]:

$$CH_2 = CH_2 + CCl_4 \longrightarrow CCl_3-CH_2-CH_2Cl$$

The most important industrial reactions of tetrachloromethane are its liquid-phase conversion with anhydrous hydrogen fluoride in the presence of antimony (III/V) fluorides or its gas-phase reaction over aluminum or chromium fluoride catalysts, both of which give the widely used and important compounds trichloromonofluoromethane (CFC 11), dichlorodifluoromethane (CFC 12), and monochlorotrifluoromethane (CFC 13).

1.3. Production

1.3.1. Theoretical Bases

The industrial preparation of chloromethane derivatives is based almost exclusively on the treatment of methane and/or monochloromethane with chlorine, whereby the chlorination products are obtained as a mixture of the individual stages of chlorination:

$$CH_4 + Cl_2 \longrightarrow CH_3Cl + HCl \qquad \Delta H = -103.5\ kJ/mol \tag{2}$$
$$CH_3Cl + Cl_2 \longrightarrow CH_2Cl_2 + HCl \qquad \Delta H = -102.5\ kJ/mol \tag{3}$$
$$CH_2Cl_2 + Cl_2 \longrightarrow CHCl_3 + HCl \qquad \Delta H = -99.2\ kJ/mol \tag{4}$$
$$CHCl_3 + Cl_2 \longrightarrow CCl_4 + HCl \qquad \Delta H = -94.8\ kJ/mol \tag{5}$$

Thermodynamic equilibrium lies entirely on the side of the chlorination products, so that the distribution of the individual products is essentially determined by kinetic parameters.

Monochloromethane can be used in place of methane as the starting material, where this in turn can be prepared from methanol by using hydrogen chloride generated in the previous processes. The corresponding reaction is:

$$CH_3OH + HCl \longrightarrow CH_3Cl + H_2O \qquad \Delta H = -33 \text{ kJ/mol} \qquad (6)$$

In this way, the unavoidable accumulation of hydrogen chloride (hydrochloric acid) can be substantially reduced and the overall process can be flexibly tailored to favor the production of individual chlorination products. Moreover, given the ease with which it can be transported and stored, methanol is a better starting material for the chloro derivatives than methane, a substance whose availability is tied to natural gas resources or appropriate petrochemical facilities. There has been a distinct trend in recent years toward replacing methane as a carbon base with methanol.

Methane Chlorination. The chlorination of methane and monochloromethane is carried out industrially by using thermal, photochemical, or catalytic methods [26]. The thermal chlorination method is preferred, and it is also the one on which the most theoretical and scientific investigations have been carried out.

Thermal chlorination of methane and its chlorine derivatives is a radical chain reaction initiated by chlorine atoms. These result from thermal dissociation at 300–350 °C, and they lead to successive substitution of the four hydrogen atoms of methane:

$$Cl_2 \longrightarrow 2\,Cl\cdot \qquad \text{(initiation step)} \qquad (7)$$

$$CH_4 + Cl\cdot \longrightarrow CH_3\cdot + HCl \qquad (8\,a)$$
$$CH_3\cdot + Cl_2 \longrightarrow CH_3Cl + Cl\cdot \qquad \text{chain propagation} \qquad (8\,b)$$

$$2\,Cl\cdot + M \longrightarrow Cl_2 + M \qquad \text{chain termination} \qquad (9)$$

(where M = walls, impurities, O_2)

The conversion to the higher stages of chlorination follows the same scheme [27]–[31]. The thermal reaction of methane and its chlorination products has been determined to be a second-order process:

$$dn(Cl_2)/dt = k \cdot p(Cl_2) \cdot p(CH_4)$$

It has further been shown that traces of oxygen strongly inhibit the reaction.

Controlling the high heat of reaction in the gas phase (which averages ca. 4200 kJ per m^3 of converted chlorine) at STP is a decisive factor in successfully carrying out the process. In industrial reactors, chlorine conversion first becomes apparent above 250 to 270 °C, but it increases exponentially with increasing temperature [32], and in the region of commercial interest — 350 to 550 °C — the reaction proceeds very rapidly. As a result, it is necessary to initiate the process at a temperature which permits the reaction to proceed by itself, but also to maintain the reaction under adiabatic conditions at the requisite temperature level of 320–550 °C dictated by both chemical and technical considerations. If a certain critical temperature is exceeded in the reaction mixture (ca. 550–700 °C, dependent both on the residence time in the hot zone and on the materials making up the reactor), decomposition of the metastable methane chlorination products occurs. In that event, the chlorination leads to formation of undesirable byproducts, including highly chlorinated or high molecular mass compounds (tetrachloroethene, hexachloroethane, etc.). Alternatively, the reaction with chlorine can get completely out of control, leading to the separation of soot and evolution of HCl (thermodynamically the most stable end product). Once such carbon formation begins it acts autocatalytically, resulting in a progressively heavier buildup of soot, which can only be halted by immediate shutdown of the reaction.

Proper temperature control of this virtually adiabatic chlorination is achieved by working with a high methane:chlorine ratio in the range of 6–4:1. Thus, a recycling system is employed in which a certain percentage of inert gas is maintained (nitrogen, recycled HCl, or even materials such as monochloromethane or tetrachloromethane derived from methane chlorination). In this way, the explosive limits of methane and chlorine are moved into a more favorable region and it becomes possible to prepare the more highly substituted chloromethanes with lower $CH_4:Cl_2$ ratios.

Figure 3 shows the explosion range of methane and chlorine and how it can be limited through the use of diluents, using the examples of nitrogen, hydrogen chloride, and tetrachloromethane.

The composition and distribution of the products resulting from chlorination is a definite function of the starting ratio of chlorine to methane, as can be seen from Figure 4 and Figure 5.

These relationships have been investigated frequently [33], [34]. The composition of the reaction product has been shown to be in excellent agreement with that predicted by calculations employing experimental relative reaction rate constants [35]–[38]. The products arising from thermal chlorination of monochloromethane and from the pyrolysis of primary products can also be predicted quantitatively [39]. The relationships among the rate constants are nearly independent of temperature in the region of technical interest. If one designates as k_1 through k_4 the successive rate constants in the chlorination process, then the following values can be assigned to the relative constants for the individual stages:

$k_1 = 1$ (methane)
$k_2 = 2.91$ (monochloromethane)
$k_3 = 2.0$ (dichloromethane)
$k_4 = 0.72$ (trichloromethane)

With this set of values, the selectivity of the chlorination can be effectively established with respect to optimal product distribution for reactors of various residence time (stream type or mixing type, cf. Fig. 4 and Fig. 5). Additional recycling into the reaction of partially chlorinated products (e.g., monochloromethane) permits further control over the ratios of the individual components [40], [41].

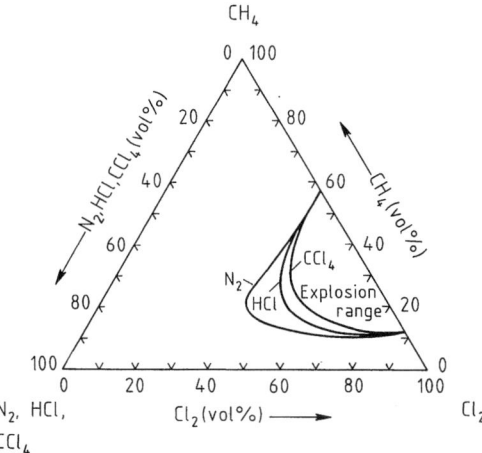

Figure 3. Explosive range of CH_4–Cl_2 mixtures containing N_2, HCl, and CCl_4
Test conditions: pressure 100 kPa; temperature 50 °C; ignition by 1-mm spark

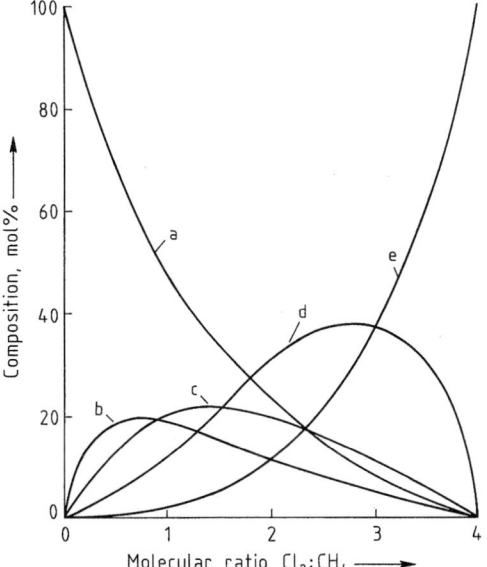

Figure 4. Product distribution in methane chlorination, plug stream reactor
a) Methane; b) Monochloromethane; c) Dichloromethane; d) Trichloromethane; e) Tetrachloromethane

It has been recognized that the yield of partially chlorinated products (e.g., dichloromethane and trichloromethane) is diminished by recycling. This factor has to be taken into account in the design of reactors for those methane chlorinations which are intended to lead exclusively to these products. If the emphasis is to lie more on the side of trichloro- and tetrachloromethane, then mixing within the reactor plays virtually no role, particularly since less-chlorinated materials can always be partially or wholly recycled. Details of reactor construction will be discussed below in the context of each of the various processes.

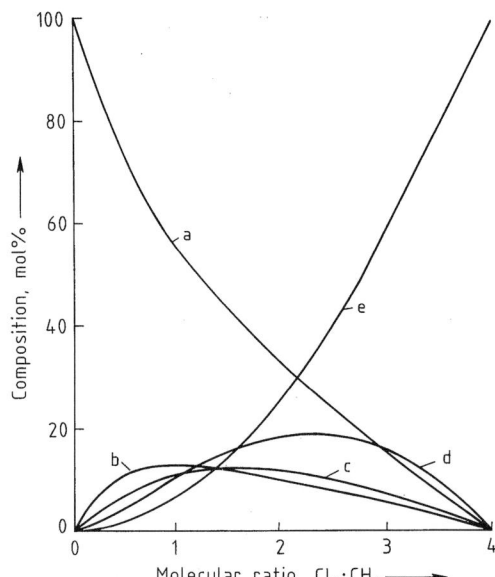

Figure 5. Product distribution in methane chlorination, ideal mixing reactor
a) Methane; b) Monochloromethane; c) Dichloromethane; d) Trichloromethane; e) Tetrachloromethane

Chlorinolysis. The technique for the production of tetrachloromethane is based on what is known as perchlorination, a method in which an excess of chlorine is used and C_1- to C_3-hydrocarbons and their chlorinated derivatives are employed as carbon sources. In this process, tetrachloroethene is generated along with tetrachloromethane, the relationship between the two being consistent with Eq. 1 in Section p. 1260 and dependent on pressure and temperature (cf. also Fig. 6).

It will be noted that at low pressure (0.1 to 1 MPa, 1 to 10 bar) and temperatures above 700 °C, conditions under which the reaction takes place at an acceptable rate, a significant amount of tetrachloroethene arises. For additional details see Chap. 3.5. Under conditions of high pressure — greater than 10 MPa (100 bar) — the reaction occurs at a temperature as low as 600 °C. As a result of the influence of pressure and by the use of a larger excess of chlorine, the equilibrium can be shifted essentially 100% to the side of tetrachloromethane. These circumstances are utilized in the Hoechst high-pressure chlorinolysis procedure (see below) [42], [43].

Methanol Hydrochlorination. Studies have been conducted for purposes of reactor design [44] on the kinetics of the gas-phase reaction of hydrogen chloride with methanol in the presence of aluminum oxide as catalyst to give monochloromethane. Aging of the catalyst has also been investigated. The reaction is first order in respect to hydrogen chloride, but nearly independent of the partial pressure of methanol. The rate constant is proportional to the specific surface of the catalyst, whereby at higher temperatures (350 – 400 °C) an inhibition due to pore diffusion becomes apparent.

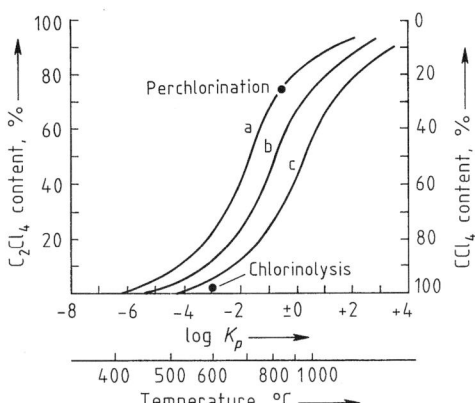

Figure 6. Thermodynamic equilibrium
$2\,CCl_4 \rightleftharpoons C_2Cl_4 + 2\,Cl_2$
a) 0.1 MPa; b) 1 MPa; c) 10 MPa

1.3.2. Production of Monochloromethane

Monochloromethane is currently produced commercially by two methods: by the hydrochlorination (esterification) of methanol using hydrogen chloride, and by chlorination of methane. Methanol hydrochlorination has become increasingly important in recent years, whereas methane chlorination as the route to monochloromethane as final product has declined. The former approach has the advantage that it utilizes, rather than generating, hydrogen chloride, a product whose disposal — generally as hydrochloric acid — has become increasingly difficult for chlorinated hydrocarbon producers. Moreover, this method leads to a single target product, monochloromethane, in contrast to methane chlorination (cf. Figs. 4 and 5). As a result of the ready and low-cost availability of methanol (via the low pressure methanol synthesis technique) and its facile transport and storage, the method also offers the advantage of avoiding the need for placing production facilities in the vicinity of a methane supply.

Since in the chlorination of methane each substitution of a chlorine atom leads to generation of an equimolar amount of hydrogen chloride — cf. Eqs. 2 – 5 in Section p. 1260 — a combination of the two methods permits a mixture of chlorinated methanes to be produced without creating large amounts of hydrogen chloride at the same time; cf. Eq. 6.

Monochloromethane production from methanol and hydrogen chloride is currently carried out catalytically in the gas phase at 0.3 – 0.6 MPa (3 – 6 bar) and temperatures of 280 – 350 °C. The usual catalyst is activated aluminum oxide. Excess hydrogen chloride is introduced in order to provide a more favorable equilibrium point (located 96 – 99 % on the side of products at 280 – 350 °C) and to reduce the formation of dimethyl ether as a side product (0.2 to 1 %).

The raw materials must be of high purity in order to prolong catalyst life as much as possible. Technically pure (99.9 %) methanol is employed, along with very clean hydrogen chloride. In the event that the latter is obtained from hydrochloric acid, it must be subjected to special purification (stripping) in order to remove interfering chlorinated hydrocarbons.

Process Description. In a typical production plant (Fig. 7), the two raw material streams, hydrogen chloride and methanol, are warmed over heat exchangers and led, after mixing and additional preheating, into the reactor, where conversion takes place at 280–350 °C and ca. 0.5 MPa (5 bar).

The reactor itself consists of a large number of relatively thin nickel tubes bundled together and filled with aluminum oxide. Removal of heat generated by the reaction (33 kJ/mol) is accomplished by using a heat conduction system. A hot spot forms in the catalyst layer as a result of the exothermic nature of the reaction, and this migrates through the catalyst packing, reaching the end as the latter's useful life expires.

The reaction products exiting the reactor are cooled with recycled hydrochloric acid (> 30%) in a subsequent quench system, resulting in separation of byproduct water, removed as ca. 20% hydrochloric acid containing small amounts of methanol. Passage through a heat exchanger effects further cooling and condensation of more water, as well as removal of most of the excess HCl. The quenching fluid is recovered and subsequently returned to the quench circulation system. The gaseous crude product is led from the separator into a 96% sulfuric acid column, where dimethyl ether and residual water (present in a quantity reflective of its partial vapor pressure) are removed, the concentration of the acid diminishing to ca. 80% during its passage through the column. In this step, dimethyl ether reacts with sulfuric acid to form "onium salts" and methyl sulfate. It can be driven out later by further dilution with water. It is advantageous to use the recovered sulfuric acid in the production of fertilizers (superphosphates) or to direct it to a sulfuric acid cleavage facility.

Dry, crude monochloromethane is subsequently condensed and worked up in a high-pressure (2 MPa, 20 bar) distillation column to give pure liquid monochloromethane. The gaseous product emerging from the head of this column (CH_3Cl + HCl), along with the liquid distillation residue — together making up ca. 5–15% of the monochloromethane product mixture — can be recovered for introduction into an associated methane chlorination facility. The overall yield of the process, calculated on the basis of methanol, is ca. 99%.

The commonly used catalyst for vapor-phase hydrochlorination of methanol is γ-aluminum oxide with an active surface area of ca. 200 m^2/g. Catalysts based on silicates have not achieved any technical significance. Catalyst aging can be ascribed largely to carbon deposition. Byproduct formation can be minimized and catalyst life considerably prolonged by doping the catalyst with various components and by introduction of specific gases (O_2) into the reaction components [45]. The life of the catalyst in a production facility ranges from about 1 to 2 years.

Liquid-Phase Hydrochlorination. The once common liquid-phase hydrochlorination of methanol using 70% zinc chloride solution at 130–150 °C and modest pressure is currently of lesser significance. Instead, new production techniques involving treat-

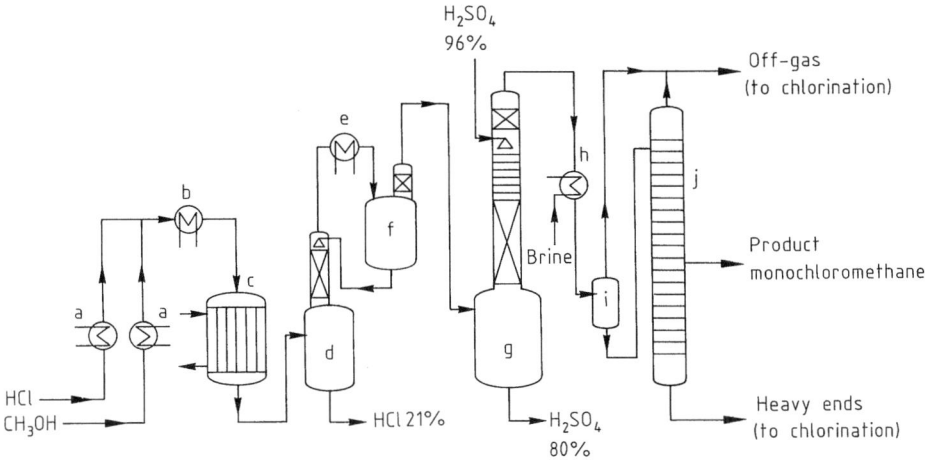

Figure 7. Production of monochloromethane by methanol hydrochlorination
a) Heat exchangers; b) Heater; c) Multiple-tube reactor; d) Quench system; e) Quench gas cooler; f) Quenching fluid tank; g) Sulfuric acid column; h) CH$_3$Cl condensation; i) Intermediate tank; j) CH$_3$Cl distillation column

ment of methanol with hydrogen chloride in the liquid phase without the addition of catalysts are becoming preeminent. The advantage of these methods, apart from circumventing the need to handle the troublesome zinc chloride solutions, is that they utilize aqueous hydrochloric acid, thus obviating the need for an energy-intensive hydrochloric acid distillation. The disadvantage of the process, which is conducted at 120–160 °C, is its relatively low yield on a space–time basis, resulting in the need for large reaction volumes [46]–[48].

Other Processes. Other techniques for producing monochloromethane are of theoretical significance, but are not applied commercially.

Monochloromethane is formed when a mixture of methane and oxygen is passed into the electrolytes of an alkali chloride electrolysis [49]. Treatment of dimethyl sulfate with aluminum chloride [50] or sodium chloride [51] results in the formation of monochloromethane. Methane reacts with phosgene at 400 °C to give CH$_3$Cl [52]. The methyl acetate–methanol mixture that arises during polyvinyl alcohol synthesis can be converted to monochloromethane with HCl at 100 °C in the presence of catalysts [53]. It has also been suggested that monochloromethane could be made by the reaction of methanol with the ammonium chloride that arises during sodium carbonate production [54].

The dimethyl ether which results from methylcellulose manufacture can be reacted with hydrochloric acid to give monochloromethane [55]. The process is carried out at 80–240 °C under sufficient pressure so that water remains as a liquid. Similarly, cleavage of dimethyl ether with antimony trichloride also leads to monochloromethane [56].

Of increasing importance today are so-called "methanolysis reactions," in which monochloromethane is recovered in the manufacture of silicones and then reintroduced into the process of silane formation [57]:

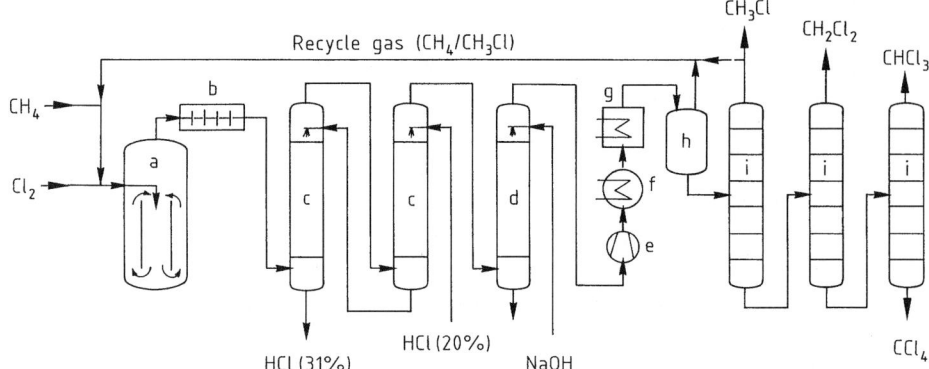

Figure 8. Methane chlorination by the Hoechst method (production of dichloromethane and trichloromethane) a) Loop reactor; b) Process gas cooler; c) HCl absorption; d) Neutralization system; e) Compressor; f) First condensation step (water); g) Gas drying system; h) Second condensation system and crude product storage vessel (brine); i) Distillation columns for CH_3Cl, CH_2Cl_2, and $CHCl_3$

$SiCl_2(CH_3)_2 + 2\ CH_3OH \longrightarrow Si(OH)_2(CH_3)_2 + 2\ CH_3Cl$ (10)
$Si + 2\ CH_3Cl \longrightarrow SiCl_2(CH_3)_2$ (11)

1.3.3. Production of Dichloromethane and Trichloromethane

The industrial synthesis of dichloromethane also leads to trichloromethane and small amounts of tetrachloromethane, as shown in Figure 4 and Figure 5. Consequently, di- and trichloromethane are prepared commercially in the same facilities. In order to achieve an optimal yield of these products and to ensure reliable temperature control, it is necessary to work with a large methane and/or monochloromethane excess relative to chlorine. Conducting the process in this way also enables the residual concentration of chlorine to be kept in the fully reacted product at an exceptionally low level (< 0.01 vol%), which in turn simplifies workup. Because of the large excess of carbon-containing components, the operation is customarily accomplished in a recycle mode.

Process Description. One of the oldest production methods is that of Hoechst, a recycle chlorination which was introduced as early as 1923 and which, apart from modifications reflecting state-of-the-art technology, continues essentially unchanged, retaining its original importance. The process is shown in Figure 8.

The gas which is circulated consists of a mixture of methane and monochloromethane. To this is added fresh methane and, as appropriate, monochloromethane obtained from methanol hydrochlorination. Chlorine is then introduced and the mixture is passed into the reactor. The latter is a loop reactor coated with nickel or highalloy steel in which internal gas circulation is constantly maintained by means of a coaxial inlet tube and a valve system. The reaction is conducted adiabatically, the necessary temperature of 350–450 °C being achieved and maintained by proper choice of the chlorine to starting material ($CH_4 + CH_3Cl$) ratio and/or by prewarming the mixture [58]. The fully reacted gas mixture is cooled in a heat exchanger and passed through an absorber cascade in which dilute hydrochloric acid and water wash out the resulting hydrogen chloride in the form of 31 % hydrochloric acid. The last traces of acid and chlorine are removed by washing with sodium hydroxide, after which the gases are compressed, dried, and cooled and the reaction products largely condensed. Any uncondensed gas — methane and to some extent monochloromethane — is returned to the reactor. The liquified condensate is separated by distillation under pressure into its pure components, monochloromethane, dichloromethane, trichloromethane (the latter two being the principal products), and small amounts of tetrachloromethane. The product composition is approximately 70 wt % dichloromethane, is approximately 27 wt % trichloromethane, and 3 wt % tetrachloromethane.

Methane chlorination is carried out in a similar way by Chem. Werke Hüls AG, whose work-up process employs prior separation of hydrogen chloride by means of an adiabatic absorption system. After the product gas has been washed to neutrality with sodium hydroxide, it is dried with sulfuric acid and compressed to ca. 0.8 MPa (8 bar), whereby the majority of the resulting chloromethanes can be condensed with relatively little cooling (at approximately – 12 to – 15 °C). Monochloromethane is recycled to the chlorination reactor. The subsequent workup to pure products is essentially analogous to that employed in the Hoechst process.

Other techniques, e.g., those of Montecatini and Asahi Glass, function similarly with respect to drying and distillation of the products.

The loop reactor used by these and other manufacturers (e.g., Stauffer Chem. Co.) [59] has been found to give safe and trouble-free service, primarily because the internal circulation in the reactor causes the inlet gases to be brought quickly to the initiation temperature, thereby excluding the possibility of formation of explosive mixtures. This benefit is achieved at the expense of reduced selectivity in the conduct of the reaction, however (cf. Figs. 4 and 5). In contrast, the use of an empty tube reactor with minimal axial mixing has unquestionable advantages for the selective preparation of dichloromethane [60], [61]. The operation of such a reactor is considerably more complex, however, especially from the standpoint of measurement and control technology, since the starting gases need to be brought up separately to the ignition temperature and then, after onset of the reaction with its high enthalpy, heat must be removed by means of a cooling system. By contrast, maintenance of constant temperature in a loop reactor is relatively simple because of the high rate of gas circulation. A system operated by Frontier Chem. Co. employs a tube reactor incorporating recycled tetrachloromethane for the purpose of temperature control [62].

Reactor Design. Various types of reactors are in use, with characteristics ranging between those of fully mixing reactors (e.g., the loop reactor) and tubular reactors.

Chem. Werke Hüls operates a reactor that permits partial mixing, thereby allowing continuous operation with little or no preheating.

Instead of having the gas circulation take place within the reactor, an external loop can also be used for temperature control, as, e.g., in the process described by Montecatini [63] and used in a facility operated by Allied Chemical Corp. In this case, chlorine is added to the reacted gases outside of the chlorination reactor, necessary preheating is undertaken, and only then is the gas mixture led into the reactor.

The space–time yield and the selectivity of the chlorination reaction can be increased by operating two reactors in series, these being separated by a condensation unit to remove high-boiling chloromethanes [64].

Solvay [65] has described an alternative means of optimizing the process in respect to selectivity, whereby methane and monochloromethane are separately chlorinated in reactors driven in parallel. The monochloromethane produced in the methane chlorination reactor is isolated and introduced into the reactor for chlorination of monochloromethane, which is also supplied with raw material from a methanol hydrochlorination system. The reaction is carried out at a pressure of 1.5 MPa (15 bar) in order to simplify the workup and separation of products.

Because of its effective heat exchange characteristics, a fluidized-bed reactor is used by Asahi Glass Co. for methane chlorination [66]. The reaction system consists of two reactors connected in series. After separation of higher boiling components, the low-boiling materials from the first reactor, including hydrogen chloride, are further treated with chlorine in a second reactor. Reactors of this kind must be constructed of special materials with high resistance to both erosion and corrosion. Special steps are required (e.g., washing with liquid chloromethanes) to remove from the reaction gas dust derived from the fluidized-bed solids.

Raw Materials. Very high purity standards must be applied to methane which is to be chlorinated. Some of this methane is derived from petrochemical facilities in the course of naphtha cleavage to ethylene and propene, whereas some comes from low-temperature distillation of natural gas (the Linde process). Components such as ethane, ethylene, and higher hydrocarbons must be reduced to a minimum. Otherwise, these would also react under the conditions of methane chlorination to give the corresponding chlorinated hydrocarbons, which would in turn cause major problems in the purification of the chloromethanes. For this reason, every effort is made to maintain the level of higher hydrocarbons below 100 mL/m^3. Inert gases such as nitrogen and carbon dioxide (but excluding oxygen) have no significant detrimental effect on the thermal chlorination reaction, apart from the fact that their presence in excessive amounts results in the need to eliminate considerable quantities of off-gas from the recycling system, thus causing a reduction in product yield calculated on the basis of methane introduced.

Chlorine with a purity of ca. 97% (residue: hydrogen, carbon dioxide, and oxygen) is compressed and utilized just as it emerges from electrolysis. Newer chlorination procedures are designed to utilize gaseous chlorine of higher purity, obtained by evaporization of previously liquified material.

Similarly, monochloromethane destined for further chlorination is a highly purified product of methanol hydrochlorination, special procedures being used to reduce the dimethyl ether content, for example, to less than 50 mL/m^3.

Depending on the level of impurities present in the starting materials, commercial processes incorporating recycling can lead to product yields of 95–99% based on chlorine or 70–85% based on methane. The relatively low methane-based yield is a consequence of the need for removal of inert gases, although the majority of this exhaust gas can be subjected to further recovery measures in the context of some associated facility.

Off-Gas Workup. The workup of off-gas from thermal methane chlorination is relatively complicated as a consequence of the methane excess employed. Older technologies accomplished the separation of the hydrogen chloride produced in the reaction through its absorption in water or azeotropic hydrochloric acid, leading to ordinary commercial 30–31% hydrochloric acid. This kind of workup requires a major outlay for materials of various sorts: on the one hand, coatings must be acid-resistant but at the same time, materials which are stable against attack by chlorinated hydrocarbons are required.

A further disadvantage frequently plagues these "wet" processes is the need to find a use for the inevitable concentrated hydrochloric acid, particularly given that the market for hydrochloric acid is in many cases limited. Hydrogen chloride can be recovered from the aqueous hydrochloric acid by distillation under pressure, permitting its use in methanol hydrochlorination; alternatively, it can be utilized for oxychlorination of ethylene to 1,2-dichloroethane. Disadvantages of this approach, however, are the relatively high energy requirement and the fact that the hydrogen chloride can only be isolated by distillation to the point of azeotrope formation (20% HCl).

Newer technologies have as their goal workup of the chlorination off-gas by dry methods. These permit use of less complicated construction materials. Apart from the reactors, in which nickel and nickel alloys are normally used, all other apparatus and components can be constructed of either ordinary steel or stainless steel.

Hydrogen chloride can be removed from the off-gas by an absorption–desorption system developed by Hoechst AG and utilizing a wash with monochloromethane, in which hydrogen chloride is very soluble [67]. A similar procedure involving HCl removal by a wash with trichloromethane and tetrachloromethane has been described by Solvay [65].

Other Processes. The relatively complicated removal of hydrogen chloride from methane can be avoided by adopting processes that begin with methanol as raw material. An integrated chlorination/hydrochlorination facility (Fig. 9) has been developed for this purpose and brought on stream on a commercial scale by Stauffer Chem. Co. [68].

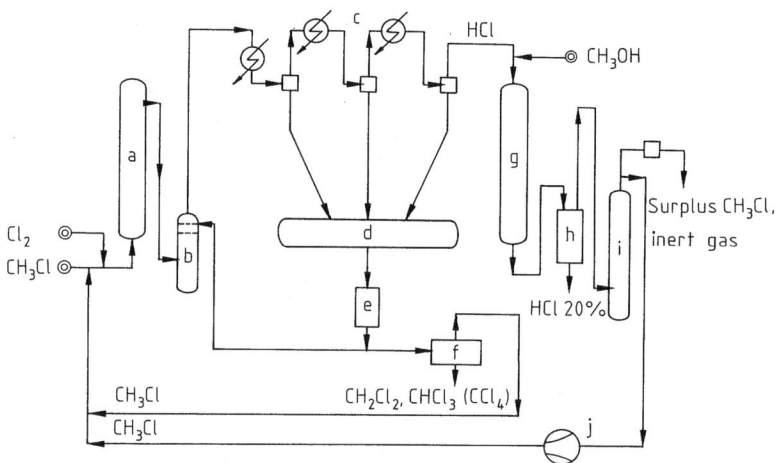

Figure 9. Chlorination of monochloromethane by the Stauffer process [69]
a) Chlorination reactor; b) Quench system; c) Multistage condensation; d) Crude product storage vessel; e) Drying; f) Distillation and purification of CH_2Cl_2 and $CHCl_3$; g) Hydrochlorination reactor; h) Quench system; i) H_2SO_4 drying column; j) Compressor

Monochloromethane is caused to react with chlorine under a pressure of 0.8–1.5 MPa (8–15 bar) at elevated temperature (350–400 °C) with subsequent cooling occurring outside of the reactor. The crude reaction products are separated in a multistage condensation unit and then worked up by distillation to give the individual pure components. Monochloromethane is returned to the reactor. After condensation, gaseous hydrogen chloride containing small amounts of monochloromethane is reacted with methanol in a hydrochlorination system corresponding to that illustrated in Figure 7 for the production of monochloromethane. Following its compression, monochloromethane is returned to the chlorination reactor. This process is distinguished by the fact that only a minimal amount of the hydrogen chloride evolved during the synthesis of dichloromethane and trichloromethane is recovered in the form of aqueous hydrochloric acid.

As a substitute for thermal chlorination at high temperature, processes have also been developed which occur by a photochemically-initiated radical pathway. According to one patent [69], monochloromethane can be chlorinated selectively to dichloromethane at −20 °C by irradiation with a UV lamp, the trichloromethane content being only 2–3 %. A corresponding reaction with methane is not possible.

Liquid-phase chlorination of monochloromethane in the presence of radical-producing agents such as azodiisobutyronitrile has been achieved by the Tokuyama Soda Co. The reaction occurs at 60–100 °C and high pressure [70]. The advantage of this low-temperature reaction is that it avoids the buildup of side products common in thermal chlorination (e.g., chlorinated C_2-compounds such as 1,1-dichloroethane, 1,2-dichloroethene, and trichloroethene). Heat generated in the reaction is removed by evaporation of the liquid phase, which is subsequently condensed. Hydrogen chloride produced during the chlorination is used for gas-phase hydrochlorination of methanol to give monochloromethane, which is in turn recycled for chlorination.

It is tempting to try to avoid the inevitable production of hydrogen chloride by carrying out the reaction in the presence of oxygen, as in the oxychlorination of ethylene or ethane. Despite intensive investigations into the prospects, however, no commercially feasible applications have resulted. The low reactivity of methane requires the use of a high reaction temperature, but this in turn leads to undesirable side products and an unacceptably high loss of methane through combustion.

In this context, the "Transcat" process of the Lummus Co. is of commercial interest [71]. In this process, methane is chlorinated and oxychlorinated in two steps in a molten salt mixture comprised of copper(II) chloride and potassium chloride. The starting materials are chlorine, air, and methane. The process leaves virtually no residue since all of its byproducts can be recycled.

Experiments involving treatment of methane with other chlorinating agents (e.g., phosgene, nitrosyl chloride, or sulfuryl chloride) have failed to yield useful results. The fluidized-bed reaction of methane with tetrachloromethane at 350 to 450 °C has also been suggested [72].

The classical synthetic route to trichloromethane proceeded from the reaction of chlorine with ethanol or acetaldehyde to give chloral, which can be cleaved with calcium hydroxide to trichloromethane and calcium formate [73]. This method is presumably still used commercially in the USSR. Trichloromethane and calcium acetate can also be produced from acetone using an aqueous solution of chlorine bleach at 60–65 °C. A description of these archaic processes can be found in [74].

1.3.4. Production of Tetrachloromethane

Chlorination of Carbon Disulfide. The chlorination of carbon disulfide was, until the late 1950s, the principal means of producing tetrachloromethane, according to the following overall reaction:

$$CS_2 + 2\,Cl_2 \longrightarrow CCl_4 + 2\,S \tag{12}$$

The resulting sulfur is recycled to a reactor for conversion with coal or methane (natural gas) to carbon disulfide. A detailed look at the reaction shows that it proceeds in stages corresponding to the following equations:

$$2\,CS_2 + 6\,Cl_2 \rightarrow 2\,CCl_4 + 2\,S_2Cl_2 \tag{13}$$
$$CS_2 + 2\,S_2Cl_2 \rightleftharpoons CCl_4 + 6\,S \tag{14}$$

The process developed at the Bitterfeld plant of I.G. Farben before World War II was improved by a number of firms in the United States, including FMC and the Stauffer Chem. Co. [75]–[77], particularly with respect to purification of the tetrachloromethane and the resulting sulfur.

In a first step, carbon disulfide dissolved in tetrachloromethane is induced to react with chlorine at temperatures of 30–100 °C. Either iron or iron(III) chloride is added as catalyst. The conversion of carbon disulfide exceeds 99% in this step. In a subsequent distillation, crude tetrachloromethane is separated at the still head. The disulfur dichloride recovered from the still pot is transferred to a second stage of the process where it is consumed by reaction with excess carbon disulfide at ca. 60 °C. The resulting sulfur is separated (with cooling) as a solid, which has the effect of shifting the equilibrium in the reactions largely to the side of tetrachloromethane. Tetrachloromethane and excess carbon disulfide are withdrawn at the head of a distillation apparatus and returned to the chlorination unit. A considerable effort is required to purify the tetrachloromethane and sulfur, entailing hydrolysis of sulfur compounds with dilute alkali and subsequent azeotropic drying and removal from the molten sulfur by air stripping of residual disulfur dichloride. Yields lie near 90% of the theoretical value based on carbon disulfide and about 80% based on chlorine. The losses, which must be recovered in appropriate cleanup facilities, result from gaseous emissions from the chlorination reaction, from the purification systems (hydrolysis), and from the molten sulfur processing.

The carbon disulfide method is still employed in isolated plants in the United States, Italy, and Spain. Its advantage is that, in contrast to chlorine substitution on methane or chlorinating cleavage reactions, no accumulation of hydrogen chloride or hydrochloric acid byproduct occurs.

Perchlorination (Chlorinolysis). Early in the 1950s commercial production of tetrachloromethane based on high-temperature chlorination of methane and chlorinating cleavage reactions of hydrocarbons ($\leq C_3$) and their chlorinated derivatives was introduced. In processes of this sort, known as perchlorinations or chlorinolyses, substitution reactions are accompanied by rupture of C–C bonds. Starting materials, in addition to ethylene, include propane, propene, dichloroethane, and dichloropropane. In recent years, increasing use has been made of chlorine-containing byproducts and the residues from other chlorination processes, such as those derived from methane chlorination, vinyl chloride production (via either direct chlorination or oxychlorination of ethylene), allyl chloride preparation, etc. The course of the reaction is governed by the position of equilibrium between tetrachloromethane and tetrachloroethene, as illustrated earlier in Figure 6, whereby the latter always arises as a byproduct. In general, these processes are employed for the production of tetrachloroethylene (see Section 3.5.3 and [78]), in which case tetrachloromethane is the byproduct. Most production facilities are sufficiently flexible such that up to 70 wt% tetrachloromethane can be achieved in the final product [79]. The product yield can be largely forced to the side of tetrachloromethane by recycling tetrachloroethylene into the chlorination reaction, although the required energy expenditure is significant. Higher pressure [80] and the use of hydrocarbons containing an odd number of carbon atoms increases the yield of tetrachloromethane. When the reaction is carried out on an industrial scale, a temperature of 500 to 700 °C and an excess of chlorine are used. The corresponding reactors either can be of the tube type, operated adiabatically by using a recycled coolant (N_2, HCl, CCl_4, or C_2Cl_4) [81]–[83], or else they can be fluidized-bed systems operated isothermally [84], [85]. Byproducts under these reaction conditions include ca. 1–7% perchlorinated compounds (hexachloroethane, hexachlorobutadiene, hexachlorobenzene), the removal of which requires an additional expenditure of effort.

Pyrolytic introduction of chlorine into chlorinated hydrocarbons has become increasingly important due to its potential for consuming chlorinated hydrocarbon wastes and residues from other processes. Even the relatively high production of hydrogen chloride can be tolerated, provided that reactors are used which operate at high pressure and which can be coupled with other processes that consume hydrogen chloride. Another advantage of the method is that it can be used for making both tetrachloromethane and tetrachloroethylene. The decrease in demand for tetrachloromethane in the late 1970s and early 1980s, a consequence of restrictions (related to the ozone hypothesis) on the use of chlorofluorocarbons prepared from it, has led to stagnation in the development of new production capacity.

Hoechst High-Pressure Chlorinolysis. The high-pressure chlorinolysis method developed and put in operation by Hoechst AG has the same goals as the process just described. It can be seen in Figure 6 that under the reaction conditions of this process — 620 °C and 10 to 15 MPa (100 to 150 bar) — the equilibrium

$$2\ CCl_4 \rightleftharpoons C_2Cl_4 + 2\ Cl_2$$

lies almost exclusively on the side of tetrachloromethane, especially in the presence of an excess of chlorine [42], [43], [86]. This method utilizes chlorine-containing residues from other processes (e.g., methane chlorination and vinyl chloride) as raw material, although these must be free of sulfur and cannot contain solid or polymerized components.

The conversion of these materials is carried out in a specially constructed high-pressure tube reactor which is equipped with a pure nickel liner to prevent corrosion. Chlorine is introduced in excess in order to prevent the formation of byproducts and in order maintain the final reaction temperature (620 °C) of this adiabatically conducted reaction. If hydrogen-deficient starting materials are to be employed, hydrogen-rich components must be added to increase the enthalpy of the reaction. In this way, even chlorine-containing residues containing modest amounts of aromatics can be utilized. Hexachlorobenzene, for example, can be converted (albeit relatively slowly) at the usual temperature of this process and in the presence of excess chlorine to tetrachloromethane according to the equilibrium reaction:

$$C_6Cl_6 + 9\ Cl_2 \rightleftharpoons 6\ CCl_4$$

The mixture exiting the reactor is comprised of tetrachloromethane, the excess chlorine, hydrogen chloride, and small amounts of hexachlorobenzene, the latter being recycled. This mixture is quenched with cold tetrachloromethane, its pressure is reduced, and it is subsequently separated into crude tetrachloromethane and chlorine and hydrogen chloride. The crude product is purified by distillation to give tetrachloromethane meeting the required specifications. This process is advantageous in those situations in which chlorine-containing residues accumulate which would otherwise be difficult to deal with (e.g., hexachloroethane from methane chlorination facilities and high-boiling residues from vinyl chloride production).

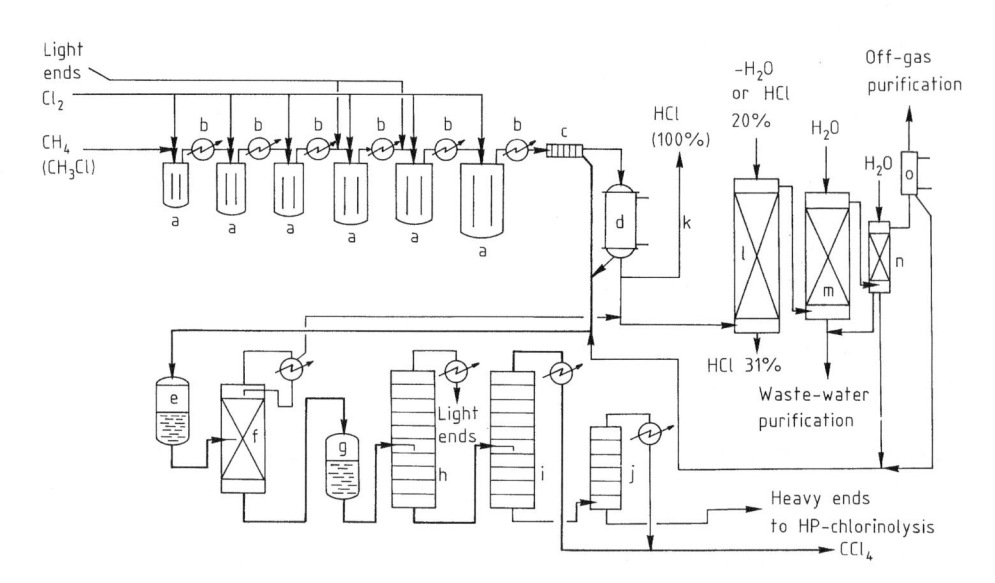

Figure 10. Production of tetrachloromethane by stepwise chlorination of methane (Hoechst process)
a) Reactor; b) Cooling; c) First condensation (air); d) Second condensation (brine); e) Crude product storage vessel; f) Degassing/dewatering column; g) Intermediate tank; h) Light-end column; i) Column for pure CCl_4; j) Heavy-end column; k) HCl stream for hydrochlorination; l) Adiabatic HCl absorption; m) Vapor condensation; n) Cooling and phase separation; o) Off-gas cooler

A number of serious technical problems had to be overcome in the development of this process, including perfection of the nickel-lined highpressure reactor, which required the design of special flange connections and armatures.

Multistep Chlorination Process. Despite the fact that its stoichiometry results in high yields of hydrogen chloride or hydrochloric acid, thermal chlorination of methane to tetrachloromethane has retained its decisive importance. Recent developments have assured that the resulting hydrogen chloride can be fed into other processes which utilize it. In principle, tetrachloromethane can be obtained as the major product simply by repeatedly returning all of the lower boiling chloromethanes to the reactor. It is not possible to employ a 1:4 mixture of the reactants methane and chlorine at the outset. This is true not only because of the risk of explosion, but also because of the impossibility of dealing with the extremely high heat of reaction. Unfortunately, the simple recycling approach is also uneconomical because it necessitates the availability of a very large workup facility. Therefore, it is most advantageous to employ several reactors coupled in series, the exit gases of each being cooled, enriched with more chlorine, and then passed into the next reactor [87]. Processes employing supplementary circulation of an inert gas (e.g., nitrogen) have also been suggested [88].

The stepwise chlorination of methane and/or monochloromethane to tetrachloromethane is based on a process developed in the late 1950s and still used by Hoechst AG (Fig. 10) [89].

The first reactor in a six-stage reactor cascade is charged with the full amount of methane and/or monochloromethane required for the entire production batch. Nearly quantitative chlorine conversion is achieved in the first reactor at 400 °C, using only a portion of the necessary overall amount of chlorine. The gas mixture leaving the first reactor is cooled and introduced into the second reactor along with additional chlorine, the mixture again being cooled after all of the added chlorine has been consumed. This stepwise addition of chlorine with intermittent cooling is continued until in the last reactor the component ratio $CH_4:Cl_2 = 1:4$ is reached. The reactors themselves are loop reactors with internal circulation, a design which, because of its efficient mixing, effectively shifts the product distribution toward more highly chlorinated materials. The gas mixture leaving the reactors is cooled in two stages to -20 °C, in the course of which the majority of the tetrachloromethane is liquified, along with the less chlorinated methane derivatives (amounting to ca. 3% of the tetrachloromethane content). This liquid mixture is then accumulated in a crude product storage vessel.

The residual gas stream is comprised largely of hydrogen chloride but contains small amounts of less highly chlorinated materials. This is subjected to adiabatic absorption of HCl using either water or azeotropic (20%) hydrochloric acid, whereby technical grade 31% hydrochloric acid is produced. Alternatively, dry hydrogen chloride can be withdrawn prior to the absorption step, which makes it available for use in other processes which consume hydrogen chloride (e.g., methanol hydrochlorination). The steam which arises during the adiabatic absorption is withdrawn from the head of the absorption column and condensed in a quench system. The majority of the chloromethanes contained in this outflow can be separated by subsequent cooling and phase separation. Wastewater exiting from the quench system is directed to a stripping column where it is purified prior to being discarded. Residual off-gas is largely freed from remaining traces of halogen compounds by low-temperature cooling and are subsequently passed through an off-gas purification system (activated charcoal) before being released into the atmosphere, by which point the gas consists mainly of nitrogen along with traces of methane.

The liquids which have been collected in the crude product containment vessel are freed of gaseous components — Cl_2, HCl, CH_3Cl — by passage through a degassing/dehydrating column, traces of water being removed by distillation. Volatile components are returned to the reaction system prior to HCl absorption. The crude product is then worked up to pure carbon tetrachloride in a multistage distillation facility. Foreruns (light ends) removed in the first column are returned to the appropriate stage of the reactor cascade. The residue in the final column (heavy ends), which constitutes 2–3 wt% of the tetrachloromethane production, is made up of hexachloroethane, tetrachloroethylene, trichloroethylene, etc. This material can be converted advantageously to tetrachloromethane in a high-pressure chlorinolysis unit.

Overall yields in the process are ca. 95% based on methane and > 98% based on chlorine.

Other Processes. Oxychlorination as a way of producing tetrachloromethane (as well as partially chlorinated compounds) has repeatedly been the subject of patent documents [90]–[92], particularly since it leads to complete utilization of chlorine without any HCl byproduct. Pilot-plant studies using fluidized-bed technology have not succeeded in solving the problem of the high rate of combustion of methane. On the other hand the Transcat process, a two-stage approach mentioned on p. 1271 and embodying fused copper salts, can be viewed more positively.

Direct chlorination of carbon to tetrachloromethane is thermodynamically possible at atmospheric pressure below 1100 K, but the rate of the reaction is very low because of the high activation energy (lattice energy of graphite). Sulfur compounds have been introduced as catalysts in these experiments. Charcoal can be chlorinated to tetrachloromethane in the absence of catalyst with a yield of 17% in one pass at 900 to 1100 K and 0.3–2.0 MPa (3–20 bar) pressure. None of these suggested processes has been successfully introduced on an industrial scale. A review of direct chlorination of carbon is found in [93].

In this context it is worth mentioning the dismutation of phosgene

$$2\ COCl_2 \longrightarrow CCl_4 + CO_2$$

another approach which avoids the formation of hydrogen chloride. This reaction has been studied by Hoechst [94] and occurs in the presence of 10 mol% tungsten hexachloride and activated charcoal at 370 to 430 °C and a pressure of 0.8 MPa. The process has not acquired commercial significance because the recovery of the WCl_6 is very expensive.

1.4. Quality Specifications

1.4.1. Purity of the Commercial Products and Their Stabilization

The standard commercial grades of all of the chloromethanes are distinguished by their high purity (> 99.9 wt%). Dichloromethane, the solvent with the broadest spectrum of applications, is also distributed in an especially pure form (> 99.99 wt%) for such special applications as the extraction of natural products.

Monochloromethane and tetrachloromethane do not require the presence of any stabilizer. Dichloromethane and trichloromethane, on the other hand, are normally protected from adverse influences of air and moisture by the addition of small amounts of efficient stabilizers. The following substances in the listed concentration ranges are the preferred additives:

Table 6. Analytical testing methods for chloromethanes

Parameter	Method	
	DIN	ASTM
Boiling range	51 751	D 1078
Density	51 757	D 2111
Refraction index	53 491	D 1218
Evaporation residue	53 172	D 2109
Color index (Hazen)	53 409	D 1209
Water content (K. Fischer)	51 777	D 1744
pH value in aqueous extract	–	D 2110

Ethanol	0.1 – 0.2 wt%
Methanol	0.1 – 0.2 wt%
Cyclohexane	0.01 – 0.03 wt%
Amylene	0.001 – 0.01 wt%

Other substances have also been described as being effective stabilizers, including phenols, amines, nitroalkanes, aliphatic and cyclic ethers, epoxides, esters, and nitriles.

Trichloromethane of a quality corresponding to that specified in the Deutsche Arzneibuch, 8th edition (D.A.B. 8), is stabilized with 0.6 – 1 wt% ethanol, the same specifications as appear in the British Pharmacopoeia (B.P. 80). Trichloromethane is no longer included as a substance in the U.S. Pharmacopoeia, it being listed only in the reagent index and there without any specifications.

1.4.2. Analysis

Table 6 lists those classical methods for testing the purity and identity of the chloromethanes that are most important to both producers and consumers. Since the majority of these are methods with universal applicability, the corresponding Deutsche Industrie Norm (DIN) and American Society for the Testing of Materials (ASTM) recommendations are also cited in the Table.

Apart from these test methods, gas chromatography is also employed for quality control both in the production and shipment of chloromethanes. Gas chromatography is especially applicable to chloromethanes due to their low boiling point. Even a relatively simple chromatograph equipped only with a thermal conductivity (TC) detector can be highly effective at detecting impurities, usually with a sensitivity limit of a few parts per million (mg/kg).

1.5. Storage, Transport, and Handling

Dry *monochloromethane* is inert with respect to most metals, thus permitting their presence during its handling. Exceptions to this generalization, however, are aluminum, zinc, and magnesium, as well as their alloys, rendering these unsuitable for use. Thus most vessels for the storage and transport of monochloromethane are preferentially constructed of iron and steel.

Since it is normally handled as a compressed gas, monochloromethane must, in the Federal Republic of Germany, be stored in accord with Accident Prevention Regulation (Unfallverhütungsvorschrift, UVV) numbers 61 and 62 bearing the title "Gases Which Are Compressed, Liquified, or Dissolved Under Pressure" ("Verdichtete, verflüssigte, oder unter Druck gelöste Gase") and issued by the Trade Federation of the Chemical Industry (Verband der Berufsgenossenschaften der chemischen Industrie). Additional guidelines are provided by general regulations governing high-pressure storage containers. Stored quantities in excess of 500 t also fall within the jurisdiction of the Emergency Regulations (Störfallverordnung) of the German Federal law governing emission protection.

Gas cylinders with a capacity of 40, 60, 300, or 700 kg are suitable for the transport of smaller quantities of monochloromethane. Shut-off valves on such cylinders must be left-threaded. Larger quantities are shipped in containers, railroad tank cars, and tank trucks, these generally being licensed for a working pressure of 1.3 MPa (13 bar).

The three *liquid chloromethanes* are also normally stored and transported in vessels constructed of iron or steel. The most suitable material for use with products of very high purity is stainless steel (material no. 1.4 571). The use in storage and transport vessels of aluminum and other light metals or their alloys is prevented by virtue of their reactivity with respect to the chloromethanes.

Storage vessels must be protected against the incursion of moisture. This can be accomplished by incorporating in their pressure release systems containers filled with drying agents such as silica gel, aluminum oxide, or calcium chloride. Alternatively, the liquids can be stored under a dry, inert gas. Because of its very low boiling point, dichloromethane is sometimes stored in containers provided either with external water cooling or with internal cooling units installed in their pressure release systems.

Strict specifications with respect to safety considerations are applied to the storage and transfer of chlorinated hydrocarbons in order to prevent spillage and overfilling. Illustrative is the document entitled "Rules Governing Facilities for the Storage, Transfer, and Preparation for Shipment of Materials Hazardous to Water Supplies" ("Verordnung für Anlagen zum Lagern, Abfüllen und Umschlagen wassergefährdender Stoffe", VAwS). Facilities for this purpose must be equipped with the means for safely recovering and disposing of any material which escapes [104].

Shipment of solvents normally entails the use of one-way containers (drums, barrels) made of steel and if necessary coated with protective paint. Where product quality

Table 7. Identification number and hazard symbols of chloromethanes

Product	Identification number	Hazard symbol
Monochloromethane	UN 1063	H (harmful)
		IG (inflammable gas)
Dichloromethane	UN 1593	H (harmful)
Trichloromethane	UN 1888	H (harmful)
Tetrachloromethane	UN 1846	P (poison)

standards are unusually high, especially as regards minimal residue on evaporation, stainless steel is the material of choice.

Larger quantities are shipped in containers, railroad tank cars, tank trucks, and tankers of both the transoceanic and inland-waterway variety. So that product specifications may be met for material long in transit, it is important during initial transfer to ensure high standards of purity and the absence of moisture.

Rules for transport by all of the various standard modes have been established on an international basis in the form of the following agreements: RID, ADR, GGVSee, GGVBinSch, IATA-DGR. The appropriate identification numbers and warning symbols for labeling as hazardous substances are collected in Table 7.

The use and handling of chloromethanes — both by producers and by consumers of the substances and mixtures containing them — are governed in the Federal Republic of Germany by regulations collected in the February 11, 1982 version of the "Rules Respecting Working Materials" ("Arbeitsstoff-Verordnung"). To some extent, at least, these have their analogy in other European countries as well. Included are stipulations regarding the labeling of the pure substances themselves as well as of preparations containing chloromethane solvents. The central authorities of the various industrial trade organizations issue informational and safety brochures for chlorinated hydrocarbons, and these should be studied with care.

The standard guidelines for handling monochloromethane as a compressed gas are the "Pressure Vessel Regulation" ("Druckbehälter-Verordnung") of February 27, 1980, with the related "Technical Rules for Gases" ("Technische Regeln Gase", TRG) and the "Technical Rules for Containers" ("Technische Regeln Behälter", TRB), as well as "Accident Prevention Guideline 29 — Gases" ("Unfallverhütungsvorschrift [UVV] 29, Gase").

For MAK values, TLV values, and considerations concerning the toxicology see Chap. 10. The ecology and the ecotoxicology of the chloromethanes are described in Chapter 10.1.5.

Table 8. Atmospheric concentration of chloromethanes (in 10^{-10} vol.%)

Compound	Continents	Oceans	Urban areas
CH_3Cl	530 ... 1040	1140 ... 1260	834
CH_2Cl_2	36	35	<20 ... 144
$CHCl_3$	9 ... 25	8 ... 40	6 ... 15 000
CCl_4	20 ... 133	111 ... 128	120 ... 18 000

1.6. Behavior of Chloromethanes in the Environment

Chloromethanes are introduced into the environment from both natural and anthropogenic sources. They are found in the lower atmosphere, and tetrachloromethane can even reach into the stratosphere. Trichloromethane and tetrachloromethane can be detected in many water supplies.

The chloromethanes, like other halogenated hydrocarbons, are viewed as water contaminants. Thus, they are found in both national and international guidelines related to water quality protection [95], [96].

There are fundamental reasons for needing to restrict chlorocarbon emissions to an absolute minimum. Proven methods for removal of chloromethanes from wastewater, off-gas, and residues are

Vapor stripping with recycling
Adsorption on activated charcoal and recycling
Recovery by distillation
Reintroduction into chlorination processes [97]
Combustion in facilities equipped with offgas cleanup

1.6.1. Presence in the Atmosphere

All four chloromethanes are emitted to the atmosphere from anthropogenic sources. In addition, large quantities of monochloromethane are released into the atmosphere by the combustion of plant residues and through the action of sunlight on algae in the oceans. Estimates of the extent of nonindustrial generation of monochloromethane range from 5×10^6 t/a [98] to 28×10^6 t/a [99].

Natural sources have also been considered for trichloromethane [100] and tetrachloromethane [101] on the basis of concentration measurements in the air and in seawater (Table 8).

The emission of chloromethanes from industry is the subject of legal restrictions in many countries. The applicable regulations in the Federal Republic of Germany are those of the TA Luft [102].

The most important sink for many volatile organic compounds is their reaction in the lower atmosphere with photochemically generated OH radicals. The reactivity of

Table 9. Velocity of decomposition of chloromethanes in the atmosphere

Compound	Reaction velocity with OH radicals $k_{OH} \times 10^{12}$ cm^3 molecule^{-1} s^{-1}	Half-life, weeks
CH$_3$Cl	0.14	12
CH$_2$Cl$_2$	0.1	15
CHCl$_3$	0.1	15
CCl$_4$	<0.001	>1000

Table 10. Chloromethane concentration in the Rhine river (µg/L)

Compound	Date	Mean value	Max. value
CH$_2$Cl$_2$	1980	not detected	
CHCl$_3$	1980	4.5	
CHCl$_3$	1982	0.4 ... 12.5	50.0
CCl$_4$	1982	<0.1 ... 3.3	44.4

monochloromethane, dichloromethane, and trichloromethane with OH radicals is so high that in the troposphere these substances are relatively rapidly destroyed.

By contrast, the residence time in the troposphere of tetrachloromethane is very long, with the result that it can pass into the stratosphere, where it is subjected to photolysis from hard ultraviolet radiation. The Cl atoms released in this process play a role in the ozone degradation which is presumed to occur in the stratosphere (Table 9) [103].

1.6.2. Presence in Water Sources

Seawater has been found to contain relatively high concentrations of monochloromethane (5.9 – 21 × 10^{-9} mL of gas/mL of water) [99], in addition to both trichloromethane (8.3 – 14 × 10^{-9} g/L) [100] and tetrachloromethane (0.17 – 0.72 × 10^{-9} g/L) [100]. Dichloromethane, on the other hand, could not be detected [98].

Chloromethanes can penetrate both surface and groundwater through the occurrence of accidents or as a result of improper handling during production, transportation, storage, or use (Table 10). Groundwater contamination by rain which has washed chlorinated hydrocarbons out of the air is not thought to be significant on the basis of current knowledge. One frequent additional cause of diffuse groundwater contamination that can be cited is defective equipment (especially leaky tanks and wastewater lines) [104].

The chloromethanes are relatively resistant to hydrolysis. Only in the case of monochloromethane in seawater is abiotic degradation of significance, this compound being subject in weakly alkaline medium to cleavage with the elimination of HCl.

The microbiological degradability of dichloromethane has recently been established [107]–[113]. This is understood to be the reason for the absence or only very low concentrations of dichloromethane in the aquatic environment [104].

Since trichloromethane and tetrachloromethane are stable compounds with respect to both biotic and abiotic processes, their disappearance is thought to be largely a consequence of transfer into the atmosphere by natural stripping phenomena.

Treatment with chlorine is a widespread technique for disinfecting drinking water. In the process, trihalomethanes result, largely trichloromethane as a result of the reaction of chlorine with traces of organic material. A level of 25 µg/L of trihalomethanes is regarded in the Federal Republic of Germany as the maximum acceptable annual median concentration in drinking water [114].

1.7. Applications of the Chloromethanes and Economic Data

As a result of very incomplete statistical records detailing production and foreign trade by individual countries, it is very difficult to describe precisely the world market for chloromethanes. The information which follows is based largely on systematic evaluation of the estimates of experts, coupled with data found in the secondary literature, as well as personal investigations and calculations.

The Western World includes about 40 producers who produce at least one of the chlorinated C_1 hydrocarbons. No authoritative information is available concerning either the production capacity or the extent of its utilization in the Comecon nations or in the People's Republic of China. It can be assumed, however, that a large part of the domestic requirements in these countries is met by imports. In reference to production capacity, see [115].

In comparing the reported individual capacities it is important to realize that a great many facilities are also capable of producing other chlorinated hydrocarbons. This situation is a result of the opportunities for flexibility both in the product spectrum (cf. Sect. 1.3.1) and in the various manufacturing techniques (e.g., tetrachloromethane/tetrachloroethene, cf. Sect. 1.3.4). If one ignores captive use for further chlorination (especially of monochloromethane), it can be concluded that the largest portion of the world use of chloromethanes (ca. 34%) can be attributed to *tetrachloromethane*. The most important market, accounting for over 90% of the material produced, is that associated with the production of the fluorochlorocarbons CFC 11 (trichloromonofluoromethane) and CFC 12 (dichlorodifluoromethane). These fluorochlorocarbons possess outstanding properties, such as nonflammability and toxicological safety, and are employed as refrigerants, foaming agents, aerosol propellants, and special solvents.

The production level of tetrachloromethane is directly determined by the market for its fluorinated reaction products CFC 11 and CFC 12. The appearance of the so-called ozone theory, which asserts that the ozone layer in the stratosphere is affected by these compounds, has resulted since 1976 in a trend toward reduced production of tetrachloromethane. This has been especially true since certain countries (United States, Canada, Sweden) have imposed a ban on aerosol use of fully halogenated fluorochlor-

Table 11. Production capacities of chloromethanes (1981) 1000 t/year

	Western Europe (FRG)	United States	Japan
Monochloromethane	265 (100)	300	70
Dichloromethane	410 (170)	370	65
Trichloromethane	140 (60)	210	65
Tetrachloromethane	250 (150)	380	70

ocarbons. However, since 1982/1983 there has been a weak recovery in demand for tetrachloromethane in the production of fluorine-containing compounds.

Outside Europe, a smaller amount of tetrachloromethane finds use as a disinfectant and as a fungicide for grain.

Monochloromethane and *dichloromethane* each account for about 25% of the world market for chloromethanes (Table 11). The demand for monochloromethane can be attributed largely (60–80%) to the production of silicones. Its use as a starting material for the production of the gasoline anti-knock additive tetramethyllead is in steep decline.

The most important use of dichloromethane, representing ca. 40–45% of the total market, is as a cleaning agent and paint remover. An additional 20–25% finds application as a pressure mediator in aerosols. One further use of dichloromethane is in extraction technology (decaffeination of coffee, extraction of hops, paraffin extraction, and the recovery of specialty pharmaceuticals).

In all of these applications, especially those related to the food and drug industries, the purity level requirements for dichloromethane are exceedingly high (> 99.99 wt%).

Trichloromethane holds the smallest market share of the chloromethane family: 16%. Its principal application, amounting to more than 90% of the total production, is in the production of monochlorodifluoromethane (CFC 22), a compound important on the one hand as a refrigerant, but also a key intermediate in the preparation of tetrafluoroethene. The latter can be polymerized to give materials with exceptional thermal and chemical properties, including PTFE, Hostaflon, Teflon, etc.

Chloroform is still used to a limited extent as an extractant for pharmaceutical products. Due to its toxicological properties, its use as an inhalatory anaesthetic is no longer significant. Small amounts are employed in the synthesis of orthoformic esters.

Table 12 provides an overview of the structure of the markets for the various chlorinated C_1 compounds, subdivided according to region.

Table 12. Demand and use pattern of chloromethanes (1983)

	Western Europe	United States	Japan
Monochloromethane	230 000 t	250 000 t	50 000 t
Silicone	52%	60%	83%
Tetramethyllead	12%	15%	–
Methylcellulose	15%	5%	1%
Other methylation reactions, e.g., tensides, pharmaceuticals	ca. 21%	ca. 20%	ca. 16%
Dichloromethane	210 000 t	270 000 t	35 000 t
Degreasing and paint remover	46%	47%	54%
Aerosols	18%	24%	19%
Foam-blowing agent	9%	4%	11%
Extraction and other uses	27%	25%	16%
Trichloromethane	90 000 t	190 000 t	45 000 t
CFC 22 production	78%	90%	90%
Other uses, e.g., pharmaceuticals, intermediate	22%	10%	10%
Tetrachloromethane	250 000 t	250 000 t	75 000 t
CFC 11/12 production	94%	92%	90%
Special solvent for chemical reactions	6%	8%	10%

2. Chloroethanes

The class of chloroethanes comprises:

Monochloroethane (ethyl chloride)
1,1-Dichloroethane
1,2-Dichloroethane (ethylene dichloride, EDC)
1,1,1-Trichloroethane
1,1,2-Trichloroethane
1,1,1,2-Tetrachloroethane (asymmetric Tetra)
1,1,2,2-Tetrachloroethane (symmetric Tetra)
Pentachloroethane
Hexachloroethane

As with other chlorinated hydrocarbons, a common characteristic of these compounds is their low solubility in water and excellent solubility in most organic solvents, their volatility with water vapor, and their tendency to form azeotropic mixtures.

Boiling point, density, viscosity, and surface tension increase with increasing chlorine substitution, whereas heat of formation, solubility, and inflammability decrease. Substitution of vicinal positions generally has a greater impact than geminal substitution, as shown in Table 13. With the exception of gaseous monochloroethane (ethyl chloride) and solid hexachloroethane, the chloroethanes are colorless liquids at ambient conditions. Figure 11 shows the vapor pressure as a function of temperature. If light and

Table 13. Physical properties of chlorinated ethanes

Compound	Boiling point (at 101 kPa), °C	Relative density, d_4^{20}
Monochloroethane	12.3	0.9240
1,1-Dichloroethane	57.3	1.1760
1,2-Dichloroethane	83.7	1.2349
1,1,1-Trichloroethane	74.1	1.3290
1,1,2-Trichloroethane	113.5	1.4432
1,1,1,2-Tetrachloroethane	130.5	1.5468
1,1,2,2-Tetrachloroethane	146.5	1.5958
Pentachloroethane	162.0	1.6780
Hexachloroethane	*mp* 186–187	2.0940

oxygen are absent, most chlorinated ethanes are fairly stable. At higher temperatures (> 300 °C), they are susceptible to the elimination of hydrogen chloride. In the presence of light and oxygen, oxidation occurs yielding phosgene, carbon oxides, and acetyl or chloroacetyl chlorides. The latter easily hydrolyze with traces of moisture forming the corresponding chloroacetic acids, which are well-known as strongly corrosive agents. To prevent this unwanted decomposition, most industrially used chlorinated hydrocarbons are stabilized with acid acceptors such as amines, unsaturated hydrocarbons, ethers, epoxides or phenols, antioxidants, and other compounds able to inhibit free radical chain reactions. Longer storage periods and use without appreciable effect on tanks and equipment is then possible.

Of all chlorinated ethanes, approximately half are of industrial importance. Monochloroethane (ethyl chloride) is an intermediate in the production of tetraethyllead and is widely used as an ethylating agent. 1,2-Dichloroethane has by far the highest production rates. It is an intermediate for the production of 1,1,1-trichloroethane and vinyl chloride (see p. 1312 and Section 3.1.3.2), but it is also used in synthetic applications (e.g., polyfunctional amines) and as a fuel additive (lead scavenger).

1,1,1-Trichloroethane, trichloroethylene, (see Section 3.4) and tetrachloroethylene (see Section 3.5) are important solvents widely used in dry cleaning, degreasing, and extraction processes.

The other chlorinated ethanes have no important end uses. They are produced as intermediates (e.g., 1,1-dichloroethane) or are formed as unwanted byproducts. Their economical conversion into useful end products is achieved either by cracking — tetrachloroethanes yield trichloroethylene — or more commonly by chlorinolysis, which converts them into carbon tetrachloride and tetrachloroethylene (see p. 1364).

Basic feedstocks for the production of chlorinated ethanes and ethylenes (see Chap. 3) are ethane or ethylene and chlorine (Fig. 12). The availability of ethylene from naphtha feedstocks has shifted the production of chlorinated C_2 hydrocarbons during the past three decades in the Western World from the old carbide–acetylene–vinyl chloride route toward the ethylene route. With the dramatic increase of naphtha prices during the past decade, the old carbide route has regained some of its attractiveness [116]. Even though a change cannot be justified presently in most

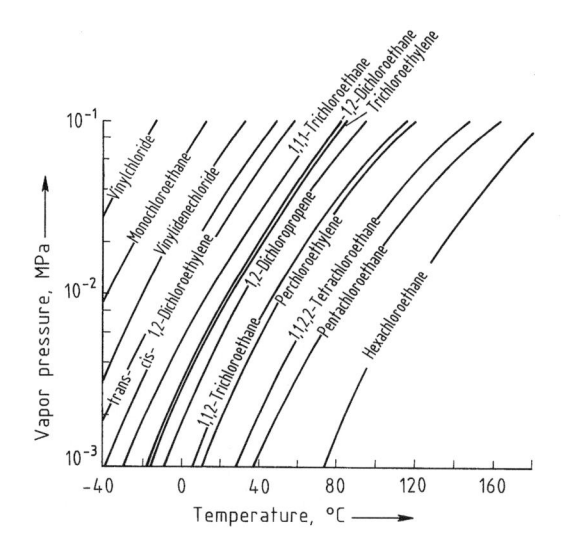

Figure 11. Vapor pressure as a function of temperature for chlorinated hydrocarbons

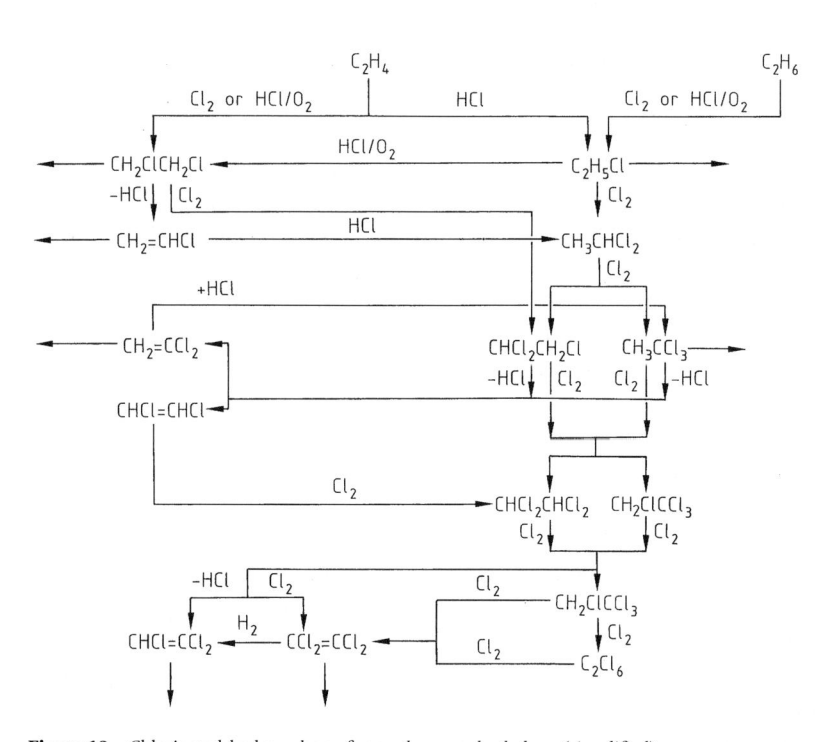

Figure 12. Chlorinated hydrocarbons from ethane and ethylene (simplified)

countries, it could offer an alternative for countries where cheap coal is readily available. The use of ethanol derived from biomass as a starting material could likewise also be considered [117], [118].

In a few cases, ethane is used directly as a hydrocarbon feedstock. This 'direct' ethane route could offer an attractive alternative in some cases, because of the substantial cost differences between ethane and ethylene. It becomes evident why numerous patents on ethane-based processes have been filed in recent years. However, the major cost advantage of such processes is the reduced capital investment for cracker capacity. The direct ethane route must certainly be considered for future grass-root-plants, but at present, the conversions and selectivities obtained seem not to justify the conversion of existing plants if cracker capacity is available.

Less is known about the situation in Eastern block countries. The available information indicates, however, that in some Eastern European countries the acetylene route is still used.

Because chlorine is needed as a second feedstock, most plants producing chlorinated hydrocarbons are connected to a chlor-alkali electrolysis unit. The hydrocarbon feedstock is either supplied from a nearby cracker, — typical for U.S. gulf coast, — or via pipelines and bulk ship transports. The chlorine value of the hydrogen chloride produced as a byproduct in most chlorination processes can be recovered by oxychlorination techniques, hydrochlorination reactions (for synthesis of methyl and ethyl chloride) or, — less economically — by aqueous HCl electrolysis. A minor but highly valuable outlet is ultrapure-grade anhydrous HCl used for etching in the electronic industry.

Although most unwanted byproducts can be used as feed for the chlorinolysis process [119] (see p. 1364), the byproducts of this process, mostly hexachloroethane, hexachlorobutadiene, and hexachlorobenzene together with residual tars from spent catalysts and vinyl chloride production, represent a major disposal problem. The optimal ecological solution is the incineration of these residues at a temperature above 1200 °C, which guarantees almost complete degradation. Presently, incineration is performed at sea on special ships [120] without HCl scrubbing or on site with subsequent HCl or chlorine recovery. The aqueous HCl recovered can then be used for pH adjustment in biological effluent treatment or brine electrolysis.

Due to their unique properties, the market for chlorinated C_2 hydrocarbons has shown excellent growth over the past 30 years and reached its maximum in the late 1970s. With increasing environmental consciousness, the production rate of some chlorinated hydrocarbons such as ethyl chloride, trichloroethylene (see p. 1359), and tetrachloroethylene (see 3.5.4) will in the long run decrease due to the use of unleaded gasoline, solvent recovery systems, and partial replacement by other solvent and extraction chemicals. However, new formulations for growing markets such as the electronic industry, the availability of ecologically safe handling systems, know-how in residue incineration, and the difficulty in finding superior replacements — causing fewer problems — guarantee chlorinated ethanes and ethylenes a long-term and at least constant market share.

2.1. Monochloroethane

Monochloroethane (ethyl chloride) [75-00-3] is thought to be the first synthesized chlorinated hydrocarbon. It was produced in 1440 by VALENTINE by reacting ethanol with hydrochloric acid. GLAUBER obtained it in 1648 by reacting ethanol (spirit of wine) with zinc chloride. Because of the growing automotive industry in the early 1920s, monochloroethane became an important bulk chemical. Its use as a starting material for the production of tetraethyl-lead initiated a significant increase in ethyl chloride production and is still its major consumer. The trend toward unleaded gasoline in most countries, however, will in the long run lead to a significant decrease in production.

2.1.1. Physical Properties

M_r	64.52
mp	− 138.3 °C
bp at 101.3 kPa	12.3 °C
ϱ of the liquid at 0 °C	0.924 g/cm^3
ϱ of the vapor at 20 °C	2.76 kg/m^3
n_D^{20}	1.3798
Vapor pressure at	
− 50 °C	4.480 kPa
− 20 °C	25.090 kPa
− 10 °C	40.350 kPa
0 °C	62.330 kPa
+ 10 °C	92.940 kPa
+ 20 °C	134.200 kPa
+ 30 °C	188.700 kPa
+ 60 °C	456.660 kPa
+ 80 °C	761.100 kPa
Heat of formation (liquid) $\Delta H°_{298}$	− 133.94 kJ/mol
Specific heat at 0 °C	1.57 kJ kg^{-1} K^{-1}
Heat of evaporation at 298 K	24.7 kJ/mol
Critical temperature	456 K
Critical pressure	5270 kPa
Viscosity (liquid, 10 °C)	2.79×10^{-4} Pa s
Viscosity (vapor, bp)	9.3×10^{-5} Pa s
Thermal conductivity (vapor)	1.09×10^{-3} W m^{-1} K^{-1}
Surface tension (air, 5 °C)	21.18×10^{-3} N/m
Dielectric constant (vapor, 23.5 °C)	1.0129
Flash point (open cup)	− 43 °C
Ignition temperature	519 °C
Explosive limits in air	3.16 – 15 vol % monochloroethane
Solubility in water at 0 °C	0.455 wt %
Solubility of water in monochloroethane at 0 °C	0.07 wt %

At ambient temperature, monochloroethane is a gas with an etheral odor.
Monochloroethane burns with a green-edged flame.
Combustion products are hydrogen chloride, carbon dioxide, and water.

Binary azeotropic mixtures of monochloroethane have been reported [121]. The data, however, have not been validated.

2.1.2. Chemical Properties

Monochloroethane has considerable thermal stability. Only at temperatures above 400 °C, considerable amounts of ethylene and hydrogen chloride are formed due to dehydrochlorination [121a]. This decomposition can be catalyzed by a variety of transition metals (e.g. Pt), transition-metal salts, and high-surface area oxides such as alumina and silica. Catalyzed decomposition is complete at temperatures slightly above 300 °C according to the thermodynamic equilibrium.

At ambient atmospheric conditions, both, hydrolysis (to ethanol) and oxidation (to acetaldehyde) are moderate.

At temperatures up to 100 °C, monochloroethane shows no detrimental effect on most structural materials if kept dry. Contact with aluminum, however, should be avoided under all circumstances for safety reasons.

Monochloroethane has the highest reactivity of all chlorinated ethanes. It is mainly used as an ethylating agent in Grignard- and Friedel-Crafts-type reactions, for ether, thioether, and amine synthesis. Halogene exchange [121b] and fluorination is also possible [121c].

2.1.3. Production

Monochloroethane can be produced by a variety of reactions. Only two are of industrial importance: the hydrochlorination of ethylene and the thermal chlorination of ethane.

Hydrochlorination of Ethylene. Exothermic hydrochlorination of ethylene can be carried out in either the liquid or gas phase.

$$C_2H_4 + HCl \longrightarrow C_2H_5-Cl \qquad \Delta H = -98 \text{ kJ/mol}$$

The liquid-phase reaction is carried out mostly at near ambient temperatures (10–50 °C) and moderate pressure (0.1–0.5 MPa) in a boiling-bed type reactor. The heat of reaction is used to vaporize part of the monochloroethane formed, which in turn is then cooled down, purified, or partially recycled. The reactor temperature is controlled by the recycle ratio and the feed rate of the reactants. Unconverted ethylene and hydrogen chloride from reflux condensers and overhead light end columns are recycled back to the reactor. Sufficient mixing and catalyst contact time is achieved through recirculation of the reactor sump phase. Aluminum chloride in a 0.5–5 wt% concentration is mostly used as a catalyst. A part of it is continuously or intermittently removed via a recirculation slip stream, together with unwanted high boiling impurities

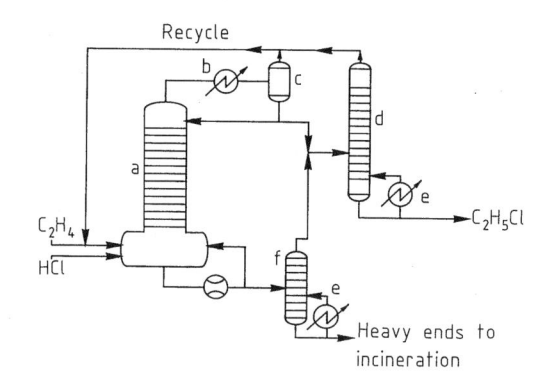

Figure 13. Schematic diagram (simplified) of an ethylene hydrochlorination process
a) Reactor; b) Cooler; c) Knock-out drum; d) Light-end columns; e) Reboiler; f) Stripper column (heavy ends)

consisting mostly of low molecular mass ethylene oligomers formed in a Ziegler-type reaction of the catalyst with the ethylene feed. New catalyst is added to the system either by a hopper as a solid or preferably as a solution after premixing with monochloroethane or monochloroethane/ethylene. A gaseous feed of vaporized $AlCl_3$ has also been suggested [122]. A simplified process diagram is shown in Figure 13; an optimized process has been patented [123]. In other process variations, the formed monochloroethane (sump phase) is washed with diluted NaOH to remove catalyst and acid and then dried and distilled. Excess ethylene is recycled.

Ethylene and HCl yields for hydrochlorination are almost quantitative; selectivities of 98–99% have been reported. In addition to $AlCl_3$, other Lewis-acid catalysts, such as $FeCl_3$ [124], $BiCl_3$ [125], and $GaCl_3$ [126], have been patented. Suggestions to perform the reaction in benzene or higher boiling hydrocarbons [127], in 1,1,2-trichloroethane [128] or to complex $AlCl_3$ by nitrobenzene [129] have not found industrial acceptance.

The troublesome handling of the catalyst is minimized when ethylene and hydrogen chloride are reacted in the gas phase. Although the reaction equilibrium becomes unfavorable at a temperature above 200 °C, the process is carried out at temperatures of 250–450 °C in order to achieve sufficient conversion. Ethylene and HCl are preheated, mixed, and sent across the catalyst, which can be used as fixed or fluidized bed. The chloroethane formed is separated and purified. Unreacted ethylene and HCl are recycled. Selectivities are comparable to those of the liquid-phase process, conversion per pass, however, may not exeed 50%, so that relatively high recycle rates are necessary. Because high pressure favors the formation of monochloroethane, the reaction is preferably carried out at 0.5–1.5 MPa.

Thorium oxychloride on silica [130], platinum on alumina [131], and rare-earth oxides on alumina and silica [132] have been patented as catalysts.

Chlorination of Ethane. Thermal chlorination of ethane for the production of monochloroethane can be used industrially in a tandem process developed by the Shell Oil Company (Fig. 14) [133]. This process was especially designed for a plant in which sufficient ethylene feedstock could only be supplied by increasing the cracker capacity. Ethane and chlorine were available, but not hydrogen chloride. For this feedstock constellation, the tandem process seems advantageous.

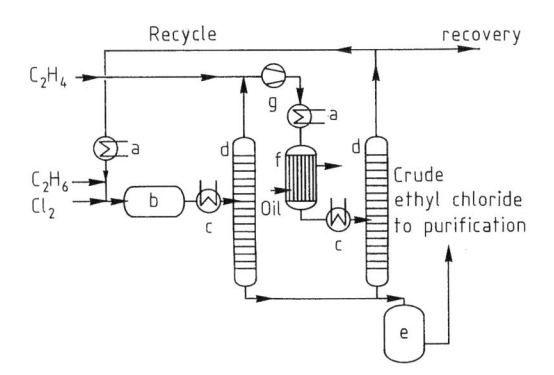

Figure 14. Production of monochloroethane by the Shell process [133]
a) Preheater; b) Ethane chlorinator; c) Cooler; d) Lightend tower; e) Crude chloroethane storage; f) Hydrochlorinator; g) Compressor

In the first stage, ethane and chlorine are reacted noncatalytically after sufficient preheating at 400–450 °C in an adiabatic reactor. The reaction gases are separated after cooling in a first monochloroethane distillation tower. The heavy bottoms of this tower containing chloroethane and more higly chlorinated products (mostly 1,1-dichloroethane and 1,2-dichloroethane) are sent to the purification stage. The overheads consisting mainly of unconverted ethane, hydrogen chloride, and ethylene are sent to a second isothermal fixed-bed reactor. Before entering this reactor, fresh ethylene is added to achieve a 1:1 ethylene to HCl feed ratio. Even though the type of catalyst used in the isothermal section is not described, any of the catalysts mentioned for gas-phase hydrochlorination in the previous section can be used. Conversion at this stage is 50–80 %. The products are then separated in a second tower. Unconverted ethane, ethylene, and hydrogen chloride are recycled to the first reactor. The monochloroethane formed by hydrochlorination is drawn off and purified together with the stream from the first tower.

Even though the recycled ethylene from the hydrochlorination step is present during thermal chlorination, the formation of 1,2-dichloroethane is insignificant. Because the first reaction is carried out at high temperatures, chlorine addition to the ethylene double bond is suppressed.

The process is balanced by the overall reaction equation:

$C_2H_6 + Cl_2 \longrightarrow C_2H_5Cl + HCl$
$HCl + C_2H_4 \longrightarrow C_2H_5Cl$

A 90 % overall yield for ethane and ethylene and a 95 % chlorine yield to monochloroethane are reported.

Monochlorination of ethane is favored because ethane chlorination is four times faster than the consecutive chlorination of monochloroethane to dichloroethanes.

Major byproducts from the chlorination step are 1,1-, 1,2-dichloroethane and vinyl chloride. To achieve a high selectivity for monochloroethane, a high ethane surplus — preferably a 3–5-fold excess over chlorine [134], [135] — and good mixing is required. Insufficient heat dissipation may enhance cracking and coking. A thermal chlorination reactor providing thorough premixing and optimal heat transfer by means of a fluidized bed has been described in [135]. Other patents claim contact of the reaction gases with metal chlorides [136] or graphite [137].

The photochemical chlorination of ethane described in several patents [138] is less important, because it is difficult to implement in large volume plants and offers no major advantages over the thermal process.

Monochloroethane as a Byproduct of the Oxy-EDC Process. Monochloroethane is a major byproduct in the Oxy-EDC process (see p. 1300), in which it is formed by direct hydrochlorination of ethylene. It can be condensed or scrubbed from the light vent gases and recovered after further purification.

Monochloroethane from Ethanol. The esterification of ethanol with HCl is possible in the liquid phase by using $ZnCl_2$ or similar Lewis-acid catalysts at 110–140 °C [139]. Similar to the production of monochloromethane (see 1.3.2), the reaction can also be carried out in the gas phase by using γ-alumina [140], $ZnCl_2$ and rare earth chlorides on carbon [141] or zeolites [142] as catalysts. At the present ethanol prices, these procedures are prohibitive. With some modification, however, they can offer outlets for surplus byproducts such as ethyl acetate from PVA production which can be converted to monochloroethane by HCl using a $ZnCl_2$/silica catalyst [143].

Other Synthetic Routes to Monochloroethane. Non-commercial routes to monochloroethane consist of electrolytic chlorination of ethane in melts [144], reactions with diethyl sulfates [145], metathesis of 1,2-dichloroethane [146], hydrogenation of vinyl chloride [147], and conversion of diethyl ether [148]. The oxychlorination of ethane is discussed later in this Chapter.

Small amounts of monochloroethane are formed during the reaction of synthesis gas–chlorine mixtures over Pt/alumina [149] and methane–chlorine mixtures in the presence of cation-exchange resins complexed with TaF_5 [150].

2.1.4. Uses and Economic Aspects

Monochloroethane became industrially significant as a result of the developing automotive industry. It is the starting material for tetraethyllead, the most commonly used octane booster. In the United States, about 80–90% and in Europe ca. 60% of the monochloroethane production is used for the production of tetraethyl lead.

Production has already been cut significantly due to the increased use of unleaded fuel for environmental reasons. U.S. projections indicate an average annual decline of ca. 10% per year. With some delay, the same trend can also be predicted for Western Europe.

Minor areas of use for monochloroethane are the production of ethyl cellulose, ethylating processes for fine chemical production, use as a blowing agent and solvent for extraction processes for the isolation of sensitive natural fragrances.

Production in 1984 in the Western World was about 300 000 t. Almost all processes in use at present are ethylene based.

2.2. 1,1-Dichloroethane

1,1-Dichloroethane [75-34-3] is the less important of the two dichloroethane isomers. It occurs — often as an unwanted byproduct — in many chlorination and oxychlorination processes of C_2 hydrocarbons.

The most important role of 1,1-dichloroethane is as an intermediate in the production of 1,1,1-trichloroethane.

Other uses are negligible.

2.2.1. Physical Properties

M_r	98.97
mp	– 96.6 °C
bp at 101.3 kPa	57.3 °C
ϱ at 20 °C	1.176 g/cm³
n_D^{20}	1.4164
Vapor pressure at	
0 °C	9.340 kPa
10 °C	15.370 kPa
20 °C	24.270 kPa
30 °C	36.950 kPa
Heat of formation (liquid) $\Delta H°_{298}$	– 160.0 kJ/mol
Specific heat at 20 °C	1.38 kJ kg⁻¹ K⁻¹
Heat of evaporation at 298 K	30.8 kJ/mol
Critical temperature	523 K
Critical pressure	5070 kPa
Viscosity at 20 °C	0.38×10^{-3} Pa s
Surface tension at 20 °C	23.5×10^{-3} N/m
Dielectric constant at 20 °C	10.9
Flash point (closed cup)	– 12 °C
Ignition temperature	458 °C
Explosive limits in air at 25 °C	5.4 – 11.4 vol% 1,1-dichloroethane
Solubility in water at 20 °C	0.55 wt%
Solubility of water in 1,1-dichloroethane at 20 °C	0.97 wt%

1,1-Dichloroethane is a colorless liquid. It is readily soluble in all liquid chlorinated hydrocarbons and in a large variety of other organic solvents (ethers, alcohols).

Binary azeotropes are formed with water and ethanol: with 1.9% water, bp 53.3 °C (97 kPa) and with 11.5% ethanol, bp 54.6 °C (101 kPa).

2.2.2. Chemical Properties

At room temperature, 1,1-dichloroethane is adequately stable. Cracking to vinyl chloride and hydrogen chloride takes place at elevated temperatures. However, compared to other chlorinated C_2 hydrocarbons, the observed cracking rates are moderate. This reaction can be promoted by traces of chlorine and iron [151]. 2,3-Dichlorobutane is often found as a dimeric byproduct of decomposition.

1,1-Dichloroethane was also found to enhance 1,2-dichloroethane cracking when added in lower concentrations ($\leq$ 10 wt%) [152].

Corrosion rates for dry 1,1-dichloroethane are marginal, increase however, with water content and temperature. Aluminum is easily attacked.

In the presence of water or in alkaline solution, acetaldehyde is formed by hydrolysis.

2.2.3. Production

Theoretically 1,1-dichloroethane can be produced by three routes:

1) Addition of HCl to acetylene:

$$2\,HCl + C_2H_2 \xrightarrow{Cat.} CH_3-CHCl_2$$

2) Thermal or photochemical chlorination of monochloroethane:

$$C_2H_5-Cl + Cl_2 \xrightarrow[\text{or heat}]{h\nu} CH_3-CHCl_2 + HCl$$

3) Addition of HCl to vinyl chloride:

$$CH_2=CHCl + HCl \xrightarrow{Cat.} CH_3-CHCl_2$$

For the synthesis of 1,1-dichloroethane as an intermediate in the production of 1,1,1-trichloroethane only the latter route is important and industrially used.

1,1-Dichloroethane via the 1,2-Dichloroethane – Vinyl Chloride Route. Hydrogen chloride and vinyl chloride obtained from 1,2-dichloroethane cracking see p. 1335) are reacted in a boiling-bed-type reactor [153] in the presence of a Friedel-Crafts catalyst, preferably ferric chloride ($FeCl_3$). 1,1-Dichloroethane is used as solvent and the temperature ranges from 30 to 70 °C.

Depending on the process design, hydrogen chloride can be used in excess to achieve complete conversion of the vinyl chloride. The heat of reaction, which differs only slightly from the heat required for 1,2-dichloroethane cracking, can be used to distill the 1,1-dichloroethane and recover part of the energy input. Downstream hydrogen chloride and unconverted vinyl chloride are separated and recycled. If necessary, the 1,1-dichloroethane can then be further purified by distillation. Due to the formation of heavy byproducts (vinyl chloride polymers) and deactivation of the catalyst, a slip-stream from the reactor bottom must be withdrawn and new catalyst added.

Improved processes use column-type reactors with optimized height [154] (hydrostatic pressure to avoid flashing of vinyl chloride!) and recycled 1,1-dichloroethane with intermittent cooling stages. In this case, the stoichiometric ratio of hydrogen chloride to vinyl chloride, as obtained from 1,2-dichloroethane cracking, can often be used. In such a process, the downstream distillation equipment can be less complex and expensive, because almost complete conversion is achieved and because no excess hydrogen chloride or the entrained vinyl chloride must be separated. However, the energy requirements may be higher because most of the heat of formation must be dissipated by cooling.

Both process variations yield between ca. 95 and 98%. Yield losses result through polymerization of vinyl chloride. The concentration as well as the nature of the catalyst determine this side reaction. Zinc chloride ($ZnCl_2$) and aluminum chloride ($AlCl_3$), which also can be used as catalysts, promote the formation of high molecular mass byproducts more than ferric chloride ($FeCl_3$) [129], [155]. The removed spent catalyst can be burned together with the heavy byproducts in an incinerator, if the vent gases are subsequently scrubbed and the wash liquor appropriately treated. Environmental problems caused by the residues are thereby almost eliminated.

1,1-Dichloroethane via the Acetylene Route. As with the synthesis of vinyl chloride (see 3.1.3.1), 1,1-dichloroethane can be produced from acetylene by adding 2 mol of hydrogen chloride. For the first reaction sequence — the formation of vinyl chloride — mercury catalyst is required [156].

Because ethylene has become the major feedstock for chlorinated C_2 hydrocarbons, this process has lost its importance.

1,1-Dichloroethane from Ethane. 1,1-Dichloroethane may also be obtained by ethane or chloroethane chlorination. This chlorination can be carried out as thermal chlorination [157], photochlorination, or oxychlorination [158]. These processes, however, are impaired by a lack of selectivity and are not used industrially.

2.2.4. Uses and Economic Aspects

As mentioned earlier, 1,1-dichloroethane is primarily used as a feedstock for the production of 1,1,1-trichloroethane.

Although several other applications have been patented [159], currently 1,1-dichloroethane is rarely used for extraction purposes or as a solvent.

Based on estimated production figures of 1,1,1-trichloroethane and disregarding other uses, the total Western World production of 1,1-dichloroethane is estimated at 200 000 – 250 000 t for 1985.

2.3. 1,2-Dichloroethane

The first synthesis of 1,2-dichloroethane (ethylene dichloride, EDC) [*107-06-2*] was achieved in 1795.

Presently, 1,2-dichloroethane belongs to those chemicals with the highest production rates. Average annual growth rates of > 10% were achieved during the past 20 years.

Although these growth rates declined during the past several years, in the long run 1,2-dichloroethane will maintain its leading position among the chlorinated organic chemicals due to its use as starting material for the production of poly (vinyl chloride).

2.3.1. Physical Properties

M_r	98.97
mp	− 35.3 °C
bp at 101.3 kPa	83.7 °C
ϱ at 20 °C	1.253 g/cm^3
n_D^{20}	1.4449
Vapor pressure at	
0 °C	3.330 kPa
20 °C	8.530 kPa
30 °C	13.300 kPa
50 °C	32.000 kPa
70 °C	66.650 kPa
80 °C	93.310 kPa
Heat of formation (liquid) $\Delta H°_{298}$	− 157.3 kJ/mol
Specific heat (liquid, at 20 °C)	1.288 kJ kg^{-1} K^{-1}
Heat of evaporation at 298 K	34.7 kJ/mol
Critical temperature	563 K
Critical pressure	5360 kPa
Viscosity at 20 °C	0.84 × 10^{-3} Pa s
Surface tension at 20 °C	31.4 × 10^{-3} N/m
Coefficient of cubical expansion (0 – 30 °C)	0.00116 K^{-1}
Dielectric constant	10.5
Flash point (closed cup)	17 °C
Flash point (open cup)	21 °C
Ignition temperature (air)	413 °C
Explosive limits in air at 25 °C	6.2 – 15.6 vol% 1,2-dichloroethane
Solubility in water at 20 °C	0.86 wt%
Solubility of water in 1,2-dichloroethane at 20 °C	0.16 wt%

1,2-Dichloroethane is a clear liquid at ambient temperature, which is readily soluble in all chlorinated hydrocarbons and in most common organic solvents.

Binary azeotropes with 1,2-dichloroethane are listed in Table 14.

Table 14. Binary azeotropes formed by 1,2-dichloroethane

wt%	Component	Azeotrope boiling point (101.3 kPa), °C
18.0	2-propen-1-ol	79.9
38.0	formic acid	77.4
37.0	ethanol	70.3
19.5	1,1-dichloroethane	72.0
43.5	2-propanol	74.7
32.0	methanol	61.0
19.0	1-propanol	80.7
79.0	tetrachloromethane	75.6
18.0	trichloroethylene	82.9
8.2	water	70.5

2.3.2. Chemical Properties

Pure 1,2-dichloroethane is sufficiently stable even at elevated temperatures and in the presence of iron. Above 340 °C, decomposition begins, yielding vinyl chloride, hydrogen chloride, and trace amounts of acetylene [121a], [160]. This decomposition is catalyzed by halogens and more highly substituted chlorinated hydrocarbons [161].

Long-term decomposition at ambient temperature caused by humidity and UV light can be suppressed by addition of stabilizers, mostly amine derivatives. Oxygen deficient burning and pyrolytic and photooxidative processes convert 1,2-dichloroethane to hydrogen chloride, carbon monoxide, and phosgene.

Both chlorine atoms of 1,2-dichloroethane can undergo nucleophilic substitution reactions, which opens routes to a variety of bifunctional compounds such as glycol (by hydrolysis or reaction with alkali), succinic acid dinitrile (by reaction with cyanide), or ethylene glycol diacetate (by reaction with sodium acetate). The reaction with ammonia to ethylenediamine and use of 1,2-dichloroethane for the production of polysulfides is of industrial importance.

Iron and zinc do not corrode when dry 1,2-dichloroethane is used, whereas aluminum shows strong dissolution. Increased water content leads to increased corrosion of iron and zinc; aluminum, however, corrodes less [162].

2.3.3. Production

1,2-Dichloroethane is industrially produced by chlorination of ethylene.

This chlorination can either be carried out by using chlorine (direct chlorination) or hydrogen chloride (oxychlorination) as a chlorinating agent.

In practice, both processes are carried out together and in parallel because most EDC plants are connected to vinyl chloride (VCM) units and the oxychlorination process is used to balance the hydrogen chloride from VCM production (see p. 1341 and Fig. 22). Depending on the EDC/VCM production ratio of the integrated plants, additional surplus hydrogen chloride from other processes such as chlorinolysis (perchloroethylene and tetrachloromethane production, see p. 1274 and Chap. 3.5.3) or 1,1,1-trichloroethane (see p. 1312) can be fed to the oxychlorination stage for proper balancing and chlorine recovery.

The use of ethane as a starting material, although the subject of numerous patent claims, is still in the experimental stage. It could offer economic advantages if the problems related to catalyst selectivity, turnover, and long-term performance are solved.

Direct Chlorination in the Ethylene Liquid Phase. In the direct chlorination process, ethylene and chlorine are most commonly reacted in the liquid phase (1,2-dichloroethane for temperature control) and in the presence of a Lewis acid catalyst, primarily iron(III) chloride:

$$CH_2 = CH_2 + Cl_2 \xrightarrow{Cat.} CH_2Cl-CH_2Cl$$
$$\Delta H^{\circ}_{298} = 220 \text{ kJ/mol}$$

To avoid problems in product purification, the use of high-purity ethylene is recommended. Especially its propane/propene content must be controlled in order to minimize the formation of chloropropanes and chloropropenes, which are difficult to separate from 1,2-dichloroethane by distillation. Purified liquid chlorine is used to avoid brominated byproducts. Oxygen or air is often added to the reactants, because oxygen was found to inhibit substitution chlorination, yielding particularly 1,1,2-trichloroethane and its more highly chlorinated derivatives [163], [164]. Through this and an optimized reactor design, the use of excess ethylene is no longer required to control byproduct formation. In most cases, the reactants are added in the stoichiometric chlorine/ethylene ratio or with a slight excess of chlorine. This simplifies the processing equipment because an excess of ethylene, which was often used in the past [165], requires complicated condensor and post reactor equipment to avoid the loss of expensive ethylene in the off-gas [164], [166].

Although several other Lewis-acid catalysts with higher selectivities such as antimony, copper, bismuth, tin, and tellurium chlorides [167] have been patented, iron chloride is widely used. Because the reaction selectivities are not dependent on the catalyst concentration, it is used in a diluted concentration between ca. 100 mg/kg and 0.5 wt%. Some processes use iron filler bodies in the reactor to improve mass and heat transfer or use iron as a construction material. This equipment generates sufficient $FeCl_3$ in situ [168].

In the liquid-phase reaction, ethylene absorption was found to be the rate-controlling step [169].

In addition to the distinct process modifications with which each producer of 1,2-dichloroethane has improved his process during the past years, two fundamental process variations can be characterized:

1) low-temperature chlorination (LTC) and
2) high-temperature chlorination (HTC)

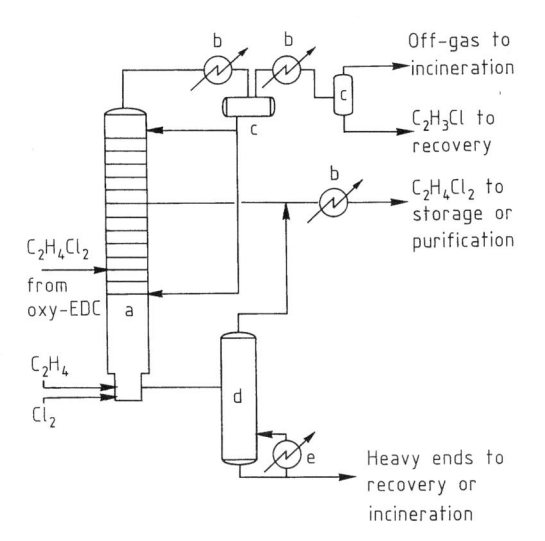

Figure 15. Simplified DC–HTC process
a) Reactor; b) Cooler; c) Knock-out drum; d) Heavy-end tower; e) Reboiler

In the *LTC process*, ethylene and chlorine react in 1,2-dichloroethane as a solvent at temperatures (ca. 20–70 °C) below the boiling point of 1,2-dichloroethane.

The heat of reaction is transferred by external cooling either by means of heat exchangers inside the reactor or by circulation through exterior heat exchangers [170].

This process has the advantage that due to the low temperature, byproduct formation is low. The energy requirements, however, are considerably higher in comparison to the HTC process, because steam is required for the rectification of 1,2-dichloroethane in the purification section. Conversions up to 100% with chlorine and ethylene selectivities of 99% are possible.

In the *HTC process*, the chlorination reaction is carried out at a temperature between 85 and 200 °C, mostly, however, at about 100 °C. The heat of reaction is used to distill the EDC. In addition, EDC from the Oxy-EDC process or unconverted EDC from the vinyl chloride section can be added, since the heat of formation equals the heat required for vaporization by a factor of ca. 6.

By sophisticated reactor design and thorough mixing conversion, and yields comparable to the LTC process may be obtained with considerably lower energy consumption for an integrated DC-Oxy-VCM process [171].

Description of the HTC Process (Fig. 15). Gaseous chlorine and ethylene are fed thoroughly mixed into a reaction tower which is also supplied with dry EDC from oxychlorination or recycled EDC from the VCM section.

The light ends are drawn off from the head section, and ethylene is condensed and recycled. In the following condensation section, vinyl chloride is separated and can then be processed with vinyl chloride from EDC cracking (see p. 1335). The remaining vent gas is incinerated. Pure EDC is taken from an appropriate section and condensed. In order to maintain a constant composition in the reactor sump phase, a slipstream is continuously withdrawn, from which the heavy byproducts are separated by rectification and sent to a recovery stage or incinerated. In some designs, the reactor is separated from the distillation tower [173]. In others, two towers are used for light ends/EDC separation. Solid adsorption has been patented for iron chloride removal [174].

For optimal heat recovery, cross exchange can be used for chlorine feed evaporation [175]. Due to the relatively low temperatures and anhydrous conditions, carbon steel equipment can be used [176].

Process developments using cracking gases instead of highly purified ethylene [177] and the use of nitrosyl chloride [178] as a chlorinating agent have not found any industrial importance.

Direct Chlorination in the Gas Phase. A catalytic gasphase process was patented by the Société Belge de l'Azote [179]. Because of the highly exothermic reaction, adequate dilution is necessary. Several catalysts have been patented [180].

The noncatalytic chlorine addition reaction has been thoroughly studied [181], but is not industrially used, as is the case for the catalytic gas-phase chlorination of chloroethane [182].

Oxychlorination of Ethylene in the Gas Phase. In the oxychlorination process, ethylene and hy-drogen chloride are reacted with oxygen in the presence of an ambivalent metal catalyst. In most cases, copper salts are used at a temperature above 200 °C. The overall reaction can be formulated as

$$C_2H_4 + 2\,HCl + 1/2\,O_2 \xrightarrow{\text{Cat.}} C_2H_4Cl_2 + H_2O$$
$$\Delta H^\circ_{298} = -295 \text{ kJ/mol}$$

The reaction sequence is similar to that of the Deacon process, although mechanistic studies indicate that ethylene is involved in the early reaction stages and the process may differ greatly from the classic Deacon process (e.g., oxidation of HCl to chlorine with subsequent addition of chlorine to ethylene by which chlorine is withdrawn from the Deacon equilibrium a high HCl conversion is achieved). The reaction sequence probably proceeds via chlorination of ethylene by cupric chloride. The copper salt is then regenerated by HCl and oxygen:

$$C_2H_4 + 2\,CuCl_2 \longrightarrow C_2H_4Cl_2 + Cu_2Cl_2$$
$$Cu_2Cl_2 + 2\,HCl + 1/2\,O_2 \longrightarrow 2\,CuCl_2 + H_2O$$
$$\overline{C_2H_4 + 2\,HCl + 1/2\,O_2 \longrightarrow C_2H_4Cl_2 + H_2O}$$

Several investigations on the reaction mechanisms have been performed [183].

Ethylene oxychlorination has attained commercial importance since about 1960, when VCM producers began to pursue the ethylene route and HCl from EDC cracking had to be recovered. Due to this historic development, several process variations are presently used by the major EDC/VCM producers and will be discussed in detail later [184] – [188]. A common characteristic of all these processes is the catalytic gas-phase oxychlorination at temperatures between 200 and 300 °C and pressures of 0.1 to 1.0 MPa, usually at 0.4 – 0.6 MPa. HCl and ethylene conversions of 93 – 97% are achieved at contact times between ca. 0.5 – 40 s with selectivities to EDC of 91 – 96% [189].

Byproducts of ethylene oxychlorination are monochloroethane, formed by direct HCl addition to ethylene, VCM from the cracking of EDC, 1,1,2-trichloroethane formed by substitution chlorination of EDC or chlorine addition to VCM, 1,1-dichloroethane formed by the addition of HCl to VCM, and other crack or substitution products such as 1,1-dichloroethylene, cis- and trans-1,2-dichloroethylene, trichloroethylene, and tetrachloroethanes. Because oxygen is present, additional oxidation products such as acetaldehyde and its chlorinated derivatives, primarily trichloroacetaldehyde (chloral), are found in the reactor effluent. Oxirane (ethylene oxide) and glycols may also be formed. The ethylene feed is partially consumed, especially at higher temperatures, by deep oxidation to yield carbon oxides (CO, CO_2) and formic acid.

In some plants, major byproducts such as chloroethane and 1,1,2-trichloroethane are recovered and sold or used as feedstock for other chlorinated hydrocarbon (CHC) processes such as 1,1-dichloroethylene production (see Chap. 3.2.3) and chlorinolysis (see p. 1364).

Reactor Feed. Polymerization grade ethylene is used to minimize byproduct formation and purification problems.

In most cases, HCl from the EDC cracking section (see p. 1335) is used as a source of chlorine. The acetylene content (derived from VCM cracking) of this hydrogen chloride may be critical and should be controlled, because acetylene tends to form more highly chlorinated byproducts and tars, which can lead to catalyst deactivation by coking (pore plugging) and may also influence down-stream operations. Selective hydrogenation to ethylene is often used to remove acetylene from this HCl [190]. An other method proposes catalytic hydrochlorination with subsequent adsorption of the vinyl chloride formed [191]. In addition to HCl from EDC cracking, HCl from other CHC processes like 1,1,1-trichloroethane-, tri-, and tetrachloroethylene production can be used without problems, if kept free of such well-known catalyst poisons as fluorine and sulfur compounds.

In most processes, air is used as an oxygen source. Microfiltration prior to compression is employed to exclude particulate matter. In oxygen-based processes, pure oxygen is supplied by a nearby air liquefaction and separation process and is used without additional processing.

Catalyst. Copper(II) salts, usually cupric chlorides, are used as standard catalysts [192], [193], [194]. In many cases, alkali, alkaline earth or aluminum chloride are added to reduce volatilization of the cupric salt. These salts form eutectic mixtures, which reduce the melting point. The reduction of the melting point, on the other hand, seems to be beneficial to the reaction rates. Furthermore, the addition of alkali salts suppresses direct addition reactions such as monochloroethane formation.

Some patents claim rare-earth salts (didymium salts) as promoters [194], [195] or use sodium/ammonium hydrogen sulfates [196] or tellurium salts [197].

High-surface-area alumina (150–300 m^2/g) is preferred as a support, because its production process allows the control of such important parameters as surface area, pore volume, and pore size distribution. Its high attrition resistance makes it very suitable for fluidized-bed reactors. Other support materials like graphite, silica, pumice, or kieselguhr are of minor importance.

For fluidized-bed reactors, alumina powder or microspheres (ca. 10–200 μm diameter) are used [198], whereas for fixed-bed reactors, catalyst tablets, extrudates, or spheres with a narrow size distribution (ca. 1/8–1/4″ diameter) are applied.

The catalyst is prepared to the support by the imbibition method using aqueous solutions of the catalyst salts followed by drying steps, or by special spray techniques [186].

Cupric chloride is usually added in concentrations of 3–12 wt% (of the total catalyst). Alkali salts are added in nearly double amounts to obtain molar alkali/copper ratios of 2:1 [199] and rare-earth salts in concentrations of 1–10 wt%.

The fine adjustment of the catalyst composition as well as the selection of the appropriate support material and preparation procedure is a well kept secret of the individual technology and closely related to the reactor design.

Reactor Design. Theoretically, two basic reactor designs are in service:

1) fixed-bed reactors
2) fluidized-bed reactors

Due to the highly exothermic oxychlorination reaction, temperature control is a problem in *fixed-bed systems*. It is achieved by proper dilution of the catalyst with inactive diluents such as undoped alumina [193], graphite [193], [200], silicon carbide [193], [201], or nickel [202]. Thus, catalyst activity at the reactor inlet is normally low and increases to its maximum at the outlet.

In order to make catalyst charging not too complicated, several blends of active catalysts and inactive diluent are prepared and sequentially charged. One patent [200] claims to use four different catalyst zones containing from the reactor inlet to the outlet active catalyst concentrations of 7, 15, 40, and 100 vol%, respectively. The active catalyst consists of 8.5 wt% $CuCl_2$ on alumina. The inert diluent is made of graphite. In another process [203], highly concentrated active catalyst is deposited at the top to start the reaction followed by an inactive zone for temperature control. In the following zones, the active catalyst concentration increases and reaches 100% in the last zone near the outlet.

Catalyst dilution requires exact mixing techniques and appropriate charging procedures in order to avoid demixing, i.e., segregation of diluent from active catalyst when different materials are used. This can lead to a rapid pressure drop buildup across the reactor.

Another approach is to vary catalyst activity through the catalyst particle size, which is not very practical, however [204]. Recent developments [203] favor staged catalysts, consisting of three to four different catalysts with varying amounts of $CuCl_2$ and KCl. The use of such catalyst systems, as offered by some manufacturers, does not require mixing and may offer advantages in some cases.

Fixed-bed technology is used by Dow Chemical, Stauffer, Toyo Soda and Vulcan. The size of the tubular reactors varies from 2 to 5 m in diameter and 4 to > 10 m in length.

They may comprise several thousand tubes for the catalyst with diameters up to 2″. Dow Chemical and Vulcan usually use one reactor, whereas Stauffer and Toyo Soda prefer successive oxychlorination systems with up to three reactors and split addition of oxygen. This latter method allows the formation of explosive mixtures at the reactor inlet to be more easily avoided, and it is claimed that fewer oxidation products are formed.

Nickel alloys are used for the construction of the tube section. Because of hot-spot formation, Alloy 200 may be prone to intergranular embrittlement, so that higher resistance may be obtained with Alloy 201 with a lower carbon content [176]. The tube sheet and the reactor head are lined with nickel on steel. For the reactor shell, carbon steel is primarily used.

Proper heat tracting for the interconnecting piping to the quench or absorber system is required to avoid corrosion.

The equipment for further processing such as the absorber – stripper – phase separator is lined with either bricks (towers) or teflon (pipes) to withstand the corrosion caused by aqueous HCl.

The heat of reaction is either used to generate steam at the side of the reactor shell or is transferred to a hot oil system, which may supply other plants.

Fluidized-bed reactors have the advantage of improved heat transfer and almost isothermal operation. However, backmixing, which influences conversion and selectivity, cannot be avoided. Nevertheless, HCl conversions of >98% have been reported [205]. This is achieved by feeding stoichiometric excesses of air or oxygen (10–80%) and ethylene (up to 60%) [206]. The temperature range between ca. 220–240 °C is somewhat lower than in fixed beds. Elevated pressure (0.2–0.5 MPa) is used to increase conversion. High-surface alumina powder (ca. 200 m^2/g) [198] or fuller's earth [207] are the preferred catalyst supports. The particle size distribution for representative samples shows a maximum at about 40–80 μm diameter [198]. Cupric chloride concentration on the catalyst varies from ca. 7–20 wt% $CuCl_2$. Higher concentrations are of no advantage, because the reaction rate will not improve and the catalyst will cake in the reactor.

Because of the lower temperature range, the reactor can be made of stainless steel if condensation (formation of aqueous HCl) can be avoided by means of proper shutdown procedures. Sparging equipment at the entrance of the reactor requires pipes, nozzles, and fittings of nickel alloys (Alloy 600 and 825) because they are more resistant to chloride stress corrosion [176].

Heat from the reaction is used to generate steam or is transferred to a hot oil system by internal cooling coils positioned in the fluidized bed.

One major advantage of the fluidized-bed reactor is that the reaction can be carried out within the explosive limit, which makes feed control less critical.

Reactor-integrated cyclones are used at the outlet to retain catalyst fines and to return them to the reaction zone.

Time–space yields may average 150–200 kg of EDC m^{-3} h^{-1} [208].

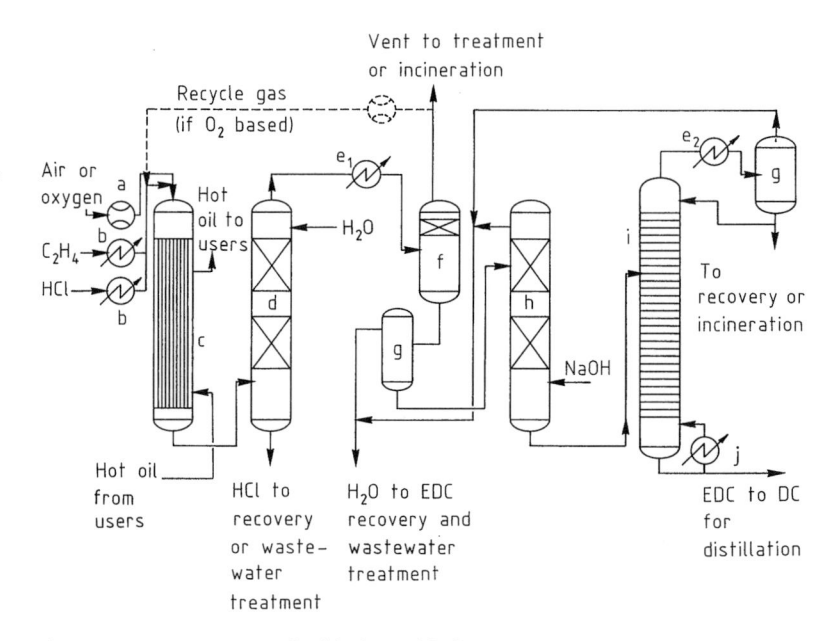

Figure 16. Oxy–EDC process (fixed bed, simplified)
a) Compressor; b) Preheater; c) Fixed-bed reactor; d) Quench tower; e) Cooler; f) Degasser; g) Separator; h) Wash tower; i) Azeotropic drying tower; j) Reboiler

Fluidized-bed reactors are more widely used than fixed-bed systems. Companies using fluidized-bed technology are B. F. Goodrich, Hoechst, Pittsburgh Plate Glass (PPG), Ethyl Corp., Solvay, ICI, and Mitsui Toatsu Chemical.

Tokoyama Soda [187] and Pechiney [209] have combined the advantages of both processes by first reacting the gases in an isothermal fluidized bed and then passing them across a fixed bed for optimal yields and conversions.

Process Description (Fig. 16, 17 and 18). Ethylene and hydrogen chloride are preheated and fed with air or oxygen to the reactor. The hot reaction gases are quenched in a brick-lined tower and the resulting aqueous HCl is either treated together with the combined wastewaters or cleaned separately by stripping for further use, e.g., in a chlor-alkali process. The gases leaving the quench tower are cooled in a heat exchanger, and the organic phase is washed with dilute NaOH in order to remove chloral [210].

The off-gas is either vented after additional condensation and/or scrubbing or adsorption steps (for air-based systems) or compressed and recycled (if pure oxygen is used).

In some process modifications, heat exchangers and separators f and g are placed behind the NaOH wash. In other processes, the quench step is performed without addition of water, and a NaOH wash tower is not always required.

The wet EDC is dried by azeotropic destillation.

The bottoms from the azeotropic distillation are sent to the DC section for final purification. The light head products are submitted for further treatment together with the azeotrope for product recovery (ethylene, monochloroethane, EDC, chlorinated

Figure 17. Oxy–EDC fluidized-bed reactor

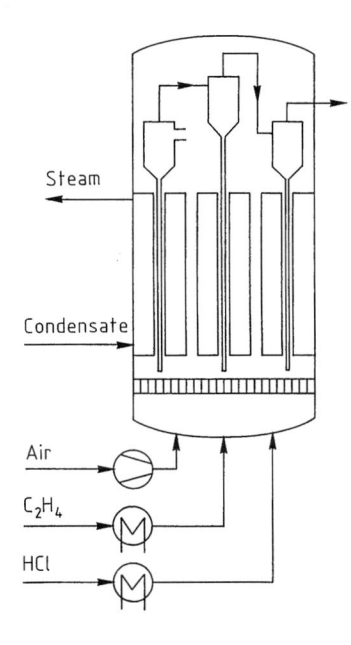

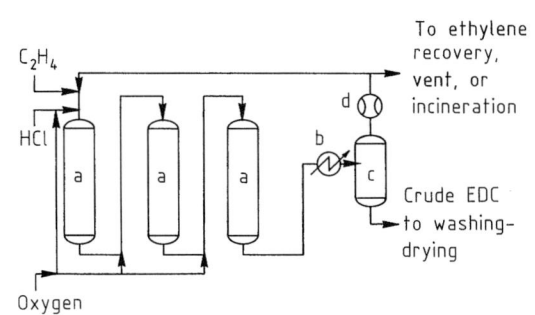

Figure 18. Stauffer oxygen-based Oxy-EDC process
a) Reactor; b) Cooler; c) Separator; d) Compressor

methanes) or incinerated. Care must be taken to remain outside of the flammable range during all process steps [211].

Oxygen-based Oxychlorination. Vent gas from air-based oxychlorination processes is one of the major emission sources of CHC plants. In spite of intensive cooling and sophisticated absorber–stripper, adsorber–desorber, and post-reaction systems [212], the restrictions on many plants have become very stringent. The large amounts of nitrogen in the vent gas, however, makes the final treatment by incineration prohibitively expensive.

If oxygen is used instead of air [213], the vent stream becomes 20–100 times smaller, allowing vent incineration or catalytic oxidation [214].

Airbased systems are more manageable than others because the nitrogen from the air acts as diluent and removes heat.

In an oxygen-based system, this function is achieved with an excess of ethylene [215], which is then recycled. Only a small quantity of the recycled stream must be drawn off in order to control the concentration of carbon oxides and other low-boiling byproducts. This slipstream is either burned or fed to the DC process to recover ethylene.

Since the heat capacity of ethylene in comparison to nitrogen is considerably higher, oxygen-based systems can be operated at lower temperatures or at higher throughput rates. This capacity increase together with the considerable savings for incineration [216] may offset the higher costs for oxygen and recycle compression energy.

During the past years, the conversion of many existing air-based facilities has proven to be feasible.

Oxychlorination of Ethylene in the Liquid Phase. An aqueous liquid-phase process for oxychlorination has been developed by the Kellog Co. [188], [217]. Ethylene, oxygen, and hydrogen chloride are fed to an aqueous solution of copper(II) salts (5 – 10 M) at 170 – 185 °C and 1.7 – 1.9 MPa. The 1,2-dichloroethane formed is stripped together with the steam generated by the heat of reaction. The gaseous products are quenched with water and further treated in a manner similar to gas-phase processes.

Although time – space yields and selectivities are comparable to the gas-phase process and feed impurities can be tolerated, the liquid-phase process is not industrially used. The main reason may be the troublesome handling of highly corrosive aqueous solutions at an elevated temperature and high pressure, even though similar problems have not been restrictive for the liquid-phase hydrochlorination of methanol (see p. 1266). Wastewater treatment may also pose more problems compared to gas-phase processes, because heavy metal contamination occurs. More information on homogeneously catalyzed oxychlorination may be found in the literature [218].

1,2-Dichloroethane from Ethane. The substantial cost margin between ethane and ethylene has prompted considerable research on direct ethane oxychlorination. This oxychlorination reaction is theoretically possible and proceeds via the sequence ethane – monochloroethane – dichloroethane.

Several processes comprising ethane and chloroethane oxychlorination or reacting mixtures of both components have been patented [158], [219].

A process developed by the Monsanto Company [220] comprises the direct thermal chlorination of ethane, yielding monochloroethane. In the next step, the reaction gases are oxychlorinated to 1,2-dichloroethane. In another variation [221], ethane is oxidized by oxygen in the presence of HCl at 400 – 600 °C to give ethylene. The resulting mixture is again oxychlorinated in a conventional manner.

This process has similarities with the autothermic cracking process [222], where ethylene, chlorine, and oxygen are reacted at 850 – 950 °C to form mainly ethylene and hydrogen chloride. In a further process step, the gases are oxychlorinated to 1,2-dichloroethane.

In both cases, chlorine balancing (HCl) in an integrated VCM process seems feasible. However, none of these processes have been implemented on an industrial scale. Compared to the oxychlorination of ethylene, ethanebased processes are frequently affected by poor conversion and selectivity. This necessitates high recycling rates, thereby increasing costs. The lack of selectivity also requires additional outlets for major byproducts (1,1-dichloroethane and trichloroethane), which may not always exist.

The cost advantage of ethane based processes further diminishes if cracker capacity for the ethylene supply and the infrastructure (loading stations, storage tanks, etc.) are already available.

Current research is directed toward the development of more specific catalysts, e.g. zeolites, and a direct route for producing VCM from ethane without isolating EDC followed by cracking (see Chap.

3.1.3.4). Such processes may offer true cost advantages for plants with easy access to ethane (U.S. gulf coast) and good integration into other CHC plants for economical byproduct recovery.

Other Processes. The production of 1,2-dichloroethane from ethanol [223] is not industrially used. It might be of interest if the cost for ethanol derived from biomass were to become competitive [117], [118].

1,2-Dichloroethane is also a byproduct of oxirane (ethylene oxide) production via the old chlorohydrine route. The EDC yield can be improved up to 50% by process modifications [224].

Because oxirane (ethylene oxide) is mostly produced by direct oxidation, this process is not important for EDC production.

2.3.4. Uses and Economic Aspects

Based on U.S. figures for 1981, ca. 85% of the total EDC production is used for the production of vinyl chloride. 10% is used in the production of chlorinated solvents such as 1,1,1-trichloroethane and tri- and tetrachloroethylene. The rest goes into various processes mainly for the synthesis of ethylenediamines. Its use as a solvent (dewaxing, deparaffinizing petroleum fractions, and coating remover) is marginal. EDC is further used in leaded gasoline as a lead scavenger. With the increasing trend toward unleaded fuel, however, this market will decline in future.

In Europe, market figures seem to be comparable to those in the US, if not shifted even more toward vinyl chloride production, because almost all European EDC plants are forward integrated to VCM units.

Production in 1985 is estimated at ca. 7×10^6 t for US, 8×10^6 t in Europe, and 2.5×10^6 t in Japan.

Installed capacity is ca. 10×10^6 t in North America, 10×10^6 t in Western Europe, and 3.5×10^6 t in Japan.

The future average growth rate is difficult to predict, since EDC production depends heavily on the big PVC consumers, the automotive industry and the construction business, and has, therefore, been subjected to severe fluctuations in the past. The growth rate may be estimated at ca. 2–5% for the decade 1985–1995.

New EDC plants in construction or in the planning phase will preferentially be located in developing countries to increase autonomy from imports and in oil-producing countries to forward integrate refineries and basic chemicals already produced.

2.4. 1,1,1-Trichloroethane

1,1,1-Trichloroethane [71-55-6] was first synthesized in the mid-19th century. It was not used industrially for more than 100 years and was frequently found as an unwanted byproduct in chlorinated hydrocarbon processes.

The Dow Chemical Co. began commercial production in the early 1950's. With the development of effective stabilizer systems, 1,1,1-trichloroethane has become one of the major solvents for cold and vapor degreasing as well as several other applications. 1,1,1-Trichloroethane is in strong competition with trichloroethylene (see 3.4.4) and has replaced this solvent in many fields.

2.4.1. Physical Properties

M_r	133.41
mp	−33 °C
bp at 101.325 kPa	74.1 °C
ϱ at 20 °C	1.325 g/cm^3
n_D^{20}	1.4377
Vapor pressure at	
0 °C	4.900 kPa
20 °C	13.300 kPa
40 °C	32.000 kPa
60 °C	62.700 kPa
70 °C	88.000 kPa
80 °C	120.000 kPa
Heat of formation (liquid) $\Delta H°_{298}$	−170 kJ/mol
Specific heat (liquid 20 °C)	1.004 kJ kg^{-1} K^{-1}
Heat of evaporation at 298 K	32 kJ/mol
Critical temperature	585 K
Critical pressure	4500 kPa
Viscosity at 20 °C	0.86 × 10^{-3} Pa s
Surface tension at 20 °C	25.6 × 10^{-3} N/m
Coefficient of cubical expansion (0 – 30 °C)	0.0013 K^{-1}
Dielectric constant at 20 °C	7.5
Flash point (closed cup)	none
Ignition temperature (air)	537 °C
Explosive limits in air at 25 °C	8.0 – 10.5 vol% 1,1,1-trichloroethane
Solubility in water at 20 °C	0.095 wt%
Solubility of water in 1,1,1-trichloroethane at 20 °C	0.034 wt%

1,1,1-Trichloroethane is a clear liquid at ambient temperature with a characteristic ethereal odor. It is soluble in all common organic solvents and is a very good solvent for fats, paraffins, and other organic compounds.

Some binary azeotropes are shown in Table 15.

2.4.2. Chemical Properties

Pure 1,1,1-trichloroethane is very unstable and tends to undergo dehydrochlorination. Noncatalyzed pyrolytic decomposition is almost complete at 300 – 400 °C [225]. When catalyzed by metal salts [226], aluminum fluoride [227], alumina [228], or others [229], this reaction proceeds at considerably lower temperatures. Dehydrochlorination yields dichloroethylenes and hydrogen chloride. High molecular mass products may

Table 15. Binary azeotropes formed by 1,1,1-trichloroethane

wt %	Component	Azeotrope boiling point (101.3 kPa), °C
4.3	water	65.0
23.0	methanol	55.5
17.4	ethanol	64.4
17.0	isopropanol	68.2
17.2	*tert*-butanol	70.2

also be obtained by polymerization of dichloroethylene [228]. Phosgene formation at elevated temperature in the presence of air is marginal [230]. Compared to olefinic chlorination solvents (tri- and tetrachloroethylene) used in similar applications, 1,1,1-trichloroethane shows better stability against oxidation [231]. Photochemical oxidation yields phosgene, carbon monoxide, and hydrogen chloride [232]. Hydrolysis with water and aqueous acid yields acetyl chloride and acetic acid [233]. Under normal conditions, this reaction proceeds slowly. Dehydrochlorination to 1,1-dichloroethylene takes place in alkaline solutions.

1,1,1-Trichloroethane is extremely corrosive to aluminum. Inhibitors must be used inevitably. Dry 1,1,1-trichloroethane moderately corrodes iron and zinc. Corrosion, however, increases with the water content.

2.4.3. Production

For the industrial production of 1,1,1-trichloroethane, three different routes are in use:

1) From 1,1-dichloroethane by thermal or photochemical chlorination
2) From 1,1,2-trichloroethane via 1,1-dichloroethylene and consecutive hydrochlorination
3) From ethane by direct chlorination

In the United States, more than 70% of the 1,1,1-trichloroethane is produced by the 1,1-dichloroethane process. An additional 20% is based on the 1,1-dichloroethane route, and ca. 10% is made by direct ethane chlorination. In Europe too, the 1,1-dichloroethane route is used by the largest producers.

Compared to the latter process, the production from 1,1-dichloroethylene has the disadvantage that more expensive chlorine is required, because one-fourth of the total chlorine required is lost as inorganic chloride. In addition, this route requires an aqueous system because dilute NaOH is used for the 1,1,2-trichloroethane dehydrochlorination (see p. 1350), which may cause environmental problems. Furthermore, 1,2-dichloroethane as feedstock for the 1,1-dichloroethane route is more readily available than 1,1,2-trichloroethane. The HCl generated by 1,1-dichloroethane chlorination can be used in other processes such as Oxy-EDC (see p. 1302) or methanol hydrochlorination (see p. 1264).

In other words, the first route is a source of HCl, whereas the second route consumes HCl. These aspects may have been also decisive for the implementation of the various processes in integrated CHC plants.

Even though photochemical reactions are rarely used for industrial purposes because reactor design and operation is somewhat troublesome, the photochemical reaction is preferred for 1,1,1-trichloroethane preparation from 1,1-dichloroethane because of its higher selectivity compared to thermal chlorination.

The direct chlorination of ethane is the least used route because of its lack of selectivity.

Besides the three routes mentioned above, several other processes have been proposed, but are not used on an industrial scale.

1,1,1-Trichloroethane from 1,1-Dichloroethane. This process uses 1,2-dichloroethane (EDC) as feedstock which is rearranged to 1,1-dichloroethane via cracking to vinyl chloride (see Chap. 3.1.3.2) followed by the addition of HCl in the presence of a catalyst.

During the final step, 1,1-dichloroethane is thermally or photochemically chlorinated.

$$CH_2Cl-CH_2Cl \xrightarrow{\Delta} CH_2=CHCl + HCl$$
$$CH_2=CHCl + HCl \xrightarrow{Cat.} CH_3CHCl_2$$
$$CH_3CHCl_2 + Cl_2 \xrightarrow{h\nu\ or\ \Delta} CH_3-CCl_3 + HCl$$
$$\Delta H^0_{298} = -103\ kJ/mol$$

Photochemical Chlorination. Photochemical chlorination is used mostly, because this reaction can be carried out at a lower temperature, which increases the selectivity toward 1,1,1-trichloroethane [234].

Preferred temperatures range between 80 and 160 °C and the reactor pressure may average 0.1–0.4 MPa. Reactor design for photochemical chlorination is a compromise: although selectivity is increased in a plug-flow reactor [235], the systems in use resemble back-mixed tank reactors (CSTR-characteristic) due to the need for sufficient actinic light radiation.

Typical byproducts are 1,1,2-trichloroethane, 1,1,1,2- and 1,1,2,2-tetrachloroethane, and pentachloroethane, which may represent up to 30% of the yield.

To minimize the formation of tetra- and more highly chlorinated byproducts and to dissipate the heat of formation, excess 1,1-dichloroethane is fed (3–10 M) to chlorine.

In order to maintain the formation of unwanted 1,1,2-trichloroethane as low as possible, the photochemical reaction is preferably carried out in the vapor phase, because liquid-phase chlorination favors the synthesis of 1,1,2-trichloroethane. Maximum selectivity toward 1,1,1-trichloroethane is ca. 90% [236], which may, however, not be achieved in industrial processes. Catalytic traces of iodine or iodine-containing compounds were also found to increase the selectivity [237].

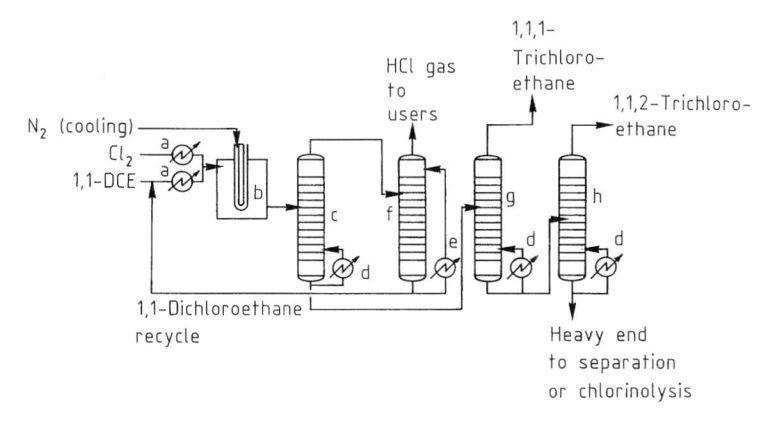

Figure 19. 1,1,1-Trichloroethane process (photochlorination)
a) Preheater; b) Photoreactor; c) Lights–heavies separation; d) Reboiler; e) Cooler; f) HCl tower; g) 1,1,1-Trichloroethane tower; h) 1,1,2-Trichloroethane tower

After separation from 1,1,1-trichloroethane, 1,1,2-trichloroethane can be used for the production of 1,1-dichloroethylene (vinylidene chloride see Chap. 3.2). The tetrachlorinated products may be used either in the production of trichloroethylene (see Chap. 3.4.3) or, without separation together with pentachloroethane, as feed for the perchloroethylene process.

Care must be taken to exclude traces of iron, which is a well-known promoter for the formation of 1,1,2-trichloroethane through a cracking–addition sequence. Monel or other copper–nickel alloys are the preferred construction materials for reactor and distillation equipment.

Special distillation processes using stabilizers [238] or thin-film evaporations [239] have been patented to avoid decomposition of 1,1,1-trichloroethane during cleanup.

Double-well UV lamps with external cooling [240] are used to prevent coking and to keep the shutdown frequency tolerable.

Process Description (Fig. 19). Finely dispersed 1,1-dichloroethane is fed together with chlorine into an adiabatic reactor equipped with an actinic light source (300–550 nm). Dichloroethane is fed in excess up to 10 M to dissipate the heat of reaction through evaporation and to suppress consecutive chlorination reactions. Reactor temperature averages between 80–100 °C at pressures of 0.1–1.0 MPa.

The reaction products are separated by distillation. In the first step, 1,1-dichloroethane is distilled together with hydrogen chloride and unreacted chlorine. HCl is then separated in a second tower. Dichloroethane and chlorine are recycled to the reactor.

The high-boiling components of the sump phase from the first tower are separated in at least two more steps yielding the crude products, which are then further purified.

Thermal Chlorination [241]–[243]. Fluidized-bed reactors offer the best technical solution for thermal chlorination because their uniform temperature profile minimizes the cracking reactions of 1,1,1-trichloroethane and reactor coking. Sand or silica [243], which must be free of iron, is used as a bed material.

Preheating of the reactants is required to maintain a reaction temperature of 350–450 °C, which is necessary to start the radical chain reaction.

Excess 1,1-dichloroethane is again added in proper dilution to avoid product losses (heavies formation).

Equipment and processing is quite similar to the photochemical system.

Although 1,1,1-trichloroethane yields of up to 82 % have been reported [242], actual yields may be considerably lower or poor conversions with higher energy requirements for dichloroethane recycling must be accepted. As far as product yield, selectivity, and specific energy consumption are concerned, this process is inferior to photochemical chlorination.

Other Processes. Catalytic liquid-phase chlorination of 1,1-dichloroethane using phosphorus catalysts (PCl$_5$) [244] as well as the highly selective chlorination by chlorine monoxide [245] are not industrially used. Either selectivity (PCl$_5$) or technical problems (Cl$_2$O) do not allow large-scale production.

1,1,1-Trichloroethane from 1,1,2-Trichloroethane. The overall reaction sequence again begins with 1,2-dichloroethane, which is chlorinated to form 1,1,2-trichloroethane. 1,1-Dichloroethylene is obtained by dehydrochlorination and is then hydrochlorinated to 1,1,1-trichloroethane:

$$CH_2Cl-CH_2Cl + Cl_2 \longrightarrow CHCl_2-CH_2Cl + HCl$$
$$CHCl_2-CH_2Cl + NaOH \longrightarrow CCl_2=CH_2 + NaCl + H_2O$$
$$CCl_2=CH_2 + HCl \xrightarrow{Cat.} CH_3-CCl_3$$
$$\Delta H^\circ_{298} = -80 \text{ kJ/mol}$$

As with the 1,1-dichloroethane route, an interim rearrangement via dehydrochlorination is again required. However, pyrolytic gas-phase dehydrochlorination of 1,1,2-trichloroethane does not have the required selectivity toward 1,1-dichloroethylene — the favored product is 1,2-dichloroethylene (see Chap. 3.3) — to be industrially attractive [246]. A yield increase to more than 90 % is only possible in aqueous systems using calcium hydroxide [247] or ammonia [248] and some of its derivatives [249]. Aqueous NaOH is primarily used for dehydrochlorination [250].

The NaCl-containing NaOH (8–10 wt % NaOH, 15–20 wt % NaCl) from diaphragm cells can be used directly without further evaporation. The use of an aqueous system however, results in loss of chlorine, which is discarded as a salt (CaCl$_2$ and NaCl).

The reaction is carried out at 80–120 °C in a packed tower or recirculation reactors. Because alkaline brine is used, nickel is applied as a construction material. Crude 1,1-dichloroethylene is withdrawn by live steam injection or flash evaporation and distilled.

In order to avoid polymerization of 1,1-dichloroethylene, all feed streams should be free of oxygen (< 1 mg/kg), or stabilizers (radical scavengers like phenols or amines) should be used.

The hydrochlorination of 1,1-dichloroethylene is carried out in a manner similar to the production of 1,1-dichloroethane from vinyl chloride at temperatures between 40–80 °C and in the presence of a Lewis acid catalyst (FeCl$_3$) [251]. Reactant ratios are almost stoichiometric or with a slight excess of HCl. The 1,1,1-trichloroethane formed can be used as solvent, but others such as 1,1,2-trichloroethane and perchloroethylene have also been mentioned [252]. Care must be taken to avoid entrainment of catalyst traces during purification. Remaining traces of catalyst and hydrogen chloride can be removed by addition of ammonia [253] and distillation or by careful filtration of the distilled 1,1,1-trichloroethane over partially deactivated NaOH flake beds or weak ion-exchange resins. Overall yields of more than 90% are obtainable.

In a special process carried out by Atochem (France), 1,1-dichloroethylene is produced as byproduct from the high-temperature chlorination of ethylene for the production of vinyl chloride [254] (see 3.1.3.4). After separation from other byproducts, it can be hydrochlorinated as usual to give 1,1,1-trichloroethane as a valuable byproduct.

A similar process has been patented by the FMC Corp. [252]. In this case ethylene is fed to the hydrochlorination stage forming monochloroethane, which is then subjected to high temperature chlorination to give 1,1-dichloroethylene.

1,1,1-Trichloroethane from Ethane. The direct synthesis of 1,1,1-trichloroethane has been patented by the Vulcan Materials Company and is mainly used in the United States:

$$CH_3-CH_3 + 3\,Cl_2 \longrightarrow CH_3-CCl_3 + 3\,HCl \qquad \Delta H^0_{298} = -330\ \text{kJ/mol}$$

The highly exothermic reaction can be controlled by recycling the chloroethane byproducts (monochloroethane, 1,1-, and 1,2-dichloroethane), which consume some of the reaction heat by endothermic dehydrochlorination reactions, so that an adiabatic reactor can be used. Hot spot temperatures of ca. 440 °C are obtained. Mean residence times of 10–20 s at reactor pressures of 0.3–0.5 MPa are found in the patent literature [255].

Due to the rigorous reaction conditions and the long reaction sequence, numerous byproducts are formed which require extensive equipment for postprocessing after quenching of the reactor gas. Even extractive distillation steps have been considered [256].

Both 1,1- and 1,2-dichloroethane are recycled. Vinyl chloride and vinylidene chloride are hydrochlorinated. 1,1-Dichloroethane from vinyl chloride hydrochlorination is added to the recycle and 1,1,1-trichloroethane resulting from the vinylidene chloride is drawn off.

At optimal conditions, the reaction product contains 60–70 mol% 1,1,1-trichloroethane and 20 mol% vinylidene chloride. Overall ethane yields of 60% and chlorine yields of 93% are obtained.

The advantage of the cheaper raw material ethane is, at least partially, offset by the higher equipment costs and the lower overall yields.

Other Processes. Other patents use monochloroethane as feedstock for the thermal chlorination [240], [257] or monochloroethane is formed in situ by feeding additional ethylene to the reactor under mild conditions [258].

The photochemical chlorination of monochloroethane has also been described [259].

All of these processes have no distinct advantages over the two basic methods that begin with 1,2-dichloroethane because of a lack of selectivity. They are, therefore, currently of no importance for the industrial production of 1,1,1-trichloroethane.

2.4.4. Uses and Economic Aspects

1,1,1-Trichloroethane is used as a solvent in numerous industrial applications such as cold and hot cleaning and vapor degreasing. Formulations are used as solvent for adhesives and metal cutting fluids [260].

New applications have been found in textile processing and finishing and in dry cleaning, where 1,1,1-trichloroethane can replace the widely used perchloroethylene.

Special grades are used for the development of photoresists in the production of printed circuit boards. Because of its lower toxicity, 1,1,1-trichloroethane has replaced trichloroethylene in many fields, especially in the United States. In Europe, however, trichloroethylene has maintained its leading position.

Further advantages of 1,1,1-trichloroethane are its graduated solvency, which allows it to be used even in very sensitive areas, its good evaporation rate, and the fact that it has no fire or flash point.

Because it readily reacts with aluminum and other metals, inhibitors must be added prior to industrial use. The inhibitor systems, which may account for 3–8% of the formulation, mainly comprise such acid acceptors as epoxides, ethers, amines, and alcohols as well as such metal stabilizers as nitro- and cyano-organo compounds, which build complexes, thereby deactivating metal surfaces or catalytic salt traces. Several formulations for proprietary grades of 1,1,1-trichloroethane have been filed and are in use [261].

U.S. producers are The Dow Chemical Co., Pittsburgh Plate Glass Inc., and the Vulcan Materials Co. In Europe, 1,1,1-trichloroethane is produced by The DOW Chemical Co., ICI Ltd., Atochem, and Solvay.

European and U.S. capacities together amount to ca. 600 000 t. Production in 1984 was estimated to be ca. 450 000 t (Europe ca. 150 000 t). In Europe, 1,1,1-trichloroethane strongly competes with trichloroethylene and may replace this solvent in several applications, which could affect the future growth rate.

However, more stringent regulations in many industrialized countries requiring the reduction of losses from vapor degreasing units and other equipment may adversely affect future demand.

Present capacity seems to be sufficient to supply the market for the next decade. No new plants have been ordered or constructed recently.

2.5. 1,1,2-Trichloroethane

1,1,2-Trichloroethane [79-00-5] is primarily an unwanted byproduct of several chlorination processes such as the production of 1,2-dichloroethane and the chlorination of ethane or 1,1-dichloroethane to 1,1,1-trichloroethane.

It has a very high solvency, but the relatively high toxicity limits its uses.

1,1,2-Trichloroethane is only important as an intermediate in the production of 1,1-dichloroethylene and to some extent for the synthesis of tetrachloroethanes.

2.5.1. Physical Properties

M_r	133.41
mp	− 37 °C
bp at 101.325 kPa	113.5 °C
ϱ at 20 °C	1.4432 g/cm^3
n_D^{20}	1.4711
Vapor pressure at	
30 °C	4.800 kPa
90 °C	49.200 kPa
100 °C	67.300 kPa
110 °C	90.600 kPa
114 °C	101.800 kPa
Heat of formation (liquid) $\Delta H°_{298}$	− 188 kJ/mol
Specific heat (liquid, 20 °C)	1.113 kJ kg^{-1} K^{-1}
Heat of evaporation at 298 K	39.1 kJ/mol^{-1}
Viscosity at 20 °C	1.20×10^{-3} Pa s
Surface tension at 20 °C	32.5×10^{-3} N/m
Coefficient of cubical expansion (0 – 25 °C)	0.001 K^{-1}
Autoignition temperature (air)	460 °C
Solubility in water at 20 °C	0.45 wt%
Solubility of water in 1,1,2-trichloroethane at 20 °C	0.05 wt%

1,1,2-Trichloroethane is a clear liquid at ambient temperature with a sweet smell. It is not flammable and easily miscible with most organic solvents.

For some binary azeotropes, see Table 16.

2.5.2. Chemical Properties

At an elevated temperature (400 – 500 °C), 1,1,2-trichloroethane is easily dehydrochlorinated to give a mixture of *cis*- and *trans*-1,2-dichloroethylene, 1,1-dichloroethylene, and hydrogen chloride [262].

Addition of alumina catalysts or an increase in temperature favors 1,2-dichloroethylene formation. In aqueous alkaline solution, 1,1,2-trichloroethane is selectively dehydrochlorinated to 1,1-dichloroethylene [263]. This reaction proceeds faster than that

Table 16. Binary azeotropes formed by 1,1,2-trichloroethane

wt%	Component	Azeotrope boiling point (101.3 kPa), °C
97	methanol	64.5
57	perchloroethylene	112.0
70	ethanol	77.8
15	water	85.3 (at 97.300 kPa)

with 1,1,1-trichloroethane. In water, hydrolysis takes place, especially under reflux conditions. Hydrolysis, however, proceeds slower than with the 1,1,1-isomer.

1,1,2-Trichloroethane is highly corrosive to aluminum, iron, and zinc. Addition of water increases the rate of corrosion.

If chlorinated, a mixture of the isomeric tetrachloroethanes is formed.

2.5.3. Production

The industrial production of 1,1,2-trichloroethane proceeds by two routes:

1) Selective chlorination of 1,2-dichloroethane:

$$\text{CH}_2\text{-CH}_2 + \text{Cl}_2 \longrightarrow \text{CHCl}_2\text{-CH}_2\text{Cl} + \text{HCl}$$
$$\text{Cl} \quad \text{Cl}$$
$$\Delta H^0_{298} = -116 \text{ kJ/mol}$$

2) Addition of chlorine to vinyl chloride:

$$\text{CH}_2 = \text{CHCl} + \text{Cl}_2 \longrightarrow \text{CH}_2\text{Cl}-\text{CHCl}_2 \quad \Delta H^0_{298} = -224 \text{ kJ/mol}$$

The liquid-phase 1,2-dichloroethane route is most often used in industrial processes when 1,1,2-trichloroethane is needed for the production of 1,1-dichloroethylene (see Chap. 3.2.3) and 1,1,1-trichloroethane (see Chap. 2.4.3).

The vinyl chloride route plays a minor role because of the more expensive feedstock and overall higher energy requirements.

A large portion of the current demand for 1,1,2-trichloroethane can be satisfied by the use of 1,1,2-trichloroethane obtained as a byproduct from the production of 1,1,1-trichloroethane. Because the latter is mostly produced by photochemical chlorination of 1,1-dichloroethane (see p. 1312), substantial amounts of 1,1,2-trichloroethane are obtained as a byproduct.

1,1,2-Trichloroethane is also one of the major byproducts in the production of 1,2-dichloroethane (see Chap. 2.3.3) and can be distilled from the heavy ends of these processes.

Thus, 1,1,2-trichloroethane is deliberately produced only when 1,1,1-tri- or 1,2-dichloroethane sources are unavailable or for the balancing of feedstocks.

1,1,2-Trichloroethane from 1,2-Dichloroethane. This process was patented by the Pittsburgh Plate Glass Co. [264] and by the Toa Gosei Chemical Industries [265]. It

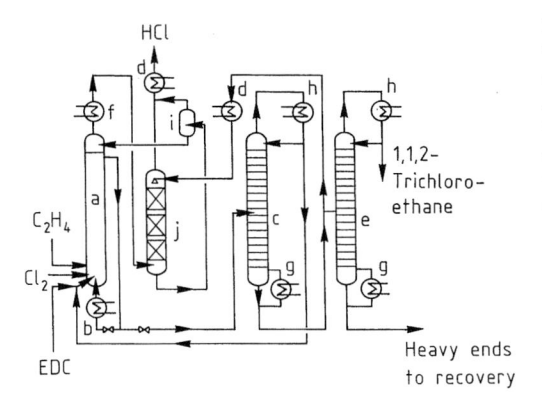

Figure 20. Process for the production of 1,1,2-trichloroethane by ethylene-induced liquid-phase chlorination [265]
a) Reactor; b) Preheater for priming; c) Light-ends tower (1,2-dichloroethane); d) Cooler (brine); e) 1,1,2-Trichloroethane finishing tower; f) Cooler (water); g) Reboiler; h) Condenser; i) Knock-out drum; j) 1,2-Dichloroethane wash tower

comprises the noncatalytic chlorination of 1,2-dichloroethane. The reaction is carried out in the liquid phase at temperatures between 100 – 140 °C. The addition of ethylene induces the reaction. The reaction mechanism has been intensively studied but is not yet clearly and fully understood [266]. Radical chlorination is very likely because metal salts, which support ionic reactions, must be excluded. The reactants should be free of radical-scavenging oxygen. Because consecutive chlorination to tetrachloroethanes and pentachloroethane takes place, the conversion per pass must be kept at 10 – 20 % for an optimum trichloroethane yield [265]. Since backmixing also favors the formation of higher chlorinated products, adiabatic plug-flow reactors are preferred.

Nickel alloys or nickel-clad steel is preferred as a structural material. As with most chlorination reactions, water must be removed to minimize corrosion.

Process Description (Fig. 20). Ethylene, chlorine, and fresh and recycled 1,2-dichloroethane are fed to a tubular reactor.

Even at low conversion rates, the heat of reaction is sufficient to maintain the reaction. Preheating of the recycled stream is, therefore, only required for startup.

The liquid phase from the reactor is first distilled to separate the excess 1,2-dichloroethane, which is then recycled to the reactor.

The crude 1,1,2-trichloroethane is then separated from the higher chlorinated by-products, mainly tetrachloroethanes, by further distillation.

The gaseous reaction phase is cooled and high-boiling components are condensed. The remaining 1,2-dichloroethane in the HCl gas is washed out with crude 1,1,2-trichloroethane and recycled to the reactor.

Other catalysts such as azodiisobutyronitrile (ADIB) or peroxides [267], actinic light [168], [265], [268] (liquid-phase photochlorination of 1,2-dichloroethane favors 1,1,2-trichloroethane, whereas by gas-phase photochlorination, 1,1,1-trichloroethane is preferentially obtained and phosphoros chloride have been proposed instead of ethylene [269]. Gas-phase chlorination catalyzed by metal chlorides has also been patented [270].

All these processes, however, are commercially more demanding and offer no real advantage over the ethylene-induced reaction.

Instead of chlorinating 1,2-dichloroethane, ethylene can be used as a starting material. Either liquid-phase chlorination with chlorine [271] (see DC–EDC p. 1300) or catalytic gas-phase oxychlorination with HCl in fixed- or fluidized-bed reactors [272] (see Oxy–EDC, p. 1302) can be used. The selectivity toward 1,1,2-trichloroethane is generally lower with this process. Especially in gas-phase oxychlorination, substantial yield losses by "deep oxidation" (CO and CO_2 formation) may occur.

1,1,2-Trichloroethane from Vinyl Chloride. The reaction is carried out similar to the DC-EDC process (see p. 1300) in liquid phase (trichloroethane) at temperatures between 50 and 90 °C. As with other addition reactions, Lewis-acid catalysts like $FeCl_3$, $AlCl_3$ or $SbCl_5$ are used, but actinic light catalysis also seems possible [273]. Yields of more than 90% are obtainable.

Catalytic oxychlorination of vinyl chloride with hydrogen chloride [274] or oxychlorination of mixtures of ethylene and vinyl chloride in fixed or fluidized beds [275] is also possible. Conditions are very similar to those for the Oxy–EDC process (see p. 1302). Instead of vinyl chloride, acetylene can also be used as a feedstock [276]. In this case, the catalyst must be doped with mercury salts in order to achieve adequate conversion.

Hydrochlorination of 1,2-dichloroethylene [277] may be of interest only for the recovery of an unusable byproduct.

2.5.4. Uses and Economic Aspects

As previously mentioned, 1,1,2-trichloroethane only plays a role as an intermediate for the production of 1,1,1-trichloroethane and 1,1-dichloroethylene.

The relatively high toxicity, which is typical for all 1,2-substituted chloroethanes, does not allow general use as a solvent.

Based on production figures for 1,1,1-trichloroethanes and 1,1-dichloroethylene, Western World production of 1,1,2-trichloroethane is estimated to be at 200 000 – 220 000 t/year for 1984.

In Western Europe, approx. 40 000 t was produced in 1984.

2.6. 1,1,1,2-Tetrachloroethane

1,1,1,2-Tetrachloroethane [630-20-6] was first synthesized by A. MOUNEYRAT in 1898. Today it is a byproduct in many industrial chlorination reactions of C_2 hydrocarbons. It is, however, not produced on an industrial scale.

If recovered from such industrial processes as the production of 1,1,1- and 1,1,2-trichloroethane, it can be used as feedstock for the production of trichloroethylene (see 3.4.3) and perchloroethylene (see p. 1364).

Because of its high toxicity, it is not used as a solvent.

2.6.1. Physical Properties

M_r	167.86
mp	– 68.7 °C
bp at 101.325 kPa	130.5 °C
ϱ at 20 °C	1.5468 g/cm^3
n_D^{20}	1.4822
Viscosity at 20 °C	1.50×10^{-3} Pa s
Surface tension at 20 °C	32.1×10^{-3} N/m
Solubility of water in 1,1,1,2-tetrachloroethane at 20 °C	0.06 wt%

1,1,1,2-Tetrachloroethane is a colorless, non-flammable heavy liquid.

2.6.2. Chemical Properties

In general, 1,1,1,2-tetrachloroethane is more stable than its symmetrically substituted isomer. Thermal decomposition at 500 – 600 °C yields trichloroethylene and hydrogen chloride. Tetrachloroethylene can be formed by disproportionation [278]. The thermal decomposition is catalyzed by numerous compounds [279], mainly Lewis acids such as FeCl$_3$ and AlCl$_3$.

Dichloroacetyl chloride is obtained through oxidation [280].

2.6.3. Production

1,1,1,2-Tetrachloroethane is not produced on an industrial scale. It is an undesired byproduct mainly from the production of 1,1,1-trichloroethane from 1,1-dichloroethane, 1,1,2-trichloroethane and 1,1,2,2-tetrachloroethane from 1,2-dichloroethane. The most economical use is its conversion to tetrachloroethylene in the chlorinolysis process (see p. 1364).

It can be prepared in highly purified form by isomerization of 1,1,2,2-tetrachloroethane or by chlorination of 1,1-dichloroethylene at approx. 40 °C in the liquid phase. Aluminum chloride is used in both reactions as a Lewis-acid catalyst [281].

2.7. 1,1,2,2-Tetrachloroethane

1,1,2,2-Tetrachloroethane [79-34-5] was first synthesized by M. BERTHELOT and E. JUNG-FLEISCH in 1869. Based on experiments by A. MOUNEYRAT, the first industrial scale production process was developed by A. WACKER in 1903. Thus, 1,1,2,2-tetrachloroethane became the first chloroethane to be produced in large quantities. For almost 70 years this process, which consists of the catalytic chlorination of acetylene,

Table 17. Binary azeotropes formed by 1,1,2,2-tetrachloroethane

wt%	Component	Azeotrope boiling point (101.3 kPa), °C
68.0	formic acid	99.3
55.0	cyclohexanone	159.1
9.0	ethylene glycol	145.1
7.0	isobutyric acid	144.8
1.8	monochloroacetic acid	146.3
60.0	propionic acid	140.4
45.0	styrene	143.5
31.1	water	93.2 (at 97.300 kPa)

was the basis for the production of such important solvents as trichloroethylene (Tri) and tetrachloroethylene (Per).

However, with the continuing replacement of trichloroethylene by 1,1,1-trichloroethane and the development of more economical processes for the production of perchloroethylene, 1,1,2,2-tetrachloroethane has become less important for the production of chlorinated solvents.

2.7.1. Physical Properties

M_r	167.86
mp	− 42.5 °C
bp at 101.325 kPa	146.5 °C
ϱ at 20 °C	1.5958 g/cm^3
n_D^{20}	1.4942
Vapor pressure at	
0 °C	0.180 kPa
20 °C	0.680 kPa
60 °C	5.330 kPa
91 °C	18.700 kPa
118 °C	46.700 kPa
138 °C	82.700 kPa
Heat of formation (liquid) $\Delta H°_{298}$	− 195 kJ/mol^{-1}
Specific heat (liquid, 20 °C)	1.122 kJ kg^{-1} K^{-1}
Specific heat (vapor, 146.5 °C)	0.92 kJ kg^{-1} K^{-1}
Heat of evaporation at 298 K	45.2 kJ/mol^{-1}
Critical temperature	688 K
Critical pressure	4000 kPa
Viscosity at 20 °C	1.77×10^{-3} Pa s
Surface tension at 20 °C	35.0×10^{-3} N/m
Coefficient of cubical expansion	0.00098 K^{-1}
Dielectric constant at 20 °C	8.00
Solubility in water at 20 °C	0.3 wt%
Solubility of water in 1,1,2,2-tetrachloroethane at 20 °C	0.03 wt%

1,1,2,2-Tetrachloroethane is a clear heavy, nonflammable liquid with a sweetish odor. It is well miscible with all common organic solvents and exhibits the highest solvency of all aliphatic chlorohydrocarbons.

1,1,2,2-Tetrachloroethane does not form explosive mixtures with air. Some binary azeotropes are shown in Table 17.

2.7.2. Chemical Properties

If moisture, air, and light are excluded, 1,1,2,2-tetrachloroethane is sufficiently stable and can be stored without adding stabilizers. At elevated temperatures (> 400 °C), it is cracked to trichloroethylene and hydrogen chloride. Tetrachloroethylene may also be formed via disproportionation [278], [279], [283]. The thermal cracking reaction can be promoted by a variety of catalysts [279].

To avoid cracking during distillation, vacuum distillation is recommended. One patent claims soft, nondecompositional evaporation by means of a fluidized-bed evaporator [284].

In weak alkali solutions, dehydrochlorination to trichloroethylene occurs. In strong alkali solutions, explosive dichloroacetylene is formed. Decomposition in the presence of air can lead to small quantities of phosgene.

Chlorination under mild conditions (eventually induced by UV light or catalysts) yields hexachloroethane via the pentasubstituted intermediate [285]. Under more rigorous conditions and more thermodynamic control (chlorinolysis reaction, see 3.5.3), tetrachloroethylene and tetrachloromethane are formed as main products. Chlorination at 400 °C in the presence of charcoal favors cleavage which gives primarily tetrachloromethane and hydrogen chloride [286]. Strong acids may hydrolyze 1,1,2,2-tetrachloroethane to glyoxal.

With hydrogen, hydrodechlorination to 1,2-dichloroethylenes occurs.

Oxidation in air yields dichloroacetyl chloride.

2.7.3. Production

Industrial processes for the production of 1,1,2,2-tetrachloroethane consist of two main routes:

1) Addition of chlorine to acetylene:

$$C_2H_2 + 2\,Cl_2 \xrightarrow{Cat.} CHCl_2-CHCl_2 \quad \Delta H^0_{298} = -422\ kJ/mol$$

2) Liquid-phase chlorination of ethylene or 1,2-dichloroethane

$$C_2H_4 + 3\,Cl_2 \longrightarrow CHCl_2-CHCl_2 + 2\,HCl \quad \Delta H^0_{298} = -436\ kJ/mol$$
$$CH_2Cl-CH_2Cl + 2\,Cl_2 \longrightarrow CHCl_2-CHCl_2 + 2\,HCl \quad \Delta H^0_{298} = -216\ kJ/mol$$

The acetylene route was used primarily in the past. The ethylene-based process was developed during the late 1960s, when the hydrocarbon feedstocks shifted toward ethylene. However, the acetylene process is still in use — mainly in the Federal Republic of Germany — where acetylene is readily available as a byproduct form naphtha crackers. Furthermore, the acetylene route has the distinct advantage of preferentially yielding the 1,1,2,2-isomer, which can easily be cracked to trichloroethylene. The ethylene-based process produces both isomers in an approximately equimolar ratio because of its radical nature.

1,1,2,2-Tetrachloroethane is also an incidental byproduct of other production processes for chlorinated hydrocarbons, such as the production of 1,1,1- and 1,1,2-trichloroethane. If necessary, it is separated together with the unsymmetric isomer and used for the production of trichloroethylene.

In the Atochem process [254] (see Figure 25), 1,1,2,2-tetrachloroethane is produced from 1,2-dichloroethylenes by chlorination as an intermediate for trichloroethylene synthesis.

Several other processes have been patented, but all of them are only of minor importance.

1,1,2,2-Tetrachloroethane from Acetylene. The technical principle of carrying out this reaction has not changed much since the development of this process [287]. It is similar to the DC–EDC process (see p. 1300 and other common liquid-phase chlorination or hydrochlorination reactions).

The reaction is carried out in the liquid phase (tetrachloroethane) at 60–90 °C. The reactor pressure is reduced to ca. 20 kPa in order to prevent the explosion of the chlorine–acetylene mixtures. Lewis-acid catalysts [287], primarily $FeCl_3$, are dissolved in the tetrachloroethane solvent from the reactor sump phase. Gaseous acetylene and chlorine are fed to the reactor sump. The highly exothermic reaction provides enough heat to distill the tetrachloroethane [288].

The upper part of the reactor is therefore a distillation tower. Tetrachloroethane is withdrawn from appropriate trays, and the overheads consisting mainly of 1,2-dichloroethylenes and acetylene can be recycled to the reactor sump. To control the heavies concentration in the sump phase, a slipstream is withdrawn. Partial evaporation for tetrachloroethane recovery followed by high-temperature incineration with subsequent flue gas scrubbing is the best treatment for this stream containing the spent catalyst.

Carbon steel can be used as a construction material. Water and moisture should be strictly avoided in order to minimize corrosion and rapid deactivation of the catalyst.

Chlorine and acetylene yields of 90–98 % have been reported.

A similar process, using crude crack gas instead of pure acetylene, has been patented [289]. Besides 1,1,2,2-tetrachloroethane, this process also produces 1,2-dichloroethane from the ethylene fraction of the crack gas feedstream.

1,1,2,2-Tetrachloroethane from Ethylene and from 1,2-Dichloroethane. Liquid-phase chlorination of ethylene or the ethylene-induced chlorination of 1,2-dichloroethane is the same process as that used for the production of 1,1,2-trichloroethane (see p. 1318). By increasing the chlorine : ethylene or chlorine : dichloroethane ratios and optimizing the residence time, an almost equimolar mixture of 1,1,2,2- and 1,1,1,2-tetrachloroethane is obtained as the main product [265], [290].

Several kinetic studies have been performed to determine the individual relative rate constants and to optimize the yield [290], [291].

At a temperature between 80 and 130 °C, chlorine conversion as high as 100% and maximum ethylene conversions of 95 – 98% can be achieved. Low-substituted products such as 1,2-dichloroethane and 1,1,2-trichloroethane can be recycled [292], so that yield losses occur only through the formation of penta- and hexachloroethane. With some minor modifications, the process is carried out as described earlier (see p. 1318). Instead of 1,2-dichloroethane, 1,2-dichloroethylenes may be fed to the reactor, which favors the formation of the symmetric isomer [293].

Similarly, the liquid-phase chlorination of mixtures containing a variety of chloroethanes and chloroethylenes has been patented [294].

Other Processes. The liquid-phase chlorination of vinyl chloride or 1,1,2-trichloroethane in the presence of $AlCl_3$ as catalyst yields 1,1,2,2-tetrachloroethane with high selectivity [295].

Specific catalysts made from graphite-intercalated copper or iron salts, alumina, and organopolysiloxanes specifically yield 1,1,2,2-tetrachloroethane by gas-phase chlorination (ca. 200 °C, 0.1 – 1.0 MPa) of mixtures comprising monochloroethane, 1,1- and 1,2-dichloroethane, and 1,1,2-trichloroethane [296].

The catalytic gas phase oxychlorination of 1,2-dichloroethane, ethylene, vinyl chloride, and 1,2-dichloroethenes has also been described [297].

Other processes use gas phase chlorination of 1,2-dichloroethane in a fluidized-bed [298] or liquid-phase photochlorination of 1,2-dichloroethane [299] or 1,2-dichloroethylenes [300].

2.7.4. Uses and Economic Aspects

1,1,2,2-Tetrachloroethane is almost always used as an intermediate in the production of trichloroethylene. Although it has a high solvency (e.g., as solvent for the production of chlorinated PVS [301]), it is very rarely used as a solvent because of its high toxicity.

Production figures for 1,1,2,2-tetrachloroethane cannot be estimated.

2.8. Pentachloroethane

Pentachloroethane [76-01-7] was first synthesized by V. REGNAULT in 1839 – 1840 by chlorination of monochloroethane.

In the past, pentachloroethane was produced as an intermediate for the tetrachloroethylene process (pentachloroethane pyrolysis). However, it is currently an unwanted byproduct of many production processes for chlorinated hydrocarbons and is mostly converted to tetrachloroethylene and tetrachloromethanes by chlorinolysis because other uses have almost disappeared.

Table 18. Binary azeotropes formed by pentachloroethane

wt%	Component	Azeotrope boiling point (101.3 kPa), °C
3	acetamide	160.5
36	cyclohexanol	157.9
28	cyclohexanone	165.4
15	glycol	154.5
43	isobutyric acid	152.9
9.5	phenol	160.9
43.4	water	95.1 (at 97.300 kPa)

2.8.1. Physical Properties

M_r	202.31
mp	− 29.0 °C
bp at 101.325 kPa	162 °C
ϱ at 20 °C	1.678 g/cm^3
n_D^{20}	1.5035
Vapor pressure at	
20 °C	0.470 kPa
60 °C	3.470 kPa
80 °C	7.860 kPa
100 °C	17.330 kPa
120 °C	33.330 kPa
140 °C	60.000 kPa
Heat of formation (liquid) $\Delta H°_{298}$	− 188.4 kJ/mol
Density of vapor (162 °C, 101.325 kPa)	5.68 g/L
Specific heat (liquid, 20 °C)	0.9 kJ kg^{-1} K^{-1}
Heat of evaporation at 298 K	45.6 kJ/mol
Viscosity at 20 °C	2.49 × 10^{-3} Pa s
Surface tension at 20 °C	34.7 × 10^{-3} N/m
Coefficient of cubical expansion	0.0009 K^{-1}
Dielectric constant at 20 °C	3.6
Solubility in water at 20 °C	0.05 wt%
Solubility of water in pentachloroethane at 20 °C	0.03 wt%

Pentachloroethane is a colorless, heavy, nonflammable liquid with a sweetish odor. It is miscible in most common organic solvents and does not form explosive mixtures with air.

Some binary azeotropes are shown in Table 18.

2.8.2. Chemical Properties

If moisture and air are eliminated, pentachloroethane shows good stability even at elevated temperatures (> 100 °C). Pyrolysis at a temperature above 350 °C yields tetrachloroethylene and hydrogen chloride [302]. The dehydrochlorination reaction is catalyzed by Lewis acids and activated alumina. Dehydrochlorination also occurs in the presence of weak alkali solution. Chlorination in the liquid phase in the presence of a catalyst or induced by ethylene (see p. 1318) yields hexachloroethylene. Dry pentachloroethane does not corrode iron; if it is stored over longer periods of time, however, the addition of amine stabilizers is recommended.

Dichloroacetyl chloride is formed with fuming sulfuric acid. Air oxidation in the presence of UV light gives trichloroacetyl chloride.

In the presence of hydrogen fluoride and Lewis acids such as $SbCl_5$– chlorine, substitution occurs.

2.8.3. Production

Since tetrachloroethylene is more economically produced by the chlorinolysis process (see p. 1364), industrial production of pentachloroethane has become unimportant and is presently no longer used.

If required, the synthesis can be performed by two routes:

1) Chlorination of trichloroethylene:

$$CHCl=CCl_2 + Cl_2 \xrightarrow[\text{or } h\nu]{\text{Cat.}} CHCl_2-CCl_3$$

The reaction is best carried out in the liquid phase. $FeCl_3$ is used as a catalyst, but UV irradiation can also be used. Stepwise chlorination in a cascade system is also possible [303].

2) Ethylene-induced chlorination of 1,2-dichloroethane [265]:

$$\underset{\underset{Cl}{|}}{CH_2}-\underset{\underset{Cl}{|}}{CH_2} + 3\,Cl_2 \longrightarrow CHCl_2-CCl_3 + 3\,HCl$$

This reaction is similar to that of 1,1,2-trichloroethane (see p. 1318). Pentachloroethane is obtained with lighter chlorinated products, which can be rechlorinated. To avoid decomposition, pentachloroethane should be distilled at reduced pressure.

By oxychlorination of ethylene, 1,2-dichloroethane, or other chlorinated C_2 hydrocarbons, pentachloroethane is also obtained.

The main industrial source, however, is the photochemical production of 1,1,1-trichloroethane (see p. 1312) and the liquid-phase chlorination process for 1,1,2-trichloroethane production (see p. 1318). Pentachloroethane formed in these processes is frequently not isolated but fed together with the tetrachloroethanes to the chlorinolysis process.

2.8.4. Uses and Economic Aspects

Because of its low stability and toxicity, uses for pentachloroethane as a solvent (cellulose derivatives, rubbers, and resins) are insignificant.

About 10 000 – 20 000 t/a (1984) of pentachloroethane may be produced as a by-product in the Western World. Most of it is used for the production of tetrachloroethylene and carbon tetrachloride.

2.9. Hexachloroethane

Hexachloroethane [67-72-1] is at ambient temperature the only solid compound of all chlorinated ethanes and ethylenes. Because it has specific properties, such as a tendency to sublime and a very high chlorine content, it has some specific applications, which are limited, however, for toxicological and ecological reasons.

2.9.1. Physical Properties

M_r	236.74
mp (closed capillary, as sublimes)	185 °C
bp at 101.325 kPa	185 °C
Crystal structure: rhombic	< 46 °C
triclinic	46 – 71 °C
cubic	> 71 °C
Specific density at 20 °C	2.094 g/cm^3
Vapor density at 185 °C	6.3 g/L
Vapor pressure at	
20 °C	0.290 kPa
40 °C	1.330 kPa
80 °C	2.400 kPa
120 °C	11.600 kPa
160 °C	45.320 kPa
180 °C	86.650 kPa
Heat of formation (liquid) $\Delta H°_{298}$	– 203.4 kJ/mol
Specific heat at 20 °C	0.615 kJ kg^{-1} K^{-1}
Heat of sublimation at 298 K	59 kJ/mol
Cryoscopic constant	5.6
Solubility in water at 22 °C	50 mg/kg

Hexachloroethane forms white crystals with a camphor-like odor. It is not flammable.

Some binary azeotropes of hexachloroethane are shown in Table 19.

2.9.2. Chemical Properties

Hexachloroethane is fairly stable and sublimes without decomposition. At temperatures above 250 °C, especially at 400 – 500 °C, it is cracked (disproportionated) [304]:

$$2\ C_2Cl_6 \longrightarrow C_2Cl_4 + 2\ CCl_4$$

Table 19. Binary azeotropes formed by hexachloroethane

wt%	Component	Azeotropic boiling point (101.3 kPa), °C
34	aniline	176.8
12	benzylic alcohol	182.0
25	monochloroacetic acid	171.2
28	o-cresol	181.3
30	phenol	173.7
15	trichloroacetic acid	181.0

Small amounts of chlorine are also formed. With metals like iron, zinc, and aluminum, chlorination reactions start at a higher temperature, forming metal chlorides and tetrachloroethylene. This reaction can be used for the synthesis of pure metal chlorides and for ultrapurification of metals.

At moderate temperature hexachloroethane is stable against aqueous alkali and acids. At temperatures above 200 °C, hydrolysis to oxalic acid occurs.

2.9.3. Production

Limited industrial uses of hexachloroethane do not justify large-scale production processes.

A primary source for hexachloroethane is from the production of tetrachloroethylene and carbon tetrachloride by chlorinolysis of hydrocarbons and chlorinated hydrocarbon residues (see p. 1364). It can be separated from the residues by distillation and fractionated crystallization.

For the intentional production of hexachloroethane, tetrachloroethylene is chlorinated batchwise in presence of iron chloride. The hexachloroethane crystallizes from the mother liquor and is isolated. The mother liquor is recycled and again chlorinated [305].

The photochemical chlorination of tetrachloroethylene is performed similarily [306].

2.9.4. Uses and Economic Aspects

Industrial uses of hexachloroethane are diminishing, so that most hexachloroethane is either recycled or incinerated with HCl — or chlorine recovery depending on the individual technology applied.

Smaller quantities are used for the synthesis of metal chlorides or for the production of fluorocarbons.

Hexachloroethane is one of the more toxic chloroethanes. Its use in plasticizer or rubber formulations is, therefore, decreasing.

Since no significant applications exist, production figures cannot be estimated.

Analysis, quality control, storage, and transportation of the chloroethanes are treated with the chloroethylenes in Sections 3.6 and 3.7

3. Chloroethylenes

The class of chloroethylene comprises:

Vinyl chloride (monochloroethylene, VCM)
1,1-Dichloroethylene (vinylidene chloride)
cis- and trans-1,2-Dichloroethylene
Trichloroethylene (Tri)
Tetrachloroethylene (perchloroethylene, Per, Perc).

3.1. Vinyl Chloride (VCM)

In addition to ethylene and NaOH, vinyl chloride [75-01-4] is one of the world's most important commodity chemicals. The 1984 worldwide consumption averaged 12–15 million t/a. About 25% of the world's total chlorine production is required for its production.

The importance of vinyl chloride results from the widespread use of poly(vinyl chloride), one of the most important polymers.

The first synthesis of vinyl chloride dates back to 1830–1834 when V. REGNAULT obtained it by dehydrochlorinating 1,2-dichloroethane with alcoholic potash. In 1902, it was obtained by BILTZ during thermal cracking of the same compound.

However, at that time, the state of the art in polymer science and technology was not very sophisticated, and this discovery did not lead to industrial or commercial consequences.

The basic work of F. KLATTE [307] on the polymerization of vinylic compounds gave rise to the industrial production of vinyl chloride in the 1930s.

Vinyl chloride was obtained by KLATTE in 1912 through catalytic hydrochlorination of acetylene [308]. This route was almost exclusively used for nearly 30 years. Because of the high energy requirements for acetylene production, its replacement by a cheaper substitute was a challenge for a long time.

From 1940–1950 on, acetylene could be partially replaced by ethylene, from which vinyl chloride was produced by direct chlorination to 1,2-dichloroethane and subsequent thermal cracking. The first large production units for this route were first constructed by Dow Chemical Co., Monsanto Chemical Co. and the Shell Oil Co. In these plants, the balance of HCl generated by dichloroethane cracking, however, was still achieved by acetylene hydrochlorination.

The complete changeover to the exclusive use of ethylene as a feedstock became possible when the large-scale oxychlorination of ethylene to 1,2-dichloroethane (see p. 1302) had been proven to be technically feasible (Dow Chemical, 1955–1958).

Since then, most plants use integrated, balanced DC – EDC – Oxy – EDC – VCM processes and more than 90% of the vinyl chloride presently produced in the Western World is based exclusively on ethylene.

In addition to its use as an intermediate in the production of trichloroethane (1,1,1- and 1,1,2-trichloroethane), most vinyl chloride is used for polymerization to PVC.

With the use of plasticizers and because of its high energy efficiency, PVC has become one of the most important industrial polymers. Even though it is one of the oldest polymers, its ready availability, relatively inexpensive production by large plants, and the continuing development of new formulations [309] with widespread uses secure its attractiveness in the future.

Several VCM plants were under construction in 1986. Due to the feedstock and market situation, the new plants will be preferentially located either in oil-producing or in developing countries.

3.1.1. Physical Properties

M_r	62.5
mp	– 153.8 °C
bp at 101.325 kPa	– 13.4 °C
ϱ at – 14.2 °C	0.969 g/cm³
at 20 °C	0.910 g/cm³
n_D^{20}	1.445
Vapor pressure at	
– 30 °C	51.000 kPa
– 20 °C	78.000 kPa
– 10 °C	115.000 kPa
0 °C	165.000 kPa
10 °C	243.000 kPa
20 °C	333.000 kPa
30 °C	451.000 kPa
40 °C	600.000 kPa
50 °C	756.000 kPa
Heat of formation (gaseous) ΔH^0_{298}	+ 35.2 kJ/mol
Specific heat (liquid, 20 °C)	1.352 kJ kg^{-1} K^{-1}
(vapor, 20 °C)	0.86 kJ kg^{-1} K^{-1}
Heat of evaporation (259.8 K)	20.6 kJ/mol
Critical temperature	429.8 K
Critical pressure	5600 kPa
Viscosity at – 40 °C	0.34 × 10^{-3} Pa s
at – 10 °C	0.25 × 10^{-3} Pa s
at 20 °C	0.19 × 10^{-3} Pa s
Dielectric constant at 17.2 °C	6.26
Flash point (open cup)	– 78 °C
Autoignition temperature	472 °C
Explosive limits in air	4 – 22 vol%
Solubility in water at 20 °C	0.11 wt%
Solubility of water in vinylchloride at – 15 °C	300 mg/kg

Vinyl chloride is a colorless, flammable gas at ambient temperature with a sweetish odor. It is soluble in most common organic liquids and solvents.

3.1.2. Chemical Properties

If oxygen and air are excluded, dry, purified vinyl chloride is highly stable and noncorrosive.

Above 450 °C, partial decomposition occurs yielding acetylene and hydrogen chloride. Trace amounts of 2-chloro-1,3-butadiene (chloroprene) may also be formed by acetylene-precursor dimerization.

Air combustion products of vinyl chloride are carbon dioxide and hydrogen chloride. Under oxygen deficient combustion, traces of phosgene may be formed. In oxidation reactions sensitized by chloride, monochloroacetaldehyde and carbon monoxide are obtainable from vinyl chloride [310].

In the presence of water, hydrochloric acid, which attacks most metals and alloys is formed. This hydrolysis most probably proceeds via a peroxide intermediate [311].

With air and oxygen, very explosive peroxides can be formed.

Because of the vinylic double bond, the most important reactions are polymerization reactions (co- and homopolymerization) and electrophilic or radicalic addition reactions — mainly chlorination or hydrochlorination — to yield 1,1,2-trichloroethane or 1,1-dichloroethane .

The substitution of the chlorine atom is more difficult to achieve. Vinyl anion addition reactions offering interesting synthetic routes are possible via the vinylmagnesium [312] and vinyllithium compounds [313].

Catalytic halogen exchange by hydrogen fluoride gives vinyl fluoride [314].

3.1.3. Production

The industrial production of vinyl chloride is based on only two reactions:

1) Hydrochlorination of acetylene:

$$C_2H_2 + HCl \longrightarrow CH_2 = CHCl \qquad \Delta H°_{298} = -99.2 \text{ kJ/mol}$$

2) Thermal cracking of 1,2-dichloroethane:

$$Cl-CH_2-CH_2-Cl \longrightarrow CH_2 = CHCl + HCl \qquad \Delta H^0_{298} = 100.2 \text{ kJ/mol}$$

All other reactions yielding vinyl chloride are industrially unimportant at present.

Acetylene hydrochlorination was mainly used in the past, when acetylene — produced via calcium carbide from coal — was one of the most important basic feedstocks for the chemical industry.

With the large-scale production of ethylene-derived polymers, such as polyethylene and polystyrene, and the general trend toward natural gas (United States), naphtha, and gas oil (Europe) as basic feedstocks, the cracker capacity increased substantially and ethylene became readily available at very competitive prices.

Besides the economic disadvantage of the higher priced hydrocarbon feed, the acetylene hydrochlorination has the drawback of not being balanced on the chlorine side because it requires only hydrogen chloride as a chlorine source.

With increasing demand for vinyl chloride and technical progress, the first balanced processes were established in the 1940s and 1950s, when acetylene was partially replaced by ethylene, which was converted to vinyl chloride by direct chlorination to 1,2-dichloroethane and subsequent thermal cracking. The hydrogen chloride from cracking could then be used for acetylene hydrochlorination:

$$\begin{aligned} C_2H_4 + Cl_2 &\longrightarrow C_2H_4Cl_2 \\ C_2H_4Cl_2 &\longrightarrow C_2H_3Cl + HCl \\ C_2H_2 + HCl &\longrightarrow C_2H_3Cl \\ \hline C_2H_2 + C_2H_4 + Cl_2 &\longrightarrow 2\,C_2H_3Cl \end{aligned}$$

By direct use of crack gas, without separation of ethylene and acetylene, this process is still currently pursued with some modifications.

With the introduction of the first large-scale Oxy–EDC plant (see p. 1302) by The Dow Chemical Co. in 1958, a balanced process based only on inexpensive ethylene became available and found rapid acceptance within the chemical industry.

Using this balanced process, vinyl chloride is made only by thermal cracking of 1,2-dichloroethane, which in turn is produced by direct chlorination or oxychlorination of ethylene. The latter process balances hereby the hydrogen chloride formed during cracking:

$$\begin{aligned} CH_2=CH_2 + Cl_2 &\longrightarrow C_2H_4Cl_2 \quad \text{DC-EDC} \\ CH_2=CH_2 + 2\,HCl + 1/2\,O_2 &\longrightarrow C_2H_4Cl_2 + H_2O \quad \text{Oxy-EDC} \\ 2\,C_2H_4Cl_2 &\longrightarrow 2\,C_2H_3Cl + 2\,HCl \quad \text{Cracking} \\ \hline 2\,CH_2=CH_2 + Cl_2 + 1/2\,O_2 &\longrightarrow 2\,C_2H_3Cl + H_2O \end{aligned}$$

Presently, more than 90% of the vinyl chloride produced is based on this route.

Since ca. 1960, considerable efforts were undertaken to replace ethylene by ethane as the basic feedstock. All ethane-based processes developed so far, however, lack selectivity, which causes increased recycle rates and losses by side reactions and requires higher capital expenditures, so that the cost advantage even for grass-root plants is marginal.

The ethylene-based processes have been improved considerably during recent years by numerous process modifications which resulted in higher yields and lower energy requirements.

However, with the development and availability of new catalysts, the ethane route may become more attractive in the future.

Due to sharply increasing energy costs, chlorine has also become a very important cost factor in vinyl chloride production. This explains the research and development on an electrolysis-free route to vinyl chloride.

3.1.3.1. Vinyl Chloride from Acetylene

The catalytic hydrochlorination of acetylene is possible in either the gaseous or the liquid phase. The gas-phase reaction dominates in industrial processes.

In this process the gaseous reactants are brought into contact with the catalyst at slightly increased pressure (0.1–0.3 MPa) and 100–250 °C (contact time 0.1–1 s) and then quenched and partially liquified. The reaction products are separated, recycled, or submitted to final purification.

The molar feed ratios, varying from almost equimolar to a 10-fold excess of HCl, depend heavily on catalyst performance. Acetylene conversions of 95–100% at almost quantitative yields are achieved.

The acetylene fed to the reactor has to be free of common catalyst poisons such as sulfur, phosphorus and arsenic compounds [315]. Unsaturated hydrocarbons must also be minimized in the feed because they may clog and inactivate the catalyst upon polymerization [316]. The hydrogen chloride must be free of chlorine to avoid explosion and should not contain chlorinated hydrocarbons, which could also act as catalyst poisons.

Water must be entirely excluded to avoid corrosion on structural materials. Because of gas-phase reaction and anhydrous conditions, most equipment is made from carbon steel. If water is used for quenching or HCl absorption, either brick-lined or polymer-made equipment is used for these parts.

Although fluidized-bed reactors have been patented [317], fixed-bed, multitubular reactors (tube size 1″–4″ inside diameter, 10–20 ft) are almost exclusively used. To avoid volatilization of the catalyst (tube side), temperature control and near to isothermal operation of the reactor is achieved by external cooling using a hot oil system or water. The heat of reaction can be transferred and used in the reboilers of the downstream purification equipment [318]. A special reactor design has been patented for use of diluted hydrogen chloride from incineration processes [319]. Mercury(II) chloride on activated carbon is used primarily as a catalyst in concentrations of 2–10 wt%. Several other metallic catalysts [320] as well as 1–3 vol% chlorine [321] have been proposed or patented. However, the mercury salt had proven to be the most effective. The reaction rate is first order with respect to acetylene [320], [322].

Activated carbon with specific properties is preferentially used as support [323]. Several patents deal with appropriate pretreatment procedures, such as oxidation and thermal activation, to improve these properties [324].

Zeolites and molecular sieves were also found to be suitable support materials [325].

Because the volatility of mercury is a very limiting factor for reactor operation and throughput, additives such as cerium chloride [326], thorium [327] and copper chloride [328], as well as polymers [329], have been proposed to reduce volatility. Mercury chloride–graphiteintercalated catalyst is also thought to possess only a very low sublimation tendency [330]. Because volatization of the mercury catalyst cannot be completely avoided, the loss of catalyst activity by moving hot spots must be minimized by operational means e.g. reversal of the flow through the reactor. Other possibilities

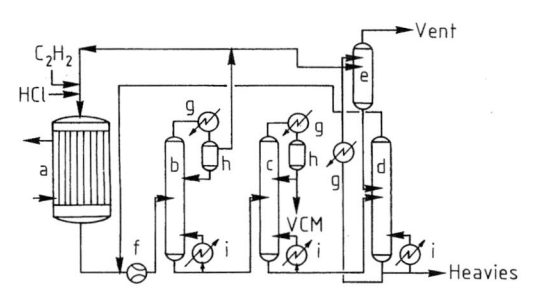

Figure 21. Production of vinyl chloride from acetylene and hydrogen chloride (schematic)
a) Reactor; b) Lights column; c) VCM column; d) Heavies stripper; e) Vent wash tower; g) Cooler; h) Knock-out drum; i) Reboiler

are the adjustment of the reactor load depending on the catalyst acitivity [331], operating two reactors in series, using fresh catalyst in the second to complete conversion from the less active first reactor at reduced heat load [332], or the fine tuning of catalyst activity and heat transfer liquid flow in a two-reactor system [333].

With fixed-bed reactor systems, time–space yields up to 300 kg m^{-3} h^{-1} are possible. An average of 70–80 kg m^{-3} h^{-1} is achieved over the lifetime of the catalyst.

The mercury is removed from the spent catalyst by either thermal (pyrolytic) treatment [334] or steam desorption [335]. The mercury-free carbon can be incinerated or reactivated.

Process Description (Fig. 21). Acetylene and hydrogen chloride are mixed and fed with recycle gas to the reactor. The gases leaving the reactor are compressed and fed to a first tower, where most of the vinyl chloride is withdrawn as a liquid from the bottom. Most of the overhead product (HCl, C$_2$H$_2$ and C$_2$H$_3$Cl) is recycled to the reactor. For removal of inert matter, a small part of this recycle stream is drawn off and washed with heavies — preferably 1,1-dichloroethane formed by competitive addition of HCl to vinyl chloride — to recover vinyl chloride and acetylene.

In the second tower, the crude vinyl chloride is purified and withdrawn at the head section. The heavy bottoms are submitted to a final stripping in the heavies column with the underflow of the washing tower and removed at the bottom for further use or for incineration. The overheads from the heavies treatment (acetylene and vinyl chloride) are recirculated to the compressor suction for optimal product recovery.

Other processes separate the heavy byproducts during the first distillation stage [336] or use water for HCl scrubbing [337], [338]. The resultant concentrated hydrochloric acid obtained can then be used for the final drying of the acetylene feed, which can contain significant amounts of moisture when produced by calcium carbide hydrolysis [338].

All process modifications attempt to achieve maximum recovery of the expensive acetylene and to avoid high pressures and temperatures, which may cause losses through polymerization [336], [339].

Vinyl Chloride from Crack Gases (Fig. 22). In the crack gas processes for vinyl chloride manufacture, unpurified acetylene produced by hightemperature cracking of naphtha [316], [340], [341] or methane [342], [343] is used. These processes are

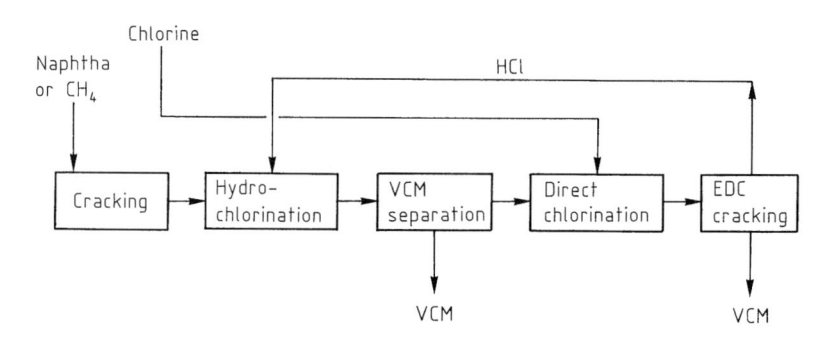

Figure 22. Balanced process for the production of vinyl chloride from crack gases

adventageous in that they do not require the cost-intensive separation of acetylene–ethylene mixtures [344].

The crack gas is fed directly to the hydrochlorinator, and the acetylene is converted to vinyl chloride which is then separated from the remaining constituents. Because all of the acetylene is consumed, the remaining ethylene is more easily separated or it can be introduced to a direct chlorination stage, where it is chlorinated to 1,2-dichloroethane, which is subsequently cracked to vinyl chloride [316], [340].

Since almost equimolar amounts of ethylene and acetylene can be achieved in the crack gas, the process can be balanced for chlorine.

Higher pressures (1.0 – 3.0 MPa) must be applied for the hydrochlorination stage in order to keep the reactor size reasonable. Because the acetylene is very diluted, hot spots are a smaller problem than with the pure acetylene process. For the chlorination and the cracking stage, standard technology can be used.

In a process developed by Solvay [341], hydrochlorination and chlorination are carried out together. The patent claims high yields without substantial formation of 1,1,2-trichloroethane, which can be formed by the addition of chlorine to vinyl chloride.

Another process modification [343] uses the quenching of the crack gases with chlorine at ca. 400 °C, which leads directly to vinyl chloride in yields up to 60%. An acetylene-based process has recently been patented, which uses HCl generated from magnesium chloride hydrate by pyrolysis [345].

All acetylene-based processes, however, have the distinct drawback of using, at least partially, the more expensive hydrocarbon feed. Completely ethylene-based processes are economically superior and in only a few cases is the acetylene route still competitively pursued.

3.1.3.2. Vinyl Chloride from 1,2-Dichloroethane

The cracking reaction of 1,2-dichloroethane can be carried out in the liquid or gas phase.

The *liquid-phase* dehydrochlorination of 1,2-dichloroethane is industrially unimportant because expensive chlorine is lost as a salt when 1,2-dichloroethane is treated with alkaline solutions:

$$CH_2Cl-CH_2Cl + NaOH \longrightarrow CH_2=CHCl + NaCl + H_2O$$

In addition, the aqueous process stream to be discarded poses severe environmental problems or requires extensive pretreatment. Even though a good dehydrochlorination reaction can be achieved by recent developments using phasetransfer catalysts [346], this process is not suitable or economical for large-scale production.

The *gas-phase* dehydrochlorination is the most important route and industrially used for the production of vinyl chloride.

It can be carried out as a pure pyrolytic reaction or in the presence of catalysts.

The noncatalyzed process is used by the majority of the vinyl chloride producers (e.g., Dow Chemical, Ethyl, B. F. Goodrich, Hoechst, ICI, Mitsui Toatsu, Monsanto, Stauffer), whereas only a few producers (e.g., Wacker) use catalytic cracking.

Improved furnace designs for the non-catalytic reaction have made conversions and yields comparable to those obtained by catalytic cracking.

Because of the time-consuming catalyst removal, shutdown periods are considerably longer for catalytic furnaces and the catalyst is an additional cost factor, so that pure thermal cracking may be currently the more economical process.

Noncatalytic Gas-Phase Reaction. The reaction occurs via a first-order free radical chain mechanism [262], [347], which starts with the homolytic cleavage of a C–Cl bond

1) $ClCH_2-CH_2Cl \longrightarrow ClCH_2-C\cdot H_2 + Cl\cdot$
2) $Cl\cdot + ClCH_2-CH_2Cl \longrightarrow ClCH_2-C\cdot HCl + HCl$
3) $ClCH_2-C\cdot HCl \longrightarrow CH_2=CHCl + Cl\cdot$
   $Cl\cdot + ClCH_2-CH_2Cl \longrightarrow ClCH_2-C\cdot HCl + HCl$
   etc.

The intermediate dichloroethane radical is stabilized by elimination of a chlorine radical, which propagates the chain.

The radical chain is terminated by recombination (reverse reaction to initiation) or wall collisions, as it is usual for this type of reaction.

Since chlorine or other radical species are important for the chain propagation, chlorine, [348]–[350] or chlorine delivering compounds such as tetrachloromethane [351] or hexachloroethane [349], as well as other radicals like oxygen [349] and nitrous oxide [349], [352] or other halogens (bromine and iodine) [350] can be added as initiators and promoters. The use of oxygen, however, is controversial because oxygen was also found to enhance coking of the furnace walls [353]. Because chlorine is readily available in vinyl chloride plants and because of its minimal interference, chlorine is primarily used as a promoter. Promoter concentration in the 1,2-dichloroethane feed may vary between a few hundred mg/kg and up to 5%. Good results were achieved when chlorine was fed at different points to the reaction zone [348], which may, however, be difficult to realize. When nitromethane was used as a promoter, high yields have been reported [354]. The addition of 1,1,2-trichloroethane was found to inhibit coke formation [355].

Even though 1,1-dichloroethane is more difficult to crack, good conversions are obtainable if the 1,1-dichloroethane concentration does not exceed 10% in the feed [152]. The crack reaction is industrially carried out at temperatures between 400 and 650 °C, preferably, however, between 500 and 550 °C. Reactor pressure may vary from 0.1 to 4.0 MPa. However, high-pressure processes (2.0–3.0 MPa) are preferred because high pressure reduces furnace size, improves heat transfer, and makes the downstream separation easier, due to increased boiling points. Mean residence time is about 10–20 s. The 1,2-dichloroethane conversion is kept at 50–60% per pass to control byproduct formation and coking, which significantly increases at higher conversion rates and causes yield losses. At these conversion rates, vinyl chloride yields of 95–99% are obtainable. High-purity 1,2-dichloroethane should be used because most impure technical-grade dichloroethane reduces conversion.

The crack furnace has a plug-flow reactor design with one or more tubes (1–10″ diameter, up to 4000 feet long) being placed in the convection zone of the furnace. The furnace may be equipped with a single burner or have multi-burner design. In most cases, natural gas is used as a burner feed; however, some plants use hydrogen-driven furnaces, using hydrogen from on-site chlor-alkali plants. Feed evaporation at ca. 200 °C and cracking at a much higher temperature is often carried out in the same furnace to make the best use of the fuel gas. Evaporation is preferably carried out in the upper, cooler part of the convection zone, whereas the cracking must take place in the lower, hotter part. Chromium–nickel alloys are the best construction materials [176].

Although several furnace designs have been patented [356] and the basic principles are quite similar, most vinyl chloride producers have developed their proprietary furnace technology for optimal yield and low shutdown frequency for pipe decoking. After leaving the reactors, the gases must be cooled down immediately to avoid yield losses by formation of heavy products. In most processes, this is achieved by a quench tower, where condensed and cooled 1,2-dichloroethane is recirculated at high rates. The heat withdrawn from the quench tower can be used for the reboilers of downstream distillation stages [357]. It is also possible to quench in two stages, first by indirect cooling (transfer line heat exchanger) and then by direct quenching [358]. Thus, substantial heat recovery is possible, which can be used by a hot oil system in other process stages. Only indirect cooling at the furnace tail pipe bears an increased risk of plugging the heat exchanger with coke and heavy byproducts.

For the downstream separation of the main constituents of the reaction gases, vinyl chloride, hydrogen chloride, and 1,2-dichloroethane, many processing possibilities have been patented [359]. However, a common principle is to first separate hydrogen chloride and then vinyl chloride from the reaction mixture by distillation. 1,2-Dichloroethane is then distilled from the remaining heavies or the whole stream is sent without separation to the DC – EDC section [360] (see p. 1300), where it can be economically purified.

The byproducts of 1,2-dichloroethane cracking can be theoretically divided into two groups:

1) *Volatile impurities* such as ethylene, acetylene, vinylacetylene, 1,3-butadiene, 2-chloro-1,3-butadiene, benzene, chlorobenzene, 1,2-and 1,1-dichloroethylene, 1,1-dichloroethane, 1,1,1- and 1,1,2-trichloroethane, methyl and methylene chloride, chloroform, and tetrachloromethane.
2) *Tars and coke.* To remove these, special filters are used [361].

2-Chloro-1,3-butadiene (chloroprene) forms tarry polymerization products, which plug the equipment when separated from 1,2-dichloroethane together with other light products. If this separation is not performed with the DC – EDC process, chlorination of the heavies is used to improve separation and to avoid excessive plugging of the equipment [362], [363].

Other volatile impurities must also be removed by posttreatment because they cannot be completely separated from the main products by distillation.

Acetylene, which codistills with the hydrogen chloride, is either converted to vinyl chloride by catalytic hydrochlorination [364] or selectively hydrogenated to ethylene [365], which does not interfere when the hydrogen chloride is used in the Oxy-EDC-process (see p. 1302. 1,3-Butadiene, which mostly contaminates the vinyl chloride fraction, can be removed by polymerization during extended residence time [366] or with Lewis-acid catalysts [367] by chlorination [362], [368], hydrochlorination [369], hydrogenation [370], or by reaction with chlorosulfuric acid [371].

The vinyl chloride obtained by distillation is suitable for polymerization. If necessary, remaining impurities can be removed by extractive distillation with acetonitrile [372], distillation in the presence of alcohols [373], or orthoesters [374] as acid scavengers, or treatment with calcium oxide [375] or zinc [376]. Removal of ionic species by an electrostatic field was also proposed [377]. Remaining traces of 1,3-butadiene can be removed by clay adsorbents [378].

Process Description (Fig. 23). Pure 1,2-dichloroethane is fed to the evaporator in the upper part of the cracking furnace. The gas phase is separated from the remaining liquids and fed to the cracking zone. After having passed the cracking zone in the furnace, the gases are cooled and quenched.

Hydrogen chloride is removed from the reaction mixture in the first distillation tower and sent back to the Oxy – EDC process or used for other purposes (e.g., methanol hydrochlorination). Vinyl chloride is distilled in the second tower and drawn off as a head product. It can be washed with diluted caustic in order to remove the last traces of hydrogen chloride and 1,2-dichloroethane.

The bottoms of the vinyl chloride column are purified in two more distillation stages. First, the low-boiling impurities are removed in the light ends column, followed by 1,2-dichloroethane separation from the heavy ends in the last tower. The purified 1,2-dichloroethane is recycled to the cracking furnace.

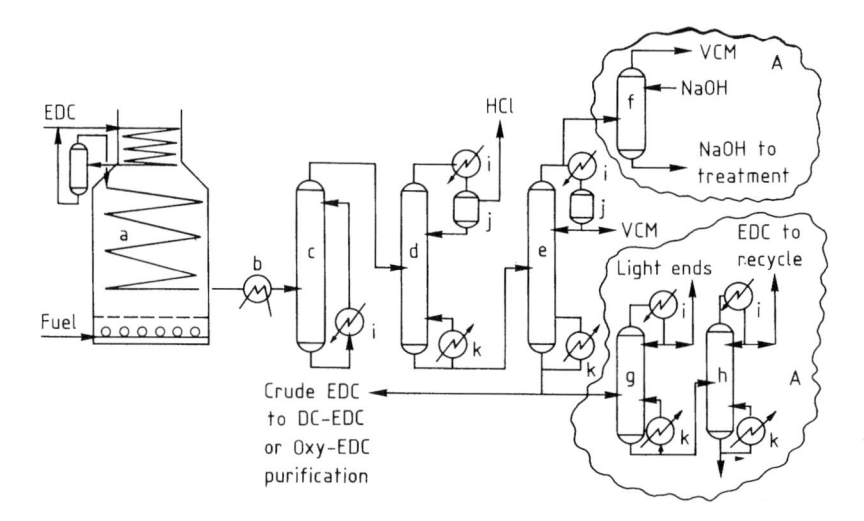

Figure 23. Schematic flow sheet for production of vinyl chloride by thermal cracking of 1,2-dichloroethane
a) Crack furnace; b) Transfer pipe heat exchanger; c) Quench tower; d) HCl distillation tower; e) VCM purification tower; f) VCM wash tower; g) Light-end tower; h) EDC – heavy end tower; i) Cooler; j) Knock-out drum; k) Reboiler A Process Modification

In integrated processes (see p. 1341), the last two stages are more economically combined with the purification of 1,2-dichloroethane from the DC and Oxy train.

If high temperature DC processes are used (see p. 1300), 1,2-dichloroethane can be purified in the DC reactor.

Catalytic Gas-Phase Reaction. The catalytic gas-phase dehydrochlorination is only used by a minority of vinyl chloride producers [315], [316], [379]. Higher selectivities toward vinyl chloride and less formation of coke, which is mainly due to the lower temperatures (200–450 °C), are often claimed as advantages.

1,2-Dichloroethane conversions, however, are not much improved compared to the noncatalytic process. On the average, 60–80% but mostly 60–70% conversion per pass is obtained.

In addition to activated carbon [380], which can be doped with ammonium salt promoters [381], a variety of other materials has been patented as catalysts, consisting of silicates [382], metal-promoted alumina [383], sodium chloride [384], and zeolites [385]. The dehydrochlorination of 1,2-dichloroethane by melts containing copper or other metals has also been described [386] (see also Transcat process, Chap. 3.1.3.4).

With the development of improved noncatalytic gas-phase processes, the catalytic route has lost most of its economic attractiveness. The higher costs of catalytic processes for catalyst and extended shutdown periods no longer compensate the slightly higher energy requirements of modern yield- and energy-optimized noncatalytic processes.

Photochemically Induced Gas-Phase Dehydrochlorination. Recently, considerable improvements in conversion and product quality were obtained by combining the thermal noncatalytic gas-phase reaction with a photochemical postreaction [387]. Either polychromatic actinic light from mercury, thallium, or tungsten lamps or, preferably, monochromatic light from suitable lasers is used as a light source for the excitation of 1,2-dichloroethane. The excited molecules then liberate chlorine atoms, which in turn start the free radical chain reaction:

$$C_2H_4Cl_2 + h\nu \longrightarrow (C_2H_4Cl_2)^* \quad \text{excited state}$$
$$(C_2H_4Cl_2)^* \longrightarrow C_2H_4Cl \cdot + Cl \cdot$$
$$Cl \cdot + C_2H_4Cl_2 \longrightarrow C_2H_3Cl_2 \cdot + HCl$$
$$C_2H_3Cl_2 \cdot \longrightarrow C_2H_3Cl + Cl \cdot$$

A photochemical postreactor implemented in existing thermal processes allows higher conversion rates at increased selectivity and decreased energy consumption.

This, however, has not yet been proven on an industrial scale.

Combined Process (Fig. 24). Most of the vinyl chloride is presently produced in so-called integrated, balanced processes comprising three units [188], [388].

1) direct ethylene chlorination
2) ethylene oxychlorination
3) 1,2-dichloroethane cracking

Ethylene and chlorine are basic feedstocks which are reacted in the direct chlorination unit to yield 1,2-dichloroethane. Additional 1,2-dichloroethane is produced in the oxychlorination process. The combined streams are fed to the cracking train, where vinyl chloride is obtained. The hydrogen chloride formed during cracking is recycled and consumed in the oxychlorination process. Thus, the process is balanced on hydrogen chloride. Integrated processes are not only advantageous due to lower energy requirements, due to the fact that energy-consuming steps can be combined with exothermic reactions, but they also allow variations in the chlorine distribution of manufacturing sites producing other chlorinated hydrocarbons and convert the largely unusable byproduct hydrogen chloride into a valuable product.

Differences in process technology used by the individual vinyl chloride producers are mainly due to the different technologies applied in the processes, such as high- or low-temperature direct chlorination, fixed- or fluidized-bed oxychlorination, which determine the needs for purification equipment and the energy requirements.

The most important processes presently used are described in detail elsewhere [184]. The major unit ratios for a balanced, air-based process are given in Table 20.

New Developments. Recently, a new route was developed [390] which allows chlorine-free production of vinyl chloride by bringing ethylene in contact with an aqueous solution containing copper(II) chloride and iodine:

$$4\ CuCl_2 + I_2 + 3\ C_2H_4 \longrightarrow 3\ C_2H_4Cl_2 + Cu_2Cl_2 + Cu_2I_2$$

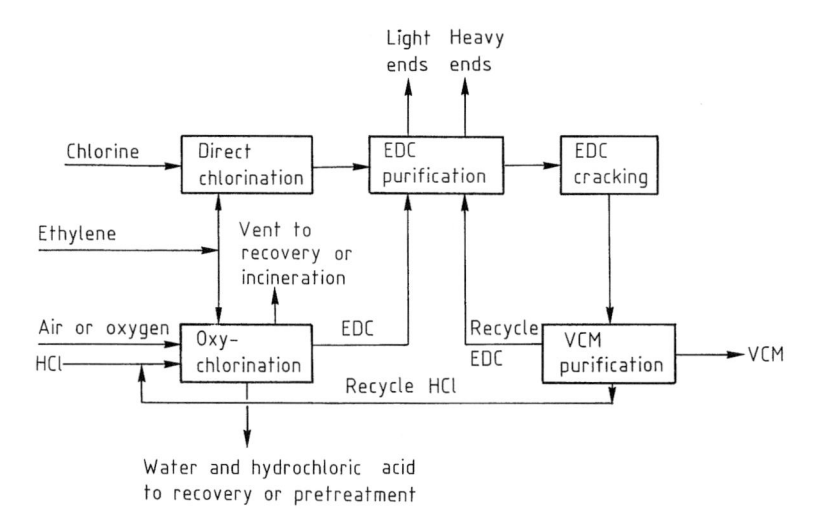

Figure 24. Ethylene-based integrated balanced process for the production of vinyl chloride

The copper salts are reoxidized after dichloroethane stripping by oxygen and an amine hydrochloride acting as a chlorine source:

$3/2\ O_2 + 6\ (CH_3)_3N \cdot HCl + Cu_2Cl_2 + Cu_2I_2 \longrightarrow 4\ CuCl_2 + I_2 + 6\ (CH_3)_3N + 3\ H_2O$

The amine hydrochloride is regenerated by sodium chloride and carbon dioxide.

$(CH_3)_3N + CO_2 + NaCl + H_2O \longrightarrow (CH_3)_3N \cdot HCl + NaHCO_3$

The driving force for this reaction is — quite similar to the Solvay soda ash process — the poor solubility of the hydrogen carbonate, which can be removed and calcined, whereby some carbon dioxide is recovered.

$2\ NaHCO_3 \longrightarrow Na_2CO_3 + CO_2 + H_2O$

The overall process yields 1,2-dichloroethane and soda ash from ethylene, oxygen, carbon dioxide, and sodium chloride. The dichloroethane must then be conventionally converted to vinyl chloride by cracking:

$C_2H_4 + 1/2\ O_2 + 2\ NaCl + CO_2 \longrightarrow C_2H_4Cl_2 + Na_2CO_3 C_2H_4Cl_2 \longrightarrow C_2H_3Cl + HCl$

Hydrogen chloride from the cracking reaction can be used to form aminohydrochloride so that balancing with sodium chloride or hydrogen chloride is possible.

$(CH_3)_3N + HCl \longrightarrow (CH_3)_3N \cdot HCl$

Whether this process will be able to replace the present route will depend primarily on the feasibility of its overall industrial verification. Financial requirements for a grass root plant, however, may be considerably lower than for a conventional plant if the required chlorine capacity is added. The requirement for inexpensive energy instead of expensive electrical power is a further advantage. The byproduct soda ash, even though

Table 20. Major unit ratios for an integrated, balanced vinyl chloride process

Component	Unit ratio (kg/kg VCM)
Raw materials:	
Ethene	0.4656
Chlorine	0.5871
Air	0.7322
Water	0.0171
Byproducts:	
Lights	0.0029
Heavies	0.0023
Vents	0.6727 (0.5779 of this is N_2)
Aqueous streams	0.1218

readily available from natural sources, should not be very limiting if one considers the fact that the conventional process also produces NaOH byproduct during brine electrolysis. However, plans by Akzo to built a commercial scale plant have been abandoned in 1986 [391].

3.1.3.3. Vinyl Chloride from Ethylene by Direct Routes

Because ethylene chlorination and oxychlorination are both highly exothermic reactions, numerous attempts have been made to combine one or both reactions with the endothermic cracking reaction for 1,2-dichloroethane, i.e., not isolating the intermediate, but directly producing vinyl chloride by high-temperature chlorination or oxychlorination of ethylene. Several processes have been patented which claim direct synthesis of vinyl chloride from ethylene and chlorine or hydrogen chloride at temperatures between 300 and 600 °C.

In *direct chlorination processes* an excess of ethylene is often used to minimize byproduct formation [392]. Other processes use two reaction zones [393] or make use of an inert fluidized bed for heat transfer [393]. The hydrogen chloride formed can be consumed in a separate oxychlorination unit [394]. Additional processes have been proposed [395].

If the *oxychlorination of ethylene* is carried out at a temperature above 350 °C, substantial amounts of vinyl chloride are obtained. As with the Oxy-EDC process (p. 1302), polyvalent metals are used as catalyst. However, low-surface area supports (e.g., α-alumina) are preferred [396], [397] because high-surface-area catalysts tend to ward rapid coking and deactivation by polymer formation at higher temperatures. The high temperature required can also cause considerable yield losses by "deep" oxidation of ethylene to CO and CO_2. Oxygen feed below the stoichiometric ratio may be required to control these unwanted side reactions [397].

Feeding the excess ethylene to the high-temperature oxychlorination reactor and converting the surplus to 1,2-dichloroethane in a second DC stage, which is then recycled and cracked in the Oxy reactor may be another possibility [398]. Further

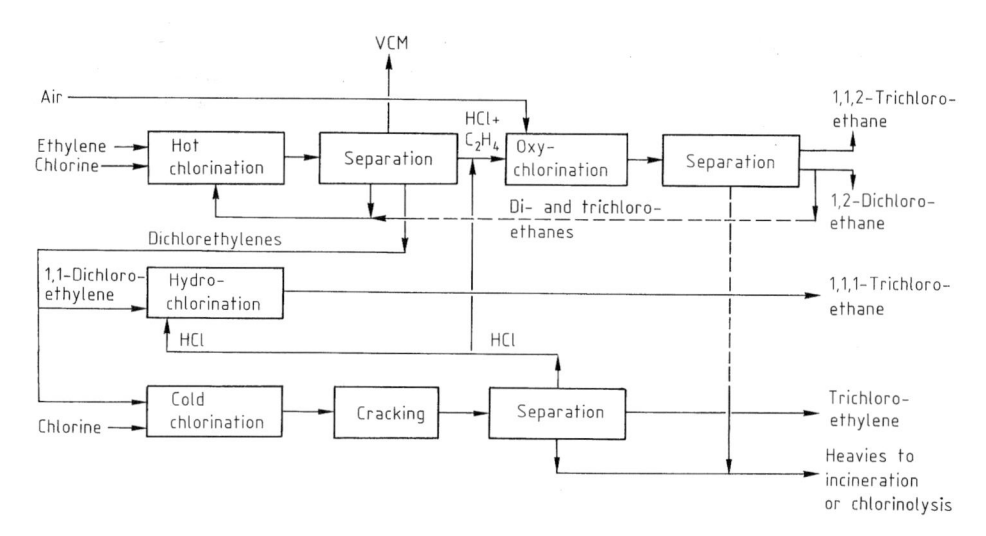

Figure 25. Schematic principle of the Atochem process for the production of vinyl chloride and other chlorinated hydrocyrbons

possibilities comprise contacting ethylene with melts containing copper(II) chloride. Vinyl chloride is formed and the reduced copper salt can be regenerated by chlorination or oxychlorination and then be recycled [399].

Common to all direct routes is the fact that the processes are difficult to control and operate and are characterized by poor selectivities because ethylene, vinyl chloride, chlorine, and hydrogen chloride undergo considerable addition and elimination reactions at elevated temperatures. Typical byproducts of direct high-temperature processes are dichloroethylenes and tri- and tetrachloroethylene. The low yield of vinyl chloride together with the need to dispose of high quantities of byproducts has considerably limited the industrial implementation of direct processes.

In a process, however, which is carried out on an industrial scale (approx. 150 000 t/a vinyl chloride) by Atochem, France, the byproducts are integrated and several other usable hydrocarbons in addition to vinyl chloride are intentionally produced (Fig. 25) [254], [400].

The process consists of high temperature chlorination of ethylene. The reaction products are separated, yielding vinyl chloride, dichloroethylenes, chloroethanes, and hydrogen chloride. Excess ethylene together with hydrogen chloride is oxychlorinated in a fluidized-bed reactor to give primarily 1,2-di- and 1,1,2-trichloroethane, which can be drawn off or recycled to the hot chlorination reactor.

Dichloroethylenes can be treated differently. 1,1-Dichloroethylene can be hydrochlorinated to give 1,1,1-trichloroethane (see p. 1315). Trichloroethylene is produced via noncatalytic cold chlorination of dichloroethylenes and subsequent cracking of the tetrachloroethanes obtained. Heavy byproducts accounting for ca. 3% of the total production are incinerated and aqueous HCl is recovered. However, they can also be used for perchloroethylene synthesis. The process is said to allow considerable fluctuations in production ratios, ranging from 2.5 : 1 to 0.8 : 1 (VCM : chlorinated solvents) allowing good responsiveness to market demands. Overall carbon yields of 94.5% are obtained.

3.1.3.4. Vinyl Chloride from Ethane

Numerous attempts have been made to convert ethane directly to vinyl chloride because this would save the processing costs for ethylene.

Ethane is readily available, particularly on the U.S. golf coast, and used as a feedstock for ethylene crackers. The direct feed of ethane to VCM plants could, thus, considerably decrease the raw material costs and make the plants less dependent on cracker capacity.

Conversion of ethane to vinyl chloride can be performed by various routes:

1) High-temperature chlorination:

$$C_2H_6 + 2\ Cl_2 \longrightarrow C_2H_3Cl + 3\ HCl$$

2) High-temperature oxychlorination:

$$C_2H_6 + HCl + O_2 \longrightarrow C_2H_3Cl + 2\ H_2O$$

3) High-temperature oxidative chlorination (combines 1 + 2)

$$2\ C_2H_6 + 3/2\ O_2 + Cl_2 \longrightarrow C_2H_3Cl + 3\ H_2O$$

A major drawback of ethane, however, is its lack of molecular functionality. In contrast to ethylene, which easily undergoes chlorine addition, ethane must first be functionalized by substitution reactions, which gives rise to a variety of consecutive and side-chain reactions (Fig. 26).

The reaction must, therefore, be kinetically controlled in order to obtain a maximal vinyl chloride yield. Conversion must be sacrificed because thermodynamic conditions would lead to stable products like tetrachloroethylene. Consequently, however, high recycle rates of unconverted ethane and byproducts such as ethyl chloride and dichloroethanes must be accepted. With special catalysts and at optimized conditions, however, ethane conversions of > 96% have been reported from oxychlorination reactions [401]. Vinyl chloride yields average 20–50% per pass. Ethylene, ethyl chloride, and 1,2-dichloroethane are obtained as major byproducts. The formation of carbon oxides can be controlled with carbon yield losses of 3–10%. The ethylene formed can either be recycled or oxychlorinated and cracked in a conventional manner. Some special processes have been suggested to purify and concentrate the aqueous hydrochloric acid obtained [402].

Another patent comprises the chlorination of ethylene – ethane mixtures with staged addition of chlorine to avoid an explosive reaction. Addition of ethylene is thought to suppress the formation of higher chlorinated byproducts [403]. Conversion and yields are comparable to the oxychlorination reactions mentioned above.

Figure 26. Ethane chlorination pathways

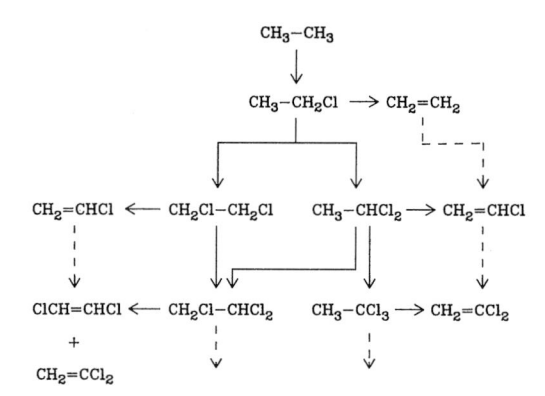

```
                    CH₃–CH₃
                       ↓
                   CH₃–CH₂Cl  →  CH₂=CH₂
            ┌──────────┤              ╎
            │          ↓              ╎
            │          ↓              ↓
CH₂=CHCl ← CH₂Cl–CH₂Cl    CH₃–CHCl₂ → CH₂=CHCl
   ╎          │              ╎           ╎
   ↓         ↓ ↓             ↓           ↓
ClCH=CHCl ← CH₂Cl–CHCl₂   CH₃–CCl₃ → CH₂=CCl₂
   +          ╎              ╎
CH₂=CCl₂      ↓              ↓
```

Some balanced ethane-based processes have been developed according to the following reaction schemes:

1) Hot ethane chlorination → VCM separation → oxychlorination of residual ethylene and chloroethane to yield additional VCM [404].
2) Thermal chlorination of ethane to ethyl chloride → oxychlorination without separation from hydrogen chloride to vinyl chloride [405].
3) Ethane chlorodehydrogenation to ethylene and hydrogen chloride → oxychlorination to 1,2-dichloroethane → thermal cracking to vinyl chloride [406].

Another balanced ethane-based process was developed by The Lummus Co. [172], [407]. In the final version of this so-called Transcat process, ethane and chlorine, as well as the recycle products ethylene, ethyl chloride, and hydrogen chloride are fed to a melt of copper(II) chloride and potassium chloride. Vinyl chloride is formed and separated. The reduced melt is transferred by an airlift system and regenerated with air, chlorine, or hydrogen chloride. It is then fed back together with the recycle products, which may also contain 1,2- and 1,1-dichloroethane, to the oxychlorinator. Even though the process has been operated on a pilot-plant scale, it has not been accepted by the vinyl chloride producers.

More ethane-based processes for chlorination [408] and oxychlorination [409] can be found in the literature. Because 1,1-dichloroethane is preferably formed by ethane chlorination or oxychlorination, its thermal cracking reaction has been intensively studied [410]. The photochemical chlorination of ethane at 250–400 °C yields ethyl chloride and the dichloroethanes (preferably the 1,1-isomer) as major products [411]. Only small amounts of vinyl chloride are formed.

3.1.3.5. Vinyl Chloride by Other Routes

Vinyl chloride can be obtained as a valuable byproduct in the synthesis of such important fluorocarbons as tetrafluoroethylene (F-1114) and chlorotrifluoroethylene

(F-1113) when saturated chlorofluorocarbons are catalytically dechlorinated by ethylene [412].

Oxidative condensation of chloromethane derived from methane or methanol can also form vinyl chloride [413].

The catalytic dehydrochlorination of 1,1,2-trichloroethane [414] or its catalytic dechlorination with ethylene [415] both yield vinyl chloride. Although both processes are not suitable for large-scale production, they could be used to recover vinyl chloride from a major byproduct when there is no demand for 1,1,2-trichloroethane. The electrochemical dechlorination of 1,1,2-trichloroethane is also possible [416].

Some ethylene-based processes comprise the production of vinyl chloride during brine electrolysis in the presence of ethylene [417], ethylene oxychlorination by nitrosyl chloride (NOCl) [418], and a bromine-based process which converts vinyl bromide into vinyl chloride by reaction with hydrogen chloride [419]. Ethane sulfochlorination has been proposed as a very exotic route similar to oxychlorination, but using sulfur instead of oxygen [420].

3.1.4. Uses and Economic Aspects

About 95% of the world production of vinyl chloride is used for the production of poly(vinyl chloride) (PVC). Thus, vinyl chloride is very dependent on the major processors of PVC as well as the housing and automotive industry with its frequent fluctuations. The rest of the vinyl chloride production goes into the production of chlorinated solvents, primarily 1,1,1-trichloroethane.

Total world capacity of vinyl chloride is about 17 million t/a. As shown in Table 21, more than half of the world's total capacity (64%) is concentrated in Western Europe and the United States. The annual growth rate is estimated between 1 and 5%, depending on the economic situation. However, capacity utilization presently averages only 70–80% [422], which may not be sufficient to make production of vinyl chloride profitable. Several new vinyl chloride plants are being planned or are under construction in Eastern Europe and in developing and oil-producing countries (see Table 22) [423]. This significant increase in capacity outside of the traditional VCM–PVC countries and its consequences may in the longrun cause a geographical shift of VCM production.

Table 21. World wide vinyl chloride capacity and production (1985, estimated)

	Capacity, 10^3 t/a	Production, 10^3 t/a
North America	4 900	4 000
South America	500	400
Western Europe	5 700	5 000
Eastern Europe/USSR	2 100	1 500
Middle/Far East	3 000	2 500
Rest	350	180
	16 550	**13 580**

Table 22. New vinyl chloride plants planned or under construction

Country	Capacity, 10^3 t/year	Technology source
Argentina	130	B. F. Goodrich
Brazil	150	B. F. Goodrich
China (P.R.)	200	Mitsui Toatsu
China (P.R.)	200	Mitsui Toatsu
Egypt	100	Mitsui Toatsu
Egypt	100	–
India	103	–
Iran	150	Toyo Soda
Nigeria	145	–
Poland	205	PPG Ind.
Portugal	110	Mitsui Toatsu
Saudi Arabia	300	B. F. Goodrich
Turkey	117	Solvay-ICI
USSR	270	Hoechst-B.F. Goodrich

3.2. 1,1-Dichloroethylene (Vinylidene Chloride, VDC)

1,1-Dichloroethylene [75-35-4] is important for upgrading 1,1,2-trichloroethane, which is very often an unwanted byproduct. Thus, 1,1-dichloroethylene is an intermediate in the production of 1,1,1-trichloroethane from 1,1,2-trichloroethane (see p. 1314). It is also used as a monomer for the production of poly(vinylidene chloride) (PVDC) and its copolymers, which are important barrier materials in the food packing industry. Of all important chloroethanes and -ethylenes, vinylidene chloride has presently the smallest sales volume. Because of its unique applications in polymers for food containers, longterm demand will grow, however.

3.2.1. Physical Properties

M_r	96.94
mp	− 122.6 °C
bp at 101.325 kPa	31.6 °C
ϱ at 20 °C	1.214 g/cm³
n_D^{20}	1.42468
Vapor pressure at	
− 60 °C	0.782 kPa
− 40 °C	3.320 kPa
− 20 °C	10.850 kPa
0 °C	28.920 kPa
20 °C	66.340 kPa
Heat of formation (liquid) $\Delta H°_{298}$	− 24.5 kJ/mol
Specific heat (liquid, 25 °C)	1.15 kJ kg⁻¹ K⁻¹
(gas, 25 °C)	0.69 kJ kg⁻¹ K⁻¹
Heat of evaporation at 25 °C	26.5 kJ/mol
Heat of fusion at − 122.6 °C	6.51 kJ/mol
Heat of polymerization at 25 °C	− 75.4 kJ/mol
Critical temperature	494 K
Critical pressure	5200 kPa
Viscosity at 20 °C	3.3×10^{-4} Pa · s
Dielectric constant at 16 °C	4.67
Flash point (open cup)	− 30 °C
Autoignition temperature	+ 460 °C
Explosive limits in air	6 − 16 vol%
Solubility in water at 20 °C	2200 mg/kg
Solubility of water in vinylidene chloride at 20 °C	320 mg/kg

1,1-Dichloroethylene is a colorless clear liquid with a sweetish odor. It is soluble in most organic solvents.

3.2.2. Chemical Properties

Vinylidene chloride belongs to the less stable chloroethylenes because it is very susceptible to both oxidation and polymerization. To avoid these reactions, oxygen scavengers such as amino and sulfur compounds or phenol derivatives must be added as stabilizers [424]. Most stabilizers in current use prevent autoxidative polymerization. They must not be removed, however, before vinylidene chloride is industrially polymerized. When pyrolyzed above 400 – 450 °C, chloroacetylene and hydrogen chloride are obtained.

With copper and other heavy metals or their salts, highly explosive acetylenes are formed. Therefore, copper and its alloys should not be used as a construction material if contact with vinylidene chloride is anticipated. Combustion with an excess of air yields carbon dioxide and hydrogen chloride. Traces of phosgene may be formed under oxygen deficient conditions.

Vinylidene chloride can be easily chlorinated at a slightly elevated temperature to give 1,1,1,2-tetrachloroethane.

The most important reaction, however, is hydrochlorination in the presence of a Lewis-acid catalyst for the production of 1,1,1-trichloroethane (see p. 1314).

3.2.3. Production

1,1-Dichloroethylene is almost exclusively produced from 1,1,2-trichloroethane. This allows the recovery of valuable hydrocarbon and chlorine from a byproduct, which is obtained in large quantities during the production of 1,2-dichloroethane and 1,1,1-trichloroethane (see 2.3.3 and 2.4.3).

1,1,2-Trichloroethane is converted to vinylidene chloride by dehydrochlorination, which can be carried out by two routes:

1) Liquid-phase dehydrochlorination in the presence of alkali, e.g., NaOH:

$$CHCl_2-CH_2Cl + NaOH \longrightarrow CCl_2 = CH_2 + NaCl + H_2O$$

2) Pyrolytic gas-phase cracking at elevated temperatures:

$$2\ CHCl_2-CH_2Cl \xrightarrow{\Delta} CCl_2 = CH_2 + ClCH = CHCl + 2\ HCl$$

The latter route has the advantage that valuable chlorine is recovered as hydrogen chloride, which can be used again for oxychlorination processes. By this route, however, vinylidene chloride selectivity is low, since the formation of 1,2-dichloroethylenes is favored [425].

In the liquid phase reaction, vinylidene chloride selectivity is well above 90%; however, hydrogen chloride is lost as a salt.

At present, the liquid-phase reaction is dominant. The development of new catalysts with increased selectivity and high stability could change this situation in the future.

Liquid-Phase Reaction. The liquid-phase reaction is carried out with aqueous solutions of alkali or alkaline earth hydroxides. As with the low solubility of alkaline earth hydroxides, the free concentration is small, NaOH (10–15 wt%) is widely preferred to increase the reaction rate. At optimum conditions [426], vinylidene chloride yields of 94–96% are obtained. The higher alkali concentration, however, bears the risk of formation of chloroacetylenes, which tend toward explosive decomposition. Different methods have been patented and are used to minimize this reaction. These include thorough mixing [427], adjustment of proper feed ratios [428], the addition of amines [429], the addition of calcium and magnesium hydroxides as emulsifiers and buffers [430], and the use of sodium chloride containing NaOH [431] (cell effluent from brine electrolysis can be used directly). Monochloroacetylene in crude vinylidene chloride can also be removed by hydrochlorination in aqueous Cu_2Cl_2–HCl solutions [432].

1,1,2-Trichloroethane can be crude [428], even heavy ends from the Oxy–EDC process (see p. 1302) can be used [433], [434]. However, washing of the feed with water is beneficial to the finished product quality [435]. To avoid polymerization during purification, the feed streams should be free of oxygen. Stabilizers are added during distillation or even to the reaction mixture [426] to inhibit polymerization.

The reaction proceeds at temperatures between 80 and 100 °C at an acceptable rate. With promoters such as charcoal, alumina, and silica [437] or quaternary ammonium salts [434], high selectivity is obtainable at even lower temperatures (≤ 40 °C).

This process is carried out continuously in packed-bed reactors to allow thorough contact. Life steam is injected to distill the vinylidene chloride. Unconverted 1,1,2-trichloroethane is recycled from the appropriate sections. A plug-flow type reactor with consecutive flash distillation has also been described [427].

Due to the alkaline conditions, nickel and some of its alloys are the best suited construction materials. This minimizes the risk of acetylide formation.

In order to obtain polymer grade vinylidene chloride, the stripped product is washed with alkali and water, dried, and fractionally distilled. Azeotropic distillation with methanol (6 wt %) and subsequent washing with water is also possible [438]. Even if purified, vinylidene chloride should be used directly and not stored for more than two days.

Recently, the use of aprotic polar solvents like dimethylsulfoxide and dimethylformamide instead of aqueous alkali has been patented [439]. The high cost of solvents, however, may not justify the large-scale production.

Gas-Phase Reaction. The noncatalyzed thermal decomposition yields *cis*- and *trans*-1,2-dichloroethylene together with 1,1-dichloroethylene in molar ratios of ca. 0.7 (1,1-/1,2-isomer). Radical chain as well as unimolecular mechanisms have been proposed for the decomposition reaction [425]. Radical chain sequences are very likely because the reaction rate can be increased by chain initiators such as chlorine and chlorine-releasing compounds [440].

By photochemically induced reactions, the formation of the 1,1-disubstituted isomer is slightly improved [441].

For industrial purposes, two routes are used to overcome the problems related to the formation of unusable 1,2-substituted isomers.

The first route comprises pure, noncatalytic, thermal cracking of 1,1,2-trichloroethane with selectivity of ca. 30 – 40 % for vinylidene chloride. The 1,2-dichloroethylenes are separated and further chlorinated to tetrachloroethane, which can be recycled and cracked in the same reactor to yield trichloroethylene. This process can be performed by feeding 1,1,2-trichloroethane and chlorine together with the recycled 1,2-dichloroethylene [442]. It can be combined with the production of 1,1,1-trichloroethane by direct ethane chlorination [443] or with a direct chlorination processes for ethylene to yield vinyl chloride [254], [444], as is the case in the Atochem process (see Chap. 3.1.3.3).

The second route makes use of specific catalysts to increase selectivity. Besides simple catalysts like sodium chloride [384], barium chloride [445], and alumina or silica, which can be activated by steam treatment [446], [447], numerous other catalysts have been patented. These catalysts mainly consist of alkali metal salts [448]–[450], alkali metal hydroxides [451], metal fluorides [452], and nitrogen-containing compounds [453] on appropriate supports such as alumina and silica gels.

The use of alkaline catalysts seems to be important because basic centers on the catalysts are mandatory for high vinylidene chloride selectivity [451], whereas acidic centers favor the formation of the 1,2-isomers. Because the activity increases with the increasing atomic mass of the metal atom [451], cesium salts are preferred [448]–[450].

Additional doping with other metals may be beneficial for further selectivity and prolonged catalyst lifetime [449], [450]. The role of the supports pore structive has also been investigated [454].

For increased selectivity, the feed is frequently diluted with inert gases such as nitrogen. Industrially more important, however, is the possibility of using vinyl chloride as a diluent [450]. When methanol is added to trichloroethane, the hydrogen chloride is consumed by the alcohol and vinylidene chloride is obtained with methyl chloride [455].

Although excellent vinylidene chloride selectivities have been reported, catalytic gas-phase dehydrochlorination is still in the developmental stage. A major drawback is the marked tendency of vinylidene chloride to polymerize on catalyst surfaces [456] which requires frequent shut downs and catalyst turnarounds. This offsets its advantages over the noncatalytic gas phase or liquid-phase reaction. The latter two methods are both used in industrial scale processes.

Other Methods. If required, vinylidene chloride can be obtained from thermal cracking of 1,1,1-trichloroethane [457], which is, however, not always economical. Other routes use vinyl chloride oxychlorination [458] or tetrachloroethane dehydrochlorination [459] and high temperature reaction of methane with chlorinating agents [460]. All of these methods are presently of little interest, because the basic feedstock for the conventional route, 1,1,2-trichloroethane, is easily available.

3.2.4. Uses and Economic Aspects

Vinylidene chloride (VDC) is often captively used for the production of 1,1,1-trichloroethane. Apart from this, VDC is a basic material for poly(vinylidene chloride) (PVDC) or its copolymers with vinyl chloride, acrylonitrile, methacrylonitrile, and methacrylate. With these materials, barrier layers for food packaging are formed as well as laminated and polymer sandwich type films.

The annual production rate for the Western World amounts to about 150 000–200 000 t, of which ca. 120 000 t are used for PVDC and its copolymers. The rest is converted to 1,1,1-trichloroethane.

Because of the unique properties of PVDC, the longterm demand will probably increase. It could easily be satisfied because the 1,1,2-trichloroethane feedstock is available from other chlorinated hydrocarbon processes.

3.3. 1,2-Dichloroethylene

Dichloroethylenes (*cis:* [156-59-2]; *trans:* [156-60-5]) often occur as an isomeric mixture during the production of chlorinated hydrocarbons, where they are produced by sidereactions, e.g., by thermal decomposition of 1,1,2-trichloroethane or from acetylene by chlorine addition.

Because there are scarcely any industrial uses for these two compounds, they are often converted to trichloro- and tetrachloroethylene.

3.3.1. Physical Properties

trans-1,2-Dichloroethylene

M_r	96.94
mp	– 49.44 °C
bp at 101.325 kPa	48.5 °C
ϱ at 20 °C	1.260 g/cm³
n_D^{20}	1.4462
Vapor pressure at	
– 20 °C	5.300 kPa
– 10 °C	8.500 kPa
0 °C	15.100 kPa
10 °C	24.700 kPa
20 °C	35.300 kPa
30 °C	54.700 kPa
40 °C	76.700 kPa
Heat of formation (liquid) $\Delta H°_{298}$	– 24.3 kJ/mol
Specific heat (liquid, 20 °C)	1.158 kJ kg^{-1} K^{-1}
Heat of evaporation (boiling point)	28.9 kJ/mol
Critical temperature	516.5 K
Critical pressure	5510 kPa
Viscosity at 20 °C	0.404×10^{-3} Pa · s
Surface tension at 20 °C	25×10^{-3} N/m
Coefficient of cubical expansion (15 – 45 °C)	0.00136 K^{-1}
Dielectric constant at 20 °C	2.15
Dipole moment	0 esu
Flash point	4 °C
Autoignition temperature	460 °C
Solubility in water at 25 °C	0.63 wt%
Solubility of water in *trans*-1,2-dichloroethylene at 25 °C	0.55 wt%

trans-1,2-Dichloroethylene is a colorless, light liquid with a sweetish odor. It forms explosive mixtures with air (9.7 – 12.8 vol% 1,2-dichloroethylene).

trans-1,2-Dichloroethylene forms azeotropic mixtures with ethanol (6 wt% ethanol, *bp* 46.5 °C) and water (1.9 wt% water, *bp* 45.3 °C). A ternary aezotrope of all three components (1.4 wt% ethanol, 1.1 wt% water) has a *bp* of 44.5 °C.

cis-1,2-Dichloroethylene

M_r	96.94
mp	– 81.47 °C
bp at 101.325 kPa	60.2 °C
ϱ at 20 °C	1.282 g/cm^3
n_D^{20}	1.4490
Vapor pressure at	
– 20 °C	2.700 kPa
– 10 °C	5.100 kPa
0 °C	8.700 kPa
10 °C	14.700 kPa
20 °C	24.000 kPa
30 °C	33.300 kPa
40 °C	46.700 kPa
Heat of formation (liquid) $\Delta H°_{298}$	– 26.8 kJ/mol
Specific heat (liquid, 20 °C)	1.176 kJ kg^{-1} K^{-1}
Heat of evaporation (boiling point)	30.2 kJ/mol
Critical temperature	544.2 K
Critical pressure	6030 kPa
Viscosity at 20 °C	0.467×10^{-3} Pa · s
Surface tension at 20 °C	28×10^{-3} N/m
Coefficient of cubical expansion (15 – 45 °C)	0.00127 K^{-1}
Dielectric constant at 20 °C	9.31
Dipole moment	0.185 esu
Flash point	6 °C
For autoignition temperature and explosive limits in air, see *trans*-1,2-dichloroethylene	
Solubility in water at 25 °C	0.35 wt%
Solubility of water in *cis*-1,2-dichloroethylene at 25 °C	0.55 wt%

cis-1,2-Dichloroethylene is a colorless, light liquid with a sweetish odor. It forms an azeotropic mixture with ethanol (9.8 wt% ethanol, *bp* 57.7 °C), methanol (13 wt% methanol, *bp* 51.5 °C) and water (3.35 wt% water, *bp* 55.3 °C). A ternary azeotrope with ethanol/water (6.55/ 2.85 wt%) has a *bp* of 53.8 °C.

The industrial product always contains both isomers and has a boiling range of 45 – 60 °C. If required, both isomers can be separated by fractional distillation.

3.3.2. Chemical Properties

Of the two isomers, the trans isomer is more reactive than the cis isomer. At higher temperatures and in the presence of bromine or alumina, isomerization is possible. Thermodynamically, the cis isomer is more stable.

If oxygen and moisture are excluded, 1,2-dichloroethylenes are sufficiently stable. With oxygen or peroxides, dimerization to tetrachlorobutene occurs. Upon oxidation, an intermediate epoxide is formed, which then undergoes rearrangement to give chloroacetyl chloride [461]. Combustion with air yields carbon oxides and hydrogen chloride. Under oxygen deficient conditions, phosgene may be formed.

In the presence of water, hydrolysis occurs to yield hydrochloric acid. Corrosion of construction material can be avoided by such stabilizers as amines and epoxides.

With weak alkali, 1,2-dichloroethylene is not attacked; concentrated alkali, however, induces dehydrochlorination to explosive monochloroacetylene. With copper or its compounds, explosive acetylides can be formed.

In the presence of Lewis-acid catalysts, 1,2-dichloroethylene can be chlorinated to 1,1,2,2-tetrachloroethane or hydrochlorinated to 1,1,2-trichloroethane.

Polymerization is difficult because very high pressures are required. It is not carried out industrially.

3.3.3. Production

Because 1,2-dichloroethylenes are industrially unimportant, they are not deliberately produced in large quantities. They occur as byproducts in some processes, such as the production of vinyl chloride and trichloroethylene, and can be withdrawn and purified if required.

Synthetic routes are possible via

1) thermal cracking of 1,1,2-trichloroethane
2) chlorination of acetylene.

In the thermal dehydrochlorination of 1,1,2-trichloroethane, the 1,2-dichloroethylenes are obtained together with the 1,1-isomer. With increasing temperature, formation of the 1,2-isomers increases. The trans isomer is preferentially formed. With catalysts, the individual ratios (1,2/1,1 and trans/cis) can be varied to some extent.

The chlorination of acetylene with activated carbon catalyst yields almost exclusively the cis isomer. An excess of acetylene is required to suppress the formation of tetrachloroethane. Instead of carbon, mercury and iron salts can be used [462].

Other routes use liquid-phase acetylene oxychlorination [463] or synthesis from 1,1,2,2-tetrachloroethylene which can be dehydrochlorinated and dehydrochlorinated by steam and iron in one reaction.

3.3.4. Uses and Economic Aspects

The 1,2-dichloroethylenes are commercially unimportant, because they do not polymerize, have relatively low boiling points, and can form explosive mixtures with air.

In applications where dichloroethylenes could be used as solvents and for low temperature extraction processes, they have been replaced by methylene chloride, which has a higher solvency, is readily available, and is based on less expensive feedstocks.

1,2-Dichloroethylenes obtained as byproducts from manufacturing processes for other chlorinated hydrocarbons are often used as feed stock for the synthesis of tri- or perchloroethylene.

3.4. Trichloroethylene

E. Fischer first obtained trichloroethylene [79-01-6] in 1864 from hexachloroethane by reductive dehalogenation with hydrogen. An acetylene-based process was developed in Austria, and the first plant became operational in Jajce/Yugoslavia in 1908, a plant still producing tri- and tetrachloroethylene [464]. Because of its high solvency and a growing demand for degreasing solvents, trichloroethylene achieved rapid growth rates in the past. Since the late 1960s, however, the production rates have strongly declined as more stringent environmental regulations became effective. Trichloroethylene is also in strong competition with other solvents such as 1,1,1-trichloroethane.

The acetylene-based process has been partially replaced mainly in the United States by ethylene chlorination and oxychlorination routes. A considerable amount of trichloroethylene is still produced from acetylene, which, however, is not made from carbide, but is obtained from ethylene crackers as a byproduct.

3.4.1. Physical Properties

M_r	131.4
mp	– 87.1 °C
bp at 101.325 kPa	86.7 °C
ϱ at 20 °C	1.465 g/cm^3
n_D^{20}	1.4782
Vapor pressure at	
– 20 °C	0.720 kPa
0 °C	2.680 kPa
20 °C	5.780 kPa
40 °C	7.700 kPa
60 °C	42.500 kPa
80 °C	82.800 kPa
Heat of formation (liquid) $\Delta H°_{298}$	– 42.0 kJ/mol
Specific heat (liquid, 20 °C)	1.01 kJ kg^{-1} K^{-1}
Heat of evaporation (boiling point)	31.5 kJ/mol
Vapor density (boiling point)	4.45 g/L
Critical temperature	544.2 K
Critical pressure	5020 kPa
Thermal conductivity (liquid)	0.14 W m^{-1} K^{-1}
Surface tension (20 °C)	26.4 × 10^{-3} N/m
Viscosity (20 °C)	0.58 × 10^{-3} Pa · s
Coefficient of cubical expansion (0 – 40 °C)	0.001185 K^{-1}
Dielectric constant (20 °C)	3.41
Dipole moment	0.9 × 10^{-18} esu

Ignition temperature	410 °C
Explosive limits in air at 25 °C	7.9 – 10.5 vol %
at 100 °C	8.0 – 52 vol %
Solubility in water at 20 °C	0.107 wt %
Solubility of water in trichloroethylene at 20 °C	0.025 wt %

Trichloroethylene is a light, colorless liquid with a sweetish smell. It is miscible with most organic solvents and has a high solvency for natural and synthetic rubbers and various other polymers. Some binary and ternary azeotropes formed by trichloroethylene are shown in Table 23.

The ternary azeotropes contain 23.8 wt % ethanol and 6.8 wt % water, *bp* 67.4 °C; or 12 wt % propanol and 7 wt % water, *bp* 71.7 °C.

3.4.2. Chemical Properties

Trichloroethylene decomposes slowly to yield hydrogen chloride, carbon oxides, phosgene, and dichloroacetyl chloride. This decomposition is enhanced by elevated temperatures (> 100 °C), air or oxygen, sunlight, and moisture and causes corrosion on construction materials.

Trichloroethylene further reacts with aluminum to form pentachlorobutadiene and higher molecular mass polymers.

Atmospheric photooxidative degradation has also been studied [465]. Hydrolysis is less pronounced. With diluted hydroxides, glycolic acid is formed. Strong hydroxides eliminate hydrogen chloride to give highly explosive dichloroacetylene. Acidic hydrolysis with sulfuric acid gives monochloroacetic acid.

Trichloroethylene can be chlorinated to pentachloroethane or hydrochlorinated to give 1,1,2,2-tetrachloroethane [466].

Although trichloroethylene can be copolymerized with a variety of other monomers, it is used in commercial polymer applications only in the production of poly(vinyl chloride), where it allows the control of molecular mass distribution.

3.4.3. Production

For the production of trichloroethylene either acetylene or ethylene is used as a feedstock.

The *acetylene route* which is still used in Europe — the entire production of trichloroethylene in the Federal Republic of Germany is based on acetylene — comprises acetylene chlorination to 1,1,2,2-tetrachloroethane followed by dehydrochlorination to trichloroethylene:

Table 23. Azeotropes formed by trichloroethylene

wt%	Component	Azeotropic boiling point (101.3 kPa), °C
18	1,2-dichloroethane	82.9
36	methanol	60.2
27	ethanol	70.9
17	1-propanol	81.8
30	2-propanol	75.5
2.5	1-butanol	86.9
33	*tert*-butanol	75.8
6.6	water	72.9
3.8	acetic acid	87.0

$$C_2H_2 + 2\,Cl_2 \longrightarrow CHCl_2-CHCl_2$$
$$CHCl_2-CHCl_2 \xrightarrow{\Delta} CHCl=CCl_2 + HCl$$

In the *ethylene-based processes*, which are widely used in the United States and Japan, ethylene or ethylene based chlorohydrocarbons, preferably 1,2-dichloroethane, are chlorinated or oxychlorinated and dehydrochlorinated in the same reactor. Perchloroethylene is obtained as a byproduct in substantial amounts.

Instead of using pure starting materials, these processes can also be carried out very economically with residues from other chlorinated hydrocarbon processes, e.g., from the production of vinyl chloride.

Trichloroethylene from Tetrachloroethane. Because the chlorination of acetylene yields 1,1,2,2-tetrachloroethane, this isomer is preferably used in the production of trichloroethylene. It dehydrochlorinates also more easily than the 1,1,1,2-substituted isomer.

Dehydrochlorination can be carried out in the liquid and gas-phase.

The *liquid-phase process* uses diluted aqueous calcium hydroxide (10–20%) for cracking [467]. The use of NaOH is not recommended because explosive dichloroacetylene could be formed. The heat of the highly exothermic reaction can be used for overhead distillation of the trichloroethylene as an aqueous azeotrope. The calcium chloride solution is continuously withdrawn from the bottom of the reactor and can be further purified from the remaining organics by steam or vacuum stripping. Although this process can be carried out with high selectivity, it is rarely used because hydrogen chloride is lost by salt formation. In carbide-derived acetylene processes, however, it offers an outlet for the calcium oxide obtained from carbide decomposition.

Gas-phase dehydrochlorination of 1,1,2,2-tetrachloroethane is an endothermic reaction.

$$CHCl_2-CHCl_2 \longrightarrow CHCl=CCl_2 + HCl \qquad \Delta H°_{298} = +61 \text{ kJ/mol}$$

It can be carried out as a pure thermal reaction at temperatures between 300–600 °C in tubular reactors. However, because this reaction forms substantial amounts of heavy byproducts, catalytic dehydrochlorination is industrially preferred.

Since catalyst activated carbon silica or porcelain are used [468], barium chloride has been patented as promoter [469]. The feed material must be thoroughly cleaned from iron chloride traces (catalyst from acetylene chlorination) to avoid poisoning of the catalyst [470]. The reaction can be carried out in either fixed or fluidized bed reactors [471] at temperatures between 250 and 400 °C. The trichloroethylene yield ranges between 90 and 95%. Catalytic traces of chloride were found to promote the reaction [472].

In addition to the acetylene chlorination–tetrachloroethane dehydrochlorination sequence, a direct synthesis by means of acetylene oxychlorination to trichloroethylene, is also possible [473].

A pure anhydrous liquid-phase process (170–200 °C), (0.3–0.6 MPa) for the dehydrochlorination of mixtures also containing the 1,1,1,2-tetrachloroethane isomer has been patented [474]. Iron chloride formed in situ from the construction material acts as a catalyst, but activated carbon can also be used [475]. The latter reaction, however, is important for ethylene-derived tetrachloroethanes, which are obtained by chlorination of 1,2-dichloroethane [476] (see p. 1318).

Trichloroethylene from Ethylene or 1,2-Dichloroethane. The synthesis of trichloroethylene from ethylene or 1,2-dichloroethane is possible by various routes, either by ethylene or 1,2-dichloroethane chlorination and subsequent dehydrochlorination or by oxychlorination of 1,2-dichloroethane.

The *chlorination–dehydrochlorination reaction* is either carried out in sequence [476], [477] or, preferably, performed in one reactor.

Although ethylene can be used as a starting material [477], 1,2-dichloroethane is the preferred feedstock because selectivities and yields can be increased.

The highly exothermic reaction is carried out at temperatures between 200 and 500 °C. Numerous catalysts such as activated carbon, silicates, graphite, and others have been patented [478]. For optimum temperature control, fluidized-bed reactors are used [478]. Even at the optimum chlorine : dichloroethane ratio of 2:1, substantial amounts of perchloroethylene are formed. This causes problems in the purification section because tetrachloromethane formed from perchloroethylene is difficult to separate from trichloroethylene. To solve this problem, a tandem process has been suggested [479]. However, the chlorination — dehydrochlorination process has the principal disadvantage of producing large amounts of hydrogen chloride, which may not fit into site balances.

Aside from ethylene and 1,2-dichloroethanes, other chlorinated ethane residues may also be used as feed [480].

The *oxychlorination process* for the production of trichloroethylene was developed by PPG Ind. [481]. It has the advantage of consuming hydrogen chloride formed during chlorination during the Deacon reaction, and only small amounts of aqueous hydrochloric acid are obtained.

In the oxychlorination process, ethylene, 1,2-dichloroethane, or chloroethane mixtures — which can be residues from other processes — are fed together with oxygen and chlorine to a fluidized-bed reactor.

The catalyst used contains potassium chloride and cupric chloride on fuller's or diatoma-ceous earth or silica. At reaction temperatures of 420–460 °C, the feed is converted by a series of substitution, crack, and Deacon reactions to trichloroethylene and tetrachloroethylene:

$C_2H_4 + Cl_2 \longrightarrow C_2H_4Cl_2$
$C_2H_4Cl_2 + 2\,Cl_2 \longrightarrow C_2H_2Cl_4 + 2\,HCl$
$C_2H_2Cl_4 + Cl_2 \longrightarrow C_2HCl_5 + HCl$
$C_2HCl_5 + Cl_2 \longrightarrow C_2Cl_6 + HCl$
$C_2H_4Cl_2 \longrightarrow C_2H_3Cl + HCl$
$C_2H_3Cl + Cl_2 \longrightarrow C_2H_3Cl_3$
$C_2H_3Cl_3 \longrightarrow C_2H_2Cl_2 + HCl$
$C_2H_2Cl_2 + Cl_2 \longrightarrow C_2H_2Cl_4$
$C_2H_2Cl_4 \longrightarrow C_2HCl_3 + HCl$
$C_2HCl_5 \longrightarrow C_2Cl_4 + HCl$
$C_2Cl_6 \longrightarrow C_2Cl_4 + Cl_2$
$4\,HCl + O_2 \longrightarrow 2\,Cl_2 + 2\,H_2O$

The chlorine yields average 90–98%, carbon yields range between 85–90%. Carbon losses occur by oxidation to carbon oxides and formation of tarry byproducts which cannot be recycled. Temperature control is very important because at a too low temperature (< 420 °C), the cracking reactions diminish whereas at a too high temperature (> 480 °C), the oxidation to carbon oxides increases.

The products are separated and purified by distillation and azeotropic distillation. Tri- and perchloroethylene are withdrawn, the light fractions and high boiling products are recycled to the reactor, and tarry byproducts can be incinerated.

Variation of the trichloroethylene : tetrachloroethylene ratio within a wide range (1.4–0.25) is possible by changing the feed ratios.

Instead of using a fluidized bed, the oxychlorination of C_2 residues in a melt of cupric iron and alkali metal chlorides has been patented [482].

Other Processes. Trichloroethylene is one of the major byproducts of the Atochem process (see 3.1.3.3), where it is obtained from dichloroethylene chlorination and subsequent cracking [254]. Other routes not industrially used are ethane chlorination [483], the pyrolysis of tri- and tetrachloromethane mixtures [484], and the hydrodehalogenation of tetrachloroethylene [485].

3.4.4. Uses and Economic Aspects

The major use for trichloroethylene is as a solvent for vapor degreasing in the metal industry. Because it can undergo hydrolysis, decomposition, and reaction with metals, it is stabilized with acid acceptors such as amines, alcohols [486], epoxides, and metal stabilizers.

Trichloroethylene is further used for degreasing in the textile industry, as an extraction solvent, in solvent formulations for rubbers, elastomers [487], paintstrippers, and industrial paints. In the production of poly(vinyl chloride), it serves as a chain-transfer agent to control the molecular mass distribution.

Since it was first produced on an industrial scale, trichloroethylene production rates have steadily increased with a peak in 1970, when 280 000 t was produced in the United States and 130 000 t in the Federal Republic of Germany. Since then, however, the production rate of trichloroethylene has declined not only because of reduced losses by improved degreasing systems, but also because of strong competition and replacement by 1,1,1-trichloroethane.

Production for 1984 is estimated at approx. 110 000 t for the United States, 80 000 t for Japan, and ca. 200 000 t for Western Europe (FRG: 30 000 t).

The annual decline of 5–7% observed in 1983 and 1984 will probably continue because the more stringent environmental regulations in most countries will further reduce emissions from degreasing units and enforce reclaiming [488].

3.5. Tetrachloroethylene

Tetrachloroethylene [*127-18-4*] (perchloroethylene, Perc) was first obtained by M. FARADAY by the thermal decomposition of hexachloroethane.

Industrial acetylene-based production began during the first decade of this century. In the 1950s, perchloroethylene became the most important drycleaning solvent.

Most producers have replaced the old acetylene route by ethylene or 1,2-dichloroethane feedstocks or by the chlorinolysis process, which uses chlorinated hydrocarbon residues as starting material.

3.5.1. Physical Properties

M_r	165.8
mp	− 22.7 °C
bp at 101.325 kPa	121.2 °C
ϱ at 20 °C	1.623 g/cm^3
ϱ at 120 °C	1.448 g/cm^3
n_D^{20}	1.5055

Vapor pressure at	
0 °C	0.590 kPa
20 °C	1.900 kPa
40 °C	5.470 kPa
60 °C	13.870 kPa
80 °C	30.130 kPa
100 °C	58.500 kPa
120 °C	100.000 kPa
Heat of formation (liquid) $\Delta H°_{298}$	− 51.1 kJ/mol
Specific heat (20 °C)	0.86 kJ kg^{-1} K^{-1}
Heat of evaporation (boiling point)	34.7 kJ/mol
Vapor density (boiling point)	5.8 kg/m^3
Critical temperature	620.3 K
Critical pressure	9740 kPa
Thermal conductivity (liquid)	0.13 W K^{-1} m^{-1}
Surface tension at 20 °C	32.1 × 10^{-3} N/m
Viscosity at 20 °C	0.88 × 10^{-3} Pa · s
at 80 °C	0.54 × 10^{-3} Pa · s
Coefficient of cubical expansion (0 – 40 °C)	0.00102 K^{-1}
Dielectric constant at 20 °C	2.20
Solubility in water at 25 °C	150 mg/kg
Solubility of water in tetrachloroethylene at 25 °C	80 mg/kg

Tetrachloroethylene is a colorless heavy liquid with a mild odor. It is soluble with most organic solvents and exhibits high solvency for organic compounds. Tetrachloroethylene is neither flammable nor does it form explosive mixtures with air.

Some azeotropes formed by tetrachloroethylene are shown in Table 24.

3.5.2. Chemical Properties

Perchloroethylene is the most stable derivative of all chlorinated ethanes and ethylenes. It is stable against hydrolysis and corrosion on construction materials is less pronounced than with other chlorinated solvents.

Tetrachloroethylene reacts with oxygen or air and light to give trichloroacetyl chloride and phosgene. This autoxidation can be suppressed by such stabilizers as amines or phenols. Liquid-phase oxidation with oxygen, however, can be used for the deliberate synthesis of trichloroacetyl chloride [489].

Hexachloroethane is obtained on chlorination. The atmospheric degradation of tetrachloroethylene has been thoroughly investigated, since it is often found during air sampling [465], [490].

Chlorine substitution by fluorine has been studied [491]. Due to the deactivating effect of the chlorine atoms, perchloroethylene cannot be polymerized under normal conditions.

Table 24. Azeotropes formed by tetrachloroethylene

wt%	Component	Azeotropic boiling point (101.3 kPa), °C
15.9	water	87.1
63.5	methanol	63.8
63.0	ethanol	76.8
48.0	1-propanol	94.1
70.0	2-propanol	81.7
29.0	1-butanol	109.0
40.0	2-butanol	103.1
50.0	formic acid	88.2
38.5	acetic acid	107.4
8.5	propionic acid	119.2
3.0	isobutyric acid	120.5
2.6	acetamide	120.5
19.5	pyrrole	113.4
43.0	1,1,2-trichloroethane	112.0
51.5	1-chloro-2,3-epoxypropane	110.1
6.0	glycol	119.1

3.5.3. Production

The production of tetrachloroethylene is theoretically possible by high temperature chlorination of chlorinated lower molecular mass hydrocarbons.

For industrial purposes, three processes are important:

1) Production from acetylene via trichloroethylene:

$$C_2H_2 + 2\,Cl_2 \longrightarrow C_2H_2Cl_4$$
$$C_2H_2Cl_4 \longrightarrow C_2HCl_3 + HCl$$
$$C_2HCl_3 + Cl_2 \longrightarrow C_2HCl_5$$
$$C_2HCl_5 \longrightarrow C_2Cl_4 + HCl$$

2) Production from ethylene or 1,2-dichloroethane through oxychlorination:

$$CH_2=CH_2 + CH_2Cl-CH_2Cl + 2.5\,Cl_2 + 1.75\,O_2$$
$$\longrightarrow CHCl=CCl_2 + CCl_2=CCl_2 + 3.5\,H_2O$$

3) Production from C_1–C_3 hydrocarbons or chlorinated hydrocarbons through high temperature chlorination

The synthesis from acetylene, which is similar to the production of trichloroethylene from acetylene, was for many years the most important production process.

With increasing prices for the acetylene feedstock, however, this route has become unimportant.

The first processes based on the high temperature chlorination of propene–propane mixtures were developed in the 1940s and early 1950s. These so-called chlorinolysis processes (*chlorin*ating pyr*olysis*) have been further developed and are currently the major source of tetrachloroethylene. Instead of propene–propane mixtures, ethane or

C_1–C_3 chlorinated hydrocarbon residues are nowadays used as feed. The chlorinolysis process has become an important step in recovering hydrocarbons and valuable chlorine from residues of other processes (e.g., from vinyl chloride and 1,1,1-trichloroethane production).

With the development of oxychlorination techniques, ethylene or 1,2-dichloroethane oxychlorination has become the second most important route. This process also allows the use of residues instead of pure feed material.

The basic difference between both processes is that tri- and tetrachloroethylene are obtained primarily from oxychlorination, whereas by the chlorinolysis route, tetrachloromethane is generated as a byproduct. Furthermore, the oxychlorination process is the most balanced on hydrogen chloride. The chlorinolysis process is a net producer of hydrogen chloride, which must be consumed by other processes.

Depending on the individual site demands and on the proprietary technology of the producers, both processes presently play a key role in modern tetrachloroethylene production.

Tetrachloroethylene from Acetylene. Even though the direct chlorination of acetylene to tetrachloroethylene is possible [492], most industrial processes use trichloroethylene as an intermediate.

Chlorination of trichloroethylene in the liquid phase (70–110 °C) and in the presence of a Lewis-acid catalyst (0.1–1 wt% $FeCl_3$) gives pentachloroethane (pentachloroethane can also be obtained from ethylene induced liquid phase chlorination of 1,2-dichloroethane (see 2.8.3)). Perchloroethylene is then produced from pentachloroethane by either liquid-phase (80–120 °C, $Ca(OH)_2$) or catalytic thermal cracking (170–330 °C, activated carbon). Overall yields (based on acetylene) of 90–94% are possible.

Because of the long production sequence of four reaction steps and the higher costs for the starting material acetylene, this process has lost its importance during the past 20 years.

Tetrachloroethylene by Oxychlorination of Ethylene, 1,2-Dichloroethane, or Chlorinated C_2 Hydrocarbon Residues. The production of tetrachloroethylene by this route has been described earlier (see p. 1359). This process produces mainly tri- and tetrachloroethylene. Heavy byproducts such as hexachloroethane, hexachlorobutadiene, and chlorinated benzenes must be withdrawn and disposed of or incinerated. The light products can be recycled, which is important for tetrachloromethane, a major byproduct [493]. For further literature, see [494].

Tetrachloroethylene by Chlorination of Hydrocarbons and Chlorinated Hydrocarbons. Theoretically, three process modifications must be distinguished for this route:

1) High temperature chlorination of ethylene, 1,2-dichloroethane, or chlorinated C_2 hydrocarbons
2) Low pressure chlorinolysis
3) High pressure chlorinolysis

The *high temperature chlorination* based on ethylene or chlorinated C_2 hydrocarbons has been mainly developed by the Diamond Alkali Co. [495] and the Donau Chemie AG [496]. The feed is reacted with chlorine at an elevated temperature (200 – 550 °C) in either fluidized (Diamond Alkali) or fixed (Donau Chemie) catalyst beds. Silica and alumina for fluidized beds and activated carbon for fixed beds have been patented as catalysts.

After quenching, hydrogen chloride and tetrachloroethylene are withdrawn and purified by distillation. The light ends can be recycled to the reactor; heavies like hexachloroethane and hexachlorobenzene must be withdrawn. Major recyclable by-products are dichloroethylenes, tetrachloroethanes, and trichloroethylene. Trichloroethylene can be converted to tetrachloroethylene by separate chlorination and recycled to the reactor, where the pentachloroethane formed is cracked [496], [497]. Because the pentachloroethane cracking is an endothermic reaction, the reactor temperature can be controlled by the addition of externally formed pentachloroethane. The carbon yield for tetrachloroethylene from the high temperature chlorination is about 90 – 92%. Yield losses result from the formation of heavies. Chlorine conversions range between 95 and 98%.

Because separation of tetrachloroethylene from 1,1,1,2-tetrachloroethane is difficult to achieve, ethylene derivatives may be added to the quench tower, which are more easily hydrochlorinated as trichloroethylene, the tetrachloroethane precursor [498].

In the *chlorinolysis process,* hydrocarbons or chlorinated hydrocarbons are chlorinated and pyrolyzed to give mainly tetrachloromethane and tetrachloroethylene.

Kinetically, the reaction consists of a whole series of radical crack and substitution reactions which lead to the most stable products.

Thermodynamically, the reaction is governed by two basic equilibria:

$$2\ CCl_4 \rightleftharpoons C_2Cl_4 + 2\ Cl_2$$
$$C_2Cl_6 \rightleftharpoons C_2Cl_4 + Cl_2$$

The thermodynamic equilibrium constants of this reaction are plotted as a function of the reciprocal temperature in Figure 27 [499]. The formation of tetrachloroethylene is favored by an increased temperature and reduced chlorine surplus and pressure. However, because industrial processes are very rarely thermodynamically controlled, the product mix can be widely varied in a range of ca. 5:1 (tetrachloroethylene:tetrachloromethane), depending on the feed products and ratios and on such physical conditions as temperature and pressure. Besides tetrachloroethylene and -methane, hexachloroethane, -butadiene, and -benzene are obtained because of their high stability. The latter three products may account for up to 10% of the carbon yield. Hexachloroethane is almost exclusively and hexachlorobutadiene frequently recycled, whereas the

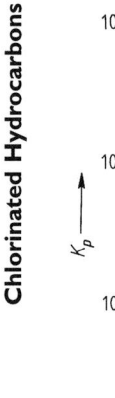

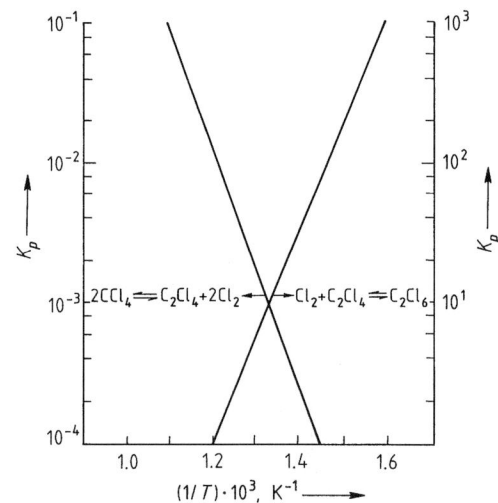

Figure 27. Thermodynamic equilibrium constants for the systems $CCl_4 - C_2Cl_4 + Cl_2$ and $C_2Cl_6 - C_2Cl_4 + Cl_2$ as function of temperature [499]

hexachlorobenzene recycle is technically more difficult and not so often practiced. It is withdrawn with some hexachlorobutadiene and disposed or incinerated for the generation of hydrogen chloride or chlorine.

Presently, two modifications of the chlorinolysis process are in use: the low pressure chlorinolysis and the high pressure chlorinolysis.

The *low pressure chlorinolysis process* is used by most producers. Feedstock for this process are C_1-C_3, preferably C_2 and C_3 hydrocarbons and chlorinated hydrocarbons.

Historically, this process dates back to the 1940s. It was first used by Dow Chemical [500] and somewhat later by Stauffer [501]. Originally designed for substitution of acetylene by cheaper feedstocks such as ethane and propane, it was increasingly used for the conversion of unwanted byproducts, mainly from chlorinated hydrocarbon (vinyl chloride, allyl chloride, 1,1,1-trichloroethane) and chlorohydrin (propylene oxide, epichlorohydrin) processes, into more valuable products.

With the shift of the traditional feedstock from hydrocarbons toward such byproducts as 1,2-dichloropropane, tetrachloroethanes, pentachloroethane, dichloroethylenes, and chlorinated propanes and propenes, the chlorinolysis process fulfills an ecologically and economically important function for integrated chlorinated hydrocarbon sites. If the demand for perchloroethylene or carbontetrachloride exceeds the available residual feedstock capacity, the process can be carried out with ethylene, ethane, propene, and propane, of which the latter three products are preferred because of the cost advantage.

The reaction is carried out at a reactor temperatures between 600 and 800 °C and a pressure between 0.2 and 1.0 MPa. The slightly increased pressure makes the anhydrous purification of the formed hydrogen chloride easier. Adiabatic as well as isothermal reactors are used [500]–[502]. Most processes use empty tubular or backmixed tank reactors, but fluidized-bed reactors have also been patented [503]. In this

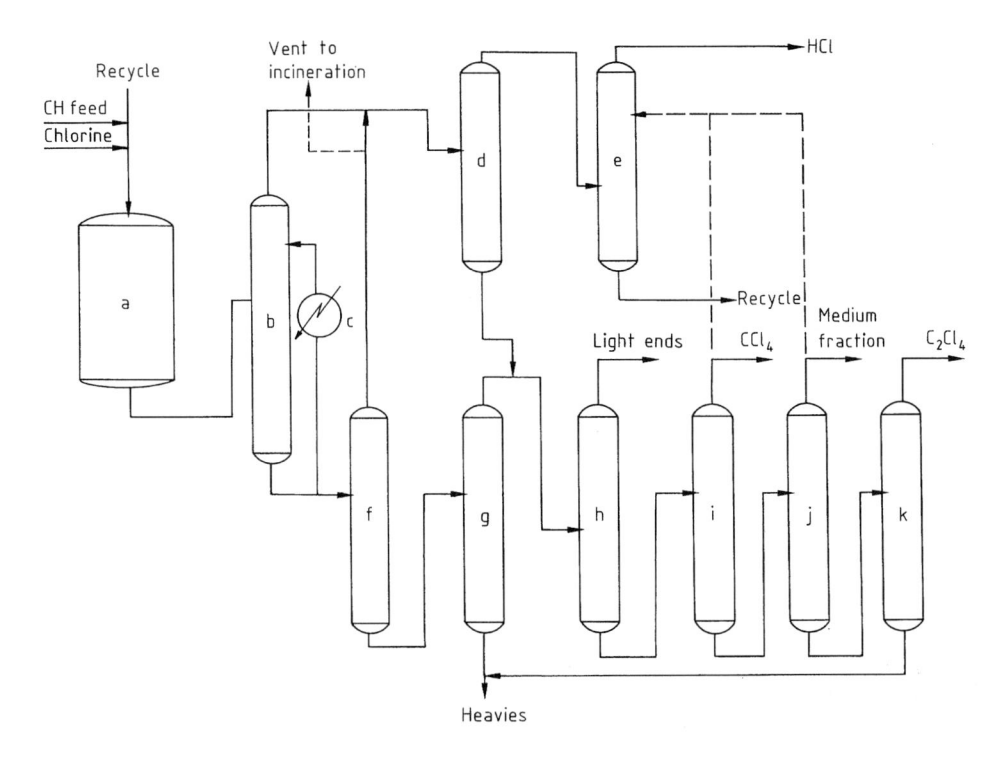

Figure 28. Schematic flow diagram of the chlorinolysis process
a) Reactor; b) Quench tower; c) Cooler; d) and e) HCl tower; f) Degasser; g) Heavies tower; h) Light-end tower; i) Carbon tetrachloride tower; j) Medium-boilers tower; k) Perchloroethylene tower

reactor type, reaction temperatures are about 500 °C, which results in higher hexachloroethane formation and increased recycle. Mean residence times range between 1 and 10 s.

If chlorohydrin residues containing oxygenated compounds are used, these feedstreams must be pretreated (water wash) because most chlorinolysis processes are very sensitive to oxygen. Oxygen containing feed can lead to the formation of phosgene, carbon oxides, and water, which may contaminate the products or cause corrosion. A process designed and carried out by Chemische Werke Hüls, however, is capable of also handling oxygenated compounds in the feed [504].

Rapid quenching of the gases is important to avoid excessive formation of hexachloroethane. Modern processes avoid the aqueous quench systems used in old plants [505] because corrosion is difficult to control and complicated drying systems are required to recover excess chlorine. In most plants, therefore, quenching is achieved by a high recirculation rate of condensed reaction gases. The heat of reaction is removed by air coolers and heat exchangers or can be used for product distillation [506]. Excess chlorine is either removed by washing or absorption–desorption with tetrachloromethane [504] or it can be used for ethylene chlorination to 1,2-dichloroethane which is either recycled or consumed for the production of vinyl chloride [507].

Process Description (Fig. 28). Hydrocarbons (ethane, propylene) or chlorinated hydrocarbon residues are preheated and fed together with vaporized chlorine to the reactor (material: Ni-alloys or brick lined carbon steel). After the reaction, the hot reaction gases are quenched and chlorine and hydrogen chloride are distilled overhead. Hydrogen chloride is purified by fractional distillation, and the remaining chlorine is removed by absorption with carbon tetrachloride or other light ends and recycled to the reactor. If carbon tetrachloride is used for absorption, the chlorine can be stripped in a second tower to avoid recycling of the solvent [504]. After degassing, the quench bottoms are submitted to fractional distillation. Light ends and medium boilers are recycled, and perchloroethylene and carbon tetrachloride are withdrawn and can be further purified. The heavy byproducts are further treated to recover hexachloroethane and hexachlorobutadiene, both of which can be recycled to the reactor. Hexachlorobenzene is either disposed or incinerated to generate hydrogen chloride or chlorine.

Maximum yield may be as much as 95% tetrachloroethylene [508]; in industrial scale processes, however, ca. 90% is achieved [509].

In a process modification by Progil Electrochimie, a reactor cascade with two reactors is used instead of one reactor [510]. The reaction temperature is kept below 600 °C by external cooling.

Another modification consists of liquid phase chlorination at 160–200 °C followed by catalytic gas phase chlorinolysis at 450–600 °C. Molybdenum pentachloride was patented as a catalyst [511].

The *high pressure chlorinolysis process* was developed by Hoechst AG [512]. The chlorinolysis reaction is non-catalytic at a pressure up to 20.0 MPa and a temperature of about 600 °C. It is claimed that this process can also use higher molecular mass feed, which may contain aromatic and alicyclic compounds. A nickel-plated steel reactor is used.

A 50 000 t/a plant was installed at the Frankfurt site.

However, apart from the tolerance against higher molecular mass feed and the easier sepa-ration of hydrogen chloride and chlorine due to the increased pressure, this process offers no major advantages over the low pressure process.

Other Processes. A process to produce tetrachloroethylene from carbon tetrachloride and carbon monoxide has been developed [513], but has not gained any importance. The conventional chlorinolysis process allows broad variations between tetrachloroethylene and carbon tetrachloride production to accommodate varying market demands without being limited by the second product.

3.5.4. Uses and Economic Aspects

The major use for tetrachloroethylene is as a solvent for dry cleaning (ca. 60% of the total consumption). It has replaced almost all other solvents in this field because it is non-flammable and allows safe operation of drycleaning units without special precautions. Because tetrachloroethylene is very stable, it contains only low concentrations of stabilizers, preferably alkylamines and morpholine derivatives. Because of its high stability, it is also used in addition to trichloroethylene and 1,1,1-trichloroethane for metal degreasing. Particularly for aluminum parts, it is superior to other degreasing formulations.

Other uses are textile finishing and dyeing and extraction processes.

In smaller quantities, tetrachloroethylene is used as an intermediate for the production of trichloroacetic acid and some fluorocarbons.

Because more than half of the Western World's tetrachloroethylene production is based on the chlorinolysis process, which coproduces carbon tetrachloride in varying ratios, capacity and output are difficult to estimate.

U.S. production capacity was estimated at 380 000 t/a for 1985. In Europe the installed capacity is ca. 450 000 t/a. Total capacity of the Western World may average 1 000 000 t/a.

Even though the consumption of tetrachloroethylene has been declining since the late seventies, the annual rate of decline is very moderate [514] compared to that of trichloroethylene because a replacement of tetrachloroethylene in its main use in dry cleaning is difficult to achieve without sacrificing safety. Thus, the reduced consumption is mainly due to improved dry-cleaning units with reduced solvent losses to the atmosphere. This trend will continue for many years because more stringent environmental regulations have been passed in most countries, which may cause an even steeper decline in the future. The production of tetrachloroethylene in 1985 is estimated to be 220 000 t in the United States and 110 000 t in the Federal Republic of Germany. Assuming a similar capacity utilization for the entire Western World, the 1985 production may be 600 000 – 700 000 t. With the poor capacity utilization and decreased consumption, new plants cannot be justified. The construction of a new unit however, was announced in Poland [515].

3.6. Analysis and Quality Control of Chloroethanes and Chloroethylenes

Standard methods are used for the analysis of chlorinated ethanes and ethylenes. A typical analysis comprises:

purity
water content
acidity/alkalinity
free chlorine content
nonvolatile residues
physical parameters such as density, refractive index, boiling point, and color

Because of the high volatility of all chlorinated ethanes and ethylenes, gas chromatographic analysis is the method of choice for purity control. Capillary columns are

Table 25. Typical degree of purity of some chlorinated ethanes and ethylenes

Compound	Purity	Acidity as HCl	Water content	Free chlorine	Residues
Chloroethane C_2H_5Cl	>99%	<200 ppm	< 15 ppm	n.d.*	<100 ppm
1,2-Dichloroethane $C_2H_4Cl_2$	>99%	< 20 ppm	<100 ppm	n.d.*	< 50 ppm
1,1,1-Trichloroethane CH_3CCl_3	>99%**	< 10 ppm	<100 ppm	n.d.*	< 50 ppm
Trichloroethylene C_2HCl_3	>99%**	< 10 ppm	<100 ppm	n.d.*	< 50 ppm
Tetrachloroethylene C_2Cl_4	>99%**	< 10 ppm	<100 ppm	n.d.*	< 50 ppm

* n.d. = not detectable;
** unstabilized

widely used for high resolution. Excellent separation is achieved with non-polar to medium polar stationary phases such as OV-1, OV-101, OV-17, OV-1701, and FFAP. The flame ionization detector (FID) is used for detection in concentrated substances. For low concentrations, the electron capture detector (ECD), together with capillary columns and direct on-column injection, offers excellent sensitivity down to the pg-level.

Unequivocal identification of chlorinated compounds in mixtures is possible by GC/MS analysis. For very diluted aqueous samples, head space, purge-and-trap or closed loop stripping techniques may be used to further enhance the sensitivity.

Analysis of air samples can be achieved by adsorption on activated carbon or other convenient materials, followed by thermal or liquid (CS_2) desorption and consecutive analysis. To eliminate matrix effects, vapor distillation or hexane extraction has been proven to be versatile for aqueous or solid samples. The water content is best determined by the Karl-Fischer method.

For the determination of the acidity and alkalinity, titrations with methanolic or ethanolic sodium hydroxide or hydrogen chloride can be used.

Free chlorine can be detected by iodometric analysis.

Some typical values are given in Table 25.

For such typical solvents as 1,1,1-trichloroethane, trichloroethylene, and tetrachloroethylene, the specifications are standardized in some countries, and standard methods for analysis (ISO, ASTM) have been developed.

3.7. Storage and Transportation of Chloroethanes and Chloroethylenes

Before being stored or transported over longer periods of time, chlorinated ethanes and ethylenes should be carefully analyzed for water, free acid, and stabilizers because decomposition may lead to excessive corrosion.

Chlorinated ethanes and ethylenes should not be brought into contact with tanks, containers, valves, etc. made of aluminum.

Contact with copper should be avoided under all circumstances because dichloroethylenes could form explosive acetylides.

Since 1,1,1-trichloroethane decomposes very easily, tanks coated with highly solvent resistant phenolic resin should be used.

Care must be taken with such high solvency compounds as 1,1,1-trichloroethane and 1,1,2,2-tetrachloroethane because they may attack coated or painted equipment.

Some transportation regulations are shown in Table 26.

3.8. Environmental Aspects in the Production of Chloroethanes and Chloroethylenes

The environmental impact of production facilities can be caused by waste (byproducts), emissions to the air, and emissions to the water. In modern, integrated plants, however, waste problems are minimized by proper balancing and a proper choice of production processes.

Hydrogen chloride generated by crack or substitution reactions is consumed by oxychlorination, hydrochlorination, or HCl electrolysis processes.

Organic byproducts can be used as feedstock for another process and converted into valuable products. Examples are the conversion of 1,1,2-trichloroethane to 1,1-dichloroethylene, the conversion of tetrachloroethanes to trichloroethylene, and the consumption of a wide range of residues in the production of tri- and tetrachloroethylene by oxychlorination or in the chlorinolysis process for the production of carbon tetrachloride and tetrachloroethylene. Recyclable products can be recovered from *tar residues* by especially designed equipment such as rotating double screw heat exchangers [518] or by vapor distillation [519]. Ferric chloride, often used as a catalyst, can be removed from heavies by extraction with hydrochloric acid [520]. The main disposal method for non-recyclable products is high temperature incineration [521] or catalytic incineration [522]. Either hydrochloric acid or hydrogen chloride is obtained, which can be recycled to oxychlorination or used for other purposes. If incineration is carried out with pure oxygen, chlorine is formed by the Deacon equilibrium and can be separated and reused [523]. With the highly developed equipment presently in use, incineration is carried out with very high destruction efficiencies and the remaining emissions do not cause environmental problems.

Other techniques which have been proposed for recovery of chlorinated hydrocarbon residues are destruction by a potassium chloride melt to give hydrogen chloride [524] and use for the production of polysulfides [525] and metal chlorides [526]. If appropriately scrubbed, it is also possible to incinerate mercury-containing catalysts from acetylene-based processes [527].

Table 26. Transportation codes and classification for some chlorinated ethanes and ethylenes

Compound	UN-No.	HAZ CHEM	IMDG	RID	Pollution category [517]
Chloroethane	1037	3 W E	class 2 D 2154 E 2057	class 2 201 3bt	–
1,1-Dichloroethane	2362	2 Y E	class 3.2 D 3268 E 3069-1	class 3 301 3 b	B
1,2-Dichloroethane	1184	2 Y E	class 3.2 D 3303 E 3079	class 3 301 16b	B
1,1,1-Trichloroethane	2831	2 Z	class 6.1 D 6387 E 6178-2	class 6.1 601 15 C	B
1,1,2,2-Tetrachloroethane	1702	2 X E	class 6.1 D 6374 E 6173	class 6.1 601 15b	B
Pentachloroethane	1669	2 Z	class 6.1 D 6322 E 6143	class 6.1 601 15b	–
Vinyl chloride	1086	2 W E	class 2 D 2244 E 2125	class 2 201 3 ct	–
1,1-Dichloroethylene	1303	3 Y E	class 3.1 D 3170 E 3050	class 3 301 1a	B
1,2-Dichloroethylene	1150	3 YE	class 2 D 3269 E 3069	class 3 301 1a	D
Trichloroethylene	1710	2 Z	class 6.1 E 6179	class 6.1 601 12b	B
Tetrachloroethylene	1897	2 Z	class 6.1 D 6375 E 6173-1	class 6.1 601 15c	B

Air emissions by tank and process vents can be collected and incinerated. When incineration facilities are unavailable, removal of chlorinated hydrocarbon impurities in vent streams is possi-ble by adsorption units, since they have often been described for the removal of vinyl chloride [528].

In the past, emissions by the Oxy–EDC vent posed the largest problems. With the conversion to oxygen, however (see p. 1302), this vent has become manageable and can be eliminated.

For the Lummus Transcat-Process (see 3.1.3.4), a vent treatment system comprising CO_2 absorption followed by ethylene stripping has been proposed [529].

Emissions into water were drastically reduced in the past since most modern processes for the production of chlorinated ethanes and ethylenes avoid the quenching of reaction gases with water to separate hydrogen chloride. By dry distillation and the use of refrigeration units, this potential source for contamination could be excluded by an improved process design.

In processes in which water is formed during the reaction (e.g., Oxy–EDC) or chlorinated hydrocarbons must be brought into contact with water for other reasons (e.g., production of vinylidene chloride from 1,1,2-trichloroethane), the wastewater in modern plants is stripped with air or steam, depending on the nature of contaminants. The gas can be either directly incinerated or condensed and separated. The aqueous condensate is recycled to the stripper, and the organics can be reused or incinerated. Thus, only minor concentrations of chlorinated hydrocarbons are discharged to the biological treatment facilities, from which the emissions are negligible, and excessive, environmentally problematic discharge of chlorinated hydrocarbons to rivers and to the sea is avoided.

4. Chloropropanes

1,2-Dichloropropane is used both as a chemical intermediate and as an industrial solvent. 2-Chloropropane and 1,2,3-trichloropropane are of lesser significance.

4.1. 2-Chloropropane

Physical Properties. 2-Chloropropane [*75-29-6*], isopropyl chloride, $CH_3-CHCl-CH_3$, M_r 78.54, is a colorless liquid.

mp	$-117.2\ °C$
bp	$36.5\ °C$
Liquid density ϱ at 20 °C	$0.8588\ g/cm^3$
Refractive index n_D^{20}	1.3776
Dynamic viscosity η at 20 °C	3.2×10^{-4} Pa s
Dielectric constant at 25 °C	9.52
Azeotropes with 1.2 wt% water, *bp*	$33.6\ °C$
with 6.0 wt% methanol, *bp*	$33.4\ °C$
with 43 wt% pentane, *bp*	$31\ °C$

Chemical Properties. Prolonged exposure to heat and light can cause dehydrochlorination, especially in the presence of aluminum, zinc, and iron. Reaction with NaOH or amines occurs only at elevated temperature.

Production. 2-Chloropropane can be synthesized by addition of hydrogen chloride to propene using, e.g., alumina, aluminum chloride, calcium chloride, or tellurium compounds as catalysts [530]–[534]. 2-Chloropropane is also formed by reaction of 2-propanol with hydrogen chloride and either zinc chloride or tellurium compounds as catalysts [535], [536].

Uses. 2-Chloropropane, like other chlorinated hydrocarbons, is employed as a solvent.

4.2. 1,2-Dichloropropane

Physical Properties. 1,2-Dichloropropane [78-87-5], propylene dichloride, $CH_2Cl-CHCl-CH_3$, M_r 112.99, is a colorless liquid.

mp	– 100.4 °C
bp	96.5 °C
Refractive index n_D^{20}	1.4394
Specific heat c_p at 30 °C	1.38 kJ kg^{-1} K^{-1}
Heat of vaporization	34.8 kJ/mol
Critical temperature T_c	577 K
Critical pressure p_c	4.43 MPa (44.3 bar)
Surface tension at 20 °C	3.0×10^{-2} N/m
Dielectric constant	8.96
Solubility in water at 20 °C	0.25 wt%
Solubility of water in 1,2-dichloropropane at 20 °C	0.16 wt%
Azeotropes with 12 wt% water, bp	78 °C
with 53 wt% methanol, bp	62.9 °C
with 57.74 wt% ethanol, bp	74.7 °C
with 84 wt% cyclohexane, bp	80.4 °C
with 84 wt% tetrachloromethane, bp	76.6 °C
Flash point (Abel-Pensky, DIN 51 755)	13 °C
Autoignition temperature (DIN 51 794)	557 °C
Explosion limits in air:	
lower limit	3.4 vol%
upper limit	14.5 vol%

Further properties are given in Table 27

1,2-Dichloropropane is miscible with most organic solvents, such as alcohols, esters, and ketones, as well as with aromatic, aliphatic, and chlorinated hydrocarbons.

Chemical Properties. 1,2-Dichloropropane is stable at room temperature but is dehydrochlorinated by thermal or catalytic cracking to allyl chloride [107-05-1] and 1-chloro-1-propene [590-21-6] [537], [538]. It is incompatible with strong oxidizers and

Table 27. Vapor pressure (p), liquid density (ϱ), and dynamic viscosity (η) as a function of temperature

	t, °C			
	0	20	50	80
p, kPa	1.8	5.1	19.8	59.9
ϱ, g/cm^3	1.1815	1.1554	1.1161	1.0751
η, Pa s×10^{-4}	12	8.5	5.8	4.4

strong acids. It is dehydrochlorinated by NaOH to give mainly 1-chloro-1-propene (45% cis and 55% trans isomer) [539].

Production. 1,2-Dichloropropane is a byproduct in the synthesis by the chlorohydrin process of the important chemical intermediate propylene oxide (1,2-epoxypropane) [75-56-9].

1,2-Dichloropropane is obtained in smaller amounts as a byproduct in the industrial synthesis of allyl chloride. Direct synthesis (e.g., by the addition of chlorine to propene) is not currently carried out [540], [541].

Quality Specifications and Chemical Analysis. 1,2-Dichloropropane is used both crude and as a commercial grade product. The crude product is used mainly as a chemical intermediate for the production of perchlorohydrocarbons (see below). The commercial grade product has a boiling range of 95–99 °C at 101.3 kPa (1013 mbar) (DIN 53 171, ASTM D 1078) and a water content below 0.1 wt% (DIN 51 777, ASTM D 1744). Quality analysis is carried out by gas chromatography.

Storage and Transportation. The usual precautions for a highly flammable liquid must be observed with 1,2-dichloropropane. The compound can be stored for several months, but it should be protected from heat, light, moisture and air. Therefore, it is recommended that the product be blanketed with nitrogen. Carbon steel is a suitable material for storage containers provided that the acid and water concentration of the product is low. Rust may increase the color number; however, this is a problem which can be avoided by using stainless steel. Light metals, such as aluminum, magnesium, and their alloys, can react violently with 1,2-dichloropropane. As material for gaskets Teflon, Hostaflon, or IT 400-C (DIN 3754) can be used but materials like PVC, Perbunane, polyethylene, polypropylene, and rubber are not suitable.

Railway tankcars, tanktrucks, and barrels are used for transportation.

The EC directive on dangerous substances must be observed in labeling any containers intended for transporting and handling 1,2-dichloropropane (July 14, 1976, No. 67/548/EC).

Hazard symbols F + Xn; R paragraphs 11–20; S paragraphs 9-16-29-33; EC no. 602-020-00-0. UN no. 1279; Rail, RID: 3.1a; Road, ADR: 3.1a; Code no. for railcars and tank cars: 33/1279; Sea, IMDG code: 3.2; Air, IATA article no.: 1507.

Legal Aspects in the Federal Republic of Germany and in the European Community. The "Verordnung über gefährliche Arbeitsstoffe (ArbStoffV) vom 11.02.1982, Anhang I Nr. 1," and the "EG-Richtlinie vom 10.06.1982" must be observed.

The MAK value (FRG, 1984) is 75 ppm (350 mg/m^3).

According to the "Verordnung über brennbare Flüssigkeiten (VbF)", 1,2-dichloropropane belongs in the "Gefahrenklasse AI". The legal aspects cited in the "Vorschriften des Grundwasserschutzes (VLwF)" must also be considered, as well as the "Unfallverhütungsvorschriften der Berufsgenossenschaften". For safety recommendations, see [542] – [546].

Uses. The most important use of 1,2-dichloropropane is as an intermediate in the synthesis of perchloroethylene [127-18-4] and tetrachloromethane [56-23-5] [538]. 1,2-Dichloropropane is a good solvent for fats, oils, resins, and lacquers. It is suitable for extraction, cleaning, degreasing, and dewaxing operations in the chemical and technical industry. Since it forms an azeotrope with water at 78 °C, it can be used for removing water from organic solutions.

1,2-Dichloropropane dissolves bitumen and tar asphalt. It is used to promote the adhesion of asphalt layers, and it is suitable for the production of roofing paper, insulation material, and shoe-polish. In combination with 1,3-dichloropropene [542-75-6] it can be used as a soil fumigant for nematodes.

1,2-Dichloropropane is applicable as a lead scavenger for antiknock fluids [546], and it is used in petroleum refineries in platforming processes to adjust the catalyst activity [547].

4.3. 1,2,3-Trichloropropane

Physical Properties. 1,2,3-Trichloropropane [96-18-4], trichlorohydrin, CH$_2$Cl – CHCl – CH$_2$Cl, M_r 147.44, is a colorless liquid.

mp	– 14.7 °C
bp	156 °C
Refractive index n_D^{20}	1.4834
Liquid density ϱ at 20 °C	1.388 g/cm^3
Dynamic viscosity η at 20 °C	2.5 × 10^{-4} Pa s
Surface tension at 20 °C	3.77 × 10^{-2} N/m
Azeotrope with 31 wt%	
cyclohexanol, bp	154.9 °C
with 39 wt% cylcohexanone, bp	160.0 °C
with ca. 14 wt% ethylene glycol, bp	ca. 152.2 °C
with 65 wt% propionic acid, bp	139.5 °C
with 7.5 wt% acetamide, bp	154.4 °C
Flash point	74.0 °C
Autoignition temperature	304 °C

Explosion limits in air:
lower limit   3.2 vol%
upper limit [548]   12.6 vol%

Chemical Properties. 1,2,3-Trichloropropane forms glycerol upon treatment with steam at 550–850 °C, and heating it with aqueous caustic gives rise to 2,3-dichloropropene, 2-chloroallyl alcohol, and the ether of the latter [549]. Heating with NaOH may also produce dangerous chloropropynes. The MAK value of 1,2,3-trichloropropane is 50 ppm (300 mg/cm^3) (FRG 1984).

Production. 1,2,3-Trichloropropane is produced by addition of chlorine to allyl chloride [550].

The selectivity of the chlorination of 1,2-dichloropropane to 1,2,3-trichloropropane is not good [551], [552].

Uses. 1,2,3-Trichloropropane is a solvent for oils, fats, waxes, chlorinated rubber, and resins. The reaction with water at high temperature to give glycerol is not employed commercially. 1,2,3-Trichloropropane is used for the synthesis of thiokol polysulfide elastomers if some branching in the polymer structure is required [553].

5. Chlorobutanes

Chlorobutanes can be prepared from butane by liquid-phase or thermal chlorination, from the butenes by addition of elemental chlorine or hydrogen chloride, and from butanol by esterification. The chlorination of butane with chlorine proceeds just as with the lower hydrocarbons, although currently the process has no commercial significance. The reaction has been found to yield primarily the 1- and 2-chlorobutanes along with the 1,3- and 1,4-dichlorobutanes (see Fig. 29) [554]–[556]. For the influence of initiators and inhibitors on the chlorination of n-butane, see [557], [558].

Photochemical chlorination in the gas phase at 15–20 °C leads to a product distribution similar to that obtained from the strictly thermal process [559].

Table 28 provides an overview of the physical properties of industrially significant chlorine derivatives of butane. 2-Chlorobutane and isobutyl chloride are used primarily as starting materials in Friedel-Crafts reactions. They are made exclusively by hydrochlorination of the corresponding alcohols, a process analogous to that used for the preparation of isopropyl chloride (see p. 1374).

Table 28. Physical properties of industrially important chlorinated derivatives of butane

	1-Chlorobutane	1-Chloro-2-methylpropane	2-Chlorobutane (optically active form)	2-Chloro-2-methylpropane	1,4-Dichlorobutane
M_r	92.57	92.57	92.57	92.57	127.02
mp, °C	−123.1	−131.2	−140.5	−25.2	−37.3
bp at 101.3 kPa, °C	78.6	68.4	68.3	50.7	155.0
Density at 20 °C, g/cm^3	0.8865	0.8780	0.8721	0.8435	1.1408
n_D^{20}	1.4025	1.3982	1.3966	1.3856	1.4550
Vapor pressure, kPa	2.0 (−10.5 °C)	16.8 (0 °C)	6.5 (0 °C)	2.7 (−32 °C)	1.3 (44 °C)
	4.0 (1 °C)		10.7 (10 °C)	13.3 (−1 °C)	2.3 (50 °C)
	10.7 (19.7 °C)		17.0 (20 °C)	53.2 (32.6 °C)	5.8 (72 °C)
	27.1 (41.1 °C)		25.9 (30 °C)	87.4 (46.6 °C)	
	66.5 (65.5 °C)		38.0 (40 °C)		
Heat of formation $\Delta H°$ of liquid at 298 K, kJ/mol	−180	−191	−193	−212	
Heat of vaporization $\Delta H_v°$ (298 K), kJ/mol	33.1	31.8	31.4	28.9	
Viscosity at 20 °C, 10^{-4} Pa s	4.5	4.6	3.6 (30 °C)	5.1	14.3
Flash point (closed cup), °C	−12		−20	−27	52
Autoignition temperature, °C	460			570	

5.1. 1-Chlorobutane

Physical Properties. 1-Chlorobutane, *n*-butyl chloride, CH$_3$(CH$_2$)$_3$Cl, is a colorless, flammable liquid with excellent solvent properties for fats, oils, and waxes. For additional physical properties, see Table 28.

Binary azeotropes are listed in Table 29.

Chemical Properties. 1-Chlorobutane is stable if kept dry. In the presenc of water, it will hydrolyze, but more slowly than secondary or tertiary chlorides. Of industrial significance is its reaction with magnesium to give the corresponding Grignard reagent.

Production. 1-Chlorobutane is obtained by esterification of *n*-butanol with hydrogen chloride or hydrochloric acid at 100 °C either without a catalyst [560] or utilizing the accelerating effect of zinc chloride [561], [562], tripentylamine hydrochloride [563], or phosphorus pentachloride [564]. *n*-Butyl chloride is also obtained, along with 2-chlorobutane, by the chlorination of butane over aluminum oxide at 200 °C [565].

Uses. *n*-Butyl chloride is used as a solvent and alkylating agent in reactions of the Friedel-Crafts type. It is also the starting material for the synthesis of bis (tributyltin) oxide, (C$_4$H$_9$)$_3$Sn−O−Sn(C$_4$H$_9$)$_3$ (TBTO), used as an antifouling agent in marine coatings and as a general fungicide.

Table 29. Azeotropes of 1-chlorobutane

wt%	Component	Boiling point of the azeotrope, °C (101.3 kPa)
25.0	formic acid	69.4
21.5	ethanol	66.2
8.0	2-butanol	77.7
28.5	methanol	57.2
38.0	methyl ethyl ketone	77.0
16.0	n-propanol	75.6
6.6	water	68.1

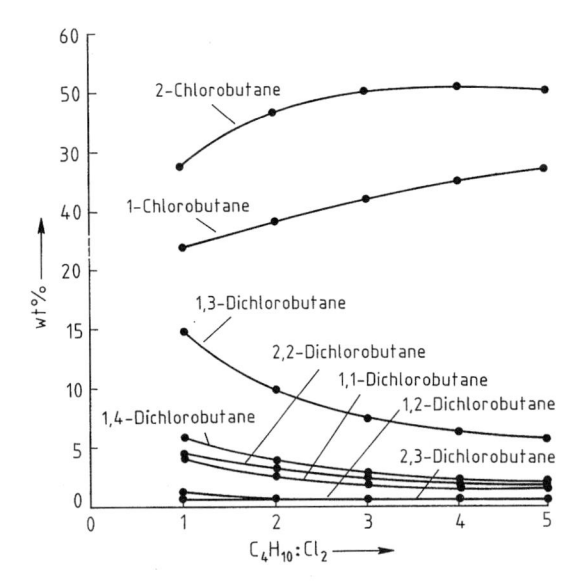

Figure 29. Product distribution in the chlorination of n-butane as a function of the n-butane : chlorine ratio (conditions: reaction temperature 390 °C, residence time 2 s, pressure 101.3 kPa)

5.2. *tert*-Butyl Chloride

Physical Properties. 2-Chloro-2-methylpropane, *tert*-butyl chloride, $(CH_3)_3CCl$, is a colorless liquid. It is miscible with common organic solvents and forms constant-boiling mixtures with a number of substances, including methanol, ethanol, and acetone.

Additional physical properties of 2-chloro-2-butane are listed in Table 28.

Chemical Properties. *tert*-Butyl chloride is very rapidly hydrolyzed to *tert*-butyl alcohol when warmed in aqueous alkaline medium. Complete decomposition to isobutylene and hydrogen chloride commences above 160 °C, particularly in the presence of such catalytic oxides as thorium oxide [566].

Production. *tert*-Butyl chloride is generally prepared by reaction of *tert*-butyl alcohol with hydrogen chloride. The hydrogen chloride is passed into the alcohol until the high acid content of the resulting aqueous phase indicates that reaction is complete. The aqueous phase is then removed and the residual product layer is washed to neutrality, dried, and, if necessary, distilled.

tert-Butyl chloride can also be prepared by gas-phase reaction of isobutylene with hydrogen chloride over catalytic chlorides and oxides such as Al_2O_3 at a temperature below 100 °C [567], as well as by passing hydrogen chloride into isobutylene at a low temperature [568], [569].

Uses. *tert*-Butyl chloride is used in Friedel-Crafts reactions (e.g., in the preparation of *tert*-butylbenzene or *tert*-butylphenol) as well as for the synthesis of 4-chloro-2,2-dimethylbutane (neohexyl chloride), used as a fragrance base.

5.3. 1,4-Dichlorobutane

Physical Properties. The compound is a liquid which is flammable when hot and miscible with numerous organic solvents. For additional physical properties of 1,4-dichlorobutane, $Cl(CH_2)_4Cl$, see Table 28.

Chemical Properties. The compound is converted to 1,3-butadiene either thermally at 600 – 700 °C or over alkaline catalysts at 350 °C. Both chlorine atoms are exchangeable. Reaction with sodium sulfide gives tetrahydrothiophene, and diacetyl peroxide converts 1,4-dichlorobutane to 1,4,5,8-tetrachlorooctane.

Production. 1,4-Dichlorobutane is obtained from butane-1,4-diol and hydrogen chloride in the presence of aqueous sulfuric acid at 165 °C [570].

It can be prepared in good yield by treatment of tetrahydrofuran either with hydrogen chloride or with aqueous hydrochloric acid at 100 – 200 °C and 1 – 2 MPa (10 – 20 bar) [571] – [574]; dehydrating agents such as sulfuric acid or zinc chloride favor the reaction [573], [575] – [577]. Other suitable reagents for chlorinating tetrahydrofuran include phosgene or phosphorus oxychloride in the presence of dimethylformamide [578].

Uses. 1,4-Dichlorobutane is used as a synthetic intermediate, as, for example, in the production of nylon.

6. Chlorobutenes

The chlorobutenes that have acquired substantial industrial and economic importance include 2-chloro-1,3-butadiene (chloroprene) and also 1,4-dichloro-2-butene, 3,4-dichloro-1-butene, 3-chloro-2-methyl-1-propene, 2,3-dichloro-1,3-butadiene, and hexachlorobutadiene.

6.1. 1,4-Dichloro-2-butene

$CH_2Cl-CH=CH-CH_2Cl$; M_r 124.96. 1,4-Dichloro-2-butene occurs as a low boiling cis form [1476-11-5] and as a higher boiling trans form [110-57-6].

Physical Properties
cis form:

mp	− 48 °C
bp at 101.3 kPa	154.3 °C
d_4^{25}	1.188
n_D^{25}	1.4887
Vapor pressure at 20 °C	0.44 kPa (4.4 mbar)

trans form:

mp	+ 1 °C
bp at 101.3 kPa	156.8 °C
d_4^{25}	1.183
n_D^{25}	1.4861
Vapor pressure at 20 °C	0.31 kPa (3.1 mbar)

Chemical Properties. 1,4-Dichloro-2-butene is stable at room temperature. If prolonged storage is envisaged, it is advisable to arrange to do so under nitrogen.

Dehydrochlorination of the compound with NaOH in the presence of a phase transfer catalyst gives 1-chloro-1,3-butadiene (cis: [10033-99-5], trans: [16503-25-6]) [579], as does heating it with KOH to 90 °C.

Production. 1,4-Dichloro-2-butene is obtained in a yield of 93% by chlorination of butadiene in the vapor phase at 300–350 °C, short residence times being possible [580], [581]. At lower temperatures about 40% of the product is 3,4-dichloro-1-butene, and higher boiling products also are formed [582]–[587].

3,4-Dichloro-1-butene [760-23-6] can be converted to 1,4-dichloro-2-butene by isomerization in the presence of copper(I) chloride and/or zirconium phosphate [588]–[591].

The two dichlorobutene isomers are also formed in the oxychlorination of C_4 cracking fractions containing butadiene and isobutene [592]–[596].

Uses. 1,4-Dichloro-2-butene occurs as an intermediate in the production of chloroprene (see 6.4.3), as does 3,4-dichloro-1-butene.

1,4-Dichloro-2-butene is a starting material in the production of adiponitrile [111-69-3] (adipic acid dinitrile), butane-1,4-diol [110-63-4], and tetrahydrofuran [109-99-9]. Adipic acid and hexamethylenediamine (starting materials in the synthesis of nylon-6,6) are obtained from adiponitrile by saponification and hydrogenation, respectively.

Butane-1,4-diol and tetrahydrofuran can be produced according to a process developed by Toyo Soda [597] which entails hydrolysis of 1,4-dichloro-2-butene in the presence of sodium formate, followed by hydrogenation to butane-1,4-diol. Tetrahydrofuran is obtained by additional elimination of water. Tetrachlorobutane (R,R: [*14499-87-7*], R,S: [*28507-96-2*]) can also be produced from 1,4-dichloro-2-butene [598].

6.2. 3,4-Dichloro-1-butene

$CH_2Cl-CHCl-CH=CH_2$; M_r 124.96; [*760-23-6*]

Physical Properties

bp at 101.3 kPa	118.6 °C
d_4^{25}	1.153
n_D^{20}	1.4630
Vapor pressure at 20 °C	1.74 kPa (17.4 mbar)

Chemical Properties. 3,4-Dichloro-1-butene is stable at room temperature. At elevated temperatures in the presence of glass or metals, however, it undergoes isomerization to 1,4-dichloro-2-butene. 3,4-Dichloro-1-butene is the starting point in the production of chloroprene (dehydrochlorination in the presence of aqueous sodium hydroxide solution (see 6.4.3).

Production. 3,4-Dichloro-1-butene is formed, together with 1,4-dichloro-2-butene (ratio 40:60), in the chlorination of butadiene [580]–[587]. It is also obtained by isomerization of 1,4-dichloro-2-butene in the presence of either copper naphthenate [599], a $PdCl_2$– benzonitrile or CuCl–adiponitrile complex [600], or a catalyst complex consisting of copper(I) chloride and an organic quaternary ammonium chloride [601].
Inhibitors such as nitriles [602], [603], thiols [604], acid amides, sulfoxides [605], and phosphines [606] are added to prevent isomerization and polymerization.

Uses. The only industrial use for 3,4-dichlorobutene is in the production of chloroprene.

6.3. 2,3,4-Trichloro-1-butene

$CH_2Cl-CHCl-CHCl=CH_2$; M_r 159.44 [*2431-50-7*].

Physical Properties

bp at 101.3 kPa	155 °C
d_4^{20}	1.3430
n_D^{20}	1.4944
Vapor pressure at 20 °C	0.25 kPa (2.5 mbar)
Flash point	63 °C
Specific heat at 20 °C	1.088 J/g

Chemical Properties. 2,3,4-Trichloro-1-butene is a colorless, stable liquid at room temperature. Dehydrochlorination with NaOH in the presence of a phase transfer catalyst converts it to 2,3-dichloro-1,3-butadiene [1653-19-6] [607].

Production. 2,3,4-Trichloro-1-butene is produced from vinylacetylene or chloroprene by addition of HCl in the presence of a CuCl complex [608]–[612], followed by chlorination of the resulting 1,3-dichloro-2-butene (cis: [10075-38-4], trans: [7415-31-8]), during which HCl is eliminated [613].

Uses. 2,3,4-Trichloro-1-butene is used exclusively for the production of 2,3-dichloro-1,3-butadiene (see Section 6.5) [607], [614]–[617].

6.4. 2-Chloro-1,3-butadiene

$CH_2 = CCl - CH = CH_2$; M_r 88.54, [126-99-8].

2-Chloro-1,3-butadiene, generally known as chloroprene, was discovered in 1930 by CAROTHERS and COLLINS [618], [619] during work on the synthesis of vinylacetylene. Chloroprene is obtained from vinylacetylene by the addition of HCl (see 6.4.3.2).

The discovery of chloroprene by CAROTHERS and COLLINS was based on the chemistry of acetylenes by Father J. NIEUWLANDS, in 1925.

Chloroprene exists in four isomeric forms: 1-Chloro-1,3-butadiene (α-chloroprene or 1-chloroprene; two isomers: cis [10033-99-5] and trans [16503-25-6]), 2-chloro-1,3-butadiene [126-99-8] (β-chloroprene, also known as 2-chloroprene or simply chloroprene), and 4-chloro-1,2-butadiene [25790-55-0] (isochloroprene). Only 2-chloroprene is economically important – as a monomer used to produce polychloroprene, also known as neoprene (the generic term for polymers of this type), e.g., Baypren, (Bayer) Butaclor (Distigugil), Denka chloroprene (Denki-Kagaku).

6.4.1. Physical Properties

Chloroprene is a colorless liquid with a characteristic ethereal odor. It is soluble in most organic solvents. Its solubility in water at 25 °C is 250 ppm.

2-Chloro-1,3-butadiene (β-chloroprene)

mp	– 130 °C
bp at 101.3 kPa	59.4 °C
d_{20}^{20}	0.9583
n_D^{20}	1.4583
Flash point	– 20 °C (ASTM)
Autoignition temperature	320 °C

| Limits of inflammability in air | 2.1–11.5 vol% |
| MAK value | 10 ppm (36 mg/m^3) |

Vapor pressure vs. temperature:

°C	0	10	20	30	40	50	59.4
kPa	10.4	16.3	25.0	37.3	53.3	73.9	101.3

Odor threshold	1 mg/L [620]
Specific heat at 20 °C	1.314 J g^{-1} K^{-1}
Latent heat of vaporization at 20 °C	325 J/g
at 60 °C	302 J/g
Reaction enthalpy (3,4-dichloro-1-butene → chloroprene)	– 83.7 kJ/mol
Heat of polymerization	– 711 J/g [621]
Reaction enthalpy of uninhibited chloroprene	– 900 J/g

1-Chloro-1,3-butadiene (α-chloroprene)

| mp at 101.3 kPa | 66–67 °C |
| d_4^{20} | 0.954 |

4-Chloro-1,2-butadiene (isochloroprene)

bp at 101.3 kPa	88 °C
d_4^{20}	0.9891
n_D^{20}	1.4775

6.4.2. Chemical Properties

Chloroprene is a colorless, highly reactive liquid. It undergoes substantial dimerization, even at room temperature, from which the following compounds have been isolated:

1,2-Dichloro-1,2-divinylcyclobutane (cis: [33817-64-0], trans: [33817-63-9]),
1,4-dichloro-4-vinylcyclohexene [65122-21-6], 1-chloro-4-(1-chlorovinyl)cyclohexene [13547-06-3], and
2-chloro-4-(1-chlorovinyl)-cyclohexene [28933-81-5].

1,6-Dichloro-1,5-cyclooctadiene [29480-42-0] is formed from the cis-cyclobutane isomer through rearrangement at elevated temperatures.

In the presence of air, uninhibited chloroprene accumulates peroxides, which may cause it to undergo spontaneous ω-polymerization to give the so-called "popcorn polymer." This process is different from the well-known α-polymerization which leads to rubber-like products. α-Polymerization inhibitors include phenothiazione and p-tert-butylcatechol. ω-Polymerization can be inhibited with nitric oxide, N-nitrosodiphenylamine, N-nitrosophenylhydroxylamine, and other nitroso compounds [622]–[626].

Due to its polymerization tendency, chloroprene is stored and transported under inert gas at temperatures below – 10 °C, with further protection by the addition of sufficient amounts of inhibitor.

6.4.3. Production

6.4.3.1. Chloroprene from Butadiene

The production of chloroprene (2-chloro-1,3-butadiene) from butadiene comprises three steps:

1) Vapor phase chlorination of butadiene to a mixture of 3,4-dichloro-1-butene and 1,4-dichloro-2-butene:

$$Cl_2 + CH_2 = CH - CH = CH_2$$
$$\longrightarrow ClCH_2 - CHCl - CH = CH_2 + ClCH_2 - CH = CH - CH_2Cl$$

2) Catalytic isomerization of 1,4-dichloro-2-butene to 3,4-dichloro-1-butene:

$$ClCH_2 - CH = CH - CH_2Cl \underset{}{\overset{cat.}{\rightleftharpoons}} ClCH_2 - CHCl - CH = CH_2$$

3) Dehydrochlorination of 3,4-dichloro-1-butene with alkali to chloroprene (2-chloro-1,3-butadiene):

$$ClCH_2 - CHCl - CH = CH_2 + NaOH$$
$$\longrightarrow CH_2 = CCl - CH = CH_2 + NaCl + H_2O$$

Chlorination of Butadiene. The reaction of chlorine with butadiene can be carried out in an adiabatically operating reactor (see Fig. 30) at temperatures not exceeding 250 °C and at pressures of 0.1 – 0.7 MPa (1 – 7 bar). Intensive and rapid mixing of the reactants is of decisive importance in preventing soot formation. In addition, temperature control is ensured by conducting the reaction in such a way that there is always an excess of butadiene present (the molar ratio of Cl_2 to butadiene should be 1:5 – 1:50). The hot gaseous reaction mixture, consisting mainly of dichlorobutenes, butadiene, and highly chlorinated $C_4 - C_{12}$ products, is condensed with liquid dichlorobutene in a scrubber-cooler. The excess butadiene is removed at the top of the cooler and returned to the chlorinator together with fresh butadiene. The HCl gas which is formed must be removed in a scrubber. The liquid reaction products are removed at the bottom of the cooler and purified by distillation. The proportion of dichlorobutenes in the reaction mixture is about 92 % [587].

Chlorination processes which utilize high reactor temperatures (280 – 400 °C) and very short residence times but avoid the need for an excess of butadiene are known [580], [581].

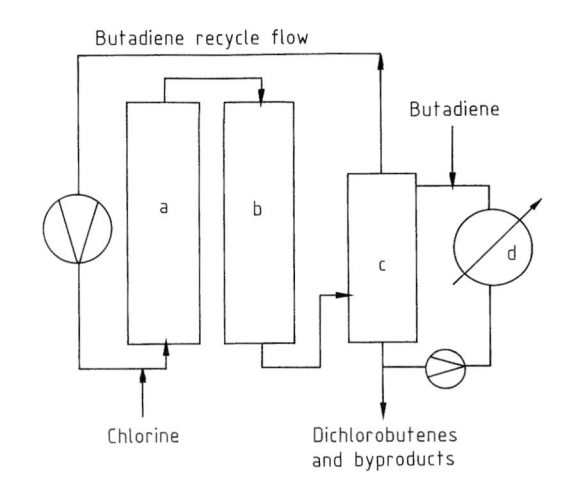

Figure 30. Flow diagram for the chlorination of butadiene
a) and b) Chlorination reactors; c) Scrubber cooler; d) Heat exchanger for carrying away heat generated by the reaction

Isomerization of the Dichlorobutenes. The distillate from the preceding chlorination step consists of a mixture of 1,4-dichloro-2-butene and 3,4-dichloro-1-butene. This mixture is subsequently isomerized to pure 3,4-dichloro-1-butene by heating to temperatures of 60–120 °C in the presence of a catalyst [599]–[601]. A mixture which is enriched in 3,4-dichloro-1-butene can be removed continuously from the reactor. The desired 3,4-dichloro-1-butene, whose boiling point is lower than that of 1,4-dichloro-2-butene, is separated from the latter in a subsequent distillation step. Unreacted 1,4-dichloro-2-butene is then returned to the isomerization reactor.

Dehydrochlorination of 3,4-Dichloro-1-butene to Chloroprene. The dehydrochlorination of 3,4-dichloro-1-butene with dilute NaOH in the presence of inhibitors gives 2-chloro-1,3-butadiene. 1-Chloro-1,3-butadiene is also formed as a byproduct [627].

The reaction is carried out at a temperature of 40–80 °C. Addition of phase transfer catalysts, such as quaternary phosphonium or ammonium compounds, accelerates the reaction considerably and gives higher yields per unit time, e.g., 99% in 30 min [628]–[630].

The chloroprene and sodium chloride solution leaving the dehydrochlorination reactor is separated in a distillation column or by decantation [631]. The salt in the wastewater is either recovered by evaporation, or else the wastewater is recycled to a chlor–alkali electrolysis plant permitting recovery of chlorine and sodium hydroxide solution (thereby decreasing the consumption of these raw materials) [632].

Processes have also been described for the dehydrochlorination of 3,4-dichloro-1-butene using an aqueous mixture of alcohols and NaOH [633]–[636]. These processes have the advantage that the resulting sodium chloride is obtained as a solid, permitting its direct use in a chlor–alkali electrolysis plant.

Chloroprene can be dried by the freezing method: moist chloroprene is cooled to <0 °C and passed through a refrigerator, on the walls of which ice crystals form. These ice crystals are separated from the liquid chloroprene in a recrystallizer, the water content of the resulting chloroprene being < 150 ppm [637].

Purification. The crude chloroprene obtained at the end of the dehydrochlorination step is purified by distillation. Chloroprene should be made prior to polymerization, as free from 1-chloroprene as possible. Accomplishing this requires a column with a large number of plates. Sieve plate or packed columns are generally used.

To prevent polymerization in the column, inhibitors are added and the crude chloroprene is distilled under reduced pressure.

Inhibited chloroprene can be stored at a temperature of −10 to −20 °C under nitrogen for fairly long periods without risk.

The purity of chloroprene (target value 99.9%) is checked by gas chromatography. Impurities can be identified by IR spectroscopy. The Fischer method can be used to determine water.

6.4.3.2. Chloroprene from Acetylene

Production of Monovinylacetylene. Acetylene dimerizes to monovinylacetylene in the presence of aqueous or anhydrous copper(I) chloride and a catalyst solution containing either an alkali metal salt or an ammonium salt.

$$2\ C_2H_2 \xrightarrow[70\ °C]{Cu_2Cl_2/NH_4Cl} CH_2=CH-C\equiv CH$$

The acetylene is led in under conditions that assure a short contact time, and the products are rapidly removed from the catalyst solution. Byproducts of the reaction are divinylacetylene, acetaldehyde, and vinyl chloride [638]. Yields of monovinylacetylene range from 75 to 95% depending on the catalyst system. After concentration and purification, monovinylacetylene is obtained as a colorless liquid boiling at 5.5 °C and 101.3 kPa. Because of its tendency to decompose, monovinylacetylene is diluted for handling, hydrocarbons commonly being used as the diluent.

Production of Chloroprene. The first step in the reaction of hydrogen chloride with monovinylacetylene is a 1,4-addition to give 4-chloro-1,2-butadiene [639].

$$CH_2=CH-C\equiv CH + HCl \longrightarrow ClCH_2-CH=C=CH_2$$

The chlorine in this compound is very reactive and the substance rearranges in the presence of the Cu_2Cl_2-containing catalyst solution to give 2-chloro-1,3-butadiene:

$$ClCH_2-CH=C=CH_2 \xrightarrow[60-70\ °C]{Cu^+} CH_2=CCl-CH=CH_2$$

Under conditions where yields of 95% are achieved the principal byproduct is 1,3-dichloro-2-butene, resulting from further HCl addition:

$$CH_2=CCl-CH=CH_2 \xrightarrow{HCl} CH_3-CCl=CH-CH_2Cl$$

The latter can be used as a starting material for 2,3-dichlorobutadiene, a comonomer of chloroprene.

Chlorides of mercury, magnesium, calcium, gold, and copper are all used in the production of chloroprene, as are ammonium chloride, ammonium bromide, pyridinium chloride, and methylammonium chloride. The most commonly used material, however, is the system $Cu_2Cl_2/NH_4Cl/HCl$. Two reactors for chloroprene production have been described [640].

6.4.3.3. Other Processes

Other processes described in the literature for the production of chloroprene include oxychlorination of butenes [592]–[597], chlorination of butadiene in organic solvents [641]–[645] and electrochemical chlorination of butadiene [646], [647].

The direct chlorination of butadiene to chloroprene has been described in patents [648], [649]. To date, however, none of these processes has become commercially important.

6.4.4. Economic Importance

Chloroprene is the starting monomer for the specialized rubber known as polychloroprene.

The vulcanizates of polychloroprene have favorable physical properties and excellent resistance to weathering and ozone. Articles made with this rubber include electrical insulating and sheathing materials, hoses, conveyor belts, flexible bellows, transmission belts, sealing materials, diving suits, and other protective suits.

Adhesive grades of polychloroprene are used mainly in the footwear industry. Polychloroprene latexes have found application for dipped goods (balloons, gloves), latex foam, fiber binders, adhesives, and rug backing.

Chloroprene and polychloroprene are produced:

– in the United States: by Du Pont and Petrotex
– in Western Europe: by Bayer in the Federal Republic of Germany, Distugil S.A. in France, and Du Pont in Northern Ireland
– in Japan: by Showa-Denko, Denki Kagaku, and Toyo Soda
– and in the Soviet Union

World polychloroprene capacity, 1983, in 1000 t [650]:

United States	213
Federal Republic of Germany	60
France	40
United Kingdom	30
Japan	85
Central planned economy countries	220

6.5. Dichlorobutadiene

6.5.1. 2,3-Dichloro-1,3-butadiene

$CH_2 = CCl - CCl = CH_2$; M_r 122.95 [1653-19-6]

Physical Properties

bp at 101.3 kPa	98 °C
d_4^{20}	1.1829
n_D^{20}	1.4890

Chemical Properties. 2,3-Dichlorobutadiene is more reactive than chloroprene. Therefore, it is not known to have been produced industrially in the pure form. Inhibition with an inert gas containing 0.1 vol% of NO has been reported [651].

Production. 2,3-Dichlorobutadiene is produced by reacting butyne-1,4-diol with phosgene, followed by isomerization of the intermediate 1,4-dichloro-1-butyne [62519-07-7] in the presence of copper(I) chloride and amines [652]–[654]. The dehydrochlorination of both 2,3,4-trichlorobutene (see 6.3) and 1,2,3,4-tetrachlorobutane with sodium hydroxide solution [607], [614]–[617], [655]–[657] has acquired industrial importance.

2,3-Dichlorobutadiene is an important monomer for special grades of polychloroprene in which the tendency to crystallize has been reduced.

6.5.2. Other Dichlorobutadienes

1,1-Dichloro-1,3-butadiene [6061-06-9] and 1,3-dichloro-1,3-butadiene [41601-60-9] are obtained by chlorination of 1-chloro-1,3-butadiene, followed by dehydrochlorination [658], [659].

1,1-Dichloro-1,3-butadiene is also obtained by reduction of 4,4-dichloro-3-buten-2-one with $NaBH_4$, followed by dehydrogenation with Al_2O_3 [660]. This compound is used as a comonomer for rubber and as an insecticide.

1,4-Dichloro-1,3-butadiene (trans,trans: [3588-12-3]; trans,cis: [3588-13-4]; cis,cis: [3588-11-2]) is manufactured by dehydrochlorination of tetrachlorobutane with zinc powder (yield: 60%) [661].

6.6. 3-Chloro-2-methyl-1-propene

3-Chloro-2-methyl-1-propene, [563-47-3] $CH_2 = C(CH_3) - CH_2Cl$, methallyl chloride, was first prepared in 1884 by M. SHESHUKOV by the chlorination of isobutylene at room temperature. Investigation of the compound was taken up again in the late 1930s by the research group of A. P. A. GROLL at the Shell Development Co. leading to its production at the pilot plant scale. It was this group which gave methallyl chloride its common name on the basis of the analogy of methacrylic acid. The IG Farben chemical works at Leuna in Germany prepared methallyl chloride on a semi-works scale a short time later (from 1939 to 1942) in order to carry out investigations in the polyamide field. Methallyl chloride production by chlorination of isobutylene resumed in Germany after World War II (about 1960) at Chemische Werke Hüls AG, which currently is the sole producer of the compound in the Federal Republic of Germany.

6.6.1. Physical Properties

M_r	90.5
mp	< -80 °C
bp at 101.3 kPa	72.0 °C
Density at 20 °C	0.9251 g/cm^3
n_D^{20}	1.4274

Vapor pressure vs. temperature:

°C	-13	18.8	36	53
kPa	2.66	13.3	26.6	53.2

Heat of formation ΔH_{298}^0 (liquid)	-105.1 kJ/mol
Heat of vaporization (calculated from the vapor pressure curve)	31.8 kJ/mol
Viscosity at 20 °C	4.2×10^{-4} Pa s
Dielectric constant	7.0
Flash point	-16 °C (Pensky-Martens)
Autoignition temperature	540 °C
Limits of inflammability in air at 20 °C	2.2 and 10.4 vol%

Methallyl chloride is a colorless liquid of low viscosity, irritating to the mucous membranes, and with a pungent odor. It is miscible with all of the common organic solvents.

Solubility in water at 20 °C is ca. 0.05 wt%; water solubility in methallyl chloride is 0.04 wt%

Methallyl chloride forms an azeotrope with water (5.8 wt%), which boils at 64.6 °C.

6.6.2. Chemical Properties

Methallyl chloride shows chemical behavior very similar to that of allyl chloride and undergoes essentially the same reactions as the latter. The presence of a β-methyl group increases its reactivity considerably, however. Under the influence of light and/or heat, the formation of low concentrations of dimeric methallyl chloride (2-methyl-4,4-bis (chloromethyl)-1-pentene) occurs. Moisture causes samples of methallyl chloride to become markedly acidic as a result of hydrolysis. A similar phenomenon is induced by either iron or iron chloride through dehydrochlorination of dimeric and polymeric material.

The chlorine atom of methallyl chloride is readily exchanged, as occurs, for example, in its conversion to alcohols, ethers, amines, and esters. Furthermore, the presence of the methyl group activates the double bond with respect to reactions such as hydration, addition of halogen or halogen halides, sulfonation, dimerization, and polymerization. Of greatest commercial significance are its reaction with sodium sulfite to give sodium methallyl sulfonate and the production of 2-methylepichlorohydrin. For further chemical reactions of methallyl chloride, see [662]–[666].

6.6.3. Production

Methallyl chloride is produced exclusively by chlorination of pure isobutylene in the gas phase. Yields of 80–85% are achieved at atmospheric pressure and a relatively low temperature (not exceeding 100 °C).

Chlorination Mechanism. The chlorination of isobutylene is somewhat unusual as compared to most olefin chlorinations. Treatment of olefins with elemental chlorine at a low temperature normally leads to addition as the dominant reaction. The competing substitution reaction, in which the double bond remains intact, increases in importance as the temperature is raised until a critical temperature is reached, above which one obtains almost exclusively the substitution product. The temperature range of this transition from addition to substitution lies highest for simple straight-chain olefins. Compounds with branching at the double bond, on the other hand, including isobutylene or 3-methyl-2-butene, react principally by substitution even at a very low temperature (−40 °C), as shown in Table 30.

Presumably, chlorination of isobutylene follows an ionic mechanism, and solid metallic and liquid surfaces catalyze the reaction [668]. According to [669]–[671], the following mechanism is likely: isobutylene reacts with a chlorine molecule such that the equivalent of a positive chlorine adds to the terminal carbon of the double bond, thereby producing a tertiary carbonium ion. This tertiary carbonium ion can react in one of three ways: either it can eliminate a proton from C_1 or C_3, both of which possibilities lead to restoration of a double bond, or else it can react further by addition of the chloride anion. All three possible products are obtained in the chlorination of

Table 30. Critical temperature ranges for selected olefins for the transition from chlorine addition to chlorine substitution, based on

Olefin	Critical temperature range, °C
Isobutylene (and other olefins branched at the double bond)	< –40
2-Pentene	125–200
2-Butene	150–225
Propylene	200–350
Ethylene	250–350

isobutylene: methallyl chloride (the favored product), the isomeric 1-chloro-2-methyl-1-propylene (isocrotyl chloride), and 1,2-dichloroisobutane.

Process Description. In carrying out the reaction on the industrial scale, gaseous isobutylene (99% pure) and chlorine are mixed at room temperature by using a dual-component jet, and the mixture is allowed to react in a cooled tube reactor at atmospheric pressure. The reaction rate is exceedingly high for such a low reaction temperature, all of the chlorine being converted within ca. 0.5 s. Proper mixing of the components by the inlet jet is a prerequisite to obtaining a high yield of methallyl chloride; reaction is so rapid that high local concentrations of chlorine cannot be dissipated, leading instead to the production of more highly chlorinated products.

In order to minimize further chlorination of methallyl chloride, an excess of isobutylene must be used, the amount necessary being dependent on the extent of reverse mixing of the reaction products in the reactor. An excess of isobutylene greater than ca. 25% leads to no further significant increase in the yield of methallyl chloride with reactors in which minimal axial mixing occurs; greater excesses would simply increase the effort required to recover unreacted isobutylene. Low reaction temperatures are advantageous in respect to the product distribution: 100 °C is generally regarded as the limit below which the reaction should be kept.

The use of liquid isobutylene as starting material would seem a logical alternative, since one could then take advantage of the cooling effect resulting from the compound's heat of vaporization. It has been shown, however, that this approach increases the yield of high boiling products, presumably due to the decrease in effectiveness of mixing of the components relative to the gas phase reaction.

A typical analysis of a methallyl chloride production run is shown in Table 31.

The *tert*-butyl chloride which is reported as being formed is a secondary product from addition of hydrogen chloride to isobutylene. For information on the influence of other variables on the chlorination, see [672].

The work-up of the crude reaction product is, in principle, quite simple. The only difficulty would be removal by distillation of isocrotyl chloride, given its very similar boiling point (see Table 31). The vinylic placement of its chlorine, however, makes this substance very unreactive relative to methallyl chloride, so that its presence is not detrimental in many applications. For this reason, it is normally separated only incompletely, if at all.

Table 31. Product distribution from isobutene chlorination, with physical properties of the substances

Compound	Concentration range, wt%	M_r	bp at 101.3 kPa, °C	d_4^{20}	n_D^{20}
tert-Butyl chloride	2–4	92.5	50.8	0.8410	1.3860
Isocrotyl chloride	4–5	90.5	68.1	0.9186	1.4221
Methallyl chloride	83–86	90.5	72.0	0.9250	1.4274
1,2-Dichloroisobutane	6–8	127.0	108.0	1.089	1.436
3-Chloro-2-chloromethyl-1-propene		125	138.1	1.1782	
1-Chloro-2-chloromethyl-1-propene	2–3	125	132.0 (cis) 130.0 (trans)	1.1659	1.4702 (25 °C)
1,2,3-Trichloroisobutane		161.5	163.9		
2-Methyl-4,4-bis(chloromethyl)-1-pentene		181	84.4 (1.3 kPa)	1.0711 (d_{20}^{20})	1.4773

The effluent gas stream is led out of the reactor into a water scrubber-cooler which absorbs the hydrogen chloride and causes the majority of the chlorinated hydrocarbons to condense. Remaining isobutylene exits from the top of the scrubber column. It is then collected, dried, compressed, and returned to the reactor. The crude methallyl chloride is separated from hydrochloric acid in separatory flasks, dried by azeotropic distillation, and freed by a preliminary distillation from more volatile byproducts (largely tert-butyl chloride and part of the isocrotyl chloride). Methallyl chloride and residual isocrotyl chloride are finally taken off at the head of a second column. The more highly chlorinated distillation residue is either worked up in a chlorinolysis unit or else it is burned under conditions permitting recovery of hydrogen chloride.

The use of glass apparatus is recommended both because of the fact that it is necessary to work to some extent with the crude product while it is still moist, and also because even the anhydrous crude product can evolve hydrogen chloride at elevated temperatures and in the presence of ferrous materials.

In contrast to other hydrocarbon chlorination procedures (those employed with methane or propylene, for example), an aqueous workup cannot in this case be avoided if isobutylene is to be recycled. This is because isobutylene reacts with hydrogen chloride to form tert-butyl chloride at the low temperature required for the separation of the chlorinated hydrocarbons. If one is willing to accept the presence of tert-butyl chloride as a byproduct, however, one could, perhaps with the aid of catalysts, avoid the complications of isobutylene recycling.

Other Production Methods. A thorough investigation has been made of the use in place of pure isobutylene in the methallyl chloride synthesis of a butane–butene (B–B) mixture remaining after extraction of butadiene from the C_4 fraction of petroleum cracking gases [673]. The C_4 fraction can also be chlorinated directly by using the liquid products as a reaction medium, in which case any straight-chain alkenes that are present react only to an insignificant extent [674]. As far as can be determined, however, these interesting possibilities have not yet seen useful commercial application.

Oxychlorination of isobutene is possible by using tellurium compounds as catalysts [675]. High yields of methallyl chloride are achieved on a small scale. Solution chlorination of isobutene in the presence of $TeCl_4$ or $SeCl_4$ is also known [676].

6.6.4. Quality Specifications and Chemical Analysis

The commercial product usually has a purity of 96–98% and contains isocrotyl chloride as its chief impurity, along with traces of 1,2-dichloroisobutane. It is stabilized with substituted phenols or marketed without a stabilizer. The degree of purity is established by gas chromatography. In many of its applications — such as copolymerizations to produce synthetic fibers — even traces of iron are detrimental. The presence of iron is best determined spectroscopically, using *o*-phenanthroline, for example. Additional information which should accompany any shipment would include the material's color, water content, pH value, residue on evaporation, and boiling range.

6.6.5. Storage and Shipment

Methallyl chloride is best stored under cool conditions in porcelain enameled vessels. Provided the material is kept very dry, it can also be stored in tightly-sealed stainless steel storage tanks. Porcelain enameled or baked enameled tankers are suitable for its transport, as are, for small shipments, special types of baked enameled drums.

6.6.6. Uses

The potential uses for methallyl chloride as a starting material in syntheses are, just as in the case of allyl chloride, extremely numerous [662]. Its use in pure form as a fumigant and disinfectant has been suggested, as has its application as a fumigating agent for seed grains [677], [678]. Methallyl chloride and other methallyl derivatives prepared from it are exceptionally well suited to copolymerization [679], [680]. The copolymerization of methallyl sulfonate with acrylonitrile has gained particular industrial significance [681]. Since 1970, 2-methylepichlorohydrin for use in the production of special epoxy resins has been manufactured from methallyl chloride at a production facility in Japan with an annual capacity of 6000 t. Methallyl chloride is used, together with allyl chloride, in the manufacture of Allethrin, a synthetic pyrethrum with applications as a pesticide.

6.7. Hexachlorobutadiene

$CCl_2 = CCl - CCl = CCl_2$; M_r 260.8 [87-68-3].

Physical Properties

mp	− 18 °C
bp at 101.3 kPa	212 °C
Density at 20 °C	1.680 g/cm^3
n_D^{20}	1.5663

Vapor pressure vs. temperature:

°C	20	34	60	64	111
kPa	0.036	0.097	0.497	0.623	5.87

Specific heat at 22 °C	0.85 kJ kg^{-1} K^{-1}
Heat of vaporization	48 kJ/mol
Coefficient of thermal conductivity	0.101 W K^{-1} m^{-1}
Viscosity at 15 °C	9.22 × 10^{-3} Pa s
at 21 °C	3.68 × 10^{-3} Pa s
at 50 °C	2.40 × 10^{-3} Pa s
Surface tension at 20 °C	3.14 × 10^{-2} N/m
Dielectric constant at 12 °C	2.56
Dielectric strength when freshly prepared	ca. 200 kV/cm

Hexachlorobutadiene is a colorless, oily liquid with a faint terpene-like odor. Its solubility in water at 20 °C is 4 mg/kg, that of water in hexachlorobutadiene at 20 °C 10 mg/kg.

Chemical Properties. Chemically, hexachlorobutadiene is very stable to acids and alkali and has no tendency to polymerize even under high pressure (10 MPa). Hexachlorobutadiene reacts with chlorine only under severe reaction conditions (e.g., under pressure in an autoclave at 230–250 °C), and then generally with cleavage of the carbon skeleton and formation of hexachloroethane and perchloroethylene [682]. Octachlorobutene and decachlorobutane are also produced in the temperature range 60–150 °C [683], [684]. For further information about the properties, reactions, and application possibilities of hexachlorobutadiene, see also the compilations [685].

Production. Hexachlorobutadiene occurs as a byproduct in all chlorinolysis processes for the production of perchloroethylene or carbon tetrachloride. Depending on the method employed, the crude product contains ca. 5 % or even more of hexachlorobutadiene, and this material can be recovered in pure form rather than being recycled into the reactor. Since chlorolysis plants are of large capacity, the demand for hexachlorobutadiene is generally met without the need for separate production facilities.

If hexachlorobutadiene is to be prepared directly, the preferred starting materials are chlorinated derivatives of butane. These are chlorinated at 400–500 °C and atmo-

Table 32. Types of chlorinated paraffins

Chlorinated paraffin type	CAS registry no
$C_{10}-C_{13}$ chlorinated paraffins	[085535-84-8]
$C_{14}-C_{17}$ chlorinated paraffins	[085535-85-9]
$C_{18}-C_{20}$ chlorinated paraffins	[063449-39-8]
$C_{20}-C_{30}$ chlorinated paraffin waxes (liquid and solid products)	[063449-39-8]

spheric pressure with a 4-fold excess of chlorine, giving a 75 % yield of hexachlorobutadiene [686].

The compound can also be prepared by chlorination of butadiene in a fluidized-bed reactor using a large excess of chlorine at 400–500 °C [687].

According to a method of the Consortium für Elektrochemische Industrie, hexachloro-1,3-butadiene can be made from 1,1,2,3,4,4-hexachlorobutane, which comprises about 60 % of the residue from the production of tetrachloroethane. The material is chlorinated repeatedly at 70–80 °C in the presence of iron(III) chloride and subsequently dehydrochlorinated at 170 °C.

Uses. As a consequence of its low vapor pressure at room temperature, hexachlorobutadiene is suitable as an absorbant for the removal of impurities in gases, e.g., removing carbon tetrachloride and other volatile compounds from hydrogen chloride [688]. Recently, a mixture composed of hexachlorobutadiene (50–70 %) and trichloroethylene has found use as a coolant in transformers [689]. The production of electrically conductive polymers with hexachlorobutadiene has been patented [690]. The compound's herbicidal properties have led to its use in the prevention of algal buildup in industrial water reservoirs, cooling towers, and cooling water systems [691]. Finally, it has been repeatedly suggested as a hydraulic fluid, as a synthetic lubricant, and as a nonflammable insulating oil.

7. Chlorinated Paraffins

"Chlorinated paraffins" is the collective name given to industrial products prepared by chlorination of straight-chain paraffins or wax fractions. The carbon chain length of commerical products is usually between $C_{10}-C_{30}$ and the chlorine content between 20–70 wt %. Four main types of chlorinated paraffins are in regular use today (see Table 32).

Compounds of this sort were first detected in the middle of the 19th century by J. B. A. DUMAS. It was he who showed that long-chain paraffins, known up to this date only as relatively inert substances, are in fact subject to certain chemical reactions, and that the influence of artificial light promotes the processes. The first systematic study on chlorinated paraffins was conducted some years later, between 1856 and 1858, by P. A. BOLLEY [692].

Chlorinated paraffins acquired industrial importance in the early 1930s and consumption grew rapidly, particularly during World War II, with increasing use of the substances as flameproofing and rot-preventing agents. Further rapid expansion subsequently took place during the 1960s when low-cost normal paraffin became freely available as feedstocks and chlorinated paraffins were recognized as effective plasticizers for PVC. Recent developments in this area have seen the introduction of a range of chlorinated α-olefins and aqueous based processes for producing solid chlorinated paraffin waxes [693], [694].

Currently, more than 300 000 t of chlorinated paraffins are consumed in the world each year (excluding China), and they are regarded as important substances with a wide range of applications. The largest single market is the United States accounting for some 41 000 t, significant consumption also exists in Europe and the Far East. The vast majority of the market comprises of liquid chlorinated paraffins with only some 16 000 t of solid chlorinated paraffin produced (excluding China).

7.1. Physical Properties

The chlorinated paraffins are homogeneous, neutral, colorless to pale yellow liquids. Their viscosities, densities and refractive indices rise with increasing chlorine content for a given carbon chain length and also with chain length at constant chlorine content (see Fig. 31). Volatilities, on the other hand, decrease with an increase in chain length and degree of chlorination (see Fig. 32). Physical properties of selected commercially available chlorinated paraffins are given in Table 33.

Chlorinated paraffins are practically insoluble in water and lower alcohols. They are soluble, however, in chlorinated aliphatic and aromatic hydrocarbons, esters, ethers, ketones, and mineral or vegetable oils. Their solubility in unchlorinated aliphatic and aromatic hydrocarbons is only moderate. Different types of chlorinated paraffins are completely miscible with one another.

At temperatures above 120 °C, some decomposition of chlorinated paraffins may occur accompanied by elimination of hydrogen chloride. The resulting polyalkenes once becoming conjugated cause a deepening of color. Accumulated hydrogen chloride operates as a catalyst for further dehydrochlorination.

Elimination of hydrogen chloride becomes significant at temperatures above 220 °C and for most applications this marks the upper limit of chlorinated paraffin use. For most commercially available grades volatility also becomes unacceptably high above this temperature.

The thermal stability of a chlorinated paraffin is defined by the extent to which it undergoes degradation at 175 °C over a 4 h period. The determination of the thermal stability requires use of a standardized piece of apparatus. There are a number of efficient stabilization systems, which increase the thermal stability of chlorinated paraffins, facilitating their use at the upper end of the temperature range noted above.

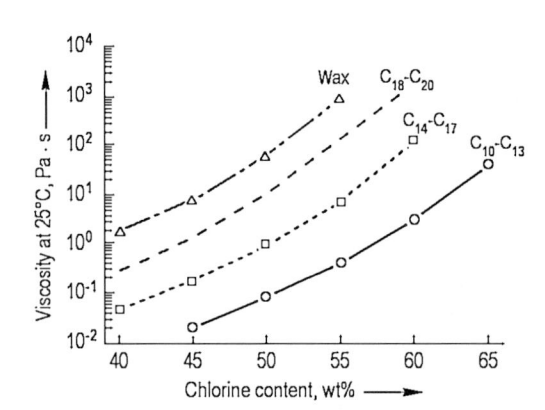

Figure 31. Viscosities of chlorinated paraffins

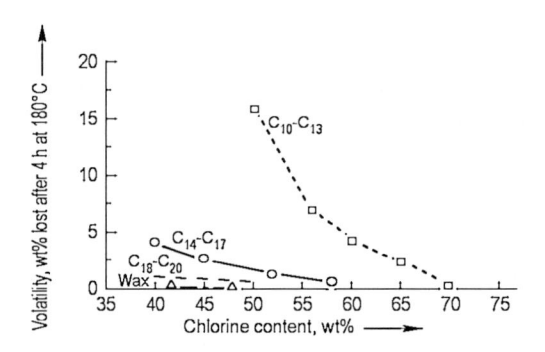

Figure 32. Densities of chlorinated paraffins

Typically these involve the addition of small amounts of epoxide-containing compounds, antioxidants and organic phosphites. Likewise, stabilizers are known which prevent darkening of chlorinated paraffins caused by the adverse effects of ultraviolet light.

Apart from chlorine content and chain length, no convenient physical or physicochemical properties are available for the characterization of individual types of chlorinated paraffins.

7.2. Chemical Properties and Structure

Commercially available chlorinated paraffins are not simple well defined chemical compounds. Instead they are complex mixtures of many molecular species differing in the lengths of their carbon chains and in the number and relative positions of chlorine atoms present on each carbon chain. Their chemical formula can be stated as follows:

$C_nH_{2n+2-m}Cl_m$

where n usually lies between 10 and 30 and m between 1 and 22. In the less chlorinated products, unchlorinated paraffin molecules may also be present. Studies

Table 33. Physical properties of selected commercial chlorinated paraffins

Trade name	Paraffin carbon chain length	Nominal chlorine contents, wt%	Color hazen, APHA	Viscosity [a], mPas	Density [a], g/mL	Thermal stability [b], wt% HCl	Volatility [c], wt%	Refractive index
Cereclor 50LV	$C_{10}-C_{13}$	50	100	80	1.19	0.15	16.0	1.493
Cereclor 56L		56	100	800	1.30	0.15	7.0	1.508
Cereclor 60L		60	135	3 500	1.36	0.15	4.4	1.516
Cereclor 63L		63	125	11 000	1.41	0.15	4.3	1.522
Cereclor 65L		65	150	30 000	1.44	0.20	2.5	1.525
Cereclor 70L		70	200	800 [d]	1.50 [d]	0.20	0.5	1.537
Cereclor S40	$C_{14}-C_{17}$	40	80	70	1.10	0.20	4.2	1.488
Cereclor S45		45	80	200	1.16	0.20	2.8	1.498
Cereclor S52		52	100	1 600	1.25	0.20	1.4	1.508
Cereclor S58		58	150	40 000	1.36	0.20	0.7	1.522
Cereclor M40	$C_{18}-C_{20}$	40	150	300	1.13	0.20	1.2	1.491
Cereclor M47		47	150	1 700	1.21	0.20	0.8	1.506
Cereclor M50		50	250	18 000	1.27	0.20	0.7	1.512
Cereclor 42	wax > C_{20}	42	250	2 500	1.16	0.20	0.4	1.506
Cereclor 48		48	300	28 000	1.26	0.20	0.3	1.516
Solid chlorinated paraffin wax [f]		70	100 [e]		1.63	0.20		

[a] At 25 °C unless otherwise stated.
[b] Measured in a standard test for 4 h at 175 °C.
[c] Measured in a standard test for 4 h at 180 °C.
[d] At 50 °C.
[e] 10 g in 100 mL toluene.
[f] Solid, softening point 95–100 °C.

have been conducted on the distribution of chlorine atoms within the molecule [695]. The results show that it is rare to find two chlorine atoms bound to one and the same carbon atom.

During paraffin chlorination, which is essentially a free radical process, tertiary carbon atoms react faster than those which are secondary; secondary carbon atoms in turn react faster than primary ones. The stability of chlorinated paraffins toward dehydrochlorination follows the reverse pattern, with tertiary chlorides being the least stable and primary chlorides the most stable. Therefore, it can be concluded that most of the chlorine atoms present in the commercial product are secondary with no more than about one in twelve primary hydrogens substituted by chlorine atoms. Methyl side-chains chains where they exist are usually unchlorinated.

Because of their very limited reactivity, chlorinated paraffins are of little significance as chemical intermediates, although their conversion to alcohols [696] and the substitution of various chlorine atoms by sulfuric acid have been studied [697].

7.3. Production

Raw Materials. Because of the known instability of compounds containing tertiary chlorine atoms, only straight-chain paraffins with a minimum content of branched isomers can be used as raw material for the industrial production of useful chlorinated paraffins. The quality of the technical products underwent a significant increase when the petrochemical industry succeeded in producing paraffin fractions enriched in straight-chain components through urea adduction and later by treatment with molecular sieves.

The paraffin fractions used most frequently by producers of chlorinated paraffins are the three straight-chain mixtures $C_{10}-C_{13}$, $C_{14}-C_{17}$, $C_{18}-C_{20}$ and waxes in the range of $C_{20}-C_{28}$. The feedstocks used most frequently to make chlorinated olefins are C_{14} or C_{16} linear α-olefins or mixtures of the two. The ranges employed can vary, depending on the regional availability of suitable raw materials.

Preparation. Chlorinated paraffins are prepared by reacting pure gaseous chlorine with the starting paraffins in the absence of any solvents at temperatures between 80 °C and 100 °C.

$$C_nH_{2n+2} + m\,Cl_2 \rightarrow C_nH_{2n+2-m}Cl_m + m\,HCl$$

Small amounts of oxygen are frequently used to catalyze the chlorination process. Temperatures above 120 °C must be avoided as these may cause dark or black products. Ultraviolet light is used by some manufacturers for initiating the reaction at relatively low temperatures. Once the reaction has started, the light source may be reduced in intensity or eliminated.

The reaction between chlorine and paraffin is exothermic with an enthalpy of reaction of ca. −150 kJ/mol; therefore, the reaction once initiated must be carefully controlled by cooling. Other important considerations are the need for efficient mixing to prevent localized overheating and control of the chlorine flow rate in order to minimize the amount of unreacted chlorine in the off gas. These requirements become increasingly important as the reaction proceeds and the viscosity of the product increases. For these reasons it is nearly impossible to prepare chlorinated paraffins containing more than 71 wt % chlorine by this method.

The reaction is terminated by stopping the chlorine flow once the required degree of chlorination is reached. The end point is assessed by a variety of methods including refractive index and viscosity. The product is then blown with nitrogen gas to remove any unreacted chlorine and residual HCl. In most case a small amount of a storage stabilizer, usually an epoxidized vegetable oil, is then added prior to sending the finished product for storage or drumming.

Other important considerations when producing chlorinated paraffins are the ability to deal effectively with HCl which is co-produced in large amounts.

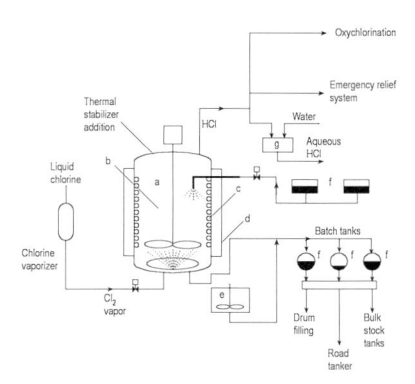

Figure 33. Commercial production of chlorinated paraffins
a) Reactor; b) Agitator; c) Heating/cooling coils; d) Jackets;
e) Addition of special stabilizers; f) Batch tanks; g) Absorption of hydrogen chloride

Both batch and continuous processes are known for the preparation of chlorinated paraffins on an industrial scale. *Batch processes* are preferred, because of the large variety of special products synthesized. Figure 33 gives a general outline of the process employed in the commercial production of chlorinated paraffins.

The same basic principle is used in *continuous processes*. In this case, however, groups of three or four reactors are arranged in series.

The production of solid chlorinated paraffin wax generally employs a two-stage process, the first of which involves production of an intermediate liquid chlorinated paraffin wax with between 40 – 50 wt % chlorine. This is then dissolved in carbon tetrachloride and further chlorinated to around 70 wt % at which point the carbon tetrachloride is stripped off and the resulting cake ground to a fine powder. Since the beginning of the 1990s producers of the solid chlorinated paraffin wax have started to use water as a reaction medium rather than carbon tetrachloride [693]. This is largely a result of the restrictions on the use and emissions of carbon tetrachloride arising from the Montreal Protocol.

Construction Materials. Older reactors are generally lead, ceramic- or occasionally silver-lined, whereas more modern reactors are generally constructed of glass-lined steel. Ferrous metal surfaces must be avoided to prevent discoloring or blackening of the finished product. Other equipment such as pumps, stirrers, pipework and valves should be constructed of corrosion-resistant materials and be resistant to hydrochloric acid.

Environmental Protection. The production process of chlorinated paraffin has minimal environmental impact because little contaminated water or other wastes are formed in the course of the reaction. Either after absorption or while it is still in the gaseous phase, the hydrogen chloride leaving the reactors as a byproduct must be cleaned in accordance with its future use. Hydrochloric acid is usually separated from any organic liquids and subsequently treated with activated charcoal. Purification of gaseous hydrogen chloride can be carried out after refrigeration by use of appropriate separators. Thus, the only wastes are those to be expected from cleaning procedures in

Table 34. Producers of chlorinated paraffins and trade names

Producer	Country	Trade name
ICI	UK (Canada, Thailand, France, Australia)	Cereclor
Caffaro	Italy	Cloparin, Cloparol
Hoechst AG *	Germany	Hordalub, Hordaflam, Hordaflex, Chlorparaffin
Occidental	United States	Chlorowax
Dover **	United States	Chlorez, Paroil, Chloroflo
Keil (Ferro Corp)	United States	"CW"
Novaky	Czech Republic	Chlorparaffin
Handy	Taiwan	Paroil
Rhône Poulenc	France	Alaiflex
NCP	South Africa	Plasticlor
Polifin	South Africa	
Koruma	Turkey	Chlorinated paraffin
Tosoh	Japan	Toyoparax
Asahi Denka	Japan	ADKciser

* Hoechst AG have announced their intention to close their plant at the end of 1998. ** Largest producer of solid chlorinated paraffin waxes.

the plant, in addition to activated charcoal. Such wastes and other byproducts should be burned in approved incinerators at a temperature above 1200 °C to prevent formation of cyclic or polycyclic chlorinated compounds.

Contamination of water caused by spillages must also be avoided. Asphalt surfaces of roads may be damaged by chlorinated paraffins, particularly at elevated temperatures.

Producers and Trade Names. Commercial grades of chlorinated paraffins are always mixtures of different alkanes chlorinated to varying degrees. Producers normally supply a range of materials intended to meet the differing needs of their customers.

In order to provide sufficient definition for individual products, numbers are usually appended to the trade names. These indicate the approximate chlorine content of a given material and in some cases provide additional information regarding the alkanes used as starting material (see Table 33).

Table 34 gives the major producers of chlorinated paraffins together with the corresponding trade names. The producers cited account for approximately 290 000 t capacity out of a total world capacity of ca. 380 000 t. The balance is made up of small plants producing products to meet local market demands in China, India and Russia. The largest producer of chlorinated paraffins is ICI who operate 5 plants worldwide.

7.4. Analysis and Quality Control

The properties of chlorinated paraffins may be influenced by the process used for their preparation. Products prepared in a continuous process sometimes have a lower thermal stability and a higher volatility than products prepared in a batch process [698]. A less homogeneous distribution of the introduced chlorine seems to be responsible for this difference. Moreover, batch products are said to be more compatible with poly(vinyl chloride).

Even though chlorinated paraffins are well-known and very important substances with many applications, analysis for determining their composition or their presence in environmental samples is particularly problematic. Techniques employing gas chromatographic investigations are very difficult, both because chlorinated paraffins are complex mixtures and because dehydrochlorination may occur during separation in a GC column. Furthermore, the sensitivity of the usual GC detection systems is often insufficient. Recent work has shown that capillary gas chromatography combined with negative ion chemical ionization mass spectrometry represents a suitable approach as does on-column reduction followed by either GC or GC–MS [699]. Resolution of groups of isomers and homologues has been obtained by a GC–MS method [700].

For routine quality control of chlorinated paraffins, usually the density, the viscosity, color, thermal stability and the chlorine content of specific products are determined. Approved standard methods are available for determination of densities, viscosities, and color. In order to determine their chlorine content, chlorinated paraffin samples are combusted under well-defined conditions, after which chloride is determined argentometrically, microcoulometrically, or by use of ion-selective electrodes. Good results have been obtained by the Wurtzschmitt method, in which a small sample of chlorinated paraffin is oxidized in a nickel bomb in the presence of an excess of sodium peroxide. Routine chlorine content determinations of production batches are usually carried out by direct correlation with other physical properties such as density or refractive index where well established relationships exist for each feedstock.

7.5. Storage and Transportation

Chlorinated paraffins are non-corrosive substances at ordinary temperatures. Therefore, their storage for several months without deterioration of either products or containers is possible in mild steels drums. Nevertheless, temperatures of storage should not exceed 40 °C and prolonged storage should not be in direct sunlight, since this may cause discoloration. Reconditioned drums should be internally lacquered before use. In the case of bulk storage, stainless or mild steel tanks are recommended. If mild steel is used as the construction material, the tanks must be internally lined with epoxy or phenol–formaldehyde resins. Appropriate producers of resins should be contacted with respect to the suitability of their coatings. Many of the chlorinated paraffin types are very difficult to handle at low temperature because of their high viscosities. Installation of storage tanks in heated rooms is, therefore, advantageous in order to keep the products in a usable state. Outdoor tanks must be equipped with some sort of moderate warming system. Circulated warm water is recommended as the preferred heating agent, since it does not produce local hot spots where deterioration of the stored product might occur.

If steam is used for warming, its pressure must be reduced to ensure that the temperature will not exceed 105 °C at the inlet. It is also recommended that external pipe work is electrically trace-heated to prevent the formation of cold plugs of product. Chlorinated paraffins swell most types of rubber, therefore, any gaskets, etc. should be made of poly(tetrafluoroethylene).

Positive displacement type pumps, preferably ones equipped with an automatically operated security device, are preferred for material transfer at end use locations. It is particularly important that adequate flow rates be maintained on the suction side of such pumps.

7.6. Health and Safety Regulations

Chlorinated paraffins are generally regarded as low toxicity products. They have very low acute toxicity, are not absorbed through the skin and exhibit extremely low vapor pressures. In longer term studies, short-chain ($C_{10}-C_{13}$) chlorinated paraffins were found to cause target organ toxicity leading to tumors in various organs of exposed rodents [701]. Subsequent studies with appropriate chlorinated paraffins have elucidated the mechanism of carcinogenesis in rodents and have shown that the mechanisms are not operative in guinea pigs [702]. Chlorinated paraffins have been shown to be non-genotoxic [703]. The relevance of the rodent tumors for human health hazard assessment can therefore be questioned [702]. In general, the highest exposure levels under industrial conditions of use are likely to be far less than those which would elicit toxicological effects. There are no requirements within the European Union to classify or label chlorinated paraffins for health effects.

Laboratory studies have demonstrated that chlorinated paraffins based on $C_{10}-C_{13}$ paraffins exhibit a toxic effect on some varieties of aquatic life. Consequently these products are classified as Severe Marine Pollutants under IMO regulations and have been provisionally classified by producers as "Dangerous For the Environment" according to EEC Directive 93/21/EEC (7th Amendment to Dangerous Substances Directive). Other types of chlorinated paraffins exhibit lower aquatic toxicity and are not classified for either supply or transport purposes. In Germany the same $C_{10}-C_{13}$ based chlorinated paraffins are classified as WGK III under the criteria laid down by the Kommission zur Bewertung Wassergefährdender Stoffe. Chlorinated paraffins biodegrade only slowly, with products containing > 50 wt % chlorine biodegrading most slowly.

7.7. Uses

Chlorinated paraffins find widespread industrial use as plasticizers, flame retardants, solvents, extreme-pressure additives and to lesser extent, as resin extenders. Choice of grade is frequently a compromise between the physical properties outlined above, in particular chlorine content, viscosity and volatility, together with a consideration of compatibility with the host polymer or base oil.

Plasticizers. This represents by far the largest use of chlorinated paraffins with the single largest application being the partial replacement of phthalate plasticizers in flexible PVC. The main reason for their use is reduction in formulation costs plus the additional benefits of flame retardancy, improved water and chemical resistance and better viscosity aging stability (i.e., a lesser increase in viscosity with time) of plastisols. Typical end use applications include cables, flooring, wall covering and general extrusion and injection molding. The products most frequently used in these applications are Cereclor S45 and S52 (Table 33).

Chlorinated paraffins also act as plasticizers in polyurethane and liquid polysulfide sealants where they also replace phthalates. Again cost reduction is an important feature as is their very low solubility in water and stability towards biodegradation, hence their use in sealants for aggressive biological environments such as sewage works. Very low volatility also makes them useful as plasticizers for insulating glass sealants. Typical grades used in these applications are Cereclor 56L, 63L, S52 and M50 (Table 33). Adhesive systems such as certain types of hot melt, pressure sensitive and poly-(vinyl acetate) emulsion adhesives are also effectively plasticized and tackified by chlorinated paraffins.

Finally their very low volatility, inert nature and low water solubility makes them effective plasticizers for a range of paint systems. In general use is focused in heavy duty industrial applications with typical host paint resins being chlorinated rubbers, chlorosulfonyl polyethylene (Hypalon), styrene–butadiene rubbers (Pliolite), Plioway and other modified acrylics. Typical grades used for paint applications are those based on wax feedstocks such as Cereclor 42 (Table 33). Solid 70 wt % chlorinated paraffin waxes also find widespread use in this application where their low plasticizing action allows them to act as a paint resin extender as well as conferring additional fire retardancy.

Fire Retardants. Chlorinated paraffins are excellent cost effective flame retardants [704], however their plasticizing action and limit on upper processing temperature restrict their use somewhat. Important areas of use include PVC where they act as a fire retardant plasticizer and are used to partially replace expensive phosphate plasticizers in applications such as mine belting and safety flooring. Chlorinated paraffins are often used in combination with a synergist, such as antimony trioxide, to enhance the fire retardancy of phthalate-based PVC formulations, such as those found in fire retardant wire and cable. Other fire retardant applications include their use in a range of rubbers including natural, nitrile, styrene–butadiene and Hypalon rubbers. Also of importance is their use in polyurethanes, in particular rigid foams and one-component foams (OCF) and unsaturated polyesters. The grades used vary considerably depending on the host polymer but in general they contain between 60–70% chlorine. Solid 70 wt% chlorinated paraffin wax also find widespread use as a flame retardant and is used in polyethylene, polypropylene and high-impact polystyrene compounds in addition to those polymers already mentioned.

Extreme Pressure Additives in Metal Working. Chlorinated paraffins are used as extreme-pressure additives to enhance lubrication and surface finish in demanding metal working and forming applications where hydrodynamic lubrication cannot be maintained. They function by providing a convenient source of chlorine which is liberated by frictional heat and then forms a chloride layer on the metal surface. This film has a lower shear strength than the metal itself, so the friction between the metals in sliding contact is reduced. Chlorinated paraffins are used predominantly in neat oils and to a lesser extent in soluble oil emulsions. They are frequently used in combination with other extreme-pressure additives including fatty acids, phosphorous- and sulfur-containing compounds, with which they display a synergistic action. Typical end use applications are stamping, forming, drawing and a range of cutting operations such as broaching. Historically, grades used in this application were based on $C_{10}-C_{13}$ paraffins, however, for environmental reasons there is now a move toward products based on higher molecular mass paraffins. In North America chlorinated α-olefins also find widespread use as extreme pressure additives.

Solvents. Chlorinated paraffins have found use as solvents for the color formers used in carbonless copy paper. The main benefits are good solvating power for the color formers, low volatility, stability in the encapsulation process and fast color development especially with blue color formers and clay-coated color fronts (CF). Chlorinated paraffins also tolerate higher levels of the diluents widely used by the industry compared to competitive products such as dialkyl naphthalenes.

8. Nucleus-Chlorinated Aromatic Hydrocarbons

The term "nucleus-chlorinated aromatic compound" as used here refers to a substance containing a mesomeric π-electron system in a carbocyclic framework in which at least one of the ring carbons bears a chlorine substituent rather than hydrogen.

Laboratory work on the compounds making up this class began long ago. For example, A. LAURENT reported in 1833 that he had obtained waxlike compounds in the course of chlorinating naphthalene. Nevertheless, their industrial manufacture and use was delayed until the first third of the 20th century.

Currently, chlorinated aromatic hydrocarbons are of substantial economic significance. This is particularly true of the chlorinated benzenes, the most important being monochlorobenzene, and the chlorinated toluenes. The compounds are now recognized as important starting materials and additives in the production of high-quality insecticides, fungicides, herbicides, dyes, pharmaceuticals, disinfectants, rubbers, plastics, textiles, and electrical goods.

In general, the environmental degradability of heavily chlorinated organic compounds, whether by biotic or by abiotic mechanisms, is low. This persistence has led in recent years to such drastic measures as prohibitions, restrictions on production and use, and legislation regulating waste disposal. Some highly chlorinated aromatics have been affected as well.

8.1. Chlorinated Benzenes

8.1.1. Physical Properties

Monochlorobenzene, C_6H_5Cl, is a colorless liquid which is volatile with steam and is a good solvent. It is miscible with all commonly used organic solvents and forms many azeotropes [705], [706]. Monochlorobenzene is flammable and has an aromatic odor. Upon addition of 24% benzene, it forms a eutectic mixture with a solidification point of − 60.5 °C.

Important physical data for monochlorobenzene are compiled in Table 35.

Dichlorobenzene, $C_6H_4Cl_2$, occurs in three isomeric forms. 1,2-Dichlorobenzene is a colorless, mobile liquid which is miscible with the commonly used organic solvents. It is difficult to ignite and has an unpleasant odor. The properties of 1,3-dichlorobenzene are similar to those of 1,2-dichlorobenzene. 1,4-Dichlorobenzene is a white, volatile, crystalline compound soluble in many organic solvents. It has a strong camphor odor. 1,4-Dichlorobenzene occurs as a stable monoclinic α-modification, which is transformed into the triclinic β-form at 30.8 °C.

Eutectic mixtures of the dichlorobenzenes are: 86.0% 1,2-dichlorobenzene, 14.0% 1,4-dichlorobenzene, solidification point − 23.7 °C; 85.3% 1,3-dichlorobenzene, 14.7% 1,4-dichlorobenzene, solidification point − 30.8 °C. Further physical data are given in Table 35.

Trichlorobenzene $C_6H_3Cl_3$, occurs in three isomeric forms. 1,2,3-Trichlorobenzene forms colorless tabular crystals. 1,2,4-trichlorobenzene is a colorless liquid of low flammability, and 1,3,5-trichlorobenzene forms long, colorless needles.

The trichlorobenzenes are insoluble in water, slightly soluble in alcohol, and very soluble in benzene and solvents like petroleum ether, carbon disulfide, chlorinated aliphatic and aromatic hydrocarbons.

A combination of 34% of 1,2,3-trichlorobenzene and 66% of 1,2,4-trichlorobenzene forms a eutectic mixture with a solidification point of + 1.5 °C. Further physical data are given in Table 35.

Chlorinated Hydrocarbons

Table 35. Physical properties of mono- to trichlorobenzenes

	Monochlorobenzene	1,2-Dichlorobenzene	1,3-Dichlorobenzene	1,4-Dichlorobenzene	1,2,3-Trichlorobenzene	1,2,4-Trichlorobenzene	1,3,5-Trichlorobenzene
	[108-90-7]	[95-50-1] [25321-22-6]	[541-73-1]	[106-46-7]	[87-61-6] [12002-48-1]	[120-82-1]	[108-70-3]
M_r	112.56	147	147	147	181.45	181.45	181.45
Melting point, °C	−45.2	−16.7	−26.3	53.5 (α-mod.) 54 (β-mod.)	53.5	17.0	63.5
Solidification point, °C	−45.58	−17	−24.8	53.08	52.4	17.2	62.6
Boiling point at 101.3 kPa, °C	132.2	179.0	173	173.9	218.5	213.5	208.4
Heat of fusion, J/g	84.9	87.8	85.7	123.5	95.9	85.3	95.9
Heat of vaporization, J/g							
at 50 °C	355.5						
100 °C	329.5	308.1	295.2	299.0	291.0	285.5	278.8
150 °C	307.7	286.0	273.4	277.4	271.7	265.9	259.2
200 °C	276.3	262.5	251.2	253.3	251.6	246.0	239.1
	315.0	273.0	262.9	264.6	242.4	241.2	236.1
bp at 101.3 kPa Temperature °C, corresponding, to vapor pressure							
0.13 kPa	−13.0	20.0	12.1	16.0	40.0	38.4	37.2
0.67 kPa	+10.6	46.0	39.0		70.0	67.3	63.8
1.33 kPa	22.2	59.1	52.0	54.8	85.6	81.7	78.0
2.67 kPa	35.3	73.4	66.2	69.2	101.8	97.2	93.7
5.33 kPa	49.7	89.4	82.0	84.8	119.8	114.8	110.8
8.0 kPa	58.3	99.5	92.2	95.2	131.5	125.7	221.8
13.3 kPa	70.7	112.9	105.0	108.4	146.0	140.0	136.0
26.7 kPa	89.4	133.4	125.9	128.3	168.2	162.0	157.7
53.3 kPa	110.0	155.8	149.0	150.2	193.5	187.7	183.0
101.3 kPa	132.2	179.0	173.0	173.9	218.5	213.5	208.4
Heat capacity, $J g^{-1} K^{-1}$	1.338 (20 °C) 1.462 (100 °C)	$1.131 + 0.0013 \delta$ (−37 to 104 °C)	$1.131 + 0.0013 \delta$ (−38 to 104 °C)	$0.992 + 0.0029 \delta$ (−78 to 52 °C) 1.088 (55 °C)		1.01	
Thermal conductivity, $Wm^{-1} K^{-1}$							
at −40 °C	0.141						
−20 °C	0.137						
0 °C	0.133						
20 °C	0.128	0.124				0.114	
50 °C	0.124						
100 °C	0.115			0.105 (60 °C)			

Table 35. continued

Property					
Standard heat of formation, kJ/mol	-10.65 ± 0.8	-17.57 ± 0.7	-20.46	-42.34 ± 1.1	-47.7
Density of the liquid, g/cm³					
at 20 °C	1.106	1.306	1.288	1.504	1.456
25 °C	1.101	1.3	1.282		
30 °C	1.096	1.295	1.277	1.25 (55 °C)	
70 °C	1.052	1.249	1.234	1.231	
100 °C					
Viscosity, mPa·s	0.806 (20 °C)	1.4 (20 °C)	1.11 (20 °C)		1.358
				1.381	2.08 (20 °C)
				1.68 (50 °C)	0.74 (100 °C)
				0.91 (100 °C)	
Heat of combustion at 25 °C, kJ/mol	-3100	-2962	-2955	-2934	
Surface tension at 20 °C, mN/m	33.5	36.6	36.2	29.9 (55 °C)	39.1
Dielectric constant					
at 20 °C	5.641	9.82	4.9	2.67	3.98
58 °C	5.02	8.5	4.63	2.62	
Refractive index, n_D^{20}	1.5248	1.5505	1.5464	1.5284 (55 °C)	1.5717 (25 °C)
Coefficient of expansion of liquid, K⁻¹	95×10^{-5}	110×10^{-5}	111×10^{-5}	116×10^{-5}	
Critical temperature, °C	359.2	424.1	410.8	407.5	461.8	470.8
Critical pressure, mPa	4.52	4.1	3.78	4.11	3.99	3.01
Critical density, g/cm³	0.365	0.408	0.41	0.395		
Flash point, °C	28	67	65	65	110	107
Ignition temperature, °C	590	640	>500	>500	>500	>500
Classification of danger (Verordnung über brennbare Flüssigkeiten – Flammable Liquids Order of the FRG)	A II	A III	A III	A III		
Explosive limits in air						
lower: vol% (g/cm³)	1.3 (60)	2.2 (130)				
upper: vol% (g/cm³)	11 (520)	12 (750)		ca. 18		
Maximum vapor concentration, g/cm³						
at 20 °C	54	7.8		10	2.7	
30 °C	89	15		19	5.8	
50 °C	222	45		60	18	
MAK (FRG), ppm	50	50		75	5	
mg/m³	230	300		450	40	
TLV-TWA (USA), ppm	75	50		75	5	
mg/m³	350	300		450	40	
Solubility in water, g/100 g						
at 20 °C	0.05	0.015	0.11	0.007		
60 °C		0.023	0.20	0.016		

Table 36. Physical properties of tetra-, penta-, and hexachlorobenzenes

	1,2,3,4-Tetra-chlorobenzene	1,2,3,5-Tetra-chlorobenzene	1,2,4,5-Tetra-chlorobenzene	Pentachloro-benzene	Hexachloro-benzene
	[634-66-2]	[634-90-2]	[95-94-3]	[608-93-5]	[118-74-1]
		[12408-10-5]			
M_r	215.90	215.90	215.90	250.34	284.78
Melting point, °C	47.5	51	141	85	229
Solidification point, °C	46.5	49.6	139.6	83.7	228.2
Boiling point at 101.3 kPa, °C	254	246	245	276	322
Heat of fusion, J/g				70.3	79.4
Heat of vaporization, J/g					
at 150 °C	254.6 (100 °C)				
200 °C	241.3		249		
bp at 101.3 kPa	216.6	217.9	224		191
Temperature °C, corresponding to vapor pressure,					
0.31 kPa	68.5	58.2		98.6	20.0
0.67 kPa	99.6	89		129.7	130.0
1.33 kPa	114.7	104.1		144.3	174.3
2.67 kPa	131.2	121.6		160.0	
5.33 kPa	149.2	140.0	146	178.5	209.5
8.0 kPa	160.0	152.0	157.7	190.1	
13.3 kPa	175.7	168.0	173.5	205.5	336.9
26.7 kPa	198.0	193.7	196	227.0	
53.3 kPa	225.3	220.0	220.3	251.6	290.5
101.3 kPa	254.0	246.0	245	276.0	322.0
Heat capacity, $J g^{-1} K^{-1}$					0.7 (25 °C)
Standard heat of formation, kJ/mol					−131
Liquid density, g/cm³	1.539 (100 °C)	1.523 (100 °C)	1.454 (150 °C)	1.609 (84 °C)	2.044 (23 °C)
				1.625 (100 °C)	1.462 (306 °C)
Heat of combustion at 25 °C, kJ/mol					−2372
Critical pressure, MPa	3.38	2.8	3.38		
Critical temperature, °C	498.5	526.1	486.6		
Maximum vapor concentration at 20 °C, g/cm³			2.5		0.00018

Tetrachlorobenzene $C_6H_2Cl_4$, also occurs in three isomeric forms: 1,2,3,4-Tetrachlorobenzene and 1,2,3,5-tetrachlorobenzene crystallize as colorless needles. 1,2,4,5-Tetrachlorobenzene forms colorless, sublimable needles with a strong, unpleasant odor.

All of the tetrachlorobenzenes are insoluble in water, but they are soluble in many organic solvents, particularly at an elevated temperature. Further physical data are given in Table 36.

Pentachlorobenzene C_6HCl_5, forms colorless needles and is insoluble in water.

Hexachlorobenzene C_6Cl_6, forms colorless and sublimable prismatic crystals. It is insoluble in water, but soluble at an elevated temperature in several organic solvents (e.g., benzene, chloroform, and ether).

8.1.2. Chemical Properties

The chlorobenzenes are neutral, thermally stable compounds. Reactions may occur to replace hydrogen at unsubstituted positions on the ring (e.g., halogenations, sulfonations, alkylations, nitrations), by substitution of the chlorine (e.g., hydrolysis), and with de-aromatization (e.g., chlorine addition).

Monochlorobenzene. In this compound the chlorine is firmly bound to the aromatic ring and can only be substituted under energetic conditions. Chlorobenzene can be hydrolyzed to phenol with aqueous sodium hydroxide at 360–390 °C under high pressure [707]–[709] or with steam at 400–450 °C over calcium phosphate. Monochlorobenzene reacts with ammonium hydroxide at high temperature and in the presence of copper catalysts to give aniline [710].

In electrophilic substitution, e.g., nitration, the directing influence of the chlorine atom leads to the formation of derivatives in which the add-ed substituent is found predominantly in the ortho or para position.

Light-induced addition of chlorine produces heptachlorocyclohexane [721].

Dichlorobenzenes. The reactivity to further substitution on the ring increases from 1,4- via 1,2- to 1,3-dichlorobenzene.

The position at which a third substituent is introduced into the ring depends on the directing influence of the two chlorine atoms. Thus, for example, no 1,3,5-derivatives can be formed in this way. Electrophilic substitution of 1,2-dichlorobenzene leads to 4-derivatives as main products and 3-derivatives as byproducts. Electrophilic substitution of 1,3-dichlorobenzene gives 4-derivatives as main product and 2-derivatives as byproduct, whereas 1,4-dichlorobenzene as starting material yields 2-derivatives.

The action of alkaline solutions or alcoholic ammonia solution on dichlorobenzene at 200 °C under pressure gives chlorophenols and dihydroxybenzenes, or chloroanilines and phenylenediamines, respectively. The addition of chlorine leads to octachlorocyclohexane [721].

Trichlorobenzenes. The reactivity of these compounds toward chlorine decreases in the order 1,3,5- > 1,2,3- > 1,2,4-trichlorobenzene. As expected, electrophilic substitution occurs preferentially at certain positions on the aromatic ring: With 1,2,3-trichlorobenzene an electrophilic substituent is led into the 4-position, with 1,2,4-trichlorobenzene into the 5-position (main product) and 3-position (byproduct), and with 1,3,5-trichlorobenzene into the 2-position.

The trichlorobenzenes can be hydrolyzed to dichlorophenol.

Tetrachlorobenzenes. Like the lower chlorinated benzenes, the tetrachlorobenzenes can be chlorinated and nitrated. The reactivity toward chlorine decreases from 1,2,3,5- via 1,2,3,4- to 1,2,4,5-tetrachlorobenzene. At a temperature above 160–180 °C, the chlorine substituents can be hydrolyzed with sodium hydroxide in methanol. In the preparation of trichlorophenol from tetrachlorobenzene, the toxic polychlorinated dibenzo-*p*-dioxins – and also the extremely toxic compound 2,3,7,8,-tetrachlorodibenzo-*p*-dioxin – may be formed if the very narrowly defined reaction conditions are not precisely maintained.

Pentachlorobenzene. Pentachlorobenzene can be chlorinated to hexachlorobenzene and nitrated to pentachloronitrobenzene. Hydrolysis to 2,3,4,5- and 2,3,5,6-tetrachlorophenol is also possible.

Hexachlorobenzene. Hexachlorobenzene, like the other polychlorobenzenes, can be dehalogenated to lower chlorinated benzenes with hydrogen or steam at a temperature above 500 °C in the presence of catalysts [722]. At a high temperature a mixture of hexachlorobenzene, chlorine, and ferric chloride gives carbon tetrachloride in high yield [723]. Reaction with sodium hydroxide and methanol leads to pentachlorophenol.

8.1.3. Production

Benzene Chlorination in the Liquid Phase. Chlorobenzenes are prepared industrially by reaction of liquid benzene with gaseous chlorine in the presence of a catalyst at moderate temperature and atmospheric pressure. Hydrogen chloride is formed as a byproduct. Generally, mixtures of isomers and compounds with varying degrees of chlorination are obtained, because any given chlorobenzene can be further chlorinated up to the stage of hexachlorobenzene. Because of the directing influence exerted by chlorine, the unfavored products 1,3-dichlorobenzene, 1,3,5-trichlorobenzene, and 1,2,3,5-tetrachlorobenzene are formed to only a small extent if at all. The velocity of chlorination for an individual chlorine compound depends on the compound's structure and, because of this, both the degree of chlorination and also the isomer ratio change continuously during the course of a reaction. Sets of data on the composition of products from different reactions are only comparable with one another if they refer to identical reaction conditions and materials having the same degree of chlorination. By altering the reaction conditions and changing the catalyst, one can vary the ratios of the different chlorinated products within certain limits. Lewis acids ($FeCl_3$, $AlCl_3$, $SbCl_3$, $MnCl_2$, $MoCl_3$, $SnCl_4$, $TiCl_4$) are used as principal catalysts (Table 37). Elevated temperatures in substitution reactions favor the introduction of a second chlorine atom in the ortho and meta positions, whereas para substitution is favored if cocatalysts and lower temperatures are used. As a further example of the influence of catalysts on the composition of the product, attention is drawn to the formation of 1,2,4,5-tetrachlorobenzene (Table 38). The optimal reaction temperature depends on the desired degree of chlorination. Mono- and dichlorination are carried out at 20–80 °C. In the production of hexachlorobenzene by chlorination of benzene, however, temperatures of about 250 °C are needed toward the end of the reaction.

Table 37. Influence of catalysts on the ratio 1,4-:1,2-dichlorobenzene

Catalyst	Proportion of 1,4-dichlorobenzene (in %) in the dichlorobenzene fraction	Ratio 1,4- : 1,2- dichlorobenzene	References
$MnCl_2 + H_2O$	ca. 50	1.03	[724]
$SbCl_5$		1.5	[725]
$FeCl_3$ or Fe	ca. 59	1.49 – 1.55	[726], [730], [733]
Metallosilicon organic compounds	61 – 74	1.56 – 2.8	[727]
$AlCl_3 – SnCl_4$		2.21	[728]
$AlCl_3 – TiCl_4$		2.25	[728]
Fe – S – PbO	ca. 70		[729]
$FeCl_3$ – diethyl ether		2.38	[730]
Aluminum silicate – hexamethylene-diamine		2.7	[731]
$FeCl_3 – S_2Cl_2$	ca. 76		[732]
$FeCl_3$ – divalent organic sulfur compounds	ca. 77	3.3	[733]
L-type zeolite	ca. 88	8.0	[734]
$TiCl_4$ (chlorinating agent is $FeCl_3$)		20 – 30	[735]

Table 38. Influence of catalysts on the ratio 1,2,4,5-:1,2,3,4-tetrachlorobenzene

Catalyst	Substrate	Ratio 1,2,4,5- : 1,2,3,4- tetrachlorobenzene	References
$FeCl_3$	1,2,4-trichlorobenzene	2.3 – 2.4	[726], [733]
Lewis acids – divalent organic sulfur compounds	1,2,4-trichlorobenzene	3	[733]
$FeCl_3$ – aromatic iodine compounds	1,2,4-trichlorobenzene	2.7 – 4.6	[736]
$SbCl_3 – I_2$	benzene	10 – 17	[737]

The usual catalyst employed in large scale production is ferric chloride, with or without the addition of sulfur compounds. Ferric chloride complexed with 1 mol of water is claimed to have the best catalytic effect [738]. Benzene and chlorine of technical purity always contain some water, however, thus, it may be that this hydrate compound is always present in iron-catalyzed industrial reactions.

The ratio of resulting chlorobenzenes to one another is also influenced by the benzene : chlorine ratio. For this reason, the highest selectivity is achieved in batch processes. If the same monochlorobenzene : dichlorobenzene ratio expected from a batch reactor is to result from continuous operation in a single-stage reactor, then a far lower degree of benzene conversion must be accepted (as a consequence of a low benzene : chlorine starting ratio). The selectivity of a continuous reactor approaches that of a discontinuous reactor as the number of reaction stages is increased [739]. Mathematical models analyzing and interpreting benzene chlorination will be found in [739] – [742] and elsewhere.

Solvents can also influence the chlorination rate as well as the selectivity of the reaction, although solvents are not used in industrial chlorination.

Continuous Chlorination. Benzene or a chlorobenzene derivative is treated with chlorine gas in a suitable reactor in the presence of dissolved ferric chloride. The reactants must be mixed as intensively as possible. The catalyst can be introduced along with the substrate or it can be allowed to form during the reaction on the surface of iron rings in the reactor. The reaction is highly exothermic.

$C_6H_6 + Cl_2 \longrightarrow C_6H_5Cl + HCl \qquad \Delta H = -131.5$ kJ/mol
$C_6H_5Cl + Cl_2 \longrightarrow C_6H_4Cl_2 + HCl \qquad \Delta H = -124.4$ kJ/mol
$C_6H_4Cl_2 + Cl_2 \longrightarrow C_6H_3Cl_3 + HCl \qquad \Delta H = -122.7$ kJ/mol
$C_6H_3Cl_3 + Cl_2 \longrightarrow C_6H_2Cl_4 + HCl \qquad \Delta H = -115.1$ kJ/mol

Unwanted heat of reaction can be dissipated either by circulating some of the reactor liquid through an external heat exchanger or by permitting evaporative cooling to occur at the boiling temperature. Circulation cooling has the advantage of enabling the reaction temperature to be varied in accordance with the requirements of a given situation. Evaporative cooling is more economical, however.

The reactor must be designed to ensure that the liquid within it has a suitable residence spectrum, since this favors high chlorination selectiv-ity. As noted above, the quantity ratio of the chlorobenzenes to one another is determined by the benzene : chlorine starting ratio.

Almost quantitative conversion of chlorine is achieved in continuously operated plants for the manufacture of mono- and dichlorobenzenes under normal operating conditions.

Cast iron, steel, nickel, and glass-lined steel can be used as construction materials. However, all starting materials must be substantially free from water; otherwise, severe corrosion is caused by the hydrochloric acid formed. Intrusion of water is dangerous for another reason: water causes the ferric chloride catalyst to be inactivated. If this occurs, chlorine collects in the reactor and exhaust system, and local overheating may ensue, causing spontaneous combustion of the aromatic hydrocarbon with chlorine in a highly exothermic reaction to form carbon and hydrogen chloride (2 mol of HCl/mol of Cl_2). Furthermore, above 280 °C metallic iron begins to burn in the chlorine stream. It is, therefore, advisable to monitor the reaction continuously by observing the heat production rate and the chlorine content of the waste gas.

On leaving the reactor, the liquid and gas portions of the reaction mixture are separated. The waste gas contains hydrogen chloride and — in proportions corresponding to their vapor pressures at the temperature of the waste gas — benzene and chlorobenzenes. If the chlorine conversion is incomplete, chlorine may be present as well. Concerning the treatment of the waste gas, see Section 8.5.

The liquid phase contains benzene, chlorobenzenes, hydrogen chloride, and iron catalyst. Production processes exist in which the product mixture is neutralized with sodium hydroxide solution or soda before it is subjected to fractional distillation. In

Table 39. Solubility (in wt%) of chlorine in benzene and chlorobenzenes

Solvent	Temperature			
	20 °C	50 °C	75 °C	100 °C
Benzene	27	6	0.8	
Monochlorobenzene	15	5		1.2
1,2-Dichlorobenzene	9	4		1.1
1,2,4-Trichlorobenzene	6	3.5		1.0

modern continuous distillation trains, the mixture of products can be distilled without preliminary treatment, however. The separated fractions consist of benzene, monochlorobenzene, dichlorobenzenes, trichlorobenzenes, and higher chlorobenzenes. Iron catalyst is removed along with the distillation residue, disposal of which is discussed in Section 8.5. Dissolved hydrogen chloride is removed during the benzene distillation and combined with the waste gas. Unreacted benzene is recycled to the reactor.

Whereas chlorobenzene and 1,2-dichlorobenzene are obtainable as pure distillates, the other fractions are mixtures of close-boiling polychlorobenzene isomers. A further separation, insofar as this is economically justifiable, would consist of combined crystallization/distillation processes.

Discontinuous Chlorination. Batch chlorination of liquid, molten, or dissolved aromatics is carried out industrially in agitator vessels equipped with external or internal cooling. The agitator must provide maximum exchange between the liquid and gas phases. Gaseous chlorine is introduced through a valve at the bottom of the vessel or through an ascension pipe beneath the agitator. The vessel may be constructed of glass-lined steel, cast iron, steel, or nickel. Absence of water must be ensured as a precaution against corrosion. The degree of chlorination can be ascertained from density determinations. The amount of chlorine which can be introduced in unit time depends on the heat output of the vessel and on the chlorine conversion rate. It is obviously desirable to keep the chlorine content of the off-gas as low as possible. It is also important that the reaction begins as soon as chlorine is introduced (the beginning of the reaction is indicated by a temperature increase and by the formation of hydrogen chloride). Chlorine is soluble in many hydrocarbons (see Table 39); therefore, if the onset of reaction is delayed, hydrogen chloride may be formed very rapidly, resulting in a substantial increase in the temperature of the reactants in the vessel.

Other Benzene Chlorination Processes. The following additional benzene chlorination processes are known:

1) Chlorination in the vapor phase with chlorine
2) Chlorination in the vapor or liquid phase with hydrogen chloride and air (oxychlorination)
3) Chlorination with chlorine-containing compounds
4) Electrolysis of benzene and hydrochloric acid

Apart from oxychlorination these processes are not industrially important. In oxychlorination, benzene vapor and a mixture of hydrogen chloride and air are reacted at about 240 °C in the presence of catalysts (e.g., $CuCl_2$–$FeCl_3$/Al_2O_3 or CuO–CoO/Al_2O_3). The main product is monochlorobenzene accompanied by 6–10% dichlorobenzene [744]–[752].

$$C_6H_6 + HCl + 0.5\, O_2 \longrightarrow C_6H_5Cl + H_2O$$

This process was developed in connection with the production of phenol from chlorobenzene (Raschig-Hooker process; → Phenol). The benzene conversion must be limited to 10–15% in order to control the heat in the catalyst solid bed (222 kJ/mol). An excessively high reaction temperature favors the formation of dichlorobenzene and a side reaction, the highly exothermic oxidation of benzene to carbon dioxide and water (330 kJ/mol).

Concerning material balance and heat, see [753]. The high cost of energy and for a corrosion-resistant plant, as well as an insufficient shift of the ratio of monochlorobenzene to dichlorobenzene, makes the process uneconomical.

Gulf has developed a process for the oxychlorination of benzene in the liquid phase under pressure with aqueous hydrochloric acid, catalytic quantities of nitric acid, and air or oxygen. The process leads to a high rate of benzene conversion and good selectivity for monochlorobenzene [754], but it has not been reported whether commercial operation has commenced.

The reaction of benzene and chlorine at 400 to 500 °C in the vapor phase [755], [756] is likewise uneconomical. One interesting feature is associated with vapor phase chlorination catalyzed by non-metals, however: the isomer ratio at the dichlorobenzene stage is shifted in the direction of *m*-dichlorobenzene (10% ortho, 66% meta, 24% para isomer). This effect is a consequence of the radical chain mechanism that is followed [757], [758].

Chlorine compounds are more selective for chlorination than chlorine itself, and they are thus used in the manufacture of special chlorobenzenes. The range of suitable chlorinating agents includes metal chlorides (e.g., of iron, antimony, titanium, and copper) of different valence stages [759]–[766], sulfuryl chloride [767]–[770], chlorine monoxide [771], and chlorosulfuric acid [772]. The latter three compounds are particularly suitable for the perchlorination of aromatics. In the chlorination of chlorobenzene with ferric chloride, high selectivity in favor of 1,4-dichlorobenzene (1,4 : 1,2-dichlorobenzene ratio till 25) is achieved [735], [765].

Chlorobenzenes can also be produced by electrolysis [773], [774], resulting in high selectivity for monochlorobenzene (98%) and in the exceptional yield of 94% [775].

Other Processes. The conversion of substituted aromatics to chlorobenzenes is of particular interest in connection with the preparation of not easily accessible isomers, such as 1,3-dichlorobenzene and 1,3,5-trichlorobenzene. The replacement of amino [776], [777], nitro [778]–[780], sulfonic acid or sulfonyl chloride [781]–[785], or acyl chloride groups [786] by chlorine leads to appropriate specialty chlorobenzenes.

8.1.3.1. Monochlorobenzene

Most monochlorobenzene is now produced from benzene and chlorine in continuously operated plants. Depending on the ratio of benzene to chlorine chosen, one can achieve either a low rate of benzene conversion and little dichlorobenzene formation, or almost complete conversion of the benzene with a higher degree of dichlorobenzene formation. Which of the two alternatives is favored depends on a profitability calculation, in which the distillation costs occasioned by the dichlorobenzenes need to be taken into account. The composition of a chlorination mixture containing the highest possible proportion of monochlorobenzene has been given as 4–5% unreacted benzene, 73% monochlorobenzene, and 22–23% dichlorobenzene [733].

The production of monochlorobenzene from benzene, hydrogen chloride, and air has been described briefly in "Other Benzene Chlorination Processes" 8.1.3.

Only if special chlorinating agents are used benzene can be chlorinated to monochlorobenzene without dichlorobenzene being formed simultaneously [763].

8.1.3.2. Dichlorobenzenes

Dichlorobenzenes are formed unavoidably in the production of monochlorobenzene. They arise as isomeric mixtures with a low content of the 1,3-isomer. A maximum dichlorobenzene concentration of 98% is obtainable in a batch process in which 2 mol of chlorine is used per mole of benzene in the presence of ferric chloride and sulfur monochloride at mild reaction temperatures. The remainder of the product consists of mono- and trichlorobenzene. About 75% 1,4-dichlorobenzene, 25% 1,2-dichlorobenzene, and only 0.2% 1,3-dichlorobenzene are obtained [787]. If the reaction is carried out with only ferric chloride as catalyst the maximum yield of dichlorobenzene is ca. 85%. For further information about the effects of catalysts on isomer distribution, see Table 37.

If discontinuous separation of dichlorobenzene isomer mixtures is to be carried out, a distillation column with at least 60 practical plates is needed. At a reflux ratio of about 35:1, monochlorobenzene (if present), 1,3-dichlorobenzene – 1,4-dichlorobenzene mixture, and 1,2-dichlorobenzene are removed successively at the top of the column.

The intermediate fractions are recycled, their amounts being dependent on the separation capability of the column.

The 1,4-dichlorobenzene fraction is concentrated from the melt by crystallization – in one or more steps, depending on the impurities present. It can then be separated into pure 1,4-dichlorobenzene and the eutectic of 1,3-dichlorobenzene and 1,4-dichlorobenzene (14.7% 1,4- and 85.3% 1,3-dichlorobenzene). In industry (for economic reasons), the fraction is not cooled until the eutectic is obtained. Several different techniques are employed for the melt crystallization [788], e.g., the batch tube bundle crystallizer (a new development of this type is the Proabd-Raffineur), the semi-continuous falling film crystallizer from Sulzer-MWB, or the continuous purifiers from Brodie or Kureha Chem. Ind.

Small amounts of 1,2- and 1,3-dichlorobenzene that are present as contaminants in 1,4-dichlorobenzene can be removed chemically by exploiting the fact that they are considerably more reactive than 1,4-dichlorobenzene. In sulfonation [789], [790], bromination [791], chlorination [792], and other reactions first the 1,3- and then the 1,2-dichlorobenzene are consumed; this leaves pure 1,4-dichlorobenzene, which can be separated by some suitable means.

Pure *1,3-dichlorobenzene* can be obtained by working up the unavoidable 1,3-dichlorobenzene-containing mother liquor that results from *p*-dichlorobenzene crystallization. Combined crystallization/distillation processes, extractive distillation [793], or separation on zeolites [794] are feasible. However, there also exist synthetic processes which lead directly to 1,3-dichlorobenzene:

Chlorination of benzene in the gas phase at 200–500 °C in a tube or in molten salt [755], [756], [795].

Sandmeyer reaction of 3-chloroaniline [776], [796] or direct replacement of the amino group by chlorine [797].

Reaction of 1,3,5-trialkylbenzene with chlorine in the presence of I_2 or ferric chloride to give 2,6-dichloro-1,3,5-trialkylbenzene, the alkyl groups then being cleaved in the presence of aluminum chloride [798], [799].

Chlorination of 1,3-dinitrobenzene or chloronitrobenzene at a temperature above 200 °C [778]–[780].

Two-step chlorination of benzenesulfonyl chloride or diphenyl sulfone, with sulfur dioxide being eliminated [784], [785].

Cleavage of sulfur dioxide or carbon dioxide from the corresponding disulfonyl chlorides or dicarbonic acid chlorides at temperatures above 200/300 °C in the presence of catalysts [781], [783], [786].

Catalytic dechlorination of 1,2,4-trichlorobenzene with hydrogen in the vapor phase [800], [801].

Isomerization of 1,2- and 1,4-dichlorobenzene at a temperature of 150–200 °C in the presence of catalysts containing aluminum chloride [802]–[804]. The conversion to 1,3-dichlorobenzene is said to be quantitative if the isomerization is carried out in an antimony pentafluoride–hydrogen fluoride system at 20 °C [805].

Except in the case of the chlorinolysis of 1,3-dinitrobenzene, no details are available on the extent to which these processes are employed on an industrial scale.

8.1.3.3. Trichlorobenzenes

1,2,3-Trichlorobenzene and *1,2,4-trichlorobenzene* are formed in minor quantities in the production of monochlorobenzene and dichlorobenzene. Trichlorobenzenes become the main product if the chlorine input is increased to about 3 mol of chlorine per mole of benzene.

The batch reaction of benzene with a 2.8-fold molar quantity of chlorine in the presence of fer-ric chloride at temperatures increasing to 100 °C gives a chlorination mixture consisting of 26% 1,4-dichlorobenzene, 4.5% 1,2-dichlorobenzene, 48% 1,2,4-trichlorobenzene, 8% 1,2,3-trichlorobenzene, 8% 1,2,3,4-tetrachlorobenzene, 5.5% 1,2,4,5-tetrachlorobenzene, and less than 1% pentachlorobenzene. The proportion of 1,4-dichlorobenzene, and thus also of 1,2,4-trichlorobenzene, can be raised by adding sulfur compounds as cocatalysts. After the chlorination mixture has been neutralized it can be separated by fractional distillation provided a column with more than 60 practical plates is used.

1,2,4-Trichlorobenzene can be obtained more directly by starting with pure 1,4-dichlorobenzene. Because of the directing influence of the chlorine substituents present (see 8.1.2), only 1,2,4-trichlorobenzene and (to an extent depending on the degree of conversion) higher chlorobenzenes are formed.

Another production method applicable to 1,2,4- and 1,2,3-trichlorobenzenes is based on the dehydrohalogenation of 1,2,3,4,5,6-hexachlorocyclohexane (stereoisomeric mixture), a byproduct of γ-hexachlorocyclohexane production. In the presence of aqueous alkali or alkaline earth solutions [806]–[809] or of ammonia [810], or directly through use of catalysts [811]–[817], hexachlorocyclohexane is converted at a temperature of 90–250 °C mainly to trichlorobenzene. The yield lies between 80 and 99%, with the product mixture consisting of 70–85% 1,2,4-trichlorobenzene and 13–30% 1,2,3-trichlorobenzene.

It should be pointed out that polychlorinated dibenzofurans and dibenzodioxins are formed during the reaction.

References to further literature on the decomposition of benzene hexachloride will be found in [818].

1,3,5-Trichlorobenzene is not formed in the course of liquid phase chlorination of benzene (see 8.1.3). It can be produced, however, by Sandmeyer reaction on 3,5-dichloroaniline, by reaction of benzene-1,3,5-trisulfonic acid derivatives with phosgene [783], by vapor phase chlorination of 1,3-dichlorobenzene [819], and by chlorination of 3,5-dichloronitrobenzene [820] or 1-bromo-3,5-dichlorobenzene at 300–400 °C [821]. The proportion of the 1,3,5-isomer in a mixture can be raised by isomerization of the other trichlorobenzenes with aluminum chloride [822], [823] or by reacting tetrachlorobenzenes and higher chlorobenzenes with alkali metal amides [824]. Total isomerization is reported to occur when SbF_5–HF is used as a catalyst system [805]. The three trichlorobenzenes can be separated by distillation and crystallization.

8.1.3.4. Tetrachlorobenzenes

1,2,3,4-Tetrachlorobenzene and *1,2,4,5-tetrachlorobenzene* are obtained by chlorination of benzene or of intermediate fractions obtained in the production of di- and trichlorobenzene. Antimony trichloride–iodine [825] is a catalyst combination that is particularly effective in giving 1,2,4,5-tetrachlorobenzene, the industrially preferred isomer (cf. Table 38). It is advisable to discontinue the chlorination before an appreciable amount of hexachlorobenzene has formed, since the low solubility of this compound makes separation of the chlorination mixture difficult. After the lower chlorinated benzenes have been removed by distillation, the tetrachlorobenzenes are separated from one another by crystallization with or without a solvent (the eutectic mixture consists of 12% 1,2,4,5- and 88% 1,2,3,4-tetrachlorobenzene and has a freezing point of 35 °C).

The 1,2,4,5-tetrachlorobenzene crystals which result are of high purity. The fraction containing 1,2,3,4-tetrachlorobenzene must be purified by distillation, however.

1,2,3,5-Tetrachlorobenzene, which is not available by direct chlorination of benzene, can be obtained by chlorination of 1,3,5-trichlorobenzene.

8.1.3.5. Pentachlorobenzene

Pentachlorobenzene is of no economic significance. Chlorination of benzene gives a mixture of tetra-, penta-, and hexachlorobenzenes, from which pentachlorobenzene can be isolated by combined distillation and crystallization steps.

8.1.3.6. Hexachlorobenzene

Hexachlorobenzene can be produced by exhaustive chlorination of benzene, using chlorine in the presence of such catalysts as ferric chloride. The reaction is conducted at a temperature above 230 °C in the liquid or vapor phase [826]–[828]. Because much of the chlorine remains unreacted and because hexachlorobenzene sublimes, the off-gas from the process is passed through fresh starting material, in the course of which the chlorine is reacted and the sublimate washed out. The difficulties caused by sublimed hexachlorobenzene can be eliminated by batch chlorination with liquid chlorine in an autoclave [829].

Hexachlorobenzene can also be produced by decomposition of hexachlorocyclohexane in the presence of chlorine. The resulting lower chlorinated benzenes, mainly trichlorobenzenes, are converted to hexachlorobenzene without first needing to be isolated [830]–[832].

High yields of hexachlorobenzene are obtainable under very mild reaction conditions when chlorine-containing reagents such as chlorine monoxide [771] or chlorosulfuric acid/iodine are used [772].

8.1.4. Quality and Analysis

The usual separating processes employed in industry give chlorobenzenes of high purity. Those impurities which remain, as well as their concentrations, depend on the nature of the production process and the separating technique applied.

The method now used almost exclusively to determine the quantitative compositions of commercial chlorobenzenes is gas chromatography, an approach which is both rapid and reliable. Glass capillary and specially packed columns are used. For those chlorobenzenes with solidification points above 0 °C, the temperature of solidification can serve as a criterium of purity of the main component. Typical analyses of several commercial chlorobenzenes are as follows:

Monochlorobenzene > 99.9%, < 0.02% benzene, < 0.05% dichlorobenzenes.

1,2-Dichlorobenzene. Technical grade: 70–85% 1,2-dichlorobenzene, < 0.05% chlorobenzene, < 0.5% trichlorobenzene, remainder 1,4- and 1,3-dichlorobenzene
Pure grade: > 99.8% 1,2-dichlorobenzene, < 0.05% chlorobenzene, < 0.1% trichlorobenzene, < 0.1% 1,4-dichlorobenzene.

1,3-Dichlorobenzene 85–99%, < 0.01% chlorobenzene, < 0.1% 1,2-dichlorobenzene, remainder 1,4-dichlorobenzene.

1,4-Dichlorobenzene > 99.8%, < 0.05% chlorobenzene and trichlorobenzene, < 0.1% 1,2- and 1,3-dichlorobenzene, bulk density about 0.8 kg/L.

1,2,4-Trichlorobenzene > 99%, < 0.5% dichlorobenzenes, < 0.5% 1,2,3-trichlorobenzene, < 0.5% tetrachlorobenzenes.

1,2,4,5-Tetrachlorobenzene > 98%, < 0.1% trichlorobenzenes, < 2% 1,2,3,4-tetrachlorobenzene, bulk density about 0.9 kg/L.

8.1.5. Storage and Transportation

The chlorobenzenes are all neutral, stable compounds which can be stored in the liquid state in steel vessels. The official regulations of various countries must be adhered with respect to the equipment of storage vessels, e.g., safety reservoir requirements, overflow prevention, and off-gas escape systems.

Chlorobenzenes that are liquid at ambient temperatures are shipped in drums, containers, or road/rail tankers. Solid compounds, such as 1,4-dichlorobenzene and 1,2,4,5-tetrachlorobenzene, can be transported in the molten state in heatable road/rail tankers or as granules or flakes in paper sacks and fiber drums. Steel containers are

Table 40. Freight classification

	GGVE/GGVS and RID/ADR		IMDG Code and IATA-DGR			US-DOT	
	class,	number	class	UN No.	pack. group		
Monochlorobenzene	3	3	3.3	1134	II	UN 1134	Flamm. liquid
1,2-Dichlorobenzene	3	4	6.1	1591	III	ORM. A	UN 1591
1,4-Dichlorobenzene	–		6.1	1592	III	ORM. A	UN 1592
1,3-Dichlorobenzene	3	4	6.1	1591	III	ORM. A	UN 1591
Trichlorobenzene	6.1	62	6.1	2321	III	Poison B	UN 2810
Tetrachlorobenzene	–		6.1	2811	III	Poison B	UN 2811
Hexachlorobenzene	6.1	62	6.1	2729	III	Poison B	UN 2811

suitable. Any paper or fiber materials that are used must be impermeable to vapors arising from the product.

Liquid transfer must incorporate provisions for gas compensation, as well as protection against static charge. Chlorobenzene vapors form flammable mixtures with air. The compounds are regarded as potential water pollutants and must not be allowed to enter groundwater.

Spills must be collected (proper precautions being taken to safeguard the health of the workers involved) and burned in a suitable incinerator.

It should be noted that chlorobenzenes may decompose with the release of hydrogen chloride if they are exposed to severe heat.

Freight classifications are given in Table 40.

8.1.6. Uses

The chlorobenzenes, particularly mono-, 1,2-di-, and 1,2,4-trichlorobenzene, are widely used as solvents in chemical reactions and to dissolve such special materials as oils, waxes, resins, greases, and rubber. They are also employed in pesticide formulations (the highest consumption of monochlorobenzene in the United States).

Monochlorobenzene is nitrated in large quantities, the product subsequently being converted via such intermediates as nitrophenol, nitroanisole, nitrophenetole, chloroaniline, and phenylenediamine into dyes, crop protection products, pharmaceuticals, rubber chemicals, etc. The production of phenol, aniline, and DDT from monochlorobenzene, formerly carried out on a large scale, has been almost entirely discontinued due to the introduction of new processes and legislation forbidding the use of DDT.

1,2-Dichlorobenzene, after conversion to 1,2-dichloro-4-nitrobenzene, is used mainly in the production of dyes and pesticides. It is also used to produce disinfectants and deodorants and on a small scale as a heat transfer fluid.

1,3-Dichlorobenzene has recently begun to be used in the production of various herbicides and insecticides. It is also important in the production of pharmaceuticals and dyes.

1,4-Dichlorobenzene is used mainly in the production of disinfectant blocks and room deodorants and as a moth control agent. After conversion into 2,5-dichloronitrobenzene, it finds application in the production of dyes. It is also used in the production of insecticides and, more recently, of polyphenylene-sulfide-based plastics, materials with excellent thermal stability [833], [834].

1,2,4-Trichlorobenzene is used as a dye carrier and (via 2,4,5-trichloronitrobenzene) in the production of dyes. Other uses are associated with textile auxiliaries and pesticide production (where 2,5-dichlorophenol serves as an intermediate). In the field of electrical engineering, it finds use as an additive for insulating and cooling fluids.

The hydrolysis of **1,2,4,5-tetrachlorobenzene** to 2,4,5-trichlorophenol, an intermediate for pesticides, has been almost entirely discontinued throughout the world due to the risk of formation of 2,3,7,8-tetrachlorodibenzo-p-dioxin (TCDD).

Pentachlorothiophenol, a mastication agent used in the rubber industry, is obtained from **hexachlorobenzene.** In the Federal Republic of Germany, the use of hexachlorobenzene as an active ingredient for pesticides has been prohibited since 1981.

Producers of chlorinated benzenes are: Anic SpA (Italy); Bayer AG (FRG); Hodogaya Chemical Co., Ltd. (Japan); Hoechst AG (FRG); Kureha Chemical Ind. Co., Ltd. (Japan); Mitsui Toatsu Chemicals, Inc. (Japan); Monsanto Chemical Corp. (USA); Nippon Kayaku Co., Ltd. (Japan); Produits Chimiques Ugine Kuhlmann SA (France); Rhône-Poulenc Chimie de Base (France); Standard Chlorine Chemical Co., Inc. (USA); Sumitomo Chemical Co., Ltd. (Japan).

8.2. Chlorinated Toluenes

8.2.1. Physical Properties

The chlorotoluenes occur in five chlorination stages, four of which have several isomers:

	Formula	Number of isomers	Table in which physical data are given
Monochlorotoluenes	C_7H_7Cl	3	41
Dichlorotoluenes	$C_7H_6Cl_2$	6	42
Trichlorotoluenes	$C_7H_5Cl_3$	6	43
Tetrachlorotoluenes	$C_7H_4Cl_4$	3	44
Pentachlorotoluene	$C_7H_3Cl_5$		44

Table 41. Physical properties of the monochlorotoluenes

	2-Chlorotoluene	3-Chlorotoluene	4-Chlorotoluene
	CH₃–C₆H₄–Cl (ortho)	CH₃–C₆H₄–Cl (meta)	CH₃–C₆H₄–Cl (para)
CAS reg. no.	[95-49-8]	[108-41-8]	[106-43-4]
M_r	126.59	126.59	126.59
Freezing point, °C	−35.59	−48.89	+7.6
Melting point, °C	−36.5	−47.8	+7.5
Boiling point at 101.3 kPa, °C	159.3	161.6	162.3
Density, g/cm³			
at 20 °C	1.0826	1.0722	1.0697
30 °C	1.0728	1.0625	1.0596
40 °C	1.0633	1.0530	1.0503
50 °C	1.0532	1.0433	1.0401
Viscosity, mPa s			
at 10 °C	1.188	1.0177	1.032
20 °C	1.022	0.877	0.892
60 °C	0.603	0.552	0.564
100 °C	0.439	0.392	0.397
Heat of fusion, J/g	79.99	82.95	102.46
Heat of vaporization, J/g			
at 100 °C	329.1		333.3
120 °C	319.5		323.6
140 °C	309.8		344.0
bp	299.6	327	303.5
Temperature corresponding to vapor pressure, °C			
0.13 kPa	5.4	4.8	5.5
0.67 kPa	30.6	30.3	31.0
1.33 kPa	43.2	43.3	43.8
2.67 kPa	56.9	57.4	57.8
5.33 kPa	72.0	73.0	73.5
8.00 kPa	81.0	83.2	83.3
13.3 kPa	94.7	96.3	96.6
26.7 kPa	115.0	116.6	117.1
53.3 kPa	137.1	139.7	139.8
101.3 kPa	159.3	161.6	162.3
Heat capacity, $J g^{-1} K^{-1}$	0.355		0.304
Heat of combustion at 18.8 °C, kJ/mol	−3747	−3749	−3754
Refractive index n_D^{20}	1.5267	1.5224	1.5209
Dielectric constant			
at 20 °C	4.45	5.55	6.08
58 °C	4.16	5.04	5.55
Critical temperature, °C	381.1		385.7
Flash point, °C	49	50	51
Ignition temperature, °C	> 500	> 500	> 500
Danger classification (VbF)	A II	A II	A II
Explosive limits in air			
lower, vol%	1.0		0.7
upper, vol%	12.6		12.2
TLV TWA (USA),			
ppm	50		
mg/m³	250		

Table 42. Physical properties of the dichlorotoluenes

	2,3-Dichloro-toluene	2,4-Dichloro-toluene	2,5-Dichloro-rotoluene	2,6-Dichloro-rotoluene	3,4-Dichloro-rotoluene	3,5-Dichloro-rotoluene
	[32768-54-0]	[95-73-8]	[19398-61-9]	[118-69-4]	[95-75-0]	[25186-47-4]
M_r	161.03	161.03	161.03	161.03	161.03	161.03
Freezing point, °C	+5.05	−13.35	+3.25	+2.6	−14.7	+25.1
Boiling point at 101.3 kPa, °C	208.1	201.1	201.8	200.6	209.0	202.4
Liquid density at 20 °C, g/cm³	1.266	1.25	1.254	1.266	1.254	
Refractive index n_D^{20}	1.551	1.548 (20 °C)	1.5449	1.5517	1.549 (22 °C)	1.5594
Flash point, °C	95	86	88	88	95	
Ignition temperature, °C	>500	>500	>500	>500	>500	>500
Explosive limits in air						
lower vol%		1.9				
upper vol%		4.5				
Temperature corresponding to vapor pressure, °C						
at 2.7 kPa	91.8	88.8	88.8	87.6	93.8	
8.0 kPa	120.2	115.6	115.9	114.7	121.6	
53.3 kPa	182.3	174.9	176.1	174.4	182.9	

Table 43. Physical properties of the trichlorotoluenes

	2,3,4-Tri-chlorotoluene	2,3,5-Tri-chlorotoluene	2,3,6-Tri-chlorotoluene	2,4,5-Tri-chlorotoluene	2,4,6-Tri-chlorotoluene	3,4,5-Tri-chlorotoluene
	[7359-72-0]	[56961-86-5]	[2077-46-5]	[6639-30-1]	[23749-65-7]	[21472-86-6]
M_r	195.48	195.48	195.48	195.48	195.48	195.48
Freezing point, °C	42.9	44.65	42.95	79.95	32.0	44.85
Boiling point at 101.3 kPa, °C	249.3	240.4	241.8	240.5	235.4	248.3
Liquid density at 100 °C, g/cm³	1.337	1.319	1.334	1.319	1.318	1.317

Table 44. Physical properties of tetra- and pentachlorotoluenes

	2,3,4,5-Tetra-chlorotoluene	2,3,4,6-Tetra-chlorotoluene	2,3,5,6-Tetra-chlorotoluene	Pentachloro-toluene
	[1006-32-2]	[875-40-1]	[1006-31-1]	[877-11-2]
M_r	229.92	229.92	229.92	264.37
Freezing point, °C	96.45	91.8	94.8	223.5
Boiling point at 101.3 kPa, °C	208.9	275.6	275.9	312
Liquid density at 100 °C, g/cm³	1.47	1.483	1.486	

All of the *monochlorotoluenes* are colorless, mobile, flammable liquids with a faint odor, similar to that of benzene; they form binary azeotropes with many organic compounds [705], [706].

The *dichlorotoluenes*, apart from 3,5-dichlorotoluene, are liquid at room temperature, and they are likewise colorless and flammable. The *tri, tetra-,* and *pentachlorotoluenes* are colorless crystalline compounds.

The liquid chlorotoluenes are good solvents and are miscible with most organic solvents. All chlorotoluenes are insoluble in water. The polychlorotoluenes can be dissolved in many organic solvents, particularly at elevated temperatures.

8.2.2. Chemical Properties

The chlorotoluenes are neutral and stable compounds.

Chemical reactions may occur at unsubstituted positions on the aromatic ring (e.g. halogenation, nitration, or sulfonation), by replacement of the chlorine substituent (e.g, hydrolysis), and on the methyl group (e.g., side-chain chlorination or oxidation).

The influence of the methyl group of toluene leads to electrophilic substitution at positions 2 and 4, whereas in chlorotoluenes the directing influences of the methyl and chlorine groups overlap unpredictably. Introduction of a third substituent (e.g., – Cl, – NO_2, – SO_3H) into 2-chlorotoluene can lead to all four possible isomers being formed, though position 5 is occupied preferentially. In 4-chlorotoluene, position 2 is the most preferred position. The chlorine atoms in these compounds are bound very firmly to the aromatic ring and cannot be displaced except under forcing conditions. Nevertheless, the hydrolysis of monochlorotoluene with sodium hydroxide solution is possible at 350 – 400 °C and pressures up to 30 MPa (300 bar), the result being isomeric cresol mixtures.

Hydrogenation of chlorotoluenes over noble metal catalysts leads to dechlorination, just as in the case of chlorobenzenes.

Chlorine substituents can be exchanged for amino groups [835], but this reaction has no industrial application.

Under free-radical conditions at elevated temperatures, it is possible to replace sequential-ly the three hydrogen atoms of the methyl group by halogen, leading to ring-chlorinated benzyl-, benzal-, and benzotrihalides (see Chap. 9).

Oxidation of the methyl group leads to chlorinated benzaldehydes and benzoic acids [836], [837].

Catalytic ammonoxidation with oxygen and ammonia at 350 – 550 °C converts the methyl group into a nitrile group [838].

The polychlorinated toluenes are similar to one another in their chemical behavior. Their principal industrial use is in the manufacture of side-chain-halogenated products.

Table 45. Influences of catalysts on the 2-:4-chlorotoluene ratio

Catalyst	2-:4-Chlorotoluene ratio	Dichlorotoluene in the chlorination mixture, %	Toluene conversion, %	Ref.
$TiCl_4$, $SnCl_4$, WCl_6 or $ZrCl_4$	3.3	1.5	~99	[840]
$[C_6H_5Si(OH)_2O]_4Sn$	2.2			[841]
$FeCl_3$	1.9	4.5	~75	[842]
$SbCl_3$–diethylselenide	1.9		~50	[843]
$SbCl_3$	1.6			[733]
$AlCl_3$–KCl	1.5	<1	~16	[844]
$SbCl_3$–thioglycolic acid	1.2			[733]
$FeCl_3$–S_2Cl_2	1.1	1.0	~99	[842]
Ferrocene–S_2Cl_2	1.06	0.2	~67	[845]
Lewis acids–thianthrene	0.91–1.1			[846]
$FeCl_3$–diphenylselenide	0.93		~91	[843]
PtO_2	0.89	2	~96	[847]
$SbCl_3$–phenoxathiin derivative	0.66–0.88	0.2–0.4	96–99	[848]
$SbCl_3$–tetrachlorophenoxathiin	0.85–0.87			[849]
$SbCl_3$–di- or tetrachlorothianthrene	0.7–0.9	0.1–0.2		[850]
Fe–polychlorothianthrene	0.76	1.2	~95	[851]

8.2.3. Production

Toluene Chlorination in the Liquid Phase. Monochlorotoluenes are produced on a large scale by reacting liquid toluene with gaseous chlorine at a moderate temperature and normal pressure in the presence of catalysts. Mixtures of isomers reflecting various chlorination stages are obtained. The chlorination conditions should be such as to give the highest possible yield of monochlorotoluene, because the dichlorotoluenes, all isomeric forms of which (with the exception of 3,5-dichlorotoluene) result, cannot be separated economically. The relative proportions of the various chlorotoluenes obtained can be varied within wide limits by altering the reaction conditions and catalyst. With most catalyst systems, a high reaction temperature favors ortho and meta substitution as well as further chlorination. Reducing the temperature favors substitution in the para position and increases the total yield of monochlorotoluene. Because of the directing influence of the methyl group, the 3-chlorotoluene fraction of crude monochlorotoluene is limited to between 0.2 and 2 % depending on the catalyst. The influence of catalysts on the ratio of 2-chlorotoluene to 4-chlorotoluene is apparent from the data in Table 45. Selectivity is seen to be inversely proportional to the activity of the catalyst [839].

Solvents also, influence the isomer distribution. This fact is of little significance, however, because the toluene reactant in industrial chlorinations is not diluted with solvents.

The reaction is usually conducted at a temperature between 20 and 70 °C. Chlorination at temperatures below 20 °C is uneconomical because the rate of reaction is too low.

Table 46. Comparison of catalytic chlorination of benzene and toluene

Catalyst system	Ratio of the chlorination rates of toluene to benzene	Ref.
$FeCl_3$	16	[842]
$FeCl_3$–S	200	[842]
Glacial acetic acid	345	[852]

Toluene has a higher π-basicity than benzene, however, and it therefore, shows a substantially higher rate of chlorination than the latter. As a result, it can be chlorinated at relatively low temperatures (Table 46).

This fact permits efficient conversion of toluene to monochlorotoluenes without a substantial quantity of dichlorotoluenes also being formed (see Table 45).

At a moderate reaction temperature (below 100 °C), only traces of side-chain-chlorinated products are formed, provided activation by light is prevented and effective catalysts are used (ones for which amounts of no more than several tenths of a percent are required).

Both batch and continuous processes are used commercially, with the former being more selective with respect to particular stages of chlorination. For continuous processes, it is necessary to choose a toluene:chlorine starting ratio which gives a lower degree of toluene conversion to maximize the formation of monochlorotoluene.

The remarks in 8.1.3on the continuous and batch chlorination of benzene apply correspondingly to the chlorination of toluene. Problems related to reactor design, materials, and dissipation of heat – 139 kJ is set free per mole of monochlorotoluene [853] – are also comparable. The ignition temperature of toluene in gaseous chlorine lies at 185 °C. The explosive limits of toluene in chlorine lie between 4 and 50 vol%. Conventional refinery toluene is sufficiently pure to serve as the starting material.

Traces of water that are entrained with technical chlorine and toluene have no influence on the composition of the product. They must not be so large as to reduce the efficiency of the catalyst, however.

One process for reducing the water content of the toluene below 30 ppm involves stripping with hydrogen chloride, and is described in a patent on the continuous production of monochlorotoluene [854].

As with the chlorobenzenes, the crude chlorotoluene mixture may be worked up by neutralization followed by separation in a distillation train. If the side-chain contains a detectable amount of chlorine, then this chlorine should be removed in order to prevent corrosion (e.g., pressure washing with aqueous alkaline solution may be carried out before the mixture enters the still).

Other Toluene Chlorination Processes. The following toluene chlorination processes are known in addition to liquid phase chlorination:

Chlorination in the vapor phase with chlorine [855]
Chlorination in the vapor or liquid phase with hydrogen chloride and air (oxychlorination)

Chlorination with chlorine-containing compounds
Electrolysis of toluene and hydrochloric acid

The oxidative chlorination of toluene with hydrogen chloride – air at a temperature of 150 – 500 °C in the presence of cupric chloride as both catalyst and chlorinating agent gives mono- and dichlorotoluenes along with varying amounts of side-chain-chlorinated products [751], [856] – [858]. In the Gulf process, the toluene is chlorinated with hydrochloric acid – oxygen in the liquid phase in the presence of nitric acid and an additional strong acid at a temperature of 60 – 150 °C and pressures of 350 – 1000 kPa. By adding alkanes, e.g., *n*-octanes, the extent of side-chain attack can be reduced to 0.7 %. The addition of trialkylphenols is claimed to improve the selectivity so that only one chlorine atom reacts with a given aromatic nucleus [859]. Good monochlorotoluene yields are obtained at 50 °C if ferric chloride is used as a chlorinating agent in the presence of Lewis acids, such as titanium chloride and aluminum chloride (*o*-chlorotoluene:*p*-chlorotoluene ratio 0.1 – 0.15) [735], [764], [860], [861]. Polymeric by-products are also formed but the side-chains are said to be unattacked.

Sulfuryl chloride [767], [862] and chlorine monoxide [767] are other possible chlorinating agents.

Monochlorotoluene can also be manufactured electrochemically. Thus, toluene can be chlorinated electrochemically in methanol to which water and sodium, ammonium, or lithium chloride has been added. High yields of monochlorotoluene result without dichlorotoluene being formed [863].

Other Processes. Only a few of the isomeric chlorotoluenes can be produced economically by direct chlorination and separation of the reaction mixture. The other chlorotoluenes must be produced by indirect syntheses involving such steps as the replacement of amino substituents by chlorine [776], [797] or introduction of meta-directing sulfonic acid groups, which are then removed after chlorination on the ring [864], [865]. See the following Sections for additional information.

8.2.3.1. Monochlorotoluenes

Both 2- and 4-chlorotoluene are produced by chlorination of toluene, as described in 8.2.3. The economic importance of 4-chlorotoluene is presently greater than that of 2-chlorotoluene, a circumstance which has led to the development of new catalyst systems (see Table 45), which increase the output of the para isomer. The batch chlorination of toluene at 50 °C in the presence of ferric chloride (or of ferric chloride – sulfur monochloride at 40 °C the values for which are given in parentheses for comparison) gives at most 83 % (98 %) monochlorotoluene, with 52 % (52 %) of the product being *o*-chlorotoluene, 2 % (0.3 %) *m*-chlorotoluene, and 28 % (46 %) *p*-chlorotoluene, in addition to 7 % (1 %) unreacted toluene and 11 % (1 %) dichlorotoluene [842].

2-Chlorotoluene and 4-chlorotoluene are separated by fractional distillation. Continuous separation is possible with columns having more than 200 theoretical plates, leading to a pure 2-chlorotoluene distillate and a bottom product containing about 98% 4-chlorotoluene. The boiling points of 3- and 4-chlorotoluene are so nearly alike that these two isomers cannot be separated by distillation. As in the production of pure 1,4-dichlorobenzene (see 8.1.3.2) the latter separation must be accomplished by crystallization from the melt. The eutectic mixture contains 77.5% *m*-dichlorotoluene and 22.5% *p*-dichlorotoluene and has a solidification point of − 63.5 °C. Consequently, crystallization from the melt must be conducted at low temperatures.

4-Chlorotoluene can also be separated from mixtures containing 2- and 3-chlorotoluene by adsorption on zeolites at 200 °C and 500 kPa [866].

3-Chlorotoluene can be produced from *m*-toluidine by a conventional Sandmeyer reaction [867] or by isomerization of 2-chlorotoluene on acid zeolites at 200–400 °C and 2000–4000 kPa, separation of the resulting mixture of 2- and 3-chlorotoluene being effected by distillation [866], [868].

8.2.3.2. Dichlorotoluenes

The isomers of dichlorotoluene have very similar chemical and physical properties, making it difficult to separate individual components from mixtures. Only 2,4-, 2,5-, and 3,4-dichlorotoluene can be produced economically by direct chlorination of the appropriate pure monochlorotoluene isomers.

2,3-Dichlorotoluene. In the chlorination of 2-chlorotoluene with ferric chloride, approximately 15% of the product is 2,3-dichlorotoluene, accompanied by 2,4-, 2,5-, and 2,6-dichlorotoluene; the former can be separated by distillation. Its production by a Sandmeyer reaction on 3-amino-2-chlorotoluene is more economical, however.

2,4-Dichlorotoluene. Chlorination of 4-chlorotoluene in the presence of ring chlorination catalysts (e.g. chlorides of iron, antimony, or zirconium) leads to 2,4- and 3,4-dichlorotoluene in a ratio of ca. 4:1; the two can be separated by fractional distillation. Only traces of other dichlorotoluenes are formed [869], [870]. A low reaction temperature reduces the formation of trichlorotoluene. According to [862], 2-chlorotoluene is converted primarily into 2,4-dichlorotoluene when sulfuryl chloride is used as the chlorinating agent.

2,5-Dichlorotoluene. The proportion of 2,5-dichlorotoluene in the dichlorotoluene fraction can be raised to 60% by chlorination of 2-chlorotoluene with sulfur compounds as catalysts or cocatalysts. The product is separated from the reaction mixture by crystallization [871], [872]. Chlorination with iodine catalysis is also possible [873]. In addition, 2,5-dichlorotoluene can be produced by a Sandmeyer reaction on either 2-amino-5-chlorotoluene or 2,5-diaminotoluene.

2,6-Dichlorotoluene. This isomer can be produced by chlorination of *p*-toluenesulfonyl chloride in the presence of antimonous chloride, followed by desulfonation [864]. Alternatively it can be made by chlorination of 4-*tert*-butyltoluene or 3,5-di-*tert*-butyltoluene with subsequent dealkylation [874], [875] or by means of a Sandmeyer reaction on 2-amino-6-chlorotoluene [876]. Distillatory separation of the 2,6-dichlorotoluene

Table 47. Distribution of the isomeric trichlorotoluenes in the chlorination of toluene

Catalyst	2,3,4-	2,3,6-	2,4,5-	2,4,6-
		Trichlorotoluene, %		
Ferric chloride	15	46	35	4
Ferric chloride – sulfur monochloride	11	20	67	2

from a chlorination mixture (in which it constitutes about 30% of the dichlorotoluene fraction) is prohibitively expensive. It is impossible at this time to assess the economics of its continuous separation by means of adsorption on faujasite-type zeolite [877].

3,4-Dichlorotoluene. The proportion of 3,4-dichlorotoluene in the product mixture from chlorination of 4-chlorotoluene can be raised to about 40% if sulfur compounds are used as catalysts or cocatalysts. This increase is at the expense of the 2,4-dichlorotoluene [878].

3,5-Dichlorotoluene. This isomer is not formed in the direct chlorination of toluene. Its indirect synthesis is possible via 2-amino-3,5-dichlorotoluene [879]. 3,5-Dichlorotoluene can also be produced to an extent corresponding to its equilibrium ratio by isomerization of other dichlorotoluenes.

8.2.3.3. Trichlorotoluenes

When toluene is chlorinated with 3 mol chlorine per mole of toluene, four of the six possible trichlorotoluenes are formed in the following catalyst-dependent ratios (Table 47) [880].

2,3,4-Trichlorotoluene. Chlorination of 4-chlorotoluene in the presence of iron trichloride leads to a product in which 2,3,4-trichlorotoluene accounts for 22% of the trichlorotoluene fraction. The compound can be separated from this mixture by distillation. It can also be produced from 3-amino-2,4-dichlorotoluene by using the Sandmeyer reaction [881].

2,3,6-Trichlorotoluene. The chlorination of 2-chlorotoluene with an iron powder catalyst gives a trichlorotoluene fraction containing 63% 2,3,6-trichlorotoluene [882]. As 2,3,6- and 2,4,5-trichlorotoluene have similar boiling points, their separation by distillation is impractical. The concentration of 2,3,6-trichlorotoluene can be raised to 71% by fractional crystallization in the absence of a solvent. This procedure gives a eutectic mixture containing 29% 2,4,5-trichlorotoluene and having a solidification point of 21.5 °C.

Higher yields of 2,3,6-trichlorotoluene are possible from chlorination of 2,3-dichlorotoluene (75%, in addition to 25% 2,3,4-trichlorotoluene) or 2,6-dichlorotoluene ($\sim$ 99%) [779]. 2,3,6-Trichlorotoluene can also be obtained by chlorination of *p*-toluenesulfonyl chloride or from 3-amino-2,6-dichlorotoluene by a Sandmeyer reaction [881].

2,4,5-Trichlorotoluene. Chlorination of 4-chlorotoluene at 20–50 °C gives > 80% 2,4,5-trichlorotoluene if ferrous sulfide is used as a catalyst. The product is separated from the chlorination mixture by distillation [883].

Table 48. Distribution of the isomeric tetrachlorotoluenes

Catalyst	2,3,4,5-	2,3,4,6-	2,3,5,6-
		Tetrachlorotoluene, %	
Ferric chloride	20	44	36
Ferric chloride – sulfur	25	45	30

2,4,6-Trichlorotoluene. The chlorination of 2,4-dichlorotoluene in the presence of ferric chloride gives a trichlorotoluene fraction containing 22% of the 2,4,6-isomer, which can be isolated by fractional distillation.

8.2.3.4. Tetrachlorotoluenes

The three tetrachlorotoluene isomers are all formed when toluene is chlorinated (Table 48) [880].

2,3,4,5-Tetrachlorotoluene. The proportion of the 2,3,4,5-isomer in the tetrachlorotoluene fraction can be raised to 49% by starting with 2,4,5-trichlorotoluene. Its separation by distillation is possible.

2,3,4,6-Tetrachlorotoluene. Yields of 51% and 66% of this isomer are obtainable by chlorination of 2,4,5-trichlorotoluene and 2,3,4-trichlorotoluene, respectively. 2,3,4,6-Tetrachlorotoluene is the exclusive product of chlorination of 2,4,6-trichlorotoluene [880].

2,3,5,6-Tetrachlorotoluene. This isomer is obtained by exhaustive chlorination of *p*-toluenesulfonyl chloride, followed by desulfonation [865].

The various tetrachlorotoluenes are also obtainable from the corresponding amino compounds by the Sandmeyer reaction.

8.2.3.5. Pentachlorotoluene

Exhaustive chlorination of toluene gives pentachlorotoluene. The reaction can be carried out with chlorine in either carbon tetrachloride or hexachlorobutadiene as solvent in the presence of iron powder and ferric chloride as catalysts [884], or with chlorine monoxide in carbon tetrachloride as solvent in the presence of an acid, e.g., sulfuric acid [771]. Sulfuryl chloride, used in the presence of sulfur monochloride and aluminum chloride catalysts, is also a suitable chlorinating agent, the method being that of O. SILBERRAD [767].

Further chlorination of pentachlorotoluene in the presence of a chlorination catalyst at a temperature above 350 °C leads to the formation of hexachlorobenzene and carbon tetrachloride [885].

8.2.4. Quality and Analysis

The industrially important chlorotoluenes can be produced such that they have a high degree of purity. No general agreement exists as to appropriate specifications. The following are typical analyses found for several chlorotoluenes:

2-Chlorotoluene	
Content	>99%
toluene	< 0.1%
4-chlorotoluene	< 0.9%
3-chlorotoluene	< 0.01%
4-Chlorotoluene	
Technical grade with	>98%
3-chlorotoluene	< 1%
2-chlorotoluene	< 0.5%
dichlorotoluenes	< 0.5%
Pure grade with	>99.5%
3-chlorotoluene	< 0.2%
2-chlorotoluene	< 0.2%
dichlorotoluenes	< 0.1%
2,4-Dichlorotoluene	
Content	>99%
4-chlorotoluene	< 0.2%
2,5-dichlorotoluene	< 0.3%
2,6-dichlorotoluene	< 0.1%
3,4-dichlorotoluene	< 0.4%
3,4-Dichlorotoluene	
Content	>95%
2,4-dichlorotoluene	< 5%
trichlorotoluenes	< 0.5%

Chlorotoluenes are preferably analyzed by gas chromatography. Packed columns are used to separate 2-, 3-, and 4-chlorotoluene. The more highly chlorinated toluenes can be separated by means of glass capillary columns.

8.2.5. Storage and Transportation

The mono- and dichlorotoluenes are stable, neutral liquids. They are shipped in drums, containers, and road or rail tankers. Steel is a suitable material for construction of containers.

For information on storage and handling, see Chap. 8.1.5.

Freight classifications are given in Table 49.

Table 49. Freight classification

	GGVE/GGVS and RID/ADR		IMDG Code and IATA-DGR			US-DOT
	class	number	class	UN No.	pack. group	
2- and 4-Chlorotoluene	3	3	3.3	2238	III	UN 1993 Flammable liquid
2,3-Dichlorotoluene	3	4		not restricted		
2,4-Dichlorotoluene	3	4		not restricted		
2,6-Dichlorotoluene	3	4		not restricted		
3,4-Dichlorotoluene	3	4		not restricted		

8.2.6. Uses

Isomeric mixtures of the monochlorotoluenes are hydrolyzed to cresol on a considerable scale. Chlorotoluenes are also used as solvents in reactions and to dissolve special products, e.g., dyes.

2-Chlorotoluene. 2-Chlorotoluene is a starting material in the production of 2-chlorobenzyl chloride, 2-chlorobenzaldehyde, 2-chlorobenzotrichloride, 2-chlorobenzoyl chloride and 2-chlorobenzoic acid, which are precursors for dyes, pharmaceuticals, optical brighteners, fungicides, and products of other types. 2-Chlorotoluene is also used in the production of dichlorotoluenes (chlorination), 3-chlorotoluene (isomerization), and *o*-chlorobenzonitrile (ammonoxidation).

4-Chlorotoluene. 4-Chlorotoluene is used mainly to produce *p*-chlorobenzotrichloride, from which is obtained *p*-chlorobenzotrifluoride, an important precursor of herbicides (e.g., trifluralin: α,α,α,-trifluoro-2,6-dinitro-*N,N*-dipropyl-*p*-toluidine). Other side-chain-chlorinated products or their derivatives are 4-chlorobenzyl chloride (for pharmaceuticals, rice herbicides, and pyrethrin insecticides), 4-chlorobenzaldehyde (for dyes and pharmaceuticals), 4-chlorobenzoyl chloride (for pharmaceuticals and peroxides), and 4-chlorobenzoic acid (for dyes). 4-Chlorotoluene is also a starting material in the synthesis of 2,4- and 3,4-dichlorotoluene and of 4-chlorobenzonitrile.

2,4-Dichlorotoluene. 2,4-Dichlorotoluene is used via its side-chain-chlorinated intermediates to produce fungicides, dyes, pharmaceuticals, preservatives, and peroxides (curing agents for silicones and polyesters).

2,6-Dichlorotoluene. 2,6-Dichlorotoluene is used to produce 2,6-dichlorobenzaldehyde, a dye precursor, and 2,6-dichlorobenzonitrile, a herbicide.

3,4-Dichlorotoluene. 3,4-Dichlorotoluene is used in small amounts in the production of 3,4-dichlorobenzyl chloride, 3,4-dichlorobenzaldehyde, 3,4-dichlorobenzotrichloride, and 3,4-dichlorobenzoic acid, from which disinfectants, crop protection products, and dyes are produced.

2,3,6-Trichlorotoluene. 2,3,6-Trichlorotoluene is used on a small scale, together with 2,4,5-trichlorotoluene, to produce 2,3,6-trichlorobenzoic acid, a herbicide precursor.

Producers of chlorinated toluenes are: Bayer AG (FRG); Enichem (Italy); Hodogaya Chemicals Co. Ltd. (Japan); Hoechst AG (FRG); Ihara Chemical Ind., Ltd. (Japan); Occidental Chemical Co., Ltd. (USA).

8.3. Chlorinated Biphenyls

Industrial use of the polychlorinated biphenyls first began in 1929 in the USA [886], [887]. The outstanding properties of these compounds, such as their high chemical and thermal stability, high dielectric constant, and the fact that they form only incombustible gases in an electric arc, made them appear ideally suited for use as insulating and cooling fluids for transformers and as dielectric impregnants for capacitors.

In subsequent years many other applications were found as well (see 8.3.5), particularly for isomeric mixtures containing two to six atoms of chlorine per mole of biphenyl.

In the mid-1960s, improved analytical methods revealed that polychlorinated biphenyls were accumulating in nature as a consequence of their extremely low rates of biological degradation (rates which decrease as the chlorine content rises). The compounds were detected in fresh water in all parts of the world, but also in many animals (e.g., birds, fish, and plankton). In the late 1970s, it was further discovered that at temperatures of 500 to 800 °C in the presence of oxygen, polychlorinated biphenyls can give rise to polychlorinated dibenzofurans and dibenzodioxins, including (although to a much smaller extent) the particularly toxic compound 2,3,7,8-tetrachlorodibenzodioxin [888]–[893].

In the meantime, all but a few of the well-known producers (see 8.3.5) discontinued the production of chlorinated biphenyls. Moreover, in many countries the production, sale, and use of polychlorinated biphenyls have been restricted or entirely prohibited by legislation.

For many years, o-, m- and p-terphenyl mixtures were chlorinated and then used as plasticizers, flame retardants, and fillers in thermoplastic pattern and holding waxes. This application likewise has been substantially discontinued [894], particularly in view of the persistent nature of the compounds in question and their accumulation in the environment. Possible toxicological hazards are either unknown or have not been adequately investigated (Table 50).

In view of the above, this article is devoted primarily to a review of recent patents covering methods for the disposal of polychlorinated biphenyls. For information re-

Table 50. MAK and TLV values of chlorinated biphenyls

	MAK (FRG) mg/m^3	TLV (USA) mg/m^3
Chlorinated biphenyls (42% Cl)	1 III B	1
Chlorinated biphenyls (54% Cl)	0.5 III B	0.5

garding the production of these compounds and their specific physical and chemical properties, attention is directed to earlier surveys [888], [895], [896].

8.3.1. Physical and Chemical Properties

There are 209 possible chlorinated biphenyls. The mono- and dichlorobiphenyls [*27323-18-8*], [*25512-42-9*] are colorless crystalline compounds (the melting points of the pure isomers lie between 18 and 149 °C). When burned in air, they give rise to soot and hydrogen chloride.

The most important products are mixtures whose principal components are trichlorobiphenyl [*25323-68-6*], tetrachlorobiphenyl [*26914-33-0*], pentachlorobiphenyl [*25429-29-2*], or hexachlorobiphenyl [*26601-64-9*]. Such mixtures are liquid to viscous (pour points increase with chlorine content from −22 to +18 °C), and they are fire-resistant. Further chlorination gives soft to brittle thermoplastic waxes.

Chlorinated biphenyls are soluble in many organic solvents, particularly when heated, but are soluble in water only in the ppm range. Although they are chemically very stable, including to oxygen of the air, they can be hydrolyzed to oxybiphenyls under extreme conditions, e.g., with sodium hydroxide solution at 300–400 °C and under high pressure. Toxic polychlorodibenzofurans may be formed under these conditions.

The fact that the compounds may eliminate hydrogen chloride to a small extent at a high temperature explains why hydrogen chloride acceptors are often added to transformer fluids based on polychlorinated biphenyls.

The excellent electrical property data of polychlorinated biphenyls, such as high dielectric constant, low power factor, high resistivity, favorable dielectric loss factor, and high dielectric strength, have already been mentioned.

8.3.2. Disposal

Many products containing chlorinated biphenyls are still in use throughout the world, particularly in transformers, rectifiers, and capacitors with long service lives. Industry and national governments are now faced with the need to dispose of these products without causing additional pollution of the environment. Appropriate official regulations exist in many countries [894], [897]–[908].

Attention is drawn in the following survey to recent patents concerned with the removal of polychlorinated biphenyls from electrical devices and with the disposal of these compounds. It is impossible to say which of the processes have actually reached maturity and which are already being used.

According to the present state of knowledge, polychlorinated biphenyls can be destroyed harmlessly by combustion at temperatures above 1000 °C and a residence time of 2 s, e.g., in a rotary burner equipped with a scrubbing tower for hydrogen chloride [894], [908]–[910]. Regulations in the Federal Republic of Germany specify a temperature of 1200 °C, a residence time of 0.2 s, and a residual oxygen content in the combustion gas of 6% [903].

Removal of polychlorinated biphenyls from silicone- and hydrocarbon-based transformer fluids and heat transfer media is accomplished through the formation of a separable fraction rich in polychlorinated biphenyls [911], treatment with polyalkylene glycol and alkali metal hydroxide [912], [913], treatment with sodium naphthalenide [914], or heating with a sodium dispersion to 75 °C [915].

Polychlorinated biphenyls are removed from impregnated electrical parts by irradiation with microwaves, which causes gasification of the compounds [916], or by dry distillation at 500–1000 °C followed by addition of oxygen [917].

Destruction of polychlorinated biphenyls has been reported to be possible by the following methods: treatment with sodium naphthalenide in the presence of metallic sodium [918]; treatment at 145 °C with a dehalogenating reagent prepared from an alkali metal, polyethylene glycol, and oxygen [919]; reaction with sulfur in the vapor phase [920]; adsorption on paramagnetic or ferromagnetic material and subsequent irradiation with microwaves in the presence of oxygen [921]; and irradiation with light in aqueous solution in the presence of a catalyst [922].

Thermal decomposition of polychlorinated biphenyls occurs by pyrolysis under oxidative conditions (oxygen-enriched air) on molten alkali carbonates at 900–980 °C [923]; by use of a plasma burner at 3000–4000 °C [924]; by dissolution in kerosene, followed by combustion in air and introduction of the combustion gases into a special decomposition furnace [925]; or by evaporation with hydrogen as a carrier gas, followed by combustion in oxygen [926].

8.3.3. Analysis

The analytical methods most frequently used for detecting chlorinated biphenyls are capillary column gas chromatography coupled with mass spectrometry in the MID (Multiple Ion Detection) mode and capillary column gas chromatography with ECD (Electron Capture Detector). These methods are suitable for solution of even the most difficult problems. Clean-up steps are necessary when complex matrices are concerned, such as preliminary separation by column chromatography.

HPLC (High Pressure Liquid Chromatography) and infrared spectroscopy are applicable to a limited extent.

Summary polychlorinated biphenyl determinations are also possible, though not usual. These require either exhaustive chlorination and measurement of the decachlorobiphenyl content or else dechlorination and subsequent measurement of the biphenyl content.

For literature references on the subject of analysis, see [908], [927]–[933].

8.3.4. Storage and Transportation

At a normal temperature the commercially used polychlorinated biphenyls are liquid to viscous mixtures with a comparatively low vapor pressure (trichlorobiphenyl 6.5×10^{-5} kPa at 20 °C). Steel and aluminum are suitable as container materials. The storage and shipping of these compounds are subject to a variety of national regulations. Since these compounds accumulate in the environment they must be handled so that release cannot occur.

Exposure of polychlorinated biphenyls to fire may result in the formation of toxic chlorinated dibenzofurans and dibenzodioxins [888]–[893] and in the evolution of hydrogen chloride.

Classification of polychlorinated biphenyls are:

GGVE/GGVS and RID/ADR: Class 6.1, Number 23
IMDG-Code and IATA-DGR: Class 9, UN Number 2315, Packaging group II
US D.O.T.: ORM.E, UN Number 2315

8.3.5. Uses

Use of these compounds has fallen drastically [934] as a result of the extensive discontinuation of their production, voluntary renunciation of their application, and national restrictions. No details were available concerning products in which polychlorinated biphenyls are still used, nor concerning the scale of such use. The following list of important fields of application should be regarded as retrospective:

- Cooling and insulating fluids for transformers
- Dielectric impregnating agents for capacitors
- Flame-retardant additives for resins and plastics used in the electrical industry
- Alkali- and acid-resistant plasticizers for lacquers, plastics, adhesives, fillers and sealing compositions
- Formulations for paints and printing inks
- Water-repellent additives for surface coatings
- Dye carriers for pressure-sensitive copying paper
- Additives for thermally-stable lubricants and gear oils
- Incombustible hydraulic fluids (particularly suitable for use in locations to which access is difficult, e.g., in mines)
- Heat transfer fluids of high heat stability
- Inert sealing fluids for vacuum pumps

Table 51. Trade names of chlorinated biphenyls

Apirolio	Caffaro, Italy
Aroclor	Monsanto, USA, UK
Clophen	Bayer AG, FRG
Delor	Chemco, Czechoslovakia
Fenclor	Caffaro, Italy
Inerteen	Westinghouse, USA
Kanechlor	Kanegafuchi Chem. Co., Japan
Pyralene	Prodelec, France
Pyranol	Monsanto, UK
Pyroclor	Monsanto, USA
Sovtol	USSR

Dust control agents for road construction

Mono- and dichlorobiphenyls have been used on a small scale as precursors for the corresponding oxybiphenyls.

Some registered trademarks are listed in Table 51. Most of the listed producers have discontinued production.

8.4. Chlorinated Naphthalenes

The first industrial applications of chlorinated naphthalenes took place at the beginning of the 20th century [935]. The compounds were used most extensively in the 1930s to 1950s, especially in cable and capacitor production, prompted by their dielectric, water-repellent, and flame-retardant properties.

More recently, most producers of polychlorinated naphthalenes have stopped their production, and output has been reduced drastically in all parts of the world. The reasons for this follow. First, connections have been established between highly chlorinated naphthalenes, especially pentachloronaphthalene and hexachloronaphthalene, and illness. Moreover, because of their high chemical and thermal stability, highly chlorinated naphthalenes are able to accumulate in the environment. Finally, new materials (polyesters and polycarbonate) have been introduced as substitutes for chlorinated naphthalenes in the capacitor and cable industries.

Monochloronaphthalenes, by contrast, are not considered to be problematic with regard to their effects on health and accumulation in the environment [936].

Official regulations relating to chlorinated naphthalenes differ considerably from country to country. In Japan, for example, polychlorinated naphthalenes are prohibited entirely. In the USA they may still be used without restriction, but changes in their production, importation, or use must be reported to the U.S. Environmental Protection Agency (EPA) [937], so that the effects of these changes on the environment may be monitored.

8.4.1. Physical Properties

Naphthalene has 75 chlorinated derivatives. To date, however, only a few have been synthesized and isolated in pure form. Only isomeric mixtures characterized according to their chlorine content are as a general rule commercially available. This situation arises because of the fact that the most important characteristics of the compounds are a function solely of their degree of chlorination, as a result of which there is little demand for the pure compounds. Moreover, precisely because the physical properties of the various isomers are very similar, the cost of their separation is unrealistically high.

Except for 1-monochloronaphthalene, which is a liquid at room temperature, pure chlorinated naphthalenes are colorless, crystalline compounds. Mixtures of the compounds for industrial use have different degrees of chlorination, and their softening points lie considerably below the melting points of the pure components (generally between -40 and $+190$ °C). As the degree of chlorination increases, a transition occurs from liquids, via waxes, to hard solids, which causes the vapor pressures and water solubilities to fall. In contrast, the melting points, boiling points, and densities tend to rise, and any characteristics that are dependent on these properties become more pronounced. Mono- and dichloronaphthalenes are freely soluble in most organic solvents. Highly chlorinated naphthalenes are most soluble in chlorinated aliphatic and aromatic solvents and in petroleum naphthas.

Chlorinated naphthalenes have excellent dielectric properties. The tri- to hexachloronaphthalenes have dielectric constants of $4.5-5$, a dissipation factor of 1×10^{-3} at 800 Hz and 20 °C, and a specific resistivity (100 V, 1 min) above 10^{14} Ω cm.

Chlorinated naphthalenes are compatible with many other commercial products, e.g., chlorinated paraffins, petroleum waxes, bitumen, various plasticizers (e.g., tricresyl phosphate), and polyisobutylene.

A selection of physical data obtained for pure monochloronaphthalenes and commercial mixtures is given in Tables 52 and 53. For physical data on pure polychlorinated naphthalenes, see [938].

8.4.2. Chemical Properties

Reactions of the compounds may occur on the ring (electrophilic substitution), at the chlorine substituent (e.g., hydrolysis), or with de-aromatization (chlorine addition).

1-Chloronaphthalene participates in electrophilic substitution reactions such as nitration [940], sulfonation, halogenation, and chloromethylation [941]. Reaction is especially favored at the para position relative to chlorine.

Hydrolysis with sodium hydroxide solution takes place at about 300 °C in the presence of copper catalysts [942] to give 1-naphthol and also 2-naphthol.

In the absence of a catalyst, additive chlorination of 1-chloronaphthalene yields pentachlorotrihydronaphthalene and hexachlorodihydronaphthalene [943].

Table 52. Physical data for monochloronaphthalenes *

	1-Chloronaphthalene [90-13-1]	2-Chloronaphthalene [91-58-7]
Melting point, °C	−2.3	95.5 – 60
Boiling point at 101.3 kPa, °C	260.2	258.6
Density, g/cm^3		
at 20 °C	1.194	1.178
80 °C	1.144	1.130
Temperature corresponding		
to vapor pressure, °C		
0.13 kPa	80.6	
0.67 kPa	104.6	
1.33 kPa	118.6	
2.67 kPa	134.4	
5.33 kPa	153.2	
8.00 kPa	165.6	161.2
13.30 kPa	180.4	
26.70 kPa	204.2	
53.30 kPa	230.8	
101.30 kPa	260.2	
Refractive index n_D^{20}	1.6326	
Viscosity at 25 °C, mPa s	2.94	
Flash point, °C	115	
Ignition temperature, °C	>500	

* A eutectic mixture of 75 % of 1-chloronaphthalene and 25 % of 2-chloronaphthalene has a solidification point of − 17.5 °C.

The chemical and thermal stabilities of chlorinated naphthalenes increase with the number of chlorine substituents. Highly chlorinated naphthalenes withstand acids, caustic solutions, and oxidizing agents, even at elevated temperatures. An exception is concentrated nitric acid, which forms nitro derivatives relatively easily with polychlorinated naphthalenes and which at 90 °C oxidizes octachloronaphthalene to hexachloro-1,4-naphthoquinone and tetrachlorophthalic acid [944].

If naphthalene is chlorinated beyond the stage of octachloronaphthalene at a temperature exceeding 200 °C in the presence of ferric chloride catalyst, perchloroindane and carbon tetrachloride are formed. This phenomenon is a result of chlorine addition and subsequent ring constriction [945], decachlorodihydronaphthalene being an intermediate.

8.4.3. Production

The chlorination of naphthalene proceeds less rapidly than that of benzene or toluene. Consequently, chlorinated naphthalenes are produced batchwise in agitator vessels. Molten naphthalene, initially at 80 °C, is mixed with gaseous chlorine in the presence of ferric chloride or antimony pentachloride until the desired degree of chlorination has been reached. As the degree of chlorination increases, the reaction temperature must be raised to keep the mixture above its softening point.

Chlorinated Hydrocarbons

Table 53. Physical data for commercial chlorinated naphthalene mixtures

	CAS reg. no.	Average chlorine content %	Softening point °C	Boiling point °C	Vapor pressure kPa	Density at 25 °C g/cm³	Flash point °C	MAK (FRG) mg/m³	TLV TWA (USA) mg/m³
Monochloronaphthalene	[25586-43-0]	22	−25	250–260		1.2	135		
Mono-/dichloronaphthalene	[28699-88-9]	26	−33	250–290		1.22	130		
Tri-/tetrachloronaphthalene	[1321-65-9]	50	93	304–354	8×10^{-3}	1.58	200	5	5
Tetrachloronaphthalene	[1335-88-2]	52	115	312–360	10^{-5}	1.65	210		2
Tetra-/pentachloronaphthalene	[1321-64-8]	56	120	327–371		1.67	230	0.5	0.5
Penta-/hexachloronaphthalene	[1335-87-1]	62	137	343–384		1.78	250		0.2
Heptachloronaphthalene	[32241-08-0]								
Octachloronaphthalene	[2234-13-1]	70	185–197	440 (101.3 kPa) 246 (0.067 kPa)		2.0	>430		0.1

The crude chlorination mixture is neutralized, e.g., with soda. The neutralizing agent is then separated and the crude product is fractionated by vacuum distillation.

Chlorine addition occurs if the chlorination is carried out in the absence of a catalyst, resulting in the unstable materials 1,2-dichloro-1,2-dihydro- and 1,2,3,4-tetrachloro-1,2,3,4-tetrahydronaphthalene.

Incomplete naphthalene conversion is accepted in the production of monochloronaphthalene in order to keep the proportion of polychlorinated naphthalenes small. The best possible result with ferric chloride as catalyst is 75% monochloronaphthalene and 13% polychlorinated naphthalenes in an isomer ratio of 90–94% 1-chloro- and 6–10% 2-chloronaphthalene. If peroxodisulfate and chloride ions are used as chlorinating agents, mainly monochlorinated derivatives are obtained [946]. Because of the large difference between their boiling points, the monochloronaphthalene isomer mixture is easily separated by distillation from the chlorination mixture, and it is of high purity. The isomers can be separated by crystallization [947]. Further information on the industrial chlorination of naphthalene will be found in the preceding edition of this encyclopedia [948].

The preparation of pure isomers representing the various chlorination stages is only of scientific interest. A review of the procedures is given in [938].

For information regarding the disposal of waste gas and chlorinated naphthalene wastes, see 8.5.

8.4.4. Quality and Analysis

The quality of chlorinated naphthalenes is monitored by wet analysis for chlorine content, as well as by gas chromatography employing glass capillary columns. Traces are determined by the same analytical methods that have been developed for polychlorinated biphenyls. For literature on this subject, see [938], [949]–[956].

One particular grade of monochloronaphthalene is used in wood preservatives and consists of

90–94%   1-chloronaphthalene
6–10%   2-chloronaphthalene
< 0.1%   dichloronaphthalene
< 0.01% trichloronaphthalene and a
chlorine content of 22%.

8.4.5. Storage and Transportation

The chloronaphthalenes are stable, neutral substances. They can be stored in tanks as liquids, in which case adequate heating must be provided consistent with the melting point of the material. Steel containers are suitable, although stainless steels should be used for applications in the electrical industry. Chlorinated naphthalenes that are liquid at ambient temperatures are shipped in drums, containers, or road tankers. Solid chloronaphthalenes are supplied as powder or flakes, either in fiber drums or in paper sacks. Existing national regulations must be complied with in connection with storage and transportation.

Polychlorinated naphthalenes are not expressly mentioned in the shipping regulations of either the EEC or the USA. Monochloronaphthalene has neither been allocated to a hazard class nor given a UN number.

8.4.6. Use

The use of chlorinated naphthalenes has diminished considerably during the last 30 years. Thus, except in special cases, chlorinated naphthalenes are no longer used in capacitors or electric cable coverings. Their use as lubricants has also been largely discontinued. Practice in individual countries varies, however. In the USA, for example, chlorinated naphthalenes are no longer used as wood preservatives. It is impossible to generalize concerning which of the following potential applications are permissible at present.

Monochloronaphthalenes: dye precursor; dye dispersant; fungicide and insecticide wood preservative; engine oil additive for dissolving sludges [957]; chemically and thermally stable sealing fluid; ingredient in special cleaning agents.

Polychlorinated naphthalenes: dielectric for impregnation of paper windings in automobile capacitors; insulating, waterproof, and flameretardant dipping and encapsulating compounds for special electrical parts; binder in the manufacture of ceramic elements for the electrical industry; paper coatings with water-repellent, flame-retardant, fungicidal, and insecticidal properties; plasticizers; electroplating stop-off compound.

Octachloronaphthalene: ingredient in the production of carbon elements by carbonization; additive for lubricants used under extreme conditions and for flame-retardant plastics.

Trade names for chlorinated naphthalenes. Most of the producers in the following list either have entirely discontinued the production of chlorinated naphthalenes, or else they have restricted their output and simultaneously reduced the content of highly chlorinated naphthalenes in their products.

Cerifal types, Caffaro, Italy
Clonacire types, Prodelec, France
Halowax types, Koppers Co., Inc., USA
Nibren Wax types, Bayer AG, FRG
Seekay wax types, Imperial Chem. Ind., Ltd., UK

8.5. Environmental Protection

In the production of chlorinated aromatics, organic compounds are contained in three different waste streams:

waste gas
wastewater
liquid or solid organic wastes

The correct disposal of these wastes results in no harm to the environment.

Waste Gas Treatment. It is necessary to distinguish between hydrogen chloride reaction gas from the chlorination process and substantially neutral waste gas, e.g. from distillation columns or storage containers. The second of these waste gas streams can be purified in an activated charcoal tower or incinerator. The hydrogen chloride reaction gas is processed in a complex manner to recover usable hydrochloric acid:

1) If the reaction gas contains chlorine, this is removed in a scrubbing tower (e.g., a bubble column [958]) containing an easily chlorinatable compound, preferably a raw material used in the chlorination process, and a chlorination catalyst.
2) In a second scrubbing tower, organic constituents of the reaction gas are washed out with a high-boiling solvent.
3) In addition to (or instead of) being passed through the second scrubbing tower, the hydrogen chloride gas is cooled to the lowest possible temperature in a cooler, in which further organic constituents are condensed out.
4) The hydrogen chloride is then absorbed in water in an adiabatic scrubber, from which it emerges as about 30% hydrochloric acid with < 5 ppm of organically bound carbon.

Absorption in calcium chloride solution in the presence of calcium carbonate lumps has also been described [959]. In this case, organic compounds are removed with the escaping carbon dioxide, and a 33% calcium chloride solution containing about 4 ppm of organic compounds remains.

Aqueous hydrochloric acid can be substantially freed from organic substances by extraction with dodecylbenzene [960].

The hydrogen chloride can also be liquefied in the absence of water and purified by distillation in a pressure column.

The pure hydrogen chloride and its aqueous solution are suitable for chemical processes. There are also techniques to recover the chlorine by electrolysis (e.g., the Hoechst-Uhde process) or oxidation (Shell-Deacon process or Kel Chlorine process).

Wastewater Treatment. The biological degradation rate of chlorinated aromatics decreases as their chlorine content increases. Only chlorinated aromatics with low degrees of chlorination are degradable in biological wastewater treatment plants, and then only if their concentration in the wastewater does not exceed certain levels. Therefore, wastewater streams containing chlorinated aromatics require preliminary purification. The following techniques [961] are suitable: stripping, extraction, and adsorption on activated carbon or polymeric resins [962].

Treatment of Wastes. These wastes may be distillation residues, useless fractions from separation processes, or industrial products containing chlorinated aromatics that are no longer suitable for use.

Normally these wastes are disposed of by incineration. In principle, it is also possible to convert chlorinated hydrocarbons into usable compounds by hydrodehalogenation or chlorinolysis. It is impossible to say whether these processes are already being used on an industrial scale.

Chlorine-containing aromatics are burned in special furnaces that provide reaction temperatures above 1000 °C and residence times of 1–2 s [910]. Only then is it certain that no polychlorodibenzodioxins are formed during combustion. This risk exists particularly with polychlorinated biphenyls (see Section 8.3, especially 8.3.2). In an excess of oxygen, the chlorinated compounds are converted into hydrogen chloride, carbon dioxide and water. The hydrogen chloride is removed from the flue gas by water scrubbers.

Hydrodehalogenation is effected with hydrogen on palladium, platinum, or nickel catalysts at elevated temperature and high pressure. The nucleus-bound chlorine is substituted by hydrogen, and hydrogen chloride is thus formed. Mixtures of different chlorination stages down to the chlorine-free fundamental compound are obtained [712], [963]–[965].

Exhaustive chlorination of chlorinated hydrocarbons in the vapor phase at a temperature above 600 °C and pressure up to 200 bar splits the molecules, thus giving high yields of carbon tetrachloride [713], [714].

8.6. Economic Facts

Overall, it may be said that the use of chlorinated aromatic hydrocarbons is stagnant or declining worldwide (see Tab. 54).

Monochlorobenzene output in the USA fell from 275 100 t in 1960 to 115 700 t in 1982 [715]. At the middle of 1984, Monsanto (at Muscatine, Iowa) discontinued the use of monochlorobenzene as a carrier for the herbicide Lasso [716]; this is expected to lead to a further drop in output.

As the production of monochloro-, 1,2-dichloro- and 1,4-dichlorobenzene is most economical in coproduction, it is becoming increasingly difficult to adapt the manufacturing conditions to the requirements of the market.

There are no output statistics for chlorinated toluenes. It is estimated that 30 000 t of 1- and 4-chlorotoluene was produced in Western Europe in 1983 [715].

Table 54. Production of chlorobenzenes

		Monochloro-benzene	1,2-Dichloro-benzene	1,4-Dichloro-benzene	Sum of allchloro-benzenes
USA	1970	220	30	32	
	1981	130	23	33	
FRG	1970				127
	1981	97		34	138
Japan	1981	34	9	16	50

The production of polychlorinated biphenyls has been almost completely discontinued throughout the world (cf. Section 8.3). Since the first industrial use of polychlorinated biphenyls at the beginning of the 1930s, about 1 million tons have been produced [718]; 40% of this quantity is estimated to be still in use. Consumption reached its highest level in the 1960s.

The production of chlorinated naphthalenes has also fallen greatly, with a shift toward less highly chlorinated naphthalenes. Output in the USA fell from 2270 tons in 1972 [719] to 320 tons in 1978; after the manufacture of chlorinated naphthalenes had been discontinued, the quantity imported into the USA settled at about 15 tons in 1980 and 1981 [720]. The output of the USA in 1972 consisted of 25 – 28% of mono-/di-, 65 – 66% of tri-/tetra-, and about 8% of penta- to octanaphthalenes [719]. In the Federal Republic of Germany with an output of about 1000 t of chlorinated naphthalenes in 1972 [721], the production of polychlorinated naphthalenes has likewise been discontinued; only monochloronaphthalenes are now produced, output being of the order of 100 t/a.

9. Side-Chain Chlorinated Aromatic Hydrocarbons

Alkyl aromatics are somewhat unique in their behavior with respect to chlorination reactions.

The action of elemental chlorine can lead either to addition or substitution on the aromatic ring or it can cause substitution in the aliphatic side-chain, depending on the reaction conditions.

The side-chain chlorinated alkyl aromatics, particularly those based on toluene and xylene, have an exceptional place because of their role as chemical intermediates. Indeed, they are used in the manufacture of chemical products of almost all kinds, including dyes, plastics, pharmaceuticals, flavors and perfumes, pesticides, catalysts, inhibitors, and so forth.

Those side-chain-chlorinated alkyl aromatics which are of greatest importance in industrial chemistry are the toluene derivatives benzyl chloride, benzal chloride, and benzotrichloride.

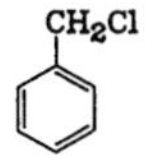
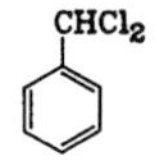
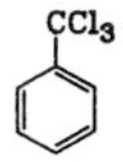

Benzyl chloride Benzal chloride Benzotri-chloride

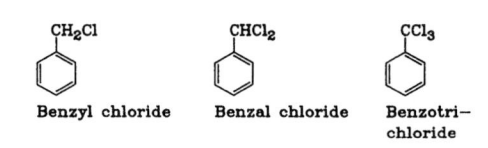

9.1. Benzyl Chloride

Benzyl chloride (chloromethylbenzene, α-chlorotoluene) [*100-44-7*] may be structurally the simplest side-chain chlorinated derivative of toluene, but economically it is the most important. Benzyl chloride is the starting material for a large number of industrial syntheses. The first preparation of it involved not the chlorination of toluene, however, but the reaction of benzyl alcohol with hydrochloric acid (S. CANNIZZARO, 1853).

9.1.1. Physical Properties

Benzyl chloride is a colorless liquid which fumes in moist air. It has a pungent odor and is irritating to the mucous membranes and the eyes (i.e., it has a powerful lachrymatory effect).

M_r	126.58
bp at 101.3 kPa	179.4 °C
mp	– 39.2 °C
ϱ at 0 °C	1.1188 g/cm^3
10 °C	1.1081 g/cm^3
20 °C	1.1004 g/cm^3
30 °C	1.0870 g/cm^3
50 °C	1.072 g/cm^3
87 °C	1.037 g/cm^3
n_D^{20}	1.5389
Dynamic viscosity η at	
15 °C	1.501 mPa s
20 °C	1.38 mPa s
25 °C	1.289 mPa s
30 °C	1.175 mPa s
Surface tension σ at	
15 °C	38.43 mN/m
20 °C	37.80 mN/m
30 °C	36.63 mN/m
88 °C	29.15 mN/m
17 °C	19.5 mN/m
Specific heat at	
0 °C	178 J mol^{-1} K^{-1} (1403 J kg^{-1} K^{-1})
20 °C	181 J mol^{-1} K^{-1} (1432 J kg^{-1} K^{-1})
25 °C	183 J mol^{-1} K^{-1} (1444 J kg^{-1} K^{-1})
50 °C	189 J mol^{-1} K^{-1} (1495 J kg^{-1} K^{-1})
100 °C	212 J mol^{-1} K^{-1} (1675 J kg^{-1} K^{-1})
Heat of vaporization at 25 °C	50.1 kJ/mol (396 kJ/kg)
Heat of combustion at constant volume	3708 kJ/mol (29.29 × 10^3 kJ/kg)

Flash point	60 °C
Ignition temperature	585 °C
Explosive limits in air,	
lower	1.1 vol%
upper	14 vol%
Explosive limits in chlorine,	
lower	ca. 6 vol%
upper	ca. 60 vol%
Specific conductivity at 20 °C	1.5×10^{-8} S/cm
Vapor pressure at	
0 °C	0.025 kPa
10 °C	0.05 kPa
20 °C	0.12 kPa
30 °C	0.37 kPa
50 °C	0.99 kPa
100 °C	7.96 kPa
130 °C	23.40 kPa
179.4 °C	101.33 kPa

Numerous binary and ternary azeotropes containing benzyl chloride are known [966]. Examples are given in Table 55.

The solubility of benzyl chloride in water is 0.33 g/L at 4 °C, 0.49 g/L at 20 °C, and 0.55 g/L at 30 °C. Benzyl chloride is freely soluble in chloroform, acetone, acetic acid esters, diethyl ether, and ethyl alcohol. The solubility of chlorine in 100 g of benzyl chloride is 8.0 g at 30 °C, 5.4 g at 50 °C, and 2.1 g at 100 °C [967].

9.1.2. Chemical Properties

Benzyl chloride can serve as a starting point for the preparation of benzal chloride and benzotrichloride, both of which are accessible by side-chain chlorination. Nuclear chlorination, on the other hand, leads to chlorobenzyl chlorides. Oxidation with sodium dichromate gives sodium carbonate in aqueous solution benzaldehyde and benzoic acid.

Metals undergo a variety of reactions with benzyl chloride. For example, magnesium in ether gives benzyl magnesium chloride (Grignard), whereas copper powder or sodium gives 1,2-diphenylethane as the main product (Wurtz synthesis). The action of Friedel-Crafts catalysts such as $FeCl_3$, $AlCl_3$, and $ZnCl_2$ gives condensation products of the $(C_7H_6)_n$ type [968], but despite the fact that the degree of condensation can be controlled by changing the reaction conditions, these polymers have no commercial significance. If benzene or toluene is added to benzyl chloride in the presence of Friedel-Crafts catalysts, one obtains diphenylmethane or the isomeric benzyltoluenes, respectively.

The action of hydrogen sulfide and sulfides of the alkali metals leads to benzyl mercaptan and dibenzyl sulfide, respectively. Reactions with sodium salts of carboxylic acids produce the corresponding benzyl esters.

Table 55. Azeotropic mixtures with benzyl chloride

Component	Boiling point, °C	Benzyl chloride, wt%
Benzaldehyde	177.9	50
Hexanoic acid	178.7	95
Isovaleric acid	171.2	38
Valeric acid	175	25
Ethyl acetoacetate	175	35
Methyl acetoacetate	167	< 80
1,3-Dichloro-2-propanol	168.9	57
2,3-Dichloro-2-propanol	171	40
Ethylene glycol	ca. 167	ca. 30

Hydrolysis with hot water is said to result in the formation of benzyl alcohol; nevertheless, this reaction has no industrial applications because the hydrochloric acid formed causes the re-formation of benzyl chloride from the product benzyl alcohol; it also catalyzes the formation of dibenzyl ether.

Hydrolysis in the presence of alkali does lead to benzyl alcohol, however.

Benzyl chloride reacts with sodium cyanide to give phenylacetonitrile (benzyl cyanide).

Reaction with ammonia or amines gives primary, secondary, tertiary amines, and quaternary ammonium salts.

The reaction with hexamethylenetetramine produces benzaldehyde (Sommelet reaction).

9.1.3. Production

Substitution of chlorine for hydrogen in the aliphatic side-chain occurs by way of a radical chain mechanism; by contrast, chlorine substitution on the aromatic ring takes place according to an electrophilic polar mechanism.

The steps in the radical chain process are as follows:

chain initiation	$Cl_2 + h\nu$	$\longrightarrow$	$2\ Cl\cdot$ (1)
chain propagation	$Cl\cdot + RH$	$\longrightarrow$	$R\cdot + HCl$ (2)
	$R\cdot + Cl_2$	$\longrightarrow$	$RCl + Cl\cdot$ (3)
chain termination	$Cl\cdot + Cl\cdot$	$\longrightarrow$	Cl_2 (4)
	$R\cdot + Cl\cdot$	$\longrightarrow$	RCl (5)
	$R\cdot + R\cdot$	$\longrightarrow$	RR (6)

This chlorination is highly exothermic (96 – 105 kJ/mol chlorine). In view of the high rate of chlorine radical formation and of hydrogen displacement, it is not surprising that, depending on the substrate and reaction conditions, radical chain lengths of 10^3 [969] to 10^6 [970] have been found. Because the mechanisms of side-chain chlorination and nuclear chlorination are fundamentally different, selectivity can be readily achieved. The prerequisites for high side-chain chlorination efficiency are as follows:

1) Achievement and maintenance of an optimal radical concentration
2) Elimination of components that might impart an electrophilic course to the reaction
3) Elimination of components capable of terminating the radical chains
4) Elimination of components conducive to other side reactions
5) In general, taking those precautions, which would encourage radical reactions or suppress electrophilic reactions

The discussion which follows considers each of these points separately.

1) Chlorine radical formation can be promoted by the addition of a radical-forming agent such as 2,2′-azobis(isobutyronitrile) (AIBN), benzoyl peroxide, or hexaphenylethane. Such compounds are consumed in the reaction, however, and thus have to be added repeatedly. For this reason, two other methods, singly or in combination, are of greater importance, particularly in industrial chlorinations: irradiation (ultraviolet light, β-*radiation) and the use of an elevated temperature (100–200 °C)*. In both cases, the effect is a result of excitation of chlorine molecules.

Irradiation is generally carried out with mercury vapor lamps that emit light in the wavelength range 300–500 nm, corresponding to the region in which the absorption bands of chlorine lie (hence, its yellowish-green color). The effects of varying the wavelength have been investigated in detail [971]–[973]. A solution containing 0.1 % chlorine absorbs 90 % of the available light energy over a distance of only 12 cm [974]. At a chlorine concentration of 3–4 mol/L, the corresponding distance is reduced to 0.01 cm [969]. Consequently, thorough mixing of the reactor contents is advantageous, as is the use of reactors of relatively small diameter. These considerations apply regardless of whether the sources of radiation are inside the reactor or external to it [975]–[977]. In the case of bubble column reactors, the height of the reactor is also important in view of the velocity of the rising gas bubbles, which is ca. 20 cm/s [978], [979].

2) Friedel-Crafts catalysts favor nuclear chlorination. For this reason, particular attention must be paid to ensure, for example, that the starting toluene is free of dissolved iron salts and that all rust particles have been filtered. By extension, steel, including stainless steel, is unsuitable as a reactor material. The preferred reactor materials are glass, enamel, and polytetrafluoroethylene.

Extensive purification of technical-grade starting materials is very expensive, however. For this reason, numerous suggestions have been made for additives that might inhibit the undesirable effects of heavy metals. Most such additives are assumed to complex with, e.g., iron cations. Some additives are reported to accelerate the side-chain chlorination, thereby inhibiting in a relative sense nuclear chlorination. The following are examples of additives which have been recommended at concentrations of 0.1–2 %: pyridine and phenylpyridines [980], alkylene polyamines [981], hexamethylenetetramine [982], phosphoramides [983], acid amides, alkyl and dialkyl acid amides [984], ureas [985], phosphorus chlorides [986], alkyl phosphites and phosphines [987], phosphorus trichloride and trialkyl phosphates [988], cyclic thioureas [989], lactams [990], red phosphorus [991], and aminoethanol [992].

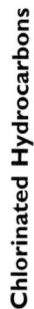

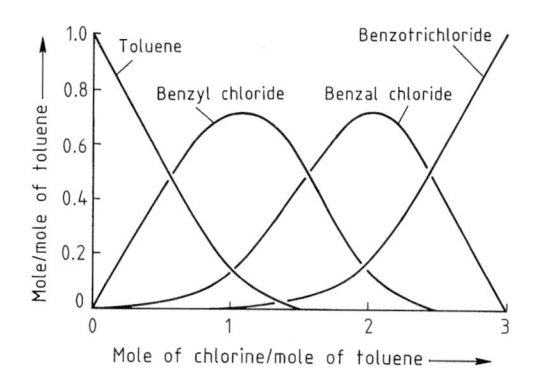

Figure 34. Progression of toluene chlorination [971], [973], [997], [1006], [1007]

3) Oxygen, a well-known radical scavenger, greatly reduces the consumption of chlorine; its presence is therefore undesirable. Toluene can be freed of oxygen, for example, puryng with an inert gas. The use of distilled chlorine, likewise, is an effective means of preventing oxygen from being present at significant concentrations [993]–[997].

It has also been reported, however, that even chlorine which contains up to 2% oxygen is suitable for side-chain chlorination at relatively low temperatures (40–50 °C) provided that phosphorus trichloride (1–3%) is present [998].

4) The presence of water leads to the formation of aqueous hydrochloric acid (indicated by cloudiness) and possibly to the hydrolysis of the chlorinated toluene; in addition, hydrolysis products which arise may be further altered by chlorination. The hydrogen chloride that is inevitably formed during chlorination is thought to be responsible for the fact that nuclear chlorination can never be entirely suppressed [999].

5) Various authors [994], [997], [1000] have drawn attention to the adverse effects caused by excessive chlorine concentrations (especially nuclear chlorination, both addition and substitution). Besides ensuring that the level of irradiation is adequate for rapid reaction, it may be useful to introduce an inert gas (N_2, HCl) into the chlorine stream. Dividing the reactor into smaller units is apparently also advantageous, particularly in the case of a continuous process [976], [977], [1001], [1002]. Much attention has also been paid to the development of kinetic models, to the design of reactors, to the effects of agitators, to the distribution of radiation, and to the effects of inhibitors [1003]–[1005].

In the course of radical chlorination of toluene, all three hydrogen atoms of the side-chain are successively replaced by chlorine. As a result, mixtures of the three expected compounds are obtained: benzyl chloride, benzal chloride, and benzotrichloride.

$$C_6H_5CH_3 \xrightarrow[k_1]{Cl_2} C_6H_5CH_2Cl \xrightarrow[k_2]{Cl_2}$$

$$C_6H_5CHCl_2 \xrightarrow[k_3]{Cl_2} C_6H_5CCl_3$$

The change in the composition of the mixture as a function of the number of moles chlorine that have reacted is shown in Figure 34 [1006]. Similar results have been reported by other authors [971], [973], [997], [1007].

Extensive investigations [971] have shown that for batch operations, it is practically impossible to alter the shape of these curves, neither by changing the intensity and wavelength of the light, by adding radical-forming agents (e.g., peroxides), by changing the quantity of chlorine introduced in unit time, by using PCl_3 or PCl_5 or other catalysts, nor by other means.

Only the rate of the reaction and the nature of the observed side reactions can be shown to depend significantly on the reaction conditions. Little is known with certainty about the absolute reaction rates. One reason for the scarcity of information is the fact that the actual concentration of the dissolved chlorine is not accurately measurable. If the chlorine concentration is assumed to be constant, the reactions can be taken to be quasi first-order, leading to the following calculated ratios of rate constants at 100 °C:

$k_1:k_2$	6.0 [971]	7.3 [995]	9.0 [997]
$k_2:k_3$	5.7 [971]	10.7 [995]	9.0 [997]

where k_1, k_2, and k_3 are defined in the equations above.

The individual components of the mixture can be isolated in pure form by fractional distillation. To prevent the decomposition of benzyl chloride during the distillation process, measures similar to those used to effect stabilization during storage and transportation (see Chap. 9.1.5) have been suggested [1008] – [1010].

Numerous attempts have been made to produce benzyl chloride that is as free as possible of the secondary products benzal chloride and benzotrichloride. One possibility is to restrict the chlorination to only 30 – 40 % of the toluene input and then to separate by distillation the resulting mixture, which still has a very high toluene content. Here, the principal disadvantage is the high cost of the distillation process.

According to [1011], pure benzyl chloride can be produced by chlorination in the vapor phase, provided the following equipment and procedure are used: The lower end of a packed column is provided with a bottom flask with a heater and an overflow. At the upper end is placed a condenser and a gas outlet. The feed point for fresh toluene and for toluene returning from the condenser is situated in the upper third of the column. The feed point for the chlorine (comprised of nozzles or frits) is located within the lower third. The temperature of the zone in which chlorination occurs is held at 125 °C during operation. This ensures that a temperature range is maintained within which benzyl chloride is a liquid whereas toluene is a gas. Under these conditions, benzyl chloride is rapidly removed from the chlorination section, flowing into the bottom flask; entrained toluene is returned from here to the chlorination section by evaporation. Hydrogen chloride formed during the chlorination, as well as toluene vapors, leave the column at its upper end. The toluene is liquefied in the condenser and returns to the chlorination section where it mixes with fresh toluene. The column is operated continuously. The bottom product is claimed to consist of 0.9 % toluene, 93.6 % benzyl chloride, and 5.5 % distillation residue. Several risks are inherent in this procedure. For example, the ignition temperature of toluene in chlorine gas is 185 °C. In addition, toluene–chlorine and benzyl chloride–chlorine mixtures are explosive over wide ranges (4–50 vol%, and 6–60 vol% respectively).

Monochlorination in the side-chain is said to be possible in the presence of alkyl or aryl sulfides or red phosphorus [991], [1012].

Today, most side-chain chlorination of toluene to produce benzyl chloride, together with benzal chloride (Fig. 34), is apparently carried out in a continuous fashion. The silver- or lead-lined reaction columns described in the past [1001] represent an obsolete technology since they are subject to considerable corrosion. For photochlorination, it is now customary to use reactors of enamel or glass, e.g., of the loop type [977]. The operating techniques have remained fundamentally unchanged, however.

Dibenzyl ether is formed as a byproduct in the alkaline hydrolysis of benzyl chloride to benzyl alcohol (see → Benzyl Alcohol). This ether can be reconverted to benzyl chloride by cleavage with hydrogen chloride at a temperature below 100 °C. Zinc chloride, iron chloride, and antimony chloride, are recommended as catalysts. Only a moderately high degree of conversion is achieved (60–70%) [1013]. Higher yields are claimed to be obtained if the metal halogenide catalysts are replaced by pyridine, pyridine derivatives, or alkylbenzylammonium chloride [1014]. Another route to benzyl chloride is the chloromethylation of benzene, although this approach is of no commercial significance [1015].

Environmental Protection. Benzyl chloride, like chlorinated hydrocarbons in general, must be handled with special care. The threshold limit value (TLV) and the Federal Republic of Germany's MAK value are both 1 ppm (5 mg/m^3). Benzyl chloride has been allocated to Category III B of the MAK list (the category comprising substances for which there is reason to suspect carcinogenic potential). Therefore, special requirements must be met concerning the sealing of production equipment and ventilation of workrooms. Since benzyl chloride is a chlorinated hydrocarbon and homologue of benzene, regular medical inspection of affected personnel is required by law.

9.1.4. Quality Specifications and Analysis

Benzyl chloride is sold in two quality grades, identified as "benzyl chloride, pure" and "benzyl chloride, pure and stabilized". Their assay is in both cases > 99%. Impurities include benzal chloride, toluene, chlorotoluene, chlorobenzene, and hydrogen chloride.

The most reliable analysis technique is gas chromatography, performed either with capillary or packed columns. The usual solid support in packed columns is Chromosorb AW-DMCS 80–100 mesh; recommended liquid phases include 4% Silicone Fluid 9.1.4DC 550 and 4% polyphenyl ether. Silicone resins are proven coating materials for capillary columns.

9.1.5. Storage and Transportation

As benzyl chloride is capable of reacting with heavy metals and their salts (Friedel-Crafts condensation reactions with the formation of HCl vapors), storage in enamel, glass, or lined vessels is essential. Suitable lining materials include bricks, lead, pure nickel, and stable synthetic resins. Drums with inserts of polyethylene or thick-walled polyethylene drums pigmented with graphite are suitable for transportation. Linings of lead, nickel, or special synthetic resins have proven to be suitable for tank cars and tank trucks. Many stabilizers have been proposed to make the storage and transportation of benzyl chloride safer. These act by neutralizing HCl and/or by forming complexes with heavy-metal ions. Examples are N,N-dimethylbenzylamine and N,N-diethylbenzylamine [1016]; pyridine and alkyl pyridines, quinoline, and bipyridyls (occasionally mixed with C_5–C_8 alcohols) [1009]; primary, secondary, and tertiary amines [1010], [1017], [1018]; phosphines [1019]; lactams [1020]; acid amides [1021]; ureas [1022]; and nitroalkanes [1023]. Aqueous sodium carbonate or sodium hydroxide solutions with a specific gravity identical with that of benzyl chloride were formerly used as stabilizers, but this practice has been largely discontinued. The emulsion these stabilizers produce lacks thermal stability and their presence makes it impossible to carry out reactions that require anhydrous benzyl chloride.

Legal Requirements. Benzyl chloride is a toxic chlorinated hydrocarbon and as such is subject to numerous regulations. The following regulations must be followed during its transport:

GGVS/ADR	Class, 6.1 no. 15 b
GGVE/RID	Class, 6.1 no. 15 b
GGV-See/IMDG	Code Class 6.1; UN no. 1738

Warning plate 6 (poison) must be displayed when benzyl chloride is transported on land. Primary label no. 6 (poison) and secondary label no. 8 (corrosive) are prescribed for marine transportation.

Benzyl chloride is additionally subject to the Arbeitsstoffverordnung of the Federal Republic of Germany and to the corresponding regulations of the European Community (EC compound no. 602-037-00-3).

9.1.6. Uses

Benzyl chloride is used mainly to produce plasticizers (e.g., benzyl butyl phthalate), benzyl alcohol, and phenylacetic acid via benzyl cyanide (used in the production of synthetic penicillin). On a smaller scale, it is used to produce quaternary ammonium salts (for disinfectants and phase-transfer catalysts), benzyl esters (benzyl benzoate and benzyl acetate for the flavors and perfumes industry), dyes of the triphenylmethane series, dibenzyl disulfide (antioxidant for lubricants), benzylphenol, and benzylamines.

9.2. Benzal Chloride

Benzal chloride (dichloromethylbenzene, α,α-dichlorotoluene, benzylidene chloride) [98-87-3] is today produced exclusively by the side-chain chlorination of toluene. It was first synthesized in 1848 by A. CAHOURS, by using the reaction of PCl$_5$ with benzaldehyde. Almost the sole application of benzal chloride is in the production of benzaldehyde.

9.2.1. Physical Properties

Benzal chloride is a liquid which fumes in moist air and which has a pungent odor and a strong irritant effect on the mucous membranes and eyes.

M_r	161.03
bp at 101.3 kPa	205.2 °C
mp	– 16.2 °C
ϱ at 0 °C	1.2691 g/cm^3
20 °C	1.2536 g/cm^3
30 °C	1.2417 g/cm^3
57 °C	1.2122 g/cm^3
79 °C	1.1877 g/cm^3
135 °C	1.1257 g/cm^3
Vapor pressure at	
45.5 °C	0.6 kPa
75.0 °C	0.8 kPa
82.0 °C	1.3 kPa
89.5 °C	1.9 kPa
105 °C	4.0 kPa
118 °C	8.0 kPa
205.2 °C	101.3 kPa
n_D^{20}	1.5503
Surface tension σ at 20 °C	40.1 mN/m
at 100 °C	31.1 mN/m
Dynamic viscosity η at 20 °C	2.104 mPa s
at 50 °C	1.327 mPa s
Specific heat at 25 °C	222 J mol^{-1} K^{-1} (1377 J kg^{-1} K^{-1})
Heat of vaporization at 72 °C	50.4 kJ/mol (313.2 kJ/kg)
Flash point	93 °C
Ignition point	525 °C
Heat of combustion at constant pressure	3.852 × 10^3 kJ/mol (23.923 × 10^3 kJ/kg)
Explosive limits in air, lower	1.1 vol%
upper	11 vol%
Specific conductivity at 20 °C	3.4 × 10^{-9} S/cm

Benzal chloride is freely soluble in alcohol, ether, chloroform, and carbon tetrachloride, but only slightly soluble in water (0.05 g/L at 5 °C; 0.25 g/L at 39 °C). The solubility of chlorine in 100 g of benzal chloride is

6.2 g   at   30 °C
4.3 g   at   50 °C
1.5 g   at 100 °C [967]

Several azeotropic mixtures are known of which benzal chloride is a component [966].

9.2.2. Chemical Properties

The action of chlorinating agents converts benzal chloride into benzotrichloride. In the presence of Lewis acids, the aromatic ring is chlorinated, with isomeric chlorobenzal chlorides being formed.
Hydrolysis under acid or alkaline conditions gives benzaldehyde.
Benzal chloride polymerizes in the presence of $AlCl_3$, $FeCl_3$, and similar compounds.
Metallic sodium converts benzal chloride into stilbene.

9.2.3. Production

Benzal chloride (together with benzyl chloride and benzotrichloride) is produced exclusively by side-chain chlorination of toluene. The preferred chlorination processes are those previously described under benzyl chloride. The pure compound is isolated by fractional distillation.

Environmental Protection. Benzal chloride is regarded as a toxic chlorinated hydrocarbon. Neither a TLV nor – in the Federal Republic of Germany – an MAK value has been established for benzal chloride. As a compound carrying a reasonable potential of being carcinogenic, benzal chloride has been allocated to Category III B of the MAK list. For this reason, stringent requirements must be met in its handling, including the sealing of production equipment and the ventilation of workrooms. Regular medical inspection is required of personnel coming in contact with the compound (chlorinated hydrocarbon and benzene homologues).

9.2.4. Quality Specifications and Analysis

The normal commercial form is "benzal chloride, pure," with an assay of > 99%. The main impurities are benzyl chloride and benzotrichloride.

Benzal chloride is analyzed by the same methods described for benzyl chloride (see p. 359).

9.2.5. Storage and Transportation

Stabilization like that employed with benzyl chloride is not absolutely necessary. It may be advisable under some conditions, however, such as storage or transportation in the tropics. Compounds used to stabilize benzyl chloride are also effective for stabilizing benzal chloride. Enameled steel, lead, and stainless steel are suitable materials for the construction of storage tanks. Stainless steel tanks or drums coated with baked enamel are suitable for transportation.

Legal Requirements. Since benzal chloride is a toxic chlorinated hydrocarbon, it is subject to the following official regulations:

GGVS/ADR	Class 6.1; 17 b
GGVE/RID	Class 6.1; 17 b
GGV-See/IMDG Code	Class 6.1; UN no. 1886

Label no. 6 (poison) must be displayed. Benzal chloride is subject to the Arbeitsstoff-Verordnung of the Federal Republic of Germany and to the corresponding EC Directive (EC compound no. 602-058-00-8).

9.2.6. Uses

Benzal chloride is used almost exclusively to produce → benzaldehyde.

Benzal chloride is hydrolyzed in the presence of water at a temperature above 100 °C by alkaline [1024] or acidic [1025]–[1028] agents. Friedel-Crafts catalysts or amines [1029] are recommended as catalysts. The latter are even recommended for mixtures of benzyl chloride and benzal chloride, whereby it is claimed that the benzyl chloride remains unchanged and that only benzaldehyde is formed. This process is unlikely to be of commercial interest; because benzyl chloride and benzaldehyde have almost identical boiling points their separation by fractional distillation would be very costly.

9.3. Benzotrichloride

Exhaustive chlorination of the side-chain of toluene leads to benzotrichloride (trichloromethyl benzene, α,α,α-trichlorotoluene, phenyl chloroform) [98-07-7]. The compound was first synthesized in 1858 by L. Schischkoff and A. Rosing, using the reaction of PCl_5 with benzoyl chloride. Benzotrichloride is now produced on a large scale, since it serves as an important intermediate in the preparation of acid chlorides (benzoyl chloride), dyes, herbicides, pesticides, and other products.

9.3.1. Physical Properties

Benzotrichloride is a colorless liquid with a pungent odor and is irritating to the eyes and mucous membranes. It fumes in moist air.

M_r	195.48
bp at 101.3 kPa	220.7 °C
mp	– 4.5 °C
ϱ at 15 °C	1.3777 g/cm^3
20 °C	1.3734 g/cm^3
30 °C	1.3624 g/cm^3
50 °C	1.342 g/cm^3
Vapor pressure at	
85 °C	1.1 kPa
95 °C	1.9 kPa
111 °C	3.1 kPa
121.5 °C	5.2 kPa
147 °C	13.3 kPa
220.7 °C	101.3 kPa
n_D^0	1.5677
n_D^{20}	1.5581
Dynamic viscosity η at 20 °C	2.40 mPa s
50 °C	1.517 mPa s
Surface tension σ at 20 °C	39.3 mN/m
100 °C	30.6 mN/m
Specific heat at 25 °C	235 J mol^{-1} K^{-1} (1206 kJ kg^{-1} K^{-1})
52 °C	248 J mol^{-1} K^{-1} (1269 kJ kg^{-1} K^{-1})
75 °C	249 J mol^{-1} K^{-1} (1273 kJ kg^{-1} K^{-1})
100 °C	250 J mol^{-1} K^{-1} (1281 kJ kg^{-1} K^{-1})
Heat of vaporization at 80 °C	52 kJ/mol (266 kJ/kg)
at 130 °C	47.5 kJ/mol (243 kJ/kg)
Flash point	108 °C
Ignition temperature	420 °C
Specific conductivity at 20 °C	6×10^{-9} S/cm
Heat of combustion at constant pressure	3684 kJ/mol (18878 kJ/kg)
Explosive limits in air, lower	2.1 vol%
upper	6.5 vol%

Benzotrichloride is freely soluble in alcohol, ether, and chloroform. It is only slightly soluble in water (0.05 g/L at 5 °C, 0.25 g/L at 39 °C). The solubility of chlorine in 100 g of benzotrichloride is

5.1 g at 30 °C
3.4 g at 50 °C
1.3 g at 100 °C [967]

Several azeotropic mixtures are known in which benzotrichloride is a component [966].

9.3.2. Chemical Properties

Acid or alkaline hydrolysis of benzotrichloride leads to benzoic acid. Partial hydrolysis gives benzoyl chloride.

Its reaction with carboxylic acids results in the corresponding acid chlorides and benzoyl chloride.

Condensation of benzotrichloride with benzene in the presence of $FeCl_3$, $AlCl_3$, or $ZnCl_2$ leads to diphenyl- and triphenylmethane.

All three chlorine atoms can be replaced by fluorine when benzotrichloride is treated with hydrofluoric acid or fluorides [1030], [1031].

Ortho-esters of benzoic acid can be prepared by reacting benzotrichloride with anhydrous alcohols.

9.3.3. Production

Exhaustive chlorination of the side-chain of toluene can be carried out in a manner analogous to that described under benzyl chloride. Photochemical chlorination in particular is widely applied for benzotrichloride production. Nevertheless, in order to prevent excessive chlorination and the appearance of ring-chlorinated materials, it is advisable, in continuous processes, to distribute the reaction over a cascade of six to ten reactors. Doing so makes it possible to introduce the chlorine at precisely the level appropriate to the progress of the reaction and results in benzotrichloride containing only a small amount of benzal chloride [976], [977].

A continuously operated plant for the production of benzotrichloride is illustrated in Figure 35 [977]. Fresh toluene flows into the first of a cascade of ten reactors. For reasons related to the removal of waste gases, the reactors can be regarded as being divided into three groups. Reactors 2–10 receive carefully metered amounts of chlorine. The off-gas from reactors 5–10 is rich in chlorine because the material in these reactors has already reached a high degree of chlorination; therefore, this gas is recycled to reactors 2 and 3.

Similarly, the off-gas from reactors 2 to 4 is introduced into reactor 1, which contains the highest proportion of toluene, so that the final traces of chlorine are removed. The off-gas from reactor 1 is thus free of chlorine.

With the condition that the chlorine and toluene are accurately metered, this technique is claimed to give practically complete conversion of toluene to benzotrichloride, and also to give a waste gas free of chlorine, i.e., consisting of pure hydrogen chloride.

Kinetic investigations of the formation of benzotrichloride have been published on several occasions [971], [995]. The yield and speed of the reaction are raised not only by exclusion of O_2 [995], but also by the use of high Reynolds numbers (35 000 – 160 000) [1032] or catalytic quantities of ammonium chloride [1033]. The chlorination of methylbenzenes using the corresponding trichlorides as solvents is claimed to give a

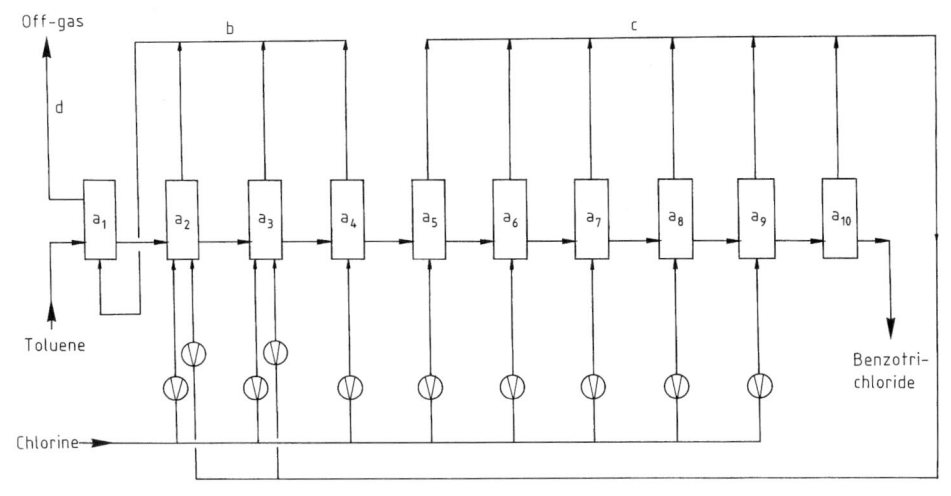

Figure 35. Continuous process for the manufacture of benzotrichloride [977]
a_1–a_{10}) Reactor cascade; b) Off-gas group 2, chlorine-containing; c) Off-gas group 3, high chlorine content; d) Off-gas group 1, chlorine-free

high yield of very pure products [1034]. According to [1035] the use of bromine in the production of benzotrichlorides increases the reaction rate and the yield. One additional manufacturing process for benzotrichloride is based on the chlorination of dibenzyl ether [1036], which is formed as a byproduct in the conversion of benzyl chloride to benzyl alcohol. This particular chlorination leads to a mixture of benzotrichloride and benzoyl chloride, which can be worked in the usual way to give pure benzoyl chloride. Indirectly, this serves as a way to improve the economics of benzyl alcohol production.

Environmental Protection. Benzotrichloride is regarded as toxic. Neither a TLV nor – in the Federal Republic of Germany – an MAK value has been established for it. Benzotrichloride has been allocated to Category III B of the MAK list (this category comprises substances reasonably suspected of having carcinogenic potential). Therefore, special requirements must be met concerning the sealing of production equipment and the ventilation of workrooms. As with other chlorinated hydrocarbons and homologues of benzene, regular medical inspection of personnel is necessary.

9.3.4. Quality Specifications and Analysis

Benzotrichloride is sold in two quality grades, known as "benzotrichloride, technical" and "benzotrichloride, pure". The corresponding assays are > 95 % and > 98 % respectively. Impurities include chlorotoluenes, benzyl chloride, chlorobenzyl chlorides, benzal chloride, chlorobenzal chlorides, and chlorobenzotrichlorides.

Gas chromatography is the preferred method of analysis. The procedure is analogous to that used for benzyl chloride (see Chap. 9.1.4).

9.3.5. Storage and Transportation

Stabilization is unnecessary for storage purposes. Enameled, lead-lined, and stainless steel vessels are suitable for storage. Stainless steel tanks and drums coated with baked enamel are suitable for transportation.

Legal Requirements. Being a corrosive chlorinated hydrocarbon, benzotrichloride is subject to various regulations:

ADR/GGVS:	Class 8, no. 66 b	
RID/GGVE:	Class 8, no. 66 b	
GGV-See/IMDG Code:	Class 6.1,	UN no. 2226

Label 8 (corrosive) must be displayed. Benzotrichloride is additionally subject to the Verordnung über gefährliche Arbeitsstoffe of the Federal Republic of Germany and to the corresponding directive of the European Community (EC compound no. 602-038-00-9).

9.3.6. Uses

Benzotrichloride is used mainly to produce benzoyl chloride, for which purpose it is either partially hydrolyzed with water or else reacted with benzoic acid. It is also of some significance in the production of pesticides (through transformation into benzotrifluoride), ultraviolet stabilizers, and dyes.

9.4. Side-Chain Chlorinated Xylenes

The side-chain-chlorinated xylenes play a less important role in the chemical industry than the corresponding toluene derivatives. In addition, substantial interest has been shown in only a few of the altogether 27 theoretically possible chloroxylenes, particularly the α-monochloro, α,α'-dichloro, and most notably, the $\alpha,\alpha,\alpha',\alpha',\alpha',\alpha'$-hexachloro derivatives.

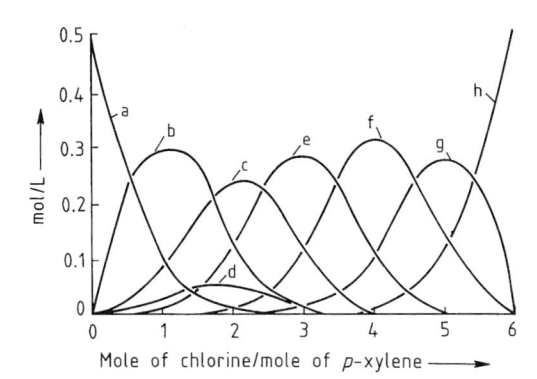

Figure 36. Progression of *p*-xylene chlorination [1043]
a) *p*-Xylene; b) α-Chloro-*p*-xylene; c) α,α'-Dichloro-*p*-xylene; d) α,α-Dichloro-*p*-xylene; e) α,α,α'-Trichloro-*p*-xylene; f) α,α,α',α'-Tetrachloro-*p*-xylene; g) α,α,α,α',α'-Pentachloro-*p*-xylene; h) α,α,α,α',α',α'-Hexachloro-*p*-xylene

9.4.1. Physical and Chemical Properties

A selection of chlorinated xylenes is listed in Table 56, together with certain physical data. The side-chain chlorinated xylenes are very similar in their chemical properties to the corresponding toluenes. They can therefore be made to undergo the same kinds of reactions as the latter. Thus, the hexachloroxylenes (*m*-, *p*-) are important in the production of carboxylic acid chlorides, and the α,α'-dichloroxylenes serve as sources of various bifunctional xylenes.

9.4.2. Production

The proven methods for the chlorination of toluene are basically suitable for the chlorination of xylene as well. Additives similar to those used in the chlorination of toluene are recommended to prevent nuclear chlorination [981]–[992]. Specialized additives include phosphoric acid esters together with sorbitol [1037], [1038] and a combination of boron trifluoride and ammonium chloride [1039]. Removal of air, moisture, and traces of metals by thorough purification of the chlorine and xylene feedstocks is also recommended [1040].

Chlorination in solvents, e.g., in carbon tetrachloride [984], [988], [996], [1041] or hexachloroxylene [1034], [1042], has been described as particularly advantageous. Since in theory, there are up to nine chlorinated derivatives of each of the xylene isomers, the course of the chlorination process is understandably very complex (Fig. 36) [1043].

The corresponding kinetics have been investigated in detail [1043]–[1048].

Thus, in the manufacture of α-chloroxylene, the chlorination must be discontinued sufficiently early to ensure that only a small amount of dichloride is formed. The product is purified by distillation [1049].

p-α,α'-Dichloroxylene can be produced analogously, whereby any xylene and α-chloroxylene recovered at the distillation stage can be returned to the chlorination reactor [1050].

Table 56. Physical data of chlorinated xylenes

Compound	Formula	M_r [CAS No.]	mp, °C	bp, °C (kPa)	ϱ, g/cm³	n_D
α-Chloro-o-xylene 1-chloromethyl-2-methylbenzene		140.61 [552-45-4]		190 (101.3) 90 (2.7) 80 (1.6)	1.083 (21 °C)	1.5410 (20 °C)
α,α′-Dichloro-o-xylene 1,2-bis(chloromethyl)benzene		175.06 [612-12-4]	55	241 (101.3)	1.393	
α,α,α′-Trichloro-o-xylene 1-chloromethyl-2-dichloromethyl-benzene		209.51 [30293-58-4]		139–141 (2.3)		
α,α,α′,α′-Tetrachloro-o-xylene 1,2-bis(dichloromethyl)benzene		243.96 [25641-99-0]	90			
α,α,α′,α′-Pentachloro-o-xylene 1-dichloromethyl-2-trichloro-methylbenzene		278.41 [2741-57-3]	53–55			
α-Chloro-m-xylene 1-chloromethyl-3-methylbenzene		140.61 [620-19-9]		195–197 (101.3) 101–102 (4.0)		1.5345 (20 °C)
α,α′-Dichloro-m-xylene 1,3-bis(chloromethyl)benzene		175.06 [626-16-4]	32–34	250 (101.3) 132 (2.1)		
α,α,α′,α′,α′-Hexachloro-m-xylene 1,3-bis(trichloromethyl)benzene		312.86 [881-99-2]	39–40	165–169 (1.6)		
α-Chloro-p-xylene 1-chloromethyl-4-methylbenzene		140.61 [104-82-5]	5	195 (101.3) 81 (2.0) 95 (2.7)	1.0512 (20 °C)	
α,α′-Dichloro-p-xylene 1,4-bis(chloromethyl)benzene		175.06 [623-25-6]	100.4	135 (2.1)		
α,α,α′-Trichloro-p-xylene 1-chloromethyl-4-dichloro-methylbenzene		209.51 [7398-44-9]	72	155–158 (3.6)		
α,α,α′,α′-Tetrachloro-p-xylene 1,4-bis(dichloromethyl)-benzene		243.96 [7398-82-5]	95			
α,α,α,α′,α′,α′-Hexachloro-p-xylene 1,4-bis(trichloromethyl)-benzene		312.86 [68-36-0]	111			

An alternative route to chloroxylenes involves the chloromethylation of toluene or benzyl chloride [1051], [1052]. This approach has the disadvantage, however, that it gives an isomer mixture, similar to that presumably formed by a double chloromethylation of benzene [1053]. This fact, together with the complexity entailed in a separation, makes its large-scale use less attractive. Isolation of various pure chloroxylenes is also possible (in some cases with high yields), not only by distillation, but also by direct crystallization from the reaction mixture [1054]–[1056].

The bis(trichloromethyl)benzenes are the chlorinated xylenes with the most commercial significance, and it is these whose manufacture has been investigated most thoroughly [988]–[990], [998], [1002], [1034], [1040], [1042], [1045], [1057].

It is worth noting that in the case of o-xylene, the exhaustive chlorination of the side-chains leads only as far as $\alpha,\alpha,\alpha,\alpha',\alpha'$-pentachloro-$o$-xylene. Steric hindrance evidently makes the hexachloro stage inaccessible. The corresponding hexafluoro derivative is known, however.

9.4.3. Storage and Transportation

Individual chloroxylenes are not subject to special regulations. The relevant regulations concerning the handling of chlorinated hydrocarbons should be appropriately applied, however.

The same is true for transportation, where, depending on the properties of the compound concerned, allocation to existing hazard categories (assimilation) is necessary.

9.4.4. Uses

In terms of output quantity the m- and p-hexachloroxylenes are the most important side-chain chlorinated xylenes. These find application particularly in the production of isophthaloyl chloride and terephthaloyl chloride, important starting materials for polyester synthesis.

The α,α'-dichloroxylenes have been used together with diamines or glycols, bisphenols, or even amino alcohols in the production of polymers.

9.5. Ring-Chlorinated Derivatives

In comparison with the toluene and xylene derivatives that are chlorinated exclusively in the side-chain, those that are also chlorinated on the ring have achieved considerably less industrial importance.

Normally, such products are made from toluenes or xylenes whose rings already bear chlorine. These are then subjected to further chlorination under the conditions described above, thereby being converted into the desired derivatives. If products are desired in which all ring positions are chlorinated, it is often possible to chlorinate both the ring and the side-chains without purification of intermediates [1058].

The uses of the ring-chlorinated compounds correspond to those of the parent series. A selection of such ring-chlorinated derivatives is compiled in Table 57.

Table 57. Physical data of chlorinated toluenes

Compound	Formula	M_r [CAS No.]	mp, °C	bp, °C (kPa)	ϱ, g/cm³	n_D
2-Chlorobenzyl chloride	C₆H₄(Cl)CH₂Cl (2-)	161.04 [611-19-8]		214 (101.3) 109 (3.4)	1.2743 (20 °C)	1.5592 (20 °C)
3-Chlorobenzyl chloride	C₆H₄(Cl)CH₂Cl (3-)	161.04 [620-20-2]		215 (100.4) 111 (3.4)	1.2695 (15 °C)	
4-Chlorobenzyl chloride	C₆H₄(Cl)CH₂Cl (4-)	161.04 [104-83-6]	31	217 (102.9) 114 (3.3) 102 (2.1)	1.241 (40 °C)	1.5500 (40 °C)
2,6-Dichlorobenzyl chloride	C₆H₃(Cl)₂CH₂Cl (2,6-)	195.49 [2014-83-7]	39–40	88 (0.4)		
3,5-Dichlorobenzyl chloride	C₆H₃(Cl)₂CH₂Cl (3,5-)	195.49 [56961-85-4]	36			
2,3,4,5,6-Pentachlorobenzyl chloride	C₆Cl₅CH₂Cl	298.84 [2136-78-9]	103			
2-Chlorobenzal chloride	C₆H₄(Cl)CHCl₂ (2-)	195.49 [88-66-4]		226–228 (99.3) 100 (1.3)	1.399 (15 °C)	1.5670 (16 °C)
3-Chlorobenzal chloride	C₆H₄(Cl)CHCl₂ (3-)	195.49 [15145-69-4]		235–237 (98.4) 105 (1.5)		
4-Chlorobenzal chloride	C₆H₄(Cl)CHCl₂ (4-)	195.49 [13940-94-8]		236 (100.6) 108 (1.3)		
2,5-Dichlorobenzal chloride	C₆H₃(Cl)₂CHCl₂ (2,5-)	229.94 [56961-83-2]	42	325 (101.3)		
2,6-Dichlorobenzal chloride	C₆H₃(Cl)₂CHCl₂ (2,6-)	229.94 [81-19-6]		250 (101.3) 124–126 (2.1)		
2,3,4,5,6-Pentachlorobenzal chloride	C₆Cl₅CHCl₂	333.29 [2136-95-0]	117	199 (1.7)		
2-Chlorobenzotrichloride	C₆H₄(Cl)CCl₃ (2-)	229.94 [2136-89-2]	30	260 (101.3) 129.5 (1.7)		
4-Chlorobenzotrichloride	C₆H₄(Cl)CCl₃ (4-)	229.94 [5216-25-1]		108–112 (1.3)	1.4947 (30 °C)	1.5690 (30 °C)
2,4-Dichlorobenzotrichloride	C₆H₃(Cl)₂CCl₃ (2,4-)	264.93 [13014-18-1]	48	147–151 (1.6) 155–159 (2.7)		

Table 58. Capacities for chlorinated toluenes, in t/a

	Europe	World
Benzyl chloride	80 000	160 000
Benzal chloride	15 000	30 000
Benzotrichloride	30 000	60 000

9.6. Economic Aspects

Production capacities for the toluene derivatives discussed above were estimated to have been as follows in 1984 (Table 58):

It is not really possible to determine the extent of utilization of these capacities, since many companies produce the products for their further own use. It is likely, however, that ca. 60% of the estimated capacity was utilized in 1984.

The 1984 price of benzyl chloride was ca. 0.90 $/kg; that of benzotrichloride was ca. 1.30 $/kg.

10. Toxicology and Occupational Health

10.1. Aliphatic Chlorinated Hydrocarbons

In this Section, chlorinated methanes, ethanes, ethylenes, propanes, and propenes of major commercial importance are discussed. A few other substances found as minor products, research chemicals, contaminants, or unwanted products are included when data are extensive or if they present an unusual or high toxicity.

The diversity of the toxic properties of chlorinated hydrocarbons is often inadequately appreciated, despite decades of use and studies proving great differences. They have all been mistakenly categorized generically as hepatotoxic, although large differences in their ability to injure the liver exist. Most of the compounds discussed are rather volatile and have a low potential for bioconcentration; nevertheless, the solvents and monomers have been incorrectly grouped with persistent chlorinated pesticides and other nonvolatile chlorinated materials.

While the substances discussed in this section show some common toxicological, chemical, and physical properties, exceptions are so common that categorization must be avoided. It is, therefore, imperative to examine the toxicity of each specific substance. Fortunately, the most common chlorinated solvents have been exhaustively studied. However, new toxicological data are being generated and the current literature and regulations should always be consulted.

Table 59. Single-dose oral toxicity of common chlorinated C_1, C_2, C_3, and C_4 aliphatic hydrocarbons

	LD_{50} (oral, rats; unless other species specified), mg/kg	Probable nature of death [a]
Chloromethane (methyl chloride)	gas	–
Dichloromethane (methylene chloride)	2 000	A
Trichloromethane (chloroform)	2 000	A, LK
Tetrachloromethane (carbon tetrachloride)	3 000	A, LK
Monochloroethane (ethyl chloride)	gas	–
1,1-Dichloroethane (ethylidene dichloride)	>2 000 [b]	–
1,2-Dichloroethane (ethylene dichloride)	700	LK
1,1,1-Trichloroethane (methyl chloroform)	10 000 – 12 000	A
1,1,2-Trichloroethane (vinyl trichloride)	100 – 200	LK
1,1,2,2-Tetrachloroethane (acetylene tetrachloride)	ca. 300 (dogs)	–
Pentachloroethane	≪1 750 (dogs)	–
Hexachloroethane	6 000	–
Monochloroethylene (vinyl chloride)	gas	–
1,1-Dichloroethylene (vinylidene chloride)	1 500	LK
1,2-Dichloroethylene (cis and trans)	1 000 – 2000	A
Trichloroethylene	4 900	A
Tetrachloroethylene (perchloroethylene)	2 000	A
Dichloroacetylene	–	–
2-Propyl chloride (isopropyl chloride)	>3 000 (guinea pigs)	A
1,2-Dichloropropane (propylene dichloride)	2 000	A
3-Chloropropene (allyl chloride)	450 – 700	LK
1,3-Dichloropropene	500 – 700	LK
2-Chloro-1,3-butadiene (chloroprene)	250	LK
Hexachlorobutadiene	200 – 350	LK

[a] A = Anesthesia, LK = Liver and kidney injury;
[b] Unpublished data, The Dow Chemical Company, Midland, Michigan, USA.

Reviews may quickly become incomplete, but a few are listed in the references [1059] – [1061]. They will be most valuable for information about effects on the skin or eyes, as well as effects of single or short, repeated exposures by ingestion or inhalation. Information on carcinogenesis, mutagenesis, and birth effects are presented, but these are active research areas and current literature must be consulted. The manufacturer or supplier of a substance is responsible for acquiring and distributing such information, and most manufacturers excercise that responsibility. Table 59 shows some data on the acute toxicity of the aliphatic chlorinated hydrocarbons.

Abbreviations and Definitions. Certain terms, names, and organizations appear in this section. Most of the abbreviations are explained in the front matter, others are defined in the text. For detailed information on general toxicology, see the corresponding articles in the B series.

Carcinogenesis, Mutagenesis, and Teratogenesis. In the current regulatory climate, mutagenic changes, reproductive effects, and particularly cancer are of prime concern. This is appropriate, but excessive concern about cancer has distracted attention from other concerns that may be of equal or more importance.

Many, if not most, chlorinated substances can be made to produce an increase of tumors in certain laboratory animals, particularly in organs in which toxicity is exhibited and high tumor rates exist normally, e.g., in the livers of mice. However, no chlorinated hydrocarbon except vinyl chloride has yet been shown to have increased cancer in human populations. Many epidemiological studies lack power due to small populations and short duration. However, a number of studies are of adequate size, duration, and power to demonstrate that cancer has not increased in the degree predicted from the studies in mice. Because of this inconsistency and a variety of other data, there has been considerable scientific discussion about interpreting animal studies in regard to human risk; see also [1062].

According to current scientific thought, certain substances, including the chlorinated hydro-carbons, increase cancer in certain organs (e.g., livers of mice) as a result of repeated stress and injury with subsequent increased cellular regeneration. Thus, preventing exposures that cause cellular changes (injury) should also prevent cancer. This nongenotoxic mechanism of induction is consistent with current human experience and other data related to most of the chlorinated hydrocarbons discussed herein except vinyl chloride.

Vinyl chloride appears to operate by a genetic mechanism and, although humans are much less responsive than rodents, the difference is quantitative and appears related to a lower rate of metabolism in humans. This indicates that exposure to all of these materials must be carefully controlled to avoid exposures that result in stress or injury. Close adherence to the occupational exposure limits (MAK or TLV) is recommended. Furthermore, good industrial hygiene practice requires that exposures to any substance be kept as low as reasonable and that careless operation should be prohibited regardless of whether the TLV or MAK is exceeded.

Reproductive effects do not appear to be of concern with any of the substances discussed in this Section, provided exposures are controlled to prevent injury to other organ systems of the mother during gestation. In other words, the reproductive system appears less sensitive than other systems [1063]. Likewise, mutagenic effects appear unlikely, based on the weight of the evidence from in vivo and in vitro studies.

Occupational Exposure Limits. Table 60 lists the 1985 TLV's and MAK's published by the American Conference of Governmental Industrial Hygienists (ACGIH) and the Deutsche Forschungsgemeinschaft (DFG), respectively [1064], [1065]. The definitions applied by these organizations must be understood in order to apply the values properly. For example, both organizations recommend that skin contact be limited if skin absorption is thought to influence the TLV or the significantly MAK. The reader must always consult the latest values published annually by these organizations and, further, must not assume that they are the legal standard. It is strongly recommended that supporting documentations be consulted when using the TLV's and MAK's.

Table 60. Summary of TLV's and MAK's for common chlorinated C_1, C_2, C_3, and C_4 aliphatic hydrocarbons

	1985 ACGIH TLV		1984 DFG MAK	
	ppm	mg/m³	ppm	mg/m³
Chloromethane	50	105	50 (III B)	105
Dichloromethane	100	350	100 (III B)	360
Trichloromethane	10 (A2)	50 (A2)	10 (III B)	50
Tetrachloromethane (skin)	5 (A2)	30 (A2)	10 (III B)	65
Monochloroethane	1000	2600	1000	2600
1,1-Dichloroethane	200	810	100	400
1,2-Dichloroethane	10	40	20 (III B)	80
1,1,1-Trichloroethane	350	1900	200	1080
1,1,2-Trichloroethane (skin)	10	45	10 (III B)	55
1,1,2,2-Tetrachloroethane (skin)	1	7	1 (III B)	7
Pentachloroethane	–	–	5	40
Hexachloroethane	10	100	1	10
Monochloroethylene (vinyl chloride)	5 (A1a)	10 (A1a)	3 (2) (TRK; III A1)	8 (3) (TRK; III A1)
1,1-Dichloroethylene	5	20	2 (III B)	8
1,2-Dichloroethylene (cis and trans)	200	790	200	790
Trichloroethylene	50	270	50 (III B)	270
Tetrachloroethylene	50	335	50	345
Dichloroacetylene	C 0.1	C 0.4	III A2	III A2
1,2-Dichloropropane	75	350	75	350
3-Chloropropene (allyl chloride)	1	3	1 (III B)	3
1,3-Dichloropropene (skin)	1	5	III A2	III A2
2-Chloro-1,3-butadiene (chloroprene) (skin)	10	36	10	36
Hexachlorobutadiene (skin)	0.02 (A2)	0.24 (A2)	(III B)	(III B)

A2 = III B = Suspected carcinogen; III A2 = Carcinogen in animal experiments; A1a = III A1 = Human carcinogen; C = Ceiling; TRK = Technical Guiding Concentration 3 ppm in existing facilities, 2 ppm in new facilities; skin = This designation is intended to suggest appropriate measures for the prevention of cutaneous absorption so that the threshold limit is not invalidated.

10.1.1. Chloromethanes

Monochloromethane. Chloromethane [74-87-3], methyl chloride, is an odorless gas and, except for freezing the skin or eyes due to evaporation, inhalation is the only significant route of exposure. It acts mainly on the central nervous system with well documented cases of excessive human exposure, leading to injury and even death [1059]. The symptoms of overexposure are similar to inebriation with alcohol (a shuffling gait, incoordination, disorientation, and change in personality), but last much longer, possibly permanent in severe exposures. According to recent experimental results, excessive exposure to methyl chloride was carcinogenic in mice and also affected the testes of male rats and fetuses of pregnant female rats [1066]. It is mutagenic in certain in vitro test systems. Available references indicate that methyl chloride may increase the rate of kidney tumors in mice in conjunction with repeated injury to this organ. The current TLV and the MAK (1985) are both 50 ppm (105 mg/m^3).

Dichloromethane. Dichloromethane [75-09-2], methylene chloride, is the least toxic of the chlorinated methanes. It is moderate in toxicity by ingestion, but the liquid is quite painful to the eyes and skin, particularly if confined on the skin [1059]–[1061]. Absorption through the skin is probably of minor consequence if exposure is controlled to avoid irritation.

Inhalation is the major route of toxic exposure. The principal effects of exposure to high concentrations (greater than 1000 ppm) are anesthesia and incoordination. Exposure to methylene chloride results in the formation of carboxyhemoglobin (COHb) caused by its metabolism to carbon monoxide. This COHb is as toxic as that derived from carbon monoxide itself. However, at acceptable levels of exposure to methylene chloride, any probable adverse effects of COHb will be limited to persons with pronounced cardiovascular or respiratory problems. Other possible toxic effects of carbon monoxide itself would not be expected.

Methylene chloride is not teratogenic in animals [1063] and has only limited mutagenic activity in *Salmonella* bacteria. It does not appear to be genotoxic in other species. Available reports of lifetime studies at high concentrations have produced inconsistent results in hamsters, rats, and mice. No tumors, benign or malignant, were increased in hamsters; rats developed only a dose related increase in commonly occurring nonmalignant mammary tumors; white mice, both sexes, had a large increase in cancers of the livers and lungs. Available epidemiological data do not indicate an increase in cancer in humans; they do indicate that the current occupational standards are protective of employee health [1064], [1065].

Trichloromethane. Trichloromethane [67-66-3], chloroform, is only moderately toxic from single exposure, but repeated exposure can result in rather severe effects [1059]–[1061]. Its use as a surgical anesthetic has become obsolete, primarily because of delayed liver toxicity and the development of anesthetics with a greater margin of safety.

Ingestion is not likely to be a problem unless large quantities are swallowed accidentally or deliberately. Chloroform has a definite solvent action on the skin and eyes and may be absorbed if exposure is excessive or repeated. Its recognized high chronic toxicity requires procedures and practices to control ingestion, skin, and eye contact, as well as inhalation exposure if liver and kidney injury, the most likely consequence of excessive exposure, is to be prevented.

In animals, chloroform is fetotoxic (toxic to the fetus of a pregnant animal) but only weakly teratogenic if at all [1063]. It does not appear to be mutagenic by common test procedures, but increases the tumor incidence in certain rats and mice. There is considerable evidence that the tumors in rat kidneys and mice livers are the result of repeated injury to these organs and that limiting exposure to levels that do not cause organ injury will also prevent cancer. It is, therefore, very important that human exposure be carefully controlled to prevent injury.

Tetrachloromethane. Tetrachloromethane [*56-23-5*], carbon tetrachloride, was once recommended as a "safety solvent." Misuse and its rather high liver toxicity, as well as the ready availability of alternate safe solvents, have eliminated its application as a solvent. Single exposures are not markedly injurious to the eyes and skin or toxic when small quantities are ingested. However, repeated exposure must be carefully controlled to avoid systemic toxicity, particularly to the liver and kidneys [1059]–[1061]. In humans, injury to the kidney appears to be the principal cause of death.

Inhalation can produce anesthesia at high concentrations, but transient liver as well as kidney injury result at much lower concentrations than those required to cause incoordination. There appears to be individual susceptibility to carbon tetrachloride, with some humans becoming nauseated at concentrations that others willingly tolerate. Ingestion of alcohol is reported to enhance the toxicity of carbon tetrachloride. Such responses should not occur, however, if exposures are properly controlled to the recommended occupational standards.

Carbon tetrachloride is not teratogenic in animals [1063] nor mutagenic in common test systems, but does increase liver tumors in mice, probably as a result of repeated injury to that organ. Therefore, it is very important that human exposure be carefully controlled to prevent liver injury.

10.1.2. Chlorinated C_2 Hydrocarbons

Monochloroethane. Chloroethane [*75-00-3*], ethyl chloride, has limited use as an industrial chemical and is most commonly recognized by its use as a local anesthetic that is sprayed on the skin for minor medical procedures. There are remarkably few published data on its toxicity, since there are very few people exposed in the few uses in industry. It appears to be low in toxicity by inhalation, the only likely route of toxic exposure. Evaporation of large quantities could freeze the skin or eyes [1059]–[1061]. The National Toxicology Program (NTP) has a cancer bioassay underway, and while it appears to be negative, complete results are not yet available.

1,1-Dichloroethane [*75-34-3*]. This flammable substance has limited use and only limited toxicological data are available. It appears to be low in toxicity by all routes including oral, dermal, and inhalation. Animals tolerated repeated exposures 7 h/d for 9 months to either 500 or 1000 ppm with no adverse effect.

It was not teratogenic when inhaled by pregnant rats. The doses fed by gavage in a carcinogenesis study of the National Cancer Institute (NCI) were so high that mortality was increased. Although no tumor increases were reported, no conclusions could be drawn concerning the induction of cancer [1059]–[1061].

1,2-Dichloroethane. 1,2-Dichloroethane [*107-06-2*], ethylene dichloride, EDC, is one of the more toxic common chlorinated substances [1059]–[1061]. It can cause depression of the central nervous system, mental confusion, dizziness, nausea, and vomiting. Liver, kidney, and adrenal injury may result from both acute overexposure and repeated overexposure at levels significantly above the recommended occupational standards. It has moderate toxicity when swallowed and is often vomited. Skin and eye irritation generally occur only if the liquid is confined. Absorption through the skin is not likely a problem from single contact, but repeated exposure should be avoided. Studies to determine the carcinogenic properties have used excessive doses and produced mixed results. According to the available data, cancer in rodents is caused by repeated injury of the organs and is not likely to occur below the current occupational standards [1067]. Particular precautions should be taken to assure that skin and inhalation exposures are carefully and appropriately controlled.

1,1,1-Trichloroethane. 1,1,1-Trichloroethane [*71-55-6*], methyl chloroform, has consistently been shown to be among the least toxic chlorinated or nonchlorinated solvents from both acute and repeated exposure [1059]–[1061]. It has been shown repeatedly to have little effect on the liver. It is low in oral toxicity, and has a typical solvent (defatting) action on the skin and eyes; hence, liquid exposure should be minimized. Exposure to more than 1000 ppm of the vapors (significantly above the occupational standards) may cause incoordination with resulting lack of judgment and possible accidents. Misuse and carelessness have resulted in unnecessary deaths, primarily while working in confined spaces. Exposure concentrations under such conditions may attain 10 000 – 30 000 ppm or more. Death is due to anesthesia or possible sensitization of the heart to endogenous adrenalin. Recovery generally has been complete and uneventful if the victim is removed from the exposure before death occurs.

Injury to the liver or other internal organs is unlikely unless severe anesthetic effects have been observed. It has been found to cause no teratogenic or reproductive effects in animals, and extensive study indicates that the vapors are not carcinogenic in rats and mice. It is metabolized only to a very slight degree in animals and humans. It is probably not mutagenic. Human experience has been favorable, as have epidemiological studies on exposed workers. Exposure to high concentrations, particularly in confined spaces, must not be permitted.

1,1,2-Trichloroethane [*79-00-5*]. This substance is relatively toxic and must not be confused with the 1,1,1-isomer discussed previously [1059]–[1061]. It has little use and human experience is limited. According to data from animals, it can cause liver injury as a result of single or repeated exposure. It has a typical solvent effect on the skin and eyes; hence, exposure should be minimized. It apparently has not been studied for its teratogenic effects on animals; it is not mutagenic in common test systems, but it increases the number of liver tumors in mice, probably as a result of organ injury and repeated regeneration. Therefore, it is very important that human exposure be carefully controlled to prevent liver injury.

Other Chloroethanes. 1,1,2,2-tetrachloroethane [79-34-5], pentachloroethane [76-01-7], and hexachloroethane [67-72-1] have limited industrial use, partially because of their recognized high toxicity on repeated exposure [1059]–[1061]. Symmetrical *tetrachloroethane*, $CHCl_2CHCl_2$, at one time was used as a solvent, but liver injury was reported among overexposed workers. It is fetotoxic in rats and increased the tumor rate in the livers of mice. The results of the study of the National Cancer Institute (NCI) on rats was inconclusive. There is limited evidence that it is mutagenic. Much less data are available on the toxicity of *pentachloroethane*, but it is assumed to be highly toxic.

Older data on *hexachloroethane* indicate a much higher toxicity than recent data [1059]. This may be due to better purity of the new sample. The recent report indicates a low to moderate oral toxicity, only slight skin and eye irritation, but a moderate to high toxicity from repeated inhalation. It was not markedly hepatotoxic and, at 15 ppm, caused no injury in dogs, rats, guinea pigs, and quail exposed 6 h/d, 5 d/week for 6 weeks. Exposure to 48 ppm caused minimal injury, primarily in the eyes and respiratory tract of the animals. It was not teratogenic in rats or mutagenic in bacteria, but it increased the rate of liver tumors in mice. It caused no cancer in rats, but both rats and mice had evidence of kidney injury. The purity of the sample needs to be verified, however, because contamination with such other substances as tetra- or pentachloroethane may have caused the reported effect.

All three of these chlorinated ethanes must be handled to control exposure and possible liver injury.

Vinyl Chloride [75-01-4]. Vinyl chloride is the only chlorinated hydrocarbon that unquestionably has caused cancer (angiosarcoma of the liver) in humans. As a result, there are numerous regulations and laws with regard to its production and use that are intended to minimize exposure. These should effectively eliminate any other toxic effects of vinyl chloride as well as the possibility of cancer.

Because vinyl chloride is a gas, ingestion is not likely in an industrial setting [1059]–[1061]. Skin and eye contact appear to be of concern only from evaporative freezing. Even at adequate protection of the respiratory tract to prevent inhalation, some vinyl chloride may be absorbed through the skin, but the total contribution seems to be slight, even at high concentrations.

When inhaled in high concentrations (10 000 ppm), anesthetic effects can occur. At even higher concentrations, the effects increase and deaths have been reported from massive exposures. Odor provides little warning of excessive exposure.

Many other effects have been alleged to occur as a result of excessive exposure to vinyl chloride, but only a few are clearly the direct result of exposure. It is hepatotoxic, possibly mutagenic, and has caused angiosarcoma of the liver. A condition known as acroosteolysis with scleroderma has been associated with cleaning autoclaves used for polymerization. Whether the disease, called *kettle cleaner's disease*, is related to vinyl chloride itself or to some other substance is not known. Likewise, other cancers alleged to be caused by vinyl chloride may or may not be the result of vinyl chloride itself, since they are not consistent throughout the industry and appear to be found only in certain populations.

Careful controls to minimize exposures and adherence to regulatory requirements are essential.

Vinylidene Chloride. Vinylidene chloride [75-35-4], 1,1-dichloroethylene, is an anesthetic at high concentrations (several thousand ppm) [1059]–[1061]. Hepatotoxicity can result from rather low exposures; therefore, low TLVs and MAK's have been recommended. It has a solvent effect on the skin and eyes but its high volatility probably precludes absorption through the skin in most situations. Exposure should be carefully controlled. The effect on the liver is quite marked on repeated exposure of animals. Although not a teratogen, it does cause injury to embryos and fetuses of exposed animals at levels causing injury to the pregnant mothers. The metabolites of vinylidene chloride are at most weakly mutagenic in bacterial test systems, and tests in mammalian systems are negative. It is probably not carcinogenic based on the weight of evidence, for only one out of 14 tests has been marginally positive. It is important that human exposure be carefully controlled to prevent liver injury.

1,2-Dichloroethylene. Most toxicological testing has been on mixed isomers [1059]. It is not clear from available data on the isomers how they compare in toxicity. Most results indicate a moderate oral toxicity to rats. A typical solvent effect on the skin and eyes is expected although data are not available to verify this conclusion. Exposure should be controlled.

Most studies on animals indicate a rather low toxicity by inhalation, with little effect on the liver. Anesthesia occurs at higher levels, but the data are inconsistent as to the actual levels required.

The effects of repeated exposure are not clear either. One reference reported no adverse effect on 7-h daily exposure for six months to 500 or 1000 ppm, but a second reported rather marked effects at 200 ppm after 14 weeks. No data were found on teratogenesis or carcinogenesis, but very limited data indicate no mutagenic effect.

Trichloroethylene [79-01-6]. There is a tremendous amount of literature on trichloroethylene because of its use as a degreasing solvent and even more so of its use as an anesthetic have resulted in considerable human exposure [1059]–[1061]. With such a vast literature, conflicting conclusions are possible. The toxicity is generally considered low to moderate. Liver and kidney injury do not appear to be a common response even after excessive exposures which cause anesthesia. Trichloroethylene has a typical solvent (defatting) action on the skin and eyes and exposure should be controlled. Absorption through the skin may occur but is not likely a significant source of exposure. When inhaled, it can have a pronounced anesthetic effect (depres-sion of the central nervous system), which may become evident as incoordination at concentrations of 400 ppm or more. Visual disturbances, mental confusion, fatigue, and sometimes nausea and vomiting are observed at higher levels. The nausea is not nearly as marked as with carbon tetrachloride or ethylene dichloride. Sensitization of the heart to adrenalin may occur, but it does not appear to be significant unless markedly anesthetic concentrations are reached (several thousand ppm).

Deliberate sniffing has been a problem, although physical dependence does not appear to be involved. A peculiar vascular dilation of the face, neck and trunk, known as *"degreaser's flush,"* occasionally occurs when alcohol is consumed during or following exposure. Although upsetting, the flush does not appear to be serious.

Urinary metabolites, trichloroacetic acid and trichloroethanol, are measured in several countries to monitor workers' exposure, but this procedure has severe disadvantages. Urinary metabolites are more related to chronic (total) exposure than they are to acute (peak) exposure. Thus, the worker may have excessive peak exposures which could result in anesthesia, incoordination and possible accidents, and yet their urinary metabolites remain within accepted limits at the end of the day.

Trichloroethylene has not shown teratogenic effects in animals. It possibly is weakly mutagenic but much of the testing is suspect due to impurities (stabilizers) present in the samples.

Trichloroethylene appears to increase the number of liver tumors in certain mice given massive doses by gavage and lung tumor in one strain of mice by inhalation, but it has generally been negative in rats and other rodent studies. The significance to human cancer is not clear [1060]. Several rather small epidemiological studies have failed to show an increase in human liver cancer of exposed workers.

Tetrachloroethylene [*127-18-4*]. Although it is not a potent anesthetic, depression of the central nervous system (incoordination) is the most common response to tetrachloroethylene at concentrations above 200 ppm [1059]–[1061]. Liver and kidney injury does not appear to be a common response even after excessive exposure. It is moderate to low in oral toxicity, has a solvent action on the skin and eyes, and is poorly absorbed through the skin. Exposure of the skin and eyes should be carefully controlled. The odor begins to be objectionable at about 400 ppm for most people. When inhaled at high concentrations, it may cause nausea and gastrointestinal upset in addition to the anesthesia and incoordination. It is much weaker in producing nausea than carbon tetrachloride and ethylene dichloride.

Reports of human injury are uncommon despite its wide usage in dry cleaning and degreasing. Sensitization of the heart to adrenalin does not appear to be a likely consequence.

Tetrachloroethylene is not extensively metabolized and most of the absorbed dose is excreted unchanged in the expired air. Analysis of metabolites in urine is therefore of even less value than with trichloroethylene.

Tetrachloroethylene is not teratogenic in standard tests in animals; it does not appear to be significantly mutagenic and increases tumors in certain strains of mice. Like many of the chlorinated hydrocarbons, the significance of the mouse liver tumor to human cancer is questionable.

Dichloroacetylene [7572-29-4]. This substance is discussed only because it is a highly toxic and pyrophoric substance formed by dehydrochlorination of trichloroethylene [1059]. This has occurred when trichloroethylene vapors (or the liquid) are passed over soda–lime or caustic soda. Much of the experience comes from the use of Hopcalite in rebreathing anesthesia machines or in closed environmental systems. Exposure to as little as 19 ppm is reported to cause the death of half of the mice (LC_{50}) in a 4-h exposure. Repeated exposure caused kidney injury and muscular paralysis. Dichloroacetylene is reported to cause headache, nausea, and nerve, liver, and kidney injury in humans or animals.

10.1.3. Chloropropanes and Chloropropenes

2-Chloropropane. 2-Chloropropane [75-29-6], isopropyl chloride, has had little use or study [1059]. Its flammability has probably discouraged its use, although it appears to be of low toxicity in animals. When fed by gavage to rats, a dose of 3 g/kg was survived. Ten repeated applications of the liquid on the skin of a rabbit was very slightly irritating if allowed to evaporate and slightly more irritating if confined under a bandage. Exposure of the skin and eyes should be prevented. Data on the effects of inhalation appear to be limited. According to one study, repeated exposure of rats to 250 ppm caused no effect, but 1000 ppm did. Another study claimed no effect in rats, mice, rabbits, guinea pigs, and monkeys exposed 7 h/d 5 d/week for 6 months to 500 or 1000 ppm. Limited data suggest that isopropyl chloride caused a mutagenic response in bacteria, but no data were found on teratogenesis or carcinogenesis.

1,2-Dichloropropane. 1,2-Dichloropropane [78-87-5], propylene dichloride, appears to be low to moderate in toxicity on single exposure, but moderately toxic on repeated exposure [1059]–[1061]. It has a solvent effect (defatting action) on the eyes and skin. Exposure should be prevented. Absorption of the liquid through the skin may occur, particularly on repeated contact. When inhaled by mice, respiratory injury rather than liver toxicity appeared to be the primary cause of death following single exposure. There appears to be little recent data on the effect of repeated inhalation by animals; hence, the recommended TLV and MAK are based on old studies.

The National Toxicology Program (NTP) has a carcinogenic study under way, but no report has been issued. No data were found in regard to teratogenic effects, but minimal data indicate it may be mutagenic in bacterial test systems.

10.1.4. Chlorobutadienes

2-Chloro-1,3-butadiene. 2-Chloro-1,3-butadiene [*126-99-8*], chloroprene, is a rather toxic, highly flammable monomer used to produce synthetic rubber [1059]. The toxicity of various samples appears to have been influenced by reaction products because chloroprene reacts with oxygen to form peroxides. It also dimerizes. It must be handled carefully to prevent these reactions. Liver injury is possible, as is hair loss. The hair loss appears to be caused by a reaction of chloroprene with the hair itself, since regrowth occurs when exposure ceases. The vapors cause respiratory irritation, as well as pain and irritation in the eyes. The pure material did not cause teratogenic effects. It is mutagenic in some bacterial test systems. Early reports of cancer among workers have not been confirmed in more carefully conducted studies, and animal studies have been negative [1060].

Hexachlorobutadiene [*87-68-3*]. Hexachlorobutadiene has been used as a pesticide, but it is a largely unwanted byproduct made during chlorination of hydrocarbons [1059]. It is highly toxic and, unlike the other substances discussed in this section, of low volatility. The literature on hexachlorobutadiene has been reviewed [1068], and a summary of the data is found in reference [1059]. There were no references to human exposure. In animals, hexachlorobutadiene has caused liver injury after single exposure and liver and kidney injury, as well as kidney tumors, after repeated ingestion. The kidneys are the primary target organ on repeated exposure.

Precautions must be taken to minimize exposure to hexachlorobutadiene by skin, ingestion, and inhalation.

10.1.5. Ecotoxicology and Environmental Degradation

Aquatic Toxicity. Most common aliphatic chlorinated hydrocarbons have been tested for their acute toxicity by both static and flow-through methods using vertebrates (fish) and invertebrates (water fleas). According to these studies the compounds tested were either nontoxic or slightly toxic as defined by the EPA (Table 61) [1072].

Presence in Water. According to analyses carried out in Western Europe, typical concentrations of 1,1,1-trichloroethane, trichloroethylene, and perchloroethylene in surface water range from 0.1 to 3 µg/L [1073]. Trace concentrations of the common chlorinated hydrocarbons have been found in drinking water. Dichloromethane has not been detected in natural waters, oceans, sediments, or fish [1074]. Trichloromethane is often formed at sub-ppm levels in the disinfection of drinking water because chlorine reacts with natural humic substances. The concentrations of common chlorinated solvents in drinking water are typically below 0.3 µg/L [1073].

Table 61. Aquatic toxicity of aliphatic chlorinated hydrocarbons (LC_{50}, mg/L)

	Fathead minnow		Daphnia magna
	96-h static	96-h flow through	48-h static
Dichloromethane	310	193	224
Trichloromethane	131*	–	28.9
Tetrachloromethane	53.2**	43.1	35.2
1,1,1-Trichloroethane	105	52.8	>530
Trichloroethylene	66.8	40.7	85.2
Tetrachloroethylene	21.4	18.4	17.7

* [1070],
** [1071]

Table 62. Tropospheric residence time of aliphatic chlorinated hydrocarbons

Compound	Time, a
Dichloromethane	0.23
Trichloromethane	0.33
Tetrachloromethane	> 30.0
1,1,1-Trichloroethane	2.7
Trichloroethylene	0.01
Tetrachloroethylene	0.18

Presence and Degradation in Air. During their use the common aliphatic chlorinated hydrocarbons escape into the atmosphere [1075]. Simple chlorinated hydrocarbons are destroyed in the troposphere, primarily by reaction of their hydrogen atoms or double bonds with hydroxyl radicals (HO·) that are naturally present in the troposphere. However, carbon tetrachloride and some of the fluorochloroalkanes (CFC 11, CFC 12, and CFC 113) resist significant attack by hydroxyl radicals; therefore, these materials have much longer half-lives than other common chlorinated solvents (Table 62) [1076]. The degradation of most commercially used chlorinated hydrocarbons in the atmosphere leads to HCl, chlorides, CO_2, and H_2O, which represent no significant threat to the general environment.

Atmospheric concentrations of chlorinated solvents are highly dependent on the site of measurement. In urban areas, the concentration of dichloromethane is $30-100 \times 10^{-6}$ mL/m^3 and the concentrations of 1,1,1-trichloroethane, trichloroethylene, or perchloroethylene are $100-600 \times 10^{-6}$ mL/m^3. In areas remote from industrial and populated regions, background concentrations are typically in the range of a few ppt (10^{-6} mL/m^3), except for 1,1,1-trichloroethane which has been reported to occur at $100-200 \times 10^{-6}$ mL/m^3 [1077].

Most of the traces of chlorinated hydrocarbons found in the atmosphere appear to be of anthropogenic origin. For methyl chloride, however, and possibly for some other halomethanes, the largest source is natural. Thus, methyl chloride results from forest fires, agricultural burning, chemical reactions in seawater, and possibly from marine plants [1078].

Table 63. Oral LD_{50}s of chlorobenzenes

Compound	LD_{50} (rat, oral), mg/kg
Monochlorobenzene	2 910
o-Dichlorobenzene	500
p-Dichlorobenzene	500
1,2,4-Trichlorobenzene	756
1,2,4,5-Tetrachlorobenzene	1 500
Pentachlorobenzene	1 080
Hexachlorobenzene	10 000

10.2. Chlorinated Aromatic Hydrocarbons

10.2.1. Chlorinated Benzenes

Acute Toxicity. Table 63 shows the LD_{50}'s of chlorinated benzenes. According to these data, the acute oral toxicity of chlorinated benzenes is low to moderate. Absorption through the skin seems to be minimal, but most of the compounds have some local irritant potency. In experimental animals, most chlorinated benzenes induce microsomal liver enzymes and cause porphyria, hypertrophy, and centrolobular necrosis of the liver. The chlorinated benzenes can induce kidney damage, changes in mucous membranes, and irritation of the upper respiratory tract, depending on the route and time of administration and on the dose applied. In addition, mono-, di-, and trichlorobenzenes are known to act as central nervous depressants, causing anesthesia and narcosis at higher doses [1059], [1060].

Chronic Effects. Hexachlorobenzene produced tumors in studies with mice and hamsters [1059]. There is no evidence for the carcinogenicity of the other chlorobenzenes. None of the compounds showed *mutagenic* activity in validated test systems. No data concerning *teratogenic* or *embryotoxic* effects of the chlorinated benzenes are available for most of the compounds. *p*-Dichlorobenzene has been tested in several species and produced no primary embryotoxic or teratogenic effects [1080].

Metabolism. Usually the chlorinated benzenes are partially hydroxylated to yield the corresponding phenols or are partially dechlorinated and then excreted in the urine and the feces. In contrast to rodents, sulfur-containing metabolites cannot be found in monkeys and humans [1059], [1081], [1082]. Only hexachlorobenzene has been shown to accumulate in animals and humans [1083], [1084].

Hexachlorobenzene has been suggested to produce *immunotoxic* effects in experimental animals, i.e., alterations of cell-mediated immune responses [1059].

Effects in Humans. Monochlorobenzene produced unconsciousness, vascular paralysis, and heart failure in a child after accidental oral uptake [1059]. *o*-Dichlorobenzene produced depression in conditioned reflex activity, demonstrating a cerebral–cortical effect. Erythropoiesis was significantly depressed. Symptoms of intoxication include headache, nausea, throat irritation, and stinging of the eyes. Skin irritation is also reported [1060].

No data are available on the effects of trichlorobenzenes in humans. Only minimal eye and throat irritation at 3–5 ppm in certain people are reported [1060]. An outbreak of cutanea tarda porphyria in Turkey in 1955 was attributed to the uptake of grain treated with hexachlorobenzene as fungicide [1059].

Regulations. The following occupational exposure limits have been established (1985):

Monochlorobenzene:	MAK 50 ppm	TLV 75 ppm
o-Dichlorobenzene:	MAK 50 ppm	TLV 50 ppm
p-Dichlorobenzene:	MAK 75 ppm	TLV 75 ppm
1,2,4-Trichlorobenzene:	MAK 5 ppm	

10.2.2. Chlorotoluenes

Little is known about the toxicity of chlorinated toluenes. The toxicity of o-chlorotoluene is considered to be relatively low and in the range of that of the chlorinated benzenes. The oral LD_{50} in rats is greater than or equal to 1600 mg/kg [1060]. Sublethal doses of o-chlorotoluene given orally to rats produce marked weakness; higher doses produce vasodilatation. Inhalation of 14 000 ppm for 6 h caused loss of coordination, vasodilatation, labored respiration, and narcosis in rats; 175 000 ppm was fatal to one of three rats [1060]. o-Chlorotoluene produces moderate skin irritation and conjunctival erythema in rabbits [1060].

p-Chlorotoluene induced no gene conversion in *Saccharomyces cerevisiae* [1085]. No data are available on chronic effects or effects on reproduction caused by chlorinated toluenes.

In *humans*, no cases of poisoning or skin irritation caused by chlorinated toluenes have been reported [1060].

Regulations. o-Chlorotoluene TLV 50 ppm

10.2.3. Polychlorinated Biphenyls

Acute Toxicity. The acute toxicity of mixtures of polychlorinated biphenyls (PCB) seems to depend on the chlorine content. Table 64 demonstrates the influence of the chlorine content in mixed isomers of PCBs, in addition to their relatively low acute oral toxicity.

The administration of acute or subacute doses results in liver enlargement, mainly due to enzyme induction; when the doses were increased, fatty degeneration and central atrophy of the liver occurred. In addition, hyperplasia and hemosiderosis of the spleen were also observed [1087], [1088]. Polychlorinated biphenyls are not likely to possess a

Table 64. Acute oral toxicity of PCBs

Chlorine content, wt %	LD_{50} (rat, oral), g/kg
21	3.98
32	4.47
42	8.65
48	11.0
62	11.3
68	10.9

substantial local irritating potential. Nevertheless, they seem to be readily absorbed through the skin, exerting such systemic effects as liver damage.

Chronic Effects. After oral application, severe liver damage (hypertrophy, fatty degeneration, and centrolobular necrosis) is most likely to be observed. The skin is also often affected (hyperplasia, hyperkeratosis, and cystic dilatation) [1065]. Polychlorinated biphenyls can interfere with heme metabolism as shown by an increased porphyrin content of the liver in rats [1065]. Hepatocellular tumors are produced in rats and mice after long-term oral application of PCBs [1062, vol. 20]. However, the tumor formation is regarded as a response to tissue damage rather than triggered by a genotoxic mechanism.

The PCBs failed to show positive response in validated *mutagenicity* test systems. The interference of polychlorinated biphenyls with reproduction could be demonstrated in numerous studies with mammals. The compounds pass through the placental barrier and exhibit embryotoxic effects [1065].

Absorption, Metabolism, and Excretion. Polychlorinated biphenyls are readily absorbed from the gastrointestinal tract after ingestion or from the lung after inhalation. The rates of metabolism and excretion decrease and the storage in body fat increases with increasing chlorine content. The compounds are generally metabolized by selective hydroxylation. In primates, most of the metabolites are excreted as conjugates in the urine, whereas excretion of free metabolites in the feces is the major route in rodents [1089].

Other Effects. Immunosuppressive action of polychlorinated biphenyls in mammals could be evidenced by a decrease in infectious resistance with atrophy of the spleen, cortical thymus atrophy, and dose-dependent decreased in the production of specific antibodies [1059], [1065]. In hens, growth retardation, high mortality, and subcutaneous edema could be observed. These findings were accompanied by focal necrosis of the liver, hydroperitoneum, and epicardial as well as lung edema (chicken edema disease) [1059], [1065].

Experience in Humans. Accidental acute intoxications with PCBs are not reported [1065]. With workers handling these compounds, acneform dermatitis was observed, in addition to liver damage with necrosis [1059], [1065].

In 1968, a subacute intoxication of more than 1000 people in Japan by contaminated rice oil was reported (Yusho disease). Initial symptoms were, for instance, swelling of the eye lids, fatigue, and gastrointestinal disturbances. Later on, discoloration of the

Table 65. Oral LD_{50}s of monochlorinated naphthalenes

Compound	Species	LD_{50} (oral), mg/kg
1-Chloronaphthalene	rats	1540
	mice	1019
2-Chloronaphthalene	rats	2078
	mice	886

skin and mucous membranes, headache, signs of sensory nerve injury, diarrhea, and jaundice were found. Cases of influence on human fetuses have been attributed to this high PCB exposure [1059], [1065].

Polychlorinated biphenyls accumulate in fat and adipose tissue. They have been demonstrated in human milk. Because the PCB level was found to be higher in infant blood but lower in umbilical blood in comparison to maternal blood, the transfer of PCBs via the milk is probably much more important than placental transfer [1059], [1090].

Regulations. The following exposure limits have been established:

Chlorine content 42%: MAK 0.1 ppm
Chlorine content 54%: MAK 0.05 ppm

Polychlorinated biphenyls are considered as possible teratogens [1091] and carcinogens [1065].

10.2.4. Chlorinated Naphthalenes

Monochlorinated naphthalenes are of low to moderate acute toxicity, as shown by their oral LD_{50} (Table 65). Subacute to subchronic uptake of mixtures of higher chlorinated naphthalenes (predominantly penta- and hexachloronaphthalene) resulted in liver injury [1059]. In general, the toxicity of chlorinated naphthalenes increases with the degree of chlorination [1060]. Chlorinated naphthalenes irritate the rabbit skin [1059]. Ingestion of lubricants containing chloronaphthalenes resulted in injury to farm animals (X disease). Marked hyperkeratosis of the skin, degenerations of the cells in pancreas, liver, and gall bladder, and damage of the renal cortex could be observed. Cattle poisoned with highly chlorinated naphthalenes show a rapid decline in vitamin A plasma levels [1059]. Octachloronaphthalene fed to rats also greatly enhances the loss of vitamin A from the liver [1092]. Mixtures of penta- and hexachloronaphthalenes can produce the so-called chicken edema disease, characterized by hydropericardium and ascites in chickens [1092].

1-Chloronaphthalene and 1,2,3,4-tetrachloronaphthalene, when tested for point mutations in the *Salmonella* assay (*Ames test*), exhibited no positive results [1093], [1094]. No data are available on the effects of chlorinated naphthalenes on reproduction.

Metabolism. Chlorinated naphthalenes are readily absorbed. Metabolism occurs by conjugation or via hydroxylation to the respective naphthols or dihydrodiols. The metabolites are excreted with the urine or the feces [1059], [1092].

Effects in Humans. The main health problem arising from use and handling of chlorinated naphthalenes is chloracne, which usually occurs from long-term contact with the compounds or exposure to hot vapors. The penta- and hexachloro derivatives are suggested to have the greatest potential to generate acne [1059].

In accidental intoxications, liver damage occurred independently from chloracne. After loss of appetite, nausea, and edema of the face and hands, abdominal pain and vomiting followed; later on jaundice developed. Autopsy in cases of fatal intoxication revealed yellow atrophy of the liver [1092].

Regulations. The following exposure limits have been established:

Trichloronaphthalene:	MAK 5 mg/m^3,
	TLV 5 mg/m^3
Tetrachloronaphthalene:	TLV 2 mg/m^3
Hexachloronaphthalene (skin):	TLV 0.2 mg/m^3
Pentachloronaphthalene:	TLV 0.5 mg/m^3
Octachloronaphthalene (skin):	TLV 0.1 mg/m^3

10.2.5. Benzyl Chloride

The acute oral toxicity (LD$_{50}$) of benzyl chloride in rats is 1231 mg/kg and in mice 1624 mg/kg [1079]. The subcutaneous LD$_{50}$ (in rats) of benzyl chloride in oil solution is 1000 mg/kg [1095]. Exposure of rats and mice to benzyl chloride concentrations of 100–1000 mg/m^3 for 2 h caused irritation of the mucous membranes and conjunctivitis [1062, vol. 11]. Benzyl chloride is a strong skin-sensitizing agent for guinea pigs [1096]. Benzyl chloride acts weakly *mutagenic* in validated test systems [1097], [1098].

Subcutaneous injection of weekly doses of 80 mg/kg for 1 year followed by a postobservation period resulted in local sarcomas with lung metastases in rats. The mean induction time was 500 d [1095]. After dermal application of benzyl chloride, skin carcinomas were observed in mice [1099].

Metabolism. Benzyl chloride is readily absorbed from the lungs and gastrointestinal tract. The compound reacts with tissue proteins after subcutaneous injection and is metabolized into *N*-acetyl-*S*-benzylcysteine [1096]. After oral administration, mercapturic acid and benzoic acid (free or conjugated with glycine) are excreted in the urine [1100].

Effects in Humans. A concentration of 16 ppm of benzyl chloride in air is reported to be intolerable to humans within 1 min. The compound is a potent lachrymator, strongly irritating to the eyes, nose, and throat and capable of causing lung edema [1060].

Regulations. The exposure limits of benzyl chloride are: MAK 1 ppm; TLV 1 ppm. Benzyl chloride should be considered as a possible carcinogen [1065].

Table 66. Acute toxicity of side-chain chlorinated xylenes

Compound	Species	Route	LD_{50}, mg/kg
m-Xylylene dichloride	mice	intravenous	100
o-Xylylene dichloride	mice	intravenous	320
p-Xylylene dichloride	rats	oral	1780

10.2.6. Benzoyl Chloride

Benzoyl chloride is of low acute oral toxicity in rats (LD_{50} 2529 mg/kg). It is more toxic by inhalation (LC_{50} 230 ppm, 4 h in male rats and 314 ppm, 4 h in female rats). The compound is irritating to skin, mucous membranes, eyes, and the respiratory tract [1101], [1102].

When benzoyl chloride or solutions of benzoyl chloride in benzene were applied to the skin of mice for up to 10 months irritation and keratinization resulted, and to some extent, ulceration and necrosis of the skin occurred. A few tumors (skin, lung) were observed in those mice [1099]. There is no clear evidence that benzoyl chloride is *mutagenic* [1097].

For *humans*, benzoyl chloride is classified as a lachrymator. It is irritating to the skin, eyes, and mucous membranes [1103]. The available data are inadequate to evaluate the carcinogenic potential of benzoyl chloride to humans [1062, vol. 29].

10.2.7. Benzotrichloride

The acute oral toxicity of benzotrichloride is 2180 mg/kg in male rats and 1590 mg/kg in female rats. The inhalative LC_{50}s are higher than 600 mg/m^3 in male rats and about 500 mg/m^3 in female rats after a 4-h exposure [1104]. Benzotrichloride irritates the skin and eyes [1105]. The compound proved to be *mutagenic* in bacterial test systems [1097]. Dermal application of benzotrichloride resulted in elevated tumor incidence in mice [1099].

In *humans*, benzotrichloride vapors are reported to be strongly irritating to the skin and mucous membranes [1103]. An increase in lung tumors has been reported in industrial plants that produce several chlorinated aromatic hydrocarbons [1106], [1107]. In the Federal Republic of Germany, and in Japan, benzotrichloride is considered as a possible carcinogen [1065], [1099].

10.2.8. Side-Chain Chlorinated Xylenes

Table 66 shows some acute toxicity data of side-chain-chlorinated xylenes. No data are available on other toxic effects in animals or humans.

11. References

General References

Ullmann, 4th ed., **9**, 404–420.
Winnacker-Küchler, 4th ed., vol. **6**; Organische Technologie II, pp. 2–11.
Kirk-Othmer, **5**, 668–714.
J. J. McKetta, W. Cunningham, *Encyl. Chem. Process. Des.* **8** (1979) 214–270.
C. R. Pearson: "C_1 and C_2-halocarbons," in: *The Handbook of Environmental Chemistry*, vol. **3**, Springer Verlag, Berlin 1982, pp. 69–88.
Chloroform, Carbon Tetrachloride and other Halomethanes: An Environmental Assessment, National Academy of Sciences, Washington, D.C., 1978.
A. Wasselle: *C_1 Chlorinated Hydrocarbons*, Private report by the Process Economic Program, SRI International, Report No. 126.
CEFIC-B.I.T. Solvants Chlorés "Methylene chloride – Use in industrial applications," 1983.
J. Schulze, M. Weiser: "Vermeidungs- und Verwertungsmöglichkeiten von Rückständen bei der Herstellung chlororganischer Produkte," Umweltforschungsplan des Bundesministers des Innern – Abfallwirtschaft. Forschungsbericht 103 01 304, UBA-FB 82-128 (1985).
Ullmann, 4th ed., **9**, pp. 420–464.
Kirk-Othmer, **5**, pp. 714–762; **23**, pp. 764–798, 865–885.
L. Scheflan: *The Handbook of Solvents*, D. van Nostrand, New York 1953.
G. Hawley: *The Condensed Chemical Dictionary*, Van Nostrand, New York 1977.
S. Patai: *The Chemistry of the C-Halogen Bond*, Part 1–**2**, J. Wiley and Sons, New York 1973.
J. L. Blackford: "Ethylene dichloride, 1,1,1-Trichloroethane, Trichloroethylene" in Chemical Economics Handbook, Stanford Research Institute, Menlo Park California 1975.
Process Economics Program, Rep. Ser. 5; Supplements 5 A – 5 C; *Vinyl Chloride*, Stanford Research Institute, Menlo Park, California 1965–1975.
J. S. Naworksi, E. S. Velez in B. E. Leach (ed.): "1,2-Dichloroethane," *Applied Industrial Catalysis*; Academic Press, New York 1983.
E. W. Flick: *Industrial Solvents Handbook*, 3rd ed., Noyes Data Corp., Park Ridge, N.J., 1985.
J. S. Sconce: *Chlorine, its Manufacture, Properties and Uses*, R. E. Krieger Publ. Co., Malabar, Fla., USA 1982.
Stanford Research Institute: *World Petrochemicals*, vol. **2**, Menlo Park, California, USA, 1984.
Stanford Research Institute: *Chem. Econ. Handbook, Marketing Research Report on PVC Resins*, Menlo Park, California, USA, 1982.
Stanford Research Institute: *Vinyl Chloride Report*, Menlo Park, California, USA, 1982.
Process Economic Program, *Chlorinated Solvents*, Rep.-No. 48; Stanford Research Institute, Menlo Park California 1969.
M. L. Neufeld et al., Market Input/Output Studies; *Vinylidene Chloride*, U.S. NTIS, PB-273205; Auerbach Corp., Philadelphia 1977.
J. A. Key et al., *Organic Chemical Manufacturing*, vol. **8**; U.S. EPA; EPA-450/3-80-028C; IT Envirosci. Inc., Knoxville 1980.
F. Asinger, *Die petrolchemische Industrie*, Akademie-Verlag, Berlin (GDR) 1971.

[1] *Beilstein*, Benzyl chloride, 5, 292, 5 (1), 151, 5 (2) 227, 5 (3), 685, 5 (4) 809.
[2] *Beilstein*, Benzalchloride 5, 297, 5 (1), 152, 5 (2), 232, 5 (3), 696, 5 (4), 817.
[3] *Beilstein*, Benzotrichloride, 5, 300, 5 (1), 152, 5 (2), 233, 5 (3), 699, 5 (4), 820.
[4] *Beilstein*, o-Xylene: 5, 364, 5 (1), 180, 5 (2), 283, 5 (3), 816, 5 (4), 926.

[5]   *Beilstein, m*-Xylene: 5, 373, 5 (1), 183, 5 (2), 291, 5 (3), 835, 5 (4), 943.
[6]   *Beilstein, p*-Xylene: 5, 384, 5 (1), 186, 5 (2), 299, 5 (3), 854, 5 (4), 966.
[7]   *Houben-Weyl,* Chlorine compounds: 5/3, 503;
[8]   *Houben-Weyl,* Photochemistry: 4/5 a, 1.
[9]   *Ullmann,* 4th ed., **3**, 305–319.
[10] *Kirk-Othmer* **5,** 828–838.

Specific References

[11] I. Mellan: *Industrial Solvents Handbook,* Noyes Data Corp., Park Ridge, N.J., 1970, p. 73.
[12] B. Kaesche-Krischer, *Chem. Ind. Tech.* **35** (1963) 856–860. H.-J. Heinrich, *Chem. Ing. Tech.* **41** (1969) 655.
[13] H. J. Schumacher, *Z. Elektrochem.* **42** (1936) 522.
[14] D. V. E. George, J. H. Thomas, *Trans Faraday Soc.* **58** (1962) 262.
[15] Du Pont, US 2 378 048, 1944 (C. W. Theobald).
[16] H. J. Schumacher, *Z. Angew. Chem.* **49** (1936) 613.
[17] A. T. Chapmann, *J. Am. Chem. Soc.* **57** (1935) 416.
[18] R. Neu, *Pharmazie* **3** (1948) 251.
[19] F. Lenze, L. Metz, *Chem. Ztg.* **56** (1932) 921.
[20] W. B. Grummet, V. A. Stenger, *Ind. Eng. Chem.* **48** (1956) 434.
[21] E. H. Lyons, R. G. Dickinson, *J. Am. Chem. Soc.* **57** (1935) 443.
[22] R. Johns, *Mitre Corp. Tech. Rep. MTR* **7144** (1976) Mc Lean, Va.; Mitre Corp.
[23] Celanese Corp., US 2 770 661, 1953 (T. Horlenko, F. B. Marcotte, O. V. Luke).
[24] United States Rubber Co., GB 627 993, 1946.
[25] R. M. Joyce, W. E. Hanford, *J. Am. Chem. Soc.* **70** (1948) 2529.
[26] W. L. Faith, D. B. Keyes, R. L. Clark: *Industrial Chemicals,* 3rd ed., J. Wiley & Sons, New York 1965, pp. 507–513.
[27] E. T. McBee et al., *Ind. Eng. Chem.* **41** (1949) 799–803.
[28] R. N. Pease et al., *J. Am. Chem. Soc.* **53** (1931) 3728–3737.
[29] W. E. Vaughan et al., *J. Org. Chem.* **5** (1940) 449–471.
[30] S. Tustev et al., *J. Phys. Chem.* **39** (1935) 859.
[31] F. Asinger: *Paraffins Chemistry and Technology,* Pergamon Press, New York 1968.
[32] T. Arai et al., *Kogyo Kagaku Zasshi* **61** (1958) 1231–1233.
[33] H. B. Hass, E. T. McBee et al., *Ind. Eng. Chem.* **34** (1942) 296–300.
[34] M. J. Wilson, A. H. Howland, *Fuel* **28** (1949) 127.
[35] G. Natta, E. Mantica, *J. Am. Chem. Soc.* **74** (1952) 3152.
[36] A. Scipioni, E. Rapsiardi, *Chim. Ind. Milan* **43** (1961) 1286–1293.
[37] S. Lippert, G. Vogel, *Chem. Tech. (Leipzig)* **21** (1969) 618–621.
[38] W. Hirschkind, *Ind. Eng. Chem.* **41** (1949) 2749–2752.
[39] B. E. Kurtz, *Ind. Eng. Chem. Process Des. Dev.* **11** (1972) 332–338.
[40] D. Kobelt, U. Troltenier, *Chem. Ing. Tech.* **38** (1966) 134.
[41] Stauffer Chem. Co., US 3 126 419, 1964 (W. M. Burks, R. P. Obrecht).
[42] H. Krekeler, W. Riemenschneider, *Chem. Ing. Tech.* **45** (1973) 1019–1021.
[43] D. Rebhahn, *Encycl. Chem. Process. Des.* **8** (1979) 206–214.
[44] E.-G. Schlosser, M. Rossberg, W. Lendle, *Chem. Ing. Tech.* **42** (1970) 1215–1219.
[45] Hoechst, EP 0 039 001, 1983 (W. Grünbein et al.).
[46] Dow Chem. Co., US 3 981 938, 1976 (J. M. Steel et al.).
[47] Dow Chem. Co., DE 2 640 852, 1978 (J. M. Steel et al.).

[48] Shinetsu Chem. Co., DE 2 447 551, 1974 (K. Habata, S. Tanaka, H. Araki).
[49] Sun Oil Co., US 3 236 754, 1962 (J. B. Bravo, R. Wynkoop).
[50] A. A. Schamschurin, *Zh. Obshch. Khim.* **9** (1939) 2207.
[51] R. F. Weinland, K. Schmidt, *Ber. Dtsch. Chem. Ges.* **38** (1905) 2327, 3696.
[52] A. Hochstetter, DE 292 089, 1914.
[53] Wacker Chemie, DE 1 211 622, 1964 (G. Künstle, H. Siegl).
[54] Olin Mathieson Chem. Corp., US 2 755 311, 1951 (H. Spencer).
[55] Hercules Powder Co., US 2 084 710, 1935 (H. M. Spurlin).
[56] Th. Goldschmidt, DE 1 150 964, 1961 (E. Ruf).
[57] Shinetsu Chem. Co., DE 3 146 526, 1981 (K. Habata, M. Shimizu, K. Ichikawa).
[58] Hoechst, DE 946 891, 1953.
[59] Stauffer Chem. Co., US 2 806 768, 1957 (H. Bender, R. P. Obrecht).
[60] Chem. Werke Hüls, DE 1 101 382, 1957 (R. Marin, H. Sauer).
[61] Hoechst, DE 2 137 499, 1973 (A. Bergdolt, H. Clasen, D. Houben).
[62] *Hydrocarbon Process. Pet. Refiner* **42** (1963) no. 11, 157.
[63] *Hydrocarbon Process. Pet. Refiner* **40** (1961) no. 11, 230–231.
[64] Stauffer Chem. Co., DE 1 443 156, 1961 (R. P. Obrecht, W. M. Burks).
[65] Solvay Cie., DE 2 106 886, 1971 (L. Forlano).
[66] M. Nagamura, H. Seya, *Chem. Econ. Eng. Rev.* **2** (1970) 34–38.
[67] Hoechst, DE 1 568 575, 1968 (M. Rossberg et al.).
[68] Stauffer Chem. Co., US 3 968 178, 1976 (R. P. Obrecht, M. J. Benett).
[69] IG Farbenind., DE 657 978, 1936 (K. Dachlauer, E. Schnitzler).
[70] S. Akliyama, T. Hisamoto, S. Mochizuku, *Hydrocarbon Process.* **60** (1981) no. 3, 76. Tokuyama Soda Co., JP Sho 55-43002, 1978 (S. Mochizuki, T. Hisamato, T. Hirashima).Tokuyama Soda Co., JP OLS Sho 56-2922, 1979 (S. Mochizuki, T. Hisamato).
[71] Lummus Co., US 3 949 010, 1976 (M. Sze).
[72] Diamond Alkali Co., US 2 792 435, 1954 (J. J. Lukes, W. J. Lightfoot, R. N. Montgomery).
[73] S. A. Miller, *Chem. Process. Eng.* **48** (1967) no. 4, 79–84.
[74] *Ullmann*, 3rd ed., **5,** 409.
[75] Food and Machinery Corp., US 3 109 866, 1963 (E. Saller et al.).
[76] Food and Machinery Corp., US 3 081 359, 1963 (E. Saller, R. Timmermann, C. J. Wenzke).
[77] Stauffer Chem. Co., US 3 859 425, 1975 (A. Wood, K. V. Darragh).
[78] *Ullmann*, 4th ed., **9,** 461–463.
[79] *Hydrocarbon Process. Pet. Refin.* **44** (1965) no. 11, 190.
[80] Halcon Chem. Co., BE 663 858, 1965 (J. W. Colton).
[81] Dow Chem. Corp., US 2 442 324, 1945 (R. G. Heitz, W. E. Brown).
[82] Stauffer Chem. Co., US 2 857 438, 1957 (R. P. Obrecht, H. Bender).
[83] Halcon Int. Co., GB 1 047 258, 1963.
[84] Soc. d'Ugine, FR 1 147 756, 1956.
[85] Solvay Cie., FR 1 373 709, 1963.
[86] Hoechst, DE 1 668 030, 1967 (H. Krekeler, H. Meidert, W. Riemenschneider, L. Hörnig). Hoechst, DE 1 668 074, 1968 (L. Hörnig, H. Meidert, W. Riemenschneider). Hoechst, DE 1 918 270, 1969 (H. Krekeler, H. Meidert, W. Riemenschneider, L. Hörnig). Hoechst, DE 2 218 414, 1972 (H. Gerstenberg, H. Krekeler, H. Osterbrink, W. Riemenschneider). Hoechst, DE 3 040 851, 1980 (D. Rebhan, H. Schmitz).
[87] Stauffer Chem. Co., US 3 126 419, 1960 (W. M. Burks, R. P. Obrecht).
[88] *Hydrocarbon Process. Pet. Refin.* **40** (1961) no. 11, 230–231.

[89] *Hydrocarbon Process. Pet. Refin.* **40** (1961) no. 11, 232.
[90] Sicedison SpA, FR 1 386 023, 1964.
[91] Pittsburgh Plate Glass Co., FR 1 355 886, 1963 (L. E. Bohl, R. M. Vancamp).
[92] B. F. Goodrich Co., US 3 314 760, 1962 (L. E. Trapasso).
[93] J. A. Thurlby, O. Stinay, *Chem. Anlagen Verfahren* **11** (1972) 65.
[94] E. Dönges, R. Kohlhaas et al., *Chem. Ing. Tech.* **38** (1966) 65.
[95] Gemeinsames Ministerialblatt 31/26, 430–452 (1980): Katalog wassergefährdender Stoffe.
[96] Amtsblatt der Europäischen Gemeinschaften C 176, 7–10 (1982): Liste von Stoffen, die gegebenenfalls in die Liste I der Richtlinie des Rates 76/464/EWG aufzunehmen sind.
[97] J. Schulze, M. Weiser, *Chem. Ind. (Düsseldorf)* **36** (1984) 347–353.
[98] C. R. Pearson: "C1 and C2 Halocarbons," in D. Hutzinger (ed.): *The Handbook of Environmental Chemistry*, vol. 3, B, Springer Verlag, Berlin 1982, pp. 69–88.
[99] J. E. Lovelock, *Nature (London)* **256** (1975) 193–194.
[100] NAS, *Chloroform, Carbon Tetrachloride and Other Halomethanes: An Environmental Assessment*, Washington, D.C., 1978.
[101] J. E. Lovelock, R. J. Maggs, R. J. Wade, *Nature (London)* **241** (1973) 194–196.
[102] Erste Allgemeine Verwaltungsvorschrift zur Reinerhaltung der Luft (TA-Luft) vom 28. 2. 1986 (GMBl. **37**, Nr. 7, S. 95–144).2. Verordnung zur Durchführung des Bundesimmissionsschutzgesetzes: Verordnung zur Begrenzung der Emission von leichtflüchtigen Halogenkohlenwasserstoffen vom 30. 4. 1986, Bundesgesetzblatt (1986) Nr. 17, S. 571–574.
[103] NAS, *Halocarbons: Environmental Effects of Chlorofluoromethane Release*, Washington, D.C., 1976.
[104] Ministerium für Ernährung, Landwirtschaft, Umwelt und Forsten, Baden-Württemberg: Leitfaden für die Beurteilung und Behandlung von Grundwasserverunreinigungen durch leichtflüchtige Chlorkohlenwasserstoffe, Heft **13** (1983) und **15** (1984) Stuttgart.
[105] Mitteilung des Landesamtes für Wasser und Abfall/Nordrhein-Westfalen, Ergebnisse der Gewässergüte-überwachung, Düsseldorf 1981.
[106] Internationale Arbeitsgemeinschaft der Wasserwerke im Rheineinzugsgebiet, Rheinbericht 1981/82, Amsterdam 1984.
[107] B. E. Rittmann, P. McCarty, *Appl. Environ. Microbiol.* **39** (1980) 1225–1226.
[108] W. Brunner, D. Staub, T. Leisinger, *Appl. Environ. Microbiol.* **40** (1980) 950–958.
[109] G. M. Klećka, *Appl. Environ. Microbiol.* **44** (1982) 701.
[110] G. Stucki et al., *Arch. Microbiol.* **130** (1981) 366–371.
[111] H. H. Tabak et al., *J. Water Pollut. Control Fed.* **53** (1981) no. 10, 1503–1518.
[112] J. Halbartschlager et al., *GWF Gas Wasserfach: Wasser/Abwasser* **125** (1984) no. 8, 380–386.
[113] T. Leisinger, *Experientica* **39** (1983) 1183–1191.
[114] Bundesgesundheitsblatt **22** (1979) 102.
[115] *SRI International*, Report no. 126: C_1-Chlorinated Hydrocarbons.
[116] K. Eisenächer, *Chem. Ing. Tech.* **55** (1983) 786.
[117] H. Giesel, *Nach. Chem. Techn. Lab.* **32** (1984) 316.
[118] W. Swodenk, *Chem. Ing. Tech.* **55** (1983) 683.
[119] J. Schulze, M. Weiser, *Chem. Ind. (Berlin)* **36** (1984) 347.
[120] J. D. Schmitt-Tegge, *Chem. Ing. Tech.* **55** (1983) 342.
[121] L. H. Horsley, *Azeotropic Data* 1 and 2, no. 6, 35; Adv. Chem. Ser., ACS, Washington, D.C., 1952, 1962.
[122] Ethyl Corp., US 2 818 447, 1953 (C. M. Neher).
[123] Halcon International Inc., US 3 345 421, 1961 (D. Brown).

[124] Shell Dev. Corp., US 2 742 511, 1950 (E. P. Franzen).
[125] Dow Chemical, US 2 031 288, 1935.
[126] Ethyl Corp., US 2 522 687, 1948 (F. L. Padgitt, G. F. Kirby).
[127] Dow Chemical, US 2 469 702, 1946 (C. C. Schwegler, F. M. Tennant).
[128] Dow Chemical, US 2 140 927, 1936 (J. E. Pierce).
[129] N. N. Lebedew, *Z. Obshch. Chim.* **24** (1954) 1959.
[130] Pure Oil Co., CA 464 069, 1950 (D. C. Bond).
[131] Marathon Oil Co., US 3 3240 902, 1966 (D. H. Olson, G. M. Bailey).
[132] Hoechst, DE-OS 2 026 248, 1970 (H. Großpietsch, H.-J. Arpe, L. Hörnig).
[133] Shell Development Co., US 2 246 082, 1939 (W. E. Vaughan, F. F. Rust).A. W. Fleer et al., *Ind. Eng. Chem.* **47** (1955) 982. *Pet. Refiner* **34** (1955) 149.*Hydrocarbon Process. Pet. Refiner* **44** (1965) 201.
[134] ICI, GB 667 185, 1949 (P. A. Hawkins, R. T. Foster).
[135] Ethyl Co., US 2 838 579, 1958 (F. Conrad, M. L. Gould, C. M. Neher).
[136] Dow Chemical, US 2 140 547, 1936 (J. H. Reilly).
[137] I.G. Farbenindustrie AG, GB 483 051, 1936 (G. W. Johnson).
[138] Standard Oil Develop. Co., US 2 393 509, 1946 (F. M. Archibald, H. O. Mottern). Ethyl Co., US 2 589 698, 1952 (H. O. Mottern, J. P. Russel).Air Reduction Co. Inc., US 3 506 553, 1970 (L. J. Governale, J. H. Huguet, C. M. Neher).
[139] Dow Chemical, US 2 516 638, 1947 (J. L. McCurdy). Société anon. des Matières Colorantes et Produits Chimiques de Saint-Denis, FR 858 724, 1940.
[140] ICI, GB 1 134 116, 1966 (S. C. Carson).
[141] H. Bremer et al., DD 85 068, 1971.
[142] Socony Mobil Oil Co. Inc., US 3 499 941, 1970 (E. N. Givens, L. A. Hamilton).
[143] Hoechst, DE-OS 1 919 725, 1970 (H. Fernholz, H. Wendt).
[144] ICI, BE 654 985, 1965.
[145] Distillers Co. Ltd., GB 566 174, 1942 (E. G. Galitzenstein). A. P. Giraitis, *Erdoel Kohle* **9** (1951) 971.
[146] Ethyl Corp., US 2 681 372, 1951 (P. W. Trotter).
[147] I.G. Farbenindustrie AG, GB 470 817, 1936 (G. W. Johnson).
[148] Hercules Powder Co., US 2 084 710, 1935 (H. M. Spurlin).
[149] Exxon, US 4 041 087, 1977 (M. A. Vannice).
[150] G. A. Olah, EP 73 673, 1983.
[151] Dow Chemical, CA 561 327, 1954 (J. H. Brown, W. E. Larson).
[152] BP Chemicals Ltd., GB 2 002 357, 1978 (D. A. Burch, E. J. Butler, C. W. Capp).
[153] Ethyl Corp., DE 1 518 766, 1965 (A. O. Wikman, L. B. Reynolds, Jr.).
[154] Dynamit Nobel, DE 1 568 371, 1966 (R. Stephan, H. Richtzenhain).
[155] R. N. Rothan, E. W. Sims, *Chem. Ind. (London)* 1970, 830.
[156] Donau Chemie, AT 163 818, 1947 (O. Fruhwirth). Donau Chemie, AT 170 262, 1950 (O. Fruhwirth).Columbia Southern Chemical, US 2 945 897, 1958 (D. H. Eisenlohr).
[157] Vulcan Materials Co., DE 1 668 842, 1963 (J. I. Jordan, H. S. Vierk).
[158] Stauffer, US 3 987 118, 1976 (M. A. Kuck). Union Carbide, US 3 427 359, 1969 (C. E. Rectenwald, G. E. Keller II, J. W. Clark).
[159] Donau Chemie, AT 163 218, 1947 (G. Gorbach).*Chem. Eng. News* **25** (1947) 1888. Donau Chemie, AT 166 251, 1949 (O. Fruhwirth).
[160] K. A. Holbrook et al., *J. Chem. Soc. B* 1971, 577.
[161] S. Inokawa et al., *Kogyo Kagaku Zasshi* **67** (1964) 1540.

[162] W. L. Archer, E. L. Simpson, *I and EC Prod. Res. Dev.* **16** (1977) 158.
[163] Jefferson Chemical Co., US 2 601 322, 1952 (R. R. Reese).
[164] Tokuyama Soda Co., JP 41 3168, 1963.
[165] Bayer, DE 1 157 592, 1961 (H. Rathjen, H. Wolz).
[166] Dynamit Nobel AG, DE-OS 2 156 190, 1971 (E. Feder, K. Deselaers).
[167] Union Carbide, US 2 929 852, 1954 (D. B. Benedict). Bayer, GB 960 083, 1962. Wacker, DE-OS 1 668 850, 1967 (R. Sieber, A. Maier); DE-OS 1 768 367, 1968 (O. Fruhwirth, L. Schmidhammer, E. Pichl).
[168] Dynamit Nobel AG, DE-OS 2 128 329, 1974 (A. Hoelle).Bayer, DE-OS 2 743 975, 1979 (S. Hartung); DE-OS 1 905 517, 1969 (R. Wesselmann, W. Eule, E. Köhler).
[169] S. N. Balasubramanian, *Ind. Eng. Chem. Fundam.* **5** (1966) 184.
[170] J. C. Vlugter et al., *Chim. Ind. (Milan)* **33** (1951) 613. *Hydrocarbon Process.* **44** (1965) 198. Knapsack AG, DE-OS 1 905 517, 1969.
[171] Società Italiana Resine S.p.A., US 3 911 036, 1975 (L. Di Fiore, B. Calcagno). Union Carbide, US 2 929 852, 1960 (D. B. Benedict).
[172] Lummus Co., US 3 917 727, 1975 (U. Tsao);US 3 985 816, 1976 (U. Tsao). Allied, US 3 941 568, 1976 (B. E. Kurtz, A. Omelian). Stauffer, DE-OS 2 652 332, 1984 (R. G. Campbell, W. E. Knoshaug).
[173] E. Lundberg, *Kem. Tidskr.* **10** (1984) 35. Hoechst, EP 80 098, 1984 (J. Hundeck, H. Hennen).
[174] B. F. Goodrich, GB 1 233 238, 1971 (R. C. Campbell). Stauffer, US 4 000 205, 1976.
[175] Dynamit Nobel AG, DE-OS 3 340 624, 1984 (H. Leuck, H.-J. Westermann).
[176] C. M. Schillmoller, *Hydrocarbon Process.* **3** (1979) 77.
[177] R. Remirez, *Chem. Eng. (N.Y.)* **75** (1968) no. 9, 142. Kureha Chemical Ind., NL 6 504 088, 1965.
[178] ICI, DE-OS 2 012 898, 1970 (N. Colebourne, P. R. Edwards).
[179] F. F. Braconier, *Hydrocarbon Process.* **43** (1964) no. 11, 140. Société Belge de l'Azote, GB 954 791, 1959 (F. F. A. Braconier, H. Le Bihan).
[180] Shell Development Co., US 2 099 231, 1935 (J. D. Ruys, J. W. Edwards).A. A. F. Maxwell, US 2 441 287, 1944.
[181] N. N. Semenov: *Some Problems in Chemical Kinetics and Reactivity,* vol. **1,** Princeton University Press, Princeton 1958, p. 211.L. F. Albright, *Chem. Eng. (N.Y.)* **74** (1967) no. 7, 123.
[182] Montecatini, IT 755 867, 1967 (C. Renato, F. Gianfranco, C. Angelo).
[183] R. V. Carrubba, *Thesis Columbia University,* 1968.R. P. Arganbright, W. F. Yates,*J. Org. Chem.* **27** (1962) 1205. H. Heinemann, *Chem. Tech. (Heidelberg)* **5** (1971) 287.
[184] R. W. McPherson et al., *Hydrocarbon Process.* **3** (1979) 75.
[185] J. A. Buckley, *Chem. Eng. (N.Y.)* **73** (1966) no. 29, 102.P. H. Spitz, *Chem. Eng. Prog.* **64** (1968) no. 3, 19.E. F. Edwards, F. Weaver, *Chem. Eng. Prog.* **61** (1965) no. 1, 21. E. M. De Forest, S. E. Penner, *Chem. Eng. (N.Y.)* **79** (1972) no. 17, 54.
[186] D. Burke, R. Miller, *Chem. Week* **5** (1964) no. 8, 93.
[187] Y. Onoue, K. Sakurayama, *Chem. Eng. Rev.* **4** (1969) no. 11, 17.
[188] L. F. Albright, *Chem. Eng. (N.Y.)* **74** (1967) no. 8, 219.
[189] E.g. The Distillers Company Ltd., DE-OS 1 443 703, 1970 (A. F. Millidge, C. W. Capp, P. E. Waight).
[190] Mitsui Toatsu Chemical, GB 1 189 815, 1970 (K. Miyauchi et al.).
[191] VEB Chemische Werke Buna, DD 157 789, 1982 (J. Koppe et al.).
[192] BASF, DE-OS 2 651 974, 1978 (P. R. Laurer, J. Langens, F. Gundel). Società Italiana Resine S.p.A., DE-OS 2 543 918, 1976 (R. Canavesi, F. Ligorati, G. Aglietti). Produits Chimiques

[193] Pechiney-Saint-Gobain, US 3 634 330, 1972 (M. Michel, G. Benaroya, R. Jacques). Vulcan Materials Co., US 3 926 847, 1975 (W. Q. Beard, Jr., P. H. Moyer, S. E. Penner). Diamond Shamrock Corp., BE 859 878, 1978 (F. C. Leitert, C. G. Vinson, Jr.). M. W. Kellog Co., US 3 114 607, 1963 (T. H. Milliken). B. F. Goodrich Co., GB 938 824, 1961. Distillers Co., GB 932 130, 1961 (C. W. Capp, D. J. Hadley, P. E. Waight); GB 971 996, 1962. Chem. Werke Hüls AG, FR 1 421 903, 1965.

[193] Shell Oil Co., US 3 892 816, 1975 (A. T. Kister).

[194] FMC, US 3 360 483, 1963 (L. H. Diamond, W. Lobunez).Shell Oil Co., US 3 210 431, 1965 (W. F. Engel). Stauffer, US 3 657 367, 1972 (R. J. Blake, G. W. Roy).

[195] Petro-Tex Chemical Corp., US 4 025 461, 1977 (L. J. Croce, L. Bajars, M. Gabliks); 4 046 821, 1977 (L. J. Croce, L. Bajars, M. Gabliks).

[196] Toyo Soda Ltd., GB 1 016 094, 1963.

[197] Hoechst, FR 1 440 450, 1965.

[198] B. F. Goodrich Co., US 3 488 398, 1970 (J. W. Harpring et al.).

[199] E. Gorin et al., *Ind. Eng. Chem.* **40** (1948) 2128.

[200] Vulcan Materials Co., GB 980 983, 1963 (US 3 184 515, 1962) (S. E. Penner, E. M. De Forest).

[201] Dow Chemical, US 2 866 830, 1956 (J. L. Dunn, Jr., B. Posey, Jr.).

[202] Pechiney, FR 1 286 839, 1961 (F. Lainé, G. Wetroff, C. Kaziz).

[203] Stauffer, US 4 206 180, 1980 (R. G. Campbell et al.).

[204] Toa Gosei Chemical Ind., US 3 699 178, 1968 (S. Yoshitaka, T. Atsushi, K. Hideo).

[205] *Chem. Week* **99** (1966) no. 20, 56.W. A. Holve et al., *Chem. Anlage + Verfahren* **11** (1972) 69.

[206] B. F. Goodrich Co., US 4 226 798, 1980 (J. A. Cowfer et al.). PPG, GB 1 220 394, 1968 (A. P. Muren, L. W. Piester, R. M. Vancamp).

[207] PPG, US 3 679 373, 1972 (R. M. Vancamp, P. S. Minor, A. P. Muren, Jr.).

[208] B. F. Goodrich Co., DE-OS 1 518 930, 1965 (J. W. Harpring et al.); 1 518 933, 1965 (J. W. Harpring, A. E. van Antwerp, R. F. Sterbenz).

[209] Pechiney, GB 959 244, 1962; US 3 190 931, 1965 (F. Lainé, C. Kaziz, G. Wetroff).

[210] Dow Chemical, US 3 966 300, 1976.

[211] PPG, GB 2 119 802, 1983 (J. S. Helfand, T. G. Taylor).

[212] Hoechst, DE-OS 2 300 844, 1974 (W. Kühn et al.). Dow Chemical, FR 2 134 845, 1972; CA 941 329, 1974 (J. E. Panzarella). BASF, DE-OS 2 400 417, 1975 (D. Lausberg). Hooker Chemical Co., NL 6 614 522, 1967.

[213] Mitsui Toatsu Chemical, JP 45-32406, 1970; JP 46-43367, 1971; JP 46-33010, 1971; GB 1 189 815, 1970 (K. Miyauchi et al.). PPG, FR 1 220 394, 1971. P. Reich, *Hydrocarbon Process.* 3(1976) 85. R. G. Markeloff, *Hydrocarbon Process.* **11** (1984) 91.

[214] Mitsubishi Chemical, DE-OS 2 422 988, 1974 (Y. Kageyama).

[215] Stauffer, US 3 892 816, 1972 (A. T. Kister).

[216] W. E. Wimer, R. E. Feathers, *Hydrocarbon Process.* **3** (1976) 83.

[217] *Chem. Eng. News* **44** (1966) 76. *Chem. Week* **99** (1966) no. 13, 93.

[218] M. L. Spektor et al., *Chem. Eng. News* **40** (1966) no. 44, 76; *Ind. Eng. Chem. Process Des. Dev.* **6** (1967) no. 3, 327.H. Heinemann, *Erdöl Kohle Erdgas Petrochem.* **20** (1967) no. 6, 400.F. Friend et al., *Adv. Chem. Ser.* **70** (1968) 168. *Chem. Ind. (Düsseldorf)* **19** (1967) no. 3, 124.

[219] Distillers Co., GB 958 458, 1962 (C. W. Capp, D. J. Hadley, P. E. Waight). Union Carbide, BE 664 903, 1965 (J. W. Clark et al.). PPG, NL 6 401 118, 1964. National Distillers and Chemical Corp., US 3 720 723, 1973 (E. G. Pritchett). Pullman Inc., US 3 159 455, 1964 (G. T. Skaperdas, W. C. Schreiner, S. C. Kurzius); 3 536 770, 1970 (G. T. Skaperdas, W. C. Schreiner). Ethyl Corp., FR 1 398 254, 1964 (M. D. Roof).

[220] Monsanto, NL 6 515 254, 1966.
[221] Monsanto, NL 6 151 253, 1966.
[222] Allied, DE-OS 2 449 563, 1975 (B. E. Kurtz et al.).
[223] Shell Development Co., US 2 442 285, 1948 (H. A. Cheney).
[224] N. Singer, DD 110 032, 1974.
[225] D. H. R. Barton, P. F. Onyon, *J. Am. Chem. Soc.* **72** (1950) 988.
[226] Ethyl Corp., US 2 765 350, 1955 (F. Conrad).
[227] S. Okazakiand, M. Komata, *Nippon Kagaku Zaishi* 1973, no. 3, 459.
[228] N. K. Taikova et al., *Zh. Org. Khim.* **4** (1968) 1880.
[229] I. Mochida, Y. Yoneda, *J. Org. Chem.* **33** (1968) 2161.
[230] W. B. Crummett, V. A. Stenger, *Ind. Eng. Chem.* **48** (1956) 434.
[231] W. L. Archer, E. L. Simpson, *I and EC Prod. Res. and Dev.* **16** (1977) no. 2, 158.
[232] L. Bertrand et al., *Int. J. Chem. Kinet.* **3** (1971) 89.
[233] Dow Chemical, US 1 870 601, 1932 (E. C. Britton, W. R. Reed).
[234] Ethyl Corp., US 3 019 175, 1959 (A. J. Haefner, F. Conrad). Pechiney-Saint Gobain, FR 1 390 398, 1964 (A. Antonini, C. Kaziz, G. Wetroff); Dynamit Nobel, DE-OS 2 026 671, 1970 (H. Richtzenhain, R. Stephan).
[235] Z. Prāsil, *Radiochem. Radioanal. Lett.* **38** (1979) 103.
[236] T. Migita et al., *Bull. Chem. Soc. Japan* **40** (1967) 920; M. Kosugi et al., *Bull. Chem. Soc. Japan* **43** (1970) 1535; T. N. Bell et al., *J. Phys. Chem.* **38** (1979) 2321.
[237] Dow Chemical, GB 2 121 416, 1983 (J. C. Stevens, J. Perettie).
[238] ICI, DE-OS 2 835 535, 1979 (C. S. Allen).
[239] A. P. Mantulo et al., DE-OS 3 033 899, 1982; FR 8 020 077, 1980.
[240] ICI, DE-OS 1 950 995, 1969 (A. Campbell, R. A. Carruthers).
[241] Ethyl Corp., GB 843 179, 1963.Monsanto, US 3 138 643, 1961 (K. M. Taylor, G. L. Wofford). PPG, US 3 059 035, 1960 (F. E. Benner, D. H. Eisenlohr, D. A. Reich). ICI, DE-OS 2 002 884, 1970 (A. Campbell, R. A. Whitelock).
[242] Ethyl Corp., GB 1 106 533, 1965; DE-OS 1 518 766, 1965 (A. O. Wikman, L. B. Reynolds).
[243] ICI, DE-OS 1 950 996, 1969 (A. Campbell, R. A. Carruthers).
[244] Montecatini Edison SpA, GB 1 170 149, 1967; FR 1 524 759, 1967 (G. Pregaglia, B. Viviani, M. Agamennone).
[245] Dow Chemical, US 3 971 730, 1976 (G. L. Kochanny, Jr., T. A. Chamberlin); 3 872 176, 1975 (G. L. Kochanny, Jr., T. A. Chamberlin).
[246] Consortium f. Elektrochem. Ind., US 1 921 879, 1933 (W. O. Herrmann, E. Baum).Saint Gobain, US 2 674 573, 1949 (M. J. L. Crauland).Distillers Comp., US 2 378 859, 1942 (M. Mugdan, D. H. R. Barton).
[247] I.G. Farbenindustrie, GB 349 872, 1930.Bataafsche Petroleum M., GB 638 117, 1948.
[248] Dow Chemical, US 2 610 214, 1949 (J. L. Amos).
[249] Ethyl Corp., US 2 989 570, 1959 (F. Conrad, M. L. Gould).
[250] PPG, DE-OS 1 443 033, 1961 (H. J. Vogt). BASF, DE-OS 1 230 418, 1961 (H. Ostermayer, W. Schweter). Dynamit Nobel, GB 997 357, 1962. Feldmühle AG, GB 893 726, 1962.
[251] Dow Chemical, US 2 209 000, 1937 (H. S. Nutting, M. E. Huscher).Solvay, BE 569 355, 1961.Dynamit Nobel, DE-OS 1 231 226, 1963 (R. Stephan, H. Richtzenhain).
[252] FMC, US 3 776 969, 1973 (W. Lobunez).
[253] Dynamit Nobel, DE-OS 1 235 878, 1963 (R. Stephan).
[254] M. D. Rosenzweig, *Chem. Eng. (N.Y.)* **10** (1971) 105.

[255] Vulcan Materials Co., DE-OS 1 518 166, 1963 (J. I. Jordan, Jr., H. S. Vierk); US 3 304 337, 1967 (J. I. Jordan, Jr., H. S. Vierk).

[256] Vulcan Materials Co., DE-OS 2 046 071, 1970 (K. F. Bursack, E. L. Johnston).

[257] Ethyl Corp., US 3 012 081, 1960 (F. Conrad, A. J. Haefner). Société d'Ugine, FR 1 514 963, 1966.

[258] Detrex Chem. Ind. Inc., CA 116 460, 1971 (C. E. Kircher).

[259] Ugine Kuhlmann, DE-OS 1 668 760, 1971 (A. Goeb, J. Vuillement).

[260] W. G. Rollo, A. O'Grady, *Can. Paint and Finish.* **10** (1973) 15. W. L. Archer, *Met. Prog.* **10** (1974) 133.R. Monahan, *Met. Finish.* **11** (1977) 26. P. Goerlich, *Ind.-Lackier-Betr.* **43** (1975) 383.J. C. Blanchet, *Surfaces* **14** (1975) 51. L. Skory et al., *Prod. Finish. (Cincinnati)* **38** (1974). H. A. Farber, G. P. Souther, *Am. Dyest. Rep.* **57** (1968) 934; G. P. Souther, *Am. Dyest. Rep.* **59** (1970) 23. J. J. Willard, *Text. Chem. Color.* **4** (1972) 62. H. Hertel, *Kunststoffe* **71** (1981) 240.

[261] Dow Chemical, US 2 838 458, 1955 (H. J. Bachtel); 2 923 747, 1958 (D. E. Rapp); 2 970 113, 1957 (H. J. Bachtel); 3 049 571, 1960 (W. E. Brown); 3 364 270, 1965 (M. J. Blankenship, R. McCarthy); 3 384 673, 1966 (M. J. Blankenship, R. McCarthy); 3 444 248, 1969 (W. L. Archer, E. L. Simpson, G. R. Graybill); 3 452 108, 1969 (W. L. Archer, E. L. Simpson, G. R. Graybill); 3 452 109, 1969 (W. L. Archer, E. L. Simpson, G. R. Graybill); 3 454 659, 1969 (W. L. Archer, E. L. Simpson); 3 467 722, 1967 (W. L. Archer, G. R. Graybill); 3 468 966, 1969 (W. L. Archer, E. L. Simpson); 3 472 903, 1969 (W. L. Archer, E. L. Simpson); 3 546 305, 1970 (W. L. Archer, E. L. Simpson); 3 681 469, 1972 (W. L. Archer, E. L. Simpson); 4 469 520, 1984 (N. Ishibe, W. F. Richey, M. S. Wing); GB 916 129, 1961; NL 6 919 176, 1969. Ethyl Corp., US 3 002 028, 1958 (A. J. Haefner, L. L. Sims); 3 060 125, 1958 (L. L. Sims); 3 074 890, 1958 (G. N. Grammer); 3 189 552, 1963 (L. L. Sims); 3 238 137, 1961 (G. N. Grammer, P. W. Trotter); 3 629 128, 1968 (J. H. Rains). Vulcan Materials Co., FR 1 369 267, 1963. Solvay Cie., BE 743 324, 1969; 755 668, 1970. PPG, US 3 000 978, 1959 (R. H. Fredenburg); 3 070 634, 1960 (D. E. Hardies, B. O. Pray); 3 128 315, 1961 (D. E. Hardies); 3 192 273, 1961 (W. E. Bissinger); 3 281 480, 1961 (D. E. Hardies); BE 613 661, 1962 (D. E. Hardies). ICI, FR 1 372 972, 1963; 1 372 973, 1963; DE 1 243 662, 1962 (P. Rathbone, Ch. W. Suckling); 1 941 007, 1969 (A. Campbell, P. Robinson, J. W. Tipping); US 3 336 234, 1964 (J. H. Speight); GB 1 261 270, 1969 (G. Marsden, J. W. Tipping). Dynamit Nobel, FR 1 371 679, 1967; 1 393 056, 1965; DE 1 246 702, 1962 (R. Stephan, H. Richtzenhain). PCUK, FR 1 555 883, 1967 (J. Vuillemenot). Diamond Shamrock Corp., DE-OS 2 102 842, 1971 (N. L. Beckers, E. A. Rowe, Jr.).

[262] D. H. R. Barton, *J. Chem. Soc.* 1949, 148.

[263] A. Suzuki et al., *Kogyo Kagaku Zasshi* **69** (1966) 1903.

[264] PPG, US 3 344 197, 1967 (Ch. R. Reiche, J. M. Jackson).

[265] Toa Gosei Chemical Ind. Co., DE-OS 1 944 212, 1969 (Y. Suzuki, R. Saito); 2 000 424, 1970 (T. Kawaguchi et al.).

[266] J. C. Martin, E. H. Drew, *J. Am. Chem. Soc.* **83** (1961) 1232; M. L. Poutsma, *J. Am. Chem. Soc.* **85** (1963) 3511. Y. H. Chua et al., *Mech. Chem. Eng. Trans.* **7** (1971) 6.

[267] ICI, GB 1 388 660, 1972 (J. S. Berrie, I. Campbell). Solvay and Cie., DE-OS 2 505 055, 1984.

[268] Bataafsche Petroleum, GB 627 119, 1947. Shell Development Co., US 2 621 153, 1947 (R. H. Meyer, F. J. F. van der Plas). Toa Gosei Chem. Ind., DE-OS 2 000 424, 1970. Dow Chemical, US 2 174 737, 1963 (G. H. Coleman, G. V. Moore).

[269] British Celanese Ltd., GB 599 288, 1945 (P. J. Thurman, J. Downing).

[270] Dow Chemical, US 2 140 549, 1937 (J. H. Reilly).

[271] PPG, US 3 065 280, 1960 (H. J. Vogt). Hoechst, FR 1 318 225, 1962. PPG, US 3 173 963, 1960 (Ch. R. Reiche, H. J. Vogt). FMCI, FR 1 491 902, 1966 (S. Berkowitz, A. R. Morgan, Jr.).

[272] Produits Chimiques Péchiney-Saint-Gobain, FR 1 552 820, 1969 (G. Benaroya, M. Long, F. Lainé); 1 555 518, 1969 (A. Antonini, P. Joffre, F. Lainé). Rhône-Poulenc Ind., US 4 057 592, 1977 (A. Antonini, P. Joffre, F. Lainé).

[273] Consortium Elektrochem. Ind., FR 690 767, 1930. I.G. Farben, FR 690 655, 1930; DE 489 454, 1927. British Celanese Ltd., GB 571 370, 1943 (P. J. Thurman, J. Downing). Hoechst, FR 1 318 225, 1962.

[274] Asahi Kasei Kogyo, FR 1 417 810, 1964.Vulcan Materials Corp., GB 980 983, 1963. Central Glass Co., JP 42-9924, 1965.

[275] Produits Chimiques Péchiney-Saint-Gobain, FR 1 552 821, 1969 (A. Antonini, G. du Crest, G. Benaroya); 1 552 824, 1969 (A. Antonini, P. Joffre, C. Vrillon).

[276] Donauchemie, DE 865 302, 1944 (O. Fruhwirth). Du Pont, US 2 461 142, 1944 (O. W. Cass).

[277] Knapsack-Griesheim AG, DE 939 324, 1954 (K. Sennewald, F. Pohl, H. Westphal).

[278] D. H. R. Barton, K. E. Howlett, *J. Chem. Soc.* 1951, 2033.

[279] Toa Gosei Chem. Ind. Ltd., FR 2 057 606, 1971; US 3 732 3422, 1973 (T. Kawaguchi). Asahi Glass Co., JP 75-34 0003, 1975.Detrex Chemical Ind., US 3 304 336, 1967 (W. A. Callahan). D. Gillotay, J. Olbregts, *Int. J. Chem. Kinet.* **8** (1976) 11.FMC, DE-OS 1 928 199, 1969 (S. Berkowitz).

[280] V. A. Poluektov et al., SU 195 445, 1976.

[281] I.G. Farben DE 530 649 (1929).

[282] L. E. Horsley, Azeotropic Data in *Adv. Chem. Ser.* **6,** Am. Chem. Soc., Washington, D.C., 1952.

[283] S. Tsuda, *Chem. Eng. (N.Y.)* **74** (May 1970).

[284] ICI, FR 2 003 816, 1969.

[285] K. S. B. Prasa, L. K. Doraiswamy,*J. Catal.* **32** (1974) 384; N. N. Lebedev et al., *Kinet. Katal.* **12** (1971) 560; J. A. Pearce, *Can. J. Res.* **24 F** (1946) 369.

[286] I.G. Farben AG, FR 836 979, 1939.

[287] G. Brundit et al., Bios Report, 1056 (1947). G. B. Carpenter, Fiat Final Report, 843 (1947).

[288] Wacker, DE 733 750, 1940.

[289] Hooker Chemical Co., NL 6 614 522, 1965.

[290] T. Kawaguchi et al., *Ind. Eng. Chem.* **62** (1970) 36.

[291] D. S. Caines et al., *Aust. J. Chem.* **22** (1969) 1177; R. G. McIver, J. S. Ratcliffe, *Trans. Inst. Chem. Eng.* **51** (1973) 68.

[292] Toa Gosei Chem. Ind. Co. Ltd., FR 2 057 605, 1971.

[293] Toa Gosei Chem. Ind. Co. Ltd., FR 2 057 604, 1971.

[294] Produits Chimiques, Péchiney-Saint-Gobain S.A., DE-OS 1 964 552, 1975 (J. C. Strini, Y. Correia); 1 964 551, 1975 (Y. Correia); US 4 148 832, 1979 (Y. Correia); DE-OS 1 817 193, 1975 (Y. Correia, J. C. Strini); 1 817 191, 1975 (Y. Correia, J. C. Strini).

[295] Kanto Denka Kogyo Ltd., DE-OS 1 568 912, 1966 (A. Suzuki et al.).

[296] Wacker, DE-OS 2 023 455, 1971 (L. Schmidhammer, O. Fruhwirth); 2 320 915, 1974 (L. Schmidhammer, D. Dempf, O. Sommer); 2 239 052, 1974 (S. Nitzsche, L. Schmidhammer).

[297] Monsanto Chem. Co., US 2 846 484, 1954 (J. E. Fox).Produits Chimiques Péchiney-Saint-Gobain, DE-OS 1 768 485, 1968 (G. du Crest, G. Benaroya, F. Lainé; 1 768 494, 1968 (A. Antonini, P. Joffre, C. Vrillon); 1 768 495, 1968 (A. Antonini, C. Kaziz, G. Wetroff); FR 1 552 825, 1969 (A. Antonini, C. Kaziz, G. Wetroff); 1 552 826, 1969 (A. Antonini, C. Kaziz, G. Wetroff).

[298] Solvay Cie., FR 1 380 970, 1964.

[299] Japan Atomic Energy Research Inst., JP 42-13842, 1967.
[300] Knapsack-Griesheim, GB 749 351, 1954.
[301] G. Kalz, *Plaste Kautsch.* **18** (1971) 500.
[302] T. J. Houser, R. B. Bernstein, *J. Am. Chem. Soc.* **80** (1958) 4439; T. J. Houser, T. Cuzcano, *Int. J. Chem. Kinet.* **7** (1975) 331.
[303] Wacker, DE 843 843, 1942 (W. Fritz, E. Schaeffer).
[304] F. S. Dainton, K. J. Ivin, *Trans. Faraday Soc.* **46** (1950) 295; J. Puyo et al., *Bull. Soc. Lorraine Sci.* **2** (1962) 75.
[305] Kali-Chemie, DE 712 784, 1938 (F. Rüsberg, E. Gruner).
[306] Du Pont, US 2 440 731, 1948 (W. H. Vining, O. W. Cass).
[307] *Nachr. Chem. Tech. Lab.* **29** (1981) 10.
[308] Chemische Fabrik Griesheim, DE 278 249, 1912.
[309] D. Hardt et al., *Angew. Chem.* **94** (1982) 159; *Int. Ed.* **21** (1982) 174.
[310] E. Sanhueza et al., *Chem. Rev.* **75** (1976) 801.
[311] M. Lederer, *Angew. Chem.* **71** (1959) 162.
[312] H. Normant, *Compt. Rend.* **239** (1954) 1510; *Bull. Soc. Chim. Fr.* 1957, 728; *Adv. Org. Chem.* **2** (1960) 1.
[313] R. West, W. H. Glaze, *J. Org. Chem.* **26** (1961) 2096.
[314] Dow Chemical, US 4 147 733, 1979 (T. R. Fiske, D. W. Baugh, Jr.).
[315] R. E. Lynn, K. A. Kobe, *Ind. Eng. Chem.* **46** (1954) 633; Société Belge de l'Azote, US 2 779 804, 1954 (F. F. A. Braconier, J. A. R. O. L. Godart).
[316] Société Belge de l'Azote, GB 954 791, 1959 (F. F. A. Braconier, H. Le Bihan).
[317] Chemische Werke Hüls AG, GB 709 604, 1957; BASF, GB 769 773, 1955; Produits Chimiques de Péchiney-Saint-Gobain, FR 1 361 884, 1963 (F. Lainé, C. Kaziz, G. Wetroff).
[318] Knapsack AG, DE-OS 2 053 337, 1972 (A. Lauke).
[319] F. J. Gattys-Verfahrenstechnik GmbH, DE-OS 2 646 129, 1979 (F. J. Gattys).
[320] D. H. R. Barton, M. Mugdan, *J. Soc. Chem. Ind.* **69** (1950) 75; F. Patat, P. Weidlich, *Helv. Chim. Acta* **32** (1949) 783.
[321] Yu. A. Pasderskii, SU 686 279, 1984.
[322] K. Washimi, Y. Wakabayashi, *Kogyo Kagaku Zasshi* **68** (1965) 113; R. D. Wesselhoft et al., *AlChEZ.* **5** (1959) 361.
[323] VEB Chemische Werke Buna, DD 159 985, 1981 (J. Glietsch et al.).
[324] VEB Chemiefaserwerk Friedrich Engels, DD 150 880, 1981 (H. E. Steglich et al.); 200 017, 1983 (H. E. Steglich et al.); VEB Chemische Werke Buna, DD 126 454, 1977 (G. Henke); 132 711, 1978 (H. Stolze et al.); 149 212, 1981 (J. Glietsch, W. Linke).
[325] Marathon Oil Co., GB 1 138 669, 1969.
[326] Monsanto, GB 600 785, 1945; 757 661, 1954.
[327] Air Reduction Co., US 2 448 110, 1946 (H. S. Miller).
[328] Gevaert Photo Production N.V., GB 655 424, 1947.
[329] Gevaert Photo Production N.V., GB 643 743, 1947.
[330] Wacker, DE-OS 1 277 845, 1968 (O. Fruhwirth, H. Kainzmeier).
[331] VEB Chemische Werke Buna, DD 139 976, 1980 (R. Adler et al.).
[332] Hoechst, DE 2 558 871, 1984 (W. Gerhardt, H. Scholz).
[333] VEB Chemische Werke Buna, DD 139 975, 1980 (K. Hartwig et al.); 143 367, 1980 (K. Hartwig et al.).
[334] Chemische Werke Hüls AG, DE 1 205 705, 1964 (W. Knepper, G. Höckele); DE-OS 2 054 102, 1970 (H. Maiwald, G. Höckele, H. Sauer).

[335] H. Bremer et al., DD 84 182, 1971.
[336] Institut für Chemieanlagen, FR 1 441 148, 1966; DD 51 850, 1968 (K. Roland, C. Gerber, G. Voigt).
[337] Institut für Chemieanlagen, GB 1 174 147, 1969; FR 1 553 573, 1968; DD 60 303, 1968 (K. Roland).
[338] Institut für Chemieanlagen, DD 50 593, 1966 (K. Roland); DE 1 260 416, 1968 (K. Roland).
[339] Grupul Industrial de Petrochimie Borzesti, FR 7 202 582, 1972 (T. Has et al.); Knapsack AG, DE-OS 1 254 143, 1967 (G. Rechmeyer, A. Jacobowsky).
[340] Kureha Kagaku Kogyo Kabushiki Kaisha, GB 1 149 798, 1969; FR 1 558 893, 1968. Japan Gas-Chemical Co. Inc., FR 1 465 296, 1967.
[341] Solvay and Cie., BE 698 555, 1967; US 3 506 727 (J. Mulders).
[342] N. L. Volodin et al., DE-OS 2 026 429, 1975;US 4 014 947, 1977; FR 2 045 822, 1971.
[343] B. J. Pope, US 3 864 409, 1975.
[344] S. Gomi, *Hydrocarbon Process.* **43** (1964) 165; K. Washimi, M. A. Kura, *Chem. Eng. (N.Y.)* **73** (1966) no. 10, 133; no. 11, 121; Y. Onoue, K. Sakurayama, *Chem. Econ. Eng. Rev.* **4** (1969) 17.
[345] F. J. Gattys Ingenieurbüro, DE-OS 2 820 776, 1980 (F. J. Gattys); DE 2 905 572, 1983 (F. J. Gattys).
[346] H. J. Pettelkau, DE 3 007 634, 1982.
[347] K. E. Howlett, *Trans. Faraday Soc.* **48** (1952) 25; L. K. Doraiswamy et al., *Br. Chem. Eng.* **5** (1960) 618;G. A. Kapralova, N. N. Semenov, *Zh. Fiz. Khim.* **37** (1963) 73; P. G. Ashmore et al., *J. Chem. Soc., Faraday Trans. 1* 1982, 657.
[348] Chemische Werke Hüls AG, DE-OS 2 130 297, 1975 (G. Scharein, J. Gaube).
[349] E. V. Sonin et al., FR 2 082 004, 1971; GB 1 225 210, 1969; DE 1 953 240, 1984.
[350] Allied, DE-OS 2 319 646, 1973 (B. E. Kurtz et al.).
[351] BASF, DE-OS 2 349 838, 1974 (G. Krome).
[352] BP Chemicals Intern. Ltd., GB 1 337 326, 1973 (N. F. Chisholm).
[353] BP Chemicals Ltd., FR 2 099 466, 1972 (D. P. Young); DE-OS 2 135 248, 1972 (D. P. Young); US 3 896 182, 1975.
[354] Mitsui Chem. Ind., JP 42 22921, 1967.
[355] BP Chemicals Ltd., GB 1 494 797, 1977 (R. W. Rae, W. F. Fry).
[356] Magyar Asvanyolaj es Földgaz Kiserleti Intezet, DE-OS 2 223 011, 1973 (L. Szepesy et al.); 2 225 656, 1973 (I. Vendel et al.); Hoechst, NL 7 503 850, 1974; Knapsack AG, DE-OS 2 313 037, 1974 (G. Rechmeier, W. Mittler, R. Wesselmann); B. F. Goodrich Co., GB 938 824, 1961.
[357] BASF, DE-OS 3 147 310, 1973 (W. Hebgen et al.).
[358] Hoechst, DE-OS 2 907 066, 1980 (A. Czekay et al.); 3 013 017, 1981 (R. Krumböck et al.); Hoechst and Uhde GmbH/Hoechst AG, EP 14 920, 1980 (A. Czekay et al.); 21 381, 1980 (G. Link et al.).
[359] Halcon Intern. Inc., FR 1 505 735, 1967 (B. J. Ozero); Knapsack AG, NL 6 612 668, 1967; Knapsack AG, DE-OS 1 250 426, 1966 (H. Krekeler et al.); 1 910 854, 1972 (G. Rechmeier, A. Jacobowsky, P. Wirtz); The Lummus Co., DE-OS 2 501 186, 1975 (R. Long, H. Unger); Hoechst AG, DE 3 024 156, 1983 (A. Czekay et al.); BASF, DE-OS 3 140 892, 1983 (W. Hebgen et al.).
[360] Hoechst, GB 2 054 574, 1981 (J. Riedl, W. Fröhlich, E. Mittermaier).
[361] PPG, NL 6 613 177, 1967; BASF, DE 3 219 352, 1984 (E. Birnbaum, E. Palme).
[362] Solvay and Cie., BE 746 270, 1970; DE-OS 1 288 594, 1969 (G. Coppens); 2 101 464, 1971 (G. Coppens).

[363] Knapsack AG, DE-OS 1 959 211, 1971 (P. Wirtz et al.); Kanegafuchi Kagaku Kogyo K.K., DE-OS 2 426 514, 1975 (T. Ohishi, N. Yoshida, T. Hino); Allied, NL 7 711 904, 1977; Wacker, DE-OS 2 754 891, 1979 (L. Schmidhammer, H. Frey); BASF, 2 307 376, 1974 (G. Krome); Dynamit Nobel AG, DE 3 135 242, 1984; (R. Stephan et al.); BASF, DE-OS 3 140 447, 1983 (W. Hebgen et al.).

[364] ICI, GB 1 405 714, 1975; Wacker, DE-OS 3 009 520, 1981 (L. Schmidhammer, R. Straßer); Solvay and Cie., EP 101 127, 1983 (A. Closon).

[365] Deutsche Gold- und Silberschmiedeanstalt, DE-OS 2 438 153, 1976 (G. Vollheim et al.).

[366] Solvay and Cie, FR 1 602 522, 1970; US 3 801 660, 1974 (G. Coppens).

[367] Dow Chemical, US 3 723 550, 1973 (R. T. McFadden).

[368] Monsanto, US 3 125 607, 1964 (H. M. Keating, P. D. Montgomery); Hoechst, DE-OS 2 903 640, 1980 (G. Rechmeier, U. Roesnik, H. Scholz).

[369] Monsanto, US 3 142 709, 1964 (E. H. Gause, P. D. Montgomery).

[370] Monsanto, US 3 125 608, 1964 (D. W. McDonald).

[371] Continental Oil Co., US 3 830 859, 1974 (R. Gordon et al.).

[372] VEB Chemische Werke Buna, DD 143 368, 1980 (H. Hauthal et al.).

[373] Rhône-Progil S.A., DE-OS 2 429 273, 1976 (Y. Correia, J.- C. Lanet).

[374] Rhône-Progil, FR 2 241 519, 1973 (Y. Correia, J.-C. Lanet).

[375] BASF, DE-OS 3 122 181, 1982 (E. Danz, G. Krome).

[376] Goodyear Fire and Rubber Co., US 4 042 637, 1977 (E. J. Glazer, E. S. Smith).

[377] M. Rätzsch et al., DD 112 603, 1975.

[378] Continental Oil Co., US 3 917 728, 1975 (R. D. Gordon).

[379] *Hydrocarbon Process.* **44** (1965) 290.

[380] Wacker, DE 1 135 451, 1960 (O. Fruhwirth).

[381] Wacker, DE-OS 2 156 943, 1973 (O. Fruhwirth, L. Schmidhammer, H. Kainzmeier).

[382] Pullman Corp., GB 1 152 021, 1969; Monsanto, FR 1 515 554, 1968 (R. L. Hartnett).

[383] Wacker, DE-OS 2 239 051, 1975 (S. Nitzsche, L. Schmidhammer); Shell Internat. Research, NL 6 610 116, 1968.

[384] FMC, DE-OS 1 928 199, 1969 (S. Berkowitz).

[385] B. F. Goodrich Co., EP 0 002 021, 1979 (A. J. Magistro).

[386] The Lummus Co., DE-OS 1 806 547, 1976 (H. D. Schindler, H. Riegel); 2 502 335, 1975 (H. D. Schindler).

[387] J. Wolfrum et al., DE 2 938 353, 1983; EP 0 027 554, 1981. K. Kleinnermanns, J. Wolfrum, *Laser Chem.* **2** (1983) 339.

[388] Kanegafuchi Kagaku Kogyo Kabushiki Kaisha, FR 1 556 912, 1969.

[389] M. Sittig: *Vinyl Chloride and PVC Manufacture,* Noyes Data Corp., Park Ridge, New York 1978.

[390] Akzo, US 4 256 719, 1981 (E. van Andel); E. van Andel, *Chem. Ind. (London)* 1983, no. 2, 139.

[391] *European Chem. News* **46** (1229) June 9, p. 4, 1986.

[392] Montecatini S.G., IT 761 015, 1967 (M. Mugero, M. Boringo).

[393] Hitachi Zosen Kabushiki Kaisha, NL 6 604 833, 1966.

[394] Badger Co. Inc, DE-OS 2 629 461, 1978 (H. R. Sheely, F. F. Oricchio, D. C. Ferrari). Pullmann Inc., GB 1 177 971, 1970 (B. E. Firnhaber).

[395] Ethyl Corp., US 2 681 372, 1951 (P. W. Trotter); Hoechst, DE 1 003 701, 1954; National Distillers, US 2 896 000, 1955 (F. D. Miller, D. P. Jenks); Société Belge de l'Azote, GB 889 177, 1959 (P. J. Leroux, F. F. A. Braconier); ESSO Research and Engin. Co., FR 1 344 322, 1962 (J.-M. Guilhaumou, M. Prillieux, P. Verrier); S. Tsutsumi, GB 956 657,

1962; Union Carbide, FR 1 385 179, 1964; Toyo Koatsu Ind., BE 637 573, 1963; 658 457, 1965 (T. Takahashi et al.).

[396] ESSO Research and Engin Co., US 3 670 037, 1972 (J. J. Dugan); ICI, DE-OS 2 837 514, 1979 (J. D. Scott); Shell Intern. Research, NL 6 611 699, 1968; Allied, US 4 039 596, 1977 (W. M. Pieters, E. J. Carlson); Du Pont, US 2 308 489, 1940 (O. W. Cass); ICI, FR 1 330 367, 1962; 1 359 016, 1963; Hoechst, BE 662 098, 1965; Osaka Kinzoku Kogyo, US 3 267 161, 1963 (R. Ukaji et al.).

[397] ICI, DE-OS 1 277 243, 1968 (B. Hancock, L. McGinty, I. McMillan).

[398] Diamond Shamrock Corp., US 4 115 323, 1978 (M. F. Lemanski, F. C. Leitert, C. G. Vinson, Jr.).

[399] Hoechst, DE-OS 1 931 393, 1971 (H. Krekeler, H. Kuckertz); British Petroleum Co. Ltd., DE-OS 1 907 764, 1972 (G. H. Ludwig); GB 1 213 402, 1970 (G. H. Ludwig); Marathon Oil Co., US 3 501 539 (D. H. Olson, G. M. Bailey).

[400] Produits Chimiques Péchiney-Saint-Gobain, FR 1 552 849, 1969 (A. Antonini, G. Stahl, C. Vrillon).

[401] British Petroleum Co. Ltd., DE-OS 2 540 067, 1976 (J. L. Barclay). B. F. Goodrich Co., DE-OS 2 613 561, 1976 (W. J. Kroenke, P. P. Nicholas); US 4 100 211, 1978 (A. J. Magistro); 4 102 935, 1978 (W. J. Kroenke, R. T. Carroll, A. J. Magistro); 4 102 936, 1978 (A. J. Magistro); 4 119 570, 1978 (W. J. Kroenke, P. P. Nicholas); CA 1 096 406, 1981 (A. J. Magistro).- Monsanto, DE-OS 2 852 036, 1979 (T. P. Li); CA 1 111 454, 1981 (T. P. Li). ICI, GB 2 095 242, 1982 (D. R. Pyke, R. Reid); 2 095 245, 1982 (D. R. Pyke, R. Reid); DE-OS 3 226 028, 1983 (D. R. Pyke, R. Reid).

[402] Owens-Illinois Inc., US 4 042 639, 1977 (T. H. Gordon, H. F. Kummerle); Princeton Chemical Research Inc., FR 1 595 619, 1970 (N. W. Frisch, R. I. Bergmann).

[403] O. A. Zaidman et al., DE-OS 2 853 008, 1980; GB 2 036 718, 1980.

[404] Princeton Chemical Research Inc., DE-OS 1 806 036, 1969 (R. I. Bergmann); Produits Chimiques Péchiney-Saint-Gobain, US 3 923 913, 1975 (A. Antonini et al.).

[405] Monsanto, NL 6 515 252, 1966.

[406] Ethyl Corp., US 3 658 933, 1972 (W. Q. Beard, Jr.); 3 658 934, 1972 (W. Q. Beard, Jr.); 3 629 354, 1971 (W. Q. Beard, Jr.).

[407] The Lummus Co., DE-OS 1 693 042, 1974 (H. Riegel); 1 812 993, 1974 (H. Riegel, H. Schindler); 1 952 780, 1970 (H. Riegel et al.); 2 230 259, 1973 (H. Riegel); 2 314 786, 1973 (U. Tsao); 2 335 949, 1974 (H. Riegel et al.); 2 336 497, 1974 (H. D. Schindler et al.); 2 509 966, 1975 (H. Riegel); 2 536 286, 1976 (U. Tsao); US 3 920 764, 1975 (H. Riegel et al.); 3 935 288, 1976 (H. Riegel); 3 937 744, 1976 (H. Riegel); 3 796 641, 1974 (H. Riegel et al.); FR 1 574 064, 1969 (M. C. Sze); 1 574 705, 1969 (H. Riegel); 1 576 909, 1969 (H. Riegel, H. D. Schindler); 1 577 105, 1969 (H. Riegel); GB 1 258 750, 1971.

[408] Ethyl Corp., US 2 838 579, 1954 (F. Conrad, M. L. Gould, C. M. Neher); Monsanto, US 3 166 601, 1961 (K. M. Taylor); Du Pont, US 3 234 295, 1961 (J. W. Sprauer, S. Heights).

[409] PPG, GB 996 323, 1962 (W. K. Snead, R. H. Chandley); 998 689, 1965; FR 1 341 711, 1962 (A. C. Ellsworth); Princeton Chemical Research Inc., DE-OS 1 929 062, 1969 (N. W. Frisch); The British Petroleum Co. Ltd., DE-OS 2 006 262, 1971 (G. H. Ludwig).

[410] Ethyl Corp., US 2 765 349–352, 1955 (F. Conrad); 2 803 677 1955 (C. M. Neher, J. H. Dunn); 2 803 678–680 1955 (F. Conrad).

[411] Air Reduction Co. Inc., US 3 506 552, 1970 (J. P. Russell).

[412] Allied, US 4 155 941, 1979 (H. E. Nychka, R. E. Eibeck).

[413] PCUK Produits Chimiques Ugine Kuhlmann, FR 2 529 883, 1982 (G. A. Olah).

[414] ICI, EP 15 665, 1980 (C. S. Allen).
[415] PPG, GB 2 121 793, 1982 (D. R. Nielsen).
[416] Central Glass Co. Ltd., DE 2 818 066, 1982 (K. Yagii, H. Oshio).
[417] Amer., US 3 558 453, 1971 (J. S. Mayell).
[418] ICI, FR 2 064 406, 1971; ZA 706 583, 1969 (C. Neville, P. R. Edwards, P. J. Craven).
[419] Gulf R a D Co., US 3 577 471, 1971 (J. G. McNulty, W. L. Walsh).
[420] Seymour C. Schuman, US 3 377 396, 1968.
[421] SRI International, *Chem. Econ. Handbook,* Marketing Research Report on PVC Resins, Menlo Park, 1982.SRI International,
Vinyl Chloride Report, Menlo Park, 1982. *Eur. Chem. News* **44** (1985) no. 1169.
[422] *Eur. Chem. News* **45** (1985) no. 1194, 4. *Chem. Week* (1985) 30.
[423] *Hydrocarbon Process., HPI Construction Boxscore, October 1984; Eur. Chem. News* **44** (1985) no. 1167.
[424] Dow Chemical, US 2 136 333, 1936 (G. H. Coleman, J. W. Zemba); 2 136 349, 1938 (R. M. Wiley); 2 160 944, 1938 (G. H. Coleman, J. W. Zemba); Ethyl Corp., FR 1 308 101, 1961 (A. J. Haefner, E. D. Hornbaker).
[425] R. J. Williams, *J. Chem. Soc.* 1953, 113; K. Feramoto et al., *Kogyo Kagaku Zasshi* **67** (1964) 50.
[426] J. Svoboda et al., *Petrochemia* **22** (1982) 21.
[427] BASF, DE-OS 1 230 418, 1966 (H. Ostermayer, W. Schweter).
[428] ICI, DE-OS 2 225 512, 1972 (G. A. Thompson, J. W. Tipping).
[429] ICI, GB 1 453 509, 1976 (I. S. McColl, A. C. P. Pugh, G. A. Thompson).
[430] Detrex Chemical Industries Inc., US 3 725 486, 1973 (W. L. McCracken et al.).
[431] Toa Gosei Chem. Ind. Co., Ltd., JP 56/104 826, 1981.
[432] PPG, DE-OS 2 651 901, 1980 (J. D. Mansell).
[433] S. A. Devinter, FR 2 131 896, 1973.
[434] Continental Oil Co., US 3 664 966, 1972 (R. D. Gordon).
[435] Kanegafuchi Chemical Industry Co., Ltd., JP 48/92312, 1973.
[436] Dow Chemical, US 3 984 489, 1976 (R. F. Mogford).
[437] Continental Oil Co., US 3 869 520, 1975 (R. D. Gordon).
[438] Dow Chemical, US 2 293 317, 1941 (F. L. Taylor, L. H. Horsley).
[439] Asahi-Dow Ltd., JP 56/57722, 1981.
[440] E. W. Sonin et al., DE-OS 1 952 770, 1973.
[441] K. Nagai, M. Katayama, *Bull. Chem. Soc. Japan* **51** (1978) 1269.
[442] Chemische Werke Hüls AG, DE-OS 2 135 908, 1974 (H. Rassaerts, G. Sticken, W. Knepper).
[443] Dow Chemical, US 4 119 678, 1978 (T. S. Boozalis, J. B. Ivy).
[444] Mitsui Toatsu Chemicals Co., Ltd., JP 47/18088, 1972.
[445] Wah Young Lee et al., *Hwahak Konghak* **14** (1976) 169; *Chem. Abstr.* **85** 142517 c.
[446] I. Mochida et al., *Sekiyu Gakkai Shi* **21** (1978) 285; *Chem. Abstr.* **90** 22226 p.
[447] I. Mochida et al., JP 51/133207, 1976.
[448] ICI, DE-OS 2 257 107, 1973 (M. H. Stacey, T. D. Tribbeck); PPG, DE-OS 2 849 469, 1979 (W. H. Rideout); US 4 144 192, 1979 (A. E. Reinhardt II).
[449] Asahi-Dow Ltd., JP 55/87730, 1980; Toa Gosei Chemical Industry Co., Ltd., JP 52/31006, 1977.
[450] Asahi-Dow Ltd., JP 55/87729, 1980.
[451] A. P. Khardin et al., *Khim. Prom-st. Moscow* 1982, 208; P. Y. Gokhberg et al., *Kinet. Katal.* **23** (1982) 50 – 53, 469 – 473;A. P. Khardin et al., *Khim. Prom-st. (Moscow)* 1981, 16.

[452] N. Yamasata, *Ibaraki Kogyo Koto Semmon Gakko Kenkyu Iho* **11** (1976) 213; *Chem. Abstr.* **89** 214 828 n.

[453] I. Mochida, *J. Mol. Catal.* **12** (1981) 359; Asahi-Dow Ltd., JP 56/40622, 1981; 58/162537, 1983; 56/150028, 1981; 56/40622, 1981.

[454] J. Kobayashi et al., *Kenkyu Hokoku – Asahi Garasu Kogyo Gijutsu Shoreikai* **37** (1980) 251; *Chem. Abstr.* **95** 6055 e.

[455] Kureha Chemical Industry Co., Ltd., JP 48/10130, 1973.

[456] L. Espada et al., *Ion* **37** (1977) 495.

[457] Wacker, DE-OS 1 925 568, 1974 (O. Fruhwirth, E. Pichl, L. Schmidhammer); Kureha Kagaku Kogyo K.K., DE-OS 2 101 463, 1971 (K. Shinoda et al.); 2 135 445, 1973 (K. Shinoda, T. Nakamura).

[458] Asahi-Dow Ltd., JP 56/57721, 1981.

[459] ICI, DE-OS 2 850 807, 1979 (J. D. Wild).

[460] Dow Chemical, US 3 726 932, 1973 (C. R. Mullin, D. J. Perettie).

[461] Dow Chemical, US 3 654 358, 1977 (G. C. Jeffrey).

[462] Donau-Chemie, AT 162 391, 1947.

[463] Knapsack-Griesheim AG, DE 969 191, 1952 (A. Jacobowsky, K. Sennewald); DE-OS 1 011 414, 1953 (A. Jacobowsky, K. Sennewald); Péchiney, US 3 197 515, 1962 (P. Chassaing, G. Clerc).

[464] B. Jakesevic, *Tekstil* **32** (1983) 321.

[465] C. J. Howard, *J. Chem. Phys.* **65** (1976) 4771; J. S. Chang, F. Kaufmann, *J. Chem. Phys.* **66** (1977) 4989.

[466] L. Bertrand et al., *J. Phys. Chem.* **72** (1968) 3926; J. Olbregts, *Int. J. Chem. Kinet.* **11** (1979) 117.

[467] Wacker, DE 901 774, 1940 (W. Fritz, J. Rambausek).

[468] Dynamit Nobel, DE 1 174 764, 1958 (E. E. Feder, K. Kienzle); Detrex Chemical Ind., US 2 912 470, 1956 (C. E. Kircher, Jr., R. J. Jones); M. Szczeszek, H. Chmielarska, *Przem. Chem.* **56** (1977) 255.

[469] Wacker, DE 846 847, 1942 (W. Fritz, E. Schaeffer); NL 6 600 884, 1966.

[470] Du Pont, US 2 894 045, 1957 (E. G. Carley); Hooker Chemical Co., US 3 100 233, 1960 (D. S. Rosenberg); Donau-Chemie AG, AT 238 699 (F. Samhaber).

[471] Donau-Chemie AG, AT 191 859, 1957 (L. Gavanda, A. F. Orlicek); ICI, GB 697 482, 1951 (R. T. Foster, S. W. Frankish).

[472] The Distillers Co., US 2 378 859, 1942 (M. Mugdan, D. H. R. Barton).

[473] Du Pont, US 3 388 176, 1968 (J. L. Sheard); 3 388 177, 1968 (L. J. Todd).

[474] Produits Chimiques Péchiney-Saint-Gobain, DE-OS 2 008 002, 1975 (R. Clair, Y. Correia); Toa Gosei Chem. Ind., DE-OS 1 943 614, 1969 (T. Kawaguchi et al.).

[475] Detrex Chemical Ind. Inc., DE-OS 1 643 872, 1967 (C. E. Kirchner, Jr., D. R. McAllister, D. L. Brothers).

[476] S. Suda, *Chem. Eng. (N.Y.)* **77** (1970) 74. T. Kawaguchi et al., *Ind. Eng. Chem.* **62** (1970) 36.

[477] Detrex Chemical Ind. Inc., US 3 691 240, 1972 (C. E. Kirchner, D. R. McAllister, D. L. Brothers).

[478] Dow Chemical, US 2 140 548, 1938 (J. H. Reilly); Diamond Alkali Corp., GB 673 565, 1952; Ethyl Corp., US 2 725 412, 1954 (F. Conrad); Donau Chemie AG, NL 6 607 204, 1966.

[479] Diamond Shamrock Corp., DE-OS 2 061 508, 1970 (J. J. Lukes, R. J. Koll).

[480] Ruthner AG, AT 305 224, 1973 (F. Samhaber); PPG, US 3 793 227, 1974 (W. K. Snead, F. Abraham).

[481] D. Jaqueau et al., *Chem. Ing. Tech.* **43** (1971) 184; J. F. Knoop, G. R. Neikirk, *Hydrocarbon Process.* **59** (1972) 109; PPG, GB 913 040, 1961 (L. E. Bohl, R. M. Vancamp); 904 084, 1961

(L. E. Bohl); 904 405, 1961 (L. E. Bohl, A. P. Muren, R. M. Vancamp); 916 684, 1961 (R. E. McGreevy, J. E. Milam, W. E. Makris); 969 416, 1962 (A. C. Ellsworth; 1 012 423, 1964 (L. E. Bohl, R. M. Vancamp); 1 027 279, 1963 (L. E. Bohl, R. M. Vancamp); 1 104 396, 1965 (L. W. Piester, R. M. Vancamp); Du Pont, GB 1 362 212, 1974, US 3 232 889, 1966; R. L. Espada, *Ion* **36** (1976) 595; L. M. Kartashov et al., *Khim. Prom-st.* (*Moscow*) **1983**, 587; Z. Trocsanyi, J. Bathory; *Kolor. Ert.* **17** (1975) 184; *Chem. Abstr.* **84** 31 534 r; L. Dubovoi et al., *Khim. Prom.-st.* (*Moscow*) **1982**, 658.

[482] Du Pont, US 3 697 608, 1972 (H. E. Bellis); 4 130 595, 1978 (H. E. Bellis); Sumitomo Chemical Co., Ltd., DE-OS 2 413 148, 1977 (K. Iida, T. Takahashi, S. Kamata); Marathon Oil Co., US 3 689 578, 1972 (D. H. Olson, G. M. Bailey).
[483] Allied, BE 841 998, 1976.
[484] Dow Chemical, US 4 105 702, 1978 (C. R. Mullin, D. J. Perettie).
[485] Wacker, DE-OS 2 819 209, 1979 (W. Mack et al.).
[486] Wacker, EP 59 251, 1984 (K. Blum, W. Mack, R. Strasser).
[487] Dow Chemical, FR 1 508 863, 1967 (S. G. Levy).
[488] *Eur. Chem. News* **44** (1985) no. 1161.
[489] Dow Chemical, US 3 959 367, 1976 (G. C. Jeffrey).
[490] G. Huybrechts et al., *Trans Faraday Soc.* **63** (1967) 1647; H. B. Singh et al., *Environ. Lett.* **10** (1970) 253;D. Lillian et al., *Environ. Sci. Technol.* **9** (1975) 1042; U. Lahl et al., *Sci. Total Environ.* **20** (1981) 171; E. P. Grimsrud, R. A. Rasmussen, *Atmos. Environ.* **9** (1975) 1014.
[491] G. Kauschka, L. Kolditz, *Z. Chem.* **16** (1976) 377.
[492] Wacker, DE 725 276, 1937 (G. Basel, E. Schäffer); Donau-Chemie AG, DE 734 024, 1940 (O. Fruhwirth, H. Walla); Société d'Ugine, FR 1 073 631, 1953.
[493] FMC, ZA 715 781, 1970 (J. S. Sproul et al.).
[494] Rhône-Progil, FR 2 260 551, 1974 (J.-C. Strini); FMC, ZA 715 780, 1970 (J. S. Sproul, B. R. Marx).
[495] Diamond Alkali Co., GB 701 244, 1951; 673 565, 1952; *Hydrocarbon Process.* **46** (1967) no. 11, 210.
[496] Donau-Chemie AG, DE-OS 1 277 245, 1965; FR 1 525 811, 1966 (F. Samhaber); NL 6 607 204, 1966.
[497] Donau Chemie AG, GB 1 143 851, 1967 (F. Samhaber).
[498] Diamond Shamrock Corp., US 3 860 666, 1976 (N. L. Beckers).
[499] K. Shinoda et al., *Kogyo Kagaku Zasshi* **70** (1967) 1482; N. A. Bhat et al., *Indian J. Technol.* **5** (1967) 255.
[500] Dow Chemical, US 2 442 323, 1944 (C. W. Davis, P. H. Dirstine, W. E. Brown); 2 442 324, 1944 (R. G. Heitz, W. E. Brown); 2 577 388, 1945 (G. W. Warren).
[501] Stauffer, US 2 857 438, 1957 (R. P. Obrecht, H. Bender).
[502] Halcon International Inc., GB 1 047 258, 1963.
[503] Société d'Ugine, FR 1 147 756, 1956; Solvay and Cie, FR 1 373 709, 1963; J. Kraft et al., *Chim Ind. (Paris)* **83** (1960) 557.
[504] Chemische Werke Hüls AG, DE 1 074 025, 1958 (F. Krüll, O. Nitzschke, W. Krumme); F. Krüll, *Chem. Ing. Tech.* **33** (1961) 228.
[505] C. H. Chilton, *Chem. Eng. (N.Y.)* **65** (1958) no. 9, 116.
[506] V. N. Tychinin et al., *Khim. Prom-st (Moscow)* 1984, no. 4, 199.
[507] Halcon International, Inc., FR 1 539 714, 1967 (I. E. Levine).
[508] K. German, J. Rakoczy, *Przem. Chem.* **63** (1984) 93;*Chem. Abstr.* **100** 17684 w.
[509] Tokuyama Soda KK, JP 8 4038 932, 1984.

[510] Progil-Electrochimie Cie., GB 819 987, 1957; *Hydrocarbon Process.* **44** (1965) no. 11, 190.
[511] Produits Chimiques Ugine Kuhlmann, US 4 211 728, 1980 (J. G. Guérin).
[512] Hoechst, DE-OS 1 793 131, 1973 (H. Krekeler et al.); 2 100 079, 1973 (H. Krekeler, W. Riemenschneider); 2 150 400, 1974 (W. Riemenschneider); 2 231 049, 1974 (R. Walburg, H. Gerstenberg, H. Osterbrink).
[513] FMC, US 3 364 272, 1968 (J. W. Ager).
[514] *Chem. Week,* July 24, 1985.
[515] *Eur. Chem. News* **44** (1985) no. 1167.
[516] G. Hommel: *Handbuch der gefährlichen Güter,* Springer Verlag, Berlin 1985.
[517] According to Annex II of Marpol 73/78, category regulation 3.
[518] Chemische Werke Hüls AG, DE-OS 2 540 178, 1976 (R. Wickbold et al.); NL 7 610 030, 1977.
[519] Solvay and Cie., EP 15 625, 1980 (R. Hembersin, R. Nicaise).
[520] Stauffer, EP 94 527, 1983 (W. Burks, Jr. et al.).
[521] BASF, DE-OS 2 827 761, 1980; Ezaki Shigeho,Chigasaki, Kanagawa, DE-OS 2 045 780, 1974; Knapsack AG, DE-OS 1 228 232, 1966; The Lummus Co., US 3 879 481, 1975; Nittetu Chemical Engineering Ltd., US 3 876 490, 1975.
[522] The B. F. Goodrich Co., DE-OS 2 531 981, 1976; 2 532 027, 1976 (J. S. Eden); 2 532 043, 1976; 2 532 052, 1976; 2 532 075, 1976.
[523] Chemische Werke Hüls AG, DE 1 246 701, 1961.
[524] BASF, DE-OS 2 261 795, 1974.
[525] Tessenderlo Chemie S.A., DE-OS 2 531 107, 1978.
[526] Dow Chemical, US 4 435 379, 1984.
[527] Hoechst, DE-OS 2 361 917, 1975 (H. Müller).
[528] Grupul Industrial de Chimie Rimnicu, DE-OS 2 148 954, 1974; FR 2 156 496, 1973; GB 1 333 650, 1973; Shell International Research, GB 1 483 276, 1977; GAF, US 3 807 138, 1974; The Goodyear Tire a. Rubber Co., DE-OS 2 842 868, 1979; Air Products, DE-OS 2 704 065, 1977; BOC Ltd., GB 2 020 566, 1979; DE-OS 2 733 745, 1978.
[529] The Lummus Co., DE-OS 2 604 239, 1976 (U. Tsao).
[530] L. E. Swabb, H. E. Hoelscher, *Chem. Eng. Prog.* **48** (1952) 564–569.
[531] N. N. Lebedev, *J. Gen. Chem. USSR (Engl. Transl.)* **24** (1954) 1925–1926.
[532] R. Letterer, H. Noller, *Z. Phys. Chem. (Munich)* **67** (1969) 317–329.
[533] M. F. Nagiev, Z. J. Kashkai, R. A. Makhumdzade, *Azerb. Khim. Zh.* 1969, no. 6, 27–32; *Chem. Abstr.* **74** (1971) 14 590 g.
[534] Hoechst, DE 1 805 805, 1968 (H. Kuckertz, H. Grospietsch, L. Hörnig).
[535] J. F. Norris, H. B. Taylor, *J. Am. Chem. Soc.* **46** (1924) 756.
[536] Hoechst, DE 1 805 809, 1968 (H. Kuckertz, H. Grospietsch, L. Hörnig).
[537] Shell Dev. Co., US 2 207 193, 1937 (H. P. A. Groll); *Chem. Abstr.* **34** (1940) 7934.
[538] E. T. McBee, H. B. Hass, T. H. Chao, Z. D. Welch et al., *Ind. Eng. Chem.* **33** (1941) 176–181.
[539] G. W. Hearne, T. W. Evans, H. L. Yale, M. C. Hoff, *J. Am. Chem. Soc.* **75** (1953) 1392–1394.
[540] H. Gerding, H. G. Haring, *Recl. Trav. Chim. Pays-Bas* **74** (1955) 841–853.
[541] H. P. A. Groll, G. Hearne, F. F. Rust, W. E. Vaughan, *Ind. Eng. Chem.* **31** (1939) 1239–1244.
[542] R. Kühn, K. Birett: *Merkblätter Gefährliche Arbeitsstoffe,* paper no. D 26, Ecomed Verlagsgesellschaft, Landsberg/Lech.
[543] G. Hommel: *Handbuch der gefährlichen Güter,* leaflet 170, Springer Verlag, Berlin-Heidelberg 1980.
[544] N. I. Sax: *Dangerous Properties of Industrial Materials,* 6th ed., Van Nostrand Reinhold Co., New York 1984.

[545] *Hazardous Chemicals Data Book*, Noyes Data Corp., Park Ridge, USA, 1980.
[546] M. Sittig: *Handbook of Toxic and Harzardous Chemicals and Carcinogens*, 2nd ed., Noyes Publ., New Jersey (1985), 329.
[547] *Ullmann*, 4th ed., **10**, 684.
[548] R. Kühn, K. Birett: *Merkblätter Gefährliche Arbeitsstoffe*, paper no. T 24, Ecomed Verlagsgesellschaft, Landsberg/Lech.
[549] A. L. Henne, F. W. Haeckl, *J. Am. Chem. Soc.* **63** (1941) 2692.
[550] A. D. Herzfelder, *Ber. Dtsch. Chem. Ges.* **26** (1893) 2432–2438.
[551] Du Pont, US 2 119 484, 1931 (A. A. Levine, O. W. Cass); *Chem. Abstr.* **32** (1938) 5413[4].
[552] T. Kleinert, *Chem. Ztg.* **65** (1941) 217–219.
[553] *Kirk-Othmer*, **18** 815.
[554] H. B. Hass, E. T. McBee, P. Weber, *Ind. Eng. Chem.* **27** (1935) 1191.
[555] Purdue Research Foundation, US 2 105 733, 1935.
[556] A. Maillard et al., *Bull. Soc. Chim. Fr.* 1961, no. 5, 1640.
[557] A. Pichler, *Peint. Pigm. Vernis* **41** (1965) no. 5, 293.
[558] A. Pichler, G. Levy, *Bull. Soc. Chim. Fr.* **1964**, no. 9, 2815; **1966**, no. 11, 3656.
[559] A. I. Gershenovich, V. M. Kostyuchenko, *Zh. Prikl. Khim.* **39** (1966) no. 5, 1160.
[560] Usines de Melle, FR 1 352 211, 1962 (J. Mercier).
[561] Sharpless Solvents Corp., US 2 122 110, 1937.
[562] E. G. Bondarenko et al., SU 172 289, 1965; *Chem. Abstr.* **63** (1965) 16212 f.
[563] Du Pont, US 2 570 495, 1946 (N. D. Scott).
[564] J. Gerrad, M. Phillips, *Chem. Ind. (London)* 1952, 540.
[565] G. Toptschijew, *J. Prakt. Chem.* **2** (1955) no. 4, 185.
[566] F. Asinger: *Chemie und Technologie der Monoolefine*, Akademie Verlag, Berlin 1956, p. 248.
[567] J. J. Leendertse et al., *Recl. Trav. Chim. Pays Bas* **52** (1933) 514–524; **53** (1934) 715.
[568] C. C. Coffin, O. Maass, *Can. J. Res.* **3** (1930) 526.
[569] C. C. Coffin, H. S. Sutherland, O. Maass, *Can. J. Res.* **2** (1930) 267.
[570] BASF, DE 1 004 155, 1954 (J. Schmidt, W. Ritter).
[571] BASF, DE 859 734, 1939.
[572] O. W. Cass, *Ind. Eng. Chem.* **40** (1948) 216.
[573] W. Reppe et al., *Justus Liebigs Ann. Chem.* **596** (1955) no. 1, 118.
[574] Du Pont, US 2 889 380, 1956 (E. E. Hamel).
[575] H. Jura, *J. Chem. Soc. Japan Ind. Chem. Sect.* **54** (1951) 433, C.A. 1238, 1954.
[576] N. Shono, S. Hachihama, *Chem. High Polym. (Japan)* **8** (1951) 70; *Chem. Abstr.* **56** (1953) 7483.
[577] G. Lutkowa, A. Kutsenko, *Zh. Prikl. Khim. (Leningrad)* **32** (1959) 2823.
[578] Chem. Werke Hüls AG, DE 1 188 570, 1961.
[579] E. M. Asatryan, G. S. Girgoryan, A. Malkhasyan, G. Martirosyan, *Arm. Khim. Zh.* **36** (1983) no. 8, 527–530.
[580] R. F. Tylor, G. H. Morsey, *Ind. Eng. Chem.* **40** (1948) 432.
[581] Distillers Co., DE 1 090 652, 1957 (C. W. Capp, H. P. Croker, F. J. Bellinger).
[582] Du Pont, FR 1 504 112, 1967 (J. L. Hatten, K. W. Otto).
[583] Knapsack, DE-OS 241 312, 1973 (K. Gehmann, A. Ghorodnik, U. Dettmeier, H. J. Berns).
[584] ICI, GB 1 119 862, 1966 (R. L. Heath).
[585] H. J. Zimmer, Verfahrenstechnik, DE 1 118 189, 1960 (N. W. Luft, K. Esser, H. Waider).
[586] Sumitomo Chem. Co., DE-AS 1 961 721, 1969.
[587] Du Pont, DE-OS 2 046 007, 1970 (O. K. Wayne, J. L. Hatten).
[588] H. Stemmler et al., DD 14 502, 1958.

[589] Showa K.K., JP 7 108 281, 1967.
[590] ICI, (W. Costain, B. W. H. Terry).
[591] I. G. Farben, US 2 242 084, 1938 (O. Nicodemus, W. Schmidt).
[592] Showa Denko K.K., JP 7 601 682, 1976 (Ogawa, Masaro; Yoshinaga, Yasuo).
[593] Agency of Ind. Science and Technology, JP 7 227 906, 1972; JP 7 334 570, 1973.
[594] Monsanto Chem. Co., FR 1 326 120, 1962.
[595] Shell Intern. Res., GB 1 007 077, 1964.
[596] BASF, DE 1 115 236, 1961 (M. Minsinger).
[597] K. Weissermel, H. J. Arpe: *Ind. Org. Chemistry,* Verlag Chemie, Weinheim 1978, p. 98.
[598] Du Pont, US 294 987, 1972 (J. H. Richards, C. A. Stewart).
[599] BP Chemicals Ltd., GB 1 296 482, 1 296 483, 1972 (P. J. N. Brown, C. W. Capp).
[600] Du Pont, USA 1 918 067, 1968 (D. D. Wild).
[601] Du Pont, DE-OS 2 248 668, 1972 (B. T. Nakata, E. D. Wilhoit).
[602] Du Pont, GB 1 260 691, 1969 (D. D. Wild).
[603] BP Chemicals Ltd., DE-OS 2 107 467, 2 107 468, 2 107 469, 1971 (C. W. E. Capp, P. J. N. Brown).
[604] Distillers Co. Ltd., GB 877 586, 1958.
[605] Electrochemical Ind., JP 6 809 729, 1967.
[606] Toyo Soda Ind., JP 6 920 330, 1968.
[607] Bayer, DE 3 208 796, 1982 (J. Heinrich, R. Casper, M. Beck).
[608] W. H. Carothers et al., *J. Am. Chem. Soc.* **53** (1931) 4203; **54** (1932) 4066.
[609] L. F. Hatch, S. G. Ballin, *J. Am. Chem. Soc.* **71** (1949) 1039.
[610] Du Pont, US 1 950 431/32, 1934; 2 102 611, 1937; 2 178 737, 1937.
[611] Distillers Co. Ltd., DE 1 193 936, 1963.
[612] Bayer, DE 2 318 115, 1973 (F. Hagedorn, R. Mayer-Mader, K. F. Wedemeyer).
[613] G. J. Beschet, W. H. Carothers, *J. Am. Chem. Soc.* **55** (1933) 2004.
[614] Du Pont, US 1 965 369, 1934.
[615] Knapsack, DE 1 149 001, 1961.
[616] Bayer, DE 2 545 341, 1975 (H. J. Pettelkau).
[617] Bayer, DE 2 717 672, 1977 (G. Scharfe, R. Wenzel, G. Rauleder).
[618] Du Pont, US 1 950 431, 1934 (W. H. Carothers, A. M. Collins).
[619] W. H. Carothers et al., *J. Am. Chem. Soc.* **53** (1931) 4203.
[620] A. E. Nyström, *Acta Med. Scand. Suppl.* **132** (1948) 219.
[621] *Chem. Abstr.* **44** (1950) 8758. S. E. Ekegren et al., *Acta Chem. Scand.* **4** (1950) 126–139.
[622] Knapsack, DE 1 148 230, 1959 (W. Vogt, H. Weiden, K. Gehrmann).
[623] Distillers Co., Ltd., US 2 926 205, 1960 (F. J. Bellringer).
[624] G. E. Ham: *High Polymers,* vol. **XVIII,** Interscience, New York 1964, p. 720.
[625] Du Pont, US 2 395 649, 1946 (F. C. Wagner).
[626] Bayer, GB 858 444, 1961 (A. R. Heinz, W. Graulich).
[627] Distillers Co. Ltd., DE 1 222 912, 1963 (D. A. Tadworth, A. B. Sutton, A. Foord, E. S. Luxon).
[628] Du Pont, DE 1 618 790, 1967 (J. B. Campbell, R. E. Tarney).
[629] Du Pont, DE 1 909 952, 1969 (J. B. Campbell, J. W. Crary, C. A. Piaseczynski, J. Stanley).
[630] Denka Chem. Corp., US 4 132 741, 1965 (A. J. Besozzi).
[631] Du Pont, US 4 418 232, 1982 (J. L. Maurin).
[632] Du Pont, US 4 215 078, 1979 (C. A. Hargreaves, A. T. Harris, R. A. Schulze).
[633] Knapsack AG, DE 2 139 729, 1971 (A. Ghrodnik, U. Dettmeier, K. Gehrmann, H. J. Berns).
[634] Bayer, DE 2 310 744, 1973 (R. Wenzel, G. Scharfe).

[635] Bayer, DE 2 460 912, 1974; 2 533 429, 1975 (G. Scharfe, R. Wenzel).
[636] BP Chem. Ltd., DE 2 707 073, 1976 (A. H. P. Hall, J. P. Merle).
[637] Bayer, DE 529 387, 1975 (G. Beilstein, B. Ehrig, D. Grenner, K. Nöthen).
[638] Du Pont, DE 588 283, 589 561, 1931.
[639] W. H. Carothers, *J. Am. Chem. Soc.* **54** (1932) 4066.
[640] Du Pont, US 2 207 784, 1937; 2 221 941, 1937.
[641] Du Pont, US 2 949 988, 1972 (J. H. Richards C. A. J. Stewart).
[642] Bayer, DE 2 234 571, 1972 (M. Weist, M. Leopold).
[643] Bayer, DE 2 357 194, 1973 (M. Weist, M. Leopold).
[644] Electrol-Chemical Ind. K.K., JP 011 135, 1977 (Ito, Akira; Watanabe, Seiichi).
[645] M. Stojanowa-Antoszczyn, A. Z. Zielinski, E. Chojnicki, *Przem. Chem.* **58** (1979) no. 3, 160–162.
[646] Y. Takasu, Y. Matsuda et al., *Chem. Lett.* **12** (1981) 1685–1686; *J. Electrochem. Soc.* **131** (1984) no. 2, 349–351.
[647] Toyo Soda, JP 81 139 686, 1981; JP 82 161 076, 1982.
[648] BASF, DE 1 115 236, 1961 (M. Minsinger).
[649] Du Pont, US 3 406 215, 1968 (H. E. Holmquist).
[650] K. Möbius, *Gummi Asbest Kunststoffe* **35** (1982) 6.
[651] Toyo Soda Mfg Co Ltd., JP 7 525 048, 1975 (H. Kisaki, K. Tsuzuki, Shimizu).
[652] Du Pont, GB 880 077, 1959.
[653] Distillers Co. Ltd., GB 825 609, 1957 (H. P. Crocker).
[654] *Chem. Eng. News* **41** (1963) 38.
[655] Denki Kagaku Kogyo, JP 80 166 970, 1980 (A. Okuda, Y. Totake, H. Matsumura).
[656] Denki Kagaku Kogyo, JP 7 700 926, 1977 (T. Kadowaki, T. Iwasaki, H. Matsumura).
[657] Toyo Soda Mfg Co Ltd., JP 7 426 207, 1974 (S. Ootsuki, H. Ookado, Y. Tamai, T. Fujii).
[658] G. M. Mkryan, R. K. Airapetyan et al., *Arm. Khim. Zh.* **34** (1981) 242–246.
[659] BP Chemicals Ind. Ltd., FR 2 148 586, 1973 (C. W. Capp, P. J. N. Brown).
[660] Bayer, DE 2 642 006, 1976, DE 2 655 007, 1976 (R. Lantzsch, E. Kysela).
[661] F. Asinger: *Petrolchemische Ind.*, vol. **II**, Akademie Verlag, Berlin (GDR) 1971, p. 1330.
[662] E. Profft, H. Oberender, *J. Prakt. Chem.* **25** (1964) no. 4–6, 225–271.
[663] M. Tamele et al., *Ind. Eng. Chem.* **33** (1941) 115.
[664] J. Burgin et al., *Ind. Eng. Chem.* **33** (1941) 385.
[665] G. Hearne et al., *Ind. Eng. Chem.* **33** (1941) 805 and 940.
[666] L. H. Gale, *J. Org. Chem.* **31** (1966) 2475–2480.
[667] *Houben-Weyl*, vol. **V/3**, 586.
[668] T. D. Stewart, D. M. Smitt, *J. Amer. Chem. Soc.* **52** (1930) 2869.
[669] W. Reeve, D. H. Chambers, *J. Amer. Chem. Soc.* **73** (1951) 4499.
[670] W. Reeve, C. S. Prikett, *J. Amer. Chem. Soc.* **74** (1952) 5369.
[671] I. V. Bodrikov et al., *Usp. Khim.* **3** (1966) no. 5, 853–880; *Chem. Abstr.* **69** (1966) 657 003.
[672] A. Striegler, *Chem. Tech.* **9** (1957) 523–529.
[673] Chiyoda Kako Kensetsu Kabushiki Kaisha, DE-AS 1 281 430, 1966.
[674] Z. S. Smoljan, I. V. Bodrikov, *Chim. Prom.* **43** (1967) no. 3, 192.
[675] Hoechst, DE-AS 1 237 557, 1965; DE-AS 1 237 555, 1965; DE-AS 1 243 670, 1965. Showa Denko KK, DE 1 817 281, 1969.
[676] Mitsubishi Petrochemical, JP 117 576, 1977.
[677] F. C. Hymas, *Food* **9** (1940) 254.
[678] C. H. Richardson, H. H. Walkden, *J. Econ. Entomol.* **38** (1945) 471.

[679] Pittsburgh Plate Glass Ind., US 2 356 871, 1940 (E. N. Moffett, R. E. Smith).
[680] Standard Oil Development Corp.US 2 440 494, 1941. United States Rubber Co., US 2 553 430, 1948;2 553 431, 1948; 2 568 872, 1946; 2 569 959, 1948; 2 576 245, 1949; 2 592 211, 1948; 2 594 825, 1947; 2 597 202, 1948; 2 643 991, 1949.
[681] Industrial Rayon Corp., US 2 601 256, 1950. Du Pont, GB 823 345, 1957. V. Gröbe, H. Reichert, W. Maschkin, *Faserforsch. Textiltech.* **15** (1964) no. 10, 463.
[682] Kureha Chemical Ind. Co., JP 6 804 321, 1965.
[683] Kureha Chemical Ind. Co., JP 6 804 326, 1965; *Chem. Abstr.* **70** (1969) 11 103 g.
[684] K. Shinoda, H. Watanabe, *Kogyo Kagaku Zasshi* **70** (1967) 2262; *Chem. Abstr.* **68** (1968) 104 303 w.
[685] L. M. Kogan, *Usp. Khim* **33** (1964) no. 4, 396–417. K. Hisashi, K. Shigeaki, *Sekiyu To Sekiyu Kagaku* **16** (1972) no. 1, 73; *Chem. Abstr.* **76** (1972) 129 588 w.
[686] E. T. McBee, R. E. Hatton, *Ind. Eng. Chem.* **41** (1949) 809.
[687] Y. G. Mamedaliev, M. M. Gusenov et al., SU 161 713, 1963.
[688] Hooker Electrochemical Co., US 2 841 243, 1957.
[689] Siemens AG, Wacker-Chemie, FR 1 564 230, 1968.
[690] Matsushita Electric Industrial Co., Ltd., JP 7 010 150, 1970; *Chem. Abstr.* **73** (1970) 360 899.
[691] L. M. Kogan et al., SU 151 156, 1962.
[692] P. A. Bolley, *Liebigs Ann. Chem.* **106** (1858) 230.
[693] Daisoh Kabushiki, JP-Kokai 6 239 774, 1994 (Y. Itaya, H. Takakata).
[694] Hoechst Aktiengesellschaft, EP-A 0 625 196, 1994 (D. Bewart, W. Freyer).
[695] A. S. Bratolyubov, *Russ. Chem. Rev. Engl. Transl.* **30** (1961) 602.
[696] A. M. Kuliev et al., *Pris. Smaz. Masl.* **116** (1967).
[697] T. T. Vasil'eva et al., *Isv. Akad. Nauk. SSSR, Ser. Khim.* (1973) 1068.
[698] V. B. Shumyanskaya et al., *Sb. Tr. Vses. Nauchno Issled. Inst. Nov. Stroit. Mater.* **32** (1972) 46.
[699] World Health Organisation, International Programme on Chemical Safety: Environmental Health Criteria 181, Chlorinated Paraffins, Geneva 1997, in press.
[700] M. D. Müller, P. P. Schmid, *J. High Resolut. Chromatogr. and Chromatogr. Commun.* **7** (1984) 33.
[701] International Agency for Research on Cancer: Chlorinated Paraffins, vol. **48**, Lyon 1990, p. 55.
[702] S. C. Hasmall, J. R. Foster, C. R. Elcombe, *Arch.Toxicol.* (1997) in press.
[703] I. Wyatt et al., *Arch. Toxicol.* (1997) in press.
[704] R. T. Cook, *Speciality Chemicals* (1995) 159.
[705] L. H. Hoorsley et al., *Azeotropic Data III,* Advances in Chemistry Series, No. 116, Am. Chem. Soc., Washington, D.C., 1973.
[706] M. Lecat, *Tables azeotropiques,* 2nd ed., Chez l'Auteur, Brussels 1949.
[707] Dow Chemical, US 1 607 618, 1926 (W. J. Hale, E. C. Britton).
[708] J. W. Hale, E. C. Britton, *Ind. Eng. Chem.* **20** (1928) 114–124.
[709] R. M. Crawford, *Chem. Eng. News* **25** (1947) no. 1, 235–236.
[710] Dow Chemical, US 2 432 551, 1947; 2 432 552 (W. H. Williams et al.).
[711] Oiwa et al., *Bl. Inst. Chem. Res. Kyoto* **29** (1952) 38–56.
[712] M. Bonnet et al., *J. Org. Chem.* **48** (1983) 4396–4397.
[713] Hoechst, EP 51 236, 1980 (D. Rebhan, H. Schmitz).
[714] *Chem. Eng.,* March 31. 1975, 62.
[715] *Chem. Ind.* 37, p. 106, February 1985.
[716] Chemical Marketing Reporter, vol. **226,** no. 13, September 24, 1984.
[717] *Japan Chem. Week,* April 15, 1982, 8.
[718] U.S. Interdepartmental Task Force on PCBs, COM – 72-10419, Washington 1972.

[719] F. D. Kover, *Environmental Protection Agency*, Washington, D.C., 1975, EPA 560/8-75-001, p. 8, p. 11.
[720] U.S. Federal Register, vol. **48**, no. 89, May 6, 1983, p. 20 668.
[721] Figure from an unpublished survey conducted by Bayer (produced in the FRG for PCN only).
[722] Dow Chemical, US 2 886 605, 1959 (H. H. McClure, J. S. Melbert, L. D. Hoblit).
[723] Hoechst, BE 660 985, 1964.
[724] PPG, US 4 017 551, 1977 (J. E. Milam, G. A. Carlson).
[725] K. Shinoda, H. Watanabe, *Kogyo Kagaku Zasshi* **73** (1970) no. 11, 2549–2551; *Chem. Abstr.* **77** (1972) 5072.
[726] S. Yamamoto et al., *Kagaku to Kogyo (Osaka)* **45** (1971) no. 9, 456–460; *Chem. Abstr.* **76** (1972) 99244.
[727] Moscow Gubkin Petrochem., SU 514 860, 1976; SU 536 185, 1976; SU 540 657, 1976; SU 594 995, 1978 (J. M. Kolesnikow et al.).
[728] Dow Chemical, US 3 636 171, 1972 (K. L. Krumel, J. R. Dewald).
[729] T. Kametani et al., *Yakugaku Zasshi* **88** (1968) no. 8, 950–953; *Chem. Abstr.* **70** (1978) 19681.
[730] PPG, US 4 235 825, 1980 (J. E. Milam).
[731] Sergeev E. V., SU 650 985, 1979.
[732] Nippon Kayaku Co., JP-Kokai 64/3821, 1961.
[733] Union Carbide, US 3 226 447, 1965 (G. H. Bing, R. A. Krieger).
[734] Ihara Chem. Ind. Co., Ltd., EP 118 851, 1984 (Y. Higuchi, T. Suzuki).
[735] R. Commandeur et al., *Nouv. J. Chim.* **3** (1979) no. 6, 385–391; PCUK, DE-OS 3 344 870, 1983 (R. Commandeur, H. Mathais).
[736] Dow Chemical, US 4 186 153, 1980; US 4 205 014, 1980 (J. W. Potts).
[737] Sanford Chem. Co., US 3 557 227, 1971 (M. M. Fooladi).
[738] Von den Berg et al., NL 72/1533, 1974; *Ind. Eng. Chem. Fundam.* **15** (1976) no. 3, 164–171.
[739] R. B. McMullin, *Chem. Eng. Prog.* **44** (1948) no. 3, 183–188.
[740] H. J. Hörig, D. Förster, *Chem. Tech. (Leipzig)* **22** (1970) no. 8, 460–461.
[741] S. Sacharjan, *Chem. Tech. (Leipzig)* **31** (1979) no. 5, 229–231.
[742] N. T. Tishchenko, V. M. Sheremet, A. I. Triklio, *Khim. Prom-st. (Moscow)* 1980, no. 7, 399–401.
[743] *Ullmanns*, 4th ed., **9**, p. 506.
[744] Raschig GmbH, DE 539 176, 1930; DE 575 765, 1931; US 1 963 761, 1934; US 1 964 768, 1933 (W. Prahl, W. Mathes).
[745] Hooker Chem. Corp., DE 1 443 988, 1962; CH 432 485, 1962; US 3 644 542, 1972 (W. Prahl et al.).
[746] Union Carbide, BE 636 194, 1962 (C. C. Diffteld, T. T. Szabo, C. A. Rivasi).
[747] Vulcan Materials Co., DE 1 518 164, 1963 (E. S. Penner, A. L. Malone).
[748] Hodogaya Chem. Ind. Co., JP-Kokai 65/12 613, 1962.
[749] Asahi Chem. Ind. Co., JP-Kokai 67/3619, 1965.
[750] W. H. Prahl, DE-OS 2 442 402, 1975.
[751] Sumitomo, JP-Kokai 73/39 449, 1971; DE 2 232 301, 1972 (E. Ichiki et al.); *Chem. Abstr.* **79** (1973) 91 746.
[752] C. A. Gorkowa, A. M. Potapow, C. R. Rafikow, SU 654 600, 1979.
[753] U. Schönemann, *Chem. Ind. (Düsseldorf)* **5** (1953) no. 7, 529–538.
[754] Gulf Res. a. Develop. Co., US 3 591 644; 3 591 645, 1971 (V. A. Notaro, C. M. Selwitz).
[755] ICI, GB 388 818, 1933 (T. S. Wheeler).
[756] Dow Chemical, US 2 123 857, 1938 (J. P. Wibaut, L. van de Lande, G. Wallagh).

[757] C. E. Kooyman, *Adv. Free Radical Chem.* **1** (1965) 137–153.
[758] W. Dorrepaal, R. Louw, *Recl. Trav. Chim. Pays-Bas* **90** (1971) 700–704; *J. Chem. Soc. Perkin Trans 2* (1976) 1815–1818; *Int. J. Chem. Kinet.* **10** (1978) no. 3, 249–275.
[759] P. Kovacic, N. O. Brace, *J. Am. Chem. Soc.* **76** (1954) 5491–5494.
[760] P. Kovacic, A. K. Sparks, *J. Am. Chem. Soc.* **82** (1960) 5740–5743; *J. Org. Chem.* **26** (1961) 1310–1213.
[761] Pullman Inc., DE 1 468 141, 1963 (H. Heinemann, K. D. Miller).
[762] National Distillers Chem. Corp., US 3 670 034, 1972 (R. E. Robinson).
[763] J. S. Grossert, G. K. Chip, *Tetrahedron Lett.* **30** (1970) 2611–2612.
[764] Nikkei Kako, Co., Ltd., Sugai Kogy Ind., Ltd., DE-OS 2 230 369, 1972 (K. Sawazaki, H. Fujii, M. Dehura).
[765] Y. Tamai, Y. Ohtsuka, *Bull. Chem. Soc. Japan* **46** (1973) no. 7, 1996–2000.
[766] Monsanto, US 3 029 296, 1962 (W. A. White, R. A. Rührwein).
[767] O. Silberrad, *J. Chem. Soc. (London)* **121** (1922) 1015–1022; US 259 329, 1925.
[768] M. Ballester, C. Molinet, J. Castaner, *J. Am. Chem. Soc.* **82** (1960) 4254–4258.
[769] M. Ballester, *Bull. Soc. Chim. Fr.* 1966, 7–15.
[770] S. Kobayashi, JP-Kokai 74/43 901, 1972; *Chem. Abstr.* **81** (1974) 35 683.
[771] Du Pont, US 4 327 036, 1982 (F. D. Marsh).
[772] R. J. Cremlyn, T. Cronje, *Phosphorus Sulfur* **6** (1979) no. 3, 495–504.
[773] N. C. Weinberg, "Tech. Elektroorg. Synth.," *Tech. Chem.* (N.Y.) 5 (1975) Pt. 2, 1–82.
[774] Texaco Inc., US 3 692 646, 1972 (W. B. Mather, H. Junction, E. R. Kerr).
[775] Sumgait Petrochem., SU 612 923, 1978.
[776] C. Galli, *Tetrahedron Lett.* **21** (1960) no. 47, 4515–4516.
[777] S. Oae, K. Shinhama, Y. H. Kim, *Bull. Chem. Soc. Japan* **53** (1980) no. 7, 2023–2026.
[778] *BIOS* 986, 151.
[779] A. A. Ponomarenko, *Zh. Obshch. Khim.* **32** (1962) no. 12, 4029–4040.
[780] N. N. Vorozhtsov et al., *Zh. Obshch. Khim.* **31** (1961) no. 4, 1222–1226.
[781] Monsanto, GB 948 281, 1962; DE 1 468 241, 1963 (E. B. McCall, W. Cummings).
[782] J. Blum, *Tetrahedron Lett.* **26** (1966) 3041–3045.
[783] Du Pont, DE 1 804 695, 1968 (C. C. Cumbo).
[784] Bayer, DE 2 326 414, 1973; US 3 897 321, 1975; US 4 012 442, 1977 (H. U. Blank, K. Wedemeyer, J. Ebersberger).
[785] Mitsui Toatsu Chem. Inc., JP-Kokai 80/104 220, 1979; *Chem. Abstr.* **94** (1974) 30 327.
[786] Monsanto, GB 957 957, 1962 (E. B. McCall, P. J. S. Bain).
[787] Dow Chemical, US 1 946 040, 1934 (W. C. Stösser, F. B. Smith).
[788] S. Rittner, S. Steiner, *Chem. Ing. Tech.* **57** (1985) no. 2, 91–102.
[789] Politechnika Slaska, PL 78 610, 1976 (W. Zielinski, J. Suwinski).
[790] Mitsui Toatsu Chem., Inc., JP-Kokai 78/44 528, 1976.
[791] Dow Chemical, US 3 170 961, 1965 (E. C. Britton, F. C. Beman).
[792] Dow Chemical, US 1 923 419, 1933 (E. C. Britton).
[793] Rhône-Progil, DE 2 332 889, 1973 (R. Chanel et al.).
[794] Mitsui Toatsu Chem., Inc., JP-Kokai 77/6229, 1975 (T. Kiyoura); *Chem. Abstr.* **87** (1972) 84 178.
[795] S. Kikkawa et al., *Kogyo Kagaku Zasshi* **73** (1970) no. 5, 964–969; *Chem. Abstr.* **73** (1970) 87 542.
[796] E. Pfeil, *Angew. Chem.* **65** (1953) no. 6, 155–158.

[797] W. Brackmann, P. J. Smit, *Recl. Trav. Chim. Pays-Bas* **85** (1966) no. 8, 857–864; Shell Int. Res., NL 6 515 451, 1967.
[798] Mobil Oil Corp., US 3 358 046, 1967 (R. D. Offenhauer, P. G. Rodewald).
[799] Sumitomo, JP-Kokai 73/86 828, 1972 (H. Suda, S. Yamamoto, Y. Sato); *Chem. Abstr.* **80** (1974) 59 655.
[800] Allied, US 2 866 828, 1958 (J. A. Crowder, E. E. Gilbert).
[801] Ethyl Corp., US 2 943 114, 1960 (H. E. Redman, P. E. Weimer).
[802] Columbia-Southern Chem. Corp., US 2 819 321, 1958 (B. O. Pray).
[803] BASF, DE 1 020 323, 1956 (E. Merkel).
[804] Union Carbide, US 2 666 085, 1954 (J. T. Fitzpatrick).
[805] Y. G. Erykalov et al., *J. Org. Chem. USSR (Engl. Transl.)* **9** (1973) 348–351; SU 392 059, 1973; SU 351 818, 1972.
[806] Kanegabuchi Chem. Ind. Co., JP-Kokai 57/510, 1957 (J. Nishida, J. Mikami, S. Kawato); *Chem. Abstr.* **52** (1958) 2905 E.
[807] H. Scholz et al., DD 9869, 1955.
[808] D. Vrzgula, Z. Stoka, J. Auerhan, CS 92 749, 1959.
[809] Mitsubishi, JP-Kokai 75/32 132, 1973; *Chem. Abstr.* **83** (1975) 96 688.
[810] BASF, DE 1 002 302, 1952 (G. Grassl).
[811] Asahi Glass Co., JP-Kokai 55/7421, 1955 (M. Fujioka et al.); *Chem. Abstr.* **51** (1957) 17 989 G.
[812] Ethyl Corp., GB 738 011, 1952 (L. Merritt).
[813] Diamond Alkali Co., US 2 914 574, 1959 (E. Zinn, J. J. Lukes).
[814] Pennsalt Chem. Corp., US 2 914 573, 1959 (G. McCoy, C. Inman).
[815] sterreichische Chem. Werke, AT 261 576, 1967 (H. Maier).
[816] Uzima Chimica Turda, BG 54 958, 1972 (J. Todea).
[817] Combinatul Chim. Riminicu-Vilcea, BG 61 469, 1976 (L. Raduly et al.).
[818] *Ullmann*, 4th ed., **9**, 508.
[819] Ishihara Sangyo Kaisha, Ltd., JP-Kokai 81/92 227, 1979; *Chem. Abstr.* **95** (1981) 203 526.
[820] Hoechst, DE-OS 2 920 173, 1979 (G. Volkwein, K. Bässler, H. Wolfram).
[821] Ishihara Sangyo Kaisha, Ltd., JP-Kokai 79/112 827, 1978; US 4 368 340, 1983; DE-OS 3 047 376, 1980 (T. R. Nishiyama et al.).
[822] BASF, DE 947 304, 1954 (F. Becke, H. Sperber).
[823] Bayer, DE-OS 3 134 726, 1981 (G.-M. Petruck, R. Wambach).
[824] Rhône Poulenc Ind., EP 43 303, 1982 (G. Soula).
[825] Sonford Chem. Co., US 3 557 227, 1971 (M. M. Fooladi).
[826] Dover Chem. Corp., US 3 274 269, 1966 (R. S. Cohen).
[827] Jefferson Chem. Co., Inc., DE-OS 2 054 072, 1971 (S. Burns).
[828] T. Magdolen, D. Vrzgula, CS 122 646, 1967.
[829] Hoechst, DE 1 028 978, 1955 (H. Klug, J. Kaupp).
[830] Nippon Soda Co., JP-Kokai 75/7572, 1975 (K. Iwabuchi); *Chem. Abstr.* **51** (1957) 17 989 F.
[831] Pechiney, US 2 852 571, 1958 (J. Frejacques, H. Guinot).
[832] C. Popa, C. Raducanu, A. Stefanescu, *Rev. Chim. (Bucharest)* **15** (1964) no. 10, 601–607.
[833] Phillips Petr. Co., US 4 096 132, 1978 (J. T. Edmonds Jr.).
[834] Bayer, EP 23 314, 1981; EP 40 747, 1981; EP 65 689, 1982 (K. Idel et al.).
[835] Swann Research Inc., US 1 954 469, 1934 (C. F. Booth).
[836] *Ullmann*, 4th ed., **8**, 344, 377–379.
[837] Ihara Chem. Ind. Co., Ltd., EP 2 749, 1978 (M. Shigeyasu et al.); Dynamit Nobel AG, EP 70 393, 1982; EP 121 684, 1984 (M. Feld).

[838] Nippon, DE-OS 2 711 332, 1977 (H. Hayami, H. Shimizu, G. Takasaki).
[839] N. N. Lebedev, J. J. Baltadzhi, *Zh. Org. Khim.* **5** (1969) no. 9, 1604–1605.
[840] Heyden Newport Chem. Co., US 3 000 975, 1961 (E. P. Di Bella).
[841] Moscow Gubkin Institute, SU 1 002 297, 1983 (N. N. Belov et al.).
[842] *Ullmann*, 4th ed., **9**, 512–513.
[843] Hooker Chem. & Plas. Corp., US 4 013 730, 1977 (J. C. Graham).
[844] Tokuyama Soda Co. Ltd., JP-Kokai 81/99 429, 1980; *Chem. Abstr.* **95** (1981) 186 820.
[845] Tenneco Chem., US 4 013 144, 1977 (E. P. Di Bella).
[846] Hooker Chem. & Plas. Corp., US 4 031 142, 1977; DE-OS 2 634 340, 1976 (J. C. Graham).
[847] Tenneco Chem., US 3 317 617, 1967 (E. P. Di Bella).
[848] Hodogaya Chem. Co., Ltd., JP-Kokai 81/110 630, 1980; *Chem. Abstr.* **96** (1982) 34 810. EP 63 384, 1982 (R. Hattori et al.).
[849] Ihara Chem. Ind. Co., Ltd., JP-Kokai 81/105 752, 1980; US 4 289 916, 1981; DE-OS 3 023 437, 1980 (J. Nahayama, C. Yazawa, K. Jamanski).
[850] Hooker Chem. & Plas. Corp., US 4 024 198, 1977 (H. E. Buckholtz); US 4 069 263, 1978; US 4 069 264, 1978 (H. C. Lin).
[851] Hooker Chem. & Plas. Corp., US 4 250 122, 1981 (H. C. Lin, S. Robots).
[852] P. B. De La Mare, P. W. Robertson, *J. Chem. Soc.* 1943, 279–281.
[853] H. G. Haring, *Chem. Eng. Monogr.* **15** (1982) 193–208, 208 A.
[854] Hoechst, Krebskosmos GmbH, DE-OS 2 536 261, 1975 (H. Horst et al.).
[855] J. M. Ballke, J. Liaskar, G. B. Lorentzen, *J. Prakt. Chem.* **324** (1982) no. 3, 488–490.
[856] A. B. Solomonov, P. P. Gertsen, A. N. Ketov, *Zh. Prikl. Khim. (Leningrad)* **42** (1969) no. 6, 1418–1420; *Zh. Prikl. Khim. (Leningrad)* **43** (1970) no. 2, 471–472.
[857] A. M. Potapov, S. R. Rafikov, *Izv. Akad. Nauk SSSR, Ser. Khim.* **31** (1982) no. 6, 1334–1338.
[858] Ajinomoto Co., Inc., JP Kokai 73/81 822, 1972 (R. Fuse, T. Inoue, T. Kato); *Chem. Abstr.* **80** (1974) 108 160.
[859] Gulf Research and Development Co., US 3 644 543, 1972 (V. A. Notaro); US 3 717 684, 1973 (V. A. Notaro, C. M. Selwitz); US 3 717 685, 1973 (H. G. Russell, V. A. Notaro, C. M. Selwitz).
[860] Int. Minerals & Chem. Corp., JP Kokai 74/76 828, 1972 (J. T. Traxler); *Chem. Abstr.* **85** (1976) 6167.
[861] H. Inoue, M. Izumi, E. Imoto, *Bull. Chem. Soc. Japan* **47** (1974) no. 7, 1712–1716.
[862] T. Tkaczynski, Z. Winiarski, W. Markowski, *Przem. Chem.* **58** (1979) no. 12, 669–670; *Chem. Abstr.* **92** (1980) 180 745.
[863] Sugai Chem. Ind. Co., Ltd., JP Kokai 83/61 285, 1981; *Chem. Abstr.* **99** (1983) 79 068.
[864] Shell Int. Res., FR 1 343 178, 1963.
[865] R. Nishiyama et al., *J. Soc. Org. Synth. Chem. Tak.* **23** (1965) 515–525.
[866] Hoechst, EP 72 008, 1982 (H.-J. Arpe, H. Litterer, N. Mayer).
[867] BIOS Final Rep. No. 1145.
[868] Toray, EP 62 261, 1982 (K. Tada et al.).
[869] Tenneco Chem. Inc., FR 1 489 686, 1967; US 3 366 698, 1968 (E. P. Di Bella).
[870] Hooker Chem. & Plas. Corp., DE-OS 2 258 747, 1972; US 4 006 195, 1977 (S. Gelfand).
[871] Bayer, DE-OS 2 523 104, 1975 (H. Rathjen).
[872] Tenneco Chem. Inc., US 4 031 146, 1977 (E. P. Di Bella).
[873] Bayer, EP 46 555, 1982 (G.-M. Petruck, R. Wambach).
[874] Shell Int. Res., GB 1 110 030, 1968 (C. F. Kohll, H. D. Scharf, R. Van Helden).
[875] Shell Int. Res., NL 6 907 390, 1970 (D. A. Was).
[876] Rhône-Poulenc Ind., FR 2 475 535, 1981 (J.-C. Lauct, J. Bourdon).

[877] Toray, EP 125 077, 1984 (K. Miwa et al.).
[878] Tenneco Chem. Inc., US 4 031 145, 1977 (E. P. Di Bella).
[879] T. Asinger, G. Loch, *Monatsh. Chem.* **62** (1933) 345.
[880] *Ullmann*, 4th ed., **9**, pp 513–514.
[881] H. C. Brimelow, R. L. Jones, T. P. Metcalfe, *J. Chem. Soc.* 1951, 1208–1212.
[882] Tenneco Chem. Inc., US 3 517 056, 1970 (J. F. De Gaetano).
[883] Tenneco Chem. Inc., US 3 692 850, 1972 (E. P. Di Bella).
[884] Kureha Chem. Ind. Co., Ltd., JP-Kokai 70/28 367, 1965 (M. Ishida); *Chem. Abstr.* **74** (1971) 3397.
[885] A. L. Englin et al., SU 1 727 736, 1965.
[886] Swann Research, Inc., US 1 836 180, 1929 (C. R. McCullough, R. C. Jenkins).
[887] C. H. Penning, *Ind. Eng. Chem.* **22** (1930) no. 11, 1180–1182.
[888] K. Soldner, G. Gollmer, *Elektrizitätswirtschaft* **81** (1982) no. 17/18, 573–583.
[889] H. R. Buser, H.-P. Bosshardt, C. Rappe, *Chemosphere* **1** (1978) 109–119.
[890] H. R. Buser, C. Rappe, *Chemosphere* **3** (1979) 157–174.
[891] R. R. Bumb et al., *Science (Washington, DC)* **210** (1980) no. 4468, 385–390.
[892] G. G. Choudhry, O. Hutzinger, *Toxicol. Environ. Chem. Rev.* **5** (1982) 1–65, 67–93, 97–151, 153–165 and 295–309.
[893] A. Schecter, *Chemosphere* **12** (1983) no. 4/5, 669–680.
[894] Environment Committee, Paris, March 1982, Report on the implementation by member countries of decision by the council on the protection of the environment by control of polychlorinated biphenyls, ENV (82) 5.
[895] *Ullmann*, 4th ed., **9**, pp. 515–518.
[896] O. Hutzinger, S. Safe, V. Zitko: *The Chemistry of PCB's*, CRC Press, Cleveland 1974.
[897] Zehnte Verordnung zur Durchführung des Bundes-Immissionsschutzgesetzes (Tenth Implementing Order Issued Pursuant to the Federal Immission Protection Act of the FRG); restrictions on PCB, PCT and VC. July 26, 1978.
[898] Council Directive of July 27, 1976, harmonizing the legal and administrative provisions of the member states restricting the sale and use of dangerous substances and preparations (76/769/EEC).
[899] Gesetz über die Beseitigung von Abfällen (Waste Disposal Act) of January 5, 1977 (FRG).
[900] Verordnung zur Bestimmung von Abfällen nach § 2 Abs. 2 des Abfallbeseitigungsgesetzes (Order Regulating the Classification of Wastes According to § 2 Para. 2 of the Waste Disposal Act), May 24, 1977 (FRG).
[901] lbeseitigung (act embodying measures to ensure the disposal of scrap oil), December 11, 1979 (FRG).
[902] Erste Allgemeine Verwaltungsvorschrift zum Bundes-Immissionsschutzgesetz (First General Administrative Provision Issued Pursuant to the Federal Immission Protection Act) – Technische Anleitung Luft (Technical Guideline – Air). August 28, 1974 (FRG).
[903] U.S. Public Law 94-469, October 11, 1976, Section 6 (e) of TSCA.
[904] U.S. Environmental Protection Agency 560/ 6-75-004, 1976 (Proceedings of the National Conference on PCBs, Chicago, November 19–21, 1975).
[905] *U.S. Environmental Protection Agency 230-03/ 79-001*, 1979, Washington, D.C., (Support Document/Voluntary Environmental Impact Statement for PCBs).
[906] NIOSH, Washington, D.C., U.S. Government Printing Office, 1977, pp. 40–49 (summary of the literature on the toxicity of polychlorinated biphenyls).
[907] C. Nels, *Elektrizitätswirtschaft* **83** (1984) no. 8, 375–379 (disposal of askarels).

[908] M. G. McGraw, *Electrical World*, *(The PCB Problem: separating fact from fiction)* February 1983, pp. 49–72.
[909] G. T. Hunt, P. Wolf, P. F. Fennelly, *Environ. Sci. Technol.* **18** (1984) 171–179.
[910] *NATO-CC MS Report II, (Disposal of Hazardous Wastes – Thermal Treatment)* March 23, 1981, pp. 84–89.
[911] General Electric Co., US 4 353 798, 1982 (S. D. Foss).
[912] General Electric Co., US 4 353 793, 1982 (D. J. Brunelle).
[913] General Electric Co., US 4 351 718, 1982 (D. J. Brunelle).
[914] Goodyear Tire & Rubber Co., US 4 284 516, 1981 (D. K. Parker, R. J. Steichen). D. K. Parker, W. L. Cox, *Plant Eng.* (1980) Aug. 21, 133–134.
[915] Sunohio, GB 2 063 908, 1981 (O. C. Norman, L. H. Handler); GB 2 081 298, 1982 (O. D. Jordan).
[916] ITO T., JP-Kokai 81/111 078, 1980.
[917] Fuji Elektric Co., Ltd., DE-OS 3 036 000, 1980 (Masuda et al.).
[918] University of Waterloo, US 4 326 090, 1982 (J. G. Schmitz, G. L. Bubbar).
[919] The Franklin Institute, DE-OS 3 033 170, 1980; EP 60 089, 1982 (L. L. Pytlewski, K. Krevitz, A. B. Smith).
[920] ITO T., JP Kokai 82/29 373, 1980; *Chem. Abstr.* **96** (1982) 222 772.
[921] Queens University, Kingston ONT, US 4 345 983, 1982 (J. K. S. Wan).
[922] Tokyo Shibaura Elec., Ltd., JP-Kokai 82/166 175, 1981; *Chem. Abstr.* **98** (1983) 59 404.
[923] Rockwell Int. Corp., WO 8 002 116; EP 26 201, 1980 (L. F. Grantham et al.).
[924] L. Bjorklund, I. Fäldt, WO 8 200 509, 1980.
[925] SOTOO Z., JP-Kokai 79/151 953, 1978; *Chem. Abstr.* **93** (1980) 31 358.
[926] ICI Australia, Ltd., WO 8 202 001, 1980 (V. Vasak, K. K. Mok, J. Sencar).
[927] K. Ballschmiter, M. Zell, *Fresenius Z. Anal. Chem.* **302** (1980) 20–31.
[928] H. M. Klimisch, D. N. Ingebrigtson, *Anal. Chem.* **52** (1980) no. 11, 1675–1678.
[929] P. W. Albro, J. T. Corbett, J. C. Schröder, *J. Chromatogr.* **205** (1981) 103–111.
[930] D. A. Newton et al., *J. Chromatogr. Sci.* **21** (1983) 161.
[931] E. Schulte, R. Malisch, *Fresenius Z. Anal. Chem.* **314** (1983) 545–551 and 319 (1984) 54–59.
[932] W. J. Dunn et al., *Anal. Chem.* **56** (1984) no. 8, 1308–1313.
[933] R. H. Lin et al., *Anal. Chem.* **56** (1984) no. 11, 1808–1812.
[934] Versar Inc., National Technical Informal Service, *Industrial Use and Environmental Distribution* (NTIS 252-012/3 W P) 1976.
[935] Fireproof Products Co., US 914 222, 1909; US 914 223, 1909; Halogen Products Co., US 1 111 289, 1913 (J. W. Aylsworth).
[936] F. D. Kover, *Environmental Hazard Assessment Report*, Chlorinated Naphthalenes, Environmental Protection Agency 560/8-75-001, 1975.
[937] *U.S. Federal Register* vol. **48,** no. 89 of May 6, 1983, pp. 20 668–20 679; vol. **49,** no. 166 of August 24, 1984, pp. 33 649–33 654.
[938] U. A. Th. Brinkman, H. G. M. Reymer, *J. Chromatogr.* **127** (1976) 203–243.
[939] AUER-Technikum, no. 10, 1982, Published by Auergesellschaft GmbH, Berlin 44 (FRG).
[940] P. Ferrero, C. Caflisch, *Helv. Chim. Acta* **11** (1928) 795–812.
[941] D. H. S. Horn, F. L. Warrer, *J. Chem. Soc.* 1946, 144.
[942] Frontier Chemical Co. (Division of Vulcan Materials Co.), US 3 413 357, 1968 (K. F. Bursack, E. L. Johnston, H. J. Moltzau).
[943] E. G. Turner, W. P. Wynne, *J. Chem. Soc.* 1941, 244–245.
[944] W. Schwemberger, W. Gordon, *Zh. Obshch. Khim.* **2** (1932) 926; **5** (1934) 695.

[945] Bayer, DE 844 143, 1951; DE 946 055, 1953 (H. Vollmann).
[946] A. Ledwith, P. J. Russel, *J. Chem. Soc. Perkin Trans.* **2** (1975) no. 14, 1503–1508.
[947] Dow Chemical Co., US 1 917 822, 1933 (E. C. Britton, W. R. Reed).
[948] *Ullmann,* 4th ed., **9,** pp. 512–520.
[949] S. Safe, O. Hutzinger, *J. C. S. Chem. Commun.* 1972, 260–262.
[950] M. P. Gulan, D. D. Bills, T. B. Putnam, *Bull. Environ. Contam. Toxicol.* **11** (1974) no. 5, 438–441.
[951] U. A. T. Brinkman, A. De Kok, *J. Chromatogr.* **129** (1976) 451.
[952] U. A. T. Brinkman et al., *J. Chromatogr.* **152** (1978) 97–104.
[953] M. Cooke et al., *J. Chromatogr.* **156** (1978) 293–299.
[954] P. W. Albro, C. E. Parker, *J. Chromatogr.* **197** (1980) 155–169.
[955] P. A. Kennedy, D. J. Roberts, M. Cooke, *J. Chromatogr.* **249** (1982) 257–265.
[956] J. J. Donkerbroek, *J. Chromatogr.* **255** (1983) 581–590.
[957] Koppers Company, Inc., U.S., Product bulletin: *Halowaxes, Chlorinated Naphthalene Oils and Waxlike Solids.*
[958] J. Todt, *Chem. Ing. Tech.* **54** (1982) no. 10, 932.
[959] Hodogaya Chem. Co., Ltd., JP-Kokai 79/78 399, 1977 (K. Imaizumi, A. Arai, G. Tsunoda); *Chem. Abstr.* **91** (1979) 142 751.
[960] Sumitomo Chem. Co., Ltd., JP-Kokai 79/43 194, 1977 (Y. Matsushita, J. Kamimura); *Chem. Abstr.* **91** (1979) 76 268.
[961] M. F. Nathan, *Chem. Eng.* Jan. 30. 1978, 93–100.
[962] Celanese Corp., US 4 276 179, 1981 (J. W. Soehngen).
[963] Hooker Chem. Corp., US 2 949 491, 1960 (J. R. Rucker).
[964] Osaka Prefecture, JP-Kokai 78/84 923, 1976 (Y. Hatano); *Chem. Abstr.* **89** (1978) 163 306.
[965] Teijin K. K., JP-Kokai 79/59 233, 1977 (T. Yamaji); *Chem. Abstr.* **91** (1979) 192 958.
[966] L. H. Horsley, *Azeotropic Data III,* Advances in Chemistry Series No. 116; Amer. Chem. Soc., Washington DC, 1973.
[967] A. V. Egunov, B. J. Konobeev, E. A. Ryabov, T. J. Gubanova, *Zh. Prikl. Khim. (Leningrad)* **46** (1973) no. 8, 1855–1856;*Chem. Abstr.* **79** (1973) 140 122 x.
[968] H. C. Haas, D. J. Livingston, M. Saunders, *J. Polym. Sci.* **15** (1955) 503.
[969] R. Pieck, J. C. Jungers, *Bull. Soc. Chim. Belg.* **60** (1951) 357–384.
[970] M. L. Poutsma in E. S. Huyser (ed.): *Methods in Free-Radical Chemistry,* Free-Radical Chlorination, vol. **I,** Marcel Dekker, New York 1969, pp. 79–193.
[971] H. G. Haring, H. W. Knol, *Chem. Process Eng.* **1964,** no. 10, 560–567; **1964,** no. 11, 619–623; **1964,** no. 12, 690–693; **1965,** no. 1, 38–40.
[972] H. G. Haring: "Chlorination of Toluene," *Chem. Eng. Monogr.* **15** (1982) 193–208, 208 A.
[973] J. Y. Yang, C. C. Thomas, Jr., H. T. Cullinan, *Ind. Eng. Chem. Process Des. Dev.* **9** (1970) no. 2, 214–222.
[974] W. Hirschkind, *Ind. Eng. Chem.* **41** (1949) no. 12, 2749–2752.
[975] BASF, DE-AS 1 245 917, 1967 (W. Beckmann). BASF, DE-AS 1 249 831, 1967 (W. Beckmann, W. L. Kengelbach, H. Metzger, M. Pape). Toyo Rayon Kabushiki Kaisha, DE-AS 1 191 342, 1965 (Y. Ito, R. Endoh).
[976] Bayer, DE-OS 2 152 608, 1973; DE-OS 2 227 337, 1974; US 3 816 287, 1974 (W. Böckmann, R. Hornung).
[977] Hoechst, DE-AS 2 530 094, 1976 (R. Lademann, F. Landauer, H. Lenzmann, K. Schmiedel, W. Schwiersch).
[978] T. Yokota, T. Iwano, A. Saito, T. Tadaki, *Int. Chem. Eng.* **23** (1983) no. 3, 494–502.

[979] S. K. Fong, J. S. Ratcliffe, *Mech. Chem. Eng. Trans.* **8** (1972) no. 1, 1–8, 9–14.

[980] Hoechst, DE-AS 2 139 779, 1972 (H. Schubert, K. Baessler).

[981] The Goodyear Tire and Rubber Co., US 2 844 635, 1958 (R. H. Mayor).

[982] The Goodyear Tire and Rubber Co., US 2 817 632, 1957 (R. H. Mayor).

[983] Nihon Nohyaku Co., Ltd., Japan, *Kokai Tokkyo Koho* **80** (1980) 113 729; *Chem. Abstr.* **94** (1981) 174 606 h. Hodogaya Chemical Co., Ltd., JP-Kokai 76 08 221, 1976 (R. Sasaki, T. Kanashiki, K. Fukazawa); *Chem. Abstr.* **84** (1976) 164 373 j.

[984] Hooker Electrochemical Co., US 2 695 873, 1954 (A. Loverde).Mitsubishi Gas Chemical Co., Inc., DE-OS 2 303 115, 1973 (T. Sano, M. Doya).

[985] Hodogaya Chemical Co., Ltd., JP-Kokai 76 08 223, 1976 (M. Fuseda, K. Ezaki); *Chem. Abstr.* **85** (1976) 32 599 y. Hodogaya Chemical Co., Ltd., JP-Kokai 76 08 222, 1976 (R. Sasaki, T. Kanashiki, K. Fukazawa); *Chem. Abstr.* **84** (1976) 164 374 k.

[986] Dow Chemical Co., Ltd., GB 2 087 377 A, 1982 (D. P. Clifford, R. A. Sewell); Diamond Alkali Co., US 3 350 467, 1967 (R. H. Lasco).G. Messina, M. Moretti, P. Ficcadenti, G. Cancellieri, *J. Org. Chem.* **44** (1979) no. 13, 2270–2274.

[987] Velsicol Chemical Co., US 3 363 013, 1968 (G. D. Kyker); The Goodyear Tire and Rubber Co., US 2 817 633, 1957; US 2 817 632, 1957 (R. H. Mayor).

[988] Diamond Alkali Co., US 3 580 854, 1971; US 3 703 473, 1972 (R. H. Lasco).

[989] Hodogaya Chemical Co., Ltd., JP-Kokai 82 98 224, 1982; *Chem. Abstr.* **97** (1982) 181 930 z.

[990] Hodogaya Chemical Co., Ltd., JP-Kokai 82 98 225, 1982; *Chem. Abstr.* **97** (1982) 162 552 m.

[991] S. Niki, JP-Kokai 77 111 520, 1977; *Chem. Abstr.* **88** (1978) 50 451 k.

[992] Shin-Etsu Chemical Industry Co., Ltd., JP-Kokai 78 77 022, 1978 (J. Hasegawa, Y. Kobayashi, T. Shimizu); *Chem. Abstr.* **89** (1978) 215 045 s. J. Besta, M. Soulek, CS 179 600, 1979; *Chem. Abstr.* **92** (1980) 58 414 d.

[993] W. E. Vaughan, F. F. Rust, *J. Org. Chem.* **5** (1940) 449–471, 472–503.

[994] M. S. Kharasch, M. G. Berkman, *J. Org. Chem.* **6** (1941) 810–817.

[995] Y. A. Serguchev, V. F. Moshkin, Y. V. Konoval, G. A. Stetsyuk, *Zh. Org. Khim.* **19** (1983) no. 5, 1020–1023; *Chem. Abstr.* **99** (1983) 69 885 d.

[996] M. Ritchie, W. I. H. Winning, *J. Chem. Soc.* 1950, 3579–3583.

[997] G. Benoy, L. de Maeyer, *R. 27e Congr. Int. Chim Ind.* **2** (1954); *Ind. Chim. Belge,* **20** (1955) 160–162.

[998] H. P. Dürkes, DE-OS 1 643 922, 1971; Du Pont de Nemours Co., US 2 446 430, 1948 (J. A. Norton). Mitsubishi Chemical Industries Co., Ltd., JP-Kokai 75 62 906, 1975 (S. Ueda, K. Zaiga); *Chem. Abstr.* **83** (1975) 113 912 f.

[999] E. Clippinger (Chevron Res. Co.) *Am. Chem. Soc., Div. Pet. Chem., Prepr.* **15** (1970) no. 1, B37–B40.

[1000] V. A. Averyanov, S. E. Kirichenko, V. F. Shvets, Y. A. Treger, *Zh. Fiz. Khim* **56** (1982) no. 5, 1136–1140; *Chem. Abstr.* **97** (1982) 38 235 w.

[1001] VEB Farbenfabrik Wolfen, DD 8523, 1954 (A. Weissenborn). *Ullmann,* 4th ed., **9**, 530.

[1002] V. R. Rozenberg et al., SU 687 061, 1979; *Chem. Abstr.* **92** (1980) 76 078 b.

[1003] J. C. André et al.a) *J. Photochem.* **18** (1982) 47–45; *Chem. Abstr.* **96** (1982) 133 058 d; **18** (1982) 57–79; *Chem. Abstr.* **96** (1982) 131 388 n; **22** (1983) 7–24; *Chem. Abstr.* **99** (1983) 37 958 p; **22** (1983) 137–155; *Chem. Abstr.* **99** (1983) 139 410 w; **22** (1983) 213–221; *Chem. Abstr.* **99** (1983) 79 916 d; **22** (1983) 223–232; *Chem. Abstr.* **99** (1983) 96 704 a; **22** (1983) 313–332; *Chem. Abstr.* **99** (1983) 141 891 x.b) *React. Kinet. Catal. Lett.* **16** (1981) no. 2–3, 171–176; *Chem. Abstr.* **95** (1981) 123 904 s; **16** (1981) no. 2–3, 177–183; *Chem. Abstr.* **95** (1981) 122 382 h; **17** (1981) no. 3–4, 433–437; *Chem. Abstr.* **96** (1982) 43 767 e; **21** (1982) no. 1–2, 1–5; *Chem. Abstr.* **98** (1983) 88 526 h.c) *J. Chim. Phys. Phys.-Chim. Biol.* **79** (1982) no. 7–8, 613–616; *Chem. Abstr.* **98** (1983) 178 410 v.d) *AIChE J.* **28** (1982) no. 1, 156–166; *Chem. Abstr.* **97** (1982) 82 599 s.e) *Entropie* **18** (1982) no. 107–108, 62–81; *Chem. Abstr.* **98** (1983) 129 139 e.f) *Oxid. Commun.* **4** (1983) no. 1–4, 13–26; *Chem. Abstr.* **101** (1984) 229 721 n.

[1004] B. Bonath, B. Foertsch, R. Saemann (Geigy AG), *Chemie. Ing.-Tech.* **38** (1966) no. 7, 739–742.

[1005] E. Stoeva, Z. Angelova, M. Gramatikova, Z. Zhelyazkov, *Khim. Ind. (Sofia)* 1982, no. 7, 294–296; *Chem. Abstr.* **98** (1983) 74 188 n.

[1006] E. Borello, D. Pepori, *Ann. Chim. (Rome)* **45** (1955) 449–466.

[1007] J. S. Ratcliffe, *Br. Chem. Eng.* **11** (1966) no. 12, 1535–1537.

[1008] Bayer, DE-OS 2 003 932, 1971; US 3 715 283, 1973 (W. Böckmann).

[1009] ICI, DE-OS 2 105 254, 1971 (E. Illingworth, A. Fleming). M. Vrana, V. Janderova, M. Danek, CS 134 477, 1969; *Chem. Abstr.* **75** (1971) 5452 g.

[1010] J. Besta, M. Soulek, CS 159 100, 1975; *Chem. Abstr.* **84** (1976) 164 371 g. Albright and Wilson Ltd., GB 1 410 474, 1975 (C. H. G. Hands).

[1011] A. Scipioni, *Ann. Chim. (Rome)* **41** (1951) 491–498.

[1012] Occidental Chemical Co., DE 2 604 276, 1984 (S. Gelfand).

[1013] Hercules Powder Co., US 2 108 937, 1938 (L. H. Fisher). Hercules Powder Co., US 2 100 822, 1937 (H. M. Spurlin).

[1014] I.G. Farbenindustrie AG, DE 659 927, 1936 (O. Leuchs). G. L. Kamalov et al., SU 872 525, 1981; *Chem. Abstr.* **96** (1982) 85 217 f.

[1015] H. C. Brown, K. L. Nelson, *J. Am. Chem. Soc.* **75** (1953) 6292. R. C. Fuson, C. H. McKeever: *Organic Reactions*, vol. 1, J. Wiley & Sons, New York 1947, p. 63. Heyden Newport Chem. Co., US 2 859 253, 1958 (J. E. Snow).

[1016] Kureha Chem. Ind. Co., JP 2173/66, 1966.

[1017] FMC, US 2 493 427, 1950 (R. M. Thomas).

[1018] Mitsui Toatsu Chemicals Co., Ltd., JP-Kokai 73 05 725, 1973 (N. Kato, Y. Sato); *Chem. Abstr.* **78** (1973) 97 293 u.

[1019] Velsicol Chemical Co., US 3 535 391, 1970 (G. D. Kyker).

[1020] Bayer, DE-AS 2 206 300, 1974 (W. Böckmann, K. A. Lipper).

[1021] Mitsui Toatsu Chemicals, Inc., JP-Kokai 73 05 726, 1972 (N. Kato, Y. Sato); *Chem. Abstr.* **84** (1976) 121 408 g;cf. *Chem. Abstr.* **78** (1973) 97 294 v.

[1022] Monsanto, US 2 542 225, 1951 (J. L. West).

[1023] Dow Chemical, DE-OS 2 410 248, 1974 (R. H. Hall, D. H. Haigh, W. L. Archer, P. West).

[1024] EKA AB, EP-A 64 486, 1981 (R. K. Rantala, G. L. F. Hag).

[1025] Bayer, DE-OS 2 752 612, 1977 (F. Brühne, K. A. Lipper).

[1026] Ciba Geigy AG, DE-AS 2 044 832, 1969 (P. Liechti, F. Blattner).

[1027] GAF, US 3 087 967, 1960 (D. E. Graham, W. C. Craig).

[1028] Tenneco Chemicals Inc., US 3 524 885, 1967 (A. J. Deinet).

[1029] Kureha Chem. Ind. Co., Ltd., JP 69 12 132, 1969 (H. Funamoto); *Chem. Abstr.* **71**(1969) 80 948 u.

[1030] J. H. Simons, C. J. Lewis, *J. Am. Chem. Soc.* **60** (1938) 492.

[1031] E. Pouterman, A. Girardet, *Helv. Chim. Acta* **30** (1947) 107.

[1032] S. F. Khokhlov et al., SU 530 019, 1976; *Chem. Abstr.* **86** (1977) 43 386 y; cf *Chem. Abstr.* **85** (1976) 142 731 t; *Chem. Abstr.* **87** (1977) 22 600 u.

[1033] Rütgerswerke AG, DE-AS 1 200 275, 1965 (G. Bison, H. Binder).

[1034] Mitsubishi Gas Chemical Co., Inc., DE-AS 2 443 179, 1977 (S. Yoshinaka, M. Doya, S. Uchiyama).
[1035] Dow Chemical, US 4 046 656, 1977 (R. A. Davis, R. G. Pews); *J. Chem. Soc., Chem. Commun.* 1978, no. 3, 105–106.
[1036] Bayer, DE-OS 1 909 523, 1970 (W. Böckmann).
[1037] Diamond Alkali Co., US 2 979 448, 1961 (G. A. Miller).
[1038] Diamond Alkali Co., US 2 994 653, 1959 (G. A. Miller).
[1039] Asahi Electro-Chemical Co., JP 28 788 ('65), 1964 (K. Harasawa); *Chem. Abstr.* **64** (1966) 9633 e.
[1040] Dynamit Nobel AG, DE-OS 2 461 479, 1976 (H. Richtzenhain, P. Riegger). Dynamit Nobel AG, DE-OS 2 535 696, 1977 (P. Riegger, H. Richtzenhain, G. Zoche).
[1041] P. G. Harvey, F. Smith, M. Stacey, J. C. Tatlow, *J. Appl. Chem. (London)* **4** (1954) 319–325.
[1042] Mitsubishi Gas Chemical Co., DE-OS 2 614 139, 1976 (S. Yoshinaka, M. Doya, S. Uchiyama, S. Nozaki).
[1043] P. Beltrame, S. Carrà, S. Mori, *J. Phys. Chem.* **70** (1966) no. 4, 1150–1158; *Tetrahedron Lett.* **44** (1965) 3909–3915.
[1044] H. Fürst, H. Thorand, J. Laukner, *Chem. Tech. (Berlin)* **20** (1968) no. 1, 38–40.
[1045] P. Riegger, K. D. Steffen, *Chem. Ztg.* **103** (1979) no. 1, 1–7.
[1046] V. I. Titov, G. S. Mironov, I. V. Budnii, *Osnovn. Org. Sint. Neftekhim* **13** (1980) 93–96; *Chem. Abstr.* **97** (1982) 162 006 m.
[1047] O. Cerny, J. Hajek, *Collect. Czecho. Chem. Commun.* **26** (1961) 478–484.
[1048] L. M. Kosheleva, V. R. Rozenberg, G. V. Motsarev, *Zh. Org. Khim.* **16** (1980) no. 9, 1890–1893; *Chem. Abstr.* **94** (1981) 14 815 x.
[1049] Bayer, EP 9787, 1982; DE-OS 2 844 270, 1978; DE-OS 2 905 081, 1979 (R. Schubart, E. Klauke, K. Naumann, R. Fuchs); ICI EP-A 28 881, 1981 (J. O. Morley).
[1050] Mitsubishi Gas Chemical Co., Inc., JP 8 225 009, 1982;*Chem. Abstr.* **97** (1982) 184 357 d.
[1051] Bergwerksverband GmbH, GB 830 052, 1960.
[1052] T. Nishi, J. Onodera, *Kogyo Kagaku Zasshi* **71** (1968) no. 6, 869–871;*Chem. Abstr.* **70** (1969) 3381 f.
[1053] H. Trautmann, W. Seidel, K. Seiffarth, DD 116 451, 1975.
[1054] Mitsui Toatsu Chemicals, Inc., JP-Kokai 78 65 830, 1978 (M. Oba, M. Kawamata, T. Shimokawa, S. Koga); *Chem. Abstr.* **90** (1979) 6063 c.
[1055] Dynamit Nobel AG, DE-OS 2 161 006, 1973 (K. Redecker, H. Richtzenhain).
[1056] BASF, DE 845 503, 1943 (W. Rohland).
[1057] Hodogaya Chem. Co., GB 1 442 122, 1974.
[1058] Dynamit Nobel AG, DE-OS 2 358 949, 1975 (G. Blumenfeld).
[1059] T. R. Torkelson, V. K. Rowe in G. D. Clayton, F. L. Clayton (eds.): *Patty's Industrial Hygiene and Toxicology*, 3rd ed., vol. 2 **B**, Wiley-Interscience, New York 1981, pp. 3433–3601.
[1060] American Conference of Governmental Industrial Hygienists Inc. Documentation of the Threshold Limit Values 1980 (With Annual Supplements) 6500 Glenway Bldg. D-5, Cincinnati, OH 45211.
[1061] Commission for Investigation of the Health Hazards of Chemical Compounds in the Work Environment: *Toxikologisch-arbeitsmedizinische Begründung von MAK-Werten*, Verlag Chemie, Weinheim 1984.
[1062] Monograph on the Evaluation of the Carcinogenic Risk of Chemicals to Humans. Chemicals, Industrial Processes and Industries Associated with Cancer in Humans. *IARC Monogr. Suppl.* **1–29** (1982) no. 4.

[1063] J. A. John, D. J. Wroblewski, B. A. Schwetz: Teratogenicity of Experimental and Occupational Exposure to Industrial Chemicals in H. Kalter, ed.: *Issues and Reviews in Teratology*, vol. **2**, Plenum, 1984, p. 267–324.

[1064] American Conference of Governmental Industrial Hygienists. TLV's® Threshold Limit Values for Chemical Substances in the Work *Environment Adopted by the ACGIH for 1985–1986*. ACGIH 6500 Glenway Bldg. D-5, Cincinnati, OH 45211.

[1065] Deutsche Forschungsgemeinschaft: MAK, Verlag Chemie, Weinheim 1984.

[1066] The Chemical Industry Institute of Toxicology: *Unpublished data on methyl chloride*. Research Triangle Park, North Carolina 27709.

[1067] B. N. Ames, P. F. Infante, R. H. Reitz, *Banbury Report 5. Ethylene Dichloride: A Potential Health Risk*, Cold Spring Harbor Laboratory, Maine, USA, 1980.

[1068] R. J. Kociba, B. A. Schwetz, D. G. Keyes, G. C. Jersey, J. J. Ballard, D. A. Dittenber, J. F. Quast, C. E. Wade, C. G. Humiston: *Environ. Health* **21** (1977) 49.

[1069] Ambient Water Quality Criteria for: Halomethanes, Chloroform, Carbon Tetrachloride, Chlorinated Ethanes, Trichloroethylene and Tetrachloroethylene. U.S. Environmental Protection Agency, Oct. 1980.

[1070] M. A. Mayes, H. C. Alexander, D. C. Dill, *Bull. Environ. Contam. Toxicol.* **31** (1983) 139–147.

[1071] Dow unpublished data.

[1072] C. Bushon, *Ecological Effects Branch Chief*, Office of Pesticides and Toxic Substances. U.S. EPA, Washington, DC, Dec. 16, 1983.

[1073] Y. Correia, G. J. Martens, F. H. van Mensch, B. P. Whim, *Atmos. Environ.* **11** (1977) 1113–1116.

[1074] P. R. Edwards, I. Campbell, G. S. Milne, *Chem. Ind.* **1** (1982) 574–578.

[1075] K. K. Beutel: *Health and Environmental Aspects of Chlorinated Solvents in Industrial Applications*, Juris Druck + Verlag, Zürich 1984.

[1076] W. L. Dilling in R. A. Conway (ed.): *Environmental Risk Analysis for Chemicals*, Van Nostrand Reinhold Co., New York 1982, pp. 166–174.

[1077] M. A. K. Khalil, R. A. Rasmussen, *Chemosphere* **7** (1984) 789–800.

[1078] National Academy of Sciences: *Nonfluorinated Halomethanes in the Environment*, Washington 1978, p. 97.

[1079] NIOSH: Registry of Toxic Effects of Chemical Substances, U.S. Dept. of Health, Education and Welfare, NIOSH (1983).

[1080] E. Loeser, M. H. Litchfield, *Fd. Chem. Toxic.* **21** (1983) 825–832.

[1081] J. Kohli et al., *Can. J. Biochem.* **54** (1976) 203.

[1082] R. D. Lingg et al., *Drug Metab. Dispos.* **10** (1982) 134.

[1083] J. H. Koeman et al., *Nature (London)* **221** (1969) 126.

[1084] L. Acker, R. Schilte, *Naturwissenschaften* **57** (1970) 497.

[1085] F. K. Zimmermann et al., *Mutat. Res.* **133** (1984) 199.

[1086] L. Fishbein, *J. Chromatogr.* **68** (1972) 345.

[1087] M. Ogawa, *Fukuoka Igaku Zasshi* **62** (1971) 74; cited in: J. R. Allen, D. H. Norback, *Science (Washington, D.C.)* **179** (1973) 498.

[1088] J. H. Miller, *Public Health Rep.* **59** (1944) 1085.

[1089] Report of the Cooperative Research Program between the Gesellschaft für Strahlen- und Umweltforschung mbH, München (GSF) and the Albany Medical College of the Union University, Albany, N.Y., 1974–1978, GSF-Bericht Ö 486.

[1090] B. Bush et al., *Arch. Environ. Contam. Toxicol.* **13** (1984) 517.

[1091] A. Hofmann et al., *Arbeitsmed. Sozialmed. Präventivmed.* **18** (1983) 181.

[1092] U. A. T. Brinkman, H. G. M. Reymer, *J. Chromatogr.* **127** (1976) 203.
[1093] G. Löfroth et al., *Mutat. Res.* **155** (1985) 91.
[1094] S. Haworth et al., *Environ. Mutagen.* **5** (1983) Suppl. 1, 3.
[1095] H. Druckrey et al., *Z. Krebsforsch.* **74** (1970) 241.
[1096] W. F. von Oettingen: U.S. Department of Health, Education and Welfare, Public Health Service Publication No. 414, U.S. Government Printing Office, Washington, D.C., 1955.
[1097] K. Yasuo et al., *Mutat. Res.* **58** (1978) 143.
[1098] J. McCann et al., *Proc Natl. Acad. Sci. U.S.A.* **72** (1975) 979.
[1099] K. Fukuda et al., *Gann* **72** (1981) 655.
[1100] H. G. Bray et al., *Biochem. J.* **70** (1958) 570.
[1101] Bayer AG, Bericht Nr. 8418 (1979), unpublished
[1102] Verordnung über gefährliche Arbeitsstoffe (ArbStoffV), Fassung vom 11. Februar 1982, (BGBl. I. S. 144).
[1103] M. Windholz (ed.): *The Merck Index,* Merck & Co., Rathway 1976.
[1104] Bayer AG, Bericht Nr. 7839 (1978), unpublished.
[1105] H. F. Smyth et al., *Arch. Ind. Hyg. Occup. Med.* **4** (1951) 119.
[1106] H. Sakabe et al.: The New York Academy of Sciences, TSCA Sec. 8(e) Submission 8EHQ-0777-0001.
[1107] H. Sakabe, K. Fukuda, *Ind. Health* **15** (1977) 173.

Chloroacetaldehydes

REINHARD JIRA, Wacker-Chemie GmbH, Burghausen, Federal Republic of Germany (Chaps. 1 and 2)
ERWIN KOPP, Wacker-Chemie GmbH, München, Federal Republic of Germany (Chaps. 1 and 2)
BLAINE C. MCKUSICK, Wilmington, Delaware 19803, United States (Chaps. 3 and 4)
GERHARD RÖDERER, Wacker-Chemie GmbH, Burghausen, Federal Republic of Germany (Section 4.1)

1.	Monochloroacetaldehyde	1522	2.3.	Production ... 1530
1.1.	Physical Properties	1522	2.3.1.	Chlorination of Acetaldehyde or Paraldehyde ... 1530
1.2.	Chemical Properties and Uses.	1523	2.3.2.	Hypochlorination of 1,2-Dichloroethylene ... 1530
1.2.1.	Reactions of the Aldehyde Group	1523	2.3.3.	Chlorination of Ethanol ... 1530
1.2.2.	Reactions at the Chlorine Atom	1524	2.4.	Availability, Storage, and Transportation ... 1530
1.2.3.	Reactions of the Chlorine and Aldehyde Groups	1525	2.5.	Polymeric Dichloroacetaldehydes ... 1531
1.2.4.	Uses of Chloroacetaldehyde and Its Derivatives as Pesticides	1525	3.	Trichloroacetaldehyde ... 1531
1.3.	Production	1525	3.1.	Physical Properties ... 1532
1.3.1.	Chlorination of Acetaldehyde and Paraldehyde	1525	3.2.	Chemical Properties ... 1532
1.3.2.	Chlorination of Vinyl Compounds in Aqueous Media and Alcohols	1526	3.3.	Production ... 1532
1.3.3.	Preparation of Anhydrous Chloroacetaldehyde	1526	3.4.	Analysis ... 1533
1.4.	Analysis	1527	3.5.	Legal Aspects ... 1533
1.5.	Transportation, Storage, and Safety	1527	3.6.	Uses ... 1534
1.5.1.	Transportation	1527	3.7.	Economic Aspects ... 1534
1.5.2.	Storage	1527	4.	Toxicology ... 1534
1.5.3.	Safety	1528	4.1.	Chloroacetaldehyde ... 1534
1.6.	Polymeric Chloroacetaldehydes	1528	4.2.	Dichloroacetaldehyde ... 1535
2.	Dichloroacetaldehyde	1529	4.3.	Trichloroacetaldehyde (Chloral) ... 1535
2.1.	Physical Properties	1529	5.	References ... 1536
2.2.	Chemical Properties and Uses	1529		

1. Monochloroacetaldehyde

Monochloroacetaldehyde [107-20-0], 2-chloroethanal, CH_2ClCHO, M_r 78.50, was first prepared in pure form by K. NATTERER in 1882 by heating chloroacetaldehyde diethyl acetal, $CH_2ClCH(OC_2H_5)_2$ [621-62-5], with anhydrous oxalic acid at 100–150 °C. Monochloroacetaldehyde was obtained from the aqueous solution as a crystalline hemihydrate.

1.1. Physical Properties

Anhydrous monochloroacetaldehyde is a colorless, mobile liquid with extremely pungent odor; *bp* 85–85.5 °C (99.7 kPa), enthalpy of formation of the liquid from the elements −256.4 kJ/mol, enthalpy of combustion −981.4 kJ/mol (at 101.3 kPa, 20 °C), dipole moment 1.99 D (at 298 K in benzene). It is soluble in water (to form a hydrate) and in common organic solvents.

Monochloroacetaldehyde hemihydrate, 1,1′-dihydroxy-2,2′-dichlorodiethyl ether, 1,1′-oxybis-(2-chloroethanol), $CH_2ClCH(OH)-O-CH(OH)CH_2Cl$, M_r 175.01, colorless crystals, *mp* 43–50 °C (elimination of water), *bp* (with decomposition into the components) 84 °C, is soluble in water (see Tables 1 and 2), ethanol, ether, benzene, methylene chloride, and chloroform.

Two CAS numbers, [34789-09-8] and [7737-02-2], are given for chloroacetaldehyde hemihydrate. The first is valid for the term "acetaldehyde, chloro dimer, hydrate" and the second for "1,1′-oxybis(2-chloroethanol)." These compounds are identical.

The hemihydrate distills from the aqueous solution as an azeotropic mixture. Both components, chloroacetaldehyde and water, are present in the vapor phase, but they cannot be separated by a simple distillation. Figure 1 shows the composition of the vapor and the liquid of a monochloroacetaldehyde–water mixture, and Figure 2 shows the boiling and condensation points as functions of the composition.

Safety Data (for the hemihydrate):

Flash point 60 °C
Ignition point 405 °C [2]
Explosive limits in air (at 20 °C, 101.3 kPa), lower 305 g/m^3
MAK 3 mg/m^3

Table 1. Solubility of monochloroacetaldehyde hemihydrate in water as a function of temperature

Temperature, °C	1	10	20	30	40
Concentration, wt%	13.35	22.2	44.3	62.7	81.5

Table 2. Density of aqueous chloroacetaldehyde hemihydrate solutions at 15°C as a function of concentration

Concentration, wt%	10	20	30	40	50	60	72.5
Density, g/cm^3	1.041	1.085	1.137	1.188	1.238	1.290	1.355

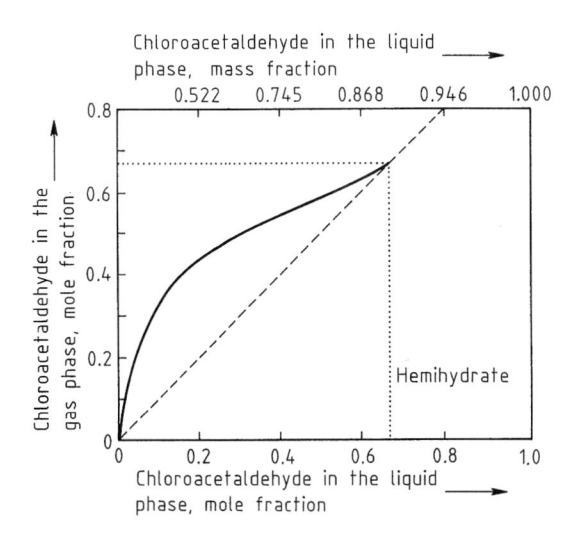

Figure 1. Vapor–liquid equilibrium curve for monochloroacetaldehyde–water [1]

1.2. Chemical Properties and Uses

Like other α-halocarbonyl compounds, chloroacetaldehyde is exceptionally reactive. The aldehyde group takes part in reactions typical of this group, e.g., addition and condensation. Other reactions involve the chloromethyl group. Nucleophilic substitution of the chlorine is facilitated by the inductive effect of the aldehyde group. The α-hydrogens are likewise strongly activated by the inductive effects of the aldehyde group and the chlorine atom.

1.2.1. Reactions of the Aldehyde Group

Like acetaldehyde, anhydrous chloroacetaldehyde self-condenses easily at room temperature to form a cyclic trimer and tetramer as well as polymers with an acetal (polyoxymethylene) structure. Cotrimers and copolymers with acetaldehyde are also known. The hemihydrate has the structure 1,1′-dihydroxy-2,2′-dichlorodiethyl ether, $ClCH_2CH(OH)-O-CH(OH)CH_2Cl$, and reacts like an aldehyde.

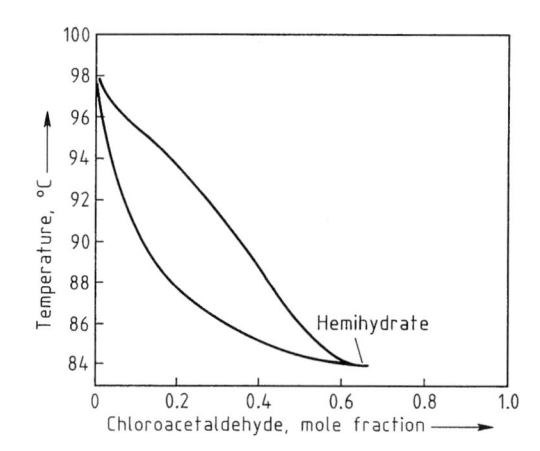

Figure 2. Boiling and condensation curve of aqueous chloroacetaldehyde solutions [1]

Among industrially important derivatives are the crystalline sodium bisulfite addition compound, sodium 2-chloro-1-hydroxyethanesulfonate [13064-50-1], the dimethyl acetal [97-97-2] (bp 127 – 128 °C), and the diethyl acetal [621-62-5] (bp 157.4 °C), which are used in preparing pharmaceutical precursors and pesticides.

Oxidation with air in the presence of cobalt salts [3], hydrogen peroxide, nitric acid [4], or other oxidants yields monochloroacetic acid. Aldol addition in an alkaline medium gives 2,4-dichloroacetaldol, which can be dehydrated to 2,4-dichlorocrotonaldehyde in an acid medium. In coaldolization with acetaldehyde, chloroacetaldehyde reacts exclusively as the methylene component to produce 2-chlorocrotonaldehyde. This aldehyde is also obtained as a byproduct of acetaldehyde production ($\rightarrow$ Acetaldehyde).

Other typical aldehyde reactions are the formation of the aldehyde ammonia, 1,3,5-tris-(chloromethyl)hexahydro-2,4,6-triazine [7038-17-7] [5], the cyanohydrin [33965-80-9], $ClCH_2CH(OH)CN$ [6], [7], and 2-chloroethyl chloroacetate, $ClCH_2COOCH_2CH_2Cl$ [3848-12-2], (Tishchenko reaction) [5].

1.2.2. Reactions at the Chlorine Atom

The chlorine atom can be removed by hydrogenation using a palladium – carbon catalyst in aqueous solution [8], and can be replaced by nucleophilic agents [9].

Mercaptoacetaldehyde is formed by the reaction of chloroacetaldehyde with sodium mercaptan. It is used as an intermediate in the synthesis of thienodiazepines, a new class of tranquilizers [10], [11], of thiophenic azo dyes, of d,l-cystein, and of certain 2-alkylthiazoline food flavors [12]. Mercaptoacetaldehyde exists as an equilibrium mixture of the dimer, 2,5-dihydroxy-1,4-dithian [40018-26-6] [13], and oligomers [14].

Substitution of the chlorine atom by the amino or methylamino groups is carried out on the acetal to protect the aldehyde group. The amino and methylamino derivatives are intermediates in the synthesis of pesticides and pharmaceuticals [15].

Reaction of chloroacetaldehyde dimethyl acetal with sodium methoxide yields 1,1,2-trimethoxyethane, which is also an intermediate in pharmaceutical syntheses, e.g., sulfamethoxydiazine.

1.2.3. Reactions of the Chlorine and Aldehyde Groups

Chloroacetaldehyde, usually in form of the hemihydrate or an acetal, is used as a starting compound for preparing numerous heterocyclic compounds. The most industrially important reaction is the synthesis of 2-aminothiazole by reaction with thiourea.

2-Aminothiazole is an intermediate in the preparation of pharmaceuticals, e.g., sulfathiazole, and dyes. Cyclocondensation of chloroacetaldehyde with methyl acetoacetate gives 2-methyl-3-carbomethoxyfuran, an intermediate in making the seed disinfectant fenfuram [16]. 5-Chlorothiadiazole, an intermediate for the cotton defoliant and plant growth regulator thidiazuron, is obtained by condensation of chloroacetaldehyde with carboethoxyhydrazine and subsequent cyclization with thionyl chloride [17].

Stilbenes are produced by the reaction of chloroacetaldehyde with aromatic hydrocarbons, phenols, or phenol ethers [18]. Vinyl dialkyl phosphates are produced by reaction with trialkyl phosphites (Perkow reaction) [19]. Reaction with sodium acetate and acetic anhydride forms 1,1,2-triacetoxyethane, which can be used as a starting material for producing *d,l*-serine [20].

1.2.4. Uses of Chloroacetaldehyde and Its Derivatives as Pesticides

Chloroacetaldehyde and its bisulfite addition compound have bactericidal, fungicidal, algicidal [21], and nematicidal [22] properties. They are also used to preserve dispersion and emulsion paints [23].

1.3. Production

1.3.1. Chlorination of Acetaldehyde and Paraldehyde

Chlorination of acetaldehyde or paraldehyde yields monochloroacetaldehyde mixed with small amounts of dichloroacetaldehyde and trichloroacetaldehyde. These three compounds are also obtained as byproducts of the production of acetaldehyde by the Wacker process (→ Acetaldehyde).

1.3.2. Chlorination of Vinyl Compounds in Aqueous Media and Alcohols

Simultaneous introduction of vinyl chloride and chlorine into water at ca. 20 °C forms chloroacetaldehyde in a nearly quantitative yield if its concentration in the reaction solution is not allowed to rise above ca. 5 wt%. At a higher concentration, increasing amounts of 1,1,2-trichloroethane are formed [24].

$$CH_2 = CHCl + HOCl \longrightarrow [ClCH_2CH(OH)Cl] \longrightarrow ClCH_2CHO + HCl$$

This method is not suitable for industrial use because it yields only a dilute solution of chloroacetaldehyde. Concentrated aqueous solutions of chloroacetaldehyde are obtained in nearly quantitative yields by the reaction of chlorine with vinyl acetate in water [25]–[27]. This reaction is preferably carried out in two steps. Chlorine is first added to vinyl acetate at room temperature with cooling to form 1,2-dichloroethyl acetate, which is then hydrolyzed at 50–60 °C.

$$CH_2 = CHOOCCH_3 + Cl_2 \longrightarrow ClCH_2CHClOOCCH_3$$
$$\xrightarrow{H_2O} ClCH_2CHO + HCl + CH_3COOH$$

The resulting solution is suitable for many further reactions without isolation of the chloroacetaldehyde.

Chloroacetaldehyde acetals are obtained in high yield and purity by carrying out the above reactions in alcohols [28], [29].

1.3.3. Preparation of Anhydrous Chloroacetaldehyde

Direct chlorination of dry acetaldehyde or paraldehyde gives only a low yield of monochloroacetaldehyde. Azeotropic dehydration of the hemihydrate with chloroform, toluene [30], or carbon tetrachloride [31] followed by distillation over anhydrous calcium chloride gives better yields. Anhydrous monochloroacetaldehyde is obtained in high yield by the depolymerization of trichloroparaldehyde or polychloroacetaldehyde at 145 °C in the presence of oxalic acid or trichloroacetic acid [32]. Monochloroacetaldehyde is also obtained in high yield by pyrolysis of chloroethylene carbonate in the presence of quaternary ammonium salts [33], [34].

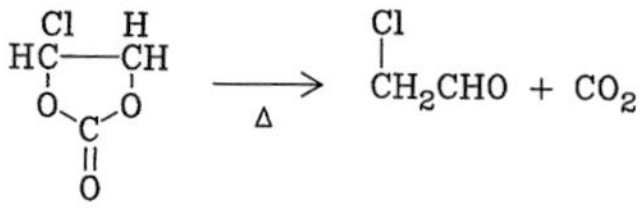

$$\underset{\underset{O}{\overset{\|}{C}}}{\overset{Cl\ \ \ \ H}{\underset{O\ \ \ \ \ \ O}{HC-CH}}} \xrightarrow{\Delta} \overset{Cl}{\underset{}{CH_2CHO}} + CO_2$$

1.4. Analysis

Chloroacetaldehyde is commercially available as a 45 wt% aqueous solution. The monochloroacetaldehyde content can be determined by reaction with excess hydrazine sulfate to form chloride ion, which can be determined by the Volhard method. Other chloroacetaldehydes and chloroacetic acids do not interfere [35]. This reaction is specific for monochloroacetaldehyde.

The usual aldehyde reactions are suitable for identification purposes. Monochloroacetaldehyde forms derivatives with characteristic boiling or melting points: oxime, *bp* 61 °C at 2.7 kPa; semicarbazone, *mp* 146–148 °C; thiosemicarbazone, *mp* 141–142 °C; 2,4-dinitrophenylhydrazone, *mp* 154–156 °C; 2,4,6-trinitrophenylhydrazone, *mp* 159–160 °C (crystallized from glacial acetic acid).

Gas chromatography can separate and determine chloroacetaldehyde and its impurities [9].

1.5. Transportation, Storage, and Safety

1.5.1. Transportation

A 45% aqueous chloroacetaldehyde solution can be shipped or stored in glass, enamel, and enamelled steel containers. Stainless steel and polyethylene are other suitable materials for transporting the product and for storing it for a short time. Polyethylene drums holding 30 L, stainless steel drums holding 205 L, and stainless steel tanks holding 800 L are suitable containers. The product is also supplied in road and rail tankers.

Transportation of chloroacetaldehyde is regulated. The GGVE (rail), GGVS (road), RID, ADR, and ADNR are all class 6.1, section 16 B; IACO/IATA: forbidden; IMDG Code is class 6.1; and the UN no. is 2232.

1.5.2. Storage

Aqueous 45% chloroacetaldehyde should be stored in an inert atmosphere, and it remains in good condition for at least 3 months if kept in the dark at room temperature. If it is exposed to sunlight or allowed to become warm, it discolors, and eventually brown, resinous products form with the liberation of hydrogen chloride. If the solution is kept at temperatures below 15 °C, the hemihydrate may crystallize out.

1.5.3. Safety

Chloroacetaldehyde must not be conveyed with compressed air. Tightly fitting goggles and gloves should be worn when chloroacetaldehyde is handled because of its strongly irritating effect. The compound should not be allowed to come in contact with the skin, eyes, or clothes, and its vapors should not be inhaled. Adequate ventilation should be provided. Areas in which chloroacetaldehyde is being handled should be provided with a safety shower or comparable means of quickly washing the skin.

A gas mask fitted with filter A (brown, for organic vapors) should be worn if there is a risk of inhaling the vapor. In the event of a spill, all potential sources of ignition must be kept far away. The spill should be mopped up with an absorbent substance, such as sawdust, fine sand, or silica granules. Disposal of chloroacetaldehyde can be effected by treating a dilute water solution with alkali at elevated temperature followed by biological treatment or incineration, always in compliance with local regulations.

Because of its high biocidal activity, chloroacetaldehyde interferes with the biological degradation of organic substances in wastewater treatment plants. Alkali treatment at elevated temperature converts it into nontoxic biodegradable compounds by eliminating chlorine.

1.6. Polymeric Chloroacetaldehydes

Trichloroparaldehyde [1129-52-8], 2,4,6-tris-(chloromethyl)-1,3,5-trioxane, $C_6H_9Cl_3O_3$, M_r 235.50, mp 88–89 °C, bp 140–144 °C (1.3 kPa), colorless crystals, soluble in common organic solvents, can be recrystallized from ethanol and forms monochloroacetaldehyde on heating in the presence of acids, e.g., oxalic acid.

Trichloroparaldehyde is formed by the reaction of concentrated sulfuric acid with the hemihydrate of monochloroacetaldehyde [36] and is also part of the residue (with tetrachlorometaldehyde) left by azeotropic distillation of the hemihydrate [5]. Trichloroparaldehyde can also be obtained by decomposing 1,2-dichloroethyl nitrate in the presence of Lewis acids, such as tin tetrachloride [37].

Tetrachlorometaldehyde [7038-25-7], 2,4,6,8-tetrakis(chloromethyl)-1,3,5,7-tetroxocane, $C_8H_{12}Cl_4O_4$, M_r 313.99, is obtained as a residue following azeotropic dehydration of chloroacetaldehyde hemihydrate [5]. It forms colorless crystals, mp 65–67 °C, bp 127–130 °C at 1.3 Pa.

Polychloroacetaldehydes [27577-42-0], $(C_2H_3ClO)_x$, are formed by the polymerization of anhydrous monochloroacetaldehyde at −40 to −78 °C, particularly in the presence of such Lewis acids as boron trifluoride in ether or chloroform, to form amorphous, elastomeric products with a polyacetal (polyoxymethylene) structure [38], [39].

Crystalline polymers are obtained at −78 °C in the presence of catalytic amounts of organometallic compounds, such as triethylaluminum, diethylzinc, or butyllithium [40].

Polymeric chloroacetaldehydes are commercially unimportant.

2. Dichloroacetaldehyde

Dichloroacetaldehyde [79-02-7], 2,2-dichloroethanal, $CHCl_2CHO$, M_r 112.94, was first obtained in 1868 by F. PATERNO by distillation of dichlorodiethyl acetal [619-33-0], $CHCl_2CH(OC_2H_5)_2$, with sulfuric acid.

2.1. Physical Properties

Anhydrous dichloroacetaldehyde is a colorless liquid with a strong odor, mp −37.6 to −37.4 °C, bp 89.2 °C, d_{25} 1.4113, dipole moment 2.36 D at 30 °C, soluble in water with formation of a hydrate and in common organic solvents.

Dichloroacetaldehyde monohydrate [16086-14-9], $CHCl_2CH(OH)_2$, 2,2-dichloro-1,1-ethanediol, crystallizes from water, M_r 130.96, mp 35–50 °C, bp 85–95 °C at 101.3 kPa, d_4^{20} 1.53–1.54, vapor pressure ca. 6.5 kPa at 20 °C and ca. 25 kPa at 50 °C, soluble in polar organic solvents, insoluble in nonpolar organic solvents. The flash point is ca. 95 °C and the ignition temperature ca. 605 °C.

2.2. Chemical Properties and Uses

On standing, dichloroacetaldehyde forms a solid, colorless polymer which depolymerizes if heated to 120 °C [41]. Dichloroacetaldehyde shows all of the typical aldehyde reactions.

Oxidation of dichloroacetaldehyde with chromic acid yields dichloroacetic acid [42]. Reduction with ethylaluminum gives 2,2-dichloroethanol [43]. Aldol condensation yields 2,2,4,4-tetrachloroacetaldol [44].

Condensation with ethylbenzene yields p,p′-diethyl-1,1-diphenyl-2,2-dichloroethane [72-56-0], which is known as Perthane and is used as an insecticide. Condensation with chlorobenzene produces p,p′-dichloro-1,1-diphenyl-2,2-dichloroethane [72-54-8], which is also an insecticide and is known as TDE or DDD.

$$CHCl_2CHO + 2\ C_6H_5Cl \longrightarrow CHCl_2CH(C_6H_4Cl)_2 + H_2O$$

The most important current use for dichloroacetaldehyde and its acetals is in the synthesis of the diuretic trichloromethiazid and the cytostatic mitotane [53-19-0], 1,1-dichloro-2-(o-chlorophenyl)-2-(p-chlorophenyl)ethane.

2.3. Production

The most important industrial production process is the chlorination of acetaldehyde or paraldehyde. Hypochlorination of 1,2-dichloroethylene yields pure dichloroacetaldehyde.

2.3.1. Chlorination of Acetaldehyde or Paraldehyde

Dichloroacetaldehyde is formed, along with mono- and trichloroacetaldehyde, by the chlorination of acetaldehyde or paraldehyde [45]. High yields are obtained in the presence of antimony trichloride (83% yield) [46] or phosphoric acid (>90% yield) [47].

Dichloroacetaldehyde is also formed, together with mono- and trichloroacetaldehyde, by oxychlorination in the preparation of acetaldehyde by the Wacker process (→ Acetaldehyde).

2.3.2. Hypochlorination of 1,2-Dichloroethylene

Dichloroacetaldehyde is obtained without impurities such as chloral and monochloroacetaldehyde, by treating 1,2-dichloroethylene with 1 mol of chlorine at 0–20 °C in aqueous dioxane [48] or in water alone with vigorous mixing [49].

2.3.3. Chlorination of Ethanol

Chlorination of ethanol can be controlled in the presence of nickel(II) chloride to obtain principally dichloroacetaldehyde along with some chloral and monochloroacetaldehyde. Dichloroacetaldehyde forms a hemiacetal, which can be used in the same manner as the pure aldehyde or the hydrate [50].

2.4. Availability, Storage, and Transportation

Dichloroacetaldehyde is available in a polyethylene container either as the solid hydrate (minimum 82% anhydrous product) or as an aqueous solution (minimum 65% anhydrous product).

Transportation of dichloroacetaldehyde is regulated. The GGVE (rail), GGVS (road), RID, and ADR are all class 6.1, section 16 B; IATA-DGR (air) is class 6.1, section II, PAC 609, CAC 611 for the 65% aqueous solution and class 8, section III, PAC 822,

CAC 823 for the hydrate. The IMDG Code is class 6.1 for the 65% aqueous solution and class 8 for the hydrate. The UN no. is 1610 for the 65% aqueous solution; it is 1759 for the hydrate.

2.5. Polymeric Dichloroacetaldehydes

Hexachloroparaldehyde [17352-16-8], 2,4,6-tris(dichloromethyl)-1,3,5-trioxane, $C_6H_6Cl_6O_3$, M_r 338.83, mp 131–132 °C, colorless crystals, bp 210–220 °C with the formation of small amounts of dichloroacetaldehyde.

Hexachloroparaldehyde is made by treating dichloroacetaldehyde with such Lewis acids as antimony trichloride, iron(III) chloride, aluminum trichloride, tin(IV) chloride, or boron trifluoride [51]. Hexachloroparaldehyde is also formed by the direct chlorination of paraldehyde with 6–7 mol of chlorine at 35 °C under anhydrous conditions. It forms dichloroacetaldehyde hydrate in the presence of aqueous acids [52].

Polydichloroacetaldehydes [28388-44-5], $(C_2H_2Cl_2O)_x$, are formed by polymerization of dichloroacetaldehyde. Crystalline polymers, are obtained by treating dichloroacetaldehyde at −78 °C with organometallic compounds such as triethylaluminum, diethylzinc, or butyllithium [40]. Amorphous polymers are obtained by using Lewis acids at a temperature below 0 °C [53]. These products have a polyoxymethylene (polyacetal) structure and are more or less soluble in common organic solvents, depending on the degree of polymerization. They dissolve in concentrated sulfuric acid at 100 °C with discoloration, are soluble in dimethylformamide, and form copolymers with other aldehydes.

3. Trichloroacetaldehyde

Trichloroacetaldehyde [75-87-6], chloral, 2,2,2-trichloroethanal, CCl_3CHO, M_r 147.39, was first synthesized by J. VON LIEBIG in 1832 by chlorination of ethanol. It was the first hypnotic drug (1869) and is a precursor of the famous insecticide DDT, 1,1,1-trichloro-2,2-bis(4-chlorophenyl)ethane [50-29-3], discovered in 1941. Chloral was an important chemical in the 1960s, but its importance has steadily declined since then because the use of DDT and other chlorinated insecticides has been restricted.

3.1. Physical Properties

Chloral, mp −56.5 °C, bp 97.8 °C at 101.3 kPa, d_4^{20} 1.5121, n_D^{20} 1.4557, vapor pressure 4.7 kPa at 20 °C, is a colorless, mobile, oily liquid. It has a penetrating odor and is readily soluble in water, alcohol, ether, and chloroform. It forms azeotropes with 1,2-dichloroethane, heptane, and benzene.

3.2. Chemical Properties

The strong electron-withdrawing effect of its three chlorine atoms gives chloral chemical properties different from those of most aldehydes [54]. It has a strong tendency to form addition compounds with nucleophilic reagents. For example, with alcohols (ROH, R = alkyl) it forms isolable hemiacetals, $CCl_3CH(OH)OR$, and with water isolable chloral hydrate, $CCl_3CH(OH)_2$ [302-17-0], mp 53 °C; bp 97.5 °C at 101.3 kPa; d_4^{20} 1.908. Chloral cannot undergo many of the base-catalyzed condensations typical of aldehydes because strongly basic reagents tend to cleave its carbon–carbon bond.

$CCl_3CHO + NaOH \longrightarrow CHCl_3 + HCO_2Na$
$CCl_3CHO + R_2NH \longrightarrow CHCl_3 + HCONR_2$ (R = alkyl)

Under acidic conditions chloral usually reacts as a typical aldehyde, forming acetals with alcohols and condensing with aromatic compounds to form diaryltrichloroethanes, e.g., DDT from chlorobenzene.

$CCl_3CHO + 2\ C_6H_5Cl \xrightarrow{H_2SO_4} CCl_3CH(C_6H_4Cl\ p)_2 + H_2O$

Anionic initiators such as lithium *tert*-butoxide convert chloral to an insoluble, nonflammable high polymer, $CH(CCl_3)O_n$, that has not found commercial use [55].

3.3. Production

Chloral is produced by the chlorination of acetaldehyde [56] (→ Acetaldehyde) or ethanol (→ Ethanol). Acetaldehyde is usually the more economical starting material.

$CH_3CHO + 3\ Cl_2 \longrightarrow CCl_3CHO + 3\ HCl$

The chlorination is carried out in hydrochloric acid, which speeds up the reaction, represses condensation and aldehyde-oxidation reactions, and facilitates temperature control. Antimony trichloride is sometimes used as a catalyst. The process can be performed either in batches or continuously. The rate of chlorination decreases as the stepwise chlorination proceeds; thus, the temperature is gradually increased from 0 °C to 90 °C during the process. Chloral is distilled from the reaction mixture as the hydrate. The hydrate is then mixed with concentrated sulfuric acid, the heavier acid layer is drawn off, and chloral is distilled through a fractionating column of moderate height.

Chlorination of a mole of ethanol requires 4 mol of chlorine, with the first serving to oxidize the ethanol to acetaldehyde.

$$CH_3CH_2OH + Cl_2 \longrightarrow CH_3CHO + 2\,HCl$$

To withstand hot hydrochloric acid, the reaction vessels and piping for these processes are lined with a ceramic or other acid-resistant material, and valves of nickel–molybdenum alloys such as Hastelloy B are used.

3.4. Analysis

Technical-grade chloral ranges in purity from 94 to 99 wt%, with water being the main impurity. Other impurities sometimes present are chloroform, hydrogen chloride, dichloroacetaldehyde, and phosgene.

A common analytical method is to treat the chloral for 2 min with $1N$ sodium hydroxide, which cleaves chloral to chloroform and sodium formate; the excess alkali is then titrated with acid [57]. Alternatively, chloral is treated with quinaldine ethyl iodide [606-55-3] to form a blue cyanine dye whose quantity is measured spectrophotometrically [58].

Gas chromatography can be used for quantitative analysis of chloral and its hydrate, which breaks down to chloral on vaporization.

3.5. Legal Aspects

Chloral hydrate and other forms of chloral are classified by the U.S. Bureau of Narcotics and Dangerous Drugs (BNDD) as Schedule IV drugs, i.e., drugs of relatively low potential for abuse that lead only to limited physical or psychological dependence. To possess chloral, one must have a license from the BNDD, keep stored chloral locked up, and maintain records of where it goes and for what purposes. Licensees are periodically audited by the BNDD. Similar regulations are found in other countries [59].

3.6. Uses

The principal use of chloral is in the production of the insecticide DDT. Much smaller amounts are used to make other insecticides (methoxychlor, naled, trichlorfon, dichlorvos), a herbicide (trichloroacetic acid), and hypnotic drugs (chloral hydrate, chloral betaine, α-chloralose, and triclofos sodium).

3.7. Economic Aspects

Chloral (as the hydrate) has been produced for use as a hypnotic drug in relatively low and gradually declining volume for many years. As an indication of the scale, United States production for this purpose was about 135 t in 1978. A far greater demand for chloral arose during World War II to produce DDT, which was proving highly successful in controlling the transmission of typhus and malaria. Demand for chloral for insecticide production rose rapidly after the war and reached a peak around 1963 when 40 000 t was produced in the United States. It then declined as insects developed resistance to chlorinated insecticides, some animals at the top of food chains showed adverse effects from concentrating DDT in their bodies, and public concern arose about the possibility of harmful effects in humans. The use of DDT was banned in the United States in 1972 and subsequently in many other countries, but it is still produced for use in the tropics. Consumption of chloral in the United States for other pesticides was 1400 t in 1972. Montrose Chemical Corporation was the only United States producer of chloral in 1978 [60].

4. Toxicology

4.1. Chloroacetaldehyde

Chloroacetaldehyde is toxic if swallowed and in contact with the skin and it is highly toxic by inhalation, causing instant to delayed death in experimental animals. The acute toxicities in laboratory animals are: rat, oral: $LD_{50} = 89$ mg/kg; rabbit, dermal: $LD_{50} = 267$ mg/kg [61]; and rat, inhalation: $LC_{50} = 650$ mg m^3 h^{-1} [62]. For mice exposed to flowing air containing choroacetaldehyde vapor, the LT_{50} (exposure time lethal to 50% of the test animals) was 2.57 min. A skin sensitization test with guinea pigs (maximization test) was negative. Due to the extremely irritating nature of the compound it had to be diluted to 0.002% [63]. Even ca. 0.03% aqueous solutions cause definite eye irritation, and a 0.5% solution causes skin irritation.

Chloroacetaldehyde is mutagenic in a variety of test systems, such as bacteria, molds, fish and mammalian cells, and in human lymphocytes. The compound exhibits alkylating properties and reacts with single-strand DNA [64]. In an initiation/promotion experiment in which chloroacetaldehyde was subcutaneously administered to mice, no increase in tumor incidence was observed [65].

Human experience has shown that chloroacetaldehyde vapor is extremely irritating to the skin, mucous membranes, and respiratory tract [63]. A 40% aqueous solution is corrosive to skin and it represents a serious hazard of eye injury. Upon skin contact it causes skin effects resembling burn blisters and tissue damage after a latency period of 1–2 h [66].

The recommended TLV and MAK value is 3 mg/m^3 (1 ppm) [67], this also being the maximum airborne concentration to which workers should be exposed. Adequate personal protection (goggles, gloves, protection mask) must be applied to prevent contact with this highly dangerous compound.

4.2. Dichloroacetaldehyde

Dichloroacetaldehyde is known to be irritating to breathe, but no formal testing has been reported. It should be handled carefully, as with the other two chloroacetaldehydes [61], [62].

4.3. Trichloroacetaldehyde (Chloral)

Chloral and chloral hydrate have the same physiological properties and can be discussed together. Chloral hydrate was extensively used for many years to induce sleep and to calm or sedate, but in recent decades it has been relegated to a minor role by the barbiturates, benzodiazepines, etc.

Because of chloral hydrate's long use as a drug, much is known about the effects of chloral on humans [59]. It is a central nervous system depressant. In therapeutic doses of 0.5–1.0 g, it has little effect on respiration and blood pressure, but in toxic doses it depresses both. Ethanol enhances chloral's effects. Chloral is quite irritating to skin and mucous membranes. In the body it is readily reduced to trichloroethanol, which is responsible for most of chloral's physiological action. The trichloroethanol is excreted in the urine, mainly as the glucuronide. Oral doses of 4–30 g have been lethal to adults [59].

Rat studies indicate that chloral is highly toxic by inhalation [60]. The LC$_{50}$ for 4-h exposure is 440 mg/m^3. On exposure to 78 mg/m^3 of chloral 4 h/d for 10 d, rats had severe lung injury and 6 of 10 died; similar results were obtained with 75 mg/m^3 for 2 h/d for 13 d. It is mutagenic in the Ames test. TLV and MAK values have not been assigned for chloral. However, the inhalation studies in rats indicate that chloral should be handled at least as carefully as chloroacetaldehyde [62].

5. References

[1] E. Kopp, Wacker-Chemie, unpublished result, 1967.
[2] Physikalisch-Technische Bundesanstalt, Braunschweig, Federal Republic of Germany, 1980.
[3] Soc. des Usines Chimiques Rhône-Poulenc, FR 1 423 671, 1964.
[4] B. G. Yasnitskij, A. P. Zaitsev, *Zh. Org. Khim.* **2** (1966) 1022; *Chem. Abstr.* **65** (1966) 18 450; SU 119 875, 1959; *Chem. Abstr.* **54** (1960) 2178.
[5] E. Kopp, J. Smidt, *Justus Liebigs Ann. Chem.* **693** (1966) 117.
[6] Saint-Gobain Chauny & Cirey, FR 1 120 234, 1955.
[7] R. M. Novak, *J. Org. Chem.* **28** (1963) 1182.
[8] Consortium für elektrochemische Industrie, DE 1 141 631, 1961.
[9] Wacker Chemie company publication, *"Chloracetaldehyd"* 1983.
[10] Hoffmann La Roche Inc, US 3 872 089, 1975; *Chem. Abstr.* **83** (1975) 79 293.
[11] Hoffmann La Roche Inc, DE-OS 2 221 623, 1972 (O. Hromatka and B. Binder); *Chem. Abstr.* **83** (1975) 486.
[12] L. Givaudan & Ci, S.A., DE-OS 2 226 780, 1973 (P. DubsM. Pesaro); *Chem. Abstr.* **78** (1973) 481.
[13] O. Hromatka, R. Haberl, *Monatsh. Chem.* 1954, no. 5, 1088.
[14] G. Künstle and Ch. Solbrig, Wacker-Chemie, unpublished results.
[15] Velsicol Chem. Corp., CH 617 932, 1980 (J. Kreuzer); *Chem. Abstr.* **94** (1981) 15 741; A. Wohl et al., *Ber. Dtsch. Chem. Ges.* **22** (1889) 1354.
[16] Shell Int. Res. Maatschapij Ger., DE-OS 1 914 954, 1969.
[17] Schering AG, DE-OS 2 636 994, 1978; DE-OS 2 214 632, 1973.
[18] R. H. Sieber, Wacker-Chemie, *Justus Liebigs Ann. Chem.* **730** (1969) 31; DE 1 593 718, 1966.
[19] W. Perkow, E. W. Krockow, K. Knoevenagel, *Chem. Ber.* **88** (1955) 662.
[20] Degussa, DE-OS 2 239 278, 1972.
[21] Wacker-Chemie, DE 1 284 152, 1962.
[22] Directie van de Staatsmijnen I. Limburg, NL 77 330, 1955.
[23] Consortium für elektrochemische Industrie, DE 1 208 039, 1964. J. Smidt, W. Hafner, R. Jira, *Angew. Chem.* **74** (1962) 93. Consortium für elektrochemische Industrie, DE 1 130 426, 1961; DE 1 147 211, 1961.
[24] *Houben-Weyl*, **7/1**, p. 369.
[25] T. Tsukamoto, Mitsubishi Chemical Ind. Co., JP 8027, 1954; *Chem.Abstr.* **50**, (1956) 13 998.
[26] V. Valenta, CS 122 819, 1967; *Chem.Abstr.* **68** (1968) 49 067.
[27] L. L. Shukovskaya, S. N. Ushakov, N. K. Galanina, *Izv. Akad. Nauk USSR, Otd. Khim. Nauk* 1962, 1692; *Chem.Abstr.* **58** (1963) 8891.
[28] Dow Chemical Corp., US 2 803 668, 1956.
[29] Dynamit-Nobel, FR 1 592 333, 1968; DE-OS 1 643 899, 1967.
[30] Pechiney Compagnie de Produits Chimiques et Electrometallurgiques, FR 1 033 574, 1951.
[31] B. G. Yasnitskij, A. P. Zaitsev, SU 162 136, 1964; *Chem. Abstr.* **61** (1964) 13 195.
[32] Chemische Werke Hüls, DE 1 130 425, 1961.
[33] Chemische Werke Hüls, DE 1 136 996, 1961.
[34] H. Gross, *J. Prakt. Chem.* **21** (1963) 99.
[35] B. G. Yasnitskij, Ts. I. Satanovskaya, *Med. Prom. SSSR,* **14** (1960) 36; *Chem. Abstr.* **55** (1961) 7167.
[36] K. Natterer, *Monatsh. Chem.* **3** (1882) 447.

[37] W. Fink, *Angew. Chem.* **73** (1961) 466; Monsanto Co., US 3 164 612, 1962.
[38] T. Iwata, G. Wasai, T. Saegusa, J. Furukawa, *Makromol. Chem.* **77** (1964) 229.
[39] Consortium für elektrochemische Industrie, DE 1 189 714, 1961.
[40] G. Wasai, T. Iwata, K. Hirano, M. Suragano, T. Saegusa, J. Furukawa, *Kogyo Kagaku Zasshi* **67** (1964) 1920; *Chem. Abstr.* **62** (1965) 13 249.
[41] M. N. Shchukina, *Zh. Obshch. Khim.* **18** (1948) 1653; *Chem. Abstr.* **43** (1949) 2575.
[42] J. Rocek, *Tetrahedron Lett.* 1959, no. 5, 1.
[43] J. Boeseken, *Rec. Trav. Chim. Pays-Bas.* **57** (1938) 75.
[44] W. Deinhammer, *Consortium für elektrochemische Industrie*, unpublished results.
[45] Shawinigan Chem. Ltd., GB 644 914, 1947.
[46] FMC Corp., FR 1 337 054, 1962. FMC Corp., US 3 499 935, 1965.
[47] FMC Corp., US 3 253 041, 1963; US 3 562 334, 1964.
[48] Olin Mathieson, US 3 381 036, 1966.
[49] Consortium für elektrochemische Industrie, DE-OS 2 306 335, 1973.
[50] K. Akashi, *Bull. Inst. Phys. Chem. Res. Tokyo* **12** (1933) 329; *Chem. Abstr.* **27** (1933) 3447.
[51] FMC Corp., US 3 322 792, 1963.
[52] Michigan Chem. Corp., US 2 768 173, 1952.
[53] Diamond Alkali Co., NL 6 408 665, 1963.
[54] F. I. Luknitskii, *Chem. Rev.* **75** (1975) 259–289.
[55] P. Kubisa, L. S. Corley, T. Kondo, O. Vogl et al., *Polym. Eng. Sci.* **21** (1981) 829–838.
[56] W. T. Cave, *Ind. Eng. Chem.* **45** (1953) 1853–1857.
[57] *United States Pharmacopeia*, Mack, Easton, Pa., 1980, p. 129.
[58] Association of Official Analytical Chemists: *Official Methods of Analysis*, Washington 1980, p. 623.
[59] L. S. Goodman, A. G. Gilman, A. Gilman: *The Pharmacological Basis of Therapeutics*, MacMillan, New York 1980, pp. 361–363, 1671.
[60] SRI International: *Chemical Economics Handbook*, Menlo Park, Cal., 1979, p. 565, 52 000.
[61] Registry of Toxic Effects of Chemical Substances (RTECS), NIOSH, Cincinnati, Ohio, 1997.
[62] Wacker-Chemie, Report No. V 87.094/261236, 1987.
[63] W. H. Lawrence et al., *J. Pharm. Sci.* **61** (1972) 19–25.
[64] J. S. Jacobson et al., *Genetics* **121** (1989) 213–222.
[65] F. Zajdela et al., *Cancer Res.* **40** (1980) 352–356.
[66] Wacker-Chemie, personal communication, 1996.
[67] Deutsche Forschungsgemeinschaft, MAK-und BAT-Werte Liste, 1995.

Chloroacetic Acids

GÜNTER KOENIG, Hoechst Aktiengesellschaft, Augsburg, Federal Republic of Germany (Chaps. 2.6–4.6)

ELMAR LOHMAR, Hoechst Aktiengesellschaft, Köln, Federal Republic of Germany (Chaps. 2–2.5 and 5–8)

NORBERT RUPPRICH, Bundesanstalt für Arbeitsschutz, Dortmund, Federal Republic of Germany (Chap. 9)

1.	Introduction	1539
2.	**Chloroacetic Acid**	1540
2.1.	Physical Properties	1540
2.2.	Chemical Properties	1541
2.3.	Production	1543
2.3.1.	Hydrolysis of Trichloroethylene	1543
2.3.2.	Chlorination of Acetic Acid	1544
2.4.	Quality Specifications	1545
2.5.	Uses	1546
2.6.	Derivatives	1547
2.6.1.	Sodium Chloroacetate	1547
2.6.2.	Chloroacetyl Chloride	1548
2.6.3.	Chloroacetic Acid Esters	1548
2.6.4.	Chloroacetamide	1549
3.	**Dichloroacetic Acid**	1549
3.1.	Physical Properties	1549
3.2.	Chemical Properties	1550
3.3.	Production	1550
3.4.	Quality Specifications	1550
3.5.	Uses	1551
3.6.	Derivatives	1551
3.6.1.	Dichloroacetyl Chloride	1551
3.6.2.	Dichloroacetic Acid Esters	1552
4.	**Trichloroacetic Acid**	1552
4.1.	Physical Properties	1552
4.2.	Chemical Properties	1553
4.3.	Production	1553
4.4.	Quality Specifications	1553
4.5.	Uses	1553
4.6.	Derivatives	1554
4.6.1.	Trichloroacetyl Chloride	1554
4.6.2.	Trichloroacetic Acid Esters	1555
4.6.3.	Trichloroacetic Acid Salts	1555
5.	**Environmental Protection**	1556
6.	**Chemical Analysis**	1557
7.	**Containment Materials, Storage, and Transportation**	1558
8.	**Economic Aspects**	1559
9.	**Toxicology and Occupational Health**	1560
10.	**References**	1561

1. Introduction

Chloroacetic acid and its sodium salt are the most industrially and economically important of the three chlorination products of acetic acid discussed in this review. The sections on physical and chemical properties, production, quality specifications, uses, and derivatives are reported separately for each of these three acids, whereas those on environmental protection, chemical analysis, containment materials, storage, transportation, and economic aspects are considered together.

2. Chloroacetic Acid

Chlorinated acetic acids have become important intermediates in organic synthesis because of the ease of substitution of the Cl atoms. Chloroacetic acid [*79-11-8*], ClCH$_2$COOH, M_r 94.50, monochloroacetic acid, chloroethanoic acid, is the most industrially significant [1]. It does not occur in nature and was first discovered as a chlorination product of acetic acid by N. LEBLANC in 1841. It was synthesized in 1857 by R. HOFFMANN, who chlorinated acetic acid by using sunlight to initiate the reaction. Discovery of other reaction accelerators, such as phosphorus, iodine, sulfur, or acetic anhydride, followed rapidly. Development of commercial processes, based mainly on acetic acid chlorination and later on acid hydrolysis of trichloroethylene, followed.

2.1. Physical Properties

Pure chloroacetic acid is a colorless, hygroscopic, crystalline solid, which occurs in monoclinically prismatic structures existing in the α-, β-, γ-, and also possibly the δ-form. Of these the α-form is the most stable and the most important industrially.

Published physical data vary widely [1]. Some of the most common values are as follows:

Modification:	α	β	γ	δ
fp, °C:	62–63	55–56	50–51	43.8
Latent heat of fusion ΔH_f, kJ/mol:	19.38	18.63	15.87	–

Density: d_4^{65} 1.3703 (liquid)
d_{20}^{20} 1.58 (solid)
Refractive index: n_D^{65} 1.4297
Surface tension: σ 35.17 mN/m (100 °C)
Viscosity: η 2.16 mPa · s (70 °C)
1.32 mPa · s (100 °C)
1.30 mPa · s (130 °C)
Degree of dissociation in water (potentiometric): 1.52×10^{-3} (25 °C)
Dielectric constant: 16.8 (100 °C)
Electrical conductivity, lowest value measured (70 °C): 3.1 µS/cm, (rises steeply if traces of water present)
Specific heat capacity: c_p 144.02 J mol^{-1} K^{-1} (solid, 45–15 °C)
180.45 J mol^{-1} K^{-1} (liquid, 70 °C)
187.11 J mol^{-1} K^{-1} (liquid, 130 °C)
Heat of combustion: ΔH_c 715.9 kJ/mol
Heat of evaporation: ΔH_v 50.09 kJ/mol
Heat of formation (100 °C): ΔH_f –490.1 kJ/mol
Heat of sublimation (25 °C): ΔH_{subl} 88.1 kJ/mol

Heat of solution in H_2O (16 °C):
  solid $\Delta H_{solv.}$ − 14.0 kJ/mol
  liquid $\Delta H_{solv.}$ 1.12 kJ/mol
Flash point (DIN 51 758): 126 °C
Ignition temperature (DIN 51 794): 470 °C
Lower explosion limit in air (101.3 kPa): 8 vol%

Vapor pressure

t, °C	189	160	150	140	130	100	90	80
p, kPa	101.3	40.0	28.0	19.0	13.0	4.3	2.6	1.1

Solubility in water, α-form

t, °C	0	10	20	30	40	50	60
g/100 g soln.	71.0	76.0	80.8	85.8	90.8	95.0	99
g/100 g H_2O	245	317	421	604	987	1900	–

Chloroacetic acid has excellent solubility in water and good solubility in methanol, acetone, diethyl ether, and ethanol, but is only sparingly soluble in hydrocarbons and chlorinated hydrocarbons. Chloroacetic acid forms azeotropes with a number of organic compounds [2]. The freezing points of various binary mixtures of chloroacetic acids are shown in Figure 1.

2.2. Chemical Properties

The high reactivity of the carboxylic acid group and the ease of substitution of the α-Cl atom are directly related. As a result, chloroacetic acid is a common synthetic organic intermediate, either as the acid itself or as an acid derivative (e.g., salt, ester, amide, hydrazide, etc.). Some important reactions that are used for industrial applications are as follows.

Reaction with inorganic bases, oxides, and carbonates or with organic bases gives salts; some salts form adducts with chloroacetic acid. Sodium chloroacetate [3926-62-3] is an important commercial product.

Chloroacetic acid esters are obtained by reaction with alcohols or olefins; methyl chloroacetate [96-34-4] and ethyl chloroacetate [105-39-5] are also important industrially.

Chloroacetyl chloride [79-04-9] is produced from the acid by reaction with $POCl_3$, PCl_3, PCl_5, thionyl chloride ($SOCl_2$), phosgene ($COCl_2$), etc. (see Section 2.6.2.).

Chloroacetic acid reacts with chloroacetyl chloride to form bis(chloroacetic)anhydride [541-88-8], which can also be obtained by dehydration of chloroacetic acid with P_2O_5 or by reaction of chloroacetic acid with acetic anhydride. Chloroacetyl chloride forms mixed anhydrides with other carboxylic acids, e.g., acetic chloroacetic anhydride [4015-58-1].

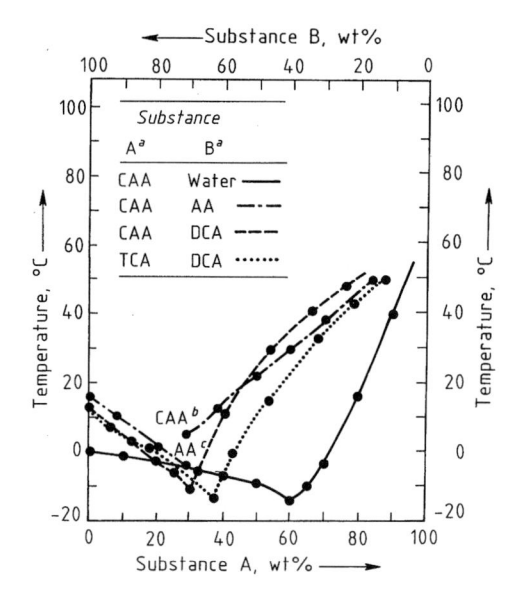

Figure 1. Freezing points of binary mixtures
[a] Acetic acid (AA), chloroacetic acid (CAA), dichloroacetic acid (DCA), trichloroacetic acid (TCA).
[b] Crystalline phase CAA.
[c] Crystalline phase AA.

Nucleophilic substitution of the chlorine atom is an important reaction when the product is used as an intermediate in organic syntheses. For example, heating neutral or basic aqueous solutions hydrolyzes the chlorine atom. This is an industrial method of producing glycolic acid [79-14-1](hydroxyacetic acid) and diglycolic acid [110-99-6] (2,2′-oxydiacetic acid).

Heating the salts gives glycolide, 1,4-dioxine-2,5-dione [502-97-6]. Reaction with sodium or potassium hydrogen sulfide forms thioglycolic acid [68-11-1] and thiodiglycolic acid [123-93-3].

Reaction with ammonia gives either aminoacetic acid [56-40-6] (glycine) as the main product or, depending on reaction conditions, nitrilotriacetic acid [139-13-9]. If methyl chloroacetate reacts with ammonia at low temperature, chloroacetamide [79-07-2] is obtained.

Aromatic compounds, such as naphthalene, undergo electrophilic substitution with chloroacetic acid over suitable catalysts to form arylacetic acids.

Reaction with potassium cyanide in a neutral solution gives the commercially important cyanoacetic acid [372-09-8]. Reaction with potassium iodide forms iodoacetic acid [64-69-7].

The corresponding phenoxyacetic acids, some of which are of industrial importance, are made by phenol etherification in the presence of sodium hydroxide. Another industrially significant ether formation process gives carboxymethyl derivatives with a relatively high degree of etherification by reacting polysaccharides, such as cellulose, starch, guar, etc., in a strongly alkaline sodium hydroxide medium.

2.3. Production

A multitude of methods have been proposed and patented for production of chloroacetic acid [1], [3]–[15]. However, of all the methods described, only two are used on an industrial scale: the hydrolysis of 1,1,2-trichloroethylene [79-01-6] catalyzed with sulfuric acid (Eq. 1), and the catalyzed chlorination of acetic acid with chlorine (Eq. 2).

$$ClCH=CCl_2 + 2\,H_2O \xrightarrow[\text{Heat}]{H_2SO_4} ClCH_2COOH + 2\,HCl \qquad (1)$$

$$CH_3COOH + Cl_2 \xrightarrow[\text{Heat}]{\text{Cat.}} ClCH_2COOH + HCl \qquad (2)$$

The first method (Eq. 1) has become of technological significance; the second (Eq. 2) is older, historically and commercially, and currently is the more important process.

2.3.1. Hydrolysis of Trichloroethylene [13]–[15]

Equal amounts of trichloroethylene and 75% sulfuric acid are reacted at 130–140 °C in a continuous process so that with complete trichloroethylene conversion, the resultant reaction mixture contains about 50% chloroacetic acid and 1–2% water. This blend is vacuum distilled to give pure chloroacetic acid. During this process the vapors are washed with water, which is returned to the sulfuric acid as a diluent. The resultant hydrogen chloride gas is washed with the fresh trichloroethylene and then purified by freezing and absorbing in water. Trichloroethylene (1500–1850 kg) and H_2SO_4 (600 kg, 95%) gives 1000 kg of finished product and 700–750 kg of HCl gas as byproduct.

The trichloroethylene method produces highly pure chloroacetic acid free of di- or trichloroacetic acid. The purification procedure consists of separation from trichloroethylene, sulfuric acid, and water. Despite the purity of the chloroacetic acid formed, this method has fallen into disuse because of the high cost of trichloroethylene and the large amount of HCl produced. It is, however, still used on an industrial scale in several countries (Spain, USSR, Romania, Yugoslavia, Japan, German Democratic Republic, etc.).

2.3.2. Chlorination of Acetic Acid

Synthesis. In this method the object is to convert acetic acid to chloroacetic acid with high selectivity [1]. This is achieved by using suitable catalysts [13], [16]–[33]. When acetic anhydride is the catalyst, the reaction mechanism is as follows [22]:

$$(CH_3CO)_2O + Cl_2 \longrightarrow ClCH_2\overset{O}{\overset{\|}{C}}-O-\overset{O}{\overset{\|}{C}}CH_3 + HCl$$

$$ClCH_2\overset{O}{\overset{\|}{C}}-O-\overset{O}{\overset{\|}{C}}CH_3 + CH_3COOH \longrightarrow ClCH_2COOH + (CH_3CO)_2O$$

$$(CH_3CO)_2O + HCl \longrightarrow CH_3COCl + CH_3COOH$$

$$CH_3COCl + Cl_2 \longrightarrow ClCH_2COCl + HCl$$

$$ClCH_2COCl + CH_3COOH \longrightarrow ClCH_2COOH + CH_3COCl$$

Various inhibitors have also been proposed to suppress formation of dichloroacetic acid, which results from chlorination of chloroacetic acid in the crude mixture. This eliminates or reduces the purification process for technical grades [34]–[38].

Purification. The high degree of purity required for many products can only be obtained by separating the di- and trichloroacetic acids. Fractional distillation is unsuitable because the boiling points of the three chlorinated acetic acids are so close. Azeotropic distillation [39] and extractive distillation [40], [41] have been suggested for separating dichloroacetic acid; it is doubtful, however, that these methods are used.

One of the most widely practiced purification processes is *crystallization* without use of a solvent. It is based on the higher melting point of the α-modification of chloroacetic acid. The di- and trichloroacetic acids are removed in the mother-liquor.

Crystallization is carried out either in stationary finger crystallizers [42] or in agitated stirrer crystallizers. With the latter, the mother-liquor is separated from the crystal slurry after crystallization by using a centrifuge [43], [44]. The product is washed with water or acetic acid and discharged. The chloroacetic acid is usually melted and converted to flakes. In stationary machines, crystallization is carried out by using cold fingers. When all the chloroacetic acid has crystallized, the mother-liquor is drained; the pure crystalline product is then melted and flaked.

Another purification method that has been described is a thin-layer crystallization process with the raw material [45]. Use of a water content of 5–25% without organic solvent is also possible [46].

Solvents have also been used for crystallization. Solvents, such as carbon tetrachloride [44], [47], dichloromethane [48], or hydrocarbons with three chlorine atoms (e.g., trichloroethylene), give crystals that are easy to filter [49].

Common to all of these crystallization methods is obtaining a mother-liquor consisting of acetic acid, chloroacetic acid, and di- and trichloroacetic acids. In the most favorable cases, this mixture is further chlorinated to form the industrially useful trichloroacetic acid (see Section 4.3).

In many instances, di- and trichloroacetic acids in either the mother-liquors, resulting from crystallization, or in the total chlorinated crude acid can be dechlorinated by catalytic hydrogenation at elevated temperature to form chloroacetic acid or acetic acid. Palladium on a carrier, such as carbon or silica gel, is normally employed.

When the reaction is carried out in the vapor phase, dichloroacetic acid is dechlorinated primarily to acetic acid [50]. However, when Pd is used on a finely dispersed, inert carrier in the liquid phase at 130–150 °C, dichloroacetic acid is dechlorinated selectively to form chloroacetic acid [51]. Modifications of this procedure, such as spraying the crude acid with the hydrogen gas under vacuum [52] or trickling the acid over the catalyst in the fixed bed [53], [54] have also been described.

Selectivity is increased if HCl is mixed with the crude acid before it and the circulating hydrogen contact the catalyst in the fixed bed [55]. Acid chlorides and anhydrides are saponified before dechlorination [56]. A particularly active and selective catalyst is Pd on silica gel (particle size 40–200 µm) [57]. Especially good results are obtained by employing a cocatalytically effective additive, such as sodium acetate [58]. The catalyst can be made more effective by enrichment of the nobel metal on its surface [59], [60]. Spent catalyst can be reactivated by treatment with chlorine [61], [62].

An industrial-scale chlorination process is shown in Figure 2. The mixture of acetic acid, acetic anhydride, and recycled acetyl chloride is chlorinated at 90–140 °C in reactor (a) or in several cascade reactors. Only traces of chlorine are still present in the HCl gas formed. Chloroacetic acid, acetic acid, and acetic anhydride are condensed by using water-cooled condensers (b) and then returned to the reactor. Acetyl chloride entrained in the HCl gas is condensed (c) in a subsequent low-temperature process and recycled. The HCl gas is further purified and usually converted to concentrated aqueous hydrochloric acid.

The crude acetic acids are vacuum distilled (d). Di- and trichloroacetic acids in the crude distilled material are dechlorinated (f) to chloroacetic acid at 120–150 °C, using a palladium catalyst and a large excess of hydrogen. Acetic acid is taken overhead from the vacuum fractionation column (i); the bottom product is pure chloroacetic acid.

A total of 660–780 kg of acetic acid and 780–1020 kg of chlorine are required per 1000 kg of pure acid, depending on the method used (in crystallization processes, the mother-liquors are regarded as lost). The process also gives 400–420 kg of HCl.

2.4. Quality Specifications

Chloroacetic acid is usually marketed to the following specifications:

Chloroacetic acid	min. 99.0 wt%
Dichloroacetic acid	max. 0.5 wt%
Acetic acid	max. 0.2 wt%
Sulfate	max. 0.5 wt%
Water	max. 0.2 wt%
Iron	max. 5 mg/kg
Lead	max. 1 mg/kg

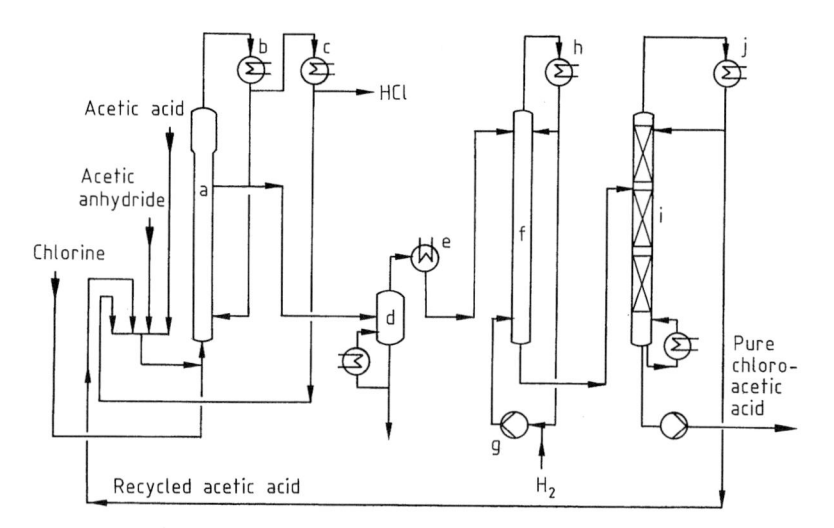

Figure 2. Chloroacetic acid obtained by the chlorination–hydrogenation process
a) Chlorinating reactor; b) Condenser for acetic acid, acetic anhydride, chloroacetic acid; c) Condenser for acetyl chloride; d) Evaporator; e) Condenser for chloroacetic acid; f) Hydrogenation reactor; g) Hydrogen compressor; h) Condenser; i) Distillation column; j) Condenser for acetic acid

Acid manufactured by the trichloroethylene method is virtually free of dichloroacetic acid; specially purified acid from the chlorination of acetic acid is marketed with an acetic acid content of max. 0.1% and max. 0.3% dichloroacetic acid. Technical grades contain up to 5% dichloroacetic acid.

2.5. Uses

Most of the chloroacetic acid produced is used to manufacture several hundred thousand tons annually of *carboxymethyl cellulose* [9004-32-4], CMC (→ Cellulose Ethers). Starch can be reacted with chloroacetic acid to give carboxymethyl starch, which is as widely used as CMC (→ Starch and other Polysaccharides). Other polysaccharides modified with chloroacetic acid are less important.

Another major application is the production of *herbicides* based on arylhydroxyacetic acids. These herbicides are some of the most widely used. Chloroacetic acid and methyl chloroacetate are also employed for making the insecticide dimethoate and the herbicides benazoline and methyl β-naphthyloxyacetate.

A third important outlet for chloroacetic acid is the manufacture of *thioglycolic acid* (mercaptoacetic acid [68-11-1]), obtained fro sodium or potassium hydrogen sulfide or other sulfur compounds. It is used as its salt, ester, or another derivative. The largest amount is employed to produce stabilizers for poly(vinyl chloride). Moreover, thioglycolic acid and its esters are used in hair cosmetics.

A minor use of chloroacetic acid is in the production of glycolic acid by saponification with an alkali hydroxide. Glycolic acid is used as an auxiliary in textile printing, leather treating, furs finishing, and as a component for cleaners; the butyl ester is employed as a paint additive, and glycolic acid esters acylated with o-phthalic acid half-esters are used as plasticizers.

Apart from the major fields of application mentioned above, chloroacetic acid and its derivatives are used in a multitude of other organic synthetic reactions. For example, caffeine and barbiturates, which are important as sleeping tablets, can be made from cyanoacetic acid or its esters. Chloroacetic acid condenses with aromatic hydrocarbons to form arylacetic acids. Reaction with naphthalene gives 1-naphthylacetic acid as the main product and 2-naphthylacetic acid as the byproduct. Both substances promote plant growth. Chloroacetic acid also is important in the syntheses of coumarin and vitamin B_6.

2.6. Derivatives

2.6.1. Sodium Chloroacetate

Physical, Chemical Properties. The sodium salt $ClCH_2COONa$ [3926-62-3], M_r 116.5, is of particular importance. It is colorless and slightly hygroscopic, and has good storage stability. It dissolves readily in water (44 wt% at 20 °C), giving a neutral solution. It has limited solubility in other polar solvents and is insoluble in nonpolar solvents. It hydrolyzes in water, depending on temperature and time, forming glycolic acid and sodium chloride.

Production. This salt is manufactured by reacting sodium carbonate with chloroacetic acid in batches or in a continuous process [63], [64].

In exceptional cases, localized superheating (about 150 °C) may occur during the reaction. If it does, slow thermal decomposition can take place, producing sodium chloride and polyglycolide as the main products with pronounced evolution of gas.

Another method of manufacturing sodium chloroacetate is spraying molten chloroacetic acid together with 50% caustic soda solution into a spray drier [65].

The production of salts of chloroacetic acids in fluidized beds has been described [66].

Uses. The uses for sodium chloroacetate are virtually the same as those for its free acid. The amount of salt required is less and depends on whether the free chloroacetic acid is used as an 80% aqueous solution or as a melt.

2.6.2. Chloroacetyl Chloride

Physical Properties. Chloroacetyl chloride [79-04-9], ClCH$_2$COCl, M_r 112.95, is a colorless, highly corrosive liquid that has a pungent odor and fumes when exposed to moist air; bp 105 °C (101.3 kPa), d_4^{20} 1.42, n_D^{20} 1.454.

Production. Chloroacetyl chloride is usually manufactured from chloroacetic acid by reaction with phosphorus trichloride, thionyl chloride, sulfuryl chloride, or phosgene. It is also obtained by chlorination of acetyl chloride in the presence of stronger aliphatic acids, preferably chloroacetic acids, or from sodium chloroacetate and the usual chlorinating agents.

One patent describes the manufacture of chloroacetyl chloride by chlorination of a mixture of 5–50 wt% acetyl chloride in acetic anhydride at 70–100 °C [67]. Another claims reaction of chloroacetic acid and trichloroethylene in the presence of iron(III) chloride and hydrochloric acid at 150 °C and 2 MPa [68]. Chloroacetyl chloride also has been obtained in 97.1% yield and with a purity of 99.8% by reacting chloroacetic acid with phosgene in the presence of palladium chloride at 110 °C [69]. Chlorination of ketene, which must be present in an excess of at least 50%, at 100–200 °C gives chloroacetyl chloride with less than 7% dichloroacetyl chloride [70]. Chloroacetyl chloride can also be made from 1,2-dichloroethylene and oxygen by using catalysts, such as bromine.

Uses. Chloroacetyl chloride is used for many syntheses, e.g., to make adrenaline, chloroacetic acid esters, and the anhydride.

2.6.3. Chloroacetic Acid Esters

Physical Properties. The *methyl ester* [96-34-4], ClCH$_2$COOCH$_3$, M_r 108.53, is of particular importance. It is a colorless liquid with a pungent odor, bp 130–131 °C (101.3 kPa), fp −32.7 °C, d^{20} 1.236, soluble in alcohol and ether, and only sparingly soluble in water.

Also important is the *ethyl ester* [105-39-5], ClCH$_2$COOCH$_2$CH$_3$, M_r 122.55, bp 144.2 °C (101.3 kPa), fp −26.0 °C, d^{20} 1.159, insoluble in water, and readily soluble in alcohol and ether.

Production. These esters can be manufactured from chloroacetic acid and either methyl or ethyl alcohol. In another method, trichloroethylene is converted to ethyl 1,2-dichlorovinyl ether, which can be readily hydrolyzed to form chloroacetic acid ethyl ester.

Uses. The reactivity of the ester, which is greater than that of the free acid, makes it suitable for many syntheses, e.g., sarcosine, chloroacetamide, thioglycolic acid ester [71] for pharmaceuticals (vitamin A), and crop protection agents (dimethoate). Condensation with aldehydes and ketones gives glycide esters [72]. Other uses are the synthesis of heterocyclic compounds, e.g., 2-phenylimino-4-oxooxazolidine from the ester and phenyl urea [73], and the well-known condensation of chloroacetic acid and its esters with thioureas to form pseudothiohydantoins.

2.6.4. Chloroacetamide

Physical Properties. Chloroacetamide [79-07-2], $ClCH_2CONH_2$, M_r 93.52, colorless crystalline needles, bp 224 °C (101.3 kPa), fp 121 °C, is soluble in water and alcohol, and sparingly soluble in all nonpolar solvents.

Production. Chloroacetamide is obtained on an industrial scale by reaction of methylchloroacetate with ammonia at low temperature [74]. Manufacture from chloroacetic acid and cyanamide at 150 – 200 °C also has been described [75].

Uses. Chloroacetamide is a versatile intermediate. In addition, it has biocidal properties and, therefore, is used as an industrial preservative. Because of its good solubility in water, chloroacetamide is a particularly suitable biocide for protection of the aqueous phase, e.g., in drilling fluids [76], in water-containing paints [77], and as a wood preservative [78]. Its insecticidal action on aphids [79] and its use as a hardener for urea and melamine resins [80] have also been described.

Various derivatives have the same biocidal effect as chloroacetamide. For instance, N-octadecylchloroacetamide is used as an antimicrobial plasticizer [81].

3. Dichloroacetic Acid

3.1. Physical Properties

Dichloroacetic acid [79-43-6], 2,2-dichloroethanoic acid, $Cl_2CHCOOH$, M_r 128.95, bp 192 °C (101.3 kPa), 102 °C (2.7 kPa), fp 13.5 °C, d_4^{20} 1.564, n_D^{20} 1.466, vapor pressure 0.19 kPa (at 20 °C), dissociation constant 5×10^{-2} mol/L (at 18 °C), is a colorless, highly corrosive liquid that gives off acidic vapors, which irritate the mucous membranes. It is miscible with water in any proportion. Dichloroacetic acid is readily soluble in the usual organic solvents, such as alcohols, ketones, hydrocarbons, and chlorinated hydrocarbons.

3.2. Chemical Properties

The two chlorine atoms of dichloroacetic acid are susceptible to displacement. For instance, with aromatic compounds, diaryl acetic acids are formed, and with phenol, diphenoxy acetic acids are the products. However, dichloroacetic acid is less prone to hydrolysis than chloroacetic acid. In the manufacture of carboxymethyl celluloses and starches, the dichloroacetic acid impurity in the chloroacetic acid gives rise to cross-linking, which is either desirable or undesirable, depending on the use of the end product.

3.3. Production

The most cost-effective production method is the hydrolysis of dichloroacetyl chloride (see Section 3.6.1). Moreover, 98 % dichloroacetic acid can be obtained in 90 % yield by hydrolysis of pentachloroethane with 88 – 99 % sulfuric acid [82] or by oxidation of 1,1-dichloroacetone with nitric acid and air [83]. Extremely pure dichloroacetic acid can be produced by hydrolysis of the methyl ester [84], which is readily available by esterification of crude dichloroacetic acid. Furthermore, dichloroacetic acid and ethyl dichloroacetate can be obtained by catalytic dechlorination of trichloroacetic acid or ethyl trichloroacetate with hydrogen over a palladium catalyst [85].

Separation of pure dichloroacetic acid from the other chloroacetic acids cannot be carried out by physical methods, especially fractional distillation, because the differences in boiling points, particularly between di- and trichloroacetic acid, are too small. The ester mixtures, on the other hand, can be satisfactorily fractionated in efficient distillation columns. In addition, mixtures of the salts of the three chloroacetic acids can be washed with water, alcohol, or water – alcohol solutions. The dichloroacetate can be dissolved preferentially and acidified to give pure dichloroacetic acid.

Dichloroacetic acid can be produced in the laboratory by reacting chloral hydrate and potassium or sodium cyanide:

$$Cl_3CCHO + H_2O + KCN \longrightarrow HCN + KCl + Cl_2CHCOOH$$

3.4. Quality Specifications

Dichloroacetic acid is a colorless liquid that is marketed, for instance, by Hoechst, with the following technical data:

Content	ca. 99 %
bp at 101.3 kPa	192 °C
Water, max.	0.3 %
fp	> 12 °C

3.5. Uses

Dichloroacetic acid is used as a test reagent for analytical measurements during fiber manufacture [poly(ethylene terephthalate)] and as a medicinal disinfectant (substitute for formalin).

Dichloroacetic acid, particularly in the form of its esters, is an important intermediate in organic synthesis. It is a reactive starting material for the production of glyoxylic acid, dialkoxy and diaroxy acids, and sulfonamides.

3.6. Derivatives

3.6.1. Dichloroacetyl Chloride

Physical Properties. Dichloroacetyl chloride [79-36-7], $Cl_2CHCOCl$, M_r 147.40, is a colorless liquid, which has an unpleasant odor and fumes when exposed to moist air; bp 107 °C (101.3 kPa), d_4^{16} 1.5315, n_D^{16} 1.4638, vapor pressure 3.06 kPa (at 20 °C).

Chemical Properties. Dichloroacetyl chloride undergoes not only reactions typical of acid chlorides, but also displacement reactions of the chlorine atoms in the 2-position. The chemistry of dichloroacetyl chloride and its derivatives is analogous in certain respects to that of glyoxylic acid. Thus, dichloroacetyl chloride reacts with ammonia and amines to form amino acids, with alcohols to form ester acetals and acetals, with benzene to form diarylacetic acids, and with phenols to form diphenoxyacetic acids.

Production. Dichloroacetyl chloride is produced by *oxidation of trichloroethylene*. Oxidation with oxygen to form a mixture of dichloroacetyl chloride and chloral has been known since the early 1900s [86]. More recent patents describe methods to manufacture dichloroacetyl chloride of > 98% purity, e.g., at 65–200 °C and 0.2–2 MPa in the presence of free-radical initiators [87], using catalytic quantities of azo compounds and amines [88], and initiating oxidation with UV light and only adding organic nitrogen bases in quantities of 0.005–0.05% once oxidation has begun [89]. Correspondingly, chloroacetyl chloride and trichloroacetyl chloride can be produced from 1,2-di- and tetrachloroethylene, but these methods are of little industrial significance because the products can be obtained more easily by other processes.

Furthermore, dichloroacetyl chloride is manufactured from pentachloroethane and fuming sulfuric acid or from chloroform and carbon dioxide in the presence of aluminum chloride at high pressure [90]. It has also been produced from ketene and chlorine in the presence of sulfur dioxide [91].

Uses. Dichloroacetyl chloride is used to manufacture dichloroacetic acid (see Section 3.3). In addition, it can be employed for the production of esters and anhydrides. It is used as the starting material to manufacture pharmaceuticals, such as sulfonamides and antibiotics (chloramphenicol), and to produce crop protection agents.

3.6.2. Dichloroacetic Acid Esters

Physical Properties. *Methyl dichloroacetate* [116-54-1], $Cl_2CHCOOCH_3$, M_r 142.98, bp 143.3 °C (101.3 kPa), d_4^{20} 1.3808, n_D^{20} 1.4428.

Ethyl dichloroacetate [535-15-9], $Cl_2CHCOOCH_2CH_3$, M_r 157.00, bp 158.5 °C (101.3 kPa), d_4^{20} 1.2821, n_D^{20} 1.4386.

Chemical Properties. These esters are highly reactive and, for example, condense with aldehydes and ketones in the presence of dilute metal amalgams to form α-chloro-β-oxyacid esters, which can be converted to the corresponding aldehydes via glycide esters [92]. Dichloroacetic acid esters are readily saponified by boiling water.

Production. These esters are obtained by the usual esterification methods. The methyl and ethyl esters are best produced by alcoholysis of dichloroacetyl chloride.

Uses. These esters are employed as intermediates in the manufacture of chloramphenicol, dichloroacetamide, and crop protection agents, and are employed as paint resin solvents. The glyceryl and ethylene glycol esters also serve as plasticizers for cellulose derivatives.

The vapors of dichloroacetic acid esters are heavier than air and can form explosive mixtures with air.

4. Trichloroacetic Acid

4.1. Physical Properties

Pure trichloroacetic acid [76-03-9] forms hygroscopic, rhombohedral crystals that are extremely soluble in water and soluble in many organic solvents; Cl_3CCOOH, M_r 163.40, bp 197.6 °C (101.3 kPa), bp ca. 107 °C (2.8 kPa), fp 59.2 °C, d_4^{60} 1.630, n_D^{65} 1.459.

4.2. Chemical Properties

The complete substitution of the hydrogen atoms on the methyl radical by chlorine makes trichloroacetic acid a strong acid ($K = 2 \times 10^{-1}$ mol/L); it also makes the chlorine atoms less easily substituted than those of chloroacetic acid and dichloroacetic acid. Trichloroacetic acid decomposes to chloroform and carbon dioxide when its aqueous solution is exposed to heat. The decomposition is particularly fast in the presence of organic or inorganic bases. When water is absent, aniline, resorcinol, and activated carbon can catalyze the decomposition. Purely thermal decomposition takes place only when the boiling point has been exceeded, resulting in the formation of chlorinated hydrocarbons, carbon monoxide and dioxide, and phosgene. Trichloroacetic acid salts with inorganic and many organic bases are known. They decompose when heated in the presence of water to form chloroform.

4.3. Production

Trichloroacetic acid is produced on an industrial scale by chlorination of acetic acid or chloroacetic acid mother-liquors at 140–160 °C [93]. If necessary, calcium hypochlorite is added as a chlorination accelerator [94]. There are conflicting views concerning adding heavy metal salts as chlorination catalysts. Examples of catalysts that have been used are iron and copper compounds, which are precipitated with sulfuric acid or phosphoric acid if decomposition of the reaction mixture occurs [95]; 2% phosphoric acid [96]; and catalysts and UV light [97]. Trichloroacetic acid has also been produced without catalysts [98], [99].

The crude product, containing about 95% trichloroacetic acid, is best isolated by crystallizing the melt, removing the mother-liquor with most of its impurities, and increasing the purity by centrifugation or recrystallization.

4.4. Quality Specifications

Trichloroacetic acid is marketed in various degrees of purity. The specifications in Table 1 are typical.

4.5. Uses

The main application for trichloroacetic acid is production of its sodium salt, which is used as a selective herbicide and in formulations with 2,4-D and 2,4,5-T preparations as a total herbicide.

Table 1. Typical quality specifications

Property	Grade			
	Crude[a]	Industrial[a]	Ph. Eur.[b]	ACS[c]
Freezing point, min., °C	55	56	–	–
Trichloroacetic acid, % min.	96.5	98	98	99
Dichloroacetic acid, % max.	2.5	1.2	–	–
Sulfuric acid, % max.	0.5	0.3	–	0.02
Sulfated ash, % max.	–	–	0.1	0.03
Water, % max.	0.5	0.2	–	–
Heavy metals in the form of:				
Lead, ppm max.	10	5	–	20
Iron, ppm max.	20	10	–	10
Phosphate, ppm max.	–	–	–	5
Chloride, ppm max.	–	–	100	10
Nitrate, ppm max.	–	–	–	20

[a] Specifications of Hoechst.
[b] European Pharmacopeia (formerly DAB 8) [100].
[c] Specifications of ACS.

Trichloroacetic acid also is employed as an etching or pickling agent in the surface treatment of metals, as a swelling agent and solvent in the plastics industry, as an albumin precipitating agent in medicine, as an auxiliary in textile finishing, and as an additive to improve high-pressure properties in mineral lubricating oils. Because it is strongly corrosive, trichloroacetic acid is used to remove warts and hard skin and to treat various skin afflictions [101]. Trichloroacetic acid, particularly its esters, are important starting materials in organic syntheses. The acid undergoes numerous color reactions that can serve to identify a wide range of organic compounds [102].

4.6. Derivatives

4.6.1. Trichloroacetyl Chloride

Physical Properties. Trichloroacetyl chloride [76-02-8], Cl_3CCOCl, M_r 197.9, bp 118 °C (101.3 kPa), d_4^{20} 1.620, n_D^{20} 1.4695, is similar to dichloroacetyl chloride.

Chemical Properties. The acid chloride can be hydrolyzed at 75–85 °C with water to form the free acid and hydrolyzed with ammonium hydroxide or concentrated sodium carbonate solution to form the salts.

Production. The acetyl chloride is prepared from trichloroacetic acid and various inorganic acid chlorides (e.g., $SOCl_2$, PCl_3) or with P_2O_5 and HCl. More useful methods are the oxidation of tetrachloroethylene with fuming sulfuric acid, oxygen, or fuming nitric acid and sulfuric acid at 18–20 °C [103], or from the reaction of pentachloroethane and dry oxygen under UV light [104]. It has been obtained in 37% yield from carbon tetrachloride and carbon monoxide in the presence of aluminum chloride at 200 °C and high pressure [105].

The most common production method is the gas-phase, photochemical oxidation of tetrachloroethylene with oxygen. The reaction is initiated with UV light, with radioactive irradiation, or it is sensitized with chlorine or iodine [106].

Uses. Trichloroacetyl chloride can be used for the manufacture of the esters and the anhydrides of trichloroacetic acid.

4.6.2. Trichloroacetic Acid Esters

Only the methyl and ethyl esters are of industrial interest. They can be used as solvents, if necessary, and for the production of the amide and polyalcohol esters, which have been suggested for use as plasticizers.

Methyl trichloroacetate [598-99-2], $Cl_3CCOOCH_3$, M_r 177.43, bp 153 °C (101.3 kPa), d_4^{20} 1.4864, n_D^{20} 1.4572.

Ethyl trichloroacetate [515-84-4], $Cl_3CCOOCH_2CH_3$, M_r 191.45, bp 167.5 °C (101.3 kPa), d_4^{20} 1.3823, n_D^{20} 1.4505.

4.6.3. Trichloroacetic Acid Salts

The sodium salt is the only one of industrial importance.

Physical Properties. Sodium trichloroacetate [650-51-1], TCA, sodium TCA, STCA, NaTA, $Cl_3CCOONa$, M_r 185.40, is a colorless salt that decomposes at temperatures below the melting point. The salt is very soluble in water and methanol and soluble in polar solvents. The solubility in water at -5 °C is 50 wt%, at 20 °C 60 wt%, and at 40 °C 70 wt%.

Chemical Properties. The salt is stable in the dry state and at normal storage temperature. Hydrolytic cleavage takes place in aqueous solution:

$Cl_3CCOONa + H_2O \longrightarrow Cl_3CH + NaHCO_3$

At the boiling point, this reaction takes place quantitatively within an hour and is used for quantitative analysis of sodium trichloroacetate [107]. At 20 – 25 °C the aqueous solution is relatively stable; with a 50% solution only ca. 1% is hydrolyzed in 4 – 6 weeks.

In aprotic solvents, such as 1,2-dimethoxyethane, the reactive intermediate dichlorocarbene is formed as the primary product at elevated temperature (ca. 80 °C) [108]:

$Cl_3CCOONa \longrightarrow NaCl + CO_2 + :CCl_2$ \hfill (3)

When phase-transfer catalysis is used, the dichlorocarbene can add, for example, across olefinic double bonds to give dichlorocyclopropane derivatives [108]. If dichlorocarbene is formed in the absence of other acceptors, it reacts with additional sodium trichloroacetate:

$$Cl_3CCOONa + :CCl_2 \longrightarrow Cl_3CCOCl + CO + NaCl \qquad (4)$$

Consequently, the main products formed during the thermal decomposition of sodium trichloroacetate (combination of Eqs. 3 and 4) were sodium chloride, trichloroacetyl chloride, carbon monoxide, and carbon dioxide; also observed were chloroform, carbon tetrachloride, and traces of phosgene [109].

Thermal decomposition is exothermic (42 kJ/mol) and starts between 125 and 170 °C. This decomposition reaction is quenched by adding water; hydrolysis then takes place.

Production. Sodium trichloroacetate is produced industrially by neutralizing trichloroacetic acid with sodium hydroxide solution or sodium carbonate [110], [111].

Uses. Sodium trichloroacetate, together with reducing agents and traces of heavy-metal salts, is recommended as a polymerization catalyst for vinyl compounds [112]. It is useful as a dyeing auxiliary because of the formation of $NaHCO_3$ during hydrolysis [113]. It also improves absorption of disperse dye systems on polyester and cellulosic fibers [114]. Moreover, in diazo papers developed by heat, sodium trichloroacetate is used to liberate base at 100–200 °C [115].

Most important, however, is its application as a selective herbicide to control monocotyledonous annual or perennial weeds [116]–[118]. As a soil-applied herbicide its half-life is 1–2 months [119]. For use as a crop protection agent, a minimum content of 95% sodium trichloroacetate is required. Because the salt in powder form causes severe irritation, the product is usually marketed as granules [120].

5. Environmental Protection

The waste gases formed during chlorination of acetic acid depend on the degree of chlorination. When purification is done by fractional crystallization, less than 1% of unreacted acetic acid is necessary in the crude acid. In this case chlorine occurs in these waste gases; recovery of the chlorine has been described [121].

Formation of chlorinated hydrocarbon impurities also depends on the production method used. They form as a result of decarboxylation and occur particularly in the manufacture of trichloroacetic acid.

In the trichloroethylene method and in the manufacture of dichloroacetic acid by catalytic oxidation of trichloroethylene, residues of the starting materials are always present in the waste gases. Such waste gases are purified by adsorption on activated charcoal or are incinerated.

Moreover, the high sulfate salt load can be a wastewater problem in the trichloroethylene method. Chlorinated acetic acids are biodegradable by activated sludge floc, but they can cause serious interference particularly in community wastewater treatment plants that have not been bioprepared. Preventing effluent contaminated with products of acetic acid chlorination from draining into public sewer systems is essential.

Spills of chlorinated acetic acids and their derivatives, resulting, for instance, from transport accidents, should be contained with inert materials (sand or earth) and collected for safe disposal, or neutralized with agents, such as Na_2CO_3 or $NaHCO_3$. Any contaminated soil must be removed (the chlorinated acetic acids are strong pollutants to groundwater). Waste material should be taken to an approved disposal facility (special landfill disposal, incinerator, or wastewater treatment plant).

Chloroacetic acid is flammable at higher temperature and forms explosive mixtures with air (lower explosion limit 8 vol%). The two other acids are nonflammable. If the acids are involved in a fire, the decomposition gases of trichloroacetic acid and dichloroacetic acid may contain phosgene, whereas chloroacetic acid gases are free of phosgene.

6. Chemical Analysis

Rapid analysis of the purity of mono-, di-, and trichloroacetic acid is adequately provided by determining the melting and freezing points, particularly when the melting depression is caused by known contamination.

Quantitative analysis of chloroacetic acid should include total acidity and a separate determination of dichloroacetic acid, sulfuric acid, hydrochloric acid, water, and heavy metals, as well as total chlorine content by approved methods. From these determinations acetic acid and chloroacetic acid content can be calculated.

Dichloroacetic acid can be determined with a high degree of accuracy by saponification with strong base and subsequent manganometric titration. Dichloroacetic acid values are exaggerated in the presence of trichloroacetic acid. If trichloroacetic acid is present, preliminary saponification with dilute base is advisable.

Acetic acid and dichloroacetic acid can be directly separated and quantitatively determined by gas chromatography. In addition to these two acids, trichloroacetic acid can also be detected in the form of its methyl ester by gas chromatography. Dichloroacetic acid and acetic acid in the presence of chloroacetic acid can also be determined accurately by HPLC [122]. This method is not recommended for chloroacetic acid itself because titration to determine total chlorine content is more accurate.

Trichloroacetic acid is determined by hydrolysis in acid solution to chloroform and quantitative determination of the hydrogen ions needed for saponification. Formation of a deep-red complex from pyridine and chloroform can be used for a colorimetric method [123]. Nitron [1,4-diphenyl-3-(phenylamino)-1H-1,2,4-triazolium hydroxide] produces a salt with trichloroacetic acid, which is suitable for detecting the acid [124].

Dichloroacetic acid can be determined by polarography in mixtures with trichloroacetic acid [125]. Large amounts of chloroacetic acid interfere with the determination.

Determination of heavy metals in all chlorinated acetic acids and derivatives is carried out by familiar methods, such as atomic absorption or colorimetry. For lead, inverse voltametry is also used.

7. Containment Materials, Storage, and Transportation

Because of the corrosiveness of these chemicals, most metals are unsuitable for use as *container materials*. Thus, chlorination is carried out in ceramic-lined, lead-coated steel containers or in glass-lined vessels. Pipelines are made of glass-lined steel or steel lined with polytetrafluoroethylene (PTFE) or perfluoroalkoxy (PFA) polymers (e.g., tetrafluoroethylene). Silver that has been rendered inert by a silver chloride layer can be used for parts of the equipment (e.g., valves, heat exchangers, distillation column linings, stirrers). Tantalum and titanium (with certain temperature limits) are other suitable metallic materials. Carbon (Diabon) and glass are effective materials for heat exchangers. In addition to PTFE and PFA/TFA, fluorinated ethylene–propylene copolymers (FEP), poly(vinyl chloride), polyethylene, and polypropylene can be used at relatively low temperature and for aqueous solutions.

The aqueous solution can be handled in special stainless steel or rubber-lined steel containers at temperatures below 40 °C. Iron with a baked phenolic resin coating can be used below 100 °C. In the trichloroethylene method, steel (cast iron) is an adequate material in many parts of the process.

During *storage and transportation*, high corrosivity and, with chloroacetic acid, toxicity must also be considered. Chlorinated acetic acids and their derivatives are classified as dangerous substances. They are subject to nearly all national and international handling regulations. Containers must be labeled in the stipulated manner. In the case of chloroacetic acid, the most important forms that are shipped are the melt (in glass-lined railroad cars or titanium-lined tank trucks) and the 80% aqueous solution (in stainless steel or rubber-lined tank trucks and containers).

Crystallized, flaked chloroacetic acid and trichloroacetic acid are obtained by utilizing chilled rolling flakers and, more recently, with chilled conveyor-type flakers; the coarser flakes manufactured in this manner tend to agglomerate less and, therefore, are preferred. Flakes are packed in multilayer paper sacks or in polyethylene bags contained in fiberboard or iron drums. Small quantities are supplied in polyethylene containers or in glass bottles. Dichloroacetic acid and most of its liquid derivatives usually are shipped in iron drums with polyethylene inner containers.

If stored in a cool, dry place, the chlorinated acetic acids and their derivatives are stable almost indefinitely, although the crystallized products have a marked tendency to agglomerate.

8. Economic Aspects

The worldwide manufacturing capacity for chloroacetic acid is estimated at 420 000 t/a, divided as follows:

Western Europe	ca. 50%
Federal Republic of Germany (ca. 23%), Finland, France, United Kingdom, Netherlands, Sweden, Spain	
Eastern bloc	ca. 22%
Asia	ca. 12%
United States	ca. 11%
South America, South Africa	ca. 5%

Because French manufacturers have switched to the acetic acid method, 80–85% of worldwide production is now manufactured by the chlorination method. One manufacturer uses 1,1-dichloroethene/chloroacetyl chloride as the starting material, whereas the remainder are probably still using the older trichloroethylene process.

Recognized manufacturers are AKZO, Atochem, Billerud/Uddeholm, Courtaulds Acetate/UK, Cros, Dow, Hercules, Hoechst, and Mätselliito.

Worldwide production cannot be determined precisely because no statistics have been published. Worldwide consumption was estimated to be 290 000 t in 1983. The main consuming region is Western Europe (ca. 125 000 t in 1983), followed by the Eastern bloc, the United States, and Japan. An increase is expected in some areas in the next few years.

Accurate figures are not available on the amounts of chloroacetic acid and its derivatives used for different applications. However, by far the largest amount goes into the production of carboxymethyl cellulose, starch, and polysaccharides. This is followed by herbicides based on phenoxyacetic acids (2,4-D, MCPA, 2,4,5-T) and thioglycolic acid (ca. 70% of which is used as esters in industry and about 30% as salts in cosmetics). All the other applications (glycolic acid, glycine, vitamins, and derivatives of chloroacetic acid for organic syntheses) account for probably not more than 10% of annual production.

Dichloroacetic acid is of little economic importance today. More significant are its acid chloride and methyl ester, which are used as intermediates in the manufacture of agrochemicals and chloramphenicol.

In the case of trichloroacetic acid, only the sodium salt is important. About 21 000 – 23 000 t/a are used worldwide as a selective herbicide.

9. Toxicology and Occupational Health

Chloroacetic Acid and Chloroacetate. Chloroacetic acid is a corrosive solid material. The lowest published LD_{50} is 76 mg/kg for oral administration to rats [126]. Single oral administration of lethal doses resulted in apathy and loss of body mass. Death occurred several days after administration [127].

Acute inhalational exposure of rats led to eye and pulmonary irritation. The irritation threshold is 23.7 mg/m^3. The LC_{50} is 180 mg/m^3 [128].

Prolonged inhalation of chloroacetic acid resulted in reduced body mass, reduced oxygen consumption, hemoglobinemia, and inflammation of the respiratory tract. A threshold limit value for chloroacetic acid of 1 mg/m^3 is recommended if these experimental animal findings are taken as a basis [128]. However, the ACGIH has not established a TLV; neither has an MAK been set.

Repeated intake of 0.1 % chloroacetate in the diet led to depressed growth curves and an increased level of liver glycogen [129]. A major metabolic pathway for chloroacetic acid is reaction with glutathion, finally yielding thiodiacetic acid [130]. One report has shown that chloroacetic acid elevates mutant rates in the mouse lymphoma forward mutation assay [131].

Dichloroacetic Acid and Dichloroacetate. Dichloroacetic acid, a clear colorless liquid, is corrosive to the skin and mucous membranes. Its oral LD_{50} is 2820 mg/kg for rats, which is of the same order of magnitude as that for trichloroacetic acid. Dichloroacetic acid is able to penetrate rabbit skin with a dermal LD_{50} of 797 mg/kg [126]. The lowest oral LD_{50} reported for dichloroacetate is 4480 mg/kg for fasted rats [127].

Sodium dichloroacetate was examined in a 3-month oral toxicity study in rat and dogs. Both species manifested toxic effects, such as depressed appetites and increased body mass, degenerative changes of testicular germinal epithelium, hind limb muscular weakness or paralysis, and brain lesions. Cessation of administration brought about an amelioration or recovery from all effects except brain lesions. Dogs proved to be more sensitive than rats, showing toxic effects at 50 mg kg^{-1} d^{-1}. The CNS observations are extremely relevant in light of a report that described polyneuropathy of a 21-year-old man whose severe hypercholesterolemia had been treated orally with dichloroacetate (50 mg/kg) for 16 consecutive weeks. The polyneuropathy was characterized by weakness of facial, finger, and lower extremity muscles, by diminished deep tendon reflexes, and by slowing of nerve conduction velocity. The neuropathy improved when the therapy was stopped [132].

Dichloroacetate demonstrated low-grade mutagenicity in the Ames Salmonella/mammalian microsome mutagenicity test [133].

Trichloroacetic Acid and Trichloroacetate. Trichloroacetic acid is a strong protein precipitant and is corrosive to the skin and eye. The solid material is not readily absorbed through the skin [133]. The lowest reported oral LD_{50} in rats for trichloroacetic acid was 3320 mg/kg and 5060 mg/kg for trichloroacetate. The lowest lethal dose of trichloroacetate for female dogs was 1590 mg/kg with toxic effects, such as salivation, emesis, sedation, ataxia, and coma [134], [135]. A 4-h exposure to 4800 mg/kg trichloroacetate in ambient air did not result in any clinical symptoms or macroscopically detectable pathologic lesions in rats, guinea pigs, rabbits, and cats [135].

Medical reports of acute exposure effects of trichloroacetic acid showed mild to moderate skin and eye burns [136]. In humans, trichloroacetate was an irritant to skin and mucous membranes. Opaqueness of the cornea may occur. Oral ingestion led to intestinal colics, salivation, emesis, diarrhea, muscular weakness, anorexia, and apathy [135].

Little relevant chronic toxicity data exists for trichloroacetic acid. The toxicological profile of airborne trichloroacetic acid is assumed to be dominated by lesions of mucous membranes of the upper respiratory tract. A TLV-TWA of 1 ppm, 7 mg/m^3 is suggested, based largely on analogy to 2,2-dichloropropionic acid (TLV-TWA 1 ppm, 6 mg/m^3) [136].

Trichloroacetic acid strongly binds to plasma proteins. High concentration in blood is capable of displacing drugs from protein-binding sites. Because of this systemic effect of potential drug displacement from plasma proteins, a TLV of 1–2 ppm may be accepted for trichloroacetic acid [134].

For trichloroacetate, a provisionally tolerated daily intake of 0.075 mg/kg body mass has been established. This limit value takes into account the results of a subchronic dog feeding study with concentrations in food of $\geq$ 2000 mg/kg, resulting in loss of body mass, malaise, necrosis of gingival and buccal mucosae, changes in white blood count, pathologic urine findings, lesions of the liver and myocardium, atrophy of skeletal muscle, and impairment of spermiogenesis. The nontoxic effect level was 500 mg/kg in the diet [135].

On the basis of available in vitro data, trichloroacetate does not exert mutagenic effects [135].

10. References

[1] *Beilstein*, 4th ed., **1**, 194–197; **2(1)**, 87–88; **2(2)**, 187–191; **2(3)**, 428–438; **2(4)**, 474–480.
[2] "Azeotropic Data 3" in R. F. Gould (ed.): *Advances in Chem., Series 116*, Amer. Chem. Soc., Washington, D.C., 1973, pp. 97–99.
[3] Du Pont, US 2 455 405, 1943 (L. S. Burrows, M. F. Fuller).
[4] SU 110 145, 1958 (B. G. Yasnitskii, A. P. Zaĭtzev); *Chem. Abstr.* **52** (1958) 17 110 g.
[5] SU 119 875, 1959 (B. G. Yasnitskii, A. P. Zaĭtzev); *Chem. Abstr.* **54** (1960) 2178 a.
[6] Rhône-Poulenc, FR 1 423 671, 1964 (P. Mounier).
[7] Rhône-Poulenc, FR 1 413 926, 1964 (G. Boullay).
[8] SU 1 004 346, 1981 (J. N. Norikov, V. J. Kondratenko, G. S. Cibulskaja); *Chem. Abstr.* **99** (1983) 5218 f.

[9] Gulf Research, US 3 627 826, 1970 (C. M. Selwitz).
[10] Chevron Research, US 4 221 921, 1976 (S. Suzuki).
[11] Mitsubishi Gas, JP-Kokai 7 844 520, 1976 (T. Isshiki, W. Yoshino, T. Kato).
[12] Plantex, JL 41 782, 1973 (S. Roiter, Y. Fein).
[13] British Intelligence Objectives Subcommittee (BIOS), Final Report no. 929, Item no. 22, London 1947, pp. 1–4, 7–9.
[14] British Intelligence Objectives Subcommittee (BIOS), Final Report no. 1154, Item no. 22, London 1947, pp. 37, 38.
[15] Field Information Agency Technical (FIAT), Report no. 1051, London 1947.
[16] Hooker Electrochem. Co., US 2 595 899, 1949 (J. A. Sonia, C. E. Lisman).
[17] BASF, DE 1 224 298, 1960 (H. Schlecht, H. Albers, R. Oster).
[18] H. Brückner, *Angew. Chem.* **40** (1927) 973–974; **41** (1928) 226–229.
[19] Akzo, DE 2 163 849, 1971 (H. D. Rupp, G. Meyer, H. Zengel).
[20] Iron Enterprise Co., JP 528, 1959 (H. Fukuda).
[21] Solvay Cie., GB 759 450, 1956 (J. Viriot).
[22] W. Richarz, A. Mathéy, *Chimia* **21** (1967) 388–395.
[23] CS 113 592, 1962 (J. Hrabovecky, S. Regula, J. Kozacek); *Chem. Abstr.* **63** (1965) 17 908 b.
[24] Comp. Espanola de Petroleos, ES 464 673, 1977 (C. R. Huertas, A. Y. Chinchilla, J. M. Lopez Bahamonde).
[25] Politechnika Lodzka, PL 63 908, 1968 (A. Kowalski); *Chem. Abstr.* **76** (1972) 139 941 v.
[26] Celanese Corp., US 2 503 334, 1950 (A. R. Hammond, J. A. John, R. Page).
[27] Monsanto, US 2 539 238, 1949 (C. M. Eaker).
[28] Dow Chemical, US 2 826 610, 1955 (E. K. Morris).
[29] Uddeholms, GB 928 178, 1960.
[30] Hoechst, DE 1 919 476, 1969 (W. Opitz, A. Jacobowsky, W. Burkhard).
[31] Fondazione de Nora, US 4 281 184, 1978 (P. Spaziante, C. Sioli, L. Giuffre).
[32] Daicel, JP 81 127 329, 1980.
[33] GB 1 176 109, 1968 (A. L. Englin, E. V. Sergeev, E. R. Berlin); *Chem. Abstr.* **72** (1970) 110 815 p.
[34] Dow Chemical, US 3 152 174, 1960 (E. K. Morris, W. W. Bakke).
[35] Frontier Chemical Co., US 2 917 542, 1958 (J. I. Jordan).
[36] Shawinigan Chemicals, US 2 809 993, 1955 (G. M. Glavin, H. B. Stevens).
[37] New York-Ohio-Chemical, US 2 688 634, 1950 (J. T. Pinkston).
[38] Monsanto, US 3 576 860, 1967 (D. A. Zazaris).
[39] Dow Chemical, US 3 772 157, 1971 (L. H. Horsley).
[40] Mitsui Toatsu Chemical, JP 7 230 165, 1970 (R. Fujiwara, H. Ohira).
[41] Mitsui Toatsu Chemical, JP 7 229 886, 1970 (R. Fujiwara, H. Ohira).
[42] Knapsack, DE 1 919 751, 1969 (H. Baader, G. Berger, A. Jacobowsky).
[43] Knapsack-Griesheim, DE 1 033 653, 1958 (R. Wesselmann).
[44] New York-Ohio Chemical, US 2 809 214, 1957 (I. N. Haimsohn).
[45] SU 374 278, 1968 (E. V. Sergeev, V. N. Egorova, R. P. Milyukova); *Chem. Abstr.* **79** (1973) 41 946 u.
[46] Tsukishima Kikai Co., EP 32 816, 1981 (R. Sugamiya, K. Nakamaru, K. Takegami).
[47] Comp. Espanola de Petroleos, ES 468 685, 1978 (C. R. Huertas).
[48] Dow Chemical, US 3 365 493, 1964 (D. D. Deline, F. B. Wortman, A. A. Holzschuh).
[49] Uddeholms, DE 1 268 130, 1960 (S. H. Persson, P. T. Akerström, P. E. Larberg).
[50] Knapsack-Griesheim, DE 910 778, 1951 (K. Sennewald, A. Wolfram).
[51] Hooker Chemical Co., US 2 863 917, 1954 (J. T. Rucker, J. S. Scone).

[52] Knapsack-Griesheim, DE 1 201 326, 1963 (W. Opitz).
[53] Hoechst, DE 1 668 023, 1967 (W. Freyer, M. Gscheidmeier, R. Höltermann).
[54] Knapsack-Griesheim, DE 1 072 980, 1958 (W. Opitz, K. Sennewald).
[55] N. V. Koninklijke Nederlandsche Zoutindustrie, NL 109 768, 1962 (G. van Messel).
[56] N. V. Koninklijke Nederlandsche Zoutindustrie, NL 109 769, 1962 (A. Blaauw, G. van Messel).
[57] Knapsack, DE 1 816 931, 1968 (K. Sennewald, A. Ohorodnik, W. Mittler).
[58] Hoechst, DE 1 915 037, 1969 (K. Sennewald, J. Hundeck, A. Ohorodnik).
[59] Hoechst, DE 2 240 466, 1972 (A. Ohorodnik, K. Gehrmann, J. Hundeck).
[60] Hoechst, DE 2 323 777, 1973 (M. Gscheidmeier).
[61] Knapsack, DE 1 920 805, 1969 (H. Baader, W. Opitz, A. Jacobowsky).
[62] Knapsack, DE 1 920 806, 1969 (K. Sennewald, J. Hundeck, W. Opitz).
[63] Hoechst, DE 871 890, 1942 (V. Hilcken, H. Petri). Hoechst, DE 860 354, 1942 (H. Petri, P. Landmann).
[64] Hercules Powder, US 2 446 233, 1946 (B. T. Lamborn, N. J. Matawan).
[65] Svenska Cellulosa, GB 782 479, 1955.
[66] Hoechst, DE 2 432 567, 1974 (H. Scholz, G. Koenig).
[67] Merck, DE 2 120 194, 1971 (H. Hornhardt).
[68] Pechiney St. Gobain, DE 2 059 597, 1970 (J. C. Strini).
[69] Knapsack, DE 1 804 436, 1968 (W. Opitz).
[70] Montecatini, US 2 843 633, 1954 (G. Natta, E. Beati).
[71] Monsanto, US 2 262 686, 1940 (L. P. Kyrides, W. Groves, F. B. Zienty).
[72] *Bull. Soc. Chim. Fr. Mem.* **5** (1939) no. 6, 1616–1625. L. Givaudan & Cie., FR 854 038, 1938.
[73] H. Aspelund, *Soum. Kemistiseuran Tied.* **49** (1940) 49–63.
[74] Dow Chemical, US 2 321 278, 1941 (E. C. Britton, W. R. Shawver).
[75] A. E. Kretov, A. P. Momsenko, SU 141 150, 1961.
[76] Phillips Petroleum Co., US 2 815 323, 1956 (R. P. Schneider).
[77] Dow Chemical, US 2 758 103, 1952 (W. A. Henson, W. M. Westveer).
[78] M. Schlecker, DE 1 492 569, 1964 (M. Schlecker).
[79] Leek Chemicals, GB 859 714, 1959 (F. N. Morris).
[80] General Electric Co., US 2 328 424, 1940 (G. F. D'Alelio).
[81] Ciba, CH 496 843, 1967 (A. Hiestand, A. Therwil).
[82] Comp. des Prod. Chim. et Electrometallurgiques Alais, DE 610 317, 1933.
[83] Shell, US 2 051 470, 1934 (M. de Simo, C. C. Allen).
[84] Dow Chemical, US 2 495 440, 1946 (E. C. Britton, L. F. Berhenke).
[85] Hoechst, DE 1 085 518, 1957 (W. Protzer).
[86] Consortium f. Elektrochem. Ind., DE 340 872, 1919; DE 391 674, 1921; DE 531 579, 1929 (M. Mugdan, J. Wimmer).
[87] Shell Int. Research, NL 6 606 933, 1965 (G. W. Gaertner, D. E. Ramey).
[88] Shell Int. Research, DE 1 793 446, 1968 (R. K. June).
[89] Hoechst, DE 1 568 547, 1966 (K. Petz).
[90] C. A. Frank, A. T. Hallowell, C. W. Theobald, *Ind. Eng. Chem.* **41** (1949) 2061–2062.
[91] F. B. Erickson, E. J. Prill, *J. Org. Chem.* **23** (1958) 141–143. Monsanto, US 2 889 365, 1957 (E. J. Prill). Distillers Co.,US 2 862 964, 1956 (R. M. Lacey).
[92] G. Darzens, *C. R. Hebd. Acad. Sci.* **203** (1936) 1374–1376.
[93] Dow Chemical, US 1 757 100, 1924 (C. J. Strosacker).
[94] Lech-Chemie Gersthofen, DE 860 211, 1944 (P. Heisel).
[95] Hoechst, DE 1 020 619, 1957 (H. Kolb, T. Sievers, H. Hoyer).

[96] Bozel-Maltera, FR 1 156 426, 1956 (F. Peto, L. Marcheguet, M. Girard).
[97] Amer. Cyanamid, US 2 382 803, 1943 (A. A. Miller).
[98] Monsanto, US 2 613 220, 1950 (C. M. Eaker).
[99] Knapsack-Griesheim, DE 1 031 778, 1954 (K. Sennewald, A. Wolfram).
[100] European Pharmacopeia, vol. 3.
[101] E. Sidi, G. E. Goetschel, F. Casalis, *Presse Med.* **32** (1950) 570.
[102] L. Rosenthaler, *Parfuem. Kosmet.* **36** (1955) 271.
[103] V. P. Rudav'skii, *Khim. Promst. Kiev* 1965, 22–23.
[104] E. Müller, K. Ehrmann, *Ber. Dtsch. Chem. Ges.* **69** (1936) 2207–2210.
[105] Du Pont, US 2 378 048, 1944 (C. W. Theobald).
[106] Uddeholms, DE 2 118 540, 1971 (C. Östlund, J. Dahlberg, A. Moden).SU 176 286, 1964 (V. A. Poluektov).
[107] G. R. Raw (ed.): *CIPAC-Handbook: Analysis of Technical and Formulated Pesticides,* vol. **1**, W. Heffer & Sons, Cambridge, England, 1970, p. 691.
[108] E. V. Dehmlow, *Tetrahedron Lett.* 1976, no. 2, 91–94.
[109] E. V. Dehmlow, K. H. Franke, *Z. Naturforsch.* **33б** (1978) 686–687.
[110] Nobel Bozel, FR 1 215 896, 1959 (M. L.-J. Girard).
[111] Monsanto, US 2 643 220, 1950 (J. Loumiet-Lauigne).
[112] Dynamit Nobel, DE 1 203 957, 1962 (W. Trautvetter).
[113] Ciba, BE 646 398, 1964 (F. Raff).
[114] Ciba, GB 1 035 340, 1964 (F. Raff).
[115] General Anilin & Film, BE 672 553, 1966 (W. J. Welch, B. E. Tripp).
[116] G. M. Tsukermann, *Best. Sel'sk. Nauk. Min. Sel'sk. Khoz. Kaz. SSR* **8** (1965) 117–121.
[117] H. Ansorge, U. Koss, *Albrecht-Thaer-Arch.* **7** (1963) 889–899.
[118] *Z. Pflanzenkr. Pflanzenschutz (Sonderheft)* **1** (1964) 49–53.
[119] H. Neururer, *Pflanzenschutzberichte* **28** (1962) 145–180.
[120] Hoechst, DE 1 166 761, 1962 (H. Kolb, R. Hartmann, H. Buckmiller).
[121] Hoechst, DE 3 246 953, 1984 (G. Koenig, J. Maginot).
[122] N. E. Skelly, *Anal. Chem.* **54** (1982) no. 4, 712–715.
[123] *Chem. Ztg.* **77** (1953) 585.
[124] W. L. Dulière, *Bull. Soc. Chim. Biol.* **34** (1952) 991.
[125] P. J. Elving, Chang-Slang Tang, *Anal. Chem.* **23** (1950) 34–43.
[126] R. J. Lewis, S., R. L. Tatken (eds.): *Registry of Toxic Effects of Chemical Substances,* NIOSH (RTECS Online Data Base), 1984.
[127] G. Woodward et al., *J. Ind. Hyg. Toxicol.* **23** (1941) 78–81.
[128] G. G. Maksimov, O. N. Dubinina, *Gig. Tr. Prof. Zabol.* 1974, no. 9, 32–35.
[129] S. Dalgaard-Mikkelson et al., *Acta Pharmacol. Toxicol.* **11** (1955) 13.
[130] S. Yllner, *Acta Phamacol. Toxicol.* **30** (1971) 69–80.
[131] D. E. Amacher, G. N. Turner, *Mutat. Res.* **97** (1982) 49–65.
[132] R. Katz et al., *Toxicol. Appl. Pharmacol.* **57** (1981) 273–287.
[133] V. Herbert et al., *Am. J. Clin. Nutr.* **33** (1980) no. 6, 1179–1182.
[134] D. Henschler (ed.): *Gesundheitsschädliche Arbeitsstoffe. Toxikologisch-Arbeitsmedizinische Begründung von MAK-Werten,* Verlag Chemie, Weinheim 1981, Part 8.
[135] DFG Pflanzenschutzkommission, *Toxikologie der Herbizide,* Verlag Chemie, Weinheim 1982, Part 4.
[136] ACGIH (ed.): *Threshold Limit Values (TLV),* Cincinnati, Ohio, 1985–1986.

Chloroamines

YASUKAZU URA, Central Research Institute, Nissan Chemical Ind., Ltd., Chiba, Japan

GOZYO SAKATA, Central Research Institute, Nissan Chemical Ind., Ltd., Chiba, Japan

1.	Introduction	1565	4.2. Uses	1569
2.	Chemical Properties	1565	5. Other Organic	
3.	Reactions	1566	N-Chloroamines	1569
4.	N-Chloroisocyanuric Acids ...	1568	6. Toxicology	1571
4.1.	Production and Chemical Properties	1568	7. References	1572

1. Introduction

Chloroamines are inorganic or organic nitrogen compounds that contain one or more chlorine atoms attached to a nitrogen atom. A familiar example is monochloroamine (NH_2Cl), which has been known since the beginning of the 19th century.

Chloroamines are used in synthetic reactions and as bleaching agents, disinfectants, and bactericides because they function as chlorinating agents and oxidants. Since chloroamines are safer and easier to handle than chlorine gas or metal hypochlorites, they are widely used in the purification of drinking water and as sanitizing agents in swimming pools. In recent years, N-chloroisocyanuric acids and 1-bromo-3-chloro-5,5-dimethylhydantoin have become important in the sanitizing and disinfectant markets, replacing calcium hypochlorite, 1,3-dichloro-5,5-dimethylhydantoin, and chloramine-T.

2. Chemical Properties

N-Chloroamines can act as oxidizing agents by absorbing two electrons:

$$Cl^+ + 2\,e^- \longrightarrow Cl^-$$

Thus, an N-chloroamine oxidizes hydriodic acid to liberate iodine, and this reaction is used for the quantitative analysis of N-chloroamines:

Table 1. values of important N-chloroamines

N-Chloroamine	CAS registry no.	K value
Trichloroisocyanuric acid	[87-90-1]	6.7×10^{-4}
1,3-Dichloro-5,5-dimethylhydantoin	[118-52-5]	2.5×10^{-4}
N-Chlorosuccinimide	[128-09-6]	6.6×10^{-5}
Dichloramine-T	[473-34-7]	8.0×10^{-7}
Chloramine-T	[127-65-1]	4.9×10^{-8}
Monochloroamine	[10599-90-3]	2.8×10^{-10}

$$RR'NCl + 2\,HI \longrightarrow HCl + RR'NH + I_2$$

The theoretical *available chlorine content* of an N-chloroamine is expressed as twice the mass fraction of chlorine in the molecule. However, for practical purposes, the value is expressed as the equivalent of elemental chlorine as determined by the oxidizing ability of the N-chloroamine.

The quantitative hydrolysis constant (*K* value) of an N-chloroamine is expressed by the equation below, and is generally in the range $10^{-4} - 10^{-10}$ (Table 1).

$$RR'NCl + H_2O \rightleftarrows RR'NH + HOCl$$

$$K = \frac{c_{RR'NH} \cdot c_{HOCl}}{c_{RR'NCl}}$$

The *K* value is used to express the bactericidal power of N-chloroamines, which depends on their generating hypochlorous acid in water.

The N–Cl bonds of N-chloroamines are covalent and are readily hydrolyzed by water with release of hypochlorous acid. They are thermally unstable; many do not melt congruently and can explode at elevated temperature. Trichloroamine (nitrogen trichloride, NCl_3) is particularly unstable and can cause explosions when present as an impurity in other N-chloroamines. N-Chloroamines should be stored cold and protected from light, water, amines and ammonium compounds, strong acids and bases, and easily oxidized organic materials.

3. Reactions

Many N-chloroamines are useful as reagents or intermediates in organic synthesis. For example, N-chloroamides of organic carboxylic acids can be reduced by alkali to the corresponding amines in good yield. This reaction, called the Hofmann degradation, is widely used for the synthesis of aromatic, heterocyclic, and alicyclic amines [1]–[5]:

$$\text{RCONH}_2 \xrightarrow{\text{Cl}_2} \text{RCONHCl} \xrightarrow{\text{OH}^-} \text{RNH}_2$$

N-Chloroureas can be converted into hydrazines by treatment with base in an alcohol or water solution [6], [7]:

$$\text{RNHCONHCl} \xrightarrow[\text{OH}^-]{\text{R'OH}} \text{RNHNHCO}_2\text{R'} \text{ or } \text{RNHNH}_2$$

N,N-Dichlorocarbamates, Cl_2NCOOR, are useful chlorinating agents and can also be added to dienes [8], [9]:

$$\text{RO}_2\text{CNCl}_2 + \text{CH}_2 = \text{CH} - \text{CH} = \text{CH}_2 \longrightarrow \text{RO}_2\text{CNClCH}_2\text{CH} = \text{CHCH}_2\text{Cl}$$

They are also the starting materials for the synthesis of N-halo-N-metallocarbamidates, which are intermediates in the synthesis of carbamate derivatives of physiologically active substances [10]–[14]:

$$\text{C}_2\text{H}_5\text{O}_2\text{CNCl}_2 \xrightarrow{\text{C}_2\text{H}_5\text{O}_2\text{CNH}_2} \text{C}_2\text{H}_5\text{O}_2\text{CNHCl} \xrightarrow[\text{metal}]{\text{Alkali}} \text{C}_2\text{H}_5\text{O}_2\text{CNNaCl}$$

8-Azaisatoic anhydride (**3**), an intermediate to several agrichemicals, is synthesized in high yield by chlorinating 2,3-pyridinedicarboximide (**1**) to *N*-chloro-2,3-pyridinedicarboximide (**2**), followed by treatment with alkali [15], [16]

N-Chloroamidines and N-chloroguanidines, which have excellent bactericidal and fungicidal activity, are widely used as intermediates in the syntheses of heterocyclic ring systems such as imidazoles, thiazoles, and oxadiazoles [17]–[21].

where R" = alkyl

Table 2. Physical constants of TCC and DCC-Na

Property	TCC [87-90-1]	DCC-Na [2893-78-9]
Physical form	White powder	White powder
Formula	(ClNCO)$_3$	Cl$_2$Na(NCO)$_3$
M_r	232.44	219.98
mp	234 °C (decomp.)	225 °C (decomp.)
Available chlorine		
Theoretical (%)	91.5	64.5
Typical value	90.0	62.0
pH (1% aqueous solution)	2.7–3.3	6.2–6.8
Solubility (g/100 g)		
Water (25 °C)	1.0	30.0
Acetone (30 °C)	35.0	0.5

4. N-Chloroisocyanuric Acids

N-Chloroisocyanuric acids (chloroisocyanurates), such as 1,3,5-trichloro-1,3,5-triazine-2,4,6(1*H*,3*H*,5*H*)trione [87-90-1], trichloroisocyanuric acid, Symclosene, TCC (**7**), and 1,3-dichloro-1,3,5-triazine-2,4,6(1*H*,3*H*,5*H*)trione sodium salt [2893-78-9], sodium dichloroisocyanurate, DCC-Na (**8**), have become increasingly important as disinfectants because they are more stable and easier to handle than metal hypochlorites. Table 2 gives physical properties of TCC and DCC-Na.

TCC
7

DCC–Na
(DCC–Na · 2 H$_2$O [51580-86-0])
8

In particular, TCC and DCC-Na have acquired a substantial share of the market traditionally served by calcium hypochlorite, pine oils, quaternary ammonium salts, and phenols.

4.1. Production and Chemical Properties

N-Chloroisocyanuric acids are prepared by continuous reaction of chlorine with isocyanuric acid in aqueous sodium hydroxide at 0–15 °C. Careful control of pH and reaction temperature are essential to prevent the formation of explosive NCl$_3$ [23], [24].

Trichloroisocyanuric acid (TCC) is gradually decomposed in an alkaline medium, yielding NCl$_3$, NHCl$_2$, and NH$_2$Cl. Both DCC-Na and its dihydrate have high water solubility compared with TCC; the dihydrate is thermally more stable than DCC-Na. The dihydrate of DCC-Na is obtained by cooling a saturated aqueous solution of DCC-Na (45 °C) to 10 °C.

4.2. Uses

The oxidizing and bioactive properties of the N-chloroisocyanuric acids derive from the hypochlorous acid that is slowly released from them in water. These compounds are used for many purposes such as nonshrinking treatment of wool, disinfectants for swimming pools, cleaning and sterilizing of bathrooms, and laundry bleach.

Cleaners and sanitizers based on N-chloroisocyanuric acids contain the following other components: phosphates, sodium metasilicates, surfactants, and such neutral salts as sodium sulfate and sodium carbonate. They are effective in removing oil and protein on stainless steel; they are also recommended for dishwashing in hotels, hospitals, restaurants, and food factories.

TCC and DCC-Na are sold in the form of granules or tablets. In general, TCC is used to achieve an active chlorine level of 1 – 2 mg/L in swimming pools, and DCC-Na a level of 100 – 300 mg/L in food processing.

Economic Aspects. In 1985, the estimated world production of N-chloroisocyanuric acids was 85 000 t. Suppliers include Olin, Monsanto, CDF Chimie, Chlorchem, Delsa, Nissan Chem. Ind., and Shikoku Chem. Ind. The demand for these compounds is increasing by 8 – 10 % per year for swimming pool use and 3 – 5 % per year for food processing.

5. Other Organic N-Chloroamines

1,3-Dichloro-5,5-dimethylhydantoin [*118-52-5*], Dactin, Halane (BASF Wyandotte), M_r 197.03, *mp* 132 °C, theoretical chlorine content 77.6 %, water solubility 2.1 g/L, is prepared by passing chlorine through an aqueous solution of 5,5-dimethylhydantoin [25], [26]. It has been used since the 1930s as a bactericide and disinfectant, but its use and market are decreased by the use of other dominant N-chloroamines, such as TCC and Di-Halo.

1-Bromo-3-chloro-5,5-dimethylhydantoin (**9**) [*16079-88-2*], M_r 241.48, *mp* 130 °C (decomp.), pH 2.88 (0.1 % solution), water solubility 2 g/L (20 °C), is prepared by successive bromination and chlorination of 5,5-dimethylhydantoin [27], [28]. The compound is a stable white powder, and the formation of chloroamines is not observed in weakly basic aqueous solution.

It has become widely used at an active halogen level of 0.5–3 mg/L for swimming pool disinfectants because of its stability in water and in storage, as well as its broad bactericidal spectrum [29], [30]. It is marketed under the name of Di-Halo (Aquabrom/TESCO; Division of Great Lakes Chemicals). Production in 1985 was estimated at ca. 2500 t, growing at a rate of 10–20% per year.

1-Chloro-2,5-pyrrolidinedione (**10**) [*128-09-6*], *N*-chlorosuccinimide, M_r 133.54, mp 150–151 °C, is a white solid that is sparingly soluble in ether, chloroform, and carbon tetrachloride. It has 50–54% available chlorine and is used principally as a chlorinating agent in organic synthesis [31].

N-Chloroglycolurils have been studied as protective agents against poison gas and are prepared by condensation of 1,2-diketones with urea followed by chlorination. An important example is 2,4,6,8-tetrachloro-2,4,6,8-tetrazabicyclo[3.3.0]octane-3,7-dione (**11**) [*776-19-2*], M_r 279.82.

1,3,4,6-Tetrachloro-3α,6α-diphenylglycolurill [*51592-06-4*], Iodogen (**12**), M_r 431.94, is used as a bactericide and as an oxidizing agent in syntheses of peptides [32]–[34].

3-Chloro-4,4-dimethyl-2-oxazolidinone (**13**) [*58629-01-9*], M_r 149.50, is an example of a new class of N-chloroamines that are being investigated as disinfectants [35], [36].

13

Sodium *N*-chloro-*p*-toluenesulfonamide (**14**) [*127-65-1*], chloramine-T, M_r 227.67 (hydrate: *mp* 175 °C; explosive), has an available chlorine content of 25 %. It is prepared in 75 – 95 % yield by passing chlorine into a sodium hydroxide solution of *p*-toluenesulfonamide [37], [38]. It is a strong electrolyte in acid solution and a good oxidizing agent in base. It is fairly soluble in water, and practically insoluble in benzene, chloroform, and ether. The compound reacts readily with mustard gas to yield a harmless crystalline sulfimide; chloramine-T derivatives are being studied as protective agents against poison gas [14].

Other examples of this group of compounds are *N*-chlorobenzenesulfonamide (**15**) [*127-52-6*], chloramine-B, M_r 213.64 (hydrate: *mp* 180 °C; explosive), and *N*-dichlorobenzenesulfonamide (**16**) [*473-29-0*], dichloramine-B, M_r 226.08, *mp* 76 °C (decomp.).

N,N′,N″-Trichloromelamine (**17**) [*12379-38-3*], M_r 229.46, *mp* 175 °C (decomp.), theoretical chlorine content 92.8 %, is used as a sterilizing agent to a smaller extent because of its low solubility in water.

N-Chlorinated derivatives of various amino acids are being studied as bleaching agents and bactericides [39] – [41].

6. Toxicology

The toxicity of N-chloroamines is very important because of their widespread use in drinking water, swimming pools, and food processing. When they are used in water that contains organic compounds, special care must be taken to prevent the formation of halomethanes, which are weak carcinogens. The formation of halomethanes in water

Table 3. Acute toxicity of N-chloroamines

Compound	LD_{50}, mg/kg (rat, oral)	Reference
TCC	1300	[22]
DCC-Na	1420	[22]
1,3-Dichloro-5,5-dimethylhydantoin	542	[44]
N-Chlorosuccinimide	2700*	[45]

* Minimum lethal dose.

is measured by the head space method, purge trap method, or solvent extraction, followed by analysis with GC (electron capture detector) or GC–MS [42], [43].

Most N-chloroamines irritate the eyes, moist skin, and the upper respiratory tract. Solutions of N-chloroisocyanuric acids containing up to 100 ppm of available chlorine are not toxic, irritating, or sensitizing. However, ingestion of the pure solid or a fresh strong suspension attacks the stomach lining. The isocyanuric acid formed after hydrolysis has low toxicity. Table 3 gives a few acute toxicity values.

Chloramine-T is a skin irritant; it is highly toxic when absorbed in the blood and should, therefore, be used only in diluted solution. Monochloroamine is widely used as a primary disinfectant for drinking water because of the low rate at which it forms halomethanes from organic compounds in the water. It has a long duration of action, although its specific bactericidal activity is rather low. Recently, however, it has been implicated as a mutagen and as a toxic agent for aquatic life, and the U.S. Environmental Protection Agency proposes to prohibit its use in drinking water [46]. 1-Bromo-3-chloro-5,5-dimethylhydantoin is less toxic to test fish species than 1,3-dichloro-5,5-dimethylhydantoin [47].

7. References

[1] Kyowa Gas Chem., JP-Kokai 77-53860, 1977 (K. Aida, H. Segawa, K. Nakao).
[2] AKZO GmbH, DE 2 313 496, 1974 (H. Zengel, M. Bergfeld).
[3] AKZO GmbH, DE 2 710 595, 1978 (H. Zengel, M. Bergfeld).
[4] SNIA Viscosa, US 4 198 348, 1980 (C. A. Pauri).
[5] Nippon Kayaku Co., JP-Kokai 78-130648, 1978 (R. Hasegawa, S. Kawamoto).
[6] Pennwalt Corp., US 3 956 366, 1976 (C. S. Sheppard, L. E. Korczykowski).
[7] Pennwalt Corp., US 3 657 324, 1972 (C. S. Sheppard, L. E. Korczykowski).
[8] Exxon Research & Eng. Co., US 4 146 557, 1979 (F. A. Daniher, A. A. Oswald).
[9] Exxon Research & Eng. Co., US 4 171 449, 1979 (F. A. Daniher, A. A. Oswald).
[10] M. M. Campbell, G. Johnson, *J. Chem. Soc. Chem. Commun.* 1975, 479.
[11] D. H. Bremner, M. M. Campbell, G. Johnson, *Tetrahedron Lett.* 1975, 2955–2958.
[12] W. Traube, H. Gockel, *Ber. Dtsch. Chem. Ges.* **56 B** (1923) 384–391.
[13] P. Chabrier, *C. R. Acad. Sci. Ser. B* **214** (1942) 362.
[14] M. M. Campbell, G. Johnson, *Chem. Rev.* **78** (1978) 65–79.

[15] Sherwin Williams Co., US 3 734 921, 1973 (L. C. Vacek).
[16] Sherwin Williams Co., US 3 828 038, 1974 (L. C. Vacek).
[17] Ciba Geigy, DE 2 242 772, 1973 (J. C. Petitpierre, C. Weis).
[18] Ube Ind. Ltd., JP-Kokai 82-32205, 1982 (Y. Hirota, T. Yorie).
[19] Medeau Researches S.r.l., DE 2 812 304, 1978 (R. Stradi).
[20] Sankyo Co. Ltd., JP-Kokai 74-18899, 1974 (T. Konotsune, T. Yanai).
[21] Sankyo Co. Ltd., JP-Kokai 73-80560, 1973 (T. Konotsune, T. Yanai, M. Suzuki).
[22] Nissan Chem. Ind., Technical Information on Nissan Chlorinated Isocyanurates, 1985.
[23] E. Nachbaur, W. Gottardi, *Monatsh. Chem.* **97** (1966) 115–120.
[24] Shikoku Chem. Ind., FR 1 571 705, 1969; *Chem. Abstr.* **72** (1970) 121588.
[25] H. Biltz, K. H. Slotta, *J. Prakt. Chem.* **113** (1926) 248; *Chem. Abstr.* **21** (1927) 1794.
[26] O. O. Orazi, O. A. Orio, *Chem. Abstr.* **48** (1954) 13634.
[27] Drug Research Inc., US 2 779 764, 1957 (L. O. Paterson).
[28] Drug Research Inc., US 3 147 259, 1964 (L. O. Paterson).
[29] L. O. Paterson, US 3 147 219, 1964.
[30] L. M. G. Padilla, GB 1 139 188, 1969.
[31] H. Zimmer, L. F. Audrieth, *J. Am. Chem. Soc.* **76** (1954) 3856–3857.
[32] P. Palitzsch, DD 123 467, 1976.
[33] P. R. P. Salacinski, C. McLean, L. E. Sykes, V. V. Clement-Jones, *Anal. Biochem.* **117** (1981) 136–146.
[34] T. B. Schulz, R. Jorde, H. L. Waldum, P. G. Burhol, *Scand. J. Gastroenterol.* **17** (1982) 379–382.
[35] Interx Research Corp., US 4 000 293, 1976 (J. J. Kaminski, N. S. Bordor).
[36] *Chem. Week* 1985, May 8, 33.
[37] M. Bugla, J. Hok, M. Veger, CS 159 409, 1975.
[38] Toyo Chem., JP-Kokai 76-4141, 1976 (S. Masutani).
[39] Interx Research Corp., US 4 036 843, 1977 (J. J. Kaminski, N. S. Bordor).
[40] Interx Research Corp., US 4 045 578, 1977 (J. J. Kaminski, N. S. Bordor).
[41] S. H. Gerson, S. D. Worley, N. S. Bordor, J. J. Kaminski, *J. Med. Chem.* **21** (1978) 686–688.
[42] A. A. Nicholson, O. Meresz, B. Lemyk, *Anal. Chem.* **49** (1977) 814–819.
[43] W. H. Glaze, G. R. Peyton, R. Rawley, *Environ. Sci. Technol.* **11** (1977) 685–690.
[44] R. J. Lewis: *Registry of Toxic Effects of Chemical Substances 1981–2*, National Institute of Occupational Safety and Health, NIOSH.
[45] *The Merck Index – An Encyclopedia of Chemicals, Drugs and Biologicals*, 9th ed.,1983.
[46] G. S. Moore, E. J. Calabrese, *J. Environ. Pathol. Toxicol.* **4** (1980) 257–263.
[47] E. W. Wilde, R. J. Soracco, L. A. Mayack et al., *Bull. Environ. Contam. Toxicol.* **31** (1983) 309–314.

Chloroformic Esters

SIEGFRIED BÖHM, Bayer AG, Leverkusen, Federal Republic of Germany

1.	Introduction 1575	6.	Quality Specifications and Analysis 1579
2.	Physical Properties 1575	7.	Storage and Transportation . . 1579
3.	Chemical Properties. 1577	8.	Uses and Economic Aspects . . 1580
4.	Production 1577	9.	Toxicology 1581
5.	Safety 1578	10.	References 1581

1. Introduction

Chloroformic esters are also known as chloroformates, chlorocarbonates, or chlorocarbonic esters (Chemical Abstracts Service calls them carbonochloridic acid esters). They are represented by the general formula ClCOOR, wherein R can be any alkyl or aryl group [1], [2]. They were first described by DUMAS in 1833 [3]. They are highly reactive compounds and, since about 1950, have attained considerable importance as intermediates in organic syntheses [4].

2. Physical Properties

Commercially important chloroformic esters are predominantly colorless liquids having boiling points only slightly higher than the corresponding alcohols. The esters of low-boiling alcohols have pungent odors and are highly volatile.

In general, chloroformic esters have densities greater than 1.00 and are immiscible or only slightly miscible with water. They are, however, miscible with inert organic solvents. The order of thermal stability of the different types of esters is aryl > primary alkyl > secondary alkyl > tertiary alkyl. The phenyl ester decomposes only slightly at its boiling point of 189 °C, but the *tert*-butylester starts decomposing at 10 °C. Decomposition is catalyzed by metal salt impurities (iron, aluminum, copper, cobalt, magnesium, and nickel) and by tertiary amines. It can proceed in two ways and lead to two different sets of products both of which include gases, which cause increased pressure in closed containers.

Table 1. Physical properties of the commercially important chloroformic esters

Ester	Empirical formula	M_r	bp, °C/kPa	d_4^{20}	n_D^{20}	Flash point, °C	Autoignition temperature, °C
Methyl [79-22-1]	$C_2H_3ClO_2$	94.50	71/101.3	1.2298	1.3880	13	510
Ethyl [541-41-3]	$C_3H_5ClO_2$	108.53	94/101.3	1.1403	1.3947	15	500
Allyl [2937-50-0]	$C_4H_5ClO_2$	120.54	110/101.3 56.5/12.9	1.1360	1.4223	28	
n-Propyl [109-61-5]	$C_4H_7ClO_2$	122.55	115/101.3	1.09202	1.4034	34	
Isopropyl [108-23-6]	$C_4H_7ClO_2$	122.55	105/101.3 (decomp.) 66.3/26.7	1.0777	1.3996	14	535
n-Butyl [592-34-7]	$C_5H_9ClO_2$	136.58	142/101.3 35/1.7	1.0513	1.4121	40	435
Phenyl [1885-14-9]	$C_7H_5ClO_2$	156.57	189/101.3 88/2.67	1.2350	1.5160	69	500
Cyclohexyl [13248-54-9]	$C_7H_{11}ClO_2$	162.62	ca. 180/101.3 (decomp.) 72/2.67	1.1260	1.4588	65	320
2-Ethylhexyl [24468-13-1]	$C_9H_{17}ClO_2$	192.69	183/101.3 (decomp.)	0.9840	1.4330	74	235
4-tert-Butylcyclohexyl [42125-46-2] cis- [15595-62-7]* trans- [15595-61-6]*	$C_{11}H_{19}ClO_2$	218.73	>137 (decomp.) 58–62/0.13	1.0450	1.4636	64	390
Diethylene glycol bis- [106-75-2]	$C_6H_8Cl_2O_5$	231.03	95–97/0.05	1.3880	1.4536	160	
Cetyl [26272-90-2]	$C_{17}H_{33}ClO_2$	304.90	ca. 140/101.3 (decomp.)	0.9250	1.448	>140	ca. 300

* No physical data reported; identification by IR and NMR data [5].

$$ClCOOCH_2CH_2R \begin{cases} CH_2 = CHR + HCl + CO_2 \\ ClCH_2CH_2R + CO_2 \end{cases}$$

where R may be any alkyl or aryl group

The physical properties of the commercially important chloroformic esters are given in Table 1.

3. Chemical Properties

Chloroformic esters exhibit the characteristic chemical properties of acid chlorides. They react with all nucleophiles including carbon acids. They react with water to produce hydrogen chloride and their corresponding alcohol. Moist chloroformic esters are, therefore, highly corrosive. They form carbonates (→ Carbonic Acid Esters) and thiocarbonates in reactions with alcohols and mercaptans, respectively. Chloroformic esters react with carboxylic acids to form carboxylic acid anhydrides [6] and with primary and secondary amines to form carbamic acid esters [7].

Their reactions with alkali peroxides form dialkyl peroxydicarbonates [8]. Bis(chloroformic) esters obtained from glycols or dihydric phenols undergo reactions with dihydric alcohols to produce polycarbonates, with dicarboxylic acids to give polyesters, or with amino compounds to yield polyurethanes [1]. They react with nitriles to form amides [9].

4. Production

The most important method used to produce chloroformic esters is the reaction of phosgene with anhydrous alcohols.

$ROH + COCl_2 \longrightarrow ClCOOR + HCl$

where R may be any alkyl or aryl group

The reaction with aliphatic alcohols proceeds with no need for hydrogen chloride acceptors or catalysts. The conversion of phenols with phosgene occurs less readily. The reaction depends on the presence of catalysts such as alkali salts [10], tertiary amines or quaternary ammonium salts [11], carboxylic acid amides (e.g., dimethylformamide or N-methylpyrrolidone) [12], or organic phosphorus compounds (e.g., triphenylphosphine, triphenylphosphine oxide) [13] and is carried out discontinuously at high temperature. The formation of carbonates by a competing reaction depends on the nature of the catalyst. Alkali salt catalysts lead to substantially more carbonate formation than phosphorus compounds. To suppress carbonate formation, the reaction heat is rapidly carried off and an excess of phosgene is maintained in the reaction zone.

Continuous and discontinuous processes are available for the preparation of chloroformic esters. In the laboratory the synthesis is carried out by passing alcohol into liquid phosgene at low temperature. The discontinuous method is generally not applicable to commercial syntheses for reasons of safety. In special cases where this method is employed, an inert diluent is required, e.g., the ester itself [14] or aliphatic, aromatic, or chlorinated hydrocarbons, and the alcohol and phosgene enter the reaction vessel simultaneously. The solvent, containing dissolved hydrogen chloride and excess phosgene, is removed by distillation [15]. The discontinuous method is used for the production of chloroformic esters from glycols and the conversion of sterically hindered or acid-sensitive alcohols. An auxiliary base, e.g., pyridine or dimethylaniline, increases the reaction rate.

The continuous procedure is the method of choice for the commercial production of aliphatic chloroformic esters. It can be carried out in either cocurrent or countercurrent flow. The required amounts of alcohol and phosgene can be passed in cocurrent flow into a circulation reactor, cascade vessel [16], or nozzle atomizer [17], or the components can react in the gas phase [18], [19]. The countercurrent method is carried out in a packed column [20], [21] because hydrogen chloride and excess phosgene then escape from the top of the column and the liquid ester collects at the bottom. A product of 99% purity can be obtained if the temperature in the reaction zone is maintained below 10 °C for the formation of the methyl ester or below 20 °C for the ethyl ester [21]. Esters of alcohols with an intermediate chain length (C_3–C_8) are produced at 30–70 °C. Purity of 97–99% and a yield greater than 90% are achieved with esters of primary alcohols, whereas the more unstable esters of secondary and tertiary alcohols give 70–80% yields.

Chloroformic esters can also be made by reacting alcohols with carbon monoxide and chlorine under pressure [22], by chlorinating formates or thioesters, by reacting phosgene with aldehydes, or by reacting phosphorus pentachloride with cyclic carbonates [2], [4]. Chloroformic esters with β-chlorine atoms are made by reacting phosgene with the corresponding epoxide [1], by reacting alcohols with trichloroacetyl chloride [1], and by cleaving α-keto sulfoxides with hydrogen chloride [2], [23]. The crude ester is purified by distillation unless, as is the case for the aliphatic esters, thermal instability precludes this method.

5. Safety

Safety precautions necessary for the handling of phosgene apply to the production of chloroformic esters. The exhaust gas is hydrolyzed and then burned. Undesirable substances, e.g., phenol, are extracted from wastewater.

Some chloroformic esters are classified as dangerous working substances [24] and may be handled only when protective clothing and equipment are used (gas masks with filter B, rubber gloves, etc.). They must not enter the atmosphere or the water supply. If spilled or leaked, they can be hydrolyzed by alkaline substances. Vapors must be destroyed in a scrubber with alkali or with water in the presence of activated carbon or by incineration. Clothing that comes into contact with chloroformic esters must be removed immediately. Any part of the body that is touched by a chloroformic ester must be washed well with water immediately. All important chloroformic esters are flammable; they form explosive mixtures with air. Electrostatic charges must, therefore, be avoided.

6. Quality Specifications and Analysis

Commercial products have concentrations of 97–99 wt%. Along with small amounts of carbon dioxide, hydrogen chloride, phosgene, and alkyl chlorides, the main impurities are the corresponding carbonates. The amounts of carbon dioxide, hydrogen chloride, and alkyl chlorides increase with the duration of storage, depending on the stability of the ester. A purity of 97% is adequate for most applications. If especially pure product is required, 99% pure, distilled methyl chloroformate is available. The methyl ester can be distilled under normal pressure, but the ethyl ester requires vacuum conditions and continuous distillation. Esters obtained from secondary or tertiary alcohols are not sufficiently stable to be purified by distillation. They must be washed with cold water and dried with calcium chloride.

The quantitative analysis of chloroformic esters can be carried out by titration, gas chromatography, or high-performance liquid chromatography (HPLC). High-boiling or thermally labile esters are determined by titration. They are treated with aniline in methanol, and the liberated hydrochloric acid is titrated with sodium hydroxide. They can also be treated with excess dibutylamine in chlorobenzene; the remaining amine is then back-titrated with hydrochloric acid.

Bivalent esters are assayed by reacting them with an aniline derivative, e.g., methylaniline, and using HPLC to measure the dicarbamate formed.

7. Storage and Transportation

The thermal stability of chloroformic esters is important for storage and transportation. Esters of primary alcohols and phenols are stable if free of moisture and metal salts. Traces of moisture and heat can generate hydrogen chloride and increase the internal pressure. For this reason, containers should not be exposed to direct sunlight. The esters of secondary and tertiary alcohols (e.g., isopropyl ester or cyclohexyl ester) require rapid transportation and no more than brief storage because they are unstable.

The storage temperature should be maintained below 10 °C and the product used as soon as possible. Iron drums lined with polyethylene or lead-lined tank cars are used to transport chloroformic esters. The methyl and ethyl esters are mentioned in the regulations for the transportation of dangerous materials (GGVE), clause 6.1, class 4. The corrosiveness, toxicity, and flammability (primary label) of the esters must be declared on the container and EEC guidelines require that dangerous working substances be provided with an additional dangerous-materials label giving instructions for handling (R and S label).

8. Uses and Economic Aspects

Chloroformic esters are highly reactive and can be used as intermediates in the synthesis of numerous compounds. They are employed in the production of unsymmetrical carbonic acid esters (→ Carbonic Acid Esters) such as diethylene glycol bis(allyl) carbonate [25], which is an intermediate in the synthesis of plastics of excellent optical quality. Other aliphatic or aromatic bis(chloroformic) esters are also used to make plastics such as polycarbonates (Makrolon) [26], [27] and polyurethanes having a high temperature resistance [1], [28]. Alkyl and aryl esters of chloroformic acid serve as catalysts in the presence of initiator radicals in the low-pressure polymerization of ethylene [29]. The methyl and ethyl esters are used as solvents in the photographic industry. They are also employed in the production of carbamates that are used to synthesize dyes (e.g., Resolin dyes), drugs, (e.g., the tranquilizer meprobamate [57-53-4]), veterinary medicines (e.g., furazolidone [67-45-8] or rintal [58306-30-2]), herbicides, (e.g., betanal [13684-63-4] and betanex [13684-56-5]), and insecticides, (e.g., Sevin [63-25-2]).

The herbicide IPC, isopropyl N-phenylcarbamate [122-42-9], is made from isopropyl chloroformate. Another important application is the preparation of the dialkyl peroxycarbonates used as catalysts in the sodium peroxide production of PVC, poly(vinyl chloride) [78320-64-6] [30]. Such commercially important peroxycarbonates as ethyl [14666-78-5], n-butyl [16215-49-9], isopropyl [105-64-6], cetyl [26322-14-5], cyclohexyl [1561-49-5], and 4-tert-butylcyclohexyl peroxycarbonate [26523-73-9] can be made from the corresponding chloroformates. Methyl chloroformate is the starting material for the synthesis of the preservative velcorin (dimethyl pyrocarbonic ester) [4525-33-1]. Phenyl chloroformate is used in the production of Sirius dyes [31]. Benzyl chloroformate is used to substitute the carbobenzyloxy protecting group on amino functions of amino acids in peptide synthesis [32], on hydroxyl groups [33], and on hydrazino groups [34]. The protecting group is easily removed by hydrogenolysis over a catalyst [35].

Ethyl chloroformate is the most important of these esters. It reacts with ethyl xanthate [140-90-9] to produce diethylxanthogen formate [3278-35-1] and diethylxanthic anhydride [2905-52-4], which are used in the flotation of ores [8]. Ethyl chloroformate is also used as a stabilizer for PVC and in the production of modified penicillins, e.g., binotal [69-53-4]. It is used to produce carboxylic acid anhydrides from carboxylic acids [6], secondary amines from tertiary amines [36], and nitriles from amides [9] and to make heterocyclic compounds [37].

The cyclohexyl, 4-tert-butylcyclohexyl, butyl, and cetyl esters of chloroformic acid are used almost exclusively to produce peroxydicarbonates. The majority of the isopropyl ester produced is also used for this purpose (total production is several thousand metric tons per year).

Chloroformic esters are not produced on a large scale. Their economic importance is due to their usefulness as intermediates in organic syntheses. The largest producers in Western Europe are BASF, Société Nationale des Poudres et Explosifs, Hoechst, and

Table 2. LD$_{50}$ values of chloroformic esters

Ester	Oral LD$_{50}$, mg/kg	Dermal LD$_{50}$, mg/kg	Inhalation LC$_{50}$, mL/m^3
Methyl	140 (guinea pig)	7 120 (rabbit)	12–56 (rat, 1 h)
	110 (rat)	1 750 (mouse)	88 (rat, 1 h)
	67 (mouse)		
Ethyl	270 (rat)	7 120 (rabbit)	145 (rat, 1 h)
n-Propyl	650 (mouse)	10 (mouse)	319 (mouse, 1 h)
Isopropyl	1 070 (rat)	11 300 (rabbit)	299 (mouse, 1 h)
	178 (mouse)	12 (mouse)	
Allyl	178 *	1 470 *	
Cetyl	5 000 (rat)		no symptoms
Diethylene glycol bis-	813 (mouse)	3 400 (mouse)	169 (mouse, 1 h)
Phenyl	1 581 (rat)	4 800 (rabbit)	44 (rat, 4 h)

* Test animals not specified.

Bayer. In the United States, Chemetron Corp., Essex Chem. Corp., Minerec Corp., and PPG Industries are the major producers. Hodogaya Co. in Japan and Chemolimpex in Hungary are also major suppliers. The production capacity in Western Europe is ca. 20–25 kt/a.

9. Toxicology

Chloroformic esters cause acute reactions of the respiratory tract and mucous membranes. Contact with the gastrointestinal tract or absorption by the skin leads to less toxic reactions (Table 2). Chloroformic esters burn the skin, leaving its surface hard and stained brown [38]. Inhalation of short-chain chloroformic esters causes irritation, a warning of potential toxic injury [39]. Improper handling can result in pulmonary edema and cause death [40]. The inhalative toxicity of chloroformic esters decreases with increasing chain length. Cetyl chloroformate is not toxic even when saturated air is inhaled [41], [42]. The introduction of a halogen in the α-position of aliphatic esters increases their inhalative toxicity [43]. Concentrations of 5 mL/m^3 of methyl chloroformate or ethyl chloroformate cause only local irritation to the respiratory tracts of rats exposed to them 15 times for periods of 6 h. These compounds cause no symptoms at the 1 mL/m^3 level [44]. TLV values of 1 mL/m^3 for the methyl and ethyl esters and 2 mL/m^3 for the isopropyl ester have been proposed but not accepted [44].

10. References

[1] M. Matzner, P. Kurkjy, J. Cotter, *Chem. Rev.* **64** (1964) no. 6, 645–687; *Chem. Abstr.* **62** (1965) 1525a.

[2]   *Houben-Weyl,* **8**, 101–105.
      Houben-Weyl, **E4**, 15–28.
[3]   J. Dumas, *Ann. Chim. Phys., 2. Série,* **54** (1833) 225–247.
[4]   R. D. Concez, *Inf. Chem.* **106** (1972) 139.
[5]   P. Beak, J. T. Adams, J. A. Barron, *J. Am. Chem. Soc.* **96** (1974) 2494.
[6]   E. Schipper, J. Nichols, *J. Am. Chem. Soc.* **80** (1958) 5714.
[7]   *Org. Synth. Coll. Vol.* **IV** (1963) p. 780.
[8]   G. Crozier, US 4 454 051, 1984.Bosordi Vegyi Kombinat, HU 30 021, 1984 (B. Mariasi, L. Molnar, J. Scabo, P. Jubasz, A. Karcagi).
[9]   C. K. Sams, R. J. Cotter, *J. Org. Chem.* **26** (1961) 6.
[10] Bayer, DE 1 117 598, 1960 (W. Altner, E. Meisert, G. Rockstroh).
[11] Union Carbide, US 3 255 230, 1966 (R. P. Kurkjy, M. Matzner, R. J. Cotter). J. Mouralova, J. Hajicek, J. Trojanek, CS 202 454, 1983.
[12] Hoechst, DE 2 131 555, 1971 (G. Semler, G. Schaeffer). Hoechst, DE 2 213 408, 1972 (G. Semler, G. Schaeffer).
[13] Bayer, DE 3 019 526, 1980 (G. Rauchschwalbe, H.-U. Blank, K. Mannes, D. Mayer).
[14] PPG, FR 2 484 406, 1981 (A. Sathe).
[15] Synvar Associates, DE 2 144 963, 1972 (M. E. Packard, E. F. Ullmann, T. L. Burkoth).
[16] BASF, DE 2 251 206, 1972 (G. Merkel, J. Datow, J. Paetsch, H. Toussaint, H. Hoffmann, S. Winderl). Minerec Corp., DE 2 704 262, 1977 (F. S. Bell, R. D. Crozier, L. E. Strow). F. Frantiek, L. Skalicky, J. Pscheidt, F. Janda, CS 190 921, 1981.
[17] BASF, DE 2 453 284, 1974 (F. Neumayr, M. Decker, J. Paetsch). Bayer, DE 2 847 484, 1980 (W. Schulte-Huermann, E. Schellmann, J. Lahrs).
[18] Mitsubishi Chemical, JP-Kokai 51/43 719, 1974 (S. Ueda, Y. Mijagawa, H. Yoshioka, E. Hirayawa). Mitsubishi Chemical, JP-Kokai 51/43 721, 1974 (S. Ueda, Y. Mijagawa, H. Yoshioka, E. Hirayawa).
[19] Eszakmagyarorszagi Vegyimüvek, Sajobabony (Hungary), DE 2 159 967, 1971 (L. Rosza, L. Meszaros, F. Mogyorodi).
[20] ICI, FR 1 336 606, 1961.
[21] Bayer, DE 3 135 947, 1983 (D. Bauer, H. Dohm, W. Schulte-Huermann, H. Hemmerich).
[22] Ube Industries, JP-Kokai 54/61 121, 1979 (S. Nakatomi, Y. Kamitoku).
[23] T. Wakin, Y. Nakamina, S. Motoki, *Bull. Chem. Soc. Jpn.* **51** (1978) 3081.
[24] EEC Guideline 79/831/EEC.
[25] PPG, US 4 273 726, 1979 (A. Senol). PPG, EP 80 339, 1981 (C. W. Eads, J. C. Crano).
[26] Bayer, DE 971 790, 1953 (H. Schnell, L. Bottenbruch, H. Krimm).
[27] K. Johnson, *Polycarbonates Recent Developments,* Chemical Process Review No. 47, Noyes Data Corp., Park Ridge, N.J., 1970.
[28] Bayer, DE 1 720 761, 1967 (H. Krimm, G. Lenz, H. Schnell).
[29] Union Rheinische Braunkohlen Kraftstoff, DE 1 126 139, 1958 (G. Nettesheim, R. Schulze-Bentrop).
[30] Goodyear, DE 2 039 010, 1970 (F. A. Cox, D. R. Glenn). Akzona Inc., US 4 269 726, 1981 (J. R. Kolczynski, G. A. Schultz).
[31] *Ullmann,* 3rd ed., **4**, 105, 108, 109.
[32] A. C. Farthing, *J. Chem. Soc.* 1950, 3 213.
[33] L. Hough, J. E. Pridde, *J. Chem. Soc.* 1961, 3178.
[34] H. Niedrich, W. Knobloch, *J. prakt. Chem.* **17** (1962) 263, 273.

[35] R. A. Boisonnes, G. Breitner, *Helv. Chim. Acta* **36** (1953) 875. B. F. Daubert, C. G. King, *J. Am. Chem. Soc.* **61** (1939) 3328.
[36] Ciba Geigy, BE 615 026, 1962 (W. Schindler).
[37] H. Krauch, W. Kunz: *Reaktionen der Org. Chemie,* Hüthig Verlag, 4th ed., Heidelberg 1969, p. 596.
[38] G. Schreiber, Fraunhofer Institut für Toxikologie und Aerosolforschung, *Bericht vom 29. 10. 1980.*
[39] W. Hey et al., *Arch. Toxikol.* **23** (1968) 186.
[40] A. M. Thies et al., *Zentralbl. Arbeitsmed. Arbeitsschutz* **18** (1968) 141.
[41] G. Schreiber, Fraunhofer Institut für Toxikologie und Aerosolforschung, *Bericht 1 vom 20. 02. 1980.*
[42] G. Schreiber, Fraunhofer Institut für Toxikologie und Aerosolforschung, *Bericht 2 vom 20. 02. 1980.*
[43] F. Flury, F. Zernik: *Schädliche Gase,* Springer Verlag, Berlin 1931.
[44] J. C. Gage, *Br. J. Ind. Med.* **27** (1970) 1–18.
[45] R. J. Lewis, Sr. and R. L. Tatken (eds.): *Registry of Toxic Effects of Chemical Substances 1981–1982,* NIOSH, Cincinnati, Ohio, 1983.
[46] *Kirk-Othmer,* 3rd ed., **4,** 765.
[47] Safety data sheet Bayer AG, Leverkusen.
[48] Toxicological Institute, Bayer AG, report No. 9741, 1981.
[49] *Am. Ind. Hyg. Assoc. J.,* **30** (1969) 470.

Chlorohydrins

GORDON Y. T. LIU, Dow Chemical, Midland, Michigan 48640, United States

W. FRANK RICHEY, Dow Chemical, Midland, Michigan 48640, United States

JOANNE E. BETSO, Dow Chemical, Midland, Michigan 48640, United States

1.	Introduction 1585	6.	Chemical Analysis 1596
2.	Physical Properties 1585	7.	Storage and Transportation .. 1596
3.	Chemical Properties........ 1590	8.	Uses 1596
4.	Production 1590	9.	Toxicology and Occupational Health................. 1597
5.	Environmental Protection ... 1595	10.	References.............. 1600

1. Introduction

Chlorohydrins are compounds with one or more chlorine and hydroxyl groups in the nonaromatic portion of their structure. The compounds of greatest industrial importance are the propylene chlorohydrins and glyceryl chlorohydrins (see Table 1). Ethylene chlorohydrin is no longer an important intermediate for the production of ethylene oxide, one of the classic commodity chemicals.

Nomenclature. Many different names for the various chlorohydrins appear in the literature. Some of these *synonyms* for the major chlorohydrins are given in Table 1.

2. Physical Properties

Most dilute solutions of chlorohydrins in water have a somewhat sweet, pleasant odor, particularly 3-chloro-1,2-propanediol (**5**), 1,3-dichloro-2-propanol (**7**), and 3-chloro-1-propanol (**4**). The general physical properties of the more significant chlorohydrins are listed in Table 2.

Chlorohydrins

Table 1. Chlorohydrin nomenclature

Chlorohydrin	CAS registry number	Formula	M_r	Synonyms
Ethylene				
2-Chloro-1-ethanol (**1**)	[107-07-3]	C_2H_5ClO $ClCH_2CH_2OH$	80.52	ethylene chlorohydrin, 2-chloroethyl alcohol, 1-chloro-2-hydroxyethane, 2-chloro-1-hydroxyethane, β-chloroethyl alcohol, β-chloroethanol, ethylene glycol chlorohydrin, glycol chlorohydrin, glycol monochlorohydrin, glycolmonochlorohydrin, 2-monochloroethanol
Propene				
2-Chloro-1-propanol (**2**)	[78-89-7]	C_3H_7ClO $CH_3CHClCH_2OH$	94.54	2-propylene chlorohydrin, 2-chloropropyl alcohol, β-chloropropyl alcohol, propylene β-chlorohydrin, 2-chloro-1-hydroxypropane
1-Chloro-2-propanol (**3**)	[127-00-4]	$CH_3CHOHCH_2Cl$		β-chloroisopropyl alcohol, propylene α-chlorohydrin, sec-propylene chlorohydrin, 1-chloroisopropyl alcohol, propylene α-chlorohydrin, 1-propylene chlorohydrin, 1-chloro-2-hydroxypropane
3-Chloro-1-propanol (**4**)	[627-30-5]	$Cl(CH_2)_3OH$		1-chloro-3-hydroxypropane, trimethylene chlorohydrin
Glycerol				
3-Chloro-1,2-propanediol (**5**)	[96-24-2]	$C_3H_7ClO_2$ $CH_2OHCHOHCH_2Cl$	110.54	α-chlorohydrin, 1-chloropropane-2,3-diol, glycerol α-chlorohydrin, 3-chloro-1,2-dihydroxypropane, 3-chloro-1,2-propanediol, 3-chloropropylene glycol, glycerol α-monochlorohydrin, α-monochlorohydrin, 1-chlorohydrin, chlorodeoxyglycerol, 1-chloro-2,3-dihydroxypropane, 3-chloropropane-1,2-diol, 2,3-dihydroxypropyl chloride, glycerin α-monochlorohydrin, α-glycerol chlorohydrin, monochlorhydrin, monochlorohydrin
2-Chloro-1,3-propanediol (**6**)	[497-04-1]	$HOCH_2CHClCH_2OH$		2-chlorotrimethylene glycol, glyceryl β-monochlorohydrin, β-chlorohydrin, 2-chlorohydrin, chloro-1,3-propanediol, chloropropanediol-1,3
Glycerol dichlorohydrins				
1,3-Dichloro-2-propanol (**7**)	[96-23-1]	$C_3H_6Cl_2O$ $CH_2ClCHOHCH_2Cl$	128.99	1,3-dichlorohydrin, α,γ-dichlorohydrin, β,β-dichloroisopropyl alcohol, glycerin α,α′-dichlorohydrin, α-dichlorohydrin, 1,3-dichloro-2-hydroxypropane
2,3-Dichloro-1-propanol (**8**)	[616-23-9]	$CH_2ClCHClCH_2OH$		1,2-dichlorohydrin, α,β-dichlorohydrin, β,γ-dichloropropyl alcohol, allyl alcohol dichloride, asymmetric glycerin dichlorohydrin, β-dichlorohydrin, 2,3-dichloro-1-hydroxypropane
Butene				
2-Chloro-1-butanol (**9**)	[26106-95-6]	C_4H_9ClO $CH_3CH_2CHClCH_2OH$	108.57	2-chloro-n-butyl alcohol, 2-chloro-1-hydroxybutane

Table 1. (continued)

Chlorohydrin	CAS registry number	Formula	M_r	Synonyms
3-Chloro-1-butanol (10)	[2203-35-2]	$CH_3CHClCH_2CH_2OH$		3-chloro-n-butyl alcohol, 3-chloro-1-hydroxybutane
4-Chloro-1-butanol (11)	[928-51-8]	$CH_2ClCH_2CH_2CH_2OH$		4-chloro-n-butyl alcohol, 4-chloro-1-hydroxybutane, 4-chloro-1-butane-ol, 4-chlorobutanol, tetramethylene chlorohydrin
1-Chloro-2-butanol (12)	[1873-25-2]	$CH_3CH_2CHOHCH_2Cl$		1-chloro-sec-butyl alcohol, 1-chloro-2-hydroxybutane
3-Chloro-2-butanol (13)	[563-84-8]	$CH_3CHClCHOHCH_3$		3-chloro-sec-butyl alcohol, 3-chloro-2-hydroxybutane
4-Chloro-2-butanol (14)	[2203-34-1]	$CH_2ClCH_2\ CHOHCH_3$		4-chloro-sec-butyl alcohol, 4-chloro-2-hydroxybutane
Isobutene		C_4H_9ClO	108.57	
2-Chloro-2-methyl-1-propanol (15)	[558-38-3]	$(CH_3)_2CClCH_2OH$		
1-Chloro-2-methyl-2-propanol (16)	[558-42-9]	$(CH_3)_2COHCH_2Cl$		
Isobutene dichlorohydrins		$C_4H_8Cl_2O$	143.01	
1-Chloro-2-chloromethyl-2-propanol (17)	[597-32-0]	$CH_3C(OH)(CH_2Cl)_2$		
1,2-Dichloro-2-methyl-3-propanol (18)	[42151-64-4]	$HOCH_2CCl(CH_3)CH_2Cl$		
Other				
2-Chloro-1-phenyl-1-ethanol (19)	[1004-99-5]	C_8H_9ClO $C_6H_5CH(OH)CH_2Cl$	151.61	styrene chlorohydrin

Table 2. Physical properties of chlorohydrins

Chlorohydrin	mp, °C (p, kPa)	bp, °C	Flash point, °C	d_4^{20}, g/mL	η, mPa·s	Azeotrope Component	Azeotrope bp, °C	Azeotrope wt%	Water	Ethanol	Acetone	Benzene
Ethylene												
2-Chloro-1-ethanol (1)	−69 (−63 to −40 °C)	129	57	1.2133	3.43	water	97.8	42	miscible	miscible	miscible	miscible
Propylene												
2-Chloro-1-propanol (2)	–	133	–	1.1032	–	water	96.0	48	very soluble	soluble	soluble	–
1-Chloro-2-propanol (3)	–	127	52	1.1154	4.67	water	95.4	54	miscible	miscible	miscible	miscible
3-Chloro-1-propanol (4)	–	16	–	1.1318	–	–	–	–	very soluble	soluble	soluble	soluble
Glycerol												
3-Chloro-1,2-propanediol (5)	–	213	138	1.3204	159	–	–	–	soluble	miscible	miscible	miscible
2-Chloro-1,3-propanediol (6)	–	146 (2.4)	–	1.3219	300	–	–	–	high	soluble	soluble	–
Glycerol dichlorohydrins												
1,3-Dichloro-2-propanol (7)	−4	174	58	1.3645	–	water	99	23	15.6*	very soluble	soluble	–
2,3-Dichloro-1-propanol (8)	–	182	91	1.3607	–	dichlorobenzene	170.8	30	12.7*	miscible	miscible	miscible
Butene												
2-Chloro-1-butanol (9)	–	75 (3.3)	–	1.0622	–	–	–	–	–	high	–	–
3-Chloro-1-butanol (10)	–	138	–	1.0671	–	–	–	–	–	–	–	–
4-Chloro-1-butanol (11)	–	84 (2.7)	–	1.0883	–	–	–	–	–	high	–	–
1-Chloro-2-butanol (12)	–	141	–	1.0683	–	–	–	–	–	soluble	–	–
3-Chloro-2-butanol (13)	–	139	–	1.0669	–	water	94.5	59	–	high	–	–
4-Chloro-2-butanol (14)	–	70 (1.7)	–	–	–	–	–	–	–	high	–	–

Table 2. (continued)

Chlorohydrin	mp, °C	(p, kPa) bp, °C	Flash point, °C	d_4^{20}, g/mL	η, mPa·s	Azeotrope Component	bp, °C	wt%	Water	Ethanol	Acetone	Benzene
Isobutene												
2-Chloro-2-methyl-1-propanol (**15**)	–	132.5	–	1.0472	–	–	–	–	decomp.	–	–	–
1-Chloro-2-methyl-2-propanol (**16**)	–	128	–	–	–	–	93.5	66.0	–	–	–	–
Isobutene dichlorohydrins												
1-Chloro-2-chloromethyl-2-propanol (**17**)	–	174	–	–	–	–	–	–	–	–	–	–
1,2-Dichloro-2-methyl-3-propanol (**18**)	–	180	–	–	–	–	–	–	–	–	–	–
Other												
Styrene chlorohydrin (**19**)	–	110 (0.8)	–	–	–	–	–	–	soluble	soluble	soluble	soluble

* In g/100 g.

3. Chemical Properties

Chlorohydrins undergo reactions which are characteristic of both alcohols and alkyl chlorides. This combination is responsible for the most common reaction of chlorohydrins, which is *dehydrochlorination to form epoxides*:

$$\text{R-CH(OH)-CH}_2\text{Cl} + \text{OH}^- \longrightarrow \text{R-CH-CH}_2\text{(O)} + \text{H}_2\text{O} + \text{Cl}^-$$

The mechanism of dehydrochlorination of ethylene chlorohydrin has been studied by comparing the relative rates of reaction in H_2O and D_2O [1]. The kinetics of the propylene chlorohydrin conversion to propylene oxide were determined with both NaOH and $Ca(OH)_2$ [2]. The relative rates of solvolysis of ethylene chlorohydrin, propylene chlorohydrin, and butylene chlorohydrin in water at 97 °C are 1.0:0.81:5.5 [3].

Hydroxyalkyl ethers may be prepared by reacting chlorohydrins with alcohols or phenols under basic conditions [4]. *Ethers of cellulose and starch* are manufactured by reacting these materials with various chlorohydrins. *Cationic starch* is made by reacting 3-chloro-2-hydroxypropyl trimethylammonium chloride with a starch slurry at pH 11–12 [5].

Chlorohydrins react with carboxylates in the presence of base to give *hydroxyalkyl esters*. On the other hand, *β-chloro esters* are formed by reaction with carboxylic acids under acid conditions [6] or from acid chlorides [7]. *Cyclic carbonates* may be produced by the reaction of a chlorohydrin with carbon dioxide in the presence of an amine [8].

A comparison of the rate of reaction of ethylene chlorohydrin with various *amines* gives the following order: *n*-amylamine > cyclohexylamine > aniline [9]. The reaction of ethylene chlorohydrin with ammonia gives *monoethanolamine* [10]. *Quaternary ammonium compounds* result from chlorohydrins and tertiary amines [11], [12].

Other chlorohydrin reactions include formation of *nitriles* from cyanides [13], *acetals* from aldehydes [14], *oxazolidinones* from cyanates [15], and the oxidation of ethylene chlorohydrin to *monochloroacetic acid* [16].

4. Production

In 1863, Carius [17] reported that chlorohydrins could be synthesized by reacting olefins with hypochlorous acid. As early as 1904, BASF began production of ethylene chlorohydrin by introducing ethylene and CO_2 into an aqueous solution of bleaching powder [18].

Ethylene Chlorohydrin. The reaction of chlorine and water (1) produces very little hypochlorous acid because of an unfavorable equilibrium, $K = 4.2 \times 10^{-4}$ [19].

$$Cl_2 + H_2O \rightleftharpoons HOCl + HCl \qquad (1)$$

Gomberg [20] reasoned that if the reaction between ethylene and hypochlorous acid (2) is significantly more rapid than the reaction of ethylene and chlorine to form dichloroethane (3), then as the ethylene is added to the aqueous system, chlorohydrin **1** should be produced preferentially.

$$H_2C=CH_2 + HOCl \rightleftharpoons \underset{\underset{Cl}{|}}{CH_2}-\underset{\underset{OH}{|}}{CH_2} \qquad (2)$$
$$\mathbf{1}$$

$$H_2C=CH_2 + Cl_2 \longrightarrow \underset{\underset{Cl}{|}}{CH_2}-\underset{\underset{Cl}{|}}{CH_2} \qquad (3)$$

He showed that with good stirring to minimize reaction (3) in the gas phase, little dichloroethane is produced until the chlorohydrin concentration reaches 6–8%.

In reaction (1), the concentration of hypochlorous acid declines steadily as the HCl level increases. Significant quantities of dichloroethane are observed when the HCl concentration goes above 3%. Because the solubility of dichloroethane in water is low (0.869 g/100 mL at 20 °C), it quickly forms a separate phase into which both chlorine and ethylene are preferentially dissolved to produce even more dichloroethane.

In a laboratory process study of the hypochlorination of ethylene, the effects of various reaction parameters were determined with both single-column and recycle-reactor arrangements [21]. Reaction temperatures of 35–50 °C and ca. 50% excess of ethylene are preferred because these conditions allow much of the dichloroethane to be stripped from the reaction medium as it is formed.

For a continuous run with recycle at a chlorohydrin concentration of 6.4%, a reactor temperature of 35 °C, an ethylene:chlorine ratio of 1.42, and a chlorine feed rate of 71 g/h, the average yield of ethylene chlorohydrin was 88.0% with a dichloroethane yield of 10.1%. Lowering the chlorine feed rate to 42 g/h raised the chlorohydrin yield slightly to 88.9% and lowered the dichloroethane yield to 9.1%.

For many years, ethylene chlorohydrin was manufactured on a large industrial scale as a *precursor to ethylene oxide*. This process has been almost completely supplanted by the direct oxidation of ethylene to ethylene oxide using silver catalysts (→ Ethylene Oxide). However, since other commercially important epoxides, such as propylene oxide and epichlorohydrin, cannot be made by direct oxidation of the parent olefin, chlorohydrin intermediates are still important in manufacturing these products. Although it is scarcely practiced today, a review of the ethylene chlorohydrin technology is valuable because of the close similarity of all of the chlorohydrin processes.

In a *typical industrial plant*, chlorine, ethylene, and water are concurrently passed upward through packed towers [22], [23]. These reactors are designed to minimize the contact between ethylene and gaseous chlorine and to provide adequate contact between the hydrocarbon feed and the liquid phase [24]. The feed lines are positioned a sufficient distance apart so that the chlorine can be completely dissolved in the water before the ethylene is introduced into the system. In some plants, there is a separate mixing column for chlorine and water followed by an ethylene reaction tower.

The reaction product is removed at the top of the tower as a 4.5–5.0% solution of ethylene chlorohydrin, the yield of which may be as high as 85–89% of the ethylene converted. In addition to dichloroethane, small amounts of bis(2-chloroethyl) ether are formed as byproducts.

Propylene Chlorohydrin. The *hypochlorination of propene* gives two isomers:

$$CH_3-CH=CH_2 + Cl_2 + H_2O$$
$$\longrightarrow \underset{\underset{90\%}{3}}{CH_3-\underset{OH}{CH}-\underset{Cl}{CH_2}} + \underset{\underset{10\%}{2}}{CH_3-\underset{Cl}{CH}-\underset{OH}{CH_2}} + HCl$$

The major side reactions are chlorine addition to give 1,2-dichloropropane [78-87-5], and formation of bis(2-chloro-1-methylethyl) ether [108-60-1]:

$$CH_3-CH=CH_2 + Cl_2 \longrightarrow CH_3-\underset{Cl}{CH}-\underset{Cl}{CH_2}$$

$$CH_3-CH=CH_2 + CH_3-\underset{OH}{CH}-\underset{Cl}{CH_2} + Cl_2$$
$$\longrightarrow \underset{Cl}{CH_2}-\underset{CH_3}{CH}-O-\underset{CH_3}{CH}-\underset{Cl}{CH_2} + HCl$$

Propylene chlorohydrin reacts with a strong base, such as calcium hydroxide, to give *propylene oxide* [75-56-9] by dehydrochlorination (→ Propylene Oxide):

$$\left.\begin{array}{c} CH_3-\underset{}{\overset{OH}{C}H}-\underset{}{\overset{Cl}{C}H_2} \\ \text{or} \\ CH_3-\underset{Cl}{CH}-\underset{OH}{CH_2} \end{array}\right\} + 1/2\,Ca(OH)_2$$
$$\longrightarrow CH_3-\overset{O}{\overset{\diagup\diagdown}{CH-CH_2}} + 1/2\,CaCl_2 + H_2O$$

A typical flowchart for the production of propylene oxide through the chlorohydrin is shown in Figure 1 [25].

In the propylene chlorohydrin process, the separation of the feed points for chlorine and propene is even more critical than in the case with ethylene. The benefits of a recirculating-type, double-chamber reaction system have been described [26]. Chlorine was dissolved in a recycle stream of dilute aqueous propylene chlorohydrin together with makeup water. From this vessel, the effluent passed to a second chamber into which the propene was fed. This arrangement gave a propylene chlorohydrin yield of 87.5% with 11.0% dichloropropane and 1.5% dichloroisopropyl ether. In contrast, a singlereactor system produced a 69.2% chlorohydrin yield with 21.6% dichloropropane and 9.2% dichloroisopropyl ether.

Modern plants show propylene chlorohydrin yields of 87–90% with 6–9% dichloropropane. There are numerous patents that claim technology to boost chlorohydrin yields to 90–95% [27]–[33].

A novel approach to hypochlorination involves the use of *tertiary alkyl hypochlorites* [34] and is based on the following chemistry:

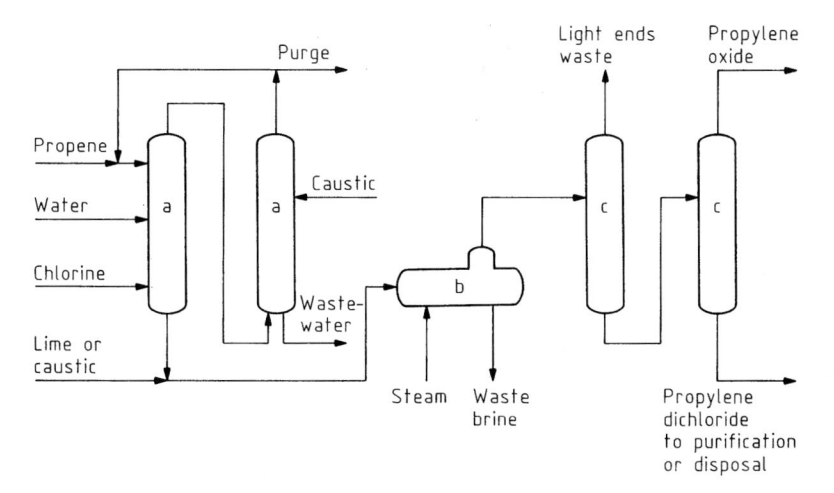

Figure 1. Typical arrangement for propylene oxide via the chlorohydrin route
a) Chlorohydrin absorber reactor; b) Chlorohydrin saponification and flashing; c) Propylene oxide purification train
Reproduced with permission from [25].

$$Cl_2 + 2\,NaOH \longrightarrow NaOCl + NaCl + H_2O$$
$$NaOCl + ROH \longrightarrow ROCl + NaOH \qquad (4)$$
$$ROCl + H_2O + C_3H_6 \longrightarrow ROH + \underset{\underset{3}{OH\ \ Cl}}{CH_3-CH-CH_2} \qquad (5)$$

The preferred alcohol in Reaction (4) is *tert*-butanol. The resulting *tert*-butyl hypochlorite has very slight water solubility and may be separated by phase. Thus, the reaction of the hypochlorite with propene and water (Eq. 5) occurs in an environment essentially free of chloride ions to reduce formation of the dichloride. A 95.8 mol% selectivity to propylene chlorohydrin is claimed [34]. A diagram of this process is shown in Figure 2 [35].

Glycerol Monochlorohydrins. *Hypochlorination of allyl alcohol* gives glycerol monochlorohydrins:

$$\underset{OH}{CH_2-CH=CH_2} + Cl_2 + H_2O \longrightarrow \left(\underset{\underset{5}{OH\ \ OH\ \ Cl}}{CH_2-CH-CH_2} + \underset{\underset{6}{OH\ \ Cl\ \ OH}}{CH_2-CH-CH_2} \right) + HCl$$

The reaction of allyl alcohol with chlorine and water at 50–60 °C is reported to give an 88% yield of monochlorohydrins and 9% dichlorohydrins [36].

Glycerol Dichlorohydrins. The *hypochlorination of allyl chloride* generates a mixture of the glycerol dichlorohydrins, 2,3-dichloro-1-propanol (**8**) and 1,3-dichloro-2-propanol (**7**), in about a 70:30 ratio.

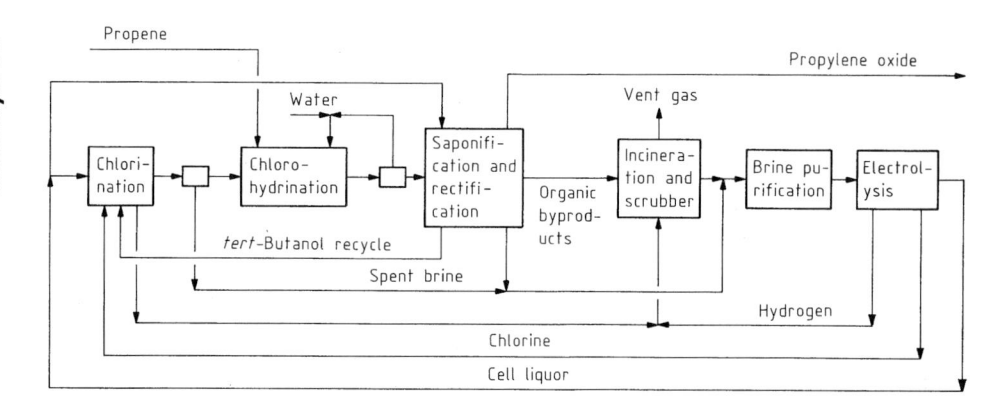

Figure 2. Hypochlorite route to propylene oxide
Reproduced with permission from [35].

$$CH_2-CH=CH_2 + Cl_2 + H_2O$$
$$\;\;\;\;|$$
$$\;\;\;\;Cl$$
$$\longrightarrow \left(\underset{8}{CH_2-CH-CH_2 \atop |\;\;\;\;\;\;|\;\;\;\;\;\;|\atop Cl\;\;\;\;Cl\;\;\;\;OH} + \underset{7}{CH_2-CH-CH_2 \atop |\;\;\;\;\;\;|\;\;\;\;\;\;|\atop Cl\;\;\;\;OH\;\;\;\;Cl} \right) + HCl$$

These may be subsequently saponified to *epichlorohydrin* [106-89-8]:

$$(7+8) + NaOH$$
$$\longrightarrow NaOH \longrightarrow CH_2-CH-CH_2 + NaCl + H_2O$$
$$\;|\;\;\;\;\backslash\;/$$
$$\;Cl\;\;\;\;O$$

Because of the low water solubility of allyl chloride, it is essential to minimize formation of an organic phase, which leads to formation of the undesired byproduct 1,2,3 trichloropropane [24], [37]. This is commonly done by maintaining the allyl chloride as a fine dispersion by vigorous agitation in an aqueous system containing 10–50 volumes of water for each volume of allyl chloride [38]–[40]. Under such conditions, dichlorohydrin selectivities of 88–93% are reported.

Another method of achieving a well-dispersed allyl chloride is by the use of a *surfactant* [41], [42]. Anionic or nonionic surfactants at a level of 0.2–0.5 wt% based on allyl chloride give reported dichlorohydrin yields of 93–95% based on allyl chloride.

An alternative way of making dichlorohydrins is by *chlorination of allyl alcohol* [107-18-6], which may be produced by the hydrogenation of acrylaldehyde [107-02-8]. This route gives reported dichlorohydrin yields of 94–98% with a 2,3-dichloro-1-propanol yield of 88–97% [43], [44].

Butene Chlorohydrins. Butylene chlorohydrin is prepared as a precursor to butene oxide. A laboratory study of the *hypochlorination of 2-butene* has been reported [45]. The dehydrochlorination of 3-chloro-2-butanol (**13**) with calcium hydroxide gave an 87% yield of 2,3-butene oxide with 11% glycol and 2% methyl ethyl ketone [46].

Table 3. Environmental data on two chlorohydrins

	Biodegradation BOD/ThOD*	Fathead minnow toxicity
Ethylene chlorohydrin	73%	LC_{50}** > 100 mg/L
1,3-Butene chlorohydrin	67%	no data

* BOD/ThOD = biochemical oxygen demand/theoretical oxygen demand.
** LC_{50} = lethal concentration to 50% of the test organisms in a specified time.

Styrene Chlorohydrin. A laboratory preparation of styrene chlorohydrin (**19**) has been described [47]. Styrene was dispersed in water with a surfactant. Calcium hypochlorite solution was added to the vigorously stirred suspension through which carbon dioxide was bubbled. A 76% yield of styrene chlorohydrin was obtained.

Long-Chain Chlorohydrins. Fatty acids containing unsaturation, such as oleic or elaidic acids, are converted to chlorohydrins as intermediates in making the epoxides or diols of these materials [48]. Long-chain olefins may be hypochlorinated with good selectivity (80–90%) by using short reaction times, which result in low olefin conversion (5–25%) [49].

Recent Process Improvements. Some process improvements and new approaches for chlorohydrin production developed in recent years are the continuous production of glycerol dichlorohydrins [50], glycerol dichlorohydrins from allyl chloride [51], electrochemical chlorohydroxylation of olefins [52], 4-halo alcohols from ring cleavage of tetrahydrofuran [53], improved continuous production of aqueous propylene chlorohydrin solutions [54], high-yield continuous preparation of propylene chlorohydrin [55], ethylene chlorohydrin from ethanol [56], and chlorohydrins from immobilized halogenating enzymes [57], [58].

5. Environmental Protection

Unpublished data [59] for two of the chlorohydrin compounds are shown in Table 3 (there were no other data). Ethylene chlorohydrin is practically nontoxic to the fathead minnow according to a categorization system used by the U.S. Environmental Protection Agency. The two compounds in Table 3 are expected to be readily biodegradable.

6. Chemical Analysis

In addition to the conventional wet quantitative method, other modern spectroscopic analyses have been used, including Raman spectroscopy [60]. Gas chromatographic analysis of styrene chlorohydrin was reported in 1980 [61]. Potentiometric determination of propylene chlorohydrins [62] and enzymatic determination of ethylene chlorohydrin [63] were also reported.

7. Storage and Transportation

Chlorohydrins generally can be stored in acid-resistant tanks with a small or low-pressure rating. Titanium metal is resistant to corrosion by most chlorohydrins [64]. Potentiometric study of the corrosion of chlorohydrins up to 120 °C on titanium and steel has been reported [65]. Chlorohydrins in general are not classified as hazardous materials by the U.S. Department of Transportation. They can be shipped in acid tank railroad cars or trucks [64].

8. Uses

Chlorohydrins find their greatest utility as intermediates in the manufacture of epoxides, especially propylene oxide and epichlorohydrin. For reviews of these processes, see the following references: [25], [66]–[68], → Epoxides, → Ethylene Oxide, → Propylene Oxide.

Adducts of ethylene chlorohydrin with tungsten or molybdenum halides plus an organoaluminum compound have been patented as catalysts for olefin metathesis and cycloalkene ring-opening polymerization [69]–[72].

Chlorohydrins have also been used in the following ways:

- intermediate for plasticizers [73]
- wet strengthening agent for paper [74], [75]
- etherizing agent of phenolic resin to produce a high-temperature heavy-metal chelating compound [76]
- preservation of biological fluids and solutions [77]
- treatment of sulfonyl halides and dimethylamine aromatic compounds for fluorescent whitening agents and photosensitizers [78]
- raw material for lubricating oil intermediates and additives [79], [80]
- reaction with 1,3-bis(dimethylamino)-2-propanol to form a flocculant [81]
- production of low molecular mass epoxy resins [82]
- raw material for flame retardants for polymers [83]

- soil-resistant yarn-treating agent [84], [85]
- hardener for polyurethane elastomers [86]

In recent years, *pharmaceutical and agricultural chemical applications* have been explored. Chlorohydrins are used as base chemicals for various drugs and chemicals, such as the following:

- carbazole derivative for treatment of anxiety [87]
- pharmaceutically acceptable salts [88]
- quaternary ammonium polymeric antimicrobial agents [89]
- substituted nitrodiphenyl ethers [90]
- phenoxyalkanol antirhinovirus agents [91]
- quaternary salts of promethazine bronchoconstriction releasing agent [92]
- substituted oxobenzothiazolines plant growth regulators [93]
- phenyl urea derivative herbicides [94]

9. Toxicology and Occupational Health

Of the various chlorohydrins, ethylene chlorohydrin and α-chlorohydrin have been the most extensively studied, the former because it is a residue produced when poly(vinyl chloride) plastics are sterilized with ethylene oxide.

Ethylene Chlorohydrin. The American Conference of Governmental Industrial Hygienists TLV-TWA is 1 ppm (3 mg/m^3) as a ceiling limit with a skin notation, which indicates that the dermal route is significant. The MAK values are the same. The Occupational Safety and Health Act PEL is 5 ppm with a skin notation.

Ethylene chlorohydrin is moderate to high in acute oral toxicity with an LD$_{50}$ of 89 mg/kg for rats [95]. It is also high in dermal toxicity as its dermal LD$_{50}$ is 84 mg/kg in rabbits [95], a value that places ethylene chlorohydrin into the class B poison category for the U.S. Department of Transportation. A volume of ca. 5 cm^3 could be lethal to the average adult male (70 kg) if it contacts the skin and is not washed off immediately. Ethylene chlorohydrin is not particularly irritating to the skin or eyes; however, it causes slight to moderate irritation, depending on the circumstances. Excessive vapor concentration of ethylene chlorohydrin is attainable, and a single exposure at elevated concentration may cause death. One fatality due to a 2-h inhalation of ethylene chlorohydrin vapor at ca. 1 mg/L (ca. 306 ppm) has been reported [96]. Two fatal and several nonfatal cases of intoxication by ethylene chlorohydrin were investigated [97]. The average concentration in the nonfatal cases was 18 ppm. Nonfatal cases showed signs of circulatory shock, lack of coordination, repeated vomiting, epigastric pain, headache, heavy urine output, cough, and skin reddening. In another case report [98], autopsy findings showed damage to the lungs, liver, kidneys, brain, and other organs.

Ethylene chlorohydrin is considered to be *noncarcinogenic* on the basis of a test in which it was given to rats by subcutaneous injection for 1 year with a 6-month hold [99], and also on the basis of a skin painting test in rats and mice [100]. Ethylene chlorohydrin did not cause birth defects in rabbits when administered intravenously in doses up to 36 mg kg^{-1} d^{-1} during the period of organogenesis. When administered to mice at higher doses, 120 mg kg^{-1} d^{-1} caused a significant increase in the number of malformed fetuses and resorptions when accompanied by maternal toxicity [101]. There have been no studies of ethylene chlorohydrin and its effect on male or female reproductive systems.

In the *Ames bacterial mutagenicity test*, ethylene chlorohydrin is sometimes positive and sometimes negative, with some of the positives only occurring after alteration by tissue homogenates (presumably creating metabolites that are known to be active) [102]–[104]. Ethylene chlorohydrin is also weakly and sporadically mutagenic when tested in animal cell systems. It is reported to have induced chromosome aberrations in rat bone marrow cells [105], but it did not produce a significant increase in the number of translocation heterozygotes (heritable translocation) when administered to male mice at 30–60 mg kg^{-1} d^{-1} for 5 weeks [106]. It was also negative in tests with fruit flies [107].

The toxicity of ethylene chlorohydrin (and presumably other activity, such as mutagenicity) is believed to be due to its *conversion to chloroacetaldehyde* in vivo. Chloroacetaldehyde and chloroacetic acid, which are metabolic intermediates, are to a great degree conjugated and inactivated by reaction with glutathione. Thus, when ethylene chlorohydrin levels are highly elevated, glutathione is depleted and chloroacetaldehyde cannot be detoxified. When chloroacetaldehyde levels at intracellular sites are elevated, other cell components are alkylated, damage to major organs occurs, and toxic symptoms are produced [108], [109].

α-Chlorohydrin. This member of the chlorohydrin series has also been extensively studied for its toxicological effects. Its chronic toxicological profile differs greatly from that of ethylene chlorohydrin. There is apparently no industrial hygiene exposure guide for α-chlorohydrin. α-Chlorohydrin is also high in acute oral toxicity, with an LD$_{50}$ of 55 mg/kg in rats [110], and is apparently also high in dermal toxicity. The rabbits placed on skin irritation tests all died with one small application, and one was examined and found to have "widespread liver damage" [59]. The material was essentially nonirritating to skin. The vast majority of toxicological information on α-chlorohydrin is from investigation of its effect on male reproduction. This antifertility effect, which α-chlorohydrin has on the male, was demonstrated in 1969 and 1970 [111], [112], and has been shown to be predictable for rats by oral, injection, and inhalation routes.

Because of the potential utility of α-chlorohydrin as a *male contraceptive*, its mechanism of action has been thoroughly studied. Only the S isomer has been determined to possess antifertility activity; it has none of the detrimental effect on the kidneys associated with the R isomer [113]. α -Chlorohydrin and epichlorohydrin produce similar effects on the male reproductive system of the rat, i.e., epididymal sperm granulomas, spermatocoeles, and an increase in the number of morphologically abnormal spermatozoa [114]. There is evidence that the profile of action may be slightly different in primates [115].

Like ethylene chlorohydrin, α-chlorohydrin is *weakly mutagenic* in some tests [116], but α-chlorohydrin may be a direct-acting mutagen.

α-Chlorohydrin was tested for *carcinogenicity* and gave ambiguous results in an oral test using rats. Toxicity (testicular effects) was noted at the two dose levels used; and although not statistically significant, there were tumors of the parathyroid that were in question in this test [117]. In mice, α-chlorohydrin produced no carcinogenic effects via skin painting and subcutaneous injection [118].

Other Chlorohydrins. Data for other chlorohydrins are primarily of the acute variety. *1-Chloro-2-propanol* is moderate in acute oral toxicity with an LD_{50} of 100–300 mg/kg in rats and moderate in acute dermal toxicity with an LD_{50} of ca. 500 mg/kg in rabbits. It is only slightly irritating to the skin, and slightly to moderately irritating to the eyes [59]. The Dow Industrial Hygiene Guide is 3 ppm as a ceiling limit with a skin notation and is based on the following: (1) acute data which showed that rats were able to survive at most a 15-min exposure to a saturated atmosphere (nominal concentration 13 000 ppm) [59]; (2) a 22-week intubation study in rats, which showed only slight liver effects (organ weight increase) at 25 mg kg^{-1} d^{-1} [119]; and (3) an inhalation test in rats, in which 14 exposures to 30 ppm caused no adverse effects, but higher concentrations caused lung congestion and lethargy [120].

A *mixture of 2-chloro-1-propanol and 1-chloro-2-propanol* was mutagenic in the Ames strains TA1535 and TA100 with and without metabolic activation but with enhanced activity after activation. The mixture was also mutagenic in a mouse lymphoma assay and in a rat cytogenetic test [121]. Such a mixture was also fed to rats and dogs for 90 days at doses ranging from 1 to 50 mg kg^{-1} d^{-1} [122]. No significant effects occurred in rats at any dose nor in dogs at 1 or 7 mg kg^{-1} d^{-1}. Only reduced food consumption, reduced body weight, and increased relative liver weight occurred at 50 mg kg^{-1} d^{-1} in dogs.

1,3-Dichloro-2-propanol is moderate in acute oral toxicity with an LD_{50} of between 126 and 252 mg/kg in rats, is moderate in acute dermal toxicity with an LD_{50} of between 500 and 1000 mg/kg in rabbits, and is moderately irritating to the eyes and moderately to severely irritating to the skin [59]. In inhalation tests, a 4-h exposure to 1000 ppm was lethal to five of five rats, while no rats died as a result of inhaling 1000 ppm for 2 h. Rats were able to survive a 4-h exposure at 300 ppm but not 7 h at 300 ppm [59]. This chlorohydrin was mutagenic toward Ames tester strains TA100 [116] and TA1535 [123] and in other such Ames tests.

2,3-Dichloro-1-propanol is moderate to high in acute oral toxicity with an LD_{50} of 90 mg/kg in rats, moderate in acute dermal toxicity with an LD_{50} of 0.2 mL/kg in rabbits, and also moderately irritating to the eyes and skin [124]. It is mutagenic toward Ames strain TA1535 [123] and in other Ames tests.

There are no toxicity data for either 3-chloro-2-butanol or 2-chloro-1-butanol (this should not be confused with chlorobutanol, which is the anesthetic having the full name trichloro-*tert*-butyl alcohol). There are also no data for the other glycerol monochlorohydrin, 2-chloro-1,3-propanediol.

Acute data indicate that *styrene chlorohydrin* is low in acute oral toxicity with an LD_{50} between 500 and 2000 mg/kg in rats [59], but there is no dermal LD_{50} available. Styrene chlorohydrin is quite irritating to the eyes and skin. Inhalation presents no hazard from exposure at room temperature; however, if the material is heated, the vapors thus generated may cause irritation and possibly organ effects (kidney and liver) [59].

10. References

[1] G. C. Swain, A. D. Ketley, R. F. W. Bader, *J. Am. Chem. Soc.* **81** (1959) 2353.
[2] S. Carra et al., *Chem. Eng. Sci.* **34** (1979) 1123.
[3] I. K. Gregor, N. V. Riggs, V. R. Stimson, *J. Chem. Soc.* 1956, 76.
[4] N. K. Bliznyuk et al., *Zh. Obshch. Khim.* **36** (1966) 480; *Chem. Abstr.* **65** (1966) 638 d.
[5] Corn Products Co., US 2 876 217, 1959 (E. F. Paschall).
[6] I. G. Rodier, *Bull. Soc. Chim. Fr.* 1948, 637.
[7] H. V. R. Iengar, P. D. Ritchie, *J. Chem. Soc.* 1957, 2556.
[8] Phillips Petroleum Co., US 3 923 842, 1975 (Y. Wu).
[9] J. Plucinski, *Rocz. Chem.* **41** (1967) 1135; *Chem. Abstr.* **68** (1968) 104 329 j.
[10] V. A. Krishnamurthy, M. R. A. Rao, *J. Indian Inst. Sci.* **40** (1958) 145; *Chem. Abstr.* **53** (1960) 19 860 g.
[11] DD 105 207, 1974 (E. Fibitz, B. Nussbuecker); *Chem. Abstr.* **81** (1974) 151 522 v.
[12] N. Parris, J. K. Weil, W. M. Linfield, *J. Am. Oil Chem. Soc.* **53** (1976) 97.
[13] I. Shinji, *J. Soc. Chem. Jpn.* **45** (1942) 359.
[14] M. F. Shostakovskii, T. T. Minakova, F. P. Sidel'kovskaya, *Izv. Akad. Nauk SSSR, Ser. Khim.* 1964, 2197; *Chem. Abstr.* **62** (1965) 8996 d.
[15] B. L. Phillips, P. A. Argabright, *J. Heterocycl. Chem.* **3** (1966) 84.
[16] Du Pont, US 2 455 405, 1948 (L. A. Burrows, M. F. Fuller).
[17] L. Carius, *Ann. Chem. Pharm.* **126** (1863) 195.
[18] J. F. Norris, *Ind. Eng. Chem.* **11** (1919) 817.
[19] F. A. Cotton, G. Wilkinson: *Advanced Inorganic Chemistry*, Wiley-Interscience, New York 1962, p. 446.
[20] M. Gomberg, *J. Am. Chem. Soc.* **41** (1919) 1414.
[21] W. G. Domask, K. A. Kobe, *Ind. Eng. Chem.* **46** (1954) 680.
[22] F. Asinger in B. J. Hazzard (trans.): *Mono-Olefins, Chemistry and Technology*, Pergamon Press, Oxford 1968, pp. 556–568.
[23] D. G. Weaver, J. L. Smart, *Ind. Eng. Chem.* **51** (1959) 894.
[24] P. W. Sherwood, *Pet. Eng.* **27** (1955) C–41.

[25] R. B. Stobaugh, V. A. Calarco, R. A. Morris, L. W. Stroud, *Hydrocarbon Process.* **52** (1973) no. 1, 99–108.
[26] P. Ferrero, L. R. Flamme, M. Fourez, *Ind. Chim. Belg.* **19** (1954) 113.
[27] Shell, BE 630 446, 1963.
[28] Bayer, FR 1 357 443, 1964.
[29] Naphthachimie, DE-OS 2 101 119, 1971 (R. Bouchet).
[30] BASF, DE-OS 2 022 819, 1971 (E. Bartholomé et al.).
[31] G. Mikula, CS 152 702, 1974; *Chem. Abstr.* **81** (1974) 63135 r.
[32] Petrocarbon Developments, FR 2 194 673, 1974.
[33] Mitsui Toatsu Chemicals, JP-Kokai 77 48 606, 1977 (Y. Watanabe et al.).
[34] Lummus, US 4 008 133, 1977 (A. P. Gelbein, J. T. Kwon).
[35] *Hydrocarbon Process.* **59** (1979) 239.
[36] J. Myszkowski, A. Z. Zielinski, *Przem. Chem.* **44** (1965) 249; *Chem. Abstr.* **63** (1965) 5516 e.
[37] Shell, US 2 605 293, 1952 (T. T. Tymstra).
[38] Shell, US 2 714 121, 1955 (J. Anderson, G. F. Johnson, W. C. Smith).
[39] T. Reis, FR 1 557 589, 1969.
[40] I. Ondrus, P. Klucovsky, CS 120 030, 1966; *Chem. Abstr.* **67** (1967) 63 741 w.
[41] Tokuyama Soda Co., JP 73 21 924, 1973 (Y. Onoue, Y. Fujii, M. Ueda).
[42] Euteco, GB 2 029 821, 1980 (C. Divo, M. Petri, M. Lazzari, A. Bigozzi).
[43] Hoechst, DE-OS 2 007 867, 1971 (D. Freudenberger, H. Fernholz).
[44] Nippon Soda Co., JP 73 18 207, 1973 (Y. Enoki et al.).
[45] A. Z. Zielinski, J. Myszkowski, A. Czubowiez, *Chem. Tech. (Leipzig)* **14** (1962) 456; *Chem. Abstr.* **58** (1963) 3305 c.
[46] J. Myszkowski, A. Z. Zielinski, *Przem. Chem.* **43** (1964) 324; *Chem. Abstr.* **61** (1964) 14 611 a.
[47] W. S. Emerson, *J. Am. Chem. Soc.* **67** (1945) 516.
[48] A. W. Ralston: *Fatty Acids and Their Derivatives,* J. Wiley & Sons, New York 1948, pp. 450–452.
[49] Procter & Gamble Co., US 3 598 847, 1971 (R. K. Kloss, G. W. Claybaugh, D. D. Whyte).
[50] G. S. Sharifov et al., DE-OS 2 951 770, 1981.
[51] Euteco SpA, GB-A 2 029 821, 1980 (G. Divo et al.).
[52] V. Kazarinov et al., *Collect. Czech. Chem. Commun.* **47** (1982) no. 11, 2849–2857.
[53] Texaco Development Corp., DE-OS 2 952 277, 1980 (K. G. Hammond).
[54] VEB Chemische Werke Buna, DD 144 907, 1980 (K. H. Kuessner et al.).
[55] BASF, DE-OS 2 910 675, 1980 (W. Koehler et al.).
[56] Financiadora de Estudos e Projetos (FINEP); Scientia-Engenharia de Sistemas, Desenvolvimento de Prototipos e Processos, Braz. Pedido PI BR 82 312 A, 1983 (A. Telles).
[57] Cetus Corp., US 4 247 641, 1981 (S. L. Neidleman et al.).
[58] J. Geigert et al., *Appl. Environ. Microbiol.* **45** (1983) no. 2, 366–374.
[59] Unpublished results, Dow Chemical.
[60] A. Gupta et al., *Spectrochim. Acta, Part A* **36 A** (1980) no. 6, 601–606.
[61] S. A. Shakun et al., *Khim. Prom.-st., Ser.: Metody Anal. Kontrolya Kach. Prod. Khim. Prom-sti.* **1** (1980) 12–14; *Chem. Abstr.* **93** (1980) 88 211 y.
[62] T. V. Shevchenko et al., *Khim. Prom.-St., Ser.: Metody Anal. Kontrolya Kach. Prod. Khim. Prom.-Sti.* **3** (1980) 43–46; *Chem. Abstr.* **93** (1980) 36 395 s.
[63] American Sterilizer Co., US 4 162 942, 1979 (D. A. Gunther).
[64] A. E. Romanushkina, *Zashch. Met.* **10** (1974) no. 1, 39–40.

[65] L. I. Komarova, *Ispol'z. Sovrem. Fiz. Khim. Metodov Issled. Protsessov. Prod. Khim. Uglekhim. Proizvod. Tedzizy Dokl. Nauchno Tekh. Konf.* 1976, 96–97; *Chem. Abstr.* **89** (1978) no. 22, 187 881 r.
[66] K. H. Simmrock, *Hydrocarbon Process.* **57** (1978) no. 11, 105–113.
[67] Lummus, US 4 008 133, 1977 (A. P. Gelbein, J. T. Kwon).
[68] "Glycerol (Epichlorohydrin)," *Pet. Refiner* **38** (1959) no. 11, 252–253.
[69] Bayer, DE-OS 2 056 178, 1972 (D. Maertens, J. Witte, M. Beck).
[70] Bayer, NL-A 7 408 469, 1975 (G. Lehnert, D. Maertens, G. Pampus, J. Witte).
[71] Bayer, DE-OS 2 332 565, 1975 (G. Pampus, G. Lehnert, J. Witte).
[72] Bayer, DE-OS 2 357 193, 1975 (J. Witte, G. Lehnert).
[73] A. Courtier, FR 2 144 084, 1973.
[74] Hercules, EP-A 416 511, 1981 (R. A. Bankert).
[75] G. Troemel, *Farben Rev. USA* 1970, no. 18, 55–63.
[76] Unitika, JP-Kokai-A2 145 719, 1983.
[77] *Chem. Mark. Rep.* 1984, 28.
[78] Mita Industrial Co., JP-Kokai 29 335, 1979 (S. Torii).
[79] Chevron Research Co., GB 2 108 485, 1983 (V. R. Small).
[80] Texaco Development Corp., FR 2 460 297, 1981 (K. G. Hammond).
[81] Kewanee Industries, US 4 188 293, 1980 (H. A. Green).
[82] Dow Chemical, US 4 408 062, 1983 (J. L. Bertram et al.).
[83] Monsanto, US 4 179 483, 1979 (G. H. Birum et al.).
[84] Allied, US 4 317 736, 1980 (R. M. Marshall).
[85] Allied, DE-OS 3 304 351, 1983 (R. H. Thomas).
[86] Ihara Chemical Industry Co., US 4 342 859, 1982 (T. Harada et al.).
[87] Burroughs Wellcome Co., US-A 4 400 383, 1983 (J. R. T. Davison).
[88] Wellcome Foundation, JP-Kokai 109 482, 1983.
[89] Kewanee Industries, US-A 4 304 910, 1981 (H. A. Green).
[90] Rohm & Haas Co., EP-A 21 692, 1981 (C. Swithenbank et al.).
[91] Richardson-Merrell, BE 887 423, 1981 (R. A. Parker).
[92] Am. Home Products Corp., US 4 183 912, 1980 (M. E. Rosenthale).
[93] Monsanto, JP-Kokai 151 966, 1979.
[94] Shell Oil Co., US 4 226 612, 1980 (K. H. Pilgram).
[95] *Gig. Sanit.* **43** (1978) no. 8, 13 (in NIOSH Registry of Toxic Effects of Chemical Substances, 1983).
[96] H. Dierker, P. G. Brown, *J. Ind. Hyg. Toxicol.* **26** (1944) 277–279.
[97] M. W. Goldblatt, W. E. Chiesman, *Br. J. Ind. Med.* **1** (1944) 207.
[98] V. Miller, R. J. Dobbs, S. I. Jacobs, *Arch. Dis. Child.* **45** (1970) 589–590.
[99] T. Balazs, *FDA By-Lines* **3** (1976) 150.
[100] National Toxicology Program (NTP) Technical Report on the Toxicology and Carcinogenesis Studies of 2-Chloroethanol (Ethylene Chlorohydrin) in F344/N Rats and Swiss CD–1 Mice (Dermal Studies), NTP TR 275, Board Draft, Jun. 1, 1984.
[101] J. B. LaBorde, C. A. Kimmel, C. Jones-Price, T. A. Marks et al., *Toxicologist* **2** (1982) no. 1, 71.
[102] H. S. Rosenkranz, T. J. Wlodkowski, *J. Agric. Food Chem.* **22** (1974) 407.
[103] J. McCann, V. Simmon, D. Streitwieser, B. N. Ames, *Proc. Natl. Acad. Sci. U.S.A.* 72 (1975) no. 8, 3190–3193.
[104] J. D. Elmore, J. L. Wong, A. D. Laumbach, U. N. Streips, *Biochim. Biophys. Acta* **442** (1976) no. 3, 405–419.

[105] V. N. Semenova, N. Y. Ramul, *Proc. Nat. Acad. Sci.* **72** (1977) no. 8, 179–181.
[106] C. W. Sheu, K. T. Cain, R. M. Grydev, W. M. Generoso, *J. Am. Coll. Toxicol.* **2** (1983) no. 2, 221–223.
[107] A. G. A. C. Knaap, C. E. Voogd, P. G. N. Kramers, *Mutat. Res.* **101** (1982) 199–208.
[108] W. Grunow, H. J. Altmann, *Arch. Toxicol.* **49** (1982) 275–284.
[109] W. H. Lawrence, K. Itoh, J. E. Turner, J. Autian, *J. Pharm. Sci.* **60** (1971) 1163–1168.
[110] R. Paul, R. P. Williams, E. Cohen, *Contraception* **9** (1974) 451–457 (in NIOSH Registry of Toxic Effects of Chemical Substances, 1981–1982).
[111] J. A. Coppola, *Life Sci.* **8** (1969) no. 1, 43–48.
[112] R. J. Ericsson, V. F. Baker, *J. Reprod. Fertil.* **21** (1970) 267–273.
[113] K. E. Porter, A. R. Jones, *Chem.-Biol. Interact.* **41** (1982) no. 1, 95–104.
[114] W. M. Kluwe, B. N. Gupta, J. C. Lamb, *Toxicol. Appl. Pharmacol.* **70** (1983) no. 1, 67–86.
[115] I. Braz, L. N. Shandilya, L. S. Ramaswami, *Andrologia* **8** (1976) no. 4, 290–296.
[116] S. J. Stolzenberg, C. H. Hine, *J. Toxicol. Environ. Health* **5** (1979) no. 6, 1149–1158.
[117] E. K. Weisburger, B. M. Ulland, J.-M. Nam, J. J. Gart et al., *J. Natl. Cancer Inst.* **67** (1981) no. 1, 75–88.
[118] B. L. Van Duuren, B. M. Goldschmidt, C. Katz, I. Seidman et al., *J. Natl. Cancer Inst.* **53** (1974) no. 3, 695–700.
[119] FDA Report, 1969.
[120] J. C. Gage, *Br. J. Ind. Med.* **27** (1970) 1–18.
[121] R. W. Biles, C. E. Piper, *Toxicologist* **1** (1981) no. 1, 41.
[122] I. W. Daly, R. D. Phillips, *Toxicologist* **1** (1981) no. 1, 13.
[123] L. A. Silhankova, F. Smid, M. Cerna, J. Davidek et al., *Mutat. Res.* **103** (1982) 77–81.
[124] H. F. Smyth, C. P. Carpenter, *J. Ind. Hyg. Toxicol.* **30** (1948) no. 1, 63–68.

Chlorophenols

FRANÇOIS MULLER, Rhône-Poulenc, Le Pont-de-Claix, France (Chaps. 2–10)
LILIANE CAILLARD, Rhône-Poulenc, Courbevoie, France (Chap. 11)

1.	Introduction	1605	4.2.	Mono-, Di-, and Trichlorophenols with Chlorine in a Meta Position ... 1610
2.	Physical Properties	1606		
3.	Chemical Properties	1607	5.	Analysis ... 1611
3.1.	Reactions of the Hydroxyl Group	1607	6.	Quality Specifications ... 1612
3.2.	Electrophilic Substitution of the Aromatic Ring	1607	7.	Economic Aspects ... 1612
			8.	Uses ... 1613
3.3.	Nucleophilic Substitution of the Aromatic Ring	1607	9.	Storage and Transportation ... 1614
3.4.	Oxidation Reactions	1608	10.	Environmental Considerations 1614
4.	Production	1608	10.1.	Toxicity to Fauna and Flora ... 1614
4.1.	Mono-, Di-, and Trichlorophenols with no Chlorine in a Meta Position; Tetrachlorophenols; and Pentachlorophenol	1609	10.2.	Persistence in the Environment ... 1615
			11.	Toxicology ... 1616
			12.	References ... 1617

1. Introduction

In chlorophenols the aromatic ring of phenol is substituted with one to five chlorine atoms. LAURENT discovered this group of compounds when he chlorinated coal tar in 1836 [1], [2].

Chlorophenols are industrially important because of their broad spectrum of antimicrobial properties and their uses as fungicides, herbicides, insecticides, ovicides, and algicides. A mixture of tetrachlorophenols and pentachlorophenol, either as such or as salts or esters, finds widespread use as a preservative for wood, glue, paint, vegetable fibers, and leather. In addition, chlorophenols are widely used as intermediates in chemical syntheses.

Table 1. Characteristics of the chlorophenols

Chlorophenol	Formula	M_r	mp, °C	bp, °C	pK_a
2-Chlorophenol [95-57-8]	C_6H_5ClO	128.56	8.7	174.5	8.52
3-Chlorophenol [108-43-0]	C_6H_5ClO	128.56	32.8	216	8.97
4-Chlorophenol [106-48-9]	C_6H_5ClO	128.56	42.8	217	9.37
2,3-Dichlorophenol [576-24-9]	$C_6H_4Cl_2O$	163.00	58	206	7.71
2,4-Dichlorophenol [120-83-2]	$C_6H_4Cl_2O$	163.00	45	210	7.90
2,5-Dichlorophenol [583-78-8]	$C_6H_4Cl_2O$	163.00	59	210	7.51
2,6-Dichlorophenol [87-65-0]	$C_6H_4Cl_2O$	163.00	66	220	6.78
3,4-Dichlorophenol [95-77-2]	$C_6H_4Cl_2O$	163.00	65	253	8.62
3,5-Dichlorophenol [591-35-5]	$C_6H_4Cl_2O$	163.00	68	233	8.25
2,3,5-Trichlorophenol [933-78-8]	$C_6H_3Cl_3O$	197.45	62	248–249	
2,4,5-Trichlorophenol [95-95-4]	$C_6H_3Cl_3O$	197.45	68	245–246	6.72
2,4,6-Trichlorophenol [88-06-2]	$C_6H_3Cl_3O$	197.45	68	246	5.99
2,3,4-Trichlorophenol [15950-66-0]	$C_6H_3Cl_3O$	197.45	83.5	subl.	6.97
3,4,5-Trichlorophenol [609-19-8]	$C_6H_3Cl_3O$	197.45	101	275	7.55
2,3,6-Trichlorophenol [933-75-5]	$C_6H_3Cl_3O$	197.45	101	272	5.80
2,3,5,6-Tetrachlorophenol [935-95-5]	$C_6H_2Cl_4O$	231.89	115	subl.	5.03
2,3,4,6-Tetrachlorophenol [58-90-2]	$C_6H_2Cl_4O$	231.89	69–70	164 (3 kPa)	5.22
2,3,4,5-Tetrachlorophenol [4901-51-3]	$C_6H_2Cl_4O$	231.89	116–117	subl.	5.64
Pentachlorophenol [87-86-5]	C_6HCl_5O	266.34	190	309–310	4.74

Table 2. Physical properties of the most important chlorophenols

Property	2-Chlorophenol	4-Chlorophenol	2,4-Dichlorophenol	2,6-Dichlorophenol	2,4,6-Trichlorophenol	Pentachlorophenol
bp, °C						
at 0.1333 kPa	12	50	53	59.5	76.5	132
at 1.333 kPa	51	92	93	101	120.2	178
at 5.333 kPa	82	125	123	131.6	152.2	
at 13.33 kPa	106	150	146	154.6	177.8	
at 53.33 kPa	150	190	187	197.7	222.5	
Specific heat at 20 °C, J · mol^{-1} K^{-1}	172	177	190			
Heat of formation at 50 °C, kJ/mol	185	198.9				295
Viscosity at 50 °C, mPa · s	2	5	2.65			
Solubility in water at 20 °C, g/L	28.5	27.1	4.5		0.8	0.014
Flash point (closed cup), °C	121	84	113			

2. Physical Properties

Important characteristics and physical properties of chlorophenols are shown in Tables 1 and 2. With the exception of 2-chlorophenol, these compounds are solid at ambient temperature; they all have strong characteristic odors. Several of them sublime readily. They can be steam-distilled. The chlorophenols are only slightly soluble in water, but highly soluble in alcohols. They impart a strongly disagreeable taste to drinking water.

3. Chemical Properties

Chlorophenols are versatile intermediates in chemical syntheses because both the hydroxyl group and the aromatic ring can react by both electrophilic and nucleophilic substitution. They are readily oxidized.

3.1. Reactions of the Hydroxyl Group

Chlorophenols are more acidic than phenol and readily form salts with alkali and alkaline-earth hydroxides or carbonates or with amines. These salts are stable and soluble in water.

Chlorophenols readily form esters of both organic and mineral acids [3]–[5]. Esters such as chlorophenyl phosphates, carbamates [6], [7], chloroformates [8], carbonates, and laurates are widely used. Pentachlorophenyl 2-ethylhexanoate is used as a textile preservative.

Chlorophenols are also readily O-alkylated, and ethers derived from aliphatic, aromatic, and heterocyclic reagents comprise a broad class of industrial products [9], [10]. One of the most widely known is 2,4-dichlorophenoxyacetic acid [95-75-7] (2,4-D), which is still used as a selective herbicide for broadleaf plants [11]–[13] (→ Chlorophenoxyalkanoic Acids).

3.2. Electrophilic Substitution of the Aromatic Ring

Electrophilic substitution is favored by the presence of chlorine atoms on the aromatic nucleus; thus, most chlorophenols partake in the standard reactions, such as alkylation [14], acylation, nitration, sulfonation, halogenation, [15], [16], carboxylation, and condensation with aldehydes, e.g., formaldehyde, chloral, glyoxylic acid [17]–[19].

3.3. Nucleophilic Substitution of the Aromatic Ring

Nucleophilic substitution for one or more of the chlorine atoms, although disfavored by the presence of the other chlorine atoms, is nevertheless used widely.

The hydrolyses of 2-chlorophenol and 4-chlorophenol were used at one time to obtain pyrocatechol [120-80-9] and hydroquinone [123-31-9], respectively. The ammonolysis of 2-chlorophenol or 4-chlorophenol produces 2-aminophenol or 4-aminophenol, respectively. This type of reaction is used primarily to prepare various substituted diphenyl ethers, which serve as efficient herbicides [20]–[22].

Chlorophenols can be converted into poly-chlorinated alicyclic ketones such as "hexachlorophenol" [599-52-0] by reaction of chlorine in aqueous acetic acid. "Hexachlorophenol" has pesticidal properties and is hydrolyzed to chloranil [2435-53-2] in the presence of an acidic catalyst such as sulfuric acid.

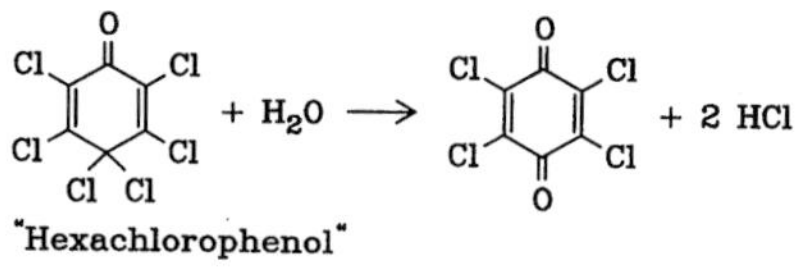

3.4. Oxidation Reactions

Like other phenols, chlorophenols can be oxidized; air oxidation forms chloroquinones, whereas oxidation with hydrogen peroxide leads to the corresponding chlorinated diphenol. Oxidation by a peroxide such as perpropionic acid in benzene at 80 °C yields chlorodiphenols (85% yield) [23]:

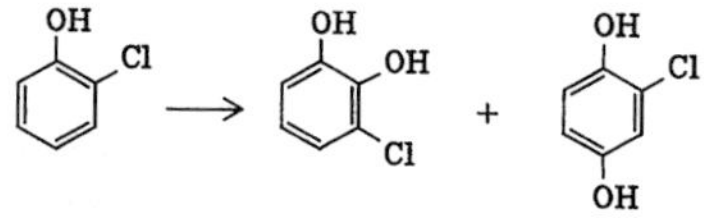

4. Production

There are two types of chlorophenols: (1) those with one to three chlorine atoms but none in a meta (3 or 5) position relative to OH and those which have four or five chlorine atoms, and (2) those with one to three chlorine atoms, at least one being in a meta position relative to OH.

4.1. Mono-, Di-, and Trichlorophenols with no Chlorine in a Meta Position; Tetrachlorophenols; and Pentachlorophenol

The mono-, di-, and trichlorophenols which have no chlorine atom in the meta (3 or 5) position relative to OH are 2-chlorophenol, 4-chlorophenol, 2,4-dichlorophenol, 2,6-dichlorophenol, and 2,4,6-trichlorophenol. The tetrachlorophenols and pentachlorophenol are also included with this type. These compounds are obtained by the direct chlorination of phenol.

The chlorination of melted phenol [108-95-2] using chlorine gas [7782-50-5] yields a mixture of ortho and para monochlorophenols with an o/p ratio of ca. 0.54 (35/65), which is scarcely dependent on temperature. The distribution between the two isomers depends especially on the nature of the chlorination reagent and the characteristics of the reaction medium. The formation of 2-chlorophenol is favored by an aprotic nonpolar solvent medium, such as hexane, carbon tetrachloride, or dichloroethane. The o/p ratio is also higher for reactions with a low concentration of phenol and a high temperature [24]. The chlorination of a 5% solution of phenol in perchloroethylene at 110 °C in the presence of 0.01 wt% diisopropylamine [108-18-9] gives a very high o/p ratio, ca. 15 [25]. Chlorinations using aqueous sodium hypochlorite [7681-52-9] at pH 10, *tert*-butyl hypochlorite [507-40-4] and alkali phenolates dispersed in an anhydrous medium [26], or chlorine in acetic acid in the presence of acetic anhydride [108-24-7] also favor the formation of 2-chlorophenol [27].

The formation of 4-chlorophenol is favored by chlorination in a polar medium, such as acetonitrile, nitrobenzene, or ether [28]–[31]. Substitution in the para position is also strongly favored by the use of sulfuryl chloride [7791-25-5] as a chlorinating agent (o/p = 0.35) [28], [31]. Certain organic sulfides, e.g., diphenyl sulfide [139-66-2], activate chlorination and favor substitution at the para position [32]. Chlorination using copper(II) chloride in aqueous hydrochloric acid in the presence of oxygen leads to 4-chlorophenol (almost 100%) [33].

Mono-, di-, and trichlorophenols are made industrially by chlorination of phenol with chlorine gas in a melt in cast-iron reactors. The isomer distribution in the product may be adjusted to market demands by the chlorination level and by recycling various intermediates. The yield, based on chlorine and phenol, is greater than 95%.

Tetrachlorophenols and pentachlorophenol are produced batchwise in nickel reactors by the chlorination of less halogenated chlorophenols in the presence of aluminum trichloride [7446-70-0] or iron trichloride [7705-08-0].

where k = relative rate constant
$k_2 = 1090$; $k_4 = 1910$; $k_{2,4} = 124$; $k'_{2,4} = 61$
$k_{2,6} = 16$; $k_{2,4,6} = 0.7$; $k'_{2,4,6} = 0.9$

4.2. Mono-, Di-, and Trichlorophenols with Chlorine in a Meta Position

The mono-, di-, or trihalogenated phenols having at least one chlorine atom in a meta position are 3-chlorophenol, 2,3-dichlorophenol, 2,5-dichlorophenol, 3,4-dichlorophenol, 3,5-dichlorophenol, 2,3,4-trichlorophenol, 3,4,5-trichlorophenol, and 2,3,6-trichlorophenol.

These compounds cannot be obtained by the chlorination of phenol, but must be made by other types of reactions, such as hydrolysis, sulfonation, hydrodechlorination, hydroxylation, and alkylation.

Hydrolysis of chlorobenzenes can be carried out in alkaline aqueous media, with or without the presence of copper salts [34]. It can be done in the vapor phase at 250–400 °C on phosphates [35] or on silica at 500–550 °C [36]. Hydrolysis is most frequently carried out in the presence of methanolic sodium or potassium at 150–200 °C and 1.0–3.0 MPa [37], [38].

Sulfonation–Desulfonation. Sulfonation of chlorobenzenes followed by the desulfonation of chlorobenzenesulfonic acids can be used to produce numerous chlorophenols [39]–[44]. The chlorobenzenes are sulfonated by using 15–20% oleum at 60–80 °C. The subsequent desulfonation is usually carried out by using 15–25% aqueous sodium hydroxide solution at 180–220 °C and 1.5–3.0 MPa.

Sandmeyer Reaction. *m*-Chlorophenols can also be made by diazotization of the corresponding chloroanilines that have been obtained either by nitration of chlorobenzenes followed by hydrogenation or by chlorination of nitrobenzenes [42].

Hydrodechlorination. Hydrodechlorination of polychlorophenols produces 3-chlorophenols. This reaction has long been carried out either in the vapor phase at 300–350 °C on an alumina catalyst [45] or in the liquid phase at 150–250 °C and 15.0–20.0 MPa, usually in the presence of sulfur [46], [47]. More recently, *catalytic* *hydrodechlorination* of polychlorophenols in organic or aqueous media with a palladium catalyst was used to obtain 3-chlorophenol and 3,5-dichlorophenol [48]. Hydrodechlorination of polychlorophenols with hydriodic acid in an aqueous or organic medium has also been reported [49].

Hydroxylation. Hydroxylation of chlorobenzenes gives a mixture of 55 wt% 4-chlorophenol, 30–35 wt% 2-chlorophenol, and 15 wt% 3-chlorophenol [50].

Hock Phenol Synthesis. The propylene alkylation of a chlorobenzene followed by oxidation and decomposition of the hydroperoxide in an acid medium produces the corresponding chlorophenol [46], [51].

The physical properties of the principal commercial chlorophenols are given in Table 2.

5. Analysis

The acidity of the phenol function in chlorophenols increases with increasing chlorination of the benzene ring. Potentiometric analysis with either soda in an aqueous medium or an amine in an organic medium is, therefore, suitable. Good sensitivity is obtained by using cyclohexylamine in ethylene glycol monoethyl ether as solvent [53], [54].

Chlorophenols are difficult to separate by gas chromatography because of their high acidity. However, the light homologs can be separated by using an acidic solid support in a glass column. The heavy homologs can be separated only after the phenol function is blocked by means of methylation, acetylation, silylation, or phthalation. Traces of chlorophenols may be determined as their trifluoroacetylated derivatives by gas chromatography with electron capture detection (limit of detection ca. 20 mg/L) [52], [54].

All phenols are correctly separated by reverse-phase HPLC on a silica column grafted with C_{18} species. Grafts of the propylamino type enable separations based on the degree of chlorination (the retention time depends on the pK_a). UV (280 nm) detection is normally used; amperometric detection is 100 times more sensitive [55]–[58].

IR spectrometry can distinguish between the different isomers, which absorb in the region 1600–400 cm^{-1} [59]. Proton NMR spectrometry can identify most of the different chlorophenols by the chemical shift and the coupling of their aromatic protons. Pentachlorophenol can be identified after the phenol group has been complexed with hexamethylphosphoric triamide [*680-31-9*]. Carbon-13 NMR is less sensitive than proton NMR, but it can provide even more useful information [60].

The individual chlorophenols can be identified and determined with great sensitivity by means of gas chromatography followed by chemical ionization mass spectrometry. The chlorophenols are first converted to their pentafluorobenzyl ether or pentafluorobenzoic ester derivatives. The analysis is then carried out on these species because of their higher electronegativity [61].

Thermal or chemical degradation of chlorophenols can produce substances that are deleterious to the environment. Certain polychlorodibenzoparadioxins, polychlorodibenzofurans, and polychlorophenoxyphenols are examples. These substances can be determined by HPLC with UV detector at a sensitivity of ca. 1 mg/kg [50], [52]. Determinations at the µg/kg level require that the sample be pretreated to remove the major portion of the chemical interferences. The sample is first extracted from an alkaline medium with hexane and then separated on an alumina liquid chromatography column.

The chosen compounds can then be collected into homologous groups [62]. The final determination is made by using GC with electron capture detection or by combined gas chromatography and mass spectrometry. The limit of detection for 2,3,7,8-tetrachlorodibenzo-p-dioxin [1746-01-6] (TCDD) is less than < 1 µg/kg [52].

6. Quality Specifications

The minimum assay for commercial chlorophenols is 98.5 wt %. More often than not, it is >99 wt % and at times even 99.5 wt %. The APHA color number is always < 100. The principal impurities are other isomers of the chlorophenol or chlorophenols with more or fewer chlorine atoms. In the majority of cases, the 2,4,5-trichlorophenol content is < 20 mg/kg. The absence of chlorobenzoparadioxins, chlorobenzofuran, and most often chlorobenzenes should be noted.

Polychlorophenoxyphenols are the major impurity (up to several percent) in the heavy chlorophenols, which are marketed as mixtures of tetrachlorophenols and pentachlorophenol. Trace quantities of chlorobenzoparadioxins and chlorobenzofuran are also found. The highly toxic compound 2,3,7,8-tetrachlorodibenzo-p-dioxin has never been detected in products made by chlorination [52]. The sodium salts of the heavy chlorophenols contain the same impurities as the parent compounds.

7. Economic Aspects

The world market, excluding the Comecon countries and China, is fairly stable and is ca. 100 000 t/a.

The production of heavy chlorophenols is 25–30 kt/a. Of this 16–20 kt/a is produced in the United States (Vulcan, Reichhold), 8–10 kt/a in Western Europe (Rhône-Poulenc, Dynamit Nobel), and 2 kt/a elsewhere in the West.

Production of light chlorophenols is ca. 60 kt/a. Of this, 18–20 kt/a is made in the United States (Dow) and 30 kt/a in Western Europe (Rhône-Poulenc, BASF, Coalite, Bayer, Chemie Linz, Esbjerg, KVK).

8. Uses

Chlorophenols are used as agricultural chemicals, pharmaceuticals, biocides, and dyes.

Agricultural Chemicals. The preparation of agricultural chemicals consumes 80–90% of chlorophenol production. The annual consumption of the herbicides 2,4-dichlorophenoxyacetic acid (2,4-D) and 2,4-dichlorophenoxy-4-butyric acid is 30–40 kt. Hoelon (Hoechst), Modown (Mobil), Ronstar (Rhône-Poulenc), and TOK E-25 (Rohm and Haas) are among the major *herbicides*. Curacron (Ciba Geigy) exemplifies the *insecticides* made from chlorophenols. Among the *fungicides* are Bayleton (Bayer), MO (Mitsui Toatsu Chemicals), and Procloraz or Sportak (Boots).

Pharmaceuticals. Pharmaceuticals derived from chlorophenols include clofibrate, ethyl 2-(4-chlorophenoxy)-2-methylpropionate (ICI), used in the treatment for high serum cholesterol, and Mervan or Aldofene, an antiinflammatory and analgesic drug (Continental Pharma).

Biocides. Biocides that are based on chlorophenols include the molluscicide Bayluscide (Bayer) and the bactericide Santophen or Chlorophen (Monsanto). Pentachlorophenol is also used to make wood and cellulose preservatives, esters for the textile and leather industries, paints, varnishes, and organic adhesives.

The sodium salt of pentachlorophenol is an active, water-soluble material. It is used as a preservative for such organic materials as freshly sawn wood (against blue rot), paints, adhesives, textiles, leather, and furs.

Dyes. Chlorophenols are used in the synthesis of anthraquinone dyes. Quinizarin, 1,4-dihydroxy-9,10-anthracenedione [*81-64-1*], is the most important example.

9. Storage and Transportation

The production, storage, transportation, and use of chlorophenols are subject to regulations and stipulations that depend on the products and on the countries concerned. Within the EEC, directive 67/548 governs classification, packing, and labeling; directive "131" 76/464 covers discharges from production plants into aquatic media; directives 77/728 and 78/631 regulate the relevant paints and pesticides, respectively.

Transportation of chlorophenols is governed by international regulations such as IMDG Code (sea), IATA-DGR (air), RID (rail), and ADR (road).

10. Environmental Considerations

Chlorophenols are hazardous to the environment and have been the subject of numerous ecological surveys of fauna and flora in aquatic, terrestrial, and aerial environments [63].

10.1. Toxicity to Fauna and Flora

Microorganisms. All chlorophenols possess bactericidal activities that increase with the degree of chlorination and attain a maximum for the trichlorophenols. Their metabolic effects include enzyme inhibitions of oxidative phosphorylation and the synthesis of adenosine triphosphate. Chlorophenols are highly toxic to algae. Trichlorophenols are at least 100 times more toxic than dichlorophenols, and pentachlorophenol is used as an algicide.

Plants. Chlorophenols possess a phytotoxicity that increases with the degree of chlorination. The absorption of chlorophenols by plants depends on the solubility of the product and pH of the environment. Terrestrial and aquatic plants can in certain cases absorb, transform, and eliminate chlorophenols without harm, but most often these plants are very sensitive to the phytotoxicity of chlorophenols.

Aquatic Organisms. Fish and other aquatic organisms absorb chlorophenols through their gills, gastrointestinal tract, or skin. Measurements over 2–4 d show that acute toxicity for invertebrates, crustacea, and freshwater and seawater fish increases with the level of chlorination. For mono-, di-, and trichlorophenols, the 2–4 d LC_{50} is 0.6–10.3 mg/L, depending on the species. For pentachlorophenol, it is 0.03–0.6 mg/L, except for certain resistant invertebrates that can survive concentrations of 5–10 mg/L.

Chlorophenols also have a long-term toxic effect at low concentrations. 2,4-Dichlorophenol and 2,4,6-trichlorophenol at concentrations of 6.2 – 10.3 mg/L affect the fertilization of sea urchin eggs, and pentachlorophenol causes renal and hepatic lesions in fish. The EPA recommends that the maximum average pentachlorophenol concentration in surface waters not exceed 0.055 mg/L. For 2,4-dichlorophenol and 2-chlorophenol, the EPA set the maximum concentrations at 2.020 and 4.380 mg/L, respectively [64]. This limitation, based on toxicity, is strengthened by the fact that chlorophenols bioaccumulate to only a small extent [65], [66].

10.2. Persistence in the Environment

Aquatic Environment. Chlorophenols may be present in the aquatic environment in many forms. They may be dissolved in free or complexed form, adsorbed on suspended inert solid or biological materials or benthic sediments, or carried in biological tissues. They are removed by biodegradation, photodecomposition, volatilization to the atmosphere, and adsorption.

Biodegradation is the principal means by which chlorophenols are removed. It must be induced because the antimicrobial activities of these products require that the bacteria adapt. Biodegradation is rapid when adapted microorganisms are already present, but pentachlorophenol is more difficult to biodegrade than the other chlorophenols.

Chlorophenols *photodecompose* when exposed to UV light, such as sunlight on the water surface.

Volatilization transfers the chlorophenol from the water to the air but does not otherwise affect it. The *adsorption* of chlorophenols on suspended or benthic materials also plays a role in the amount of chlorophenols in water. Light chlorophenols are fixed to only a small extent, but pentachlorophenol adsorbs very strongly on certain types of particles when conditions are suitable.

Terrestrial Environment. The persistence of chlorophenols in the soil depends on their adsorption – desorption characteristics. Only the adsorption of pentachlorophenol has been studied in depth. It attaches very strongly to earth and is not easily washed away by rain [67].

Atmosphere. The presence of chlorophenols in the atmosphere has been inferred from the detection of minute quantities of pentachlorophenol in rainwater and snow.

Table 3. Lethal doses (LD_{50}) of chlorophenols for rats

Chlorophenol	LD_{50}, mg/kg	
	oral	percutaneous
2-Chlorophenol	670	950
3-Chlorophenol	570	1030
4-Chlorophenol	261	1390
2,4-Dichlorophenol	580	1730
2,4,5-Trichlorophenol	820	2260
2,4,5-Trichlorophenol, sodium salt	1620	
2,4,6-Trichlorophenol	820	
2,3,4,6-Tetrachlorophenol	140	210
Pentachlorophenol	50	100
Pentachlorophenol, sodium salt	210	72

11. Toxicology

The toxicity of chlorophenols to animals depends upon the degree of chlorination, the positions of the chlorine atoms, and the purity of the sample. The LD_{50} values for rats are given in Table 3 [30]. In rats and rabbits, 50% of the administered dose of pentachlorophenol is eliminated within 24 h. In mice, 70–85% of the dose is eliminated within 4 d. The elimination of tetrachlorophenols is comparable to that of pentachlorophenol [63].

Toxic doses of chlorophenols cause convulsions, as well as other symptoms [63]. Trembling and clonic convulsions can be induced by tactile or auditory stimulation. Agitation and polypnea are observed before muscular weakness; dyspnea and coma occur and finally death [37], [66]. The toxic effects of pentachlorophenol and the tetrachlorophenols are due to the inhibition of oxidative phosphorylation.

After repeated administration, toxic doses of chlorophenols cause mainly hepatic and renal lesions. The no-effect doses were ca. 100 mg/kg for 2,4-dichlorophenol and 2,4,6-trichlorophenol and 1–5 mg/kg for pentachlorophenol [63].

Investigations of the elimination and metabolism of chlorophenols have been inconclusive. Dogs excrete 2-chlorophenol in the urine as both the free and conjugate forms (87%). However, with rabbits 2-chlorophenol is detected after the administration of chlorobenzene. 2,4-Dichlorophenol, 2,4,6-trichlorophenol, and traces of 2,4,5-trichlorophenol are found in their free and conjugate forms, following the administration of β- or γ-hexachlorocyclohexane to mice.

2,4,5-Trichlorophenol and its conjugate forms are found in the urine following the administration of 2,4,5-trichlorophenoxy-4′-butyric acid to mice; 2,3,4,5-, 2,3,4,6-, and 2,3,5,6-tetrachlorophenols, free and in their conjugate forms, are detected after the administration of pentachlorophenol [61].

Chronic effects of human exposure to low doses of chlorophenol have not been reported, but toxic doses of pentachlorophenol cause severe pathological modifications in the gastrointestinal tract, lungs, spleen, liver, kidneys, and cardiovascular system. Low levels of pentachlorophenol have been found in the urine, blood, and body fat of exposed and unexposed persons, but no explanation for their presence is available. Certain chlorophenols cause dermatoses such as photoallergic contact dermatitis, papulofollicular lesions, comedones, and sebaceous cysts [40], [41], [63], [68].

12. References

[1] A. Laurent, *Ann. Chim. Phys.* **3** (1841) 195.
[2] Grignard: *Traité de chimie organigue,* 1st ed., 1940, T6-915.
[3] Velsicol Chemical, US 4 226 813, 1979 (J. A. Albright).
[4] Dow Chemical, US 4 334 111, 1978 (R. A. Davis).
[5] Synthelabo, FR 2 475 543, 1980 (J. P. Kaplan).
[6] ANVAR, FR 2 487 343, 1980 (J. Martinez).
[7] Hitachi Manell, JP 81/10 521, 1979 (N. Ogoshi).
[8] Bayer, DE 3 019 526, 1980 (G. Rauchschwalbe).
[9] Spofa United Pharma, US 4 221 919, 1975; CS 8681, 1974.
[10] Rhône-Poulenc, EP 46 719, 1980 (S. Ratton).
[11] UBE, JP 80/62 033, 1978 (Y. Umemura).
[12] VEB Berlin Chemie, DE 2 903 343, 1979 (E. Gores).
[13] PPG, US 4 293 707, 1979 (S. B. Richter).
[14] Coalite, EP 49 126, 1980; GB 31 041, 1980 (K. Gladwin).
[15] Akad. Wiss. DDR, DD 142 333, 1979 (D. Martin).
[16] Ciba Geigy, US 4 223 166, 1977 (R. Jaeger).
[17] Nippon Synthetic Chem., JP 81/55 337, 1979 (K. Nakashima).
[18] Nippon Synthetic Chem., JP 81/55 338, 1979 (K. Gougi).
[19] Nippon Synthetic Chem., JP 81/55 339, 1979 (K. Nakashima).
[20] ICI, US 4 242 121, 1976 (A. F. Hawkins).
[21] UBE, EP 23 725, 1980; JP 79/99 914, 1979 (H. Sekioka).
[22] Dow Chemical, US 4 332 820, 1980 (D. M. Lowell).
[23] Bayer, DE 2 658 866, 1976 (H. Seifert).
[24] W. D. Watson, *J. Org. Chem.* **39** (1974) no. 8, 1161.
[25] Ciba Geigy, DE 3 318 791, 1983; CH 3234, 1982 (I. Szekely).
[26] Y. Ogata et al., *J. Chem. Soc. Perkin Trans. 2* 1984, no. 3, 451.
[27] Y. Furuya, *Chem. Ind. London* 1976, no. 15, 649.
[28] A. Campbell, D. J. Shields, *Tetrahedron* **21** (1965) 211.
[29] S. U. Zubarev et al., *J. Org. Chem. USSR (Engl. Transl.)* **4** (1968) no. 10, 1769.
[30] *Registry of Toxic Effects of Chemical Substances 1981 – 1982,* Cincinnati, Ohio, 45 226, June 1983.
[31] W. D. Watson, *J. Org. Chem.* **50** (1985) 245.
[32] W. D. Watson, *Tetrahedron Lett.* 1976, no. 30, 2591.
[33] H. P. Crocker, R. Walser, *J. Chem. Soc.* **14** (1982) 1970.
[34] Bayer, DE 2 259 433, 1974 (K. Wedemeyer).
[35] A. H. Haymob, *Zh. Obshch. Khim.* **26** (1956) 1647 – 1650.
[36] A. Galat, *J. Am. Chem. Soc.* **74** (1952) 3890 – 3891.
[37] *Rep. Gov. Chem. Ind. Res. Inst. Tokyo* **47** (1952) 327 – 336.
[38] Velsicol Chemical, US 4 094 913, 1977 (A. W. Carlson).

[39] Dow Chemical, US 2 835 707, 1956 (W. C. Stoesser).
[40] *Notices d'information médicale sur les affections figurant dans la liste européenne des maladies professionnelles* 23/7/1962 and 20/7/1966, Commission des Communautés Européennes, ed. 1969.
[41] J. C. Limasset et al., *Chlorophénols et dioxines: risques pour les travailleurs de l'industrie chimique et pour les utilisateurs*, INRS Cahier de notes documentaires 1980, note ND 1249-99-80.
[42] V. D. Simonov et al., *Zh. Prikl. Khim. (Leningrad)* **45** (1972) no. 12, 2765.
[43] Rhône-Progil, FR 2 212 827, 1972 (G. Soula).
[44] Reichhold Chemicals, GB 845 287, 1956 (F. J. Shelton).
[45] Progil, FR 2 161 861, 1971 (G. Rivier).
[46] Bayer, DE 2 443 152, 1974 (K. Wedemeyer).
[47] Bayer, DE 2 344 925, 1972 (W. Kiel).
[48] Rhône-Poulenc, EP 55 196, 1980; EP 55 197, 1980; EP 55 198, 1980 (G. Cordier).
[49] Rhône-Poulenc, EP 118 377, 1983 (G. Cordier).
[50] L. Castle et al., *J. Chem. Soc. Chem. Commun.* 1978, no. 16, 704.
[51] M. I. Faraberov et al., *Zh. Prikl. Khim.* **39** (1966) no. 9, 2094–3101.
[52] Rhône-Poulenc, internal report.
[53] W. Selig, *Mikrochim. Acta* **2** (1971) no. 3–4, 359.
[54] K. Abrahamsson, T. Min Xie, *J. Chromatogr.* **279** (1983) 199.
[55] K. Ugland, *J. Chromatogr.* **213** (1981) 83–90.
[56] J. Nair et al., *J. Liq. Chromatogr.* **6** (1983) no. 14, 2829–2837.
[57] N. G. Buckman et al., *J. Chromatogr.* **284** (1984) 441–446.
[58] R. Beauchamp, P. Boinay, *J. Chromatogr.* **204** (1981) 123–130.
[59] L. J. Bellamy: *The Infra Red Spectra of Complex Molecules,* Chapman & Hall, London 1975.
[60] M. T. Tribble, J. G. Traynham, *J. Am. Chem. Soc.* **91** (1969) no. 21, 379–388.
[61] S. S. Zhi, A. H. Duffield, *J. Chromatogr.* **284** (1984) 157–165.
[62] H. R. Buser, *J. Chromatogr.* **107** (1975) 295–310.
[63] V. P. Kozak et al., *Reviews of the Environmental Effects of Pollutants* – XI Chlorophenols – Report EPA – 600/1-179-012.
[64] K. L. E. Kaiser, *Can. J. Chem.* **60** (1982) no. 16, 2104.
[65] *Fed. Regist.* **45** (1980) no. 231, 79 318.
[66] K. Kobayashi, *Nippon Suisan Gakkaishi* **45** (1979) no. 2, 173.
[67] M. Alexander, M I. H. Aleem, *J. Agric. Food Chem.* **9** (1961) 44–47.
[68] INRS, Cahier de notes documentaires 1984, fiche toxicologique no. 196.

Chlorophenoxyalkanoic Acids

MARGUERITE L. LENG, Dow Chemical, Midland, Michigan 48640 United States

1.	Introduction 1619	5.	Toxicology and Occupational Health. 1625
2.	Production 1620	6.	Biochemical and Environmental Aspects. 1628
3.	Quality Specifications. 1623		
4.	Uses 1624	7.	References. 1629

1. Introduction

Research on chlorinated phenoxy compounds during World War II led to the discovery of the selective *herbicidal action* of 2,4-dichlorophenoxyacetic acid (2,4-D) for control of broadleaf weeds in grasses and related crops, such as cereal grains, rice, and sugarcane. Use patents issued before 1950 included several in the United States assigned to American Chemical Paint Co. (Amchem, now Union Carbide) and Dow Chemical [1].

Although all patents have expired and many new plant growth regulator type weedkillers have been developed, phenoxy compounds still comprise a large sector of the worldwide herbicide market. The estimated total *capacity* for manufacture of chlorophenoxyalkanoic acids and derivatives approached 2×10^5 t in 1982, with a value of about \$ 1.75/kg (\$ 0.80/lb.). Total production decreased about 10% worldwide in 1983 because of weak prices and lower demand.

The *principal chlorophenoxyalkanoic acids used as herbicides* are listed in Table 1. They are formulated as water-soluble salts and oil-soluble or emulsifiable esters. Major products, in terms of volume, are 2,4-D and MCPA followed by mecoprop and dichlorprop. Most production of 2,4,5-T and fenoprop has been discontinued because of the high toxicity of the dioxin contaminant formed during manufacture of the 2,4,5-trichlorophenol intermediate used for making these herbicides.

Detailed information on individual chlorophenoxyalkanoic acids and their derivatives can be found in recent editions of herbicide manuals and handbooks [1]–[4] and in extensive reviews published since 1970 [5]–[18].

Table 1. Chlorophenoxyalkanoic acids used as herbicides *

Common name	Chemical name (IUPAC)	CAS registry number	Formula	M_r
2,4-D	(2,4-dichlorophenoxy)acetic acid	[94-75-7]	$C_8H_6Cl_2O_3$	221.04
2,4-DB	4-(2,4-dichlorophenoxy)butyric acid	[94-82-6]	$C_{10}H_{10}Cl_2O_3$	249.09
Dichlorprop	2-(2,4-dichlorophenoxy)propionic acid	[120-36-5]	$C_9H_8Cl_2O_3$	235.05
MCPA **	(4-chloro-o-tolyloxy)acetic acid	[94-74-6]	$C_9H_9ClO_3$	200.63
MCPB **	4-(4-chloro-o-tolyloxy)butyric acid	[94-81-5]	$C_{11}H_{13}ClO_3$	228.68
Mecoprop **	2-(4-chloro-o-tolyloxy)propionic acid	[93-65-2]	$C_{10}H_{11}ClO_3$	214.66
2,4,5-T	(2,4,5-trichlorophenoxy)acetic acid	[93-76-5]	$C_8H_5Cl_3O_3$	225.49
Fenoprop	2-(2,4,5-trichlorophenoxy)propionic acid	[93-72-1]	$C_9H_7Cl_3O_3$	269.53

* Melting point data in Table 2.
** The trivial names MCPA, MCPB, and mecoprop are derived from former chemical nomenclature using 2-methyl-4-chlorophenoxy as the basic unit.

2. Production

Pokorny Process. Most technical phenoxy acids are produced by using the simple reaction first described in 1941 by POKORNY [19] for synthesis of 2,4-D and 2,4,5-T. The appropriate chlorophenol is reacted with a chloroalkanoic acid under alkaline conditions at about 100 °C in an aqueous medium or in an organic solvent, such as toluene, xylene, or chlorobenzene:

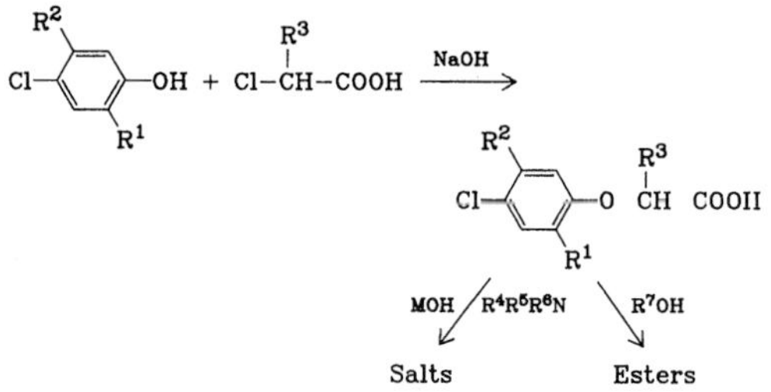

where R^1 is CH_3 or Cl, R^2 is H or Cl, R^3 is H or CH_3, $R^4R^5R^6N$ is an alkylamine, R^7OH is an alcohol, and MOH is an alkali metal base. The resultant salt is hydrolyzed to the free acid, which is purified by recrystallization from an organic solvent or by steam distillation. Chlorophenoxy acids prepared from 2-chloropropionic acid contain an asymmetric carbon atom and are isolated as racemic mixtures. A schematic diagram of the Pokorny process used to produce 2,4-D, its salts, and its esters is shown in Figure 1.

Various methods for synthesis of chlorophenoxyalkanoic acids, esters, and salts have been reviewed [10, pp. 107–123]. The nature and quantity of byproducts depend largely on the purity of the starting chlorophenols (→ Chlorophenols). For example, the 2,4-dichlorophenol used for making 2,4-D, dichlorprop, and 2,4-DB is prepared by chlorination of phenol and contains 2,6-dichlorophenol and 2,4,6-trichlorophenol as

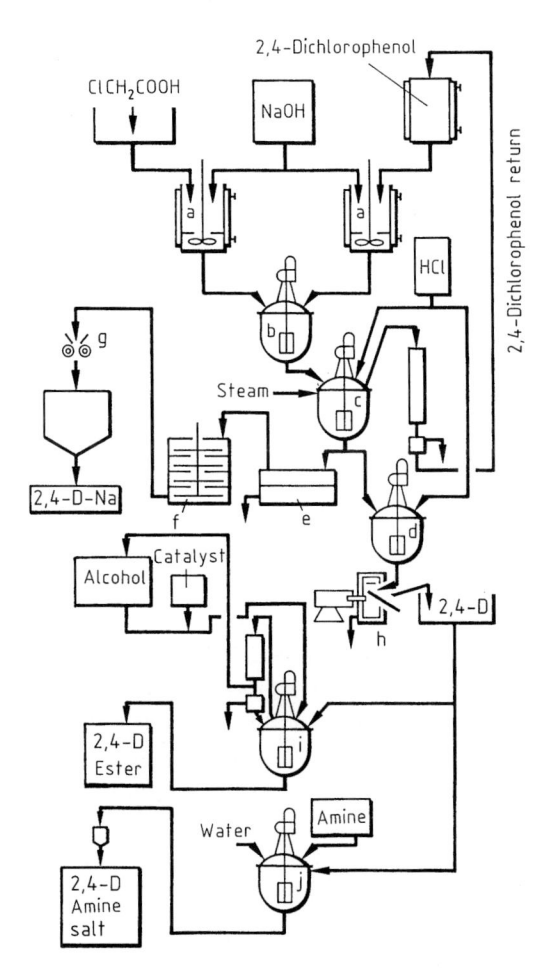

Figure 1. Production of 2,4-dichlorophenoxyacetic acid (2,4-D), as well as its salts and esters
a) Neutralization vessel; b) Reaction kettle; c) Acidification kettle; d) Precipitation vessel; e) Suction filter; f) Drier; g) Mill; h) Filter; i) Esterification kettle; j) Amination kettle

major impurities. These byproducts are not formed if the chlorination is carried out in liquid sulfur dioxide below its boiling point [10].

Similarly, the chlorophenol used to manufacture MCPA, MCPB, and mecoprop is obtained by chlorination of *o*-cresol, a starting material in abundant supply in Europe particularly in Britain, as a product of coal tar distillation. Although chlorination occurs mostly at the 4-position, large amounts of 6-chloro- and 4,6-dichloro-2-methylphenol are also produced when chlorine or alkali hypochlorites are used. A much purer product is obtained when sulfuryl chloride is used for chlorination [10].

A completely different route is used for production of 2,4,5-trichlorophenol by *hydrolysis of 1,2,4,5-tetrachlorobenzene*:

where R^1 is H or CH_3. $R^2R^3R^4N$ is an alkylamine, R^5OH is an alcohol, and MOH is an alkali metal base. One chlorine substituent is replaced by using alcoholic alkali under pressure, with careful control of temperature to avoid formation of highly toxic 2,3,7,8-tetrachlorodibenzo-*p*-dioxin (TCDD) by exothermic condensation of two molecules of 2,4,5-trichlorophenol. Trace amounts of TCDD are carried over into 2,4,5-T and, to a lesser extent, into fenoprop (also known as silvex or 2,4,5-TP). The impurities in technical-grade 2,4,5-T may vary among producers and among batches of the same product [9, p. 38].

Chlorination of phenoxyacetic acid is another important synthetic route for production of 2,4-D. However, the product may contain a high level of 2,4-dichlorophenol, which has a very strong odor [5]. Chlorination of tolyloxyacetic acid produces mainly the 4-chloro derivative, with small amounts of 6-chloro and other impurities [10].

Phenoxybutyric acid derivatives, such as 2,4-DB and MCPB, are generally prepared by using *γ-butyrolactone* to acylate the chlorophenols [5].

An alternate route to phenoxy acids is by formation of the appropriate *chlorophenolate* and reaction with an ester of the appropriate chloroalkanoic acid [10]. The resulting phenoxy ester can be saponified and acidified to release the free acid, or the phenoxy ester can be used directly for preparation of oil-soluble herbicide formulations. However, these reaction conditions could lead to condensation of chlorophenols to dioxins, and may have been responsible for the di-, tri-, and tetrachlorodibenzo-*p*-dioxins found in certain 2,4-D products in Canada [20]. The 2,3,7,8-tetrachloro isomer has not been detected in products other than 2,4,5-T and fenoprop.

Platinum-catalyzed oxidation of 2-phenoxyethanols has also been used to synthesize phenoxyacetic acids [21]. However, this method may not be commercially feasible because of cost.

Volatile products containing isopropyl or butyl esters of 2,4-D and 2,4,5-T were produced in large quantities prior to 1970, particularly for use as defoliants in the Vietnam conflict. However, most countries now permit use of only *low-volatile products* containing salts or derivatives with a long side-chain, such as isooctyl, butoxyethyl, or PGBE (polypropylene glycol butyl ether) esters of chlorophenoxyalkanoic acids [7], [9], [10].

Water-soluble salts of phenoxy acids are formed with alkali metal cations, ammonia, and organic amines, such as dimethylamine, triethylamine, or triisopropanolamine.

Magnesium and calcium salts of 2,4-D and MCPA are considerably less soluble in water than are sodium and potassium salts [10]. Oil-soluble salts are prepared from long-chain amines derived from fatty acids [1], [10].

Process *waste streams* are treated by a combination of methods to reduce levels of chlorophenolic and chlorophenoxy compounds to low ppm values prior to discharge to public waters. Methods include steam stripping, solvent extraction, carbon adsorption, resin adsorption, chlorine oxidation, biological oxidation, and incineration.

Current principal *manufacturers* of technical-grade acids and derivatives are Akzo Zout in the Netherlands, BASF and Bayer in the Federal Republic of Germany, Chemie-Linz in Austria, Dow in the United States and Brazil, Koge in Denmark, A. H. Marks and May & Baker in the United Kingdom, Rhône-Poulenc in France, and Vertac in the United States. About 25% of the world capacity in 1982 was located in East Europe, mainly in Czechoslovakia, German Democratic Republic, Hungary, Poland, Romania, and USSR. Phenoxy herbicides are also produced in Argentina, Australia, India, Japan, Mexico, New Zealand, the People's Republic of China, the Philippines, and Taiwan. Several major manufacturers have discontinued production or sold their facilities since 1982.

End-use products are formulated by most manufacturers of technical chlorophenoxyalkanoic acids and derivatives, as well as by many additional distributors worldwide. Numerous trade names have been listed for phenoxy herbicide products [1]–[4], [7], [10], [18], but many such names were discontinued or were transferred to other companies in recent years.

3. Quality Specifications

International specifications for phenoxy herbicides and other pesticides have been developed by the United Nations Food and Agriculture Organization (FAO). Methods of analysis have been tested and approved by the Collaborative International Pesticides Analytical Council (CIPAC) and the Association of Official Analytical Chemists (AOAC).

Updated specifications for phenoxyalkanoic herbicides were consolidated in a booklet issued by FAO in 1984 [22]. The values for technical acids summarized in Table 2 have "full" status for MCPA and 2,4,5-T and "provisional" status for the others. Other parameters for identification include IR spectra, GLC retention times, and R_f values for TLC analyses. The FAO also lists specifications for salts and esters, aqueous salt solutions, emulsifiable ester concentrates, and mixtures containing phenoxy salts or esters, all given in terms of the acids released by hydrolysis of these derivatives. Complete names are required of the base components of salts and the alcohol components of esters in phenoxy herbicide products.

The U.S. Environmental Protection Agency (EPA) has issued regulations and guidelines for registration of pesticide products [23]. These include chemistry requirements

Table 2. International FAO specifications for technical chlorophenoxyalkanoic acids

Common name	mp, °C	Acid (anhydrous)[a], g/kg min.	Free phenol[b], g/kg max.
2,4-D	137–141	890 ± 25	5
2,4-DB	114–119	890 ± 30	20
Dichlorprop	113–118	890 ± 25	15
MCPA	118–119[c]	840 ± 20	10
MCPB	100[c]	840 ± 30	30
Mecoprop	94–95[c]	840 ± 40	15
2,4,5-T[d]	150–156	910 ± 30	5
Fenoprop[d]	175–181	890 ± 40	15

[a] The water content of dry acids should not exceed 15 g/kg; the approximate water content should be stated for wet acids containing more than 15 g/kg.
[b] The free phenol content is expressed as the parent phenol in terms of the anhydrous phenoxy acid content.
[c] As listed in [1], [2]; may be 99–107 °C for technical-grade MCPA.
[d] Maximum content of 2,3,7,8-tetrachlorodibenzo-p-dioxin (TCDD) is 1×10^{-8} expressed as 0.01 mg/kg of the 2,4,5-T or fenoprop content on an anhydrous basis.

for the identification of all impurities and byproducts down to 0.1% in technical-grade products.

For many years, the *analytical methods* used for assessing the quality of phenoxy herbicides involved simple acid–base titration, or determination of total chlorine released by pyrolysis of technical-grade acids and derivatives [2]–[4], [18]. These nonspecific, and therefore inaccurate methods have been superseded by GLC procedures [1], [24], [25], and, recently, by isomer-specific high-pressure liquid chromatography [26]–[28]. Similar methods for other phenoxy herbicides are undergoing collaborative testing and will likely be required in the future [10].

4. Uses

Distribution, marketing, and use of pesticides are closely regulated in most countries, either through national statutes or by indirect reliance on reviews by expert committees under the auspices of international agencies, such as FAO, the World Health Organization (WHO), and the Codex Food Standards Programme of the United Nations.

Labels on containers of phenoxy herbicides are registered individually in most exporting and importing countries (and even in states within countries). These labels specify the nature and content of active ingredients; give precautions to be taken to avoid hazard to humans, livestock, wildlife, or the environment; and describe precise conditions for each approved use.

Phenoxy herbicides are used primarily for *selective control of broadleaf weeds* in cereal grains, pastures, and turf; and for removing unwanted brushy species in rangeland, forests, and noncropland. Rates of application range from as low as 0.25 kg/ha in grain crops to as high as 16 kg/ha for spot treatment of individual trees in rights-of-way. Very dilute solutions of 2,4-D and fenoprop derivatives have also been used as plant growth regulators in fruit orchards.

The choice of products depends on the species of plants to be controlled [1]–[4], [6]. For example, 2,4,5-T and fenoprop are effective on woody plants and herbaceous weeds that are resistant to 2,4-D. Dichlorprop, fenoprop, and mecoprop are effective for controlling chickweed and other unwanted plants in lawns and turf. The butyric acid derivatives 2,4-DB and MCPB can be used on sensitive crops, such as peas, peanuts, soybeans, and seedling forage legumes (alfalfa and clover), because they are inactive as herbicides until converted to the acetic acid derivatives by β-oxidation of the side-chain [6].

Phenoxy herbicides are applied alone or as mixtures with other herbicides in solutions, dispersions, or emulsions in water or oil; equipment that produces large droplets is used to avoid spray drift. Although volatile derivatives, such as isopropyl and butyl esters of 2,4-D and 2,4,5-T, are effective for controlling weeds and brush in noncropland, treatments with low-volatile long-chain esters and amine salts are less likely to cause damage to susceptible off-target crops or desirable vegetation. Granular salt formulations of 2,4-D or fenoprop have been used for aquatic weed control, but longer-acting ester formulations are more effective.

Formulated phenoxy herbicide products generally contain inert ingredients to prevent precipitation of salts when these formulations are diluted with hard water, or to help disperse or emulsify esters in oil or oil/water mixtures. Oil concentrates are less corrosive than emulsifiable concentrates or salt solutions, but are more flammable and require more restrictive labeling for safe transport, storage, and handling.

Only inert ingredients approved by the EPA are permitted in products to be used on food crops in the United States [23].

5. Toxicology and Occupational Health

Numerous toxicological studies have been conducted with phenoxy herbicides in laboratory animals, livestock, birds, and fish. In addition, pharmacokinetic studies in animals and in humans have demonstrated that these compounds do not accumulate in the body. Detailed summaries of findings are provided in several extensive reviews [7], [9], [11]–[18], [29]–[31].

Chlorophenoxyalkanoic acids, derivatives, and formulated products are *classified as moderately toxic*, based on acute studies using single large doses administered by oral, dermal, or inhalation routes in various species of animals [1]–[4]. According to World Health Organization (WHO) guidelines, 2,4-D and 2,4,5-T are in class II (moderately hazardous) with acute oral LD_{50} values of 375 and 500 mg/kg of body weight in rats [32]. Other phenoxy herbicides are in class III (slightly hazardous), with acute oral LD_{50} values in mg/kg for rats as follows: fenoprop (650), MCPB (680), 2,4-DB (700), MCPA (700), dichlorprop (800), and mecoprop (930). Acute oral toxicity values for salts and esters are comparable to those for the acids, on an acid equivalent basis [7], [9].

Dogs appear to be more sensitive to chlorophenoxyalkanoic acids than rats, whereas birds are less sensitive [7] and livestock can tolerate higher doses without apparent effect [9]. Phenoxy herbicides are not toxic to bees. The LC_{50} in fish may range from less than 1 mg/L for esters to greater than 100 mg/L for acids and salts [7].

Dry acids are only slightly irritating to eyes and skin, but concentrated solutions or wetted powders may cause burns if contact is prolonged or repeated [31]. In rabbits, no significant adverse systemic effects resulted from exposure for 3 weeks to dilute solutions of formulated products containing dimethylamine salt, isooctyl ester, or butyl ester of 2,4-D [31].

Although phenoxy compounds are not acutely toxic in dermal and inhalation studies, potential occupational exposure is greater by these routes than by ingestion. Adequate precautions should be taken to avoid contamination of skin with technical-grade materials or concentrates, and to avoid prolonged breathing of the vapors or dusts. Occupational exposure limits have been set at 10 mg/m^3 for 2,4-D and 2,4,5-T in both the United States (TWA) and the Federal Republic of Germany (MAK).

Effects of 2,4-D in humans were reviewed in 1984 by an international committee convened by WHO [17]. Signs and symptoms of acute overexposure occurred only after ingestion of large amounts of 2,4-D products in suicide cases, or where poor occupational hygiene led to pronounced dermal exposure. The acute no-observed–adverse-effect level for biological effects was estimated to be as high as 36 mg/kg of body weight in humans for 2,4-D or its equivalent in alkali salts, amine salts, or esters. On the basis of available studies on amounts of 2,4-D or 2,4,5-T absorbed through occupational exposure, the margin of safety is at least 360-fold for spraying crews and much greater than 1000-fold for the general population in areas where 2,4-D is used [17].

Little or no effects were seen in livestock or laboratory animals other than dogs given repeated oral doses up to 50 mg/kg of 2,4-D, 2,4,5-T, MCPA, or fenoprop, as acids, salts, or esters [7], [9]. Reversible embryotoxic or fetotoxic effects were seen in offspring of pregnant rats and mice given daily doses above 40 mg/kg during organogenesis [7], [9], [11]–[18]. However, true teratogenic effects, such as cleft palate, were seen only in mice given 2,4,5-T, possibly due to high levels of TCDD contaminant in some samples of technical-grade acid and esters [9, p. 115].

Dietary feeding studies have also been conducted with phenoxy acids and derivatives in rats, dogs, and mice for periods ranging from 3 months to 2.5 years to evaluate potential for causing chronic toxicity, cancer, or reproductive effects. Details are provided in extensive reviews by National Research Council of Canada [7], WHO [11]–[18], and others [30]–[37]. The major findings are as follows:

In chronic feeding studies with *2,4-D* conducted by U.S. Food and Drug Administration [33], dogs and rats were fed for 2 years on diets containing up to 500 and 1250 ppm technical-grade acid, respectively, and pregnant rats were fed up to 1500 ppm in the diet during three generations of reproduction with two litters per generation. No significant effects were seen in any of the dogs, nor in rats fed up to 625 ppm in the diet equivalent to a dosage level of 31 mg kg^{-1} d^{-1} in this species. There was no effect on fertility nor on average litter size at any dose, but at 1500 ppm the percentage of young born that survived to 21 days was sharply reduced and the weaning weight was depressed. On the basis of these studies, the acceptable daily intake (ADI) for 2,4-D in humans was estimated by the 1971 WHO Expert Committee on Pesticide Residues at 0.3 mg kg^{-1} d^{-1} with a 100-fold safety factor over the no-effect level in rats [12]. Additional studies are under way in the United States to fulfill current EPA requirements for registration of pesticides used in food crops [23].

Four long-term feeding studies have been conducted with *2,4,5-T*, each study at dosage levels of 3, 10, and 30 mg kg^{-1} d^{-1} in rats. Effects were seen at the high dose, and to a lesser extent at the middle dose, in studies conducted by Dow in the United States using purified 2,4,5-T ($<5 \times 10^{-10}$ TCDD) in the diet of rats for 2 years [34] or during three generations of reproduction [35]. On the other hand, all dose groups were comparable to controls in studies conducted in the Federal Republic of Germany using technical-grade 2,4,5-T (5×10^{-8} TCDD) in the diet of rats from before conception through weaning to age 130 weeks or through three generations of reproduction [14]. On the basis of these studies, the 1981 WHO Expert Committee estimated the ADI in humans at 0.03 mg kg^{-1} d^{-1} for 2,4,5-T containing no more than 1×10^{-8} TCDD [16].

Unpublished studies with *fenoprop* have been summarized in two reviews [30], [31]. The dosage levels causing no significant toxicological effect were 7 mg kg^{-1} d^{-1} in 90-day feeding studies with the sodium salt and PGBE esters in rats and the potassium salt in dogs, and essentially 2.6 mg kg^{-1} d^{-1} in 2-year feeding studies with the potassium salt in both rats and dogs.

Subchronic feeding studies with *MCPA* and *mecoprop* have also been reviewed [7], [9]. When diethanolamine salts of these compounds were fed to rats at up to 2500 ppm in the diet for 7 months, the only effect noted was unspecified mortality, mainly due to infection at the maximum dose [36]. On the other hand, no mortalities were reported when MCPA as the sodium salt or mecoprop as the acid was fed at levels up to 3200 ppm in the diet of rats for 3 months. The only effects noted were reduced food consumption and decreased weight gain at the highest dose [37]. Additional studies with MCPA are under way to meet current requirements in the United States for reregistration of pesticides used in food crops [23].

The *potential human carcinogenicity* of chlorophenoxyalkanoic acids and their derivatives has been evaluated by several expert groups. Committees convened by the International Agency for Research on Cancer (IARC) reviewed all of the above studies in rats, numerous studies in mice conducted by U.S. National Cancer Institute, and several epidemiological studies in humans occupationally exposed to formulated products [18]. Both *2,4-D* and *2,4,5-T* and their esters were assigned to group 3 (evidence inadequate for carcinogenicity to humans and animals, and for activity in short-term tests) [18, 1982, pp. 101–103, 235–238]. For *MCPA*, no adequate data were available to evaluate carcinogenicity to experimental animals or humans [18, 1983, pp. 255–269]. On the other hand, *TCDD* was classified in group 2B (sufficient evidence for carcinogenicity to animals and inadequate evidence in humans and in short-term tests [18, 1982, pp. 238–243], whereas *chlorophenols* and *phenoxyacetic acid herbicides* (occupational exposure) were generically classified in group 2B on the basis of only limited evidence for carcinogenicity to humans [18, 1982, pp. 88–89, 211–212].

6. Biochemical and Environmental Aspects

The following observations summarize findings reported in extensive reviews on biochemistry, metabolism, degradation, and residues of phenoxy herbicides [6], [7], [9], [11]–[18], [31], [38], [39].

The *solubility* of chlorophenoxyalkanoic acids ranges from 45 to 1300 mg/L in water and from 0.1 to 153 g/100 mL in CCl_4 and ethanol, respectively [6]. These compounds are relatively strong acids, with pK values ranging from 2.64 for 2,4-D to 6.21 for MCPB [6], and exist mainly as organic anions at the pH of living tissues. Salts dissociate in aqueous media, and esters are rapidly hydrolyzed to acids by enzymatic action in animals, plants, and soil [6].

Phenoxy acids and derivatives are rapidly *absorbed* from the gastrointestinal tract in animals and are excreted in urine, largely as free acids, with a half-life of less than 1 day [38]. Doses greater than 50 mg/kg of body weight may saturate the excretory mechanism, particularly in dogs and pigs [9]. Fenoprop is excreted more slowly than 2,4-D and 2,4,5-T, and appears partly in conjugated form in urine. In humans, small oral doses of 2,4-D or 2,4,5-T (5 mg/kg) or fenoprop (1 mg/kg) were handled much as in rats, indicating little likelihood for accumulation from repeated exposure [31]. Absorption into the body is slower from dermal exposure than from oral ingestion [17], [39].

The phenoxy ether link is stable in plants and animals, but is cleaved by bacterial action in soil and in the rumen of cattle and sheep [38]. The rate of cleavage is inhibited by substitution of a third chlorine at the meta position on the ring and is sterically hindered by the angular methyl group in the 2-propionic acid side-chain. Thus, 2,4,5-T and fenoprop accumulate in the liver and kidney of cattle and sheep ingesting high levels in the diet [38].

The first step in *degradation* of chlorophenoxy acids and chlorophenols occurs by hydroxylation at the 4-position with shift of the 4-chloro group to the 3- or 5-position (NIH shift). This reaction also is inhibited by the presence of a *m*-chloro substituent; thus, residues of 2,4,5-trichlorophenol appeared in the milk of cows and in the liver and kidney of cattle maintained on feed containing 100 or 300 ppm 2,4,5-T, respectively [38].

The rate of *microbial degradation* of chlorophenoxyalkanoic acids in the environment depends on a number of factors, including sunlight, temperature, moisture and organic matter in soil, and whether the organisms were adapted by repeated treatment [6], [7], [9]. Under field conditions favorable for degradation, 2,4-D disappears within about 2–3 weeks, whereas other phenoxy herbicides disappear more slowly [6]. The average half-lives in a laboratory study were 4, 10, 17, and 20 days for 2,4-D, dichlorprop, fenoprop, and 2,4,5-T, respectively, when added as dimethylamine salts at about 5 ppm in three soils containing partially decomposed litter obtained from two forest sites and one grassland site in Oklahoma [40].

Photodegradation is rapid when phenoxy herbicides are exposed to sunlight in water containing dissolved organic matter, and the resultant chlorophenols are photolyzed even more rapidly than the parent acids [7], [9]. The dioxin impurity TCDD is also degraded rapidly by sunlight, particularly in the presence of its herbicide carrier on the surface of leaves or in the presence of other hydrogen donors dissolved in water [7].

Plants treated with phenoxy herbicides rapidly hydrolyze esters and salts to the parent acids, which form ether-soluble *conjugates* with amino acids or water-soluble conjugates with sugars, depending on the relative susceptibility or resistance of the plants to the herbicides [6]. *Analyses for residues* in treated crops and environmental samples involve extraction with organic solvents under acid conditions [10] or with polar solvents under alkaline conditions [41]. Quantification is best accomplished by conversion to volatile esters and measurement by electron capture gas chromatography (EC/GC). Other methods include colorimetry and thin-layer chromatography [4], [10].

Tolerances have been established in the United States for residues of 2,4-D and MCPA in a number of food and feed crops, and in milk and meat [42]. Permitted levels range from 0.05 ppm (the detection limit) in strawberries, poultry, and eggs to 1000 ppm on grass immediately after treatment with 2,4-D at high rates. Similarly, international maximum residue limits (MRLs) have been recommended for 2,4-D and 2,4,5-T by FAO [11]–[13], [15], and have been accepted by many member countries attending annual meetings of the United Nations Codex Committee on Pesticide Residues [43]. However, residues of chlorophenoxyalkanoic acids have seldom been detected in monitoring studies conducted on domestic and imported foods in a number of countries [44].

Guidelines for *drinking water* quality issued in 1982 by WHO recommend a maximum concentration level of 100 µg/L for 2,4-D in water [45].

7. References

[1] *Herbicide Handbook of the Weed Science Society of America (WSSA)*, 5th ed., WSSA, Champaign, Ill., 1983, pp. 128, 141, 167, 292, 295, 297, 429, 443.

[2] D. Hartley, H. Kidd (eds.): *Agrochemicals Handbook*, Royal Society of Chemistry, Nottingham, England, 1983, updated 1984.

[3] British Crop Protection Council, C. R. Worthing (ed.): *Pesticide Manual, a World Compendium*, 7th ed., Lavenham Press, Lavenham, Suffolk, England, 1983, 655 pp.

[4] E. Y. Spencer (ed.): *Guide to the Chemicals Used in Crop Protection*, 7th ed., Publication 1093 of Research Branch of Agriculture Canada, London, Ontario, Canada, 1982, 595 pp.

[5] N. N. Melnikov: "Aryloxalkylcarboxylic Acids," Chap. 14 in "Chemistry of Pesticides," *Residue Rev.* **36** (1971) 157–176.

[6] M. A. Loos: "Phenoxyalkanoic Acids," in P. C. Kearney, D. D. Kaufman (eds.): *Herbicides: Chemistry, Degradation and Mode of Action*, 2nd ed., vol. **1**, Marcel Dekker, New York 1975, pp. 1–128.

[7] National Research Council of Canada: *Phenoxy Herbicides – Their Effects on Environmental Quality*, NRC no. 16075, NRCC/CNRC Publications, Ottawa, Canada, 1978, 440 pp.

[8] National Research Council of Canada: *2,4-D: Some Current Issues,* NRC no. 20 647, NRCC/CNRC Publications, Ottawa, Canada, 1983, 99 pp.

[9] R. W. Bovey, A. L. Young: *The Science of 2,4,5-T and Associated Phenoxy Herbicides,* Wiley-Interscience, New York 1980, 462 pp.

[10] S. S. Que Hee, R. G. Sutherland: *The Phenoxyalkanoic Herbicides, Chemistry, Analysis, and Environmental Pollution,* vol. **1**, CRC Series in Pesticide Chemistry, CRC Press, Boca Raton, Fla., 1981, 321 pp.

[11] FAO/WHO: *1970 Evaluations of Some Pesticide Residues in Food,* WHO/Food Add./71.42, FAO, Rome 1971, pp. 59–86, 459–477 (2,4-D and 2,4,5-T, respectively).

[12] FAO/WHO: *1971 Evaluations of Some Pesticide Residues in Food,* WHO Pesticide Residue Series no. 1, WHO, Geneva 1972, pp. 83–97 (2,4-D).

[13] FAO/WHO: *1974 Evaluations of Some Pesticide Residues in Food,* WHO Pesticide Residue Series no. 4, WHO, Geneva 1975, pp. 159–183 (2,4-D).

[14] FAO/WHO: *1975 Evaluations of Some Pesticide Residues in Food,* WHO Pesticide Residue Series no. 5, WHO, Geneva 1976, pp. 173–188 (2,4-D).

[15] FAO/WHO: *Pesticide Residues in Food, 1979 Evaluations,* FAO Plant Production and Protection Paper 20 Suppl., FAO, Rome 1980, pp. 469–498 (2,4,5-T).

[16] FAO/WHO: *Pesticide Residues in Food, 1981 Evaluations,* FAO Plant Production and Protection Paper 42, FAO, Rome 1982, pp. 494–501 (2,4,5-T).

[17] International Programme on Chemical Safety (IPCS): *Environmental Health Criteria 29, 2,4-Dichlorophenoxyacetic Acid (2,4-D),* WHO, Geneva 1984, 151 pp.

[18] IARC: *Monographs on the Evaluation of the Carcinogenic Risk of Chemicals to Man,* IARC, Lyon, France, vol. **15**, 1977, pp. 111–138, 273–299; vol. **1–29** suppl. 4, 1982, pp. 88–89, 101–103, 211–212, 235–243; vol. **30**, 1983, pp. 255–269.

[19] R. Pokorny: "New Compounds, Some Chlorophenoxyacetic Acids," *J. Am. Chem. Soc.* **63** (1941) 1768.

[20] W. P. Cochrane, J. Singh, W. Miles, B. Wakeford, *J. Chromatogr.* **217** (1981) 289–299.

[21] H. Fiege, K. Wedemeyer, *Angew. Chem. Int. Ed. Engl.* **20** (1981) no. 9, 783.

[22] FAO Specifications for Plant Protection Products, "Phenoxyalkanoic Herbicides," AGP:CP/100, FAO, Rome 1984.

[23] EPA, "Data Requirements for Pesticide Registrations, Final Rules," CFR 40, 1984, 158. *Fed. Regist.* **49** (1984) 42 856–42 905.

[24] Collaborative International Pesticides Analytical Council (CIPAC): *CIPAC Handbook 1, 1 A, 1 B,* Heffers Printers, Cambridge, England, 1970, 1980, 1983, respectively.

[25] Collaborative International Pesticides Analytical Council (CIPAC): *CIPAC Proceedings 1980, 1981,* Heffers Printers, Cambridge, England, 1980, 1981.

[26] N. E. Skelly, T. S. Stevens, D. A. Mapes, *J. Assoc. Off. Anal. Chem.* **60** (1977) 868–872.

[27] T. S. A. Stevens, N. E. Skelly, R. B. Grorud, *J. Assoc. Off. Anal. Chem.* **61** (1978) 1163–1165.

[28] W. P. Cochrane, M. Lanouette, J. Singh, *J. Assoc. Offic. Anal. Chem.* **66** (1983) 804–809.

[29] J. M. Way, *Residue Rev.* **26** (1969) 37–62.

[30] W. R. Mullison: "Some Toxicological Aspects of Silvex," *Proc. South. Weed Control Conf.* **19** (1966) 426–435.

[31] P. J. Gehring, J. E. Betso: "Phenoxy Acids: Effects and Fate in Mammals," in C. Ramel (ed.): "Chlorinated Phenoxy Acids and Their Dioxins," *Ecol. Bull. (Stockholm)* **27** (1978) 122–133.

[32] "WHO Recommended Classification of Pesticides by Hazard – Guidelines for Classification 1984–1985," VBC/84.2, WHO, Geneva 1984.

[33] W. H. Hansen, M. L. Quaife, R. T. Habermann, O. G. Fitzhugh, *Toxicol. Appl. Pharmacol.* **20** (1971) 122–129.
[34] R. J. Kociba, D. G. Keyes, R. W. Lisowe, R. P. Kalnins et al., *Food Cosmet. Toxicol.* **17** (1979) 205–221.
[35] F. A. Smith, F. J. Murray, J. A. John, K. D. Nitschke et al., *Food Cosmet. Toxicol.* **19** (1981) 41–45.
[36] M. R. Gurd, G. L. M. Harmer, B. Lessel, *Food Cosmet. Toxicol.* **3** (1965) 883–885.
[37] H. G. Verschuuren, R. Kroes, E. M. Den Tonkelaar, *Toxicology* **3** (1975) 349–359.
[38] M. L. Leng: "Comparative Metabolism of Phenoxy Herbicides in Animals," in G. W. Ivie, H. W. Dorough (eds.): *Fate of Pesticides in Large Animals*, Academic Press, New York 1977, pp. 53–76.
[39] M. L. Leng, J. C. Ramsey, W. H. Braun, T. L. Lavy: "Review of Studies with 2,4,5-T in Humans Including Applicators Under Field Conditions," in J. R. Plimmer (ed.): *Pesticide Residues and Exposure*, ACS Symposium Series 182, Am. Chem. Soc., Washington, D.C., 1982, pp. 133–156.
[40] J. T. Altom, J. F. Stritzke, *Weed Sci.* **21** (1973) 556–560.
[41] D. J. Jensen, R. D. Glass: "Analysis of Residues of Acidic Herbicides," in H. A. Moye (ed.): *Analysis of Pesticide Residues*, J. Wiley & Sons, New York 1981, pp. 223–261.
[42] EPA, CFR 40, "Tolerances for 2,4-D," Section 180.142; "Tolerances for MCPA," Section 180.339, Washington, D.C., 1986.
[43] Codex Alimentarius Commission (CAC), Joint FAO/WHO Food Standards Programme: "Codex Maximum Limits for Pesticide Residues," CAC, 1st ed., vol. 13, FAO, Rome 1983, p. 20-iv (2,4-D).
[44] International Group of National Assoc. of Pesticide Manufacturers (GIFAP), "Pesticide Residues in Food," GIFAP, Brussels, Belgium, Mar., 1984, 26 pp.
[45] WHO: "Recommendations," *Guidelines for Drinking Water Quality*, vol. **1**, EFP/82.39, WHO, Geneva 1982, p. 82.

Choline

YOSHIHISA SUZUKI, Central Research Laboratories, Ajinomoto Co., Inc., Kawasaki, Japan

1. Properties 1633
2. Production 1634
3. Analysis 1634
4. Salts 1634
5. Uses 1635
6. References 1636

Choline [62-49-7], $C_5H_{15}O_2N$, M_r 121.18, trimethyl(2-hydroxyethyl)ammonium hydroxide, was first isolated from hog bile by STRECKER in 1849.

$[(CH_3)_3N^+CH_2CH_2OH]\,OH^-$

Choline occurs in material derived from animals such as bile, brain, liver, and egg yolk, and in the seeds, leaves, and stems of plants. Usually, very small amounts are present as free choline; most occurs as a component of such physiologically important substances as acetylcholine, lecithin, and their metabolites.

Like the vitamins, choline plays a significant role in nutrition [2], [3], but its daily requirement seems to be hundreds of times greater than that of vitamins. Choline deficieny interferes with normal fat metabolism; excessive fat accumulates in the liver, causing cirrhosis [4] or hemorrhagic kidneys [5] in experimental animals under various conditions. Several salts of choline (chloride, citrate, and tartrate) are often added to animal feeds as a dietary supplement [6]. These salts are also used in therapy.

One of the most important derivatives of choline is *acetylcholine* [51-84-3], which occurs in the body in very small amounts whenever parasympathetic nerves are stimulated. It is essential for impulse transmission of all nerves [7].

1. Properties

Choline is a strong base (pK = 5.06) [8], which absorbs carbon dioxide and water from the air. Choline is usually a colorless, viscous liquid with a fishy odor. When heated, it decomposes to form trimethylamine and glycol [9]. Choline is easily soluble in water, methanol, and ethanol; hardly soluble in acetone and chloroform; and insoluble in ether, benzene, toluene, and carbon tetrachloride.

2. Production

Hydrolysis of lecithin [10] was once employed for the production of choline and its salts, but recently this has been replaced by the reaction of trimethylamine with ethylene oxide [11]–[14] or with chlorohydrin [14]–[18].

$$(CH_3)_3N + \overset{O}{\overset{|}{CH_2-CH_2}} + H_2O \longrightarrow [(CH_3)_3N^+CH_2CH_2OH]OH^-$$

$$(CH_3)_3N + ClCH_2CH_2OH \longrightarrow [(CH_3)_3N^+CH_2CH_2OH]Cl^-$$

Another synthesis involves the reaction of 2-(ethoxymethoxy)ethyldimethylamine with methyl formate, followed by the addition of hydrochloric acid and ethanol [19]. For continuous production, the reaction of trimethylamine, ethylene oxide, and hydrochloric acid has been suggested [20].

3. Analysis

For the gravimetric determination of choline, many nonspecific precipitants are used, such as platinum chloride, gold chloride, potassium triiodide [21], phosphotungstic acid, and phosphomolybdic acid. Choline also precipitates from an aqueous solution when ammonium reineckate [13573-16-5] is added. The resulting choline reineckate is soluble in acetone, and the light absorption of the acetone solution at 526 µm can be measured; this is presently the most widely used method for the determination of choline [2], [22]–[24].

In a microbiological assay, the growth rate of mutant ATCC 34 486 or 9277 of *Neurospora crassa* is used as an appropriate measure of choline content [2], [25].

In a physiological assay, choline is first acetylated and the acetylcholine-promoted contraction of a piece of isolated rabbit intestine is then measured (kymography) [26].

4. Salts

Choline chloride [67-48-1], $C_5H_{14}ONCl$, M_r 139.63, forms hygroscopic, white crystals or a crystalline powder, usually having a slight amine odor [27] and a strong brackish taste. It decomposes at ca. 180 °C to form dimethylaminoethanol and methyl chloride. It is very soluble in water, freely soluble in alcohol, slightly soluble in acetone and chloroform, and almost insoluble in ether and benzene. Its solution is neutral to litmus paper.

Choline hydrogen tartrate [87-67-2], $C_9H_{19}O_7N$, M_r 253.26, mp 149–153 °C, is a white crystalline powder with an acid taste and a slight amine odor. It is very soluble in water, slightly soluble in alcohol, and almost insoluble in ether, benzene, and chloroform. The pH of a 25% aqueous solution is about 3.5.

Choline dihydrogen citrate [77-91-8], $C_{11}H_{21}O_8N$, M_r 295.30, mp 105–107.5 °C, is a white, crystalline granular substance with an acid taste, which is more palatable than the chloride. This compound is very soluble in water, soluble in alcohol, and almost insoluble in ether, benzene, and chloroform. The pH of a 25% aqueous solution is about 4.3.

Tricholine citrate [546-63-4], $C_{21}H_{47}O_{10}N_3$, M_r 501.63, is sold as "concentrate," which is a clear, faint yellow-green viscous syrup containing 65% tricholine citrate and having a slight amine odor. The pH is between 9 and 10; the relative density d_4^{20} is 1.16–1.17.

Choline gluconate, $C_{11}H_{25}O_8N$, M_r 299.33, is a hygroscopic yellow mass, soluble in water and slightly soluble in ethanol.

5. Uses

Choline salts, such as the chloride, hydrogen tartrate, dihydrogen citrate, and gluconate, are used therapeutically for patients with cirrhosis or other disorders of the liver; the combination of choline with methionine and cystine is recommended in this case [4], [28], [29].

Occasionally, acetylcholine [51-84-3] is used as a vasodilator. Methacholine chloride [62-51-1], O-acetyl-β-methylcholine chloride, is used as a parasympathetic stimulant and as an antiepinephrine substance. Carbamoylcholine chloride [51-83-2] is used to increase peripheral circulation in cases of vasospasms caused by peripheral vascular disorders. Cytidine diphosphate choline [987-78-0] (CDP-choline or citicoline) is an intermediate in lipid synthesis. In neurology and neurosurgery, CDP-choline has been used extensively in various pathological conditions such as brain edema, head injury, cerebral hemorrhage, and the acute or recovery phase of cerebral ischemia [30]. Choline is sometimes combined with potent pharmaceuticals to increase their stability and solubility [31], [32].

In animal feed, choline chloride is added to chicken and turkey starting mashes [32]. Choline increases egg production and egg size [33] when it is added to a corn–soybean diet [34]. Choline is not used extensively in swine nutrition [35]; its influence on body weight gain [36] and milk fat production [37] in cattle has been studied.

Choline is used as a specific reagent for the colorimetric determination of cobalt and hexacyanoferrate(II), because a bright green color appears when hexacyanoferrate(II) is added to a mixture of cobalt salt and choline chloride.

Recently, the food and cosmetics industry has had an increasing interest in choline, brought about by the demand for harmless surface-active agents. Some choline salts of fatty acids, such as choline myristate, have been suggested for soft detergents [38], and some acylcholine compounds, such as stearoylcholine chloride, have been suggested as ingredients in hair conditioners [39].

Chlorocholine chloride [*999-81-5*], 2-chloroethyltrimethylammonium chloride, is used as a plant growth regulant to maintain the compactness of the plant [40] as opposed to the growth-accelerating effect of gibberellic acid [41], [42]; it is used for ornamental plants, especially poinsettias and azaleas. This compound is also used to prevent the lodging and flattening of wheat by wind, as well as the etiolation of greenhouse tomatoes, thus leading to earlier and larger fruit yields [43].

Toxicology. Choline has a low toxicity. An acute toxicity test of choline chloride administered orally to rats shows an LD_{50} value of 3 – 6 g/kg [44].

6. References

[1] *Beilstein* **4**, 227; **4 (2)**, 732; **4 (3)**, 726.
[2] W. H. Sebrell, R. S. Harris (eds.): *The Vitamins*, 2nd ed., vol. **3**., Academic Press, New York 1971, pp. 2 – 154.
[3] Committee on Dietary Allowances: *Recommended Dietary Allowances*, 8th ed., Food and Nutrition Board, National Research Council, National Academy of Sciences, New York, 1974, 64 pp.
[4] C. L. Connor, I. L. Chaikoff, *Proc. Soc. Exp. Biol. Med.* **39** (1938) 356.
[5] W. H. Griffith, D. J. Mulford, *J. Am. Chem. Soc.* **63** (1941) 929 – 932.
[6] H. Haertel, *Arch. Gefluegelkd.* **38** (1974) no. 5, 165 – 174.
[7] D. Nachmansohn: *Chemical and Molecular Basis of Nerve Activity*, Academic Press, New York 1959.
[8] C. W. Price, W. C. M. Lewis, *Trans. Faraday Soc.* **29** (1933) 775 – 787.
[9] E. Kahane, J. Levy: *Biochimie de la Choline et de ses Dérivés*, Hermann, Paris 1938.
[10] J. D. Riedel, DE 193 449, 1906.
[11] F. B. Moosnick, E. M. Schleicher, W. E. Peterson, *J. Clin. Invest.* **24** (1945) 278.
[12] Produits Amines, DE 665 882, 1931.
[13] F. R. Koerner, FR 736 107, 1932; GB 379 260, 1932.
[14] R. R. Renshaw, *J. Am. Chem. Soc.* **32** (1910) 128.
[15] Merck, US 2 198 629, 1940.
[16] BASF, DE 801 210, 1948.
[17] NOPCO, US 2 623 901, 1950.
[18] American Cyanamide, US 2 774 759, 1956.
[19] Du Pont, US 2 457 226, 1948.
[20] BASF, DE-OS 3 135 671, 1983.
[21] I. Sakakibara, T. Yosinaga, *J. Biochem. (Tokyo)* **23** (1936) 211 – 239.
[22] D. Glick, *J. Biol. Chem.* **156** (1944) 643.
[23] I. Hanin: *Handbook of chemical assay methods*, Raven Press Publ., New York 1971.
[24] S. E. Valdes Martinez, *Analyst (London)* **108** (1983) no. 1290, 1114 – 1119.
[25] N. H. Horowitz, G. W. Beadle, *J. Biol. Chem.* **150** (1943) 325.
[26] J. P. Fletcher, C. H. Best, D. M. Solandat, *Biochem. J.* **29** (1935) 2278.

[27] *The Pharmacopeia of the U.S.A.*, 19th rev. ed., Mack Publishing Co., Easton, Pa., 1974, p. 731.
[28] A. J. Beams, *JAMA, J. Am. Med. Assoc.* **130** (1946) 190.
[29] C. C. Otken, *Nutr. Rep. Int.* **29** (1984) no. 1, 1–10.
[30] B. S. Nilsson, *Clin. Pharmacol. Drug Epidemiol.* **2** (1979) 273–277.
[31] Nepera Chem. Co., US 2 776 287, 1957.
[32] Yakult, JP-Kokai 108 787, 1981; *Chem. Abstr.* **96** (1982) 104 576r.
[33] V. K. Tsiagbe, C. W. Kang, M. L. Sunde, *Poult. Sci.* **61** (1982) no. 10, 2060–2064.
[34] M. L. Sunde, *Proc. Cornell Nutr. Conf. Feed Manuf.* 1982, 67–73.
[35] T. H. Jukes, J. J. Oleson, A. C. Dornbush, *J. Nutr.* **30** (1945) 219–223.
[36] T. S. Rumsey, R. R. Oltjen, *J. Anim. Sci.* **41** (1975) 416.
[37] R. A. Erdman, R. D. Shaver, J. H. Vandersall, *J. Dairy Sci.* **67** (1984) no. 2, 410–415.
[38] Mitsui Toatsu,JP-Kokai 162 798, 1982; *Chem. Abstr.* **98** (1983) 181 576j.
[39] Lion, JP-Kokai 109 709, 1982; *Chem. Abstr.* **97** (1982) 168 722p.
[40] N. E. Tolbert, *J. Biol. Chem.* **235** (1960) 475.
[41] J. A. Lockhart, *Plant Physiol.* **37** (1962) 759–764.
[42] H. Kende, H. Ninnemann, A. Lang, *Naturwissenschaften* **50** (1963) 599–600.
[43] N. P. Budykina, R. I. Volkova, S. N. Drozdov, V. V. Klykova et al., *Khim. Sel'sk. Khoz.* **2** (1984) 38–40;*Chem. Abstr.* **100** (1984) 169 852y.
[44] M. W. Neuman, H. C. Hodge, *Proc. Soc. Exp. Biol. Med.* **58** (1945) 87–88.

Cinnamic Acid

Dorothea Garbe, Haarmann & Reimer GmbH, Holzminden, Federal Republic of Germany

1.	Introduction	1639	5.4.	Oxidation of Cinnamaldehyde. 1642
2.	Physical Properties	1640	5.5.	Preparation of Cinnamic Acid
3.	Chemical Properties.	1640		Esters 1642
4.	Occurrence in Nature.	1640	6.	Uses 1642
5.	Production	1641	7.	Quality Specifications. 1643
5.1.	Perkin Reaction	1641	8.	Economic Aspects 1643
5.2.	Claisen Condensation.	1641	9.	Toxicology. 1643
5.3.	Benzal Chloride Reaction	1642	10.	References. 1643

1. Introduction

Cinnamic acid [621-82-9], 3-phenylpropenoic acid, $C_9H_8O_2$, M_r 148.16, exists in cis and trans forms.

trans Isomer   cis Isomer

The more stable isomer is the trans isomer, which occurs naturally and is the usual commercial product.

Cinnamic acid was first isolated as crystals from cinnamon oil by Trommsdorf in 1780. He thought it was benzoic acid. Dumas and Péligot identified it in 1835, and in 1856 Bertagnini succeeded in synthesizing it from benzaldehyde and acetyl chloride.

2. Physical Properties

trans-Cinnamic acid [140-10-3] forms colorless crystals with a faint balsamic odor, *mp* 133 °C, *bp* 300 °C (101.3 kPa) and 173 °C (13.3 kPa), d_4^4 1.2475, d_4^{180} 1.0270. The heat of combustion is 30.5 kJ/g. *trans*-Cinnamic acid is only slightly soluble in water, but it is highly soluble in polar organic solvents.

cis-Cinnamic acid [102-94-3] exists in three crystalline forms: *cis*-allocinnamic acid, *mp* 68 °C, and two *cis*-isocinnamic acids, *mp* 58 °C and 42 °C. The lower melting *cis*-isocinnamic acid is extremely unstable and isomerizes spontaneously to *cis*-allocinnamic acid. The three *cis*-cinnamic acids can be interconverted, e.g., by inoculating their melts.

3. Chemical Properties

Cinnamic acid undergoes reactions typical of a carboxyl group and an olefinic double bond. The carboxyl group can be esterified to form cinnamates, some of which are important flavorings and fragrances. When reacted with inorganic acid chlorides, such as thionyl chloride or phosphorus chlorides, cinnamic acid gives cinnamoyl chloride [102-92-1]. When heated, cinnamic acid forms styrene [100-42-5] and carbon dioxide. With oxidizing agents or when heated with alkali, the olefinic double bond cleaves to give benzaldehyde [98-87-3].

Hydrogenation of cinnamic acid leads to 3-phenylpropionic acid [501-52-0] or 3-cyclohexylpropionic acid [701-97-3], depending on the reaction conditions. The addition of bromine to the olefinic double bond of cinnamic acid forms 2,3-dibromo-3-phenylpropionic acid [6286-30-2], an intermediate in the manufacture of the fragrant substance β-bromostyrene [103-64-0]. Hydrohalogenation gives 2-halo–3-phenylpropionic acids. Dimers, such as truxinic acid (3,4-diphenyl-1,2-cyclobutanedicarboxylic acid) [4482-52-4] and α-truxillic acid (2,4-diphenyl-1,3-cyclobutanedicarboxylic acid) [490-20-0], are formed from cinnamic acid under the influence of sunlight. Cinnamic acids with substituents on the benzene ring are seldom prepared by direct substitution reactions of cinnamic acid.

4. Occurrence in Nature

Cinnamic acid, usually the trans isomer, occurs naturally in the free state and as an ester, e.g., in essential oils. Cinnamic acid is a constituent of cassia oil, of the extract oil from *Populus balsamifera*, and of the essential oils from leaves and peels of *Citrus bigaradia*. Cinnamates, especially methyl- [1754-62-7], benzyl- [103-41-3], cinnamyl- [122-69-0], and hydrocinnamyl cinnamates [122-68-9], are found in numerous essential and balsam oils (methyl cinnamate in oils of the *Alpinia* species and varieties of *Ocimum canum*; benzyl, cinnamyl, and hydrocinnamyl cinnamates in Peru, Tolu, and storax balsam oils). Siam and Sumatra benzoin [*9000-73-1*] resins also contain cinnamates.

5. Production

Commercial synthesis of cinnamic acid almost always results in the trans isomer.

5.1. Perkin Reaction

The Perkin reaction is the oldest known method of producing cinnamic acid commercially. In this reaction benzaldehyde [*100-52-7*] is condensed with acetic anhydride in the presence of sodium acetate as catalyst.

$$\text{PhCHO} + (CH_3CO)_2O \xrightarrow{CH_3COONa} \text{PhCH=CHCOOH} + CH_3COOH$$

Other possible catalysts are potassium acetate, tertiary amines, potassium phosphate [1] and trimethyl borate [2].

5.2. Claisen Condensation

The Claisen condensation of benzaldehyde with acetic acid esters in the presence of alkali alcoholates yields cinnamic acid esters, which can be saponified to form cinnamic acid.

$$\text{PhCHO} + CH_3COOR^1 \xrightarrow{NaOR^2} \text{PhCH=CHCOOR}^1 + \text{PhCH=CHCOOR}^2$$

$$\xrightarrow{\text{Saponification}} \text{PhCH=CHCOOH} + R^1OH + R^2OH$$

where R^1 and R^2 may be the same or different alkyl groups

5.3. Benzal Chloride Reaction

Benzal chloride [98-87-3] reacts with alkali acetate in an alkaline medium to give a high yield of cinnamic acid. Cinnamic acid can be obtained by this reaction in the presence of amines such as pyridine in more than 80% yield [3].

$$\text{C}_6\text{H}_5\text{CHCl}_2 + 2\,\text{CH}_3\text{COONa} \xrightarrow{\text{pyridine}} \text{C}_6\text{H}_5\text{CH=CHCOOH} + \text{CH}_3\text{COOH} + 2\,\text{NaCl}$$

5.4. Oxidation of Cinnamaldehyde

Cinnamaldehyde [104-55-2] can be oxidized to cinnamic acid by using oxygen and a catalyst, e.g., silver [4] or palladium on charcoal [5].

$$\text{C}_6\text{H}_5\text{CH=CHCHO} \xrightarrow[\text{Catalyst}]{\text{O}_2} \text{C}_6\text{H}_5\text{CH=CHCOOH}$$

5.5. Preparation of Cinnamic Acid Esters

Cinnamic acid esters can be prepared from styrene, alcohols, carbon monoxide, and oxygen by using catalysts such as platinum [6] or palladium [7].

$$\text{C}_6\text{H}_5\text{CH=CH}_2 + \text{ROH} + \text{CO} + 1/2\,\text{O}_2 \xrightarrow{\text{Catalyst}} \text{C}_6\text{H}_5\text{CH=CHCOOR}$$

6. Uses

Cinnamic acid is an important intermediate in the preparation of its esters, which are used as fragrances, for pharmaceuticals, and for the enzymatic production of L-phenylalanine, the starting material for peptide sweeteners. Sodium cinnamate is a known corrosion inhibitor. Cinnamic acid is also used as a brightener in cyanide-free zinc electroplating baths [8], a corrosion inhibitor during removal of scale from zinc [9] and in aerosol cans [10], a low-toxicity heat stabilizer for poly(vinyl chloride) [11], a crosslinking agent for dimethyl terephthalate–ethylene glycol copolymer [12] and polyurethanes [13], a fireproofing agent for polycaprolactam [14], in laundry-resistant polyurethane adhesives for polyester fibers [15], and for improvement of the storage stability of drying-oil-modified alkyd resin coatings [16].

7. Quality Specifications

Cinnamic acid used in fragrances and foods must be 99% pure after drying. The melting point should be 133 °C, and the chlorine content not more than 0.005 wt%[17].

8. Economic Aspects

The annual worldwide production of cinnamic acid has increased to several thousand tons due to the steadily increasing demand for artificial dipeptide sweeteners. Bayer (Federal Republic of Germany), DSM (the Netherlands), and Kay Fries (United States) are among the major producers.

9. Toxicology

The acute oral LD_{50} in rats is 2.5 g/kg, and the acute dermal LD_{50} in rabbits exceeds 5 g/kg. Cinnamic acid applied neat to intact or abraded rabbit skin for 24 h was slightly irritating; a 4% solution in petrolatum produced no sensitization in man [18].

10. References

General References

    E. Gildemeister, F. Hoffmann: *Die Ätherischen Öle*, Akademie Verlag, Berlin 1966.

Specific References

[1] Agency of Industrial Sciences and Technology, JP 76 01 026, 1967 (H. Tanaka, M. Tsuda); see *Chem. Abstr.* **85** (1976) 142845h.
[2] Y. Y. Makarov-Zemlyanskii, V. V. Malyavkin, SU 175 975, 1962; see *Chem. Abstr.* **64** (1966) 8095h.
[3] Kureha Chemical Ind. Co., JP-Kokai 73 081 830, 1973 (K. Shinoda, K. Kobayashi); see *Chem. Abstr.* **80** (1974) 108194g.
[4] Union Carbide and Carbon Corp., FR 1 113 852, 1954.
[5] Rhône-Poulenc, FR 2 069 836, 1969.
[6] Ube Ind., JP-Kokai 82 070 836, 1980; see *Chem. Abstr.* **97** (1982) 109703k.
[7] Mitsubishi Chem. Ind., JP-Kokai 81 071 039, 1979; see *Chem. Abstr.* **95** (1981) 186892q.

[8] M and T Chemicals Inc., GB 2 009 790, 1977 (D. A. Arcilesi).
[9] F. Boehm, Erzeugung Chem. Produkte GmbH, DE 2 405 861, 1974 (W. Boehm).
[10] Seitetsu Kagaku Co., JP-Kokai 81 082 872, 1979; see *Chem. Abstr.* **95** (1981) 208175e.
[11] Adeka Argus Chemical Co., JP-Kokai 79 011 948, 1977 (M. Ito, T. Aoki); see *Chem. Abstr.* **90** (1979) 187987a.
[12] Teijin Ltd., JP-Kokai 79 011 997, 1977 (H. Inata, T. Morinage, M. Ogasawara); see *Chem. Abstr.* **92** (1980) 95081p.
[13] Gambro Dialysatoren GmbH & Co. KG, JP-Kokai 79 060 398, 1977; see *Chem. Abstr.* **91** (1979) 92517d.
[14] Ube Industries, JP-Kokai 78 133 256, 1977 (Y. Sasaki, T. Fujimoto, M. Yamanaka); see *Chem. Abstr.* **90** (1979) 122557r.
[15] Honny Chemicals Co., JP 79 038 131, 1971 (T. Ichinomiya, A. Imaizumi); see *Chem. Abstr.* **92** (1980) 148084a.
[16] Meidensha Co., JP 71 001 709, 1966 (T. Horiuchi, K. Yonahara, K. Tokubata); see *Chem. Abstr.* **76** (1972) 47458q.
[17] *Food Chemicals Codex 1981*, National Academy Press, Washington, D.C., pp. 364–365.
[18] D. L. J. Opdyke, *Food Cosmet. Toxicol.* **16** (1978) 687.

Citric Acid

Frank H. Verhoff, Miles Laboratories, Elkhart, Indiana 46515, United States

1.	Introduction	1645	4.3.1.	Surface Fermentation 1651
2.	Physical Properties	1647	4.3.2.	Submerged Fermentation. 1652
3.	Chemical Properties.	1648	4.3.3.	Other Fermentation Methods . . . 1652
4.	Production	1650	4.4.	Recovery 1652
4.1.	Laboratory Synthesis	1650	5.	Quality Specifications and Uses 1653
4.2.	Early Production	1651	6.	Economic Aspects 1654
4.3.	Fermentation	1651	7.	References. 1654

1. Introduction

Citric acid [77-92-9], $C_6H_8O_7$, M_r 192.13, is a tricarboxylic acid. According to IUPAC nomenclature, citric acid is 2-hydroxypropane-1,2,3-tricarboxylic acid; it is also known as β-hydroxytricarballylic acid.

$$H_2C\text{—}\underset{\underset{OH}{|}}{\overset{\overset{COOH}{|}}{C}}\text{—}CH_2$$
$$\text{COOH} \qquad\qquad \text{COOH}$$

Citric acid was isolated and crystallized in 1784 by Scheele, who precipitated calcium citrate by adding lime to lemon juice. The presence of three carboxyl and one hydroxyl groups was recognized by Liebig in 1838. Citric acid was first prepared from calcium citrate in 1860 in the United Kingdom. By 1880 citric acid was being produced from calcium citrate in France, Germany, and the United States.

Citrus fruits contain citric acid in large quantities, from 5% in the fruit [1] to 9% in the juice [2]. Citric acid is also found in many other fruits. The citric acid content (in mg/100 mL) of various common fruits is as follows [3], [4]:

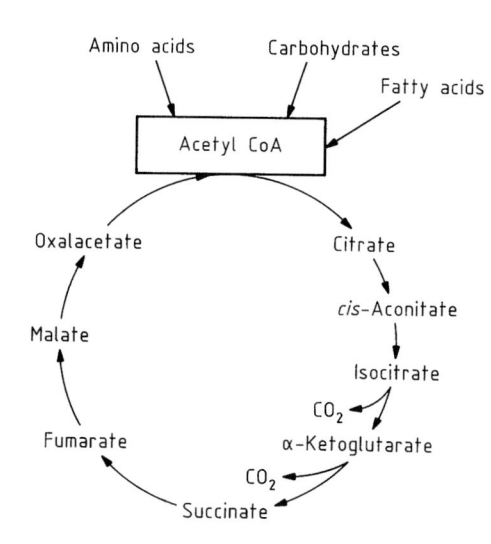

Figure 1. Citric acid cycle of plants and animals [5], [6]

Lime	7000
Lemon	5630
Raspberry	2480
Black currant	1170
Tomato	1018
Pineapple	605
Strawberry	580
Cranberry	202
Apple	14

In addition to fruits citric acid is found in almost all plant and animal species. The *citric acid cycle*, also known as the Krebs cycle or tricarboxylic acid cycle (see Fig. 1), is pivotal in the oxidation of sugars and acetates to carbon dioxide and water, releasing energy for physiological functions [5], [6].

About 100 years after citric acid had been isolated from citrus fruits, it was synthesized from glycerol. The finding, in 1893, that certain microorganisms secrete citric acid when grown on sugar media suggested the possibility of producing the acid commercially by fermentation. The first successful operation of this type began 1923 in New York; later large-scale fermentation processes were developed in Czechoslovakia, the United Kingdom, Belgium, and Germany. Submerged fermentation has been used since the middle of this century.

Until recently, citric acid fermentation, whether surface or submerged, used carbohydrate substrates with certain strains of the mold *Aspergillus niger*. Since the early 1970s, other microorganisms have been introduced, including molds, yeasts, and bacteria, and other substrates, e.g., various carbohydrates and n-alkanes from crude oil.

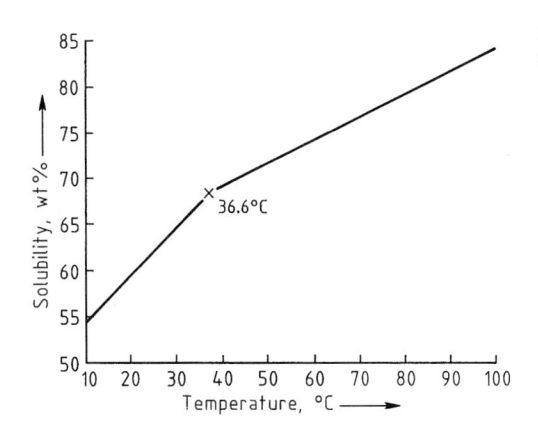

Figure 2. Solubility of citric acid in water as a function of temperature

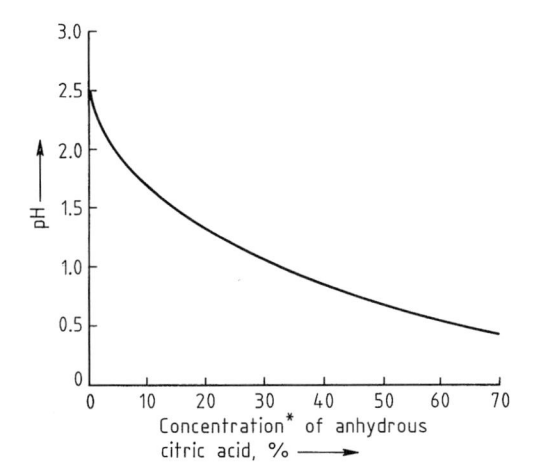

Figure 3. Concentration of citric acid in water vs. pH
* Concentration is in (grams per 100 mL solution) × 100%

2. Physical Properties

Citric acid is a colorless, odorless substance with an agreeable, though strongly acid taste. The monohydrate crystallizes from a saturated solution below 36.6 °C and the anhydrous form at higher temperature. The *monohydrate*, M_r 210.13, has a relative density of 1.542 and a heat of combustion at 30 °C of 1973 kJ/mol; it softens at 75 °C and melts at 100 °C. Its translucent, colorless crystals belong to the orthorhombic system. *Anhydrous citric acid, mp* 153 °C, has a relative density of 1.665 and a heat of combustion at 30 °C of 1990 kJ/mol. The crystals are translucent and colorless, belonging to the holohedral class of the monoclinic system. Both forms exist under conditions of normal humidity; dehydration of the monohydrate occurs in very dry air and is more rapid when it is carried out under vacuum in the presence of concentrated sulfuric acid. The anhydrous crystals gradually absorb water in moist air; both crystalline forms clump and harden in moist air.

Table 1. Freezing point depression and boiling point elevation of aqueous citric acid solutions

Concentration, mol/kg H$_2$O	Freezing point depression, °C	Boiling point elevation, °C
0.01	0.023	–
0.05	0.042	–
0.10	0.203	–
0.50	0.965	0.284
1.00	1.940	0.577
2.00	1.000	1.214
5.00	–	3.512
10.00	–	8.390
20.00	–	16.600

Table 2. Solubility of hydrated and anhydrous citric acid in organic solvents at 25°C and density of solutions

Form / Solvent	Solubility,* g/100 g	d_{25}
Monohydrate		
Amyl acetate	5.980	0.8917
Amyl alcohol	15.430	0.8774
Ethyl acetate	5.276	0.9175
Ether	2.174	0.7228
Chloroform	0.007	1.4850
Anhydrous		
Amyl acetate	4.22	0.8861
Ether (absolute)	1.05	0.7160

* Saturated solution.

The solubility of citric acid in water as a function of temperature is shown in Figure 2; the temperature of the solution drops as the acid dissolves. The pH of citric acid solutions as a function of concentration is shown in Figure 3. The freezing point depression and boiling point elevation of citric acid solutions are given in Table 1. The dissociation constants at 18 °C are K_1 8.2 × 10^{-4}, K_2 1.8 × 10^{-5}, and K_3 4.0 × 10^{-6}.

The solubility of citric acid in organic solvents is listed in Table 2, along with the relative density of the solutions. The distribution between ether and water is shown in Table 3.

3. Chemical Properties

Citric acid loses water at 175 °C to form aconitic acid, HOOCCH=C(COOH)(CH$_2$-COOH), which loses carbon dioxide to yield itaconic anhydride [2170-03-8]. Itaconic anhydride rearranges to citraconic anhydride [616-02-4] or adds water to form itaconic acid [97-65-4], (HOOCCH$_2$)(HOOC)C=CH$_2$.

Table 3. Distribution of citric acid between water and ether

Concentration, mol/L		Distribution coefficient
In water	In ether	
At 15 °C		
0.902	0.0077	117
0.460	0.0036	128
0.220	0.0017	129
0.297	0.0023	129
At 25 °C		
0.9175	0.0063	114
0.487	0.0031	155
0.241	0.00155	155
0.315	0.0020	158

$$\text{HOOCCH}_2\underset{\underset{\text{CH}_2\text{COOH}}{|}}{\overset{\overset{\text{OH}}{|}}{\text{C}}}\text{COOH} \xrightarrow[-\text{H}_2\text{O}]{\text{Heat}} \text{Itaconic anhydride} \xrightarrow{\text{Heat}} \text{Citraconic anhydride}$$

Citric acid → Itaconic anhydride → Citraconic anhydride

Addition of water to citraconic anhydride gives citraconic acid [498-23-7], cis-HOOCCH=C(CH$_3$) (COOH). Evaporation of a citraconic acid solution in the presence of nitric acid yields mesaconic acid [498-24-8], the trans isomer of citraconic acid.

Potassium permanganate oxidation of citric acid at 35 °C yields 1,3-acetonedicarboxylic acid [542-05-2], HOOCCH$_2$COCH$_2$COOH; at 85 °C the product is oxalic acid [144-62-7], HOOCCOOH. Fusion with potassium hydroxide produces oxalic and acetic [64-19-7] acids.

Citric acid forms crystalline mono-, di-, and tribasic salts with many cations. The degree of hydration of these salts varies; thus, trisodium citrate can crystallize with 2 or 5.5 molecules of water. Mixtures of metal cations give complex salts, such as ZnNa$_3$H(C$_6$H$_5$O$_7$)$_2$, ZnNa$_4$(C$_6$H$_5$O$_7$)$_2$, and Zn(Na$_3$C$_6$H$_4$O$_7$)$_2$.

Citric acid forms stable complexes with many metals; some, such as ferroammonium citrates, can be crystallized. Copper gives a complex resembling the copper–tartaric acid complex of Fehling's solution. These complexes have been studied by various classical methods.

Citric acid can chelate many metal ions in solution by forming bonds between the metal and the carboxyl or hydroxyl groups of the citric acid molecule. Sometimes two or more citric acid molecules are involved in the interaction with the metal ion. This property is valuable for preventing precipitation (e.g., bathtub ring from soap), changing chemical potential, and altering other chemical properties.

Citric acid is easily esterified with many alcohols under the usual conditions in the presence of a catalyst, such as sulfuric acid, p-toluenesulfonic acid, or an acid ion-exchange resin; esterification with alcohols boiling above 150 °C requires no catalyst.

Benzyl chloride and sodium citrate yield di- or tribenzyl esters. Trimethyl, triethyl, and tributyl citrate are used as plasticizers in food-packaging materials.

Dihydric alcohols, dihydric phenols, and polyhydric alcohols, such as mannitol, sorbitol, and glycerol, form polyesters with citric acid. In some cases the esterification reaction can be stopped before completion, leaving at least one of the carboxylic acid moieties free to form a salt; the resulting polyester may be soluble in water.

Acid chlorides and anhydrides react with the hydroxyl group of citric acid; usually the ester of citric acid is used and the allyl ester is formed. Dicarboxylic acid chlorides give bis(citric acid esters).

Epoxides, such as ethylene oxide, propylene oxide, and styrene oxide, form polymers by reacting with citric acid or its esters at the available hydroxyl and carboxyl groups.

Ammonia, amines, amides, and carbamides react with citric acid as they do with simple carboxylic acids. Aldehydes form adducts with citric acid; e.g., formaldehyde gives anhydromethylenecitric acid [*144-16-1*]5-oxo-1,3-dioxolane-4,4-diacetic acid:

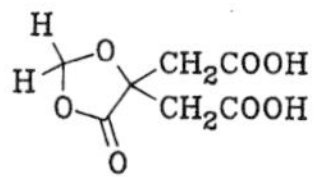

4. Production

4.1. Laboratory Synthesis

The classic laboratory synthesis of citric acid beginning with acetone has not been used on an industrial scale:

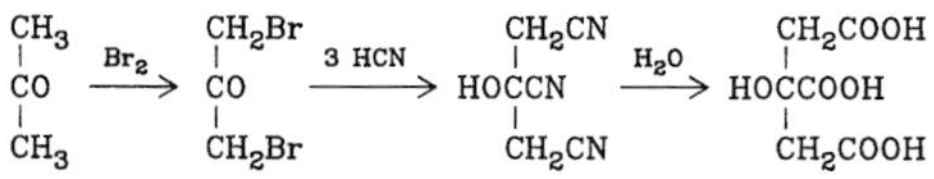

Reaction conditions at each stage are critical and use of bromine and hydrogen cyanide is hazardous. A simplified version of this synthesis has been patented [7].

Another synthesis is as follows:

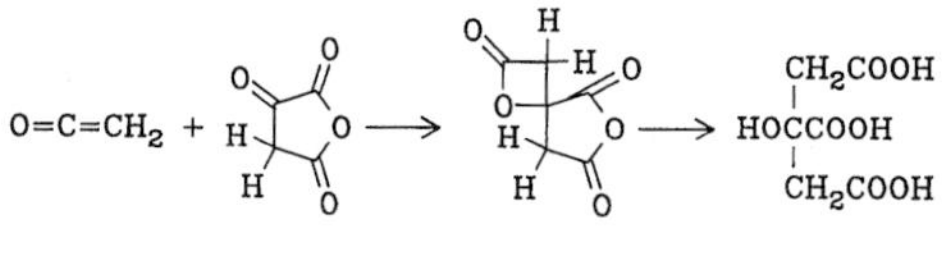

The key step in this route is the catalytic condensation of ketene with oxalacetic anhydride, which can be obtained by catalytic oxidation of malic or fumaric acid.

Other syntheses are found in the patent and scientific literature, but apparently none have been commercialized [8], [9].

4.2. Early Production

For about 50 years, citric acid was obtained exclusively from unripe citrus fruit. In southern Italy, an entire industry with orchards and processing facilities was devoted to this purpose. The juice was pressed from unripe fruit and mixed with lime (CaO) to precipitate calcium citrate. The solid was treated with sulfuric acid to precipitate calcium sulfate and form a solution from which citric acid was crystallized. The yield of purified product was about 2–3 wt% of the fruit; citrus oils were also obtained. This procedure has been displaced by fermentation.

4.3. Fermentation

The most economical method for producing citric acid since the 1930s has been fermentation, which employs a strain of *Aspergillus niger* to convert sugar to citric acid. Both surface fermentation and submerged fermentation have been used.

4.3.1. Surface Fermentation

In surface fermentation, *A. niger* is grown on a liquid substrate in pans stacked vertically in a chamber. The chamber and pans are sterilized either before or after introduction of the substrate. The pans are filled manually or automatically. The chamber is warmed by the introduction of moist, sterile air at a controlled temperature. The liquid and the surface microorganisms are removed manually or automatically from the pans. The pans are cleaned before the next batch is introduced.

The substrate for the fermentation is a carbohydrate, usually molasses or a sugar, such as raw beet, refined beet, or cane sugars, or a syrup. Glucose syrups can be prepared from wheat, corn, potato, or other starch. The sugar content of the syrup can vary from about 10 to 25 wt%. Certain inorganic nutrients, such as ammonium nitrate, potassium phosphate, magnesium sulfate, zinc sulfate, and potassium ferrocyanide, are added. The pH is adjusted to between 3 and 7, depending on the carbohydrate source. Sterilization may be batchwise or continuous; the latter uses less energy and is usually faster. After sterilization, the temperature is adjusted as required.

The surface of the sterile substrate in the pans is inoculated with *A. niger* spores, which germinate and cover the surface of the liquid with a mat of mold. After two to three days the surface is completely covered and citric acid production begins, continuing at almost a constant rate until 80–90% of the sugar is consumed. Fermentation then continues more slowly for an additional six to ten days.

The theoretical yield from 100 kg of sucrose is 123 kg of citric acid monohydrate or 112 kg of anhydrous acid. However, the *A. niger* uses some sugar for growth and respiration, and the actual yield varies between 57 and 77% of theoretical, depending on such factors as substrate purity, the particular strain of organism, and the control of fermentation.

4.3.2. Submerged Fermentation

Submerged fermentation is similar to surface fermentation, but takes place in large fermentation tanks. This method is used more frequently because labor costs are lower with large tanks than with small pans; equipment costs are also lower.

The fermentation vessel can be short and wide or tall and narrow, and equipped with mixing devices, such as top-entering or side-entering agitators of the turbine or propeller type. Agitation can be increased by use of a draft tube, a recirculation loop, or a nozzle through which air and recirculated substrate is pumped. Spargers located at the bottom of the vessel or under the stirrer supply air, which may be enriched with oxygen. Oxygen is usually recovered from the exhaust gas. The air is supplied by a compressor and passes through a sterile filter; if necessary, the air is cooled. Because the process is exothermic, the vessel must be equipped with heat-exchange surfaces, which can be the outside walls or internal coils. Ports are provided for introducing substrate, inoculum, and steam or other sterilizing agents; sampling and exhaust ports are also provided.

The substrate is prepared in a separate tank and its pH adjusted; the micronutrients may be added to this tank or directly to the fermentor. The substrate is sterilized by a batchwise or, more commonly, by a continuous operation. The fermentor is sterilized, charged with substrate, and inoculated. Fermentation requires 3–14 days. After it is completed, the air supply is stopped to prevent the microorganisms from consuming the citric acid.

4.3.3. Other Fermentation Methods

Citric acid can also be produced from hydrocarbons using yeast. The process resembles submerged fermentation, but requires more air. It is a four-phase process, i.e., air, organic liquid, aqueous liquid, and solid, and some operational details are different.

4.4. Recovery

The citric acid broth from the surface or submerged fermentation processes must be purified. First, biological solids usually are removed by filtration using a rotary vacuum filter or the more recent belt-press filter, or by centrifugation. The solids are washed to improve recovery of citric acid.

The dissolved citric acid must then be separated from residual sugars, proteins generated by the fermentation, and other soluble impurities. This has traditionally been accomplished by precipitation and crystallization. Addition of lime precipitates calcium citrate, which is filtered and stirred in dilute sulfuric acid to form a precipitate of calcium sulfate; filtration yields a purified citric acid solution. Control of pH and temperature in these operations helps to optimize the results. Citric acid is then crystallized from solution and recrystallized from water; the mother liquors are recycled to remove accumulated impurities.

Other purification techniques can be used, including ion exchange, carbon adsorption, membrane filtration, chromatography, and liquid extraction.

After filtration or centrifugation, the citric acid crystals are dried, classified, and sold in bags or in bulk. Crystals of the wrong size are recycled. Citric acid is also sold as an aqueous solution of varying degrees of purity.

Byproducts of these purification processes include biomass from the primary filtration, which can be sold for feed, and calcium sulfate, which can be used for landfill or for industrial purposes, such as manufacture of gypsum board. Soluble organic impurities are given normal wastewater treatment.

Few changes in production methods have occurred in recent years, but research has been done on the fermentation process [10] and on new separation techniques [11].

5. Quality Specifications and Uses

Specifications. The Food Chemical Code specifications for citric acid are as follows [12]:

Description: colorless, translucent crystals or white, granular to fine crystalline powder; odorless, strongly acid taste; hydrate is efflorescent in dry air
Solubility: 1 g of anhydrous acid dissolves in 0.5 mL of water, in 2 mL of alcohol, and in 30 mL of ether
Identification: 1-in-10 solution gives positive tests for citrate
Assay: not less than 99.5% $C_6H_8O_7$, calculation based on the anhydrous form
Water: not more than 0.5% (anhydrous form) and not more than 8.8% (hydrate)
Limits of impurities: arsenic (as As): not more than 3 ppm (0.0003%)
heavy metals (as Pb): not more than 10 ppm (0.001%)
oxalate: passes test; no turbidity produced
readily carbonizable substances: passes test
residue on ignition: not more than 0.05%

For specific test methods, see [12].

Recently, the number of grades of solid and liquid product has increased, which precludes a detailed discussion of each. Further information can be obtained from the manufacturers.

Uses. Citric acid and its sodium or potassium salts are used in many food products, including carbonated beverages, dry-packaged drinks, fruit drinks, jams, jellies, and canned fruits. They are also important in producing vegetable oils and in preserving the color, flavor, and vitamin content of fresh and frozen vegetables and fruit.

Citric acid is widely used in cleaning, e.g., for boilers and heat exchangers. Its chelating ability assists in removal of scale. Citric acid is employed as a detergent builder, especially in liquid formulations. Its solutions can remove sulfur dioxide from gases and chelate micronutrients in fertilizer. Citric acid esters are plasticizers for food-grade plastic containers.

6. Economic Aspects

Citric acid is manufactured worldwide. The United States and Western Europe are the major producers; however, capacity has recently increased in the Far East and in Eastern Europe. The location of new facilities depends on access to raw materials and markets.

7. References

[1] H. Moellering, W. Gruber, *Anal. Biochem.* **17** (1966) 369–376.
[2] H. Fricke, S. B. Hensen, *Food Process. Ind.* 1975, 38–44.
[3] J. J. Ryan, J. A. Du Pont, *J. Agri. Food Chem.* **21** (1973) 45–49.
[4] K. N. Paulson, M. A. Stevens, *J. Food Sci.* **39** (1974) 354–357.
[5] H. A. Krebs, *Angew. Chem.* **66** (1954) 313–319.
[6] A. L. Lehninger: *A Short Course in Biochemistry,* Worth Publ., New York 1973, p. 217.
[7] Montecatini Edison, DE-OS 2 245 892, 1972.
[8] Chevron Research Corp., DE-OS 2 225 986, 1972.
[9] Ethyl Corp., DE-OS 2 240 723, 1972.
[10] M. Rohr, C. P. Kubicek, J. Kominek: "Citric Acid" in H. Dellweg (ed.): *Biotechnology Microbiology Products, Biomass, and Primary Products,* vol. **3**, Verlag Chemie, Weinheim 1983, pp. 456–465.
[11] D. T. Friesen et al.: "Separation of Citric Acid from Fermentation Beer Using Supported Liquid Membranes", Paper no. 194, Am. Chem. Soc. Meeting, New York, April 1984.
[12] *Food Chemicals Codex,* 3rd. ed., National Academy Press, Washington, D.C., 1981.

Cresols and Xylenols

HELMUT FIEGE, Bayer AG, Leverkusen Federal Republic of Germany

1.	Cresols	1656	1.6.	Quality Specifications and Analysis ... 1685
1.1.	Physical Properties	1657	1.7.	Handling, Storage, and Transportation ... 1687
1.2.	Chemical Properties	1658		
1.3.	Formation and Isolation	1660	1.8.	Uses ... 1688
1.3.1.	Isolation from Coal Tars	1661	1.9.	Economic Aspects ... 1690
1.3.2.	Recovery from Spent Refinery Caustics	1662	2.	Xylenols ... 1691
1.4.	Production	1664	2.1.	Properties ... 1691
1.4.1.	Alkali Fusion of Toluenesulfonates	1664	2.2.	Isolation ... 1691
			2.3.	Separation ... 1693
1.4.2.	Alkaline Chlorotoluene Hydrolysis	1667	2.4.	Production ... 1694
1.4.3.	Cymene Hydroperoxide Cleavage	1668	2.5.	Quality Specifications and Analysis ... 1697
1.4.4.	Methylation of Phenol	1671		
1.4.5.	Other Routes	1676	2.6.	Handling, Storage, and Transportation ... 1698
1.5.	Separation of *m*- and *p*-Cresol	1681	2.7.	Uses ... 1698
1.5.1.	Physicochemical Processes	1681	2.8.	Economic Aspects ... 1699
1.5.2.	Separation via Addition Compounds	1683	3.	Environmental Protection ... 1699
1.5.3.	Separation via Ester or Salt Formation	1684	4.	Toxicology and Occupational Health ... 1700
1.5.4.	Separation by Nuclear Substitution	1685	5.	References ... 1702

1. Cresols [1]–[3]

Cresol, methylphenol, C_7H_8O, M_r 108.14, occurs in three isomeric forms:

o–Cresol
2-methylphenol
[95-48-7]

m–Cresol
3-methylphenol
[103-39-4]

p–Cresol
4-methylphenol
[106-44-5]

Mixtures of *m*- and *p*-cresol and of *o*-, *m*-, and *p*-cresol are occasionally referred to in the technical literature as dicresol and tri- or isocresol, respectively.

Cresylic acids [1319-77-3] are mixtures of cresols, xylenols, higher alkylated phenols, and possibly some phenol. *Tar acids* are cresylic acids obtained from tar.

STÄDELER first discovered cresol in cow's urine in 1851 [4]. FAIRLIE and WILLIAMSON found it in coal-tar creosote in 1854. (The word cresol, in which "creosote" and "phenol" are combined, is partly explained by this fact.) Cresol was first synthesized in 1866 by GRIESS, who boiled diazotized toluidine. All three cresol isomers were distinguished for the first time by ENGELHARDT and LATSCHINOFF in 1869.

Cresols and cresol derivatives (ethers and esters) are widely distributed in nature. They are formed as metabolites of various microorganisms and are found in the urine of mammals. Humans eliminate, on average, 87 mg of *p*-cresol per day in the urine [5]. Cresol in its various forms is also detectable in the extracts and water vapor distillates of many plants, e.g., in jasmine flower oil, cassia flower oil, easter lily oil, and ylang ylang oil, in the floral oil of *Yucca gloriosa*, in peppermint, eucalyptus, and camphor oil, and in the essential oils of several plants of the genus *Artemisia*, of conifers, of oak wood, and of sandalwood. Small amounts of cresol are also found in certain foods and drinks, e.g., in tomatoes and tomato ketchup, cooked asparagus, certain types of cheese, butter oil, red wine, whisky, rum, cognac, and other brandies [6], raw and roasted coffee, black tea, smoked foods, and tobacco and tobacco smoke. Considerable amounts of cresol may be formed and enter surface waters if forests are pyrolyzed by pyroclastic flow from erupting volcanoes, as happened after the eruption of Mount St. Helens in 1980 [7].

Cresols are important chemical raw materials. They were originally obtained only from coal tar; after World War II, they were also obtained from spent refinery caustics. Since the mid-1960s, they have been produced synthetically on an increasingly large scale. "Synthetic cresol" now provides ca. 60 % of the requirements of the United States, Europe, and Japan; only ca. 40 % of the requirements are now met by "natural cresol," i.e., cresol from coal tar and spent refinery caustics.

Table 1. Physical properties of pure cresols

Property	o-Cresol	m-Cresol	p-Cresol
Melting point at 101.3 kPa, °C	30.99	12.22	34.69
Depression of mp by 1 mol of impurity, °C	0.596	0.749	0.915
Boiling point at 101.3 kPa, °C	191.00	202.23	201.94
Density at 25 °C (s or l), g/cm^3	1.135	1.0302	1.154
at 50 °C (l), g/cm^3	1.0222	1.0105	1.0116
Viscosity at 50 °C, mPa · s	3.06	4.17	4.48
Refractive index at 50 °C	1.5310	1.5271	1.5269
Dielectric constant at 50 °C	6.00	9.86	10.18
Specific conductance at 50 °C, S/cm	0.43×10^{-10}	63.4×10^{-10}	60.2×10^{-10}
Surface tension at 50 °C, N/m	34.4×10^{-6}	33.4×10^{-6}	34.0×10^{-6}
Dissociation constant at 25 °C	4.8×10^{-11}	8.1×10^{-11}	5.4×10^{-11}
Critical temperature, °C	424.4	432.6	431.4
Critial pressure, MPa	5.01	4.56	5.15
Critical density, kg/m^3	384	346	391
Heat capacity c_p at 25 °C (g), J mol^{-1} K^{-1}	127.3	124.6	124.6
c_p at 25 °C (s or l), J mol^{-1} K^{-1}	154.7	225.1	150.3
c_p at 50 °C (l), J mol^{-1} K^{-1}	237.9	235.0	233.3
Enthalpy of fusion ΔH_m at mp, 101.3 kPa, kJ/mol	15.830	10.714	12.715
Enthalpy of sublimation ΔH_{sub} at 25 °C, 101.3 kPa, kJ/mol	76.07	–	73.98
Enthalpy of vaporization ΔH_v at 25 °C, 101.3 kPa, kJ/mol	–	61.76	–
ΔH_v at bp, 101.3 kPa, kJ/mol	45.222	47.429	47.581
Enthalpy of combustion $\Delta H_c°$ at 25 °C, 101.3 kPa, kJ/mol	– 3696	– 3706	– 3701
Enthalpy of formation $\Delta H_f°$ at 25 °C (s or l), 101.3 kPa, kJ/mol	– 204.8	– 194.2	– 199.5
$\Delta H_f°$ at 25 °C (g), 101.3 kPa, kJ/mol	– 128.7	– 132.4	– 125.5
Free energy of formation $\Delta G_f°$ at 25 °C (s or l), 101.3 kPa, kJ/mol	– 55.7	– 59.2	– 51.0
$\Delta G_f°$ at 25 °C (g), 101.3 kPa, kJ/mol	– 33.0	– 40.2	– 31.7
Entropy S at 25 °C (s or l), 101.3 kPa, J mol^{-1} K^{-1}	165.5	212.7	167.4
Entropy $S°$ at 25 °C (g), 101.3 kPa, J mol^{-1} K^{-1}	352.8	356.3	351.0
Cubic expansion coefficient at 25 °C		0.759×10^{-3}	
Solubility in water at 25 °C, wt%	2.6	2.3	1.9
at 100 °C, wt%	4.8	5.3	4.8
of water in cresol at 25 °C, wt%	13.4	13.1	15.8
at 100 °C, wt%	18.4	20.8	24.9
Critical solution temperature in water, °C	164.8	148.6	143.7
at % cresol, wt%	34	36	34
Vapor density (air = 1)	3.72	3.72	3.72
Flash point (closed cup), °C	81	86	86
Ignition temperature, °C	555	555	555
Flammability limit, lower, in air at 20 °C, 101.3 kPa, vol%	1.3	1.0	1.0

1.1. Physical Properties [1], [8] – [11]

As pure substances, o- and p-cresol are crystalline; m-cresol is a viscous oil at room temperature. The cresols have a phenolic odor and are colorless, but become yellow to brown after a time. Because water dissolves freely in cresols, they absorb moisture from the air.

Physical data relating to the cresols have been compiled in Tables 1 and 2. Data concerning the temperature dependence of their thermodynamic properties are found in [8] – [11]; many additional properties are contained in [10].

Table 2. Vapor pressure of pure cresols, for ≥ 260 °C (in kPa)

Temp., °C	o-Cresol,	m-Cresol,	p-Cresol,
25	0.037	0.019	0.016
50	0.25	0.12	0.11
60	0.49	0.25	0.23
80	1.57	0.86	0.83
100	4.26	2.52	2.47
120	10.10	6.36	6.29
140	21.47	14.26	14.21
160	41.72	29.01	29.07
180	75.27	54.45	54.75
200	127.6	95.52	96.24
220	204.9	158.2	159.6
260	471	378	388
300	940	774	810
340	1704	1423	1521
380	2894	2420	2659

The solubility of the cresols in phenol and in many organic solvents, e.g., aliphatic alcohols, ethers, chloroform, and glycerol, is high. They are less soluble in water than phenol. The presence of other water-soluble organic compounds (e.g., methanol) raises their solubility in water and reduces the critical solution temperature [13]. Dissolved inorganic salts lower the water solubility of cresols.

The cresols can be distilled with steam and form azeotropes with a number of compounds, e.g., decane, 1-decene, 1-undecene, dodecane, 1,2,4,5-tetramethylbenzene, divinylbenzene, ethylene glycol, diethylene glycol methyl and ethyl ether, diethylene glycol diethyl ether, ethylene glycol diacetate, benzyl formate, phenyl acetate, methyl benzoate, hexane-2,5-dione, 2-octanone, acetophenone, benzylamine, the toluidines, and numerous pyridine derivatives. Vapor–liquid equilibrium data of the cresols with numerous other substances are compiled in [14]; those of m-cresol with carbon dioxide and nitrogen under pressure at 190–390 °C are given in [15] (2–5 MPa) and [16] (2–25 MPa), respectively.

1.2. Chemical Properties [2], [17]

Acidity. The cresols are chemically similar to phenol. They are weak acids and dissolve in aqueous alkali to form water-stable salts known as cresolates. Therefore, they can be extracted into sodium hydroxide solution from solvents that are not miscible with water. Nevertheless, their acidity is so low (pK_a 10.1–10.3) that even hydrogen sulfide (pK_a 7.2) and carbon dioxide (pK_a 6.4) are able to liberate them from cresolates. Thus, they hardly dissolve at all in solutions of sodium carbonate or sodium hydrogen carbonate.

Etherification and Esterification. The hydroxyl group can be etherified with alkyl halides, dialkyl sulfates, and toluenesulfonic acid esters, and react with acyl anhydrides or acyl chlorides to give the corresponding tolyl esters, which are normally referred to as "cresyl" esters. Isocyanates react with the OH group to form urethanes.

Substitution of the Hydroxyl Group. Under drastic conditions (420 °C, Al_2O_3), the OH group is replaced by ammonia, and the corresponding toluidine is obtained. If cresols are reacted with sulfur oxytetrafluoride at 150 °C, with diphenylphosphine trichloride at 220 °C, or with phosphorus tribromide at 280 °C, the corresponding fluoro-, chloro-, or bromotoluene is obtained. Butylthiol in hydrochloric acid replaces the phenolic OH group by a thiobutyl group. Distillation with zinc powder gives toluene.

Hydrogenation. Hydrogenation in the vapor phase at 300–400 °C under pressure (up to 8 MPa) in the presence of catalysts consisting of transition metals and aluminum oxide gives toluene (hydrodeoxygenation) [18]. Hydrogenolysis over catalysts at 400–500 °C or purely thermally at 500–700 °C can be so controlled that cresol is mainly demethylated to phenol (hydrodealkylation) [19]. Hydrogenation over Raney nickel or noble-metal catalysts under less severe conditions gives methylcyclohexanols or methylcyclohexanones.

Oxidation. The cresols are sensitive to oxidation. Depending on the oxidizing agent, reaction conditions, and position of the methyl group, oxidative reactions, which occur by free radical mechanisms, lead to a large number of compounds such as hydroquinones, quinols, quinones, cyclic ketones, furans, dimeric and trimeric cresols, and tolyl ethers. Strongly oxidizing agents break down the phenol ring.

After the hydroxyl group has been protected by esterification or etherification, the methyl group can be selectively mono-, di-, or trichlorinated, or selectively oxidized to a formyl group (e.g., with manganese dioxide or oxygen) or to a carboxyl group (with acid permanganate solution). Unlike the other two isomers, *p*-cresol can be directly oxidized with oxygen to give *p*-hydroxybenzaldehyde, e.g., in methanolic sodium hydroxide in the presence of cobalt salts [20]. Alkali fusion of the cresols in the presence of lead oxide or manganese dioxide produces the corresponding hydroxybenzoic acids directly.

Electrophilic Nuclear Substitution. Cresols, like phenol, readily undergo electrophilic substitution. The substituent enters the nucleus mainly in the o- and p-positions relative to the hydroxyl group. The cresols can, therefore, be nitrated even with dilute nitric acid. Nitrosation, halogenation, sulfonation, and alkylation occur readily. If *o*- or *p*-cresol is heated with Friedel–Crafts catalysts, such as $AlCl_3$, isomerization to *m*-cresol, thermodynamically the most stable of the three isomers (see p. 1676), occurs. The heating of dry alkali cresolates with CO_2 under pressure produces hydroxy-methylbenzoic acids (Kolbe–Schmitt reaction). In the presence of alkali, formaldehyde adds onto cresols, even at room temperature, to form hydroxymethylbenzyl alcohols; under acidic conditions or at elevated temperatures these condense to form high-molecular resins. The main products of the reaction of cresols with chloroform and alkali are *o*-hydroxy-methyl-benzaldehydes (Reimer–Tiemann synthesis). Electrophilic substitutions are occasionally complicated by partial or substantial cessation of the reaction at the primary stage of the cyclohexadienone formed by addition of the electrophile. Thus, if *p*-cresol is heated with tetrachloromethane in the presence of aluminum chloride, the main product is 4-methyl-4-trichloromethyl-2,5-cyclohexadienone (Zincke–Suhl reaction).

Behavior toward Materials [21]. At elevated temperature unalloyed steel is attacked by cresols to a noticeable extent, the severity of the attack depending on the water content. Stress corrosion cracking may occur even at 100 °C. Chromium steels are scarcely more resistant than unalloyed steels. Chromium–nickel stainless steels of the V2A type (AISI 321) are corroded at elevated temperatures at a rate of 0.1–1 mm/a, whereas the rate for V4A steels (AISI 316Ti) in boiling cresol is ca. 0.1 mm/a. Nickel is more resistant, its corrosion rate in pure cresols at 190 °C being only ca. 0.001 mm/a. Tantalum is considered to be entirely resistant to cresols.

Aluminum and its alloys are attacked severely at temperatures greater than ca. 120 °C when the water content of the cresol is too low. If the cresol contains less than ca. 0.3 % water, corrosion may proceed explosively. The use of copper and brass alloys should also be avoided.

The corrosive behavior of the cresols can alter significantly, depending on the presence of other compounds. Thus, the simultaneous presence of water-soluble chlorides and basic nitrogen compounds in the distillation of cresol/xylenol fractions has synergistic effects, leading, for example, to very severe corrosion of stainless steels [22].

Extensive experience concerning the behavior both of pure cresols and of cresols mixed with other substances toward metallic and nonmetallic materials is described in [21].

1.3. Formation and Isolation [23], [24]

In addition to phenol, cresols, xylenols, and numerous other phenols are found in the wastewater of cracking processes [25], tars, and tarlike products (→ Tar and Pitch) formed in the purely thermal cracking (carbonization and coking) [23], [24], oxdizing thermal cracking [23], [24], and hydrogenating thermal cracking ([23], [24], Coal Pyrolysis [26]) of such natural materials as bituminous coal [27], brown coal [28], [29], oil shale [30], peat, wood [31], lignin [32], and other biomaterials [33].

The yields of cresols, xylenols, and other phenols and their quantity ratio depend not only on the starting material, but also largely on the process conditions (especially the temperature and the residence time), the type of reactor, and the mode of operation.

Table 3. Average yields of phenols in the tars of various cracking processes*

Phenol	Sump-phase hydrogenation, coal oil (ca. 470 °C)	Low-temperature tars (600–650 °C)	Lurgi gasifier tar (800–900 °C, 20 bar)	High-temperature coke-oven tars (1200–1300 °C)		
				USA	UK	FRG
Phenol	16	1.4	0.14	0.20	0.22	0.12
o-Cresol	6	1.4	0.16	0.08	0.12	0.06
m-Cresol	14	0.9	0.26	0.14	0.17	0.12
p-Cresol	6	0.8	0.21	0.09	0.10	0.06
Xylenols	21	6.1	0.78	0.12	0.18	0.06
Higher phenols	22	12.4	1.67	0.27	0.34	0.05
Sum of phenols	85	21.6	3.22	0.90	1.13	0.47
Sum of cresols	26	3.1	0.63	0.31	0.39	0.24

* Data in kilogram of phenol per ton of water-free bituminous coal.

Hydrogenation provides the highest yields, low-temperature carbonization gives intermediate amounts, while high-temperature coking produces the lowest yields (Table 3).

Small amounts of cresols, xylenols, and other phenol derivatives are also formed in the catalytic and thermal cracking (especially coking thermal cracking) of petroleum fractions [34].

At the present time cresols and xylenols are isolated from coal tars obtained in the high-temperature coking process, low-temperature carbonization, and Lurgi pressure gasification of coal (see Table 3), and from spent refinery caustics.

1.3.1. Isolation from Coal Tars

The main source of tar cresols is the high-temperature coke-oven tar (composition and work up → Tar and Pitch), obtained in the production of metallurgical coke by horizontal chamber coking of bituminous coal at 900–1300 °C [23]. This tar contains, on average, ca. 0.4–0.6 wt% phenol, 0.8–1 wt% cresols, and 0.2–0.5 wt% xylenols. In the United Kingdom, cresols are also produced from low-temperature coal tars obtained in the production of smokeless fuels (semicoke) according to the Coalite process, where the content of C_6–C_8 phenols is ca. 10 times higher. Another source of cresols is provided by the liquid byproducts obtained by Sasol in South Africa in pressure gasification of bituminous coal by the Lurgi process, which are similar in composition to low-temperature tars. In addition, lignite tars are used in the German Democratic Republic, Czechoslovakia, the Soviet Union, and India [28].

When the starting product is high-temperature tar, the phenols are isolated by extraction with sodium hydroxide solution or, in the Lurgi phenoraffin process (→ Tar and Pitch), from the carbolic oil that boils at 180–210 °C, from the light oil, and from the filtrate of the naphthalene oil; where other coal tars are concerned the procedures are similar. The hydrocarbons and pyridine bases still present in the crude phenolate caustic are removed by distillation with steam. The crude phenol is then set free with carbon dioxide. Frequently, the phenolate caustics from coking plant effluents, which contain primarily phenol and cresol and only very small amounts of xylenols [35], are incorporated. The composition of the resulting crude phenol may, therefore, vary extensively. For example, the composition of crude phenol obtained in this way from German high-temperature coke-oven coal tar may be as follows [27]: 15% water, 30% phenol, 12% *o*-cresol, 18% *m*-cresol, 12% *p*-cresol, 8% xylenols, and 5% trimethylphenols together with higher boiling phenols.

After the alkali content of the crude phenol has been lowered from ca. 2% to ca. 0.3% by scrubbing with water, the crude phenol is dehydrated azeotropically and rectified under vacuum into the following fractions: phenol, *o*-cresol, *m*-/*p*-cresol mixture, xylenol, and phenol tar. As the difference between the boiling points of *m*- and *p*-cresol is very small (0.3 °C), these compounds are obtained as a single fraction, in which small amounts of 2,6-xylenol and 2-ethylphenol are also present. The ratio of *m*-cresol to *p*-cresol is ca. 60:40 for high-temperature tar and ca. 50:50 for low-temperature tar.

Examples of firms that isolate cresols (and xylenols) from coal tar are the Koppers Co. and the Merichem Co. in the United States, Coalite & Chemical Products Ltd. and the Croda Synthetic Chemicals Ltd. in the United Kingdom, Rütgers-Werke AG and Chemisches Werk Lowi GmbH in the Federal Republic of Germany, CdF Chimie in France, and the Nippon Steel Chemical Co. in Japan.

The scale on which cresols and xylenols are recovered from coal tar has decreased continuously over the past 25 years for the following primary reasons: the coke consumption of the iron- and steel-producing industry, which chiefly determines the supply of tar, has fallen; processes which lead to tars with lower cresol and xylenol contents are being increasingly used in coke production; and the tar output from gasworks has become insignificant due to the changeover to natural gas.

The situation may change, however, because blast-furnace technology based on the use of high-temperature coke from bituminous coal as a reducing agent will retain at least its present importance, and because the gasification of coal to produce synthetic natural gas and syncrude will be expanded as the prices of oil and natural gas rise. This would yield more low-temperature tar, in which the content of cresols and xylenols is much greater than in high-temperature coke-oven tar.

1.3.2. Recovery from Spent Refinery Caustics

In the United States, cresols and xylenols are obtained mainly from the naphtha fractions produced by catalytic and thermal cracking in the petroleum industry, which contain, on the average, ca. 0.1% C_6-C_8 phenols [34], [36]. In the scrubbing of the sulfur compounds contained in these fractions (hydrogensulfides, alkyl- and arylthiols) with concentrated alkaline solutions, a process known as "sweetening," the acid cresols and xylenols are also extracted. The composition of the spent cresylate caustics fluctuates widely, and they contain on the average 20–25% C_6-C_8 phenols and 10–15% sulfur compounds. The caustics are collected by the reprocessing firms Merichem,

Northwest Petrochemical, and Productol, and are reprocessed in central plants by a variety of processes.

At the Northwest Petrochemical in Anacortes, Washington, the thiols in the alkaline solution are first oxidized with air to give disulfides, which are then decanted as an oily layer [37]:

$$RSNa + R'SNa + 0.5\ O_2 + H_2O \longrightarrow RSSR' + 2\ NaOH$$

The phenols are then precipitated from the aqueous alkaline phase in a packed column with a countercurrent stream of carbon dioxide (flue gas) and decanted. The phenols still present in the carbonate–hydrogen carbonate phase (2–6%) are extracted with an organic solvent, from which they are extracted again with an aqueous alkaline phase, which is then returned to the column. A typical composition of the phenol mixture obtained in this way might be as follows: 20% phenol, 18% *o*-cresol, 22% *m*-cresol, 9% *p*-cresol, 28% xylenol (all six isomers plus ethylphenol), and 3% higher phenols. This mixture is then separated by distillation into phenol, *o*-cresol, *m*-/*p*-cresol mixture, and xylenols, or mixtures in which particular constituents predominate.

The Merichem Co. of Houston, Texas, is the biggest processor of spent refinery caustics in the United States, and precipitates the phenols with hydrogen sulfide, a waste product of the refineries in Houston [38], [39], instead of using carbon dioxide. After extraction with a solvent, the sulfide solution obtained in this process is concentrated in a triple-effect evaporator to give sodium sulfide for use in the paper industry and in ore dressing [39].

The water vapor condensate formed at this stage is stripped with natural gas to remove odorous compounds, after which it is passed to a cooling tower. The natural gas is then used as a fuel in the final process air incinerator. Before being introduced into the waste air incinerator, which is operated at ca. 1370 °C, the process air from all of the other parts of the plant is scrubbed with alkaline solution. In addition, odor-neutralizing chemicals are sprayed alongside the plant area. All of the water at the plant (including rain water) is recycled in an integrated system. Water is released into the environment almost completely as water vapor from the cooling towers [39], [40].

According to the process operated until 1971 by the Pittsburgh Consolidated Chemical Co. [41], [42], the spent refinery caustics were treated directly with flue gas containing 8–14% carbon dioxide. From the crude phenol thus precipitated, a fraction boiling at 100–235 °C was removed by distillation and extracted continuously in a countercurrent of aqueous methanol–light naphtha. After removal of the final thiol and base residues with an ion-exchange resin, the cresylic acids that had entered the aqueous methanolic phase were recovered by fractional distillation.

In 1984, the "cresylics" capacities of Merichem, Northwest, and Productol were 55 000, 15 000, and 10 000 t/a, respectively [43]. However, these figures include phenol, the xylenols, and several other alkylphenols. In addition, Merichem no longer isolates cresylics exclusively from spent refinery caustics, but also works up crude phenols from coal tars of various types.

The recovery of cresols and xylenols from spent refinery caustics has so far been confined to the United States. Because different refinery techniques are used in the Federal Republic of Germany, the production of cresol-containing spent caustics is insufficient for economical reprocessing; or no spent caustics are produced at all because desulfurization is carried out differently, e.g., by hydrotreating. In the United States as well, the recovery of cresol from spent refinery caustics is stagnating or even receding because the refineries are likewise changing over either to hydrotreating or to UOP's Merox process, partly because the price of sodium hydroxide has risen very sharply in recent years. With the Merox process, substantial quantities of cresols remain in the gasoline, and considerably smaller amounts of spent caustics are obtained.

1.4. Production

Since 1965, when the recovery of cresols from coal tar and spent refinery caustics had become insufficient to meet the rising demand, these compounds have been increasingly produced by synthesis. The processes now in use are

- alkali fusion of toluenesulfonates
- alkaline chlorotoluene hydrolysis
- splitting of cymene hydroperoxide
- methylation of phenol in the vapor or liquid phase

The first three processes start from toluene and were developed from the corresponding benzene–phenol syntheses (→ Phenol); to some extent they are even carried out in converted plants that formerly produced phenol. In the following descriptions, attention will be primarily paid to the differences between these processes. The methylation of phenol is a process specifically developed to produce cresols and xylenols. Each process gives a different isomer distribution, and all are, therefore, of individual importance.

1.4.1. Alkali Fusion of Toluenesulfonates [44], [45]

The alkali fusion is used mainly to manufacture *p*-cresol and consists of four reaction steps:

$$2\,CH_3-C_6H_5 + 2\,H_2SO_4$$
$$\rightleftharpoons 2\,CH_3-C_6H_4-SO_3H + 2\,H_2O \quad (1)$$

$$2\,CH_3-C_6H_4-SO_3H + Na_2SO_3$$
$$\longrightarrow 2\,CH_3-C_6H_4-SO_3Na + H_2O + SO_2 \quad (2)$$

$$2\,CH_3-C_6H_4-SO_3Na + 4\,NaOH$$
$$\longrightarrow 2\,CH_3-C_6H_4-ONa + 2\,Na_2SO_3 + 2\,H_2O \quad (3)$$

$$2\,CH_3-C_6H_4-ONa + SO_2 + H_2O$$
$$\longrightarrow 2\,CH_3-C_6H_4-OH + Na_2SO_3 \quad (4)$$

$$2\,CH_3-C_6H_5 + 2\,H_2SO_4 + 4\,NaOH$$
$$\longrightarrow 2\,CH_3-C_6H_4-OH + 2\,Na_2SO_3 + 4\,H_2O$$

The sulfonation of toluene (1) is normally carried out with concentrated sulfuric acid at 120–130 °C and atmospheric pressure. To ensure that substantially all of the sulfuric acid has been reacted, the water of the reaction is entrained with excess toluene vapor. The sulfonic acid mixture is neutralized with sodium sulfite (and/or sodium hydroxide) (2) and then fused with excess sodium hydroxide at 330–350 °C (3). The melt is introduced into water, the sodium sulfite filtered off, and the aqueous phase acidified with sulfur dioxide from reaction step (2) and/or with sulfuric acid (4). This gives an aqueous phase containing sodium sulfite, which is returned to the neutralization unit (2), and a crude cresol phase, which is dehydrated by azeotropic distillation. The plant and process are similar to those formerly used in the synthesis of phenol by alkali fusion of sodium benzenesulfonate. The dehydrated crude cresol phase is separated by fractional distillation into phenol, *o*-cresol, *m*-/*p*-cresol mixture, and a residue containing ditolyl sulfones, xylenols, higher phenols, and other compounds. The ditolyl sulfones are formed in the sulfonation; the phenols and xylenols are formed in small quantities in the alkali fusion [46], [47]. See [46] for the mechanism of the fusion reaction.

The distribution of the cresol isomers depends mainly on the sulfonation conditions. Normally a mixture of the following composition is obtained: 6–12% *o*-cresol, 6–12% *m*-cresol, 80–85% *p*-cresol [44], [48]. After the *o*-cresol has been removed by distillation, a *p*-cresol of ca. 90% purity can be obtained by distillation. From this, *p*-cresol with a *m*-cresol content of only ca. 1% can be isolated relatively easily by melt crystallization (see Section p. 1681).

A cresol yield of 80%, based on toluene, is possible [48]. The yield falls as the *m*-cresol content of the cresol mixture increases [44], [47].

Under kinetically controlled (mild) sulfonation conditions and mild alkali fusion, it is possible to produce cresol mixtures with an even smaller content of *m*-cresol. It has been claimed that sulfonation with chlorosulfuric acid at 33–45 °C gives a product free from *m*-cresol at a yield of 90% (calculated on the basis of toluene) and with an *o*-/*p*-cresol ratio of 15:85 [49]. The sulfonation of toluene with sulfur trioxide–sulfur dioxide mixtures at 25–50 °C likewise leads to toluenesulfonic acid mixtures almost devoid of the *m*-cresol isomer, and which have an o-/p-isomer ratio as low as 3:97 [50]. In the absence of a sulfur dioxide solvent, it may even be possible to obtain an o- and m-isomer content of only 0.5% when the process is carried out in a loop reactor at 0 °C, and when gaseous SO_3, diluted with an inert gas, is introduced into the reactor at several locations [51]. The inert gas (waste gas) is not necessary if the process is carried out under vacuum [52].

Thermodynamically controlled sulfonation conditions (high temperature, long reaction time, and the use of a sulfonation mixture with a low water content) lead to an equilibrium containing ca. 5% *o*-, 54% *m*-, and 41% *p*-toluenesulfonic acid [53]. Alkali fusion of such mixtures then gives an isomer distribution of ca. 5% *o*-cresol, 56% *m*-cresol, and 39% *p*-cresol [47].

Technical-grade *m*-cresol can also be produced by toluene sulfonation according to a process used by the Honshu Chemical Industry Co. [54]. Steam at 165 °C is introduced into an toluenesulfonic acid mixture rich in m-isomer obtained under thermodynamically controlled conditions as described above (isomerization for several hours at 190–200 °C), whereupon *o*- and *p*-toluenesulfonic acid are hydrolyzed back to sulfuric acid and toluene. The toluene is distilled with the steam; *m*-toluenesulfonic acid with a purity above 90% remains, together with sulfuric acid. Side reactions during the isomerization and hydrolysis can be suppressed by adding sodium sulfate (5–10 mol%, calculated on the basis of sulfuric acid) to the sulfonation batch. In this way, removal of the sulfuric acid that is formed in the hydrolysis as a lower phase at ca. 140–150 °C is also possible [55]. It is also possible to neutralize the hydrolysis mixture with concentrated sodium hydroxide solution, filter off the resulting sodium sulfate at 80 °C, introduce the solution of sodium *m*-toluenesulfonate into a sodium hydroxide melt, containing 10–15% potassium hydroxide to facilitate stirring, at an initial temperature of 330 °C, and to fuse it at 340 °C.

The distillation of the crude cresol phase obtained after acidification gives *m*-cresol with a purity of up to 98% in a yield of ca. 65%, calculated on the basis of reacted toluene [55].

p-Cresol is produced according to the toluene sulfonation process by Sherwin–Williams (capacity 15 000 t/a) in the United States, by Croda Synthetic Chemicals Ltd. (capacity 12 000 t/a; sulfonating agent is SO_3) in the United Kingdom, and in Japan (ca. 5000 t/a). *m*-Cresol is produced in this way by the Honshu Chemical Industry Co. (ca. 2400 t/a). A plant designed to operate the Honshu process is also reported to be capable of producing 3000–5000 t/a of *m*-cresol has been built in India by Gujarat Aromatics.

The toluenesulfonic acid–cresol process is relatively simple as regards the plant required. Its primary drawback is the unavoidable formation of sodium sulfite in aqueous solution. At Croda Synthetic Chemicals, the sodium sulfite is upgraded into a product suitable for the paper industry [56]. At Sherwin–Williams, sodium sulfite is oxidized with air to form sodium sulfate [57], but *p*-cresol is still present in the aqueous sulfite solution (3000–5000 ppm); to avoid stripping, thereby causing an odor problem in the neighborhood of the plant, the solution is first passed through columns of activated carbon to adsorb the cresol. The activated carbon is regenerated every 24 h with 10% sodium hydroxide solution. The amount of *p*-cresol recovered in this way is sufficient to pay for the additional installations [58].

1.4.2. Alkaline Chlorotoluene Hydrolysis

The hydrolysis process is important in the production of cresol with a high m-cresol content, and is used in the Federal Republic of Germany by Bayer AG, the world's largest manufacturer of synthetic cresols (capacity more than 30 000 t/a).

In the first reaction step, a mixture of o- and p-chlorotoluene in a 1:1 ratio is produced by chlorinating 1 mol of toluene with 1 mol of chlorine in the presence of iron(III) chloride and disulfur dichloride ($\rightarrow$ Chlorinated Hydrocarbons).

$$CH_3-C_6H_5 + Cl_2 \longrightarrow CH_3-C_6H_4-Cl + HCl$$

In the next reaction step, this mixture is hydrolyzed with excess sodium hydroxide solution (2.5–3.5 mol/mol) at 360–390 °C and 280–300 bar (28–30 MPa) [59].

$$CH_3-C_6H_4-Cl + 2\ NaOH \longrightarrow CH_3-C_6H_4-ONa + H_2O + NaCl$$

The exothermic reaction is carried out continuously in high-pressure tubes several hundred meters long; the tubes are made of nickel steel to withstand the corrosive action of the reaction mixture. Separation of the components, which are introduced via a mixing chamber, is prevented by causing the reaction mass to flow at high velocity, with frequent changes in the direction of flow; finally, the reaction mixture is homogeneous.

The cresol in the resulting sodium cresolate solution is set free by neutralization; for this purpose the hydrochloric acid formed during chlorination can also be used. The coproduct sodium chloride can be returned to the chlor-alkali electrolysis.

For economic reasons, it is important that the starting materials are produced at locations sufficiently close to one another that conveyance through pipelines is possible. The technology of the process largely corresponds to that of the Dow–Bayer chlorobenzene–phenol hydrolysis process [60]. In the manufacture of cresols, however, the amounts of the byproducts formed are larger than in this process: these include ditolyl ethers [bis(methylphenyl) ethers], in particular, tolylcresols [methyl(methylphenyl)phenols], and small amounts of toluene, phenol, benzoic acid, and off-gas from cracking (methane and hydrogen). The removal of small amounts of phenol by azeotropic distillation, e.g., with chlorotoluene, is dealt with in [61].

Whereas it is hardly possible to control the scale on which tolylcresol is formed, the formation of ditolyl ether can be controlled via the temperature, the residence time, and the sodium hydroxide concentration; in addition, the scale on which this byproduct is formed can be minimized by recycling. This possibility arises from the fact that ditolyl ethers are hydrolyzable – though less easily than the homologous diphenyl ether. Therefore, if the reaction is optimally controlled, yields of ca. 80%, based on chlorotoluene, are attainable. However, ditolyl ether has several uses, e.g., as a heat transfer medium (Diphyl DT), as an electrical insulant (Baylectrol 4800; replacement for polychlorinated biphenyl), and as a precursor for tanning agents. Therefore, ditolyl ether is no longer recycled; indeed, the conditions may be adjusted to increase the selectivity for this compound.

The process gives very pure cresols, i.e., products that are almost free from other compounds. The o-, m-, and p-isomers are present in a ratio of ca. 1:2:1. This ratio differs from that of the chlorotoluene input because isomerization via aryne intermediates is assumed to occur [62], [63], occurs under the severe conditions of hydrolysis. After the o-cresol has been distilled, a m-/p-cresol mixture containing ca. 70% m-cresol is obtained.

The plant can also be operated on pure o- or p-chlorotoluene. The hydrolysis of p-chlorotoluene then gives a 1:1 m-/p-cresol mixture, and hydrolysis of o-chlorotoluene gives a 1:1 o-/m-cresol mixture. Technically pure m-cresol is obtainable from the latter by distilling off the o-cresol.

Hydrolysis of chlorotoluene without substantial isomerization is possible under milder conditions (200–350 °C) with alkali hydroxide [63], [64], alkali carbonates [65], or ammonium hydroxide [66], or with alkali acetate or alkali propionate solution [67] in the presence of copper or copper compounds. These processes are not used industrially.

1.4.3. Cymene Hydroperoxide Cleavage

The synthesis via cymene hydroperoxide, which is also known as the cymene–cresol process, permits the production of m- or p -cresol from the corresponding cymenes (isopropyltoluenes). It is unsuitable for the production of o-cresol. The process consists of three reaction steps:

1) Toluene propylation and cymene isomerization

2) Oxidation of cymene to cymene hydroperoxide

$$\text{cymene} + O_2 \xrightarrow[\text{(Alkali)}]{90-120\,°C} \text{cymene hydroperoxide}$$

3) Peroxide cleavage

$$\text{cymene hydroperoxide} \xrightarrow[\text{in Acetone}]{(H^+)} m/p\text{-Cresol} + CH_3-CO-CH_3$$

By comparison with the analogous Hock cumene process (→ Acetone), the cymene–cresol process has a lower space–time yield, involves considerably more costly distillation procedures, necessitates more extensive treatment of wastewater, and gives a product yield that is 20–30% lower. The reasons for this are as follows:

1) All three isomers, o-, m-, and p-cymene, are formed in the propylation of toluene. Of the three isomers, o-cymene not only is hardly oxidized but also increasingly inhibits oxidation of the other cymenes [68] as its concentration in the isomer mixture rises. For continuous oxidation, the o-cymene content of the cymene mixture must be kept below 10% at all times [69], [70]. Thus, cymene mixtures are needed that contain as little o-cymene as possible.

The lowest o-cymene content is that of a cymene mixture in thermodynamic equilibrium [70], [71]. Aluminum chloride must be used to reach this equilibrium as quickly as possible [72]–[74]. The isomer ratio obtainable with aluminum chloride under practical conditions is ca. 3% o-, 64% m-, and 33% p-cymene. As the o-cymene concentration rises while the oxidation is in progress, a large amount of the cymene must be returned to the first stage of the process after the cymene hydroperoxide and other byproducts have been separated, so that the o-cymene content can again be lowered to ca. 3% through isomerization on aluminum chloride. The diisopropyltoluenes formed as byproducts in the alkylation and separated by distillation are also returned to this isomerization step.

2) The oxidation rates of the cymenes are lower than that of cumene. In addition, the cymenes can be oxidized up to a peroxide content of only ca. 20%, whereas oxidation levels exceeding 30% are possible with cumene. The proportion of unreacted cymene that must be distilled and returned to the reaction is, therefore, greater than in the cumene–phenol process. At higher degrees of oxidation, the formation of byproducts from excessive oxidation of the isopropyl group to dimethyl(tolyl)methanol and methylacetophenone is greatly increased.

3) The methyl group of the cymenes is likewise oxidized; for details of the mechanism of p-cymene oxidation, see [75]. The ratio at which the methyl group and isopropyl group are subjected to oxidative attack may be as high as 1:4 [76]. As the primary peroxide is less stable than the tertiary peroxide, it is more easily converted into

secondary products (isopropylbenzyl alcohol, isopropylbenzaldehyde, cuminic acid). A more important drawback is that in the highly exothermic acid cleavage, in addition to isopropylphenol, 1 mol of formaldehyde is formed per mole of primary peroxide still present; because formaldehyde is able to bind up to 2 mol of cresol as a formaldehyde resin, the cresol yield is substantially diminished. Moreover, the presence of resin makes the product more difficult to work up. Various solutions to the problem have been proposed: (1) selective decomposition of the primary peroxide during the actual oxidation by adding catalytic amounts of picoline, pyrrole, thiophene [77], or triphenylphosphine [78]; (2) selective hydrogenation of the primary peroxide over a noble-metal catalyst [79] after the oxidation in a separate stage of the process (before the acid cleavage) or decomposition by passing the primary peroxide over a magnesium aluminate catalyst [80] or treatment with sodium hydroxide and heavy metals [81]; (3) extraction of the primary peroxide [82]–[85]. However, it is simpler to dissolve the products resulting from methyl group oxidation continuously and as fully as possible in excess alkali solution present during the actual oxidation itself [68], [69], [86]. The organic acids present in the spent caustics (ca. 0.5 kg of cuminic acid, toluic acid, phthalic acid, and other products per kilogram of cresol) are largely precipitated during the acidification and can be removed by filtration. Additional removal from the filtrate (wastewater) is possible by extraction [87].

According to another suggestion, hydrogen peroxide should be present during cleavage of cymene hydroperoxide to cresol [88]; alternatively, this cleavage could be effected under mildly hydrogenating conditions [89]. These measures decrease the extent of yield-lowering side reactions undergone either by the primary peroxide or by its secondary product, formaldehyde.

4) The reaction mixture after acidic peroxide cleavage contains not only cresol, acetone, and unreacted cymene, but also isopropenyltoluene, 2-tolylpropanol, isopropylbenzyl alcohol, isopropylbenzaldehyde, methylacetophenone, isopropylphenol, and numerous other compounds. The separation of these byproducts is relatively expensive because of their large number, their similar boiling points, and the formation of azeotropic mixtures [90].

Cymene–cresol plants, in which cymene mixtures are used and which have a nameplate capacity of 20 000 t/a each, have been operated in Japan by Mitsui and Sumitomo since 1969. The product has a *m*- and *p*-cresol content of more than 99.5 % and a m-/p-ratio of 60:40. The acetone coproduct is of high quality [91].

If pure *p*- or *m*-cymene is used, the high cost of recycling and isomerization necessitated by the presence of *o*-cymene is eliminated, and pure *p*- or *m*-cresol can be obtained directly. For example, until 1972 the Hercules Powder Co. operated a plant at Gibbstown in the United States for the manufacture of *p*-cresol from *p*-cymene obtained from natural terpenes [92].

Owing to the close proximity of the boiling points of the cymene isomers (*m*-cymene 175.1 °C, *p*-cymene 177.1 °C, *o*-cymene 178.3 °C, at 101.3 kPa), the *m*-cymene, at most, may possibly be isolated at reasonable cost from the cymene mixture resulting from the propylation of toluene (see above) [74]. However, both pure *m*-cymene and pure *p*-

cymene can be isolated according to UOP's Cymex process on *p*-cymene-selective 13X molecular sieves by using toluene as desorption agent [93]. The process corresponds to UOP's Parex process for the isolation of *p*-xylene from C_8 aromatics (→ Xylenes). A cymene – cresol plant, based on the UOP and Hercules Powder technologies, for the manufacture of pure *m*-cresol and *p*-cresol [for the antioxidant 2,6-di-*tert*-butyl-4-methylphenol (BHT)] is being constructed in China [94].

If in the Cymex process only one isomer, e.g., *p*-cymene, is required, the *m-/o*-cymene mixture leaving the adsorption column, which is impoverished with regard to *p*-cymene, can be reisomerized and returned to the separator [95]. Since it is less important in this case to have the lowest possible *o*-cymene content, the isomerization (and the alkylation) can, for example, be carried out on phosphoric acid – aluminum oxide catalysts at 370 °C instead of using aluminum chloride [95].

In the manufacture of *m*-cresol, *m*-cymene can also be obtained by selective cracking of the *p*-cymene present in cymene mixtures with a low ortho isomer content. Cleavage of *p*-cymene (and to a lesser extent that of *o*-cymene) to toluene and propylene is accomplished at 300 – 400 °C on shape-selective ZSM-5 zeolites [96] that are particularly selective in the MgO-modified Li^+-ZSM-5 form [97]. A further, technically less attractive way of isolating *m*-cymene consists in complexing with anhydrous hydrofluoric acid – boron trifluoride at temperatures below − 40 °C [98].

1.4.4. Methylation of Phenol

Synthetic cresol, especially *o*-cresol, and xylenol are now produced largely by methylation of phenol with methanol in the presence of catalysts. 2,6-Xylenol is produced almost entirely by this method.

The process consists of only one reaction step, but is based on phenol, which is relatively expensive; the separation of products from the reaction mixture is also costly. This is partly because the boiling points of several components are very near one another, and partly because in some cases the purity of the product must meet very strict requirements.

The reaction can be carried out either in the vapor phase or in the liquid phase.

Methylation in the Vapor Phase. Currently vapor-phase methylation is used exclusively to manufacture pure *o*-cresol and/or pure 2,6-xylenol:

$$\text{C}_6\text{H}_5\text{OH} + \text{CH}_3\text{OH} \longrightarrow \text{o-CH}_3\text{C}_6\text{H}_4\text{OH} + \text{H}_2\text{O}$$

$$\Delta H^\circ_{298} = -104 \text{ kJ/mol}$$

$$\text{o-CH}_3\text{C}_6\text{H}_4\text{OH} + \text{CH}_3\text{OH} \longrightarrow 2,6\text{-}(\text{CH}_3)_2\text{C}_6\text{H}_3\text{OH} + \text{H}_2\text{O}$$

$$\Delta H^\circ_{298} = -48 \text{ kJ/mol}$$

Side reactions:

$$\text{CH}_3\text{OH} \longrightarrow \text{CO} + 2\,\text{H}_2 \quad \Delta H^\circ_{298} = -91 \text{ kJ/mol}$$
$$\text{CO} + \text{H}_2\text{O} \longrightarrow \text{CO}_2 + \text{H}_2 \quad \Delta H^\circ_{298} = -41 \text{ kJ/mol}$$
$$\text{CO} + 3\,\text{H}_2 \longrightarrow \text{CH}_4 + \text{H}_2\text{O} \quad \Delta H^\circ_{298} = -206 \text{ kJ/mol}$$

At atmospheric or slightly elevated pressure, a superheated mixture of the vapors of phenol, methanol, and water is passed at a liquid hourly space velocity (LHSV) of 1–2 h^{-1} over a metal oxide catalyst fixed in a multitubular reactor of stainless steel at a temperature of 300–460 °C [99]–[101]. The reaction temperature required depends on the nature of the catalyst and on the desired composition of the product. The temperature is permitted to deviate from this level by only a few degrees.

The heat of the reaction is dissipated by boiling organic heat transfer media [102], by circulating salt melts [100], or through the generation of high-pressure steam [101]. The slight loss of catalytic activity that occurs in the course of time is compensated by raising the temperature accordingly. Carbon deposits are removed from time to time by burning. The presence of water in the feedstock mixture suppresses the decomposition of methanol, extends the regeneration cycle, and prolongs the lifetime of the catalyst [103], [104], which may then endure for several thousand hours.

The product mixture leaving the reactor gives up some of its heat to the feedstock mixture in a heat exchanger, and is then condensed. The uncondensable reaction products formed from the decomposition of methanol (H_2, CH_4, CO_2, and CO) are used as fuel gas in steam production. Aqueous methanol is first distilled from the liquid reaction product and then recycled. Water remaining in the product is then removed either by azeotropic distillation with toluene [102] or as an azeotropic mixture with unreacted phenol [101]. In the latter case, it is necessary to strip the phenol from the wastewater. The dehydrated mixture is then fractionated into phenol ether, phenol, *o*-cresol ($\geq 99.5\%$), and 2,6-xylenol in an additional series of highly efficient distillation columns operated continuously under vacuum. The phenol ether–phenol fraction is recycled. The distillation residue, which contains, among other compounds, 2,3-, 2,4-, and 2,5-xylenol and 2,4,6- and 2,3,6-trimethylphenol, is usually burned as a fuel oil. If the catalyst has high ortho selectivity, pure (99.5%) 2,6-xylenol can be obtained directly. If that is not the case or if the reaction conditions are too severe (high temperature, long residence time), the 2,6-xylenol fraction will contain *m*- and *p*-cresol, which must be removed by special methods (cf. Section 2.3). The ratio of *o*-cresol to 2,6-xylenol can be controlled by altering the methanol–phenol ratio, and if the *o*-cresol is recycled, 2,6-xylenol comprises the sole product. Over an extended production time, the yield (selectivity) of *o*-cresol and 2,6-xylenol, calculated on the basis of reacted phenol, is 90–95%.

With regard to the use of a fluidized-bed reactor for this reaction, see [105], [106]. Vapor–liquid equilibrium data for the alkylation product are given in [107], and details of the distillation in [108]–[110].

Magnesium oxide catalysts, in particular, are used. They require relatively high operating temperatures (420–460 °C), but are so ortho-selective that *m*- and *p*-cresol are hardly formed, even at 540 °C [111]. The activity (phenol conversion), as well as the ratio of 2,6-xylenol to *o*-cresol and the ortho selectivity depend on the conditions under which the magnesium oxide is produced [112]–[114]. These properties can be further improved by combining magnesium oxide with other oxides, such as those of manganese [113], [115], copper [113]–[115], titanium, uranium, zirconium, chromium [116], and silicon [117]. At 440 °C and at a liquid hourly space velocity (LHSV) of 2.0 h^{-1} the continuous reaction of a mixture of phenol, methanol, and water in a 1:4:2.4 molar ratio over a MgO catalyst thus gives a phenol conversion of ca. 82%, and a molar ratio of 2,6-xylenol, *o*-cresol, *p*-cresol, 2,4-xylenol, and 2,4,6-trimethylphenol of 69.5:23.0:0.05:0.8:6.6. If the catalyst also contains ca. 1% of copper, the conversion rate is 89%, and the corresponding molar ratio of the products is 73.3:18.9:0.06:0.07:7.0. The waste gas volume is in both cases 0.5 mol per mole of reacted phenol [114]. The mechanism of ortho methylation is discussed in [118].

With *manganese oxide* catalysts, the methylation can be carried out at ca. 420 °C [119]; on *chromium oxide* catalysts, at ca. 390 °C [120]–[122]; and on catalysts based on *vanadium oxide* [123], [124] or *iron oxide* [125]–[130], at ca. 350 °C, provided that these oxides are combined with smaller amounts of other oxides (e.g., of Co, Ni, Zn, Al, Ga, Si, Ge, Sn, Pb, Bi, Mg, and Na) or with one another [131]–[134]. These mixed oxide catalysts show excellent ortho selectivity and give a very high 2,6-xylenol/*o*-cresol ratio and virtually no *p*-cresol. At phenol/methanol molar ratios of ca. 1:5–6 and liquid hourly space velocities (LHSV) of ca. 1 h^{-1} they permit almost complete conversion of phenol, with 2,6-xylenol yields of ca. 95% (calculated on the basis of phenol). Combination with other metal oxides also extends the catalyst lifetime (especially if alkali-metal compounds are added) and diminishes the extensive decomposition of methanol typically seen with manganese, chromium, vanadium, and iron oxide catalysts. In spite of this measure and the addition of water vapor (generally 1–2 mol per mole of phenol), 1–2 mol of methanol are usually lost by decomposition for every mole of 2,6-xylenol manufactured. The proposed methods for reducing the loss include the addition of sulfur dioxide, hydrogen sulfide, nitric oxide, carbon monoxide, or carbon dioxide to the feedstock mixture [135], recycling a portion of the cracking gas [136], and introducing the methanol [137] or phenol at different parts of the reactor [138]. Relatively little methanol is decomposed on Fe_2O_3 catalysts doped with silicon oxide and magnesium oxide [128]. For the mechanism of the reaction on mixed oxide contacts, see [139], [140].

γ-Aluminum oxide catalyzes the methylation of phenol at a temperature as low as 300–320 °C. Under these conditions, at a methanol:phenol molar ratio of 1:2, and at a liquid hourly space velocity (LHSV) of 1 h^{-1}, *o*-cresol, 2,6-xylenol, and *m*-/*p*-cresol mixture are formed in a molar ratio of ca. 82:17:1, with a total selectivity of 97% and a phenol conversion of 33% [141]. Increasing the methanol:phenol ratio and recycling the *o*-cresol enables the *o*-cresol:2,6-xylenol ratio to be shifted toward 2,6-xylenol [142].

On γ-Al$_2$O$_3$, unlike other methylation catalysts, methanol is scarcely decomposed under the conditions described above. In addition, this catalyst must be regenerated only once in ca. 1000 h, and has a working lifetime of several years [143]. Its disadvantage is that to prevent the formation of significant quantities of other di-, tri-, and tetramethylphenols, the phenol conversion rate must not substantially exceed 60%, which means that relatively large amounts of the product must be worked up and recycled. In addition, extensive operations, e.g., extractive distillation with diethylene glycol [142], are necessary to free the 2,6-xylenol from the *p*-cresol that is unavoidably formed on this catalyst.

Below 300 °C, increasing amounts of anisole are formed on γ-Al$_2$O$_3$; however, because anisole is isomerized to cresol at elevated temperatures [144], [145], only a small amount ($\leq$ 2.5%) is present in the reaction mixture under the previously mentioned conditions. As the temperature and residence time are increased, the rate of phenol conversion on Al$_2$O$_3$ rises, but the ortho selectivity falls. This is also basically true of other catalysts. Instead of *o*-cresol and 2,6-xylenol, increasingly large amounts of *m*- and *p*-cresol, other xylenols, multimethylphenols, and methylbenzenes, especially hexamethylbenzene, are formed [146], [147]. Demethylation occurs above 500 °C [148]. Increasing the temperature also results in the formation of larger amounts of gaseous cracking products, and accelerates the carbonization of the catalyst, thus necessitating more frequent regeneration.

The reaction kinetics and the choice and calculation of an optimum multitubular reactor for phenol methylation on γ-Al$_2$O$_3$ are discussed in [149]–[151]. Industrial plants are described in [99] and [142].

Aluminum oxides with strong acid sites [152], *silica–alumina* [118], [153], *zeolites* [153]–[156], *aluminum phosphates* [147], [158], and *phosphoric acid–kieselguhr* [158] have enhanced isomerization and transmethylation properties. Therefore, they lead to products with a relatively high m/p content, especially if the temperature is simultaneously raised to ca. 450 °C. In addition, at a temperature of 250–300 °C, shape-selective zeolites [154]–[156], e.g., of the HKY type [154], and also phosphoric acid–kieselguhr [158], show remarkably high selectivity for *p*-cresol, giving a *m*-/*p*-cresol mixture with *p*-cresol contents of ca. 85% [154] or 77% [158], respectively. *Titanium dioxide* also shows para selectivity [158].

As alternatives to methanol, methylbenzenes [160], [161], dimethyl ether [162], methylamine [163], methane [164], carbon monoxide–hydrogen mixtures [165], and formaldehyde–hydrogen mixtures [166] have been proposed as methylating agents.

Methylation in the vapor phase is utilized by General Electric in the United States (capacity ca. 10 000 t/a *o*-cresol and 70 000 t/a 2,6-xylenol) and in the Netherlands (ca.

4000 t/a *o*-cresol and 16 000 t/a 2,6-xylenol); by Croda Synthetic Chemicals in the United Kingdom (ca. 8000 t/a *o*-cresol and 2,6-xylenol); and in Japan by Nippon Crenol (joint venture of Asahi Chemical and Nippon Steel Chemical; 5000 t/a *o*-cresol and 10 000 t/a 2,6-xylenol) and Gem-Polymer (joint venture of General Electric and Mitsui; 2300 t/a *o*-cresol and 18 000 t/a 2,6-xylenol). The facilities of Conoco and Koppers in the United States were closed down (temporarily) in 1983 for market reasons. Cindu, Rütag, and Midland Yorkshire Tar Distillers' (CRM) plant in the United Kingdom has been taken over by Croda. Vapor-phase methylation plants are said to be in operation in the USSR [143] and in Czechoslovakia [150].

Methylation in the Liquid Phase. Results similar to those achieved in the vapor phase are obtained if the oxide catalysts (particularly γ-Al_2O_3) are suspended in a phenol–methanol mixture and heated in an autoclave to 300–400 °C [167]–[169]. The mixture of phenol and methanol [170] or of *o*-cresol and methanol [171] can also be passed as a liquid over a γ-Al_2O_3 catalyst forming a fixed bed in a tube reactor. At 350 °C and a phenol:methanol molar ratio of 0.7:1, the pressure is ca. 35 bar [170].

According to a process operated by Chemisches Werk Lowi, Federal Republic of Germany, phenol and an equimolar amount of methanol are reacted in an autoclave at 350–400 °C in the presence of 2–3 wt% *aluminum methylate*, with the water of the reaction, together with some of the methanol, being distilled through a small column. After the catalyst residue has been removed by filtration, the product is worked up by distillation. Methanol, anisole, and phenol are recycled. The yield of cresols and xylenols is ca. 80% [172], calculated on the basis of reacted phenol (ca. 60%). The main product is *o*-cresol; *m*- and *p*-cresol, 2,6-, 2,4-, and 2,3-xylenol, and trimethylphenols are also formed. The ratio of the products can be adjusted by altering the phenol:methanol ratio, the temperature, and the residence time. The presence of bases (e.g., pyridine) increases the conversion [173], whereas that of acids raises the proportion of *p*-cresol and 2,4-xylenol in the product [174]. Aluminum phenolate can also be used as a catalyst [175]. If necessary, the process can be operated in a multipurpose plant, the capacity of which may be ca. 2000–3000 t/a.

Union Rheinische Kraftstoff AG operates a process in the Federal Republic of Germany that uses *liquid-phase catalysts* [176], [177], preferably aqueous zinc halide–hydrogen halide solutions [178], at 200–240 °C and pressures of ca. 25 bar (2.5 MPa). After the reaction, the organic components are separated from the aqueous catalyst phase by extraction with benzene or cyclohexane, washed with water several times, and worked up by distillation. The catalyst solution is united with the washing water, concentrated, and recycled [179]. The main products are *o*-cresol, *p*-cresol, 2,4-xylenol, and 2,6-xylenol. At a phenol:methanol molar ratio of 1:0.6, yields of up to 98% are obtainable, calculated on the basis of reacted phenol. Compared with other processes, this has the advantages not only that *p*-cresol and 2,4-xylenol are obtained in considerably larger quantities, but also that, provided the reaction conditions are carefully harmonized, the *p*-cresol is almost free of *m*-cresol. The extent of para substitution increases with the methanol content of the feedstock mixture, but the yield falls. The process requires materials that are highly resistant to corrosion, e.g., a reactor lined with tantalum. The plant has a capacity of ca. 16 000 t/a. The product ratio is controllable within certain limits. As an example, the *o*-cresol, *p*-cresol, 2,4-xylenol, and 2,6-xylenol ratio can be 7:5:2:1.

Transmethylation and Isomerization. After phenol has been added to the methylphenol residues either from the methylation of phenol or of different origin, or to 2,6-xylenol for which there is no other use, reaction can occur in the presence of alkylation catalysts to give a product consisting mainly of *o*-cresol, together with *m*-/*p*-cresol. The reaction may proceed either in the vapor phase or in the liquid phase; the temperature is 400–500 °C. Catalysts are aluminum oxide [180], chromium oxide–aluminum oxide [181], silica–alumina [182], metal oxide–iron oxide [183], or magnesium oxide–tungsten oxide [184]. In the case of 2,6-xylenol, this transmethylation, accompanied by isomerization, can also be accomplished without a catalyst in the vapor phase at 550–600 °C or, batchwise, in the liquid phase at 420–470 °C [185]. In this way, the *o*-cresol yield from the methylation of phenol can be raised.

Owing to the temporary lack of demand for *o*-cresol, opportunities for isomerizing it to *m*- and *p*-cresol have been investigated. The results formerly obtained with aluminum chloride [186], [187], aluminum fluoride, boron trifluoride, aluminum silicates, or aluminum oxide were unsatisfactory [187], [188] because resinification and disproportionation to phenol, dimethylphenols, etc. occurred in addition to isomerization.

Isomerization without substantial disproportionation is possible, however, on silica–alumina zeolites of the ZSM group: at a temperature of ca. 400 °C and *o*-cresol conversions of 40–50%, *m*-/*p*-cresol selectivities of 90% and an m:p ratio of 7:3 are possible [189].

1.4.5. Other Routes

The rising demand for cresol has stimulated the development of numerous syntheses, especially those that lead selectively to certain cresol isomers. It appears, however, that these processes are not yet being used industrially.

Gulf Oxychlorination. A cresol synthesis analogous to the Raschig phenol synthesis cannot be carried out satisfactorily under industrial conditions: at the temperature needed for oxychlorination in the vapor phase (5), toluene is oxidized to a considerably greater extent than benzene, and only low rates of toluene conversion are, therefore, possible.

$$CH_3-C_6H_5 + HCl + 0.5\, O_2 \longrightarrow CH_3-C_6H_4-Cl + H_2O \quad (5)$$
$$CH_3-C_6H_4-Cl + H_2O \longrightarrow CH_3-C_6H_4-OH + HCl \quad (6)$$

In a process developed by the Gulf Research Development Co., these disadvantages are avoided by oxychlorinating the toluene with aqueous hydrochloric acid and oxygen at a temperature of ca. 100 °C in the presence of catalytic amounts of nitric acid and a palladium or copper salt [190], [191]. At toluene conversion rates of 80%, primarily *o*- and *p*-chlorotoluene are obtained in a 2:1 molar ratio and with a selectivity of ≥ 95%.

The subsequent hydrolysis of the chlorotoluene mixture is carried out in the vapor phase at 400–450 °C, as in the Raschig process, but over improved catalysts, such as lanthanum phosphate that may be activated with copper [192]. At a chlorotoluene conversion of 20%, the cresol selectivity is ca. 95%. The isomer ratio differs from that given by the processes already being used industrially. A large amount of *o*-cresol is obtained in addition to *p*-cresol containing some *m*-cresol.

If the hydrochloric acid set free in the hydrolysis step (6) is returned to the oxychlorination reactor, the process becomes equivalent to the oxidation of toluene with oxygen. This fact, together with the high conversion rates and selectivities, makes the process interesting. Commercial exploitation of the process necessitates the extensive recycling and places severe demands on the resistance of the plant to corrosion.

The Gulf oxychlorination process is equally applicable to xylenes, and is, therefore, also suitable for the manufacture of xylenols.

Oxidative Decarboxylation of Methylbenzoic Acids. As in the Dow process for the manufacture of phenol from benzoic acid, methylbenzoic acids are decarboxylated to cresol when an air–steam mixture is passed through their melt at 200–240 °C in the presence of copper and magnesium salts [193].

$$\text{CH}_3\text{-C}_6\text{H}_4\text{-COOH} + 1/2\, O_2 \xrightarrow[\text{Catalyst}]{H_2O} \text{CH}_3\text{-C}_6\text{H}_4\text{-OH} + CO_2$$

By comparison with the manufacture of phenol, the cresol process is more complex and 10–15% less selective [48]. This is due partly to the fact that cresols have higher boiling points; in consequence, optimal control of the reaction is difficult, and a greater amount of tar is formed [193]. In addition, the simultaneous successive oxidation of the methyl group leads to numerous byproducts [194]. Oxidation in the presence of benzoic acid has been proposed [195] as a mean of preventing the sublimation of methylbenzoic acid.

Better results are achieved when a methylbenzoic acid–steam mixture is oxidized in the vapor phase with oxygen–nitrogen mixtures over mixed catalysts at a temperature of ca. 300 °C [196], [197]. Thus, the reaction of a mixture of *p*-methylbenzoic acid, water, oxygen, and nitrogen (molar ratio of 1:18:1:9) at 300 °C over a Mo–V–Cu–Zn–Na oxide catalyst supported on γ-Al_2O_3 is reported to give a *m*-cresol selectivity of 90% at a methylbenzoic acid conversion rate of 80%; as byproducts, 2% toluene, 5% bis-(methylphenyl)–bis(methylphenyl) ether mixture, and 3% of additional carbon dioxide are formed [197].

The process could be of interest in the manufacture of pure *m*-cresol, since only *m*-cresol is formed from both the ortho and the para forms of methylbenzoic acid [48], [193]. *m*-Methylbenzoic acid gives (in the liquid phase) *o*- and *p*-cresol in a 1:2 ratio [198], [199]. Regarding the mechanism and kinetics, see [193], [194], and [200], respectively.

The required methylbenzoic acids can be obtained by oxidation of the corresponding xylenes, which are, however, more expensive than toluene.

Baeyer–Villiger Oxidation of *p*- or *o*-Methylbenzaldehyde. This route leads to *o*-cresol or, more interestingly from the practical aspect, specifically to *p*-cresol. Hydrogen peroxide reacts with excess formic acid to give performic acid (7), which oxidizes *p*-methylbenzaldehyde to *p*-tolyl formate (8); hydrolysis (9) then leads to *p*-cresol and formic acid.

$$HCOOH + H_2O_2 \rightleftharpoons HCOOOH + H_2O \qquad (7)$$

$$\text{p-CH}_3\text{-C}_6\text{H}_4\text{-CHO} + HCOOOH \longrightarrow \text{p-CH}_3\text{-C}_6\text{H}_4\text{-O-CHO} + HCOOH \qquad (8)$$

$$\text{p-CH}_3\text{-C}_6\text{H}_4\text{-O-CHO} \xrightarrow{+H_2O} \text{p-CH}_3\text{-C}_6\text{H}_4\text{-OH} + HCOOH \qquad (9)$$

Several variants of the process have been developed [201]–[203]. The most favorable process was probably invented by Mitsubishi [204]. Hydrogen peroxide (1.2 mol; 90%) is added to a mixture consisting of 1 mol of *p*-methylbenzaldehyde, 14 mol of formic acid, and ca. 10% (4.5 mol) of water. The temperature of the mixture is maintained at 60–90 °C; the heat of the reaction is removed by reflux condensation under vacuum. Because of the presence of water, formate is hydrolyzed; after a reaction time of 1 h, the *p*-methylbenzaldehyde conversion rate is 100% and the *p*-cresol yield is 85%. If, instead of 90% hydrogen peroxide, performic acid is used, or if the process is carried out in the presence of a solid acid (zeolite or ion-exchange resin), the yield rises to 90–92%. The product is worked up by distilling the formic acid. Apart from the additional amount formed in the process, the formic acid, which has a concentration of ca. 90%, is recycled. The *p*-cresol is then isolated from the residue, which still contains 10–20% of high-boiling constituents [204].

The starting material, *p*-methylbenzaldehyde, can be produced from toluene and carbon monoxide by the so-called MGC PTAL process developed by Mitsubishi [205]. It is an intermediate in the manufacture of terephthalic acid.

Nucleus Hydroxylation. The direct hydroxylation of toluene (or xylene), e.g., with oxygen [206], hydrogen peroxide [207], [208], inorganic peroxo compounds [209], or organic peroxides [210], [211], in the presence of various catalyst systems (Friedel–Crafts catalysts, transition-metal compounds, or irradiation), has been extensively investigated.

According to Rhône–Poulenc, the most successful results so far have been obtained with 85% hydrogen peroxide in the presence of a large amount of trifluoromethanesulfonic acid and a small amount of phosphoric acid (for complexing metal ions) [212].

$$C_6H_5-CH_3 + H_2O_2 \xrightarrow[-20\ °C]{CF_3SO_3H} \text{o-cresol} + \text{p-cresol}$$

$$2 : 1$$

At −20 to −15 °C and a molar ratio of H_2O_2, toluene, CF_3SO_3H, and H_3PO_4 of ca. $1:10:17:0.04$, o- and p-cresol are obtained in a $2:1$ ratio in 80% yield (calculated on the basis of hydrogen peroxide) at a toluene conversion rate of 10.6%. One of the problems is the recycling of the very expensive trifluoromethanesulfonic acid.

Work carried out by UOP indicates that relatively good results are obtainable with hydrogen fluoride, which is less expensive and easier to handle industrially than trifluoromethanesulfonic acid, in the presence of carbon dioxide. With 30% hydrogen peroxide at 0 °C and at a molar ratio of H_2O_2, toluene, HF, and CO_2 of $1:10:70:10$, o- and p-cresol are obtained in a $2:1$ ratio in 67% yield, based on hydrogen peroxide, in addition to a small amount of m-cresol. m-Xylene gives mainly 2,4-xylenol, p-xylene mainly 2,5-xylenol, and o-xylene mainly 2,3- and 3,4-xylenol [213]. The effect of the carbon dioxide is not fully understood, but this compound acts to some extent as an internal coolant.

On titanium-modified zeolites, o-, m-, and p-cresols in a $14:11:75$ ratio are formed in a yield of at least 40%, based on H_2O_2, when a mixture of H_2O_2, water, toluene, methyl isobutyl ketone, and titanium zeolite (pretreated with H_3PO_4) is heated under reflux in a mass ratio of $1:7:17:10:2.6$ [214].

ICI has developed a cyclic process equivalent to the oxidation of toluene with peracetic acid, which gives yields of up to 95% of an o-/p-cresol mixture, free of the meta isomer, with a cresol content of 92%. The "catalyst" in the reaction is iodotoluene. This is oxidized by peracetic acid to iodosyltoluene, which without isolation reacts with toluene and sulfuric acid to form ditolyliodonium hydrogen sulfate. Hydrolysis is effected by diluting the reaction mixture. Cresol is produced; iodotoluene is simultaneously regenerated and recycled [215].

All processes involving nucleus hydroxylation are unsatisfactory for several reasons; for example, the cresols must be isolated from very dilute solutions, and considerable amounts of auxiliaries and unreacted starting materials need to be recycled with as little waste as possible.

Oxidative Methylation of Toluene. In the USSR, a pilot plant has been operating in which toluene is oxidized adiabatically in a tube reactor at 700–750 °C and with contact time of 0.2–0.5 s with oxygen in the presence of methane and steam (molar ratio $1:1-2:10-20:2-6$) [216], [217].

$$C_6H_5-CH_3 + O_2 + CH_4 \longrightarrow \begin{cases} C_6H_5-CH_2CH_3 \rightarrow C_6H_5-CH=CH_2 \\ C_6H_5OH + CH_3-C_6H_4-OH \\ C_6H_6 \end{cases}$$

At a toluene conversion of ca. 45%, 0.52 t of styrene, 0.25 t of benzene, ca. 0.11 t of phenol, and ca. 0.06 t of cresol are obtained per ton of toluene [218], [219]. Thus, if the process were to be used on the scale customary for petrochemical operations, it should be possible to produce considerable amounts of cresol by the "direct oxidation" of toluene.

Hock Reaction of Toluene Derivatives. As in the cymene–cresol process, other toluene hydrocarbons can be peroxidized and cleaved to give cresol and the corresponding carbonyl compound. Particular interest has been shown in toluene derivatives that give *p*-cresol and are more readily accessible than *p*-cymene, e.g., ethylidenebis(*p*-methylbenzene) (coproduct: *p*-methylacetophenone) [220], [221], 4-methyl-cyclohexylbenzene (coproduct: cyclohexanone) [222], and *p*-xylene (coproduct: formaldehyde) [223], [224]. Oxidation of xylene in the presence of acetic acid gives the corresponding cresyl acetate, which can be hydrolyzed to form cresol [225].

Hydrogenation of N,N'-Dialkylaminomethylphenols. The reaction of phenol with paraformaldehyde and, e.g., piperidine in methanol at a 2:1:1:100 molar ratio at ca. 10 °C gives a mixture of piperidinomethylphenols, which without isolation is directly hydrogenated at 120–130 °C and 1 MPa over a palladium–carbon catalyst:

o-Cresol and *p*-cresol in a 1:2 molar ratio are obtained in a yield of ca. 90%, based on reacted phenol, or of about 75%, based on the formaldehyde input. In addition, some 2,4-xylenol is formed (2–5%) without *m*-cresol [226]. The yield is lower (65–70%) if dimethylamine is substituted for piperidine.

Diels–Alder Ring Closure of Isoprene and Vinyl Acetate. The Diels–Alder route leads selectively to *p*-cresol in three reaction steps: (1) Isoprene (1 mol) reacts with vinyl acetate (20 mol) in the presence of 1% hydroquinone in an autoclave at 180 °C; (2) the reaction product, 1-methylcyclohexene-4-yl acetate, is saponified; and (3) finally, the 1-methylcyclohexen-4-ol is dehydrogenated catalytically to *p*-cresol [227]:

High yields are claimed to be possible when the starting materials are recycled.

Carbonylation. 2-Methallyl chloride, acetylene, and carbon monoxide in acetone as solvent give *m*-cresol in 74% yield at room temperature.

The catalyst consists of tetracarbonylnickel, sodium iodide, iron powder, and thiourea; magnesium oxide neutralizes the hydrochloric acid formed [228].

From Toluidines. Under severe conditions, toluidines in the presence of acidic catalysts (H_3PO_4, HCl, $ZnCl_2$, BF_4) can be hydrolyzed to form the corresponding cresols [229], [230]. Higher yields can be achieved by boiling the diazonium salts obtainable from toluidines [231], [232]. Continuous operation of the process is described in [233].

Hydrodealkylation of Xylenols. Xylenols can be demethylated to phenol and cresols with hydrogen under pressure, either thermally at 600–700 °C [19] or on metal oxide catalysts at 400–500 °C [234]. Simultaneous deoxygenation enables BTX aromatics (BTX = benzene–toluene–xylene) to be produced at the same time. In this way, additional cresols can be produced, if necessary, from the xylenol fractions that are formed on a large scale in the gasification and hydrogenation of coal and other processes.

1.5. Separation of *m*- and *p*-Cresol [24]

Only *o*-cresol can be separated as a pure product by distillation from mixtures of the three cresol isomers. *m*-Cresol and *p*-cresol are obtained as a single fraction because the difference between their boiling points is too small for separation by distillation, even when carried out under vacuum.

1.5.1. Physicochemical Processes

Crystallization. According to the melting point diagram for *m*-/*p*-cresol mixtures (Fig. 1), it is possible to obtain, at normal pressure, pure *p*-cresol from mixtures containing more than 58% *p*-cresol, and pure *m*-cresol from mixtures containing more than 89% *m*-cresol. Meta/para mixtures containing 42–89% of *m*-cresol cannot be separated by crystallization at normal pressure because *m*- and *p*-cresol exist as a molecular compound in a 2:1 molar ratio within this concentration range.

The range within which *p*-cresol can be isolated by crystallization can be extended by applying pressure (Fig. 1); simultaneously the crystallization temperature and also the *p*-cresol yields, relative to crystallization at normal pressure, are raised [235].

Crystallization is of practical importance as a technique for the separation of pure *p*-cresol from *m*-/*p*-cresol mixtures of high *p*-cresol content obtained from syntheses or other processes. Up to now, this method has generally been carried out at normal pressure. In Japan, a high-pressure crystallization technique known as the *Finecry process*, in which a liquid *m*-/*p*-cresol mixture is introduced into a high-pressure vessel of the piston cylinder type and crystallized adiabatically at 200 MPa, has been recently developed for industrial use by Kobe Steel. After the mother liquor has been drawn off and the plant decompressed, the *p*-cresol emerges as a pure crystalline product. The computer-controlled crystallization–decompression cycle lasts 2–5 min [236]. It has been reported that a plant for the manufacture of 500 t/a of *p*-cresol is to be constructed for Sumitomo Chemical in Japan [237].

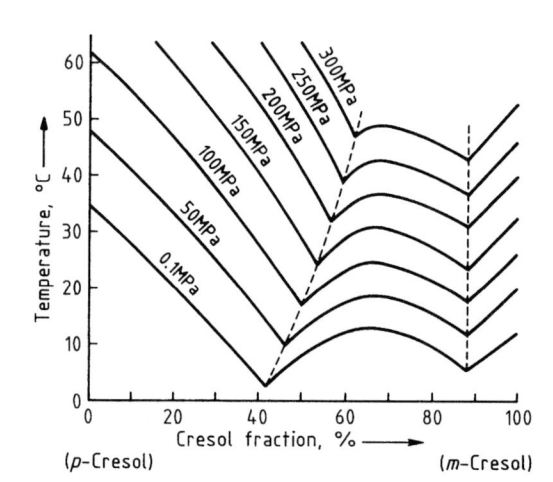

Figure 1. Liquid–solid phase diagram of *m*- and *p*-cresol mixtures under pressure [235]

Adsorption. Alkali metal-modified or alkaline earth metal-modified zeolites of type X, A [238], L, or ZMS-5 [239], and also titanium dioxide [240], adsorb *p*-cresol more strongly than *m*-cresol. Thus, one can separate *m*-/*p*-cresol mixtures in an adsorption column and dissolve them again with a suitable desorbing liquid such as an aliphatic alcohol [238] or a ketone [241]. A continuous version of this process has been developed by UOP [93]. It is known as the Cresex process and is used industrially in the United States by Merichem. The separating efficiency of the process not only depends on the adsorbent, but is also influenced to a large extent by the desorbent. With 1-hexanol, for example, only one isomer, eiher *m*- or *p*-cresol, can be isolated at a purity of ca. 99 % in a single operation (1 column); the purity of the other isomer obtained is only ca. 91 %. With 1-pentanol, however, a mixture containing ca. 69 % *m*-cresol and 31 % *p*-cresol can be separated into an extract containing 98.9 % of *p*-cresol and a raffinate containing 99.2 % of *m*-cresol in a single operation [93]. Phenolic impurities present in the cresol feed, e.g., *o*-cresol, *o*-ethylphenol, 2,4-, 2,5-, and 2,6-xylenol, migrate into the *m*-cresol during the adsorptive separation; hence, only pure *p*-cresol can be obtained in one operation if these impurities are present in significant quantities [93].

Other Methods. The separation of *m*-/*p*-cresol mixtures by azeotropic distillation with benzyl alcohol [242] or by countercurrent extraction with selective solvents (aqueous methanol–gasoline) [243] has acquired no practical importance. This also applies to dissociative extraction and distillation processes that exploit the higher dissociation constant (higher acidity) of *m*-cresol in relation to that of *p*-cresol. Dissociative extraction uses systems consisting of aqueous sodium hydroxide solution [244], [245], aqueous trisodium phosphate solution [246], or aqueous aminoethanol solution [244] and an organic solvent, e.g., benzene, hexane, or toluene, by which the *p*-cresol is isolated in a multistep extraction. In dissociative distillation, an amount of sodium hydroxide solution equivalent to the *m*-cresol content is added to the mixture and the *p*-cresol is distilled through a column with a large amount of water vapor as an azeotrope [247].

1.5.2. Separation via Addition Compounds

m-Cresol and *p*-cresol form addition compounds with anhydrous calcium bromide; the compound formed with *p*-cresol has the higher thermodynamic stability. If a *m*-/*p*-cresol mixture in toluene is heated to 100 °C with 1 mol of calcium bromide per mole of *p*-cresol, the precipitate that is filtered off after ca. 8 h consists practically only of the adduct of *p*-cresol, while the *m*-cresol content of the mother liquor is correspondingly higher. If the *p*-cresol adduct is then decomposed, e.g., by heating to 200 °C, *p*-cresol with a purity of ca. 98 % is obtained, together with reusable calcium bromide [248]. Pure *m*-cresol can be isolated from the mother liquor rich in m-isomer by selective adduct formation, e.g., with sodium acetate [249] or urea [250]. Pure *m*-cresol can also be obtained from the mother liquor after cooling to ca. 0 to −5 °C and adding calcium bromide. Under these kinetically controlled conditions, the main product is the *m*-cresol–calcium bromide adduct, the subsequent decomposition of which gives *m*-cresol [251]. In adduct formation with excess calcium bromide at 25–30 °C, the *m*-cresol content of the mother liquor is directly rendered practically free of p-isomer. The *p*-cresol from the adduct then contains correspondingly more *m*-cresol [252]. It is not known whether this process, which was developed by the Koppers Co. and has many variants, is being used industrially. Lithium bromide, magnesium bromide, or magnesium chloride can be used in place of calcium bromide, but the results are less satisfactory [248]–[252].

Many organic nitrogen compounds react with *m*- or *p*-cresol to form crystalline addition compounds that can be separated from the reaction mixture and decomposed to regenerate the nitrogen compound and the respective cresol isomer [24]. Benzidine was formerly considered particularly suitable for the separation of *m*- and *p*-cresol by virtue of the position of, and difference between, the melting points of its adducts; however, not only the carcinogenic properties of this compound but also other considerations now preclude its use. A more suitable and practical process is one in which benzylamine as the auxiliary substance forms with *m*-cresol a 1:1 adduct, which has a melting point of 39.5 °C. The adduct is centrifuged and fractionated in two distillation columns operating at different pressures (101.3 and 11.3 kPa) in such a manner that the benzylamine is recovered at the top of column 1 while pure *m*-cresol is obtained at the top of column 2. The mother liquor is distilled in the same way, then pure *p*-cresol is isolated by crystallization from the *m*-/*p*-cresol mixture rich in p-isomer obtained in this case at the top of column 2. The benzylamine is recycled [253].

The isolation of *m*-cresol via the 1:1 addition compound with urea is a process of some industrial importance. Formerly, the separated adduct was decomposed by heating with water. Now it is decomposed in the organic solvent in which the adduct has been formed. The procedure is as follows:

Urea (1.25 mol per mole of *m*-cresol) is added to a mixture of *m*-/*p*-cresol (65:35; 2 parts by weight) and toluene (0.5 parts by weight) at 30 °C. The mixture is cooled to − 10 °C, centrifuged, and washed with toluene (0.2 parts by weight). The product, consisting mainly of the urea adduct, is introduced into toluene (1.5 parts by weight), heated at 80 °C for 10 min, and filtered hot; almost all the urea is thereby held back. Distillation of the filtrate gives toluene, while that of the residue gives ($\geq$ 98%) *m*-cresol in 89% yield [254].

Advantage may be taken of adduct formation between cresols and certain other phenols with a view to effecting separation. For example, *m*-cresol forms with phenol an adduct with a melting point of 25.9 °C [255], while *p*-cresol forms an adduct with bisphenol A [isopropylidenedi(*p*-phenol)] [256]. Furthermore, *m*-cresol gives with 1,4-dioxane at 20 °C and 110 – 130 MPa an addition compound that has been proposed for the separation of *m*-cresol from *m*-/*p*-cresol mixtures in which the content of *m*-cresol is high [257].

m-Cresol can be precipitated as an addition compound with anhydrous sodium acetate from *m*-/*p*-cresol mixtures (e.g., in the ratio 65:35) in an organic solvent (gasoline, benzene, or toluene) at 20 – 40 °C. The precipitate is removed from the mother liquor and decomposed in the same solvent by heating to 80 – 95 °C; the sodium acetate is filtered. *m*-Cresol is obtained from the organic solvent as a technically pure product (96 – 99%). The yield is ca. 70% [24]. Higher yields (75%) and improved *m*-cresol purity (99.5%) are said to be possible if the adduct is decomposed at room temperature with polar solvents such as acetone [258].

1.5.3. Separation via Ester or Salt Formation

The mother liquor enriched with *p*-cresol (e.g., to a m-/p-ratio of 35:65) from the precipitation with sodium acetate can react with oxalic acid — with azeotropic dewatering — to give the semiester or diester, which is precipitated on cooling to 20 °C. The ester is separated and then hydrolyzed by heating with water at 85% in the presence of the respective nonpolar solvent. The oxalic acid may be precipitated on cooling and recycled. The *p*-cresol dissolves in the organic solvent, from which it is isolated by distillation. The alternating use of both processes separates *m*- and *p*-cresol almost entirely [259].

Separation of the cresols via the calcium cresolates, in which advantage is taken either of the relatively greater solubility of the calcium *p*-cresolate or of the easier hydrolyzability of the *m*-cresolate with superheated steam, is unsatisfactory and has no practical importance.

1.5.4. Separation by Nuclear Substitution

Partial separation of cresols is possible by taking advantage either of the lower solubility of *p*-cresolsulfonic acid in sulfuric acid [24], of the more rapid formation of *m*-cresolsulfonic acid during sulfonation [260], or of the fact that the rate of hydrolysis of *m*-cresolsulfonic acid is faster than that of *p*-cresolsulfonic acid [261]. The last-mentioned difference forms the basis of the Bruckner process, the continuous operation of which in a reaction tower is described in [262].

A large-scale industrial process, e.g., used by Sumitomo in Japan and Bayer in the Federal Republic of Germany, comprises the separation of *m*- and *p*-cresol via the di-*tert*-butyl derivatives. In this technique, the *m*-/*p*-cresol mixture reacts with up to 2 mol of isobutene per mole of cresol in the presence of acidic catalysts (sulfuric acid). The resulting alkylated product consists mainly of 2,6-di-*tert*-butyl-*p*-cresol (*bp* 147 °C at 2.67 kPa) and 2,4-di-*tert*-butyl-*m*-cresol (*bp* 167 °C at 2.67 kPa), which are easily separated by distillation, provided that the catalyst has first been entirely inactivated. The process also yields monobutylated cresols with boiling points of 124–129 °C at 2.67 kPa which are returned to the alkylation reactor, and dimers, trimers, and possibly tetramers (*bp* 125 °C at 2.67 kPa) of isobutene, which are removed from the system.

m-Cresol or *p*-cresol and isobutene are subsequently recovered by dealkylation of the di-*tert*-butylcresols at ca. 200 °C in the presence of such acidic catalysts as sulfuric acid. The individual cresols are then further purified by distillation. Their purity may exceed 99 % [263], [264]. Continuous operation of the butylation–debutylation process is described in [263] and [265]–[267]. The process is attractive not simply because it gives a very high yield of pure cresol, but also because it enables 2,6-di-*tert*-butyl-*p*-cresol, a widely used antioxidant, to be produced simultaneously.

If under suitable conditions *m*-/*p*-cresol mixtures are chlorinated with sulfuryl chloride [268], nitrosated with sodium nitrite [269], or condensed with formaldehyde [270], *m*-cresol reacts considerably faster, and *p*-cresol with a purity of 96–98 % can be isolated from the reaction mixture as a byproduct. These processes are frequently exploited in practice.

1.6. Quality Specifications and Analysis

Cresols and their mixtures are commercially available in a wide range of grades and purities. They are characterized according to the following criteria [271], [272]:

- *Content of cresols, xylenols, and other phenols*, determined preferably by GC, e.g., according to ASTM 3626-80
- *o-Cresol content* (mixtures), determined via the crystallizing point of the cineole adduct according to [271], PC 17-67; or *m-cresol content* (mixtures), determined via the crystallizing point of the urea adduct according to [271], PC 16-67

- *Crystallizing point*, directly or after drying (pure isomers), e.g., according to [271], PC 10-67
- *Distillation range* (mixtures), e.g., according to [271], PC 5–67, or DIN 52 137 and 52 139
- *Residue on evaporation*, e.g., according to [271], PC 7-67
- *Residue on distillation*, e.g., according to [271], PC 6-67
- *Specific gravity*, e.g., according to [271], PC 2-67
- *Color* (color number), e.g., according to [271], PC 1-67, or ASTM D 3627
- *Water content*, determined either by azeotropic distillation with xylene according to DEAN and STARK ([271], PC 3-67) or by the Karl Fischer method (ASTM D 1631-80)
- *Content of neutral oils*, e.g., as determined according to [271], PC 4-67, by distillation after addition of sodium hydroxide solution (cresolate formation)
- *Content of pyridine bases*, determined either according to [271], PC 4-67, by titrating the distillate from the neutral oil determination with hydrochloric acid, or by direct titration with perchloric acid in glacial acetic acid [272]
- *Total sulfur content*, e.g., according to [271], PC 22-67 (reduction method)
- *Acidity or alkalinity*, e.g., according to [271], PC 9-67, by titration against phenol red
- *Solubility in sodium hydroxide solution*, e.g., according to [271], PC 16-67

Specifications laid down in British Standards are as follows: For *o*-cresol (crystallizing point, fp, grade A 30.3–31.0 °C; grade B 29.3–30.3 °C; grade C 28.3–29.3 °C; *m*-cresol ($fp \geq 10.5$ °C), and *p*-cresol ($fp \geq 34.0$ °C) in B.S. 522:1964; for cresylic acid of specified *o*-cresol content (45 to 90%) in B.S. 517:1964; for cresylic acid of specified *m*-cresol content (60 to 39%, grades A to E) in B.S. 521:1964; and for so-called refined cresylic acid (grades A to E with altogether 0.5, 1.0, 2.0, 3.0, or 5.0% water, pyridine bases, and neutral oil content) in B.S. 524:1964.

Maximal amounts of impurities encountered in commercially available cresols are usually no more than 0.2% water, 0.15% pyridine bases, 0.1% neutral oils, and 0.01% sulfur. Synthetic cresols are practically free from pyridine bases and sulfur. Pure cresols are colorless but may have a faintly yellowish color. Their mixtures sometimes have a slightly stronger tinge, ranging from yellow to brown.

o-Cresol is produced with a purity of 99.2–99.9% ($fp \geq 30.7$ °C), e.g., for ECN resins (see Section 1.8). Commonly used technical grades have purity levels of 98.2–99% ($fp \approx 30.3$–30.6 °C), ca. 96% ($fp \approx 29.5$ °C), and ca. 94% ($fp \approx 28.5$ °C) and thus correspond to grades A, B, and C of the British Standard. For example, *m*-cresol, is sold at purities of 99% ($fp \approx 11.5$ °C), 98% ($fp \approx 11$ °C), and 96.5% ($fp \approx 10$ °C). *p*-Cresol usually has a purity of 99% ($fp \approx 34$ °C). Grades with a purity of ca. 95% ($fp \approx 31$ °C) and ca. 65% are also available. The main impurity is *m*-cresol.

Technical cresol mixtures are usually classified according to their *m*-cresol content, which is between 20 and 70%. Within this range, many different grades are supplied, depending on the manufacturer and the customer's requirements; there may also be substantial differences between them on the basis of phenols (phenol itself, *o*-cresol, *p*-cresol, xylenols, 2-ethylphenol, and other alkylphenols) that are additionally present, as well as other impurities (water, pyridine, neutral oils, and sulfur compounds). Thus, there are cresol mixtures containing up to 50% phenol, or almost 50% xylenols and other alkylphenols, or which have a total cresol content of only ca. 30%; normally the latter lies between 60 and 90%. *m-/p*-Cresol mixtures obtained by chlorotoluene hydrolysis or from the cymene–cresol process are distinguished by their high *m*-cresol content (70% and 60%, respectively) and by their high total content of cresol, normally $\geq$ 99%.

1.7. Handling, Storage, and Transportation

The cresols are corrosive, and toxic if absorbed through the skin (see Chap. 4). Work with cresols should be carried out in closed apparatus wherever possible. The rooms must be well ventilated, and precautions, such as the use of airlocks and the gas compensation technique, must ensure that cresol vapor or dust cannot be inhaled by personnel. The fundamental criterion of the effectiveness of all such precautions is the MAK value or TLV, which in 1984 was fixed at 5 ppm (22 mg/m^3) for all cresol isomers and isomer mixtures. This level corresponds to the odor threshold. At higher concentrations, e.g., resulting from operational disturbances, suitable respirators must be worn. In addition, persons handling cresol must always have adequate eye protection and, if necessary, body protection, e.g., as provided by polychloroprene or PVC [poly(vinyl chloride)] gloves, long aprons, fully protective suits, etc. Further recommendations on the handling of cresols and xylenols may be found in [273]–[276] and in the safety data sheets of cresol manufacturers.

Steel vessels are suitable for storage. The discoloration that takes place on storage is promoted by elevated temperatures, light, oxygen, water, and catalysts, e.g., by dissolved iron or copper. If the product is required to remain colorless for as long as possible, it is best to use tanks of stainless steel (V4A; AISI 316Ti). If exceptional color stability is required, the product should also be stored under nitrogen. Access of moisture must be prevented, since cresols absorb water readily. Both *o*-cresol and *p*-cresol, which are solids at room temperature, have to be handled at a temperature higher than their crystallizing points. Cresol–air mixtures reach the lower explosive limit at ca. 76 °C in the case of *o*-cresol and at ca. 83 °C in those of *m*- and *p*-cresol; therefore, if the storage temperature exceeds 70 °C or 75 °C, respectively, the gas space must be rendered inert with nitrogen. Storage practice must comply with local regulations. The storage containers must be tested regularly for soundness and freedom from leaks. As the cresols are classified as waterpollutants, measures such as the provision of tank basins must be taken to prevent cresol from entering the soil or drains.

Cresols are transported in heatable tank cars, in tank trucks, or in steel drums with a capacity of ca. 200 L. The container must bear the appropriate hazard warnings in compliance with the regulations. For unloading and sampling, see [275].

Transportation is subject to the following regulations:

International sea transport (IMDG Code) and international air transport (IATA-DGR) for *o-*, *m-*, and *p*-cresol and cresol mixtures: Class 6.1, UN-no. 2076, package group II; for cresylic acids: class 6.1, UN-no. 2022, package group II. For transport of *o-*, *m-*, and *p*-cresol, cresol mixtures, and cresylic acids on European roads (ADR) and railways (RID): class 6.1, no. 14 b, from January 1, 1986. As regards transportation on European waterways (ADNR), *o-*, *m-*, and *p*-cresol and cresol mixtures belong to class 6.1, no. 22 a, and cresylic acid to class 6.1, no. 22. The ADNR classification is to be merged with the ADR/RID classification in 1986.

National regulations: United States (DOT-CFR 49) for *o-*, *m-*, *p*-cresol and cresol mixtures: Corrosive material/UN 2076; for cresylic acids: Poison B/UN 2810. In the Federal Republic of Germany class 6.1, no. 14 b, is valid from January 1, 1986, to the transport of *o-*, *m-*, and *p*-cresol, cresol mixtures, and cresylic acids by road (GGVS) and rail (GGVE).

Cresols are liquids that are partly miscible with water and have flash points of 80–90 °C. As such, they are, therefore, classified in the Federal Republic of Germany and in West Berlin as belonging to group A, hazard class III, according to the regulations for the storage of flammable liquids (Verordnung über die Lagerung brennbarer Flüssigkeiten, VbF). In the United States, *m*- and *p*-cresol belong to Category 1 of the NFPA 704M fire hazard classification system. Water fog, foam, dry chemical extinguishers, and — for small fires — carbon dioxide are suitable for fire fighting. In the Federal Republic of Germany, VDE Specification 0165 places the cresols in Ignition Group G 1.

1.8. Uses

Most of the *o*-cresol manufactured in Europe is chlorinated to 4-chloro-*o*-cresol (PCOC), the starting material for the chlorophenoxyalkanoic acids 4-chloro-2-methylphenoxyacetic acid (MCPA), 2-(4-chloro-2-methylphenoxy)-propionic acid (MCPP), and γ-(4-chloro-2-methylphenoxy)butyric acid (MCPB), which are important as selective herbicides. A considerably smaller proportion is nitrated to 4,6-dinitro-*o*-cresol (DNCO), which has both herbicidal and insecticidal properties.

Highly pure *o*-cresol is increasingly processed, especially in Japan, to epoxy – *o*-cresol novolak resins (ECN resins), which are used as sealing materials for integrated circuits (chips). *o*-Cresol (of common quality) is also used to modify traditional phenol–formaldehyde resins.

o-Cresol is also important as a precursor of various dye intermediates, of which the most important in terms of quantity is *o*-cresotinic acid (*o*-hydroxymethylbenzoic acid) produced by the Kolbe synthesis. This acid finds further use in the manufacture of pharmaceuticals, whereas its methyl esters serve as dyeing auxiliaries.

An appreciable amount of *o*-cresol is used as a solvent, either directly or after hydrogenation to 2-methylcyclohexanol or 2-methylcyclohexanone. In the form of its carbonate ester, *o*-cresol constitutes a starting material in the synthesis of coumarin. The alkylation of *o*-cresol with propene gives carvacrol (3-isopropyl-6-methylphenol), which is used as an antiseptic and in fragrances. In addition, small amounts of *o*-cresol are used as starting materials in the production of various antioxidants.

m-Cresol, either pure or mixed with *p*-cresol, forms a starting material for such important contact insecticides as *O,O*-dimethyl-*O*-(3-methyl-4-nitrophenyl)thionophosphoric acid (fenitrothion; Folithion, Sumithion) and *O,O*-dimethyl-*O*-(3-methyl-4-methylthiophenyl)thionophosphoric acid ester (fenthion; Baytex, Lebaycid). In addition, *m*-cresol is needed in the synthesis of phenyl *m*-tolyl ether (*m*-phenoxytoluene), which after oxidation to *m*-phenoxybenzaldehyde serves as a building block in the production of insecticides of the pyrethroid type.

Pure *m*-cresol has considerable importance in the production of fragrance and flavor substances: its isopropylation gives thymol, from which (−)-menthol is obtained by hydrogenation and separation of the isomers. 6-*tert*-Butyl-*m*-cresol, obtained by alkylation of *m*-cresol with isobutene, is the starting material for the perfume fixative musk ambrette. The condensation of 6-*tert*-butyl-*m*-cresol with sulfur dichloride gives 4,4′-thiobis(6-*tert*-butyl-*m*-cresol), an important antioxidant for polyethylene and polypropylene. Further antioxidants are produced from 6-*tert*-butyl-*m*-cresol by condensation with butyraldehyde or crotonaldehyde. In Japan, *m*-toluidine is produced, on demand, by amination of *m*-cresol.

The 4-chloro-*m*-cresol obtained by selective chlorination of *m*-/*p*-cresol mixtures finds use as a disinfectant and preservative. 2,4,6-Trinitro-*m*-cresol can be used as an explosive.

p-Cresol, pure or mixed with *m*-cresol (see Section 1.5.4), is used mainly to produce 2,6-di-*tert*-butyl-*p*-cresol (BHT), a nonstaining and light-resistant antioxidant with a wide range of applications. 2-Alkyl-*p*-cresols obtained by the monoalkylation of *p*-cresol in the ortho position, e.g., *tert*-butyl-, cyclohexyl-, methylcyclohexyl-, α-methylbenzyl-, or nonyl-*p*-cresol, can be condensed with formaldehyde or sulfur dichloride to give 2,2′-methylene- or 2,2′-thiodiphenols, which are also important as antioxidants. 2,6-Dicyclopentyl-*p*-cresol, obtainable by dicyclopentylation of *p*-cresol, is a very effective antioxidant. Coupling of 2-*tert*-butyl-*p*-cresol with diazotized 4-chloro-2-nitroaniline yields Tinuvin 326, an absorber of UV light, which finds use, e.g., for polyethylene and polypropylene films and coatings. In the fragrance industry, *p*-cresol is used to obtain *p*-cresolcarboxylic acid esters and, particularly, *p*-cresol methyl ether, most of which is used in the production of anisaldehyde. Like *o*-cresol, *m*- and *p*-cresol are used as components of various dyes.

An important field of application for technical cresol mixtures is the production of modified phenolic resins by condensation with formaldehyde. The suitability and price of cresol mixtures for this purpose depends on their content of *m*-cresol, the most reactive of the three isomers. Cresol mixtures are highly important as solvents for synthetic resin coatings (wire enamels).

m-/*p*-Cresol mixtures free of the *o*-isomer are used to produce neutral phosphoric acid esters (tricresyl phosphate, diphenyl cresyl phosphate), which find use as fire-resistant hydraulic fluids, as additives in lubricants, as air filter oils, and as flame-retardant plasticizers for PVC and other plastics.

The bactericidal and fungicidal properties of cresols enable them to be used as disinfectants in soap. Synthetic tanning agents of commercial importance are obtained by the condensation of formaldehyde with cresolsulfonic acids and by the sulfonation of novolaks obtained from cresols. Crude cresols are used as wood preservatives. Cresol mixtures are also used in ore flotation and fiber treatment, as metal degreasing and cutting oils, as extracting solvents, and as agents for removing carbonization deposits from internal combustion engines. Methylcyclohexanol and methylcyclohexanone produced by hydrogenation of cresols are used in the paint and textile industries.

1.9. Economic Aspects

The combined total output of cresol of the United States, Western Europe, and Japan in 1984 was estimated to be 220 000 t:

Synthetic cresol	ca. 140 000 t
Cresol from coal tar	ca. 55 000 t
Cresol from spent refinery caustics	ca. 25 000 t
Total	ca. 220 000 t

About 65 000 t was produced in the United States, about 95 000 t in Western Europe, and about 60 000 t in Japan. Approximately half of the total cresol output is processed in the form of the pure isomers, and the rest as isomer mixtures. About 30 % of the total output is used by the manufacturers themselves.

The above figures for cresol from coal tar and spent refinery caustics refer only to the cresol content. They do not include the xylenols contained in the raw material (30 000 – 40 000 t), which to some extent are sold as though they were cresol. The tar cresols are obtained from tars produced in Eastern Europe and from crude cresols resulting from the gasification of coal in South Africa. This proportion probably accounts for ca. 25 % of the total output of tar cresols.

The total output of cresol has not increased over the last ten years. Substantial overcapacity, especially for o-cresol, has led to the (temporary) shutdown of plants in the United States (Conoco, Koppers, and Diamond Shamrock) and, thus, to a decrease in capacity of ca. 50 000 t/a [43]. Nevertheless, the capacity remaining in operation is not being used fully, partly because General Electric in the United States is now producing o-cresol in place of 2,6-xylenol on a larger scale than before and partly because new o-cresol capacity has been installed in the Netherlands (General Electric) and in Japan (Nippon Crenol and GEM-Polymer).

2. Xylenols [1]–[3]

Xylenol, dimethylphenol, $C_8H_{10}O$, M_r 122.166, occurs in six isomeric forms:

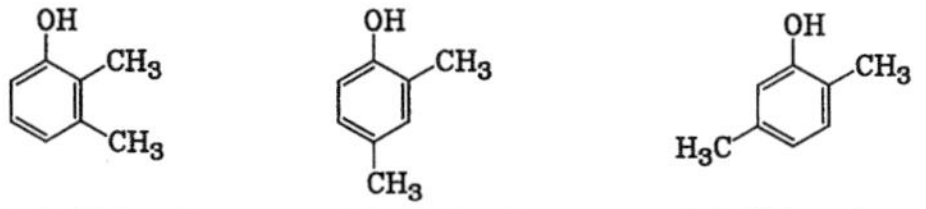

2,3-Xylenol
2,3-dimethylphenol
[526-53-0]

2,4-Xylenol
2,4-dimethylphenol
[105-67-9]

2,5-Xylenol
2,5-dimethylphenol
[95-87-4]

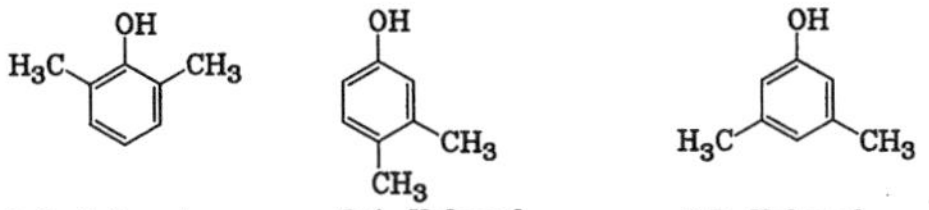

2,6-Xylenol
2,6-dimethylphenol
[576-26-1]

3,4-Xylenol
3,4-dimethylphenol
[95-65-8]

3,5-Xylenol
3,5-dimethylphenol
[108-68-9]

The numbering of the substituents normally and correctly begins from the OH group; but is occasionally (erroneously) started from the methyl group: 2,4-xylenol can be referred to as 1,3,4-xylenol, 4-hydroxy-*m*-xylene, or 4-hydroxy-1,3-dimethylbenzene.

Xylenols are present, e.g., in the essential oils of various conifers, in tea, in tobacco and tobacco smoke, in roasted coffee, and in various smoked foods. In many cases, they contribute to the flavor of these products.

From the standpoint of economics, the most important isomer is now 2,6-xylenol, of which coal tar contains only a relatively small amount and which is produced, therefore, almost entirely by synthetic means.

2.1. Properties [1]–[3]

The xylenols are colorless crystallizing compounds that are soluble in alcohol, acetone, and in many other organic solvents. They are less soluble in water than the cresols. Except for 3,5-xylenol, their acidity is likewise lower than that of the cresols. They are soluble in aqueous sodium hydroxide, but to different extents, depending on the structure. Their chemical behavior is similar to that of the cresols. Physical data are presented in Table 4. For dependence of the thermodynamic data on temperature, see [8], [277]–[279].

2.2. Isolation

With regard to the formation of xylenols in various thermal cracking processes, see Section 1.3.

Table 4. Physical properties of pure xylenols

Physical property	2,3-Xylenol	2,4-Xylenol	2,5-Xylenol	2,6-Xylenol	3,4-Xylenol	3,5-Xylenol
Melting point (101.3 kPa), °C	72.57	24.54	74.85	45.62	65.11	63.27
Depression by 1 mol% of impurities, °C	0.502	0.576	0.727	0.558	0.727	0.520
Boiling point at 101.3 kPa, °C	216.87	210.93	211.13	201.03	226.95	221.69
at 10.0 kPa, °C	141.63	137.12	136.98	125.48	152.33	147.88
Density at 25 °C, g/cm^3	1.164	1.0160 (*l*)	1.189	1.132	1.138	1.115
at (*t*, °C), g/cm^3	0.981 (85)	1.0033 (40)	0.958 (85)	0.978 (65)	0.982 (80)	0.971 (75)
Refractive index at (*t*, °C)	1.5198 (80)	1.5254 (50)	1.5092 (80)	1.5171 (60)	1.5268 (60)	1.5150 (70)
Dielectric constant at 80 °C	4.65	4.98	5.07	4.34	7.72	7.05
Dissociation constant at 25 °C	2.88×10^{-11}	2.51×10^{-11}	3.89×10^{-11}	2.34×10^{-11}	4.34×10^{-11}	6.46×10^{-11}
Critical temperature, °C	449.7	434.4	433.9	427.8	456.7	442.4
Critical pressure, MPa	4.9	4.3	4.9	4.3	5.0	3.6
Critical density, kg/m^3	340	320	350	310	350	250
Heat capacity c_p at 25 °C (g), J mol^{-1} K^{-1}	164.2	156.2	157.2	156.1	163.6	152.8
c_p at 80 °C (*l*), J mol^{-1} K^{-1}	274	281	285	246	280	296
Enthalpy of fusion ΔH_m at *mp*, 101.3 kPa, kJ/mol	19.88	12.85	13.91	15.21	13.15	17.42
Enthalpy of sublimation ΔH_{sub} at 25 °C, 101.3 kPa, kJ/mol	84.07	65.90 (*l*)	85.04	75.65	85.79	82.90
Enthalpy of vaporization ΔH_v at *bp*, 101.3 kPa, kJ/mol	47.35	47.18	46.97	44.55	49.70	49.34
Enthalpy of combustion $\Delta H_c°$ at 25 °C, 101.3 kPa, kJ/mol	– 4339	– 4351 (*l*)	– 4334	– 4343	– 4338	– 4336
Enthalpy of formation $\Delta H_f°$ at 25 °C (*s* or *l*), 101.3 kPa, kJ/mol	– 241.5	– 229.0 (*l*)	– 246.9	– 237.6	– 242.5	– 244.6
$\Delta H_f°$ at 25 °C (g), 101.3 kPa, kJ/mol	– 157.4	– 163.1	– 161.8	– 161.9	– 156.8	– 161.7
Free energy of formation $\Delta G_f°$ at 25 °C (g), 101.3 kPa, kJ/mol	– 33.37	– 41.25	– 39.70	– 39.07	– 34.30	– 39.44
Entropy $S°$ at 25 °C (g), 101.3 kPa, J mol^{-1} K^{-1}	390.4	398.0	396.0	390.0	391.4	392.0
Cubic expansion coefficient at 25 °C		0.818×10^{-3}				
Solubility in water at 25 °C, wt%	0.47	0.61	0.49	0.64	0.50	0.49
Critical solution temperature in water, °C	210	213	219	237	189	198
at % xylenol, wt%	38	34	42	45	35	35

Xylenols, with the cresols and other phenols, are isolated from high-temperature, low-temperature, and Lurgi pressure gasifier coal tars (see Section 1.3.1) and from spent refinery caustics (see Section 1.3.2). The resulting crude phenols (cresylic acids) contain all six xylenols in amounts that differ greatly according to the raw material and conditions of formation (see Table 3). Moreover, the ratio of the xylenols themselves differs. In xylenols from spent refinery caustics, 2,4-xylenol generally predominates; besides this, the content of 2,5-xylenol is high, whereas that of 3,5-xylenol is relatively small [280]. Xylenols from bituminous coal tar are characterized by their high content of 3,5-xylenol. The average isomer ratio of bituminous coal tar xylenols is given in

[281]; for example, as 31.9% 3,5-xylenol, 28.1% 2,4-xylenol, 14.1% 2,3-xylenol, 13.6% 2,5-xylenol, 6.9% 2,6-xylenol, and 6.5% 3,4-xylenol.

In the distillation of crude phenol, which is normally carried out continuously, the 2,6-xylenol present and some 2-ethylphenol enter the *m-/p*-cresol fraction. The other xylenols and the ethylphenols, as well as several C_9-phenols, are separated from the higher boiling phenols and the phenol pitch (→ Tar and Pitch) either together as a crude xylenol fraction or (in the case of xylenols from low-temperature tar) successively in the form of several differently enriched xylenol fractions.

2.3. Separation

For the isolation of individual xylenols, the crude xylenol fraction (see above) is rectified discontinuously into fractions of a narrow boiling temperature range [24].

2,4-Xylenol and 2,5-xylenol are contained in the fraction boiling at 208–212 °C at 101.3 kPa. The isomers are usually sold as a mixture. Since the difference between their boiling points (0.2 °C at 101.3 kPa) is too small for separation by distillation, other means of separation need to be utilized.

If the fraction is cooled, the greater part of the 2,5-xylenol crystallizes and is then obtained in pure form after recrystallization. After sulfonation of the centrifuged mother liquor with 92% sulfuric acid and dilution with water, the remaining 2,5-xylenol crystallizes as 2,5-xylenolsulfonic acid. Hot potassium chloride solution is then added to the filtrate, and the temperature is lowered. This precipitates the potassium salt of 2,4-xylenolsulfonic acid, from which, after recrystallization from water and acidification with sulfuric acid, the 2,4-xylenol is released by hydrolysis and stripped with steam.

It is also possible to separate 2,4- and 2,5-xylenol by butylation with isobutene; due to the steric hindrance of the phenolic OH group, the resulting 6-*tert*-butyl-2,4-dimethylphenol cannot be dissolved in 10% aqueous sodium hydroxide, whereas the 4-*tert*-butyl-2,5-dimethylphenol can. The products separated in this way can be debutylated with concentrated sulfuric acid. It has also been proposed that 2,5-xylenol be isolated by exploiting the fact that the two isomers have different butylation and debutylation rates on highly acidic ion-exchange resins [282], depending on temperature. 2,4-Xylenol and 2,5-xylenol can also be separated via adduct formation with phenols or amines [283].

2,3-Xylenol is isolated from a fraction boiling at 215–218 °C by selective dephenolation with dilute aqueous sodium hydroxide, introduction of the free phenols into milk of lime, and stripping with steam; it can then be purified by recrystallization from benzene. It is also obtained in the cooling crystallization of a very narrow distillate fraction.

3,5-Xylenol is crystallized by cooling the fraction that boils at 218–221 °C, after which it is centrifuged and recrystallized from benzene.

3,4-Xylenol is obtained in the same way as given above from the fraction that boils at 223–225 °C; it is likewise centrifuged and recrystallized from benzene.

3,5-Xylenol or 3,4-xylenol can also be separated from other xylenols and from trimethylphenols via adduct formation with anhydrous calcium bromide [284].

Faujasite zeolites modified with copper or silver cations adsorb 3,5-xylenol less readily than other xylenols, thus enabling 3,5-xylenol to be separated as a raffinate component [285].

2,6-Xylenol has a boiling point ca. 0.9 °C lower than that of *p*-cresol at 101.3 kPa; the difference is ca. 4 °C at 20 kPa [8]. Thus, it can be obtained from the first runnings of the *m*-/*p*-cresol fraction. 2,6-Xylenol can be isolated in numerous ways. In the azeotropic distillation of the *m*-/*p*-cresol fraction with steam, 2,6-xylenol crystallizes from the condensate. 2,6-Xylenol can be distilled from cresolate caustics by virtue of its lower acidity, after which it is obtained from the condensate by extraction with benzene. For this reason, the condensate from the steam distillation of the crude phenolate caustic obtained from coal tar (see Section 1.3.1) is also suitable as a starting material for the isolation of 2,6-xylenol.

In fractions highly enriched with 2,6-xylenol, e.g., those obtained from synthesis by the methylation of phenol (Section 1.4.4), the 2,6-xylenol can be separated from *m*- and *p*-cresol and other phenols by countercurrent extraction with aqueous sodium hydroxidehydrocarbon mixtures. This technique takes advantage of the lower acidity of 2,6-xylenol. The product can be isolated from the hydrocarbon phase in a purity in excess of 98% [286]. By comparison with *m*- and *p*-cresol, 2,6-xylenol has less tendency to form hydrogen bonds, and can therefore also be separated from the cresols either by countercurrent extraction with aqueous methanol–octane [287] or by extractive distillation with di- or triethylene glycol [108], [142], [288]. Furthermore, the more polar phenols (cresols) can be distilled from the 2,6-xylenol as azeotropes with $C_{10}-C_{14}$ aliphatics as entrainers [289]–[291]. Cresols are adsorbed more readily than 2,6-xylenol on zeolites, e.g., of the K–Ba–X type; therefore, 2,6-xylenol can be isolated in high purity as a raffinate stream [292], [293]. The isolation of 2,6-xylenol by the fractional crystallization of phenol methylation residues greatly enriched by distillation can be facilitated by adding phenol [294] or water [295], [296]. Good purification is achieved by crystallization under high pressure (e.g., 160 MPa at 75 °C) even in the absence of such additions [297]. Isolation of the 2,6-xylenol by distillation is possible if the *m*- and *p*-cresol impurities are selectively reacted with phosphorus oxide chloride to give phosphoric acid esters [298], or with methanol on a catalyst either of Al_2O_3 [299] or of titanium sulfate–magnesium oxide [300] to give higher boiling xylenols and trimethylphenols.

2.4. Production

Phenol Methylation. Cresols or xylenols are formed in the methylation of phenol with methanol. Details of the catalysis and industrial conditions of the reaction are found in Section 1.4.4.

In the *gas phase*, *o*-cresol and 2,6-xylenol are the preferred products. The 2,6-xylenol can be made the main product if ortho-selective catalysts and high methanol:phenol ratios are used, and/or *o*-cresol is recycled [142]. Under ortho-selective conditions, it is also possible, in principle, to produce chiefly 2,4-xylenol by methylation of *p*-cresol [301], and to produce 2,3-xylenol [302] and 2,5-xylenol by methylation of *m*-cresol [303], [304].

With *liquid-phase* catalysts used for methylation (see 1675), it is typical not only that more 2,4-xylenol than 2,6-xylenol is formed, but also that the purity of this 2,4-xylenol (ca. 90%) is higher at the outset than that of the 2,4-xylenol obtained, e.g., from coal tars or spent refinery caustics (50–75% purity).

Demethanization of Isophorone. 3,5-Xylenol, which is not normally a product of methylation processes and of which only a limited amount can be isolated in the pure form from natural xylenol mixtures, is produced in the United Kingdom by Synthetic Chemicals Ltd. by demethanization of isophorone (3,5,5-trimethyl-2-cyclohexen-1-one) according to a process developed by Shell:

$$\text{Isophorone} \xrightarrow[\text{Catalyst}]{600\,°C} \text{3,5-Xylenol} + CH_4$$

The reaction is carried out in the vapor phase in a multitubular reactor at 600 °C and 0.1 kPa and liquid hourly space velocities (LHSV) around 0.5 h^{-1}. This process, in which the isophorone conversion is practically complete, gives a 3,5-xylenol yield of ca. 70% on a cobalt–molybdenum–potassium oxide–γ-aluminum oxide catalyst [305], or of ca. 80% on a new rare-earth metal oxide–α-aluminum oxide carrier catalyst promoted with transition-metal oxides (0.1% Co) and alkali-metal oxides (0.4% K) [306]. Toluene, xylene, mesitylene, dihydroisophorone, *m*-cresol, 2,4- and 2,5-xylenol, and trimethyl-phenol are formed as byproducts. The catalyst has to be regenerated at intervals by burning off accumulated deposits of carbon. The carbon deposits are avoided in the process used industrially by Rütgerswerke AG in the Federal Republic of Germany by carrying out the reaction at 600 °C in tubes of chromium–nickel steel with a packing of the same material. The yield is stated to be 65% with an isophorone conversion of ca. 90% [307].

Homogeneous catalysis is possible at 570–600 °C if ca. 1 wt% of methyl iodide (or another halogenated hydrocarbon) is added to the isophorone. Complete conversion is achieved at a liquid hourly space velocity (LHSV) of 0.5 h^{-1}. After the 3,5-xylenol has been worked up by distillation, it is obtained as a product of ca. 99% purity in 85% yield [308]. The iodine content of the product has to be lowered by heating with ca. 0.2 wt% of metallic zinc, magnesium, or iron before the distillation [309].

Isophorone is produced by alkaline trimerization of acetone ($\rightarrow$ Ketones). The acetone can also be reacted directly on a catalyst of either magnesium oxide [310] or chromium oxide–magnesium oxide–calcium silicate [311] at 470 °C and 0.5 kPa to give a mixture consisting essentially of 3,5-xylenol, isophorone, mesityl oxide, and unreacted acetone. High space–time yields of 3,5-xylenol are only possible if acetone and mesityl oxide are recycled, and isophorone is manufactured as a coproduct [311].

Dimethylcumene Hydroperoxide Cleavage. According to the process of the Mitsubishi Gas Chemical Co. in Japan, 3,5-xylenol is obtained by oxidation of 3,5-dimethylcumene to 3,5-dimethylcumene hydroperoxide [312], which is then cleaved [313]. By comparison with the analogous cymene–cresol process (see Section 1.4.3), this method is more expensive, e.g., because the additional methyl group of the dimethylcumene leads to more side reactions (formation of more primary peroxide) and oxidation degrees of only ca. 10% are economical. The 3,5-dimethylcumene is readily accessible by propylation of m-xylene in the presence of catalytic amounts of $AlCl_3$. The ratio of 3,5- to 2,4-dimethylcumene at isomerization equilibrium is ca. 99:1 [314].

Alkaline Chloroxylene Hydrolysis. Under the conditions of the chlorotoluene–cresol process (see Section 1.4.2), chloroxylenes are hydrolyzed to xylenols. From 2,4- or 3,5-dimethylchlorobenzene, for example, one can obtain mixtures of 2,4- and 3,5-xylenol [315] that can be separated by distillation. The hydrolysis of 2,5-dimethylchlorobenzene under pressure leads to 2,5-xylenol.

Alkali Fusion of Xylenesulfonates. The toluenesulfonic acid–cresol processes (see Section 1.4.1) also give xylenols if xylene is used as the starting material instead of toluene. Thus, the sulfonation of m-xylene with 95% sulfuric acid (4 h at 150 °C) or chlorosulfuric acid (1 h at 50 °C) yields a product consisting primarily of 2,4-dimethylbenzenesulfonic acid, the fusion of which with sodium hydroxide at 320 °C leads to 2,4-xylenol in 79% yield [316]. If the sulfonation mixture is heated to a temperature of up to 220 °C, the 2,4-dimethylbenzenesulfonic acid isomerizes to the thermodynamically more stable 3,5-form, the fusion of which with excess alkali gives 3,5-xylenol [317]. If the 2,4-/3,5-dimethylbenzenesulfonic acid mixture is heated with water (steam) to 140–160 °C, the 2,4-dimethylbenzenesulfonic acid is selectively hydrolyzed; in the following alkali fusion, the residual 3,5-dimethylbenzenesulfonic acid is converted into 3,5-xylenol of 97–98% purity [318] in a yield of ca. 70%, based on reacted m-xylene [55].

The sulfonation of p-xylene with 98% sulfuric acid at 140 °C gives 2,5-dimethylbenzenesulfonic acid. Fusion of this with alkali hydroxide at 330–340 °C gives 2,5-xylenol in 80% yield, based on p-xylene [319].

3,4-Dimethylbenzenesulfonic acid, which is accessible in ca. 98% yield by sulfonation of o-xylene with 96% sulfuric acid at ca. 150 °C [320], [321], gives 3,4-xylenol in more than 90% total yield when fused with alkali hydroxide at 320–330 °C [320].

Other routes to the xylenols include the Gulf oxychlorination process, nucleus hydroxylation (see Section 1.4.5), and oxidative decarboxylation of dimethylbenzoic acids [193].

Isomerization. Xylenols for which there is inadequate demand and xylenol fractions from which certain xylenols have already been isolated can be isomerized catalytically. Aluminum oxides or silicates, the activities of which have been further increased with hydrogen fluoride, titanium dioxide, iron oxide, and other metal oxides [322]–[326], are suitable as catalysts. The equilibrium distribution of the xylenols (2,3-:2,4-:2,5-:2,6-:3,4-:3,5-isomer) after their isomerization in the vapor phase at 340 °C on aluminum silicate is given as 9:30:24:11:12:14 [326]. In the liquid phase at 105–130 °C, with aluminum chloride in combination with hydrogen chloride as catalyst, 3,5-xylenol is primarily formed [186]; however, uneconomically large amounts of $AlCl_3$ are consumed. With regard to isomerization with $HF-BF_3$, see [327]. If phenol [186] or an aromatic solvent such as benzene is present, transmethylation occurs, with the simultaneous formation of cresol or toluene (see p. 1676).

2.5. Quality Specifications and Analysis

The xylenols are characterized according to the same methods of analysis as are used for cresols (see Section 1.6).

The fraction sold as a 2,4-/2,5-xylenol mixture is a clear yellow liquid that contains ca. 50–75% 2,4-xylenol and ca. 30–10% 2,5-xylenol, depending on its origin. The main impurities are *m* - and *p*-cresol and ethylphenols. 2,4-Xylenol from the methylation of phenol on liquid-phase catalysts (UK-Wesseling) contains more than 89% 2,4-xylenol and less than 7.5% 2,5-xylenol. Pure 2,4-xylenol is offered with a content of ca. 97% (*fp* 23.0 °C). 2,3-Xylenol and 2,5-xylenol are available with purities of ca. 97% and ca. 96%, respectively. 2,6-Xylenol is traded at purities of greater than 99.5%, 98–99%, 96–98%, and 90–96%. 3,4-Xylenol is available in the grades "technically pure" ($\geq$ 95%; *fp* $\geq$ 62.2 °C) and "pure" ($\geq$ 98%; *fp* $\geq$ 64 °C). 3,5-Xylenol is commercially available in purity grades of 95% (*fp* 60.3 °C) and 99% (*fp* 62.5 °C).

Xylenol mixtures differ widely in composition, depending on the manufacturer and the requirements of the customer. For example, there are xylenol mixtures (cresylic acids) that contain almost 50% cresol or 50% higher alkyl phenols (C_9 and higher). The mixtures may also contain considerable amounts of ethylphenols in addition to the xylenols, depending on their origin. A rough classification is possible according to the distillation range (5–95%), which could be, for example, 200–220 °C, 205–215 °C, 210–225 °C, 220–240 °C, 220–255 °C, or 230–240 °C.

2.6. Handling, Storage, and Transportation

The xylenols are similar to the cresols in their physiological behavior. Their handling, storage, and transportation are, therefore, subject to the precautions applicable to cresols (see Section 1.7). This also applies to the precautions taken to prevent discoloration (storage in vessels of stainless steel, protection from light and moisture, and possible storage under nitrogen). References to the handling of xylenols are made in [273]–[276].

Xylenols are transported in ca. 200-L steel drums. Xylenol mixtures, 2,4/2,5-xylenol fractions, and 2,6-xylenol are also supplied in tank cars or tank trucks, which in the case of 2,6-xylenol must be capable of being heated. 2,6-Xylenol may have to be filled and transported under nitrogen. The shipping containers must bear the appropriate hazard warnings in accordance with the regulations.

The transportation of xylenols and xylenol mixtures is subject to the following regulations:

International sea transport (IMDG Code) and international air transport (IATA-DGR): class 6.1, UN no. 2261, packing group II. Transport on European roads (ADR) and railways (RID) from January 1986: class 6.1, no. 14 b. Transport on European waterways (ADNR): class 6.1, no. 22 b (to be merged with the ADR/RID classification in 1986).

National regulations: United States (DOT-CFR 49): ORM. A/UN 2261. With regard to their transportation on roads (GGVS) and railways (GGVE) in the Federal Republic of Germany, these products belong to class 6.1, no. 14 b.

According to "Verordnung über brennbare Flüssigkeiten" (VbF), the properties and flash points ($\geq 86\,°C$) of xylenol mixtures and xylenols place them in group A, hazard class III (inasmuch as they are liquids). Water fog, foam, dry extinguishers, and, for small fires, carbon dioxide are suitable for fire fighting.

2.7. Uses

Xylenol mixtures are used as solvents (e.g., for wire enamels), disinfectants (e.g., in sheep dips), textile auxiliaries, and in ore flotation. Mixtures, particularly if they are rich in 3,5-xylenol and xylenols with two hydrogen atoms in the o- and/or p-positions, are used in the manufacture of xylenol–formaldehyde resins. Xylenols without an *o*-methyl group are used to produce nontoxic plasticizers and trixylenyl phosphates that serve as fire-resistant hydraulic fluids.

2,4-/2,5-Xylenol mixtures constitute raw materials for antioxidants, especially those intended for gasoline (e.g., 6-*tert*-butyl-2,4-dimethylphenol) and rubber [e.g., 2-methyl-1,1-bis(2-hydroxy-3,5-dimethylphenyl)propane].

2,6-Xylenol is used primarily to produce poly(phenylene oxide) (PPO) resins, which are distinguished by high impact resistance, thermal stability, fire resistance, and dimensional stability. 2,2-Bis(4-hydroxy-3,5-dimethylphenyl)propane (tetramethylbisphenol A), obtained by the condensation of 2,6-xylenol with acetone, serves as an intermediate for polycarbonates, whose properties are similar to those of the PPO resins. 2,6-Xylenol is also reacted with ammonia to yield 2,6-dimethylaniline, which is used especially in pesticide manufacture. Smaller amounts of 2,6-xylenol are used in the formation of pesticides, and in the manufacture of disinfectants as well as the antioxidant bis(4-hydroxy-3,5-dimethylphenyl)methane.

3,5-Xylenol is used in the production of various pesticides, e.g., 3,5-dimethyl-4-methylthiophenyl methylcarbamate (methiocarb, Mesurol), a broad-spectrum insecticide, acaricide, and molluscicide. 3,5-Xylidine, obtained by amination, is used particularly for perylene pigments. After chlorination to 4-chloro- and 2,4-dichloro-3,5-dimethylphenol, 3,5-xylenol is used as a disinfectant and industrial preservative. Other xylenols or xylenol mixtures likewise serve as disinfectants after chlorination.

3,4-Xylenol is the starting material for the insecticide 3,4-dimethylphenyl methylcarbamate (MPMC, Meobal).

Pure xylenols are also used in small amounts in the synthesis of dyes, pharmaceuticals, and fragrances.

2.8. Economic Aspects

As a rough estimate, the total output of xylenols in the United States, Western Europe, and Japan in 1984 was ca. 140 000 – 150 000 t. Xylenol mixtures accounted for ca. 30 000 – 40 000 t, 2,6-xylenol for ca. 100 000 t, and the remaining individual isomers (especially 2,4- and 3,5-xylenol), obtained by synthesis or isolated from mixtures, for ca. 10 000 t.

The xylenol mixtures are obtained — mainly in the United States and Europe, especially in the United Kingdom — from spent refinery caustics and/or coal tars. The production of xylenol by synthesis is accounted for by the United States, Europe, and Japan in the approximate ratio 60:20:20.

3. Environmental Protection

Cresols and xylenols are readily degradable biologically; the xylenols are degraded somewhat slower than the cresols [328]. Biological treatment plants with suitably adapted bacteria are able to degrade wastewaters with 900 mg of cresol/L and more [329].

In the environment, cresols are degraded at low concentrations, especially by bacteria that are widely distributed in the soil and water (particularly *Pseudomonas* species). Degradation by other organisms, including yeasts, fungi, algae, and higher plants, as well as by photolysis, is also known [330]. Accordingly, these compounds do not persist in the environment [331].

However, as cresols and xylenols show high acute toxicity both toward fish (LC_{50} (96 h) 1–10 mg of cresol/L) and to aquatic invertebrates, and as they impair the taste of edible fish and drinking water even in very low concentrations [330], [332], the release of these compounds into surface waters is subject to strict limitations. Therefore, leaked or spilled cresols or xylenols must be prevented from entering surface waters or groundwater.

Liquid products can be adsorbed, e.g., with sawdust, peat, earth, kieselguhr, or other adsorbents, and then burned completely in special incineration plants. Solidified products should not be dissolved, but removed mechanically. Final residues can be neutralized with 2–3% aqueous sodium hydroxide and flushed with plenty of water into suitable drains leading to a biological wastewater treatment plant.

Cresols and xylenols can generally be removed from production wastewater by extraction, e.g., with diisopropyl ether, butyl acetate, methyl isobutyl ketone, or toluene [333], to such an extent that the residual contents can be eliminated without difficulty in a biological wastewater treatment plant. The wastewater should enter the biological treatment plant continuously, not intermittently in batches.

The phenolic constituents of wastewater can also be removed to a very large extent with ionexchange resins or such adsorbents as activated carbon [58], which can be subsequently regenerated, or with oxidizing agents such as hydrogen peroxide [334]. Stripping with steam is likewise possible, but is less effective.

Cresols and xylenols can be removed from waste gas streams by scrubbing with dilute aqueous sodium hydroxide or by incineration. Another effective method is adsorption on activated carbon, which is able to accumulate up to 0.5 kg of cresols/kg. The adsorbate is subsequently recoverable when the activated carbon is regenerated [335].

4. Toxicology and Occupational Health

Cresols and xylenols burn the skin and mucous membranes through degradation of proteins, similarly to phenols. Initially white, and later brownish black necroses appear on the skin. Certain individuals may show hypersensitivity symptoms in addition to the corrosive effect [273], [274].

Cresols and xylenols, like phenol, are toxic when absorbed through the skin. Their TLV or MAK value (5 ppm) is, therefore, supplemented by the word "skin" or the letter H (standing for the German word "Haut"). This percutaneous effect is easily overlooked because the initial burning sensation is followed by local anesthesia. The compounds are absorbed particularly rapidly when in the liquid form or dissolved (diluted). Even if only a relatively small area of the skin is exposed, e.g., a single arm or hand, there may be severe injury to health, possibly developing into a lethal paralysis of the central nervous system. Corresponding amounts of cresol or xylenol that are taken up through the respiratory tract, as vapor or as droplets, or through the stomach also cause severe poisoning in addition to corrosion. The initial or consecutive symptoms of poisoning are headache, dizziness, ringing in the ears, nausea, vomiting, muscular twitching, mental confusion, and loss of consciousness, depending on severity.

As phenols are rapidly absorbed through the skin and a causal treatment of the poisoning does not yet exist, rinsing them off the skin (not wiping them off) as soon as possible constitutes a decisively important first aid measure. A mixture of polyethylene glycol and ethanol in a 2:1 ratio has thus far proven most suitable for this purpose: it should be kept available for rapid use at the workplace in portable pressure vessels (similar to fire extinguishers, but painted a different color, e.g., green) [273]. Polyethylene glycol 400, or water, is less suitable, particularly in the case of xylenols, which are less soluble and may solidify on the skin. Water in very large quantities should only be used if no other solvents are immediately available. Eyes should be rinsed only with water, but very thoroughly, for at least 15 min. If cresols or xylenols enter the stomach, a large amount of water should be given immediately in small portions to achieve a diluting effect and, in addition, medicinal charcoal (1 tablespoon to 1 glass of water). Medical attention must be given, without delay, in every case. Further information on the prevention of accidents with cresols and xylenols, as well as the steps that should be taken if accidents occur, will be found in [273], [276].

Prolonged contact with cresols or xylenols at concentrations that, although low, are above the MAK value (5 ppm) may result in absorption through the skin, mucous membranes, and respiratory organs, and cause gastrointestinal disturbances (vomiting, loss of appetite), nervous disorders, headache, dizziness, fainting, and dermatitis.

According to epidemiological investigations [336], cresols are not carcinogenic. A carcino-genic effect would indeed be surprising, since *p*-cresol, for example, is a normal product of human metabolism. Cresols have likewise caused no carcinomas in animal experiments [330], [337]. In the *Salmonella* microsome test, all three cresol isomers have been found to be nonmutagenic toward the most widely differing *Salmonella* strains, both in the presence and in the absence of hamster or rat microsomal fractions [338]. The sister-chromatid exchange mutagenicity tests of the cresol isomers revealed no dose-dependent effects either in vitro in cultured human fibroblast or in vivo in male mice [339].

Extensive literature compilations on the toxicity of the cresols may be found in [330] and [335], which also contain some information on the xylenols, and in [337]. Toxicity data for cresols and xylenols are summarized in [340].

5. References

General References

[1] *Beilstein,* **6H,** 349–389; **6I,** 169–196; **6II,** 322–368; **6III,** 1233–1341; **6IV,** 1940–2093 (Cresols); **6H,** 480–494; **6I,** 240–245; **6II,** 453–466; **6III,** 1722–1769; **6IV,** 3006–3164 (Xylenols).
[2] *Houben-Weyl,* **6/1c,** part 1 and 2.
[3] *Ullmann,* 4th ed., **15,** 61–77.

Specific References

[4] G. Städeler, *Ann. Chem. Pharm.* **77** (1851) 17–37.
[5] *Geigy Scientific Tables,* vol. **1,** Ciba-Geigy Ltd., Basel 1984, p. 99.
[6] M. Lehtonen, *J. Assoc. Off. Anal. Chem.* **66** (1983) 62–70.
[7] D. M. Knight et al., *Org. Geochem.* **4** (1982) no. 2, 85–92.
[8] R. J. L. Andon, D. P. Biddiscombe, J. D. Cox, R. Hanley, D. Harrop, E. F. G. Herington, J. F. Martin, *J. Chem. Soc.* 1960, 5246–5254.
[9] R. J. L. Andon, J. F. Counsell, E. B. Lees, J. F. Martin, C. J. Mash, *Trans. Faraday Soc.* **63** (1967) 1115–1121.
[10] A. P. Kudchadker, S. A. Kudschadker, R. C. Wilhoit: *Key Chemicals Data Books, Cresols,* Thermodynamics Res. Center, Texas A & M University, College Station, Texas, 1978.
[11] ESDU: *Thermophysical Properties of Industrial Important Fluids on the Saturation Line: Cresols,* item 82 032, Engineering Sciences Data Unit, London 1983.
[12] J. McGarry, *Ind. Eng. Chem. Process Des. Dev.* **22** (1983) 313–322.
[13] E. Terres, F. Gebert, H. Hülsemann, H. Petereit, H. Toepsch, W. Ruppert, *Brennst. Chem.* **36** (1955) 359–372.
[14] J. Gmeling, U. Onken, W. Arlt resp. U. Weidlich: *Vapor-liquid equilibrium Data Collection* vol. **I,** part 2b (1978), vol. **I,** part 1a (1981), vol. **I,** part 2c (1982), vol. **I,** part 2d (1982), DECHEMA-Chemistry Data Series.
[15] H. M. Sebastian, H.-M. Lin, K.-C. Chao, *J. Chem. Eng. Data* **25** (1980) 381–383.
[16] H. Kim, W. Wang, H.-M. Lin, K.-C. Chao, *J. Chem. Eng. Data* **28** (1983) 216–218.
[17] *Rodd's Chemistry of Carbon Compounds,* vol. **III,** part A, Elsevier Scientific Publ. Co., Amsterdam-Oxford-New York 1983, p. 161–219.
[18] H. Weigold, *Fuel* **61** (1982) 1021–1026.
[19] Hydrocarbon Res. Inc., US 4 431 850, 1984. *Chem. & Eng. News* **59** (Nov. 1981) 32, 34.
[20] Sumitomo, EP 12 939, 1979.
[21] *DECHEMA-Werkstofftabelle: Kresole,* no. 851–852, Verlag Chemie, Weinheim 1959.
[22] A. A. Sagues, B. H. Davis, T. Johnson, *Ind. Eng. Chem. Process Des. Dev.* **22** (1983) 15–22.
[23] M. A. Elliot (ed.): *Chemistry of Coal Utilization,* 2nd suppl. vol., J. Wiley & Sons, New York-Chichester-Brisham-Toronto 1981.
[24] A. Dierichs, R. Kubička: *Phenole und Basen,* Akademie Verlag, Berlin 1958.
[25] J. Schulze, H. Sutter, M. Weiser, *Chem. Ind. (Berlin)* **35** (1983) 448–454.
[26] R. Cypres, S. Furfari, M. Ghodsi, *Erdöl Kohle* **36** (1983) 471–477.
[27] H.-G. Franck, G. Collin: *Steinkohlenteer,* Springer Verlag, Berlin-Heidelberg-New York 1968.
[28] E. Gundermann: *Chemie und Technologie des Braunkohleteers,* Akademie Verlag, Berlin (GDR) 1964.

[29] E. I. Kazakow, E. D. Vilyanskaya, G. A. Markus et al., *Khim. Tverd. Topl. Moscow* **17** (1983) no. 3, 93–96 (*Solid Fuel Chem. (Engl. Transl.)* New York (1983) 89–92).
[30] G. Bett, T. G. Harvey, T. W. Matheson, K. C. Pratt, *Fuel* **62** (1983) 1445–1454.
[31] R. E. Schirmer, T. R. Pahl, D. C. Elliot, *Fuel* **63** (1984) 368–372.
[32] H. E. Jegers, M. T. Klein, *Ind. Eng. Chem. Process Des. Dev.* **24** (1985) 173–183.
[33] J. A. Russel, P. M. Molton, S. D. Landsmann: *Alternative Energy Sources (Washington)* **3** (1980/1983) no. 3, 307–322.
[34] H. L. Lochte, E. R. Littmann: *The Petroleum Acids and Bases,* Chemical Publ. Co., New York 1955, pp. 121–125. H. L. Coonradt, B. W. Rope, *Ind. Eng. Chem. Process Des. Dev.* **2** (1963) 317–322.
[35] H.-J. Wurm, *Chem. Ing. Tech.* **48** (1976) 840–845.
[36] R. E. Maple: *Symposium on Refining Petroleum for Chemicals,* Am. Chem. Soc. New York City Meeting Sept. 7–12, 1969, D 105–113.
[37] *Chem. Eng.* **69** (1962) August 20, 66–77.
[38] R. E. Maple, A. R. Price, *Hydrocarbon Process.* **51** (1972) no. 10, 168–172.
[39] R. E. Maple, A. R. Price, *AIChE Symp. Ser.* **73** (1977) 200–206.
[40] A. R. Price, *AIChE National Conference on Complete Water Reuse,* New York, April 23–24 (1973) pp. 333–341.
[41] D. C. Jones, J. A. Kohlbeck, M. B. Neuworth, *Ind. Eng. Chem. Prod. Res. Dev.* **2** (1963) 217–220.
[42] T. P. Forbath (ed.), *Chem. Eng. (N.Y.)* **64** (1957) no. 7, 228–231.
[43] *Chemical Profile: Cresylics,* April 1, 1985, Schnell Publishing Co., Inc., New York 1985.
[44] S. W. Eglund, R. S. Aries, D. F. Othmer, *Ind. Eng. Chem.* **45** (1953) 189–197.
[45] D. McNeil, *Chem. Eng. Monogr.* **15** (1982) 209–232.
[46] W. Pritzkow, P. Grothkopf, R. Höring, H. Gross, *Z. Chem.* **5** (1965) 300.
[47] Honshu Chem. Ind. Co., JP 70 28 976, 1967; *Chem. Abstr.* **74** (1971) 12 828.
[48] G. D. Kharlampovich et al., *Khim. Promst. (Moscow)* **44** (1968) 16–20.
[49] G. D. Kharlampovich, V. F. Kollegow et al., *Khim. Tekhnol., Resp. Mezhved. Nauchno.-Tekh. Sb.* **5** (1966) 77–80.
[50] Tennessee Corp., US 2 828 333, 1956; US 2 841 612, 1956.
[51] Marni S.A., DE 24 13 344, 1974.
[52] Hoechst, DE 28 37 549, 1978.
[53] A. A. Spryskow, *Zhur. Obshei Khim.* **30** (1960) 2449–2453; *Chem. Abstr.* **57** (1962) 16 464.
[54] Honshu Chem. Ind. Co., JP 78 21 142, 1976; *Chem. Abstr.* **88** (1978) 190 393.
[55] Taoka Chem. Co., EP 80 880, 1982.
[56] P. Taffe, *Chem. Age (London)* **9** (May 1980) 18.
[57] E. Hurwitz, E. Ciabettari, R. A. Wolff, I. Bernstein, *Ind. Eng. Chem.* **51** (1959) 1301–1304.
[58] C. D. Baker, E. W. Cark, W. V. Jesernig, C. H. Huether, *AIChE Sympos. Ser.* **70** (1974) 686–692.
[59] R. N. Shreve, C. J. Marsel, *Ind. Eng. Chem.* **38** (1946) 254–261.
[60] O. Lindner, *Chem. Ing. Tech.* **36** (1964) 769–774.
[61] Bayer AG, DE 29 04 831, 1979.
[62] A. T. Bottini, J. D. Roberts, *J. Am. Chem. Soc.* **79** (1957) 1458–1462.
[63] M. Zoratti, J. F. Bunnet, *J. Org. Chem.* **45** (1980) 1769–1776.
[64] Dow Chem. Co., US 4 001 340, 1975.
[65] Dow Chem. Co., US 1 959 283, 1930; US 2 126 610, 1933.
[66] Asahi Chem. Ind., JP 69 17 372, 1967; *Chem. Abstr.* **71** (1969) 112 620.
[67] Frontier Chem. Co., US 3 413 341, 1965.
[68] Sumitomo Chem. Co., DE 18 03 036, 1968.

[69] Sumitomo Chem. Co., DE 23 25 354, 1973.
[70] Hercules Powder, GB 754 872, 1954.
[71] R. H. Allen, T. Alfrey, L. D. Yats, *J. Am. Chem. Soc.* **81** (1959) 42–46.
[72] R. H. Allen, L. D. Yats, *J. Am. Chem. Soc.* **83** (1961) 2799–2805.
[73] Rütgerswerke & Teerverwertung, DE 12 74 096, 1969.
[74] Cosden Oil & Chemical Co., DE 22 47 308, 1972.
[75] H. Boardman, *J. Am. Chem. Soc.* **84** (1962) 1376–1382.
[76] G. D. Serif et al., *Can. J. Chem.* **31** (1953)1229–1238.
[77] Sumitomo Chem. Co., JP 78 82 732, 1976; *Chem. Abstr.* **90** (1979) 6090.
[78] Sumitomo Chem. Co.,JP 78 82 731, 1976; *Chem. Abstr.* **90** (1979) 6091.
[79] Mitsui Petroch. Inds., JP 83 198 468, 1982; *Chem. Abstr.* **100** (1984) 174 433.
[80] Mitsui Petroch. Inds., JP 82 59 861, 1980; *Chem. Abstr.* **97** (1982) 162 579.
[81] Hercules Powder, US 2 728 797, 1954.
[82] Hercules Powder, US 2 779 797, 1953.
[83] G. D. Kharlampovich et al., *Zh. Prikl. Khim. (Leningrad)* **46** (1973) 118–121.
[84] Mitsui Petroch. Inds., JP 73 49 735, 1971; *Chem. Abstr.* **79** (1973) 136 803.
[85] Mitsui Petroch. Inds., US 4 408 083, 1981.
[86] Mitsui Petroch. Inds., JP 4 036 217, 1974; *Chem. Abstr.* **82** (1975) 170 367.
[87] Sumitomo Chem. Co., JP 71 23 723, 1968; *Chem. Abstr.* **75** (1971) 88 329.
[88] Mitsui Petrochem. Inds., EP 21 848, 1980.
[89] Goodyear Tire & Rubber, EP 77 749, 1982.
[90] R. S. Sushko, V. N. Kiva, *Vses. Konf. Teor. Prakt. Rektifikatsii*, [Dokl.] 3rd ed., vol. **I**, E. G. Melamed (ed.), 1973, pp. 177–180; Chem. Abstr. **88** (1978) 63 599.
[91] K. Ito, *Hydrocarbon Process.* **52** (1973) no. 8, 89–90.
[92] *Chem. Eng. News*1953, 246; 1964, 29.
[93] R. W. Neuzil, D. H. Rosback, R. H. Jensen, J. R. Teague, A. J. de Rosset, *CHEMTECH* **10** (1980) 498–503.
[94] *Eur. Chem. News,*1978, August 1, 21.
[95] UOP Inc., US 4 128 593, 1978.
[96] Mobil Oil Corp., EP 12 613, 1979.
[97] Sumitomo Chem. Co., EP 127 410, 1984.
[98] Société Chimique des Charbonnages Courbevoie, US 3 962 366, 1974.
[99] *Chem. Process. (Chicago)* 1966, Mid-November, 113.
[100] T. F. Meinhold, J. Bardzik, *Chem. Process. (Chicago)* 1968, April, 31–35.
[101] Rütgerswerke & Teerverwertung, US 3 347 936, 1965.
[102] Y. V. Churkin et al., *Khim. Promst. (Moscow)* 1972, 660–666.
[103] General Electric Co., US 4 041 085, 1975.
[104] General Electric Co., US 3 962 126, 1974.
[105] M. Tomita, *Hokkaido Kogyo Kaihatsu Shikensko Hokuku* **22** (1981) 8–23; *Chem. Abstr.* **96** (1982)70 751.
[106] Asahi Chemical Ind. Co., EP 101 138, 1983.
[107] G. A. Kirichenko, V. N. Kiva, V. S. Bogdanov, *Russ. J. Phys. Chem. (Engl. Transl.)* **52** (1978) 893–894.
[108] G. A. Kirichenko, V. N. Kiva, N. Z. Baibulatova, *Vses. Konf. Teor. Prakt, Rektifikatsii, [Dokl.]*, 3rd **1** (1973) 86–89;*Chem. Abstr.* **87** (1977) 22 649.
[109] Mitsui Toatsu Chemicals Inc., JP 81 46 829, 1979; *Chem. Abstr.* **95** (1981) 132 485.
[110] Mitsui Toatsu Chemicals Inc., JP 81 65 836, 1979; *Chem. Abstr.* **95** (1981) 186 857.

[111] General Electric Co., US 3 446 856, 1964.
[112] H. Hattori, K. Shimazu, N. Yoshi, K. Tanabe, *Bull. Chem. Soc. Japan* **49** (1976) 969–972.
[113] General Electric Co., WO 84/01 146, 1982.
[114] General Electric Co., EP 127 833, 1984.
[115] General Electric Co., DE 3 418 087, 1984.
[116] Conoco Inc., US 4 359 408, 1981.
[117] General Electric Co., US 4 201 880, 1976.
[118] K. Tanabe, T. Nishizaki, *Sixth Internat. Congress on Catalysis,* London, July 12–16, 1976, paper B 26, pp. 1–8.
[119] Mitsui Toatsu Chemicals Inc., US 4 388 478, 1980.
[120] Mitsui Toatsu Chemicals Inc., US 4 208 857, 1977.
[121] Mitsui Toatsu Chemicals Inc., JP 81 55 327, 1979; *Chem. Abstr.* **95** (1981) 186 848.
[122] Mitsui Toatsu Chemicals Inc., JP 82 28 018, 1980; *Chem. Abstr.* **96** (1982) 199 288.
[123] Asahi Chemical Ind., JP 84 194 834, 1983; *Chem. Abstr.* **102** (1985) 114 797.
[124] Asahi Chemical Ind., JP 81 133 232, 1980; *Chem. Abstr.* **96** (1982) 34 853.
[125] Mitsui Toatsu Chemicals Inc., US 4 329 517, 1979.
[126] Asahi-Dow Ltd., GB 2 089 343, 1980.
[127] Croda Synthetic Chemicals Ltd., EP 50 937, 1981.
[128] Bayer AG, EP 10 24 93, 1983.
[129] Agency of Industrial Science and Technology, JP 78 35 061, 1978; *Chem. Abstr.* **90** (1979) 6086.
[130] T. Kotanigawa, M. Yamamoto, K. Shimokawa, Y. Yoshida, *Bull. Chem. Soc. Japan,* **44** (1971)1961–1964.
[131] Union Rheinische Braunkohlen Kraftstoff AG, GB 2 072 674, 1981.
[132] Mitsubishi Gas Chemical Co., US 4 024 195, 1974.
[133] Nippon Steel Chem. Co., JP 84 181 232, 1983.
[134] Asahi-Dow Ltd., DE 3 147 026, 1981.
[135] Mitsui Toatsu Chemicals Inc., JP 81 43 229, 1979; *Chem. Abstr.* **95** (1981) 115 033.
[136] Mitsui Toatsu Chemicals Inc., JP 81 45 427, 1979; *Chem. Abstr.* **95** (1981) 97 361.
[137] Mitsui Toatsu Chemicals Inc., JP 81 49 330, 1979; *Chem. Abstr.* **95** (1981) 97 358.
[138] Mitsui Toatsu Chemicals Inc., JP 83 208 244, 1982; *Chem. Abstr.* **100** (1984) 158 561.
[139] T. Kotanigawa, *Bull. Chem. Soc. Japan* **47** (1974) 950–953.
[140] E. M. Knyazewa, N. S. Khasanova, L. M. Koval, N. N. Sudakova, *Russ. J. Phys. Chem. (Engl. Transl.)* **58** (1984) 592–594.
[141] Rütgerswerke AG, DE 27 56 461, 1977.
[142] Y. V. Churkin, V. N. Khabibullin, G. A. Kirichenko, A. D. Zotov, *Sov. Chem. Ind. (Engl. Transl.)* **4** (1972) 550–553.
[143] G. D. Kharlampovich, *Proizvod. Pereab. Plastmass Sint. Smol. (Moscow)* **7** (1978) 56–58; *Chem. Abstr.* **90** (1979) 199 365.
[144] V. N. Vinogradova, G. D. Kharlampovich, *Sov. Chem. Ind. (Engl. Transl.)* **5** (1973) 558–560.
[145] Conoco Inc., US 4 381 413, 1981.
[146] P. S. Landis, W. O. Haag, *J. Org. Chem.* **28** (1963) 585.
[147] S. E. Santos, J. M. Marinas, M. P. Alcasar, *React. Kinet. Catal. Lett.* **24** (1984) 247–251.
[148] General Electric Co., GB 1 034 500, 1965.
[149] M. F. Mazitov, A. S. Shelev, B. I. Bykow, V. Churkin, *Sov. Chem. Ind. (Engl. Transl.)* **12** (1980) 995–1002.

[150] M. Huml, J. Jona, *Sb. Pr. Vyzk. Chem. Vyuziti Uhli, Dehtu Ropy (Prague)* **15** (1978) 199–210; *Chem. Abstr.* **89** (1978) 59 735.
[151] J. Jona, CS 214 049, 1980; *Chem. Abstr.* **101** (1984) 54 706.
[152] Consolidation Coal Co., US 3 624 163, 1968.
[153] M. Janardanarao, G. S. Salvapati, R. Vaideswaran, *Proc. Natl. Symp. Catal., 4th 1978 (Bombay, India)* 1980,51–57.
[154] S. Namba, T. Yashima, Y. Itaba, N. Hara, *Stud. Surf. Sci. Catal. (Amsterdam)* **5** (1980) (Catalysis by Zeolites) 105–111.
[155] S. Balsama, P. Beltrame, P. L. Beltrame et al., *Appl. Catal.* **13** (1984) 161–170.
[156] Moscow Gubkin Petrochem. SU 1 004 342, 1981; *Chem. Abstr.* **99** (1983) 38 189. Yeda Research and Devel. Co., EP 126 245, 1984.
[157] M. Blanco, J. M. Marinas, R. Perez-Ossorio, A. Alberola, *An. Quim.* **71** (1975) 199–205.
[158] N. D. Limankina, E. A. Vdovtsova, A. S. Sultanov, *Uzb. Khim. Zh.* 1977, no. 6, 34–38; *Chem. Abstr.* **89** (1978) 129 157.
[159] Coal Tar Research Assoc., GB 1 125 087, 1966.
[160] Asahi Chem. Ind., JP 74 028 742, 1970; *Chem. Abstr.* **82** (1975) 111 767. JP 74 043 943, 1970; *Chem. Abstr.* **83** (1975) 27 884. JP 74 043 944,1970; *Chem. Abstr.* **83** (1975) 27 885.
[161] E. A. Vdovtsova, A. Kochetkova, A. S. Sultanov, *Neftekhimiya* **22** (1982) 200–206.
[162] O. V. Glukhikh, V. A. Proskuryakow, D. A. Sibarow, *Zh. Prikl. Khim. (Leningrad)* **57** (1984) 658–663.
[163] Shell Devel. Co., US 2 440 036, 1948.
[164] General Electric Co., US 4 225 732, 1977.
[165] Ashland Oil Inc., DE 2 251 553, 1972.
[166] General Electric Co., US 4 048 239, 1976.
[167] Y. V. Churkin, L. A. Rusanova, G. D. Kharlampovich et al., *Tr., Nauch.-Isled. Neftekhim. Proizvod.* **2** (1970) 7–13; *Chem. Abstr.* **74** (1971) 76 121.
[168] M. Inoue, S. Enomoto, *Chem. Pharm. Bull.* **24** (1976) 2199–2203.
[169] M. Ehsan, G. S. Salvapati, M. Janardanarao, R. Vaideswaran, *Proc. Catsympo 80, Natl. Catal. Symp., 5th 1980* (pub. 1983) 31–37, Hyderabad, India.
[170] Continental Oil Co., US 3 994 982, 1975.
[171] Continental Oil Co., US 4 022 843, 1975.
[172] Chemisches Werk Lowi GmbH, DE 1 265 755, 1965.
[173] Chemisches Werk Lowi GmbH, DE 1 543 494, 1966.
[174] Chemisches Werk Lowi GmbH, DE 1 817 342, 1968.
[175] Coal Tar Res. Assoc., GB 1 060 036, 1965.
[176] Union Rheinische Braunkohlen Kraftstoff AG, DE 1 281 448, DE 1 297 110, 1966.
[177] Union Rheinische Braunkohlen Kraftstoff AG, GB 1 160 491, 1968.
[178] Union Rheinische Braunkohlen Kraftstoff AG, GB 1 077 645, 1966; GB 1 125 077, 1967.
[179] Union Rheinische Braunkohlen Kraftstoff AG, US 3 691 238, 1967.
[180] Midland-Yorkshire Tar Distillers, GB 1 239 761, 1967.
[181] K. K. Tiwari, P. N. Mukherjee, *Indian J. Technol.* **22** (1984) 295–300.
[182] Consolidation Coal Co., GB 1 238 353, 1968.
[183] Union Rheinische Braunkohlen Kraftstoff AG, EP 119 421, 1984.
[184] Continental Oil Co., US 4 125 736, 1977.
[185] Rütgerswerke AG, US 4 149 019, 1976.
[186] L. A. Fury, jr., D. E. Pearson, *J. Org. Chem.* **30** (1965) 2301–2304.

[187]   A. Dierichs, R. Kubička: *Phenole und Basen*, Akademie Verlag, Berlin (GDR) 1958, pp. 381–385.
[188]   V. N. Vinogradova, G. D. Kharlampovich, *Khim. Promst. (Moscow)* 1975, 419–420.
[189]   Union Rheinische Braunkohlen Kraftstoff AG, US 4 283 571, 1979.
[190]   Gulf Research & Devel. Co., US 3 591 644, 1968.
[191]   Gulf Research & Devel. Co., US 3 591 645, 1968.
[192]   Gulf Research & Devel. Co., US 3 752 878, 1970.
[193]   W. W. Kaeding, R. O. Lindblom et al., *Ind. Eng. Chem. Process Des. Dev.* **4** (1965) 97–101.
[194]   N. L. D'yachenko, G. D. Kharlampovich, L. K. I'llina, *Neftekhimiya* **13** (1973) 870–875.
[195]   Imperial Chemical Ind. Ltd., GB 1 564 567, 1977.
[196]   Lummus Co., US 4 277 630, 1977.
[197]   Sumitomo Chemical Co., EP 52 839, 1981.
[198]   N. L. D'yachenko, G. D. Kharlampovich, L. K. I'llina, *Neftekhimiya* **12** (1972) 762–765.
[199]   M. P. Sharma, J. N. Chatterjea, *J. Chem. Tech. Biotechnol.* **33 A** (1983) 328–332.
[200]   G. D. Kharlampovich, N. L. D'yachenko, *Kinet. Katal.* **8** (1967) 208–210.
[201]   Rhône-Poulenc S.A., DE 2 164 004, DE 21 63 940, 1971.
[202]   Rhône-Poulenc S.A., DE 2 249 167, DE 2 256 513, 1972.
[203]   Mitsubishi Gas Chem. Co., DE 2 508 452, 1975.
[204]   Mitsubishi Gas Chem. Co., EP 74 162, 1982.
[205]   S. Fujiyama, T. Kasahara, *Hydrocarbon Process.* **57** (1978) no. 11, 147–176.
[206]   Y. Shimamura, H. Misawa et al., *Chem. Lett.* 1983, 1691–1694.
[207]   G. A. Olah, A. P. Fung, T. Keumi, *J. Org. Chem.* **46** (1981) 4305–4306.
[208]   M. A. Brooks, R. Higgins et al., *J. Chem. Soc. Perkin Trans. II* 1982, 687–692.
[209]   M. K. Eberhardt, *J. Org. Chem.* **42** (1977) 832–835. K. Tomizawa, Y. Ogata, *J. Org. Chem.* **46** (1981) 2107–2109.
[210]   Rhône-Poulenc S.A., DE 1 262 282, 1964.
[211]   S. Hashimoto, W. Koike, *Bull. Chem. Soc. Japan* **43** (1970) 293.
[212]   Rhône-Poulenc Ind., US 4 301 307, 1979.
[213]   J. A. Vesely, L. Schmerling, *J. Org. Chem.* **35** (1970) 4028–4033.
[214]   Anic S.p.A., DE 31 35 556, 1981.
[215]   D. J. Le Count, J. A. W. Reid, *J. Appl. Chem.* **18** (1968) 108–110.
[216]   K. E. Khcheyan, O. M. Revenko et al., *Nov. Metody Sint. Org. Soedin. Osn. Neftekhim. Syr'ya* 1982, 3–10; *Chem. Abstr.* **101** (1984) 74 734.
[217]   NI-Institut Sintetitscheskich Spirtow i Organitscheskich Produktow, US 3 830 853, 1971.
[218]   K. E. Khcheyan, O. M. Revenko, A. N. Shatalova, *11th World Petroleum Congress*, London 1983, Spec. Paper No. 16, pp. 1–7.
[219]   V. I. Zavorotow, G. L. Bitman, R. M. Gimronova, *Nov. Metody Sint. Org. Soedin. Osn. Neftekhim. Syr'ya* 1982, 11–19; *Chem. Abstr.* **101** (1984) 74 735.
[220]   G. N. Kiričenko, E. G. Mavljutova, T. M. Channon, *Neftekhimiya* **10** (1970) 231–235.
[221]   Sumitomo Chemical Co., JP 80 45 667, 1978; *Chem. Abstr.* **93** (1980) 204 270.
[222]   Ube Industries, DE 26 03 269, 1976.
[223]   Rhône-Poulenc S.A., DE 24 41 744, 1974.
[224]   Mitsubishi Gas Chem. Co., JP 80 51 031, 1978; *Chem. Abstr.* **93** (1980) 185 946.
[225]   M. M. Groshan et al., *Dokl. Akad. Nauk. SSSR Ser. Khim.* **204** (1972) 872–878; *Doklady Chemistry (Engl. Transl.)* **204** (1972) 469–470.
[226]   Koppers Co., US 4 215 229, 1978.
[227]   H. G. Könnecke, K. Kränke, H. Languth, DD 60 047, 1966.

[228] L. Cassar, M. Foà, G. P. Chiusolt, *Organomet. Chem. Synth.* **1** (1971) 302–304.
[229] IG-Farbenindustrie, DE 694 781, 1938.
[230] Mitsubishi Chem. Ind., JP 73 103 513, 1973; *Chem. Abstr.* **80** (1974) 70 532.
[231] T. Cohen, A. G. Dietz, J. R. Miser, *J. Org. Chem.* **42** (1977) 2053–2058.
[232] K. Pastalka et al., CS 184 670, 1976; *Chem. Abstr.* **94** (1981) 208 539. CS 193 993, 1977; *Chem. Abstr.* **98** (1983) 71 667.
[233] BASF AG, DE 14 93 492, 1965.
[234] F. P. Daly, *J. Catal.* **61** (1980) 528–532.
[235] M. Moritoki T. Fujikawa, *Process Technol. Proc.* 1984, vol. **2**, Ind. Crist, 369–372.
[236] M. Moritoki, K. Kitagawa, K. Onoe, K. Kaneko, *Process Technol. Proc.* 1984, vol. **2**, Ind. Crist., 377–380.
[237] *Process Eng. (London)* **65** (July 1984) 13.
[238] Universal Oil Prod. Co., US 3 969 422, 1966.
[239] Mitsui Petrochem. Ind., JP 82 59 824, 1980; *Chem. Abstr.* **97** (1982) 127 256.
[240] Mitsui Petrochem. Ind., JP 80 57 529, 1978; *Chem. Abstr.* **93** (1980) 149 980.
[241] Toray Ind. Inc., DE 3 327 146, 1982.
[242] Pure Oil Co., US 3 031 383, 1959.
[243] R. Rigamonti, G. Schiavino, *Chim. Ind. (Milan)* **36** (1954) 611–617.
[244] M. M. Anwar, M. W. T. Pratt, M. Y. Shaeen, *Int. Solvent Extr. Conf.* [Proc.] vol. **2** (1980), Paper 80-64.
[245] L. S. Nikolaeva, A. M. Evseev et al., *Russ. J. Phys. Chem. (Engl. Transl.)* **52** (1978) 1351–1353.
[246] M. W. Pratt, J. Spokes, *CIM spec.* vol. **21** (1979) 723–738.
[247] G. A. Markus, G. I. Mischenko, A. N. Antonova, *Koks Khim.* 1974, no. 8, 37–41; *Chem. Abstr.* **82** (1975) 33 129.
[248] Koppers Co., US 4 267 389, 1979.
[249] Koppers Co., US 4 267 390, 1979.
[250] Koppers Co., US 4 267 391, 1979.
[251] Koppers Co., US 4 267 392, 1979.
[252] Koppers Co., US 4 394 526, 1980.
[253] Bayer AG, DE 12 15 726, 1964.
[254] Sumitomo Chem. Co., DE 2 136 700, 1971.
[255] Gesellschaft für Teerverwertung GmbH, DE 1 124 046, 1958.
[256] Gesellschaft für Teerverwertung GmbH, DE 1 118 797, 1960. Phillips Petroleum Co., US 4 491 677, 1983.
[257] Mitsui Petroch. Ind./Kobe Steel, JP 77 31 039, 1975; *Chem. Abstr.* **87** (1977) 39 110.
[258] Mitsui Petrochem. Ind., JP 79 46 729, 1977; *Chem. Abstr.* **91** (1979) 123 551. JP 79 46 730, 1977; *Chem. Abstr.* **91** (1979) 74 334
[259] VEB Leuna-Werke, DD 48 625, 1964.
[260] S. N. Vyas, S. R. Patwardhan, M. M. Bhave, *Sep. Sci. Technology* **16** (1981) 377–384.
[261] Y. Muramoto, H. Asakura, *Nippon Kagaku Kaishi* 1975, 672–677; *Chem. Abstr.* **83** (1975) 113 274.
[262] Koppers Co., US 3 785 776, 1972.
[263] W. Weinrich, *Ind. Eng. Chem.* **35** (1943) 264–272.
[264] D. R. Stevens, *Ind. Eng. Chem.* **35** (1943) 665–660.
[265] Bayer AG, US 4 144 400, 1976.
[266] Bayer AG, US 4 150 243, 1976.
[267] Bayer AG, US 4 113 975, 1977.

[268] BIOS 937.
[269] Verona Chemical Co., US 1 502 849, 1922.
[270] J. Anazawa, K. Uetake, *Kogyo Kagaku Zasshi* **63** (1960) 1963–1967.
[271] Standardization of Tar Products Test Committee (S.T.P.T.C), P. V. Watkins (ed.): *Standard methods for testing tar and its Products,* 6th ed., Gomersal, Checkheaton, Yorkshire, 1967.
[272] C. Zerbe: *Mineralöle,* Part II, Springer Verlag, Berlin-Heidelberg 1969, pp. 494–497.
[273] Berufsgenossenschaft der Chemischen Industrie; Merkblatt M 018: *Phenol, Kresole und Xylenole,* Heidelberg 1984.
[274] Manufacturing Chemists Association, Chemical Safety Data Sheet SD-48: *Cresol,* Washington 1952.
[275] ASTM D 3851-83:Sampling and Handling for Phenol and Cresylic Acid, 1983.
[276] G. Hommel: *Handbuch der gefährlichen Güter,* Merkblätter **117** (Kresylsäure), **118** (o-Kresol), **118 a** (m/p-Kresol), 1980; Merkblätter **986** (2,3- und 2,4-Xylenol), **987** (2,5- und 3,5-Xylenol), **988** (2,6- und 3,4-Xylenol), Springer Verlag, Berlin-Heidelberg 1985.
[277] A. P. Kudchadker, S. A. Kudchadker: *Key chemicals data books. Xylenols.* Thermodynamics Research Center, Texas A & M University, 1978.
[278] S. A. Kudchadker, A. P. Kudchadker, R. C. Wilhoit, B. J. Zwolinski, *Hydrocarbon Process.* **58** (1979) no. 1, 169–171.
[279] ESDU, Thermophysical properties of industrial important fluids on the saturation line: *Xylenols.* Item 83 005, Engineering Science Data Unit, London 1983.
[280] I. V. Goncharov, V. I. Kulachenko, *Neftekhimiya* **19** (1979) 255–258; **18** (1978) 816–821.
[281] G. A. Makkus, *Koks Khim.* 1975, no. 9, 43–47.
[282] Conoco Inc., US 4 228 311, 4 247 719, 4 249 026, 4 292 450, 1979.
[283] ICI, GB 582 057, 1944. US 2 497 971, 1945.
[284] Koppers Co., US 4 447 658, 1982.
[285] Toray Industries Inc., DE 3 327 146, 1983.
[286] General Electric Co., US 4 001 341, 1964 (publ. 1974).
[287] A. A. Pavlova, *J. Appl. Chem. USSR (Engl. Transl.)* **48** (1975) 1394–1396.
[288] Consolidation Coal, US 3 331 755, 1964.
[289] Consolidation Coal, US 3 337 424, 1965.
[290] Koppers Co., DE 19 24 768, 1969.
[291] Mitsui Petrochem. Ind., EP 13 133, 1979.
[292] Asahi Chem. Ind., DE 2 703 777, 1977.
[293] UOP Inc., US 4 386 225, 1982.
[294] Rütgerswerke AG, DE 2 531 774, 1975.
[295] Asahi-Dow Ltd., JP 82 165 333, 1981; *Chem. Abstr.* **98** (1983) 88 963.
[296] Nippon Steel Co., JP 83 52 235, 1981; *Chem. Abstr.* **99** (1983) 22 105.
[297] Kobe Steel/Mitsui Gas Chem. Co., JP 77 38 545, 1973; *Chem. Abstr.* **87** (1977) 24 943.
[298] R. W. Maxwell, US 4 013 520, 1975.
[299] Continental Oil, US 3 996 297, 1975.
[300] Conoco Inc., US 4 258 220, 1980.
[301] Y. V. Churkin et al., *Neftepererab. Neftekhim. (Moscow)* 1974, no. 5, 32–33.
[302] Kuraray Co., JP 75 71 635, 1973; *Chem. Abstr.* **83** (1975) 205 917.
[303] Kuraray Co., JP 75 71 636, 1973; *Chem. Abstr.* **83** (1975) 205 916.
[304] Honshu Chem. Ind. Co., JP 73 78 134, 1972; *Chem. Abstr.* **80** (1974) 59 671. JP 84 84 832, 1982; *Chem. Abstr.* **101** (1984) 191 329.
[305] Shell Intern. Research, DE 25 29 773, 1975.

[306] Shell Intern. Research, EP 80 759, 1982.
[307] Rütgerswerke AG, GB 1 229 359, 1968.
[308] Shell Oil Co., US 4 086 282, 1976.
[309] Shell Intern. Research, DE 2 804 114, 1978.
[310] Atlantic Richfield Co., DE 2 316 576, 1973.
[311] Rütgerswerke AG, DE 3 015 803, 1980.
[312] Mitsubishi Gas Chemical Co., DE 2 521 324, 1974.
[313] Mitsubishi Gas Chemical Co., DE 2 650 416, 1976.
[314] E. V. Kirkland, O. P. Funderburk, F. T. Wadsworth, *J. Org. Chem.* **23** (1958) 1631–1635.
[315] Sinclair Research Inc., US 3 352 927, 1964.
[316] G. D. Kharlampovich, L. Z. Oblasova, N. A. Zhmakina, *Neftepererab. Neftekhim. (Moscow)* 1972, no. 10, 19–21.
[317] Coalite and Chemical Products Ltd., DE 17 68 679, 1968.
[318] Koppers Co., BE 844 866, 1975.
[319] L. Z. Oblasova, V. V. Moskvich, G. D. Kharlampovich, A. I. Eliseeva, *Izv. Vyssh. Uchebn. Zaved., Khim. Khim. Tekhnol.* **16** (1973) 1047–1050.
[320] G. D. Kharlampovich, L. Z. Oblasova, SU 577 202, 1975; *Chem. Abstr.* **88** (1978) 50 486.
[321] V. S. Tarnavskii, I. G. Skorina, A. N. Morozow, *Khim. Tekhnol. Polikondens. Polim.* 1977, 69–76.
[322] P. H. Given, *J. Appl. Chem.* **7** (1957) 172–193.
[323] Union Rheinische Braunkohlen Kraftstoff AG, DE 1 956 383, 1969.
[324] K. Kochloefl et al., CS 89 437, 1959.
[325] J. Jelinek, CS 95 413, 1960; *Chem. Abstr.* **55** (1961) 7357.
[326] I. Pigmen, E. Del Bel, M. B. Neuworth, *J. Am. Chem. Soc.* **76** (1954) 6169–6171.
[327] V. I. Buraev, I. S. Isaev, V. A. Koptyug, *Zh. Org. Khim.* **15** (1979) 782–789.
[328] P. Pitter, *Water Research* **10** (1976) 231–236.
[329] W. E. Olive, jr., H. D. Cobb, R. M. Atherton, *Proc. Int. Biodegradation Symp. 3rd.* 1975 (Pub. 1976), 381–388.
[330] C. T. Helmes, B. Levin, K. McCaleb et al.: *A study of industrial data on candidate chemicals for testing*, Report 1978, EPA 560/5-78-002: Cresols and cresylic acids, p. 3–63/3–128. Order No. PB-284 950, US Environmental Protection Agency, Washington D.C.
[331] R. L. Aaberg, R. A. Peloquin, D. L. Strenge, P. L. Mellinger: *An aquatic pathways model to predict the fate of phenolic compounds*, Report 1983, PNL-4202 App. A-D.; Order No. DE 83 016 661, 1984; Chem. Abstr. 100 (1984) 39 246.
[332] F. Dietz, J. Traud, *GWF Gas Wasserfach: Wasser/Abwasser* **119** (1978) 318–325.
[333] S. T. Hwang, *AIChE, Symp. Ser.* **77** (1981) no. 209, 304–315.
[334] E. J. Keating, R. A. Brown, E. S. Greenberg, *Proc. Ind. Waste Conf. 1976* **33** (publ. 1979) 464–470.
[335] J. Gordon: *Air Pollution assessment of Cresols*, U.S. NTIS, PB Report 1976, PB-256 737, 66 pp.
[336] R. L. Hervin, B. Froneberg, *Health hazard evaluation report* No. 80-020-1054, 1982, 32 pp.
[337] DHEW (NIOSH) Publication (U.S.) vol. 78–133: *NIOSH Criteria for a Recommended Standard Occupational Exposure to Cresol.* 1978, 116 pp.
[338] S. T. Haworth, K. Lawlor, W. Speck, E. Zeiger, *Environ. Mutagen.* **5** (1983) 3–142.
[339] M. Cheng, A. D. Klingerman, *Mutat. Res.* **137** (1984) 51–55.
[340] N. I. Sax: *Dangerous Properties of Industrial Materials*, 6th ed., Van Nostrand Reinhold, New York 1984, pp. 814–815 (Cresols), pp. 2741–2742 (Xylenols).

Crotonaldehyde and Crotonic Acid

WERNER BLAU, Hoechst Aktiengesellschaft, Frankfurt Federal Republic of Germany (Chap. 1)
HERBERT BALTES, Hoechst Aktiengesellschaft, Frankfurt Federal Republic of Germany (Chap. 2)
DIETER MAYER, Hoechst Aktiengesellschaft, Frankfurt Federal Republic of Germany (Chap. 3)

1.	Crotonaldehyde 1711	2.1.	Physical and Chemical Properties 1715
1.1.	Physical and Chemical Properties 1711	2.2.	Production 1717
1.2.	Production 1713	2.3.	Uses 1719
1.3.	Uses 1714	3.	Toxicology and Occupational Health................. 1719
2.	Crotonic Acid 1715	4.	References............. 1720

1. Crotonaldehyde

Crotonaldehyde [4170-30-3], C_4H_6O, M_r 70.09, 2-butenal, is a colorless liquid with strong lacrimatory properties, by which it can be easily detected. Crotonaldehyde exists in two stereoisomeric forms, *cis*-crotonaldehyde [15798-64-8] and *trans*-crotonaldehyde [123-73-9]. Commercial crotonaldehyde consists of both isomers but contains less than 5% of the cis isomer.

$$\underset{cis-}{\overset{H_3C}{\underset{}{\diagdown}}CH=CH\diagup CHO} \qquad \underset{trans-}{\overset{H_3C}{\underset{}{\diagdown}}CH=CH\diagdown CHO}$$

Crotonaldehyde

Crotonaldehyde was discovered by A. LIEBEN [1] and its structure was suggested by A. KEKULÉ [2].

1.1. Physical and Chemical Properties

Table 1 lists some physical properties of crotonaldehyde. Owing to its double bond and the aldehyde group, crotonaldehyde is very reactive. Contamination with alkali or strong acids may induce dangerous heat-generating condensation reactions [3]. The resinifying and darkening of pure crotonaldehyde can be suppressed by thorough exclusion of oxygen and by use of stainless steel containers. These measures prevent formation of peroxides and of iron salts of crotonic acid, which cause the darkening and

Table 1. Physical properties of crotonaldehyde

Boiling point	102.2 °C
Melting point	− 69 °C
Density d_4^{20}	0.852
Refractive index n_D^{20}	1.438
Heat of vaporization	515 J/g
Crotonaldehyde – water azeotrope	24.8 wt% H_2O
bp	84 °C
Solubility in water (20 °C)	18.1 g/100g
Solubility of water in crotonaldehyde (20 °C)	9.5 g/100 g
Explosion limits in air	
lower:	2.1 vol%
upper:	15.5 vol%

resinifying. When these precautions are taken, crotonaldehyde can be stored without the need for added inhibitors. Water and hydroquinone inhibit oxidation and resinification.

When exposed to strong acids, crotonaldehyde forms a dimer called dicrotonaldehyde [*13710-57-1*] (5,6-dihydro-2,6-dimethyl-2*H*-pyran-3-carboxaldehyde) [4].

Only some of the more important reactions involving crotonaldehyde are described in the following paragraphs.

Reduction. Crotonaldehyde can be reduced selectively at one or at both of its reactive groups.

When the double bond and the aldehyde group are involved, the product is butanol [5]. This can be achieved by hydrogenation at high temperature in the presence of nickel or copper catalysts. Hydrogenation at low temperature, however, leads to the formation of butyraldehyde [5]. When reducing agents such as $LiBH_4$, $NaBH_4$, or $LiAlH_4$ are used, crotonaldehyde is reduced selectively to crotyl alcohol [*6117-91-5*], $CH_3CH=CHCH_2OH$ [6], [7].

Oxidation. The oxidation of crotonaldehyde to crotonic acid by air or oxygen has been studied widely [8] – [10].

Crotonaldehyde can undergo addition reactions involving its double bond, reacting with methanol or ethanol in the presence of sodium hydroxide to form the corresponding 3-alkoxybutyraldehydes [11]. The addition of thiols to the double bond also follows the same pattern [12], [13]. For instance, addition of ethanethiol to crotonaldehyde yields 3-(ethylthio)butanal [12].

In Diels – Alder additions crotonaldehyde reacts as a dienophile [14], but commonly available derivatives of crotonaldehyde, for example, 1-acetoxybutadiene or 1-ethoxybutadiene, can react as dienes [15].

Crotonaldehyde reacts with ketene in the presence of zinc salts of organic acids to form a polyester, which then yields sorbic acid on heating or hydrolysis [16], [17]. Formation of a β-lactone by this reaction is also possible [18]:

$$CH_3CH=CH-\underset{O}{\diamond}=O$$

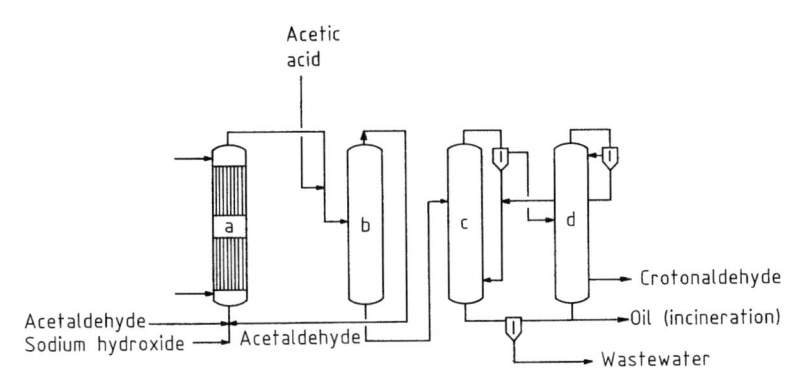

Figure 1. Crotonaldehyde production from acetaldehyde via aldol condensation
a) Aldol reactor; b) Acetaldehyde stripper; c) Azeotrope column; d) Rectification column.

1.2. Production

The aldol condensation of acetaldehyde, followed by dehydration and rectification, is the method generally used for the production of crotonaldehyde:

2 CH$_3$CHO $\longrightarrow$ CH$_3$CHOHCH$_2$CHO $\longrightarrow$ CH$_3$CH=CHCHO + H$_2$O

Past studies have concentrated on the use of different catalysts and specific conditions for this reaction, and examples are described in the literature [19]. The direct oxidation of 1,3-butadiene to crotonaldehyde with palladium catalysis (Consortium process) affords conversion rates as high as 34% [20]. The Consortium process, also known as the Wacker–Hoechst process, is normally used in the direct oxidation of ethylene to acetaldehyde. The analogous oxidation of 1,3-butadiene to crotonaldehyde has found no practical use, however.

Figure 1 shows the flow diagram of a crotonaldehyde production unit. Aldol is formed inside reactor (a) by allowing acetaldehyde to react with aqueous sodium hydroxide. The reaction is water-cooled, and the conversion rate is approximately 50%. The product is neutralized with acetic acid and fed into the acetaldehyde stripper (b), where acetaldehyde is distilled and recycled to the aldol reactor.

In the azeotrope column (c), the aldol is dehydrated and the crotonaldehyde formed is distilled as an azeotrope with water. After condensation, the distillate separates into water and an aqueous crotonaldehyde phase containing about 10% water. The aqueous crotonaldehyde is then transferred to column (d). The separated water is recycled to the azeotrope column which it leaves together with small amounts of an oily byproduct at the bottom. After separation, the oil is incinerated and the water is degraded biologically in a wastewater plant.

Crotonaldehyde is distilled in the rectification column (d) as a side cut. A crotonaldehyde–water azeotrope is distilled overhead. After separation into water and aqueous crotonaldehyde, the former is recycled to the azeotrope column (c), and the latter is fed back into the rectification column. Byproducts leaving the rectification column at the bottom are incinerated.

Analysis. The following derivates can be used for the identification of crotonaldehyde: semicarbazone, *mp* 198–199 °C [21]; 4-phenylsemicarbazone, *mp* 179 °C [22]; thiosemicarbazone, *mp* 167 °C [23]; phenylhydrazone, *mp* 56–57 °C [24]; 2,4-dinitrophenylhydrazone, *mp* 196–197 °C [25].

Several methods have been developed for the quantitative determination of crotonaldehyde in the presence of saturated aldehydes, for example, sulfitometric determination of double bonds [26] and potentiometric titration with bromine in methanol [27]. In the presence of crotonaldehyde, acetaldehyde can be determined by treatment with sodium hydrogen sulfite and subsequent distillation of the acetaldehyde with addition of sodium hydrogen carbonate [28]. Gas chromatography (ASTM E 260-69; DIN 51 405) is the method most often used for the identification and determination of crotonaldehyde. Other standardized methods include: water content (ASTM D 1364-64; DIN 51 777), acid content (ASTM D 1980-67), acetaldehyde content by oximation (ASTM D 2192-70), and color (Hazen; ASTM D 1209-69; DIN 53 409).

Storage and Transportation. Pure crotonaldehyde can be stored and shipped unstabilized without deterioration and discoloration, provided oxygen is strictly excluded. Containers made from stainless steel or aluminum are recommended for storage and transportation. Bakelite-lacquered drums have also been used. It should again be emphasized that contamination of crotonaldehyde with strong acids or alkaline substances may induce hazardous, heat-producing condensation reactions [3]. Therefore, containers must be checked carefully before filling.

Because of the lacrimatory and toxic properties of crotonaldehyde, spills or leakages must be prevented.

1.3. Uses

The largest use of crotonaldehyde is in the manufacture of sorbic acid. Crotonic acid is made commercially by oxidation of crotonaldehyde (Chap. 2). 3-Methoxybutanol [*2517-43-3*] is made by addition of methanol to crotonaldehyde and subsequent reduction of the 3-methoxybutyraldehyde formed. 3-(Ethylthio) butanal [*27205-24-9*] is manufactured by addition of ethanethiol to crotonaldehyde. Although crotonaldehyde is still used commercially, the large amounts formerly required for the production of 1-butanol have now been replaced by products from the oxo process. Because most producers use crotonaldehyde as an intermediate, the market for crotonaldehyde is small.

2. Crotonic Acid

2-Butenoic acid [3724-65-0], $C_4H_6O_2$, M_r 86.09, exists in two stereoisomeric forms:

Crotonic acid
(E-2-butenoic acid)
[107-93-7]

Isocrotonic acid
(Z-2-butenoic acid)
[503-64-0]

Only the trans form (crotonic acid) and its derivatives have industrial importance as monomers in copolymerization reactions and as intermediates in the synthesis of pharmaceuticals or fungicides. Isocrotonic acid is not a commercial product. Crotonic acid occurs naturally in crude wood distillate.

2.1. Physical and Chemical Properties

Table 3 lists the principal physical properties of crotonic acid, isocrotonic acid, and some derivatives of crotonic acid. Table 3 presents additional properties of crotonic acid.

Crotonic acid crystallizes in the monoclinic system as white needles or prisms [29]. The solubilities at 25 °C in acetone, ethanol, ethyl acetate, and toluene are 53, 52, 72, and 37 wt%, respectively. The less stable isocrotonic acid is a liquid at room temperature and is miscible with water at 25 °C.

The 70-eV mass spectra of crotonic and isocrotonic acids are shown in [30]. The ionization energy of crotonic acid is 10.08 eV (derived from the helium(I) photoelectron spectrum). The UV absorption spectra of crotonic and isocrotonic acids in isopentane have strong absorption maxima at 207 and 208 nm, respectively [31]. For proton NMR data, see [32], and for ^{13}C-NMR data, see [33]. The IR spectrum of crotonic acid (potassium bromide disk) shows bands characteristic of carboxylic acids, for example, at 1709 cm^{-1} (C=O), 1425 cm^{-1} (C-O), and 924 cm^{-1} (OH).

Isomerization and Polymerization. Heating of either crotonic or isocrotonic acid at 100–190 °C yields mixtures of crotonic acid, isocrotonic acid, and 3-butenoic acid [625-38-7] (vinylacetic acid) as well as dimeric and polymeric products [35]. After heating at 140–180 °C, the ratio of isocrotonic to crotonic acid in the equilibrium mixture is 0.16–0.18:1. Isomerization can also be achieved by acid or base catalysis or by UV irradiation. Crotonic acid forms copolymers with many common monomers by radical polymerization. The most important copolymers are those with vinyl acetate (see Section 2.3) [36].

Table 2. Physical properties of crotonic acid, isocrotonic acid, and some derivatives of crotonic acid

Compound	CAS registry number	M_r	mp, °C	bp, °C	Density d_4^t	n_D^t
Crotonic acid [(E)-2-butenoic acid]	[107-93-7]	86.09	72	189 81 (1.7 kPa)	0.9730 (72 °C) 0.9604 (77 °C)	1.4256 (72 °C) 1.4249 (77 °C)
Isocrotonic acid [(Z)-2-butenoic acid]	[503-64-0]	86.09	15	169 74 (2.0 kPa)	1.0267 (20 °C)	1.4483 (14 °C)
Crotonamide [(E)-2-butenamide]	[625-37-6]	85.11	157	140 (1.7 kPa) (subl.)		1.4420 (165 °C)
Crotonyl chloride [(E)-2-butenoyl chloride]	[625-35-4]	104.54		124 35 (2.4 kPa)	1.0905 (20 °C)	1.460 (18 °C)
Crotonic anhydride [(E,E)-2-butenoic anhydride]	[78957-07-0]	154.17		131 (2.7 kPa)	1.0397 (20 °C)	1.4745 (20 °C)
Methyl crotonate [(E)-2-butenoic acid methyl ester]	[623-43-8]	100.13		119	0.9444 (20 °C)	1.4242 (20 °C)
Ethyl crotonate [(E)-2-butenoic acid ethyl ester]	[623-70-1]	114.15		137	0.9239 (14.8 °C)	1.4245 (14.4 °C)

Table 3. Properties of crotonic acid

Density, solid (15 °C)	1.018 g/cm^3
Vapor pressure	24 Pa (20 °C), 880 Pa (70 °C)
Specific heat	3.031 J g^{-1} K^{-1} (s), 2.072 J g^{-1} K^{-1} (l)
Heat of combustion	2.00 MJ/mol
Heat of fusion	150.9 J/g
Solubility in water, g/L	41.5 (0 °C), 94 (25 °C), 656 (40 °C), 1260 (42 °C)
pK$_a$ [34]	4.817 (25 °C)

Addition Reactions. In the presence of nickel, platinum, or palladium catalysts, crotonic acid is hydrogenated to butanoic acid. Borohydrides reduce only the carboxyl group to form crotyl alcohol [504-61-0] [(E)-2-butenol]. The addition of halogen or hydrogen halide produces 2,3-dihalobutanoic and 3-halobutanoic acid, respectively. Bromination in the presence of traces of water or hydrogen bromide gives 2,3-dibromobutanoic acid [600-30-6]. Oxidation by peroxybenzoic acid or osmium tetroxide yields 2,3-dihydroxybutanoic acid [3413-97-6].

Cyclization. Crotonic acid and its esters are typical dienophiles in Diels – Alder reactions. For example, butadiene and crotonic acid form 6-methyl-3-cyclohexene-1-carboxylic acid [10479-42-2]:

Another method for the formation of ring systems is the Friedel–Crafts reaction: benzene, crotonic acid, and three equivalents of aluminum chloride yield 3-methyl-1-indanone [6072-57-7]:

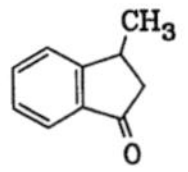

When only one equivalent of aluminum chloride is used, the product is 3-phenylbutanoic acid [4593-90-2]. More highly substituted 3-methyl-1-indanones can be synthesized in a similar manner [37], [38].

Other Reactions. Treating crotonic acid with acyl halides such as phosphorus trichloride, thionyl chloride, or benzoyl chloride produces crotonyl chloride [625-35-4] [(E)-2-butenoyl chloride]. Esters of crotonic acid are synthesized by conventional methods. Crotonamide [625-37-6] [(E)-2-butenamide] is prepared from crotonic acid or its chloride and ammonia. Crotonyl chloride and sodium crotonate [17342-77-7] or crotonic acid and acetic anhydride react to form crotonic anhydride [78957-07-0] [(E,E)-2-butenoic anhydride].

2.2. Production

Crotonic acid is prepared on an industrial scale by the oxidation of crotonaldehyde [123-73-9] [(E)-2-butenal] with oxygen or air. Initially, percrotonic acid [5813-77-4] (2-buteneperoxoic acid) is formed as an intermediate (Eq. 1), which then reacts with another mole of crotonaldehyde to form crotonic acid (Eq. 2). The second step is rate-determining:

$$CH_3-CH=CH-CHO + O_2 \rightarrow CH_3-CH=CH-C\overset{O}{\underset{OOH}{}} \quad (1)$$

$$CH_3-CH=CH-C\overset{O}{\underset{OOH}{}} + CH_3-CH=CH-CHO \rightarrow$$
$$2\ CH_3-CH=CH-COOH \quad (2)$$

The overall heat of reaction is 268 kJ/mol. The reaction is carried out below 50 °C to prevent oxidative destruction to carbon dioxide, acetic acid, and formic acid or isomerization to isocrotonic acid.

In the continuous Shawinigan process [39], water-saturated crotonaldehyde (10–15 % water) is dissolved in an organic diluent such as acetic acid, methyl acetate, ethyl acetate, acetone, benzene, or toluene and oxidized at 25–30 °C with oxygen. A mixture of cobalt and copper salts is used as catalyst. The catalyst concentration can vary from 0.02 to 2 wt% of the total batch. Yields are reported to be 70%.

In the absence of a catalyst, water-containing crotonaldehyde can also be oxidized with air in a hydrocarbon solvent at 15–25 °C and about 500 kPa [40].

In another continuous process (Hoechst), dry crotonaldehyde is oxidized with oxygen at 20 °C in the presence of small quantities of a manganese salt. No additional solvent is used. The mixture resulting from the reaction contains about 30% crotonic acid, 3% formic acid, 1–2% acetic acid, and traces of isocrotonic acid and water. This mixture is distilled at reduced pressure, and any unconverted crotonaldehyde is transferred back to the oxidation reactor.

The crude crotonic acid is fractionated at reduced pressure. Finally, it is crystallized from water in a temperature range from 40 to 5 °C to remove traces of isocrotonic acid. The overall yield is about 70%.

Other catalysts and conditions for the oxidation of crotonaldehyde to crotonic acid have been described in the literature [41]–[43]. Addition of manganese, cobalt, or copper salts catalyzes the decomposition of percrotonic acid (Eq. 2) and prevents the dangerous build-up of peroxy compounds, which can also be hindered by the addition of pyridine [44] or thallium salts [45].

Other methods of synthesis involve the oxycarbonylation of propene [46] and the carbonylation of propylene oxide [47]. In the laboratory, crotonic acid has been prepared (86% yield) from acetaldehyde and malonic acid in dry ether in the presence of pyridine [48].

Isocrotonic acid can be synthesized by concurrent thermal isomerization of crotonic acid and distillative removal from the equilibrium mixture under defined conditions [49]. Separation by gel filtration is also possible [50]. Another synthesis for isocrotonic acid starts with 2-butanone [*78-93-3*], which is brominated to 1,3-dibromo-2-butanone [*815-51-0*] and then rearranged in a stereospecific Favorskii reaction to form isocrotonic acid [51].

Analysis and Transportation. Crotonic acid may be identified as its *S*-benzylthiuronium salt, *mp* 162 °C, or 4-bromophenacyl ester, *mp* 95–96 °C. In the absence of other unsaturated compounds, quantitative determination by bromometric titration is possible. The content of isocrotonic acid can be ascertained by proton NMR spectroscopy [32] or gas chromatography.

Crotonic acid is normally shipped in polyethylene-lined fiber drums or in paper bags with an inner polyethylene bag. A typical specification for technical-grade crotonic acid is as follows [52]:

Crotonic acid	approx. 99 wt%
Hazen color index (APHA), 5% aqueous solution (DIN 53 409)	max. 20
Water content (DIN 51 777)	max. 0.5 wt%
Acid value (DIN 53 492)	approx. 645 mg KOH/g
Melting temperature	approx. 70 °C

Economic Aspects. The producers of crotonic acid are Hoechst (Federal Republic of Germany) and Tennessee Eastman Co. (United States). The total capacity probably does not exceed 2500 t/a (1985).

2.3. Uses

Copolymers of crotonic acid and vinyl acetate [25609-89-6] have a wide range of applications [36], for example, in hair sprays and wave lotions [53], hotmelt adhesives, paper coatings, sizes for paper and textiles, paints, coatings for photographic or packaging film, and in binders for pharmaceutical tablets. Examples of trade names are Mowilith (e.g., Ct 5, VP CT 6) and Aristoflex.

Crotonic acid, crotonyl chloride, and crotonic anhydride are starting materials for the manufacture of fungicides and pharmaceuticals. The 2-(1-methylheptyl)-4,6-dinitrophenyl ester of crotonic acid (dinocap) [6119-92-2] (trade names Karathane, Crotothane) and some anilides such as N-(2,6-dimethylphenyl)-N-(tetrahydro-2-oxo-3-furanyl)butenamide [75648-01-0] have fungicidal properties [54], [55]. N-Ethyl-N-(2-methylphenyl)butenamide (crotamiton) [483-63-6] is a drug used in the treatment of certain skin diseases; some trade names are Eurax, Euraxil, and Crotamitex. A mixture of N-[1-[(dimethylamino)carbonyl]propyl]-N-ethyl-butenamide (crotethamide) [6168-76-9] and N-[1-[(dimethylamino)carbonyl]propyl]-N-propyl-butenamide (cropropamide) [633-47-6] is known as prethcamide [8015-51-8] and, under the trade names Micoren or Respirot, is used as an analeptic and respiratory stimulant [56].

Some esters of crotonic acid, e.g., the pentyl [25415-76-3] and the 3,7-dimethyl-2,6-octadienyl (geranyl) [56172-46-4] esters, are components of perfumes and deodorants [57].

3. Toxicology and Occupational Health

Crotonaldehyde is already perceptible at concentrations of 0.2 ppm (0.6 mg/m^3) [58]. After exposure to 15 ppm for 30 s, no symptoms of eye irritation occur, but 45 ppm causes clear irritation [59]. The mucous membranes of the respiratory system are irritated at concentrations of 4.1 ppm [60]. Because of its warning symptoms no acute intoxication in humans has occurred and no case reports have been published.

In animal experiments, the following data on acute toxicity were determined; for additional data, see [61], [62]:

LD$_{50}$	300 mg/kg (rat, oral) [61]
LC$_{50}$	4000 mg/m^3 (rat, inhal., 30 min) [61]
LD$_{100}$	180 mg/kg (rat, subcutaneous) [62]
LDLo	100 mg/kg (dog, subcutaneous) [61]
LD$_{50}$	380 mg/kg (rabbit, dermal) [61]

Symptoms of acute intoxication in animals are dyspnea, bradypnea, excitation [63], hemorrhagic rhinitis, and congested lungs with hemorrhagic edema [59].

Crotonaldehyde does not induce point mutations in the normal Ames test [63]–[65] and is negative in the assay utilizing *Saccharomyces cerevisiae* as the indicator system [60]. When the modified methodology of the Ames test developed by RANNUG is used [66] crotonaldehyde induces point mutations [67]. Point mutations in *Salmonella typhimurium* can also be induced in the intrasanguine host-mediated assay [68]. In mice, crotonaldehyde induces chromosomal damage after intraperitoneal administration of 1 mg [69], [70], but in the micronucleus assay the oral administration of 80 mg/kg does not increase the frequency of micronucleated polychromatic erythrocytes [71].

A cell transformation assay in the I_{13} subclone of BALB/3T3 cells was negative, which indicates a lack of oncogenic properties for crotonaldehyde [72].

MAK classification: III B [73]
TLV: 2 ppm [74]

Crotonic Acid. In high concentrations, crotonic acid is corrosive and strongly irritating. Acute toxicities (LD_{50}) are 1 g/kg (rat, oral), 600 mg/kg (guinea pig, dermal) [75], and 100 mg/kg (rat, intraperitoneal) [76].

After exposure to nonirritating concentrations, no cumulative effects are likely, because crotonic acid is rapidly converted to β-hydroxybutyryl-coenzyme A by an enzyme present in the liver and other tissues. Crotonic acid is also a normal metabolite of fats [77].

4. References

[1] A. Lieben, *Liebigs Ann. Chem. Suppl.* **1** (1861) 117.
[2] A. Kekulé, *Liebigs Ann. Chem.* **162** (1872) 96.
[3] G. Hommel: *Handbuch der gefährlichen Güter* **I, 65,** Springer-Verlag, Berlin-Heidelberg-New York 1980.
[4] E. Spaeth, R. Lorenz, E. Freund, *Monatsh. Chem.* **76** (1947) 297.
[5] O. Horn, *Ullmann*, 3rd. ed., **4** (1953) 797.
[6] R. F. Nystrom, S. W. Chaikin, W. G. Brown, *J. Am. Chem. Soc.* **71** (1949) 3245.
[7] S. W. Chaikin, W. G. Brown, *J. Am. Chem. Soc.* **71** (1949) 122.
[8] L. N. Owen, *J. Chem. Soc.* 1943, 463.
[9] W. G. Young, *J. Am. Chem. Soc.* **54** (1932) 2500.
[10] Shawinigan Chemicals Inc., US 2 413 235, 1946 (D. J. Kennedy).
[11] IG Farben, DE 554 949, 1930.
[12] R. H. Hall, B. K. Howe, *J. Chem. Soc.* 1949, 2723.
[13] J. L. Szabo, E. T. Stiller, *J. Am. Chem. Soc.* **70** (1948) 3667.
[14] O. Diels, K. Alder, *Liebigs Ann. Chem.* **470** (1929) 66.
[15] W. Flaig, *Liebigs Ann. Chem.* **568** (1950) 1.
[16] H. Bestian et al., in K. Winnacker, L. Küchler (eds.): *Chemische Technologie*, vol. **4**, Hanser-Verlag, München 1972.
[17] Hoechst, DE 2 203 712, 1973 (H. Fernholz, H. Schmidt, F. Wunder).

[18] Eastman-Kodak, US 2 469 110, 1949 (H. J. Hagemeyer).
[19] F. Steinberger, *Ullmann*, 3rd ed., **5** (1954) 614.
[20] J. Smidt, W. Hafner, R. Jira, J. Sedlmeier, R. Sieber, R. Ruetlinger, M. Kojer, *Angew. Chem.* **71** (1959)176.
[21] K. von Auwers, P. Heimke, *Liebigs Ann. Chem.* **458** (1925) 203.
[22] P. Grammaticakis, *Bull. Soc. Chim. Fr.* 1949, 414.
[23] P. Grammaticakis, *Bull. Soc. Chim. Fr.* 1950, 504.
[24] K. von Auwers, A. Kreuder, *Ber. Dtsch. Chem. Ges.* **58** (1925) 1977.
[25] D. Johnson, *J. Am. Chem. Soc.* **75** (1953) 2720.
[26] B. Wurzschmidt, *Z. anal. Chem.* **128** (1948) 549.
[27] J. Kubias, S. Pilný, *Chem. Listy* **47** (1953) 672.
[28] E. Sjostrom, *Act. Chim. Scand.* **7** (1953) 1392.
[29] S. Shimizu, S. Kekka, S. Kashino, M. Haisa, *Bull. Chem. Soc. Jpn.* **47** (1974) 1627–1631.
[30] J. L. Holmes, J. K. Terlouw, P. C. Vijfhuizen, C. A'Campo, *Org. Mass. Spectrom.* **14** (1979) 204–212.
[31] T. Fueno, K. Yamaguchi, *Bull. Chem. Soc. Jpn.* **45** (1972) 3290–3293.
[32] F. H. A. Rummens, J. W. de Haan, *Org. Magn. Reson.* **2** (1970) 351–355.
[33] H. O. Kalinowski, S. Berger, S. Braun: ^{13}*C-NMRSpektroskopie*, Thieme Verlag, Stuttgart 1984, pp. 178, 440.
[34] U. N. Dash, U. K. Nayak, *Can. J. Chem.* **58** (1980) 323–327.
[35] M. B. Hocking, *Can. J. Chem.* **50** (1972) 1224–1232.
[36] L. S. Luskin, in R. H. Yokum, E. B. Nyquist (eds.): *Functional Monomers*, vol. **2**, Marcel Dekker, New York 1974, pp. 501–554.
[37] H. G. Grant, *Z. Naturforsch.* **34 b** (1979) 728–733.
[38] J. R. Merchant, R. B. Upasani, *Chem. Ind. (London)* 1983, 929.
[39] Shawinigan Chemicals, US 2 413 235, 1946 (D. J. Kennedy).
[40] Eastman-Kodak Co., US 2 945 058, 1960 (R. W. Watson, G. K. Finch).
[41] The Distillers Co., US 2 450 389, 1948 (K. H. W. Tuerck).
[42] Celanese Co., US 2 487 188, 1949 (G. W. Seymour, B. B. White, E. Barabash).
[43] BASF, DE 803 296, 1949 (H. G. Trieschmann).
[44] Consortium f. elektrochem. Ind., DE 870 846, 1949 (W. V. Herrmann, W. Haehnel).
[45] Atlantic Richfield Co., US 4 097 523, 1978 (J. Kao, J. J. Leonard).
[46] Röhm GmbH, DE-OS 2 324 132, 1974 (W. Gänzler, K. Kabs, G. Schröder).
[47] Diamond Alkali Co., US 3 024 275, 1962 (W. A. McRae, J. L. Eisenmann).
[48] R. A. Letch, R. P. Linstead, *J. Chem. Soc.*1932, 443–456.
[49] M. B. Hocking, *Synthesis* **1971,** 482–483. Dow Chemical Co., US 3 539 548, 1970 (M. B. Hocking).
[50] N. Kundu, F. Maenza, *Naturwissenschaften* **57** (1970) 544–545.
[51] A. F. Kluge, J. Meinwald, *Org. Synth.* **53** (1973) 123–127.
[52] Hoechst AG: *Crotonic acid technically pure*, product information, Hoechst AG, Frankfurt 1978.
[53] L'Oréal, DE-OS 2 513 807, 1975 (C. Papantoniou, J. C. Grognet). L'Oréal, DE-OS 2 330 956, 1974 (C. Papantoniou).
[54] Chevron Research Co., DE-OS 3 006 154, 1980 (D. C. K. Chan).
[55] Sumitomo Chemical Co., EP-A 100 165, 1983 (H. Noguchi, T. Kato et al.).
[56] Ciba-Geigy AG, US 2 447 587, 1948 (H. Martin, H. Gysin).
[57] Fritsche Brothers Inc., US 3 074 891, 1963 (K. Kulka).

[58] K. Verschuren: *Handbook of Environmental Data on Organic Chemicals*, Van Nostrand-Reinhold Co., New York 1977.
[59] W. E. Rinehart, *Am. Ind. Hyg. Assoc. J.* **28** (1967) 561.
[60] *Toxic and Hazardous Industrial Chemicals Safety Manual*, ITI, Tokyo 1975.
[61] National Institute for Occupational Safety and Health (NIOSH): *Registry of Toxic Effects of Chemicals*, Cincinnati, Ohio, 1979.
[62] E. Skog, *Acta Pharmacol. Toxicol.* **6** (1950) 299.
[63] B. N. Ames, J. McCann, E. Yamasaki, *Mutat. Res.* **31** (1975) 347.
[64] D. Gericke: Unpublished results, Hoechst AG, D-6000 Frankfurt/M. 80 (1979).
[65] V. F. Simon, K. Kauhanen, R. G. Tardiff, in D. Scott, B. A. Bridges, H. F. Sobels (eds.): *Progress of Genetic Toxicology*, Elsevier – North Holland, Amsterdam 1977.
[66] R. Rannug, R. Göthe, C. A. Wachtmeister, *Chem. Biol. Interact.* **12** (1976) 251.
[67] T. Neudecker, D. Lutz, E. Eder, D. Henschler, *Mutat. Res.* **91** (1981) 27.
[68] D. R. Jagannath, D. I. Brusick: Unpublished results, Litton Bionetics Inc., Project No. 20 998, Kensington, Md., USA (1980).
[69] C. Auerbach, M. Moutschen-Dahmen, J. Moutschen, *Mutat. Res.* **39** (1977) 317.
[70] I. Moutschen-Dahmen, M. Moutschen-Dahmen, N. Degraeve, N. Houbrechts, A. Colizzi, *Mutat. Res.* **29** (1975) 205.
[71] D. Mayer, W. Weigand, M. Kramer: Unpublished results, Hoechst AG, D-6000 Frankfurt/M. 80 (1980).
[72] J. O. Rundell, D. I. Brusick: Litton Bionetics Inc., Project No. 21 002, Kensington, Md., USA (1980).
[73] DFG (eds.): *Maximale Arbeitsplatzkonzentrationen (MAK)*, Verlag Chemie, Weinheim 1986.
[74] ACGIH, Cincinnati, Ohio (1985/1986).
[75] H. F. Smyth, C. P. Carparter, *J. Ind. Hyg. Toxicol.* **26** (1944) 269.
[76] *Patty's Industrial Hygiene and Toxicology*, 3rd. ed., vol. **2 C**, Wiley-Interscience, New York 1982, p. 4953.
[77] J. Fruton, S. Simmonds: *General Biochemistry*, 2nd ed., Wiley, New York 1958.

Crown Ethers

EDWIN WEBER, Institut für Organische Chemie und Biochemie, Universität Bonn Federal Republic of Germany

1.	Classification and Nomenclature 1723	5.	Production	1728
2.	Physical Properties 1726	6.	Uses	1730
3.	Chemical Properties........ 1726	7.	Toxicology and Occupational Health.................	1731
4.	Crown Ether Complexes..... 1726	8.	References..............	1732

1. Classification and Nomenclature

Crown ethers [1]–[13] were originally defined as macrocyclic polyethers with the following structure [7]:

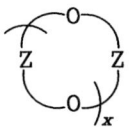

where Z = alkylene, arylene, or cycloalkylene

Typical bridging groups are 1,2-ethylene, *o*-phenylene, and 1,2-cyclohexylene, which can also be substituted. In the common crown ethers (Fig. 1, **1–9**) x runs from 3 to 6. Dioxane ($x = 1$) is not considered a crown ether (→ Dioxane).

Hetero analogues, in which oxygen is replaced either partially or completely by other heteroatoms, especially sulfur or nitrogen, are also known (**10–15**) [4], [12]. The heteroatoms can be integrated into heterocycles, as in **15**. These hetero crown ethers are also called coronands [12] or corands [13].

Occasionally, open-chain analogues such as **16** (podands) [12], [14] and bi- or oligocyclic compounds such as **17** (cryptands) [15] are also classified as crown ethers.

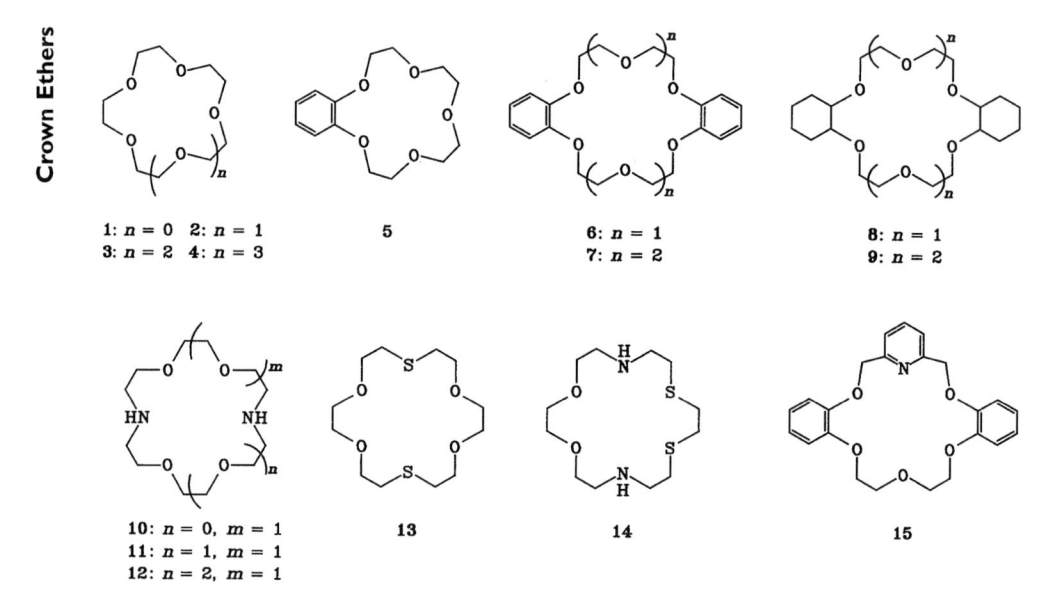

Figure 1. Structures of crown ethers

A more general term applicable in this field of substances is crown compounds.

The name crown ethers refers to the crown-like atomic arrangement. A special system of nomenclature is generally preferred to the more complicated IUPAC system (Table 1). Thus, compound **2**, which contains 5 endocyclic oxygen atoms and a total of 15 atoms in the ring, is usually called 15-crown-5 (or [15]crown-5). Additional substituents or condensation sites such as 'benzo' and 'dicyclohexano' (older nomenclature: dicyclohexyl or perhydrodibenzo) appear as a prefix. Acronyms are also used for common crown ethers (Table 1). A more recent terminology enables a more precise indication of the type and sequence of the heteroatoms and bridging units (e.g., **14**: 18 ‹$O_2NS_2N-2_6$-coronand-6›) [26]. The phane nomenclature can be used for benzo derivatives (e.g., **6**: 1,4,7,14,17,20-hexaoxa[7.7]-(1,2)benzenophane) [27].

The crown ethers were first described by C. J. PEDERSEN [7], [17] in 1967; since then they have been extensively studied by many groups [1]–[13].

Table 1. Nomenclature and physical properties of crown ethers

Compound no.	IUPAC name	CAS registry no.	Trivial name	Acronym	M_r	mp, °C	bp, °C	Ref.
1	1,4,7,10-Tetraoxacyclododecane	[294-93-9]	12-crown-4	12C4	176.21	16^a	$67-70^b$	[16]
2	1,4,7,10,13-Pentaoxacyclopentadecane	[33100-27-5]	15-crown-5	15C5	220.27	$-32^{c,d}$	78^e	[16]
3	1,4,7,10,13,16-Hexaoxacyclooctadecane	[17455-13-9]	18-crown-6	18C6	264.32	36.5–38	116^f	[17], [18], [19]
4	1,4,7,10,13,16,19-Heptaoxacycloheneicosane	[33089-36-0]	21-crown-7	21C7	308.37		oil	[20]
5	2,3,5,6,8,9,11,12-Octahydro-1,4,7,10,13-benzopentaoxacyclopentadecin	[14098-44-3]	benzo-15-crown-5	B15C5	268.31	79–79.5		[17]
6	6,7,9,10,17,18,20,21-Octahydrodibenzo [b,k][1,4,7,10,13,16]hexaoxacyclooctadecin	[14187-32-7]	dibenzo-18-crown-6	DB18C6	360.41	162.5–163.5	380	[7], [21]
7	6,7,9,10,12,13,20,21,23,24,26,27-Dodecahydrodibenzo [b,n] [1,4,7,10,13,16,19,22]-octaoxacyclotetracosin				448.52	113–114		[7]
8	2,5,8,15,18,21-Hexaoxatricyclo [20.4.0.0^{9,14}]-hexacosane	[14174-09-5]	dibenzo-24-crown-8	DB24C8	372.50	$36-56^g$		[7], [21]
9	2,5,8,11,18,21,24,27-Octaoxatricyclo[26.4.0.0^{12,17}]-dotriacontane	[16069-36-6]	dicyclohexano-18-crown-6	DCH18C6	460.61	$<26^g$		[7]
10	1,4,10-Trioxa-7,13-diazacyclopentadecane	[17455-23-1]	dicyclohexano-24-crown-8	DCH24C8	218.30	89–90		[22]
11	1,4,10,13-Tetraoxa-7,16-diazacyclooctadecane	[31249-95-3]	diaza-15-crown-5	DA15C5	262.35	115–116		[22]
12	1,4,7,13,16-Pentaoxa-10,19-diazacycloheneicosane	[23978-55-4]	diaza-18-crown-6	DA18C6	306.40	$-10^{c,h}$		[22]
13	1,4,10,13-Tetraoxa-7,16-dithiacyclooctadecane	[23978-10-1]	diaza-21-crown-7	DA21C7	296.44	90–91		[23]
14	1,4-Dioxa-10,13-dithia-7,16-diazacyclooctadecane	[296-39-9]	dithia-18-crown-6	DT18C6	294.48	47		[24]
15	19,20,22,23-Tetrahydro-12H-7,11-nitrilo-6H-dibenzo [b,k] [1,4,7,10,13] pentaoxacycloeicosin	[28843-76-7]	dithiadiaza-18-crown-6	DTDA18C6	393.44	135–137		[25]
		[59945-37-8]	dibenzopyridino-18-crown-6	DBPy18C6				

^{a} n_D^{20} 1.4621, d_4^{20} 1.109.
^{b} At 0.067 kPa.
^{c} Freezing point.
^{d} n_D^{20} 1.4615, d_4^{20} 1.113.
^{e} At 0.067 kPa.
^{f} At 0.267 kPa.
^{g} Mixture of stereoisomers (commercial product).
^{h} d_4^{20} 1.075.

Classification and Nomenclature

2. Physical Properties

Physical data for important crown ethers are given in Table 1.

Crown ethers of the ethylene oxide oligomer type (**1–4**) are colorless, odorless, viscous liquids or solids with a low melting point. They are strongly hygroscopic and readily soluble in most organic solvents and in water [4].

Crown ethers with condensed aromatic rings (**5–7**) are colorless, barely hygroscopic, crystalline compounds. At room temperature they have poor solubility in water, alcohols, and many other common solvents [4]. They are readily soluble only in halogenated hydrocarbons, pyridine, and formic acid.

Crown ethers with alicyclic bridges (**8, 9**) have much better solubility in hydrocarbons and much poorer solubility in water than the ethylene oxide oligomers [4].

The solubility of crown ethers increases markedly if salts are added [7], e.g., 25-fold for dibenzo-18-crown-6 (**6**) in MeOH after addition of KF [4]. The increase depends on the type of salt used. Conversely, presence of the crown ether increases the solubility of the salt (see Chap. 4).

3. Chemical Properties

Like simple dialkyl ethers, aliphatic and alicyclic crown ethers are chemically stable [4]. Aromatic crown ethers react like anisole or veratrole, i.e., they can be halogenated or nitrated and they react with formaldehyde [7]. Hydrolysis takes place only in special cases [28].Crown ethers are also thermally stable; dibenzo-18-crown-6 (**6**) can be distilled at 380 °C without decomposition [21]. With hydrogen ions [29] and in the presence of Lewis acids ($AlCl_3$, $TiCl_3$), oxonium compounds are formed [30]. The hetero derivatives (**10–15**) are usually more reactive than the classic crown ethers. Aza analogues are strong bases and react with acids to form salts.

4. Crown Ether Complexes

The most remarkable property of crown ethers is their ability to form stable complexes with alkali and alkaline-earth metal ions and with ammonium ions [12]. Numerous complexes of crown ethers with nonionic organic molecules are also known [31].

Crystalline complexes of crown ethers and metal ions (Table 2) are obtained by mixing the two components in a common solvent [7]. The complexes have higher melting points than the free crown ethers (see Table 1). The stoichiometry of the complexes (crown ether:salt = 1:1, 2:1, or 1:2) is largely determined by the fit of crown ether cavity and cation diameter [32], [34] (see Table 3). Structure determination

Table 2. Crystalline complexes of crown ethers with alkali or alkaline-earth metal salts

Crown ether	Salt	Complex	mp, °C (complex)	Ref.
5	NaSCN	5 · NaSCN	162–165	[32]
5	KSCN	(5)$_2$ · KSCN	176	[32]
6	KSCN	6 · KSCN	245–246	[32]
6	CsSCN	(6)$_2$ · CsSCN	146–147	[32]
6	Ba(SCN)$_2$	6 · Ba(SCN)$_2$	> 360	[17]
7	KSCN	7 · (KSCN)$_2$		[33]

Table 3. Diameters of crown ether cavities and of cations

Crown ether	Diameter, pm	Cation	Diameter, pm
1	120–150	Li$^+$	136
2, 5	170–220	Na$^+$	194
3	260–320	Ba^{2+}	268
6	260–320	K$^+$	266
8	260–320	Rb$^+$	294
4	340–430	Cs$^+$	334

by means of X-ray diffraction [34], [35] has confirmed that in the 1:1 complexes the cation is located inside the crown ether ring. For the 2:1 complexes, sandwich structures were found. In the 1:2 complexes, two cations are located inside the crown ether ring. Complexes in which the cation is surrounded by a cage-like structure have also been described. In all cases, the cations are coordinated to the ether oxygens which point into the ring. Hydrogen bridges are effective in complexes with ammonium ions. Frequently there is also contact with the counterion.

In solution, the complexes exhibit varying degrees of stability [8], [36], [37] (Table 4). They are most stable under the following conditions:

- optimum ratio of cation diameter to cavity diameter
- optimum number of ether oxygens capable of coordination
- strongly basic ether oxygens
- high charge density at the cation
- low polarity of the solvent
- lipophilic, polarizable anion

Nitrogen (e.g., **11**) and, especially, sulfur (e.g., **13**) decrease the stability of alkali and alkaline-earth ion complexes. Conversely, complexation with Ag$^+$ or other heavy-metal ions and with ammonium ions is enhanced [8], [37]. The effect of heterosubstitution on the stability of 18-crown-6 complexes is shown in Table 5. Pyridine rings (e.g., **15**) increase selectivity for sodium ions [37].

The stability constants reflect the ion selectivity of the crown ether [36]. A quantitative measure is the ratio K_{M1}/K_{M2} (M1 = cation 1, M2 = cation 2). Ion selectivity depends to a large extent on the solvent; in certain cases it can even be reversed through a change of solvent. For example, in water, 18-crown-6 (**3**) complexes of

Table 4. Stability constants (log K_s) of crown ether–metal ion complexes in water and methanol

Crown ether	Solvent	Na$^+$	K$^+$	Cs$^+$	Sr^{2+}	Ba^{2+}
15C5 (**2**)	water	0.70	0.74	0.8	1.95	1.71
	methanol	3.48	3.77	2.62	2.63	
18C6 (**3**)	water	0.80	2.03	0.99	2.72	3.87
	methanol	4.36	6.06	4.79	> 5.5	7.04
21C7 (**4**)	methanol	1.73	4.22	5.01	1.77	5.44
DB18C6 (**6**)	water	1.64	0.99	0.79	0.94	1.97
	methanol	4.36[b]	5.00[b]	3.55[b]		4.28
DB24C8 (**7**)	methanol	2.25[b]	3.60[b]	3.78[b]		
DCH18C6 (**8**)[c]	water	1.21	2.02	0.96	3.24	3.57
	methanol	4.08[b]	6.01[b]	4.61[b]		

[a] Determined by calorimetry, unless otherwise stated.
[b] Potentiometrically.
[c] Isomer A [37].

Table 5. Stability constants (log K_s) of hetero crown ether complexes

Cation	18-Crown-6 (**3**)	Diaza-18-crown-6 (**11**)	Dithia-18-crown-6 (**13**)
K$^+$ [a]	6.06	2.04	1.15
Ag$^+$ [b]	1.60	7.80	4.34

[a] In methanol.
[b] In water.

barium ions are more stable than those of potassium ions (K_{Ba}/K_K = 1.84), whereas in methanol the potassium complexes are more stable (K_{Ba}/K_K = 0.95) [37]. In multiphase systems that are suitable for selective salt extractions — e.g., H$_2$O, salts/CH$_2$Cl$_2$, crown ether — ion selectivity is determined by the relative degree of salt transfer [11]. Effective extraction requires large anions that are strongly polarizable (e.g., picrate), a highly lipophilic crown ether, and a high stability of the complex in the organic solvent.

The thermodynamic characteristics of a large number of crown ether complexes in various solvents have been determined [36], [37]. The kinetics of crown ether complexation have also been studied [37], [38].

5. Production

The most important methods for the production of crown ethers are analogous to the Williamson ether synthesis [39]. The ether rings are formed by reaction of diols with bifunctional components that possess terminal leaving groups such as chlorine or tosylate (Eq. 1):

$$\text{HO-R}^1\text{-OH} + \text{X-R}^2\text{-X} \xrightarrow{\text{Base}} \text{R}^1\underset{O}{\overset{O}{\diamond}}\text{R}^2 \qquad (1)$$

where R^1, R^2 = alkylene, arylene, or cycloalkylene

$$X = Cl, H_3C\text{-}\langle\text{-}\rangle\text{-}SO_3\text{-}$$

The reaction is carried out in the presence of a base, e.g., an alkali hydroxide or carbonate, cesium fluoride, potassium *tert*-butoxide, or sodium hydride. *n*-Butanol, *tert*-butanol, tetrahydrofuran (THF), and, less frequently, dimethylformamide (DMF) or dimethyl sulfoxide (DMSO) are used as solvents. Cyclization can take place in a single step (Eq. 1) or in several steps:

$$2\,\text{HO-R}^1\text{-OH} + \text{X-R}^2\text{-X} \xrightarrow{\text{Base}} \text{R}^1\underset{O-R^2-O}{\overset{OH\ HO}{\diamond}}\text{R}^1 \qquad (2)$$

$$\text{R}^1\underset{O-R^2-O}{\overset{OH\ HO}{\diamond}}\text{R}^1 + \text{X-R}^3\text{-X} \xrightarrow{\text{Base}} \text{R}^1\underset{O-R^2-O}{\overset{O-R^3-O}{\diamond}}\text{R}^1$$

Protecting groups may be necessary. A one-component method in which the leaving group X is produced from the diol in an intermediate step is also known [4]:

$$\text{HO-R-OH} \xrightarrow{\text{TosCl, base}} R\underset{}{\overset{}{\bigcirc}} \qquad (3)$$

$$\text{TosCl} = H_3C\text{-}\langle\text{-}\rangle\text{-}SO_2\text{-}Cl$$

A template effect, i.e., enhancement of yield and reaction rate by cations that are dimensionally compatible with the crown ether being produced, is often observed.

A different production method is based on the cyclooligomerization of oxirane [4]:

$$H_2C\text{-}CH_2 \xrightarrow{\text{Lewis acid}} \left(\text{-O-}\right)_n + \text{other products} \qquad (4)$$

This also involves template participation [40]. The oxirane method is suitable only for the production of unsubstituted aliphatic crown ethers of a limited range of ring sizes, e.g., 12-crown-4 (**1**), 15-crown-5 (**2**), and 18-crown-6 (**3**). The Williamson procedure is generally applicable and yields both aliphatic and aromatic crown ethers [39]. Crown ether sulfides are produced similarly (nucleophilic substitution by a thiol group) [39]:

$$\text{HS-R}^1\text{-SH} + \text{X-R}^2\text{-X} \longrightarrow \text{R}^1\underset{S}{\overset{S}{\diamond}}\text{R}^2 \qquad (5)$$

Synthesis of crown ether amines proceeds via a cyclic diamide or a cyclic bis(tosylamide). Crown ethers with cyclohexane bridges are obtained through catalytic hydrogenation of the appropriate aromatic compounds.

Commercial Products. Most of the crown ethers listed in Figure 1 are commercially available; the preparations are usually of > 98% or > 99% purity (specification by gas chromatography). Cyclohexano derivatives contain up to 2% of the corresponding benzo crown ether. The most important suppliers are Aldrich, Alfa, Fluka, Janssen, Merck, Nippon Soda, Parish, PCR, Sigma, Schuchardt, and Strem.

6. Uses

Crown ethers are used as complexing agents and phase-transfer catalysts in *organic syntheses* [10]–[12], [41]. They solubilize inorganic salts that are needed as reactants in organic solvents. This amounts to activation of the anion (reagent). The main advantages as compared with phase-transfer catalysts such as quaternary ammonium salts are the greater activating effect on the anion (naked ions) and the absence of foreign ions, the crown ether being neutral. In addition, phase transfer directly from the crystal into the organic phase (solid–liquid phase transfer) is usually possible.

Figure 2 illustrates the preparative usefulness of crown ethers. The industrial application is especially in the manufacture of polymers and polycondensates [42]. In principle, any reaction involving the participation of ions can be influenced by crown ethers [4], [10], [11], [41].

Another important field of application is in *chemical analysis* [43]. The ion selectivity of crown ethers permits the concentration, separation, and masking of ions and the determination of ion concentrations. Suitable methods include ion extraction, ion chromatography, membrane transport, and ion-selective electrodes [11]. It is possible in this way to separate traces of Sr^{2+} from an excess of Ca^{2+}, and even $^{44}Ca^{2+}$ from $^{40}Ca^{2+}$. Heterosubstituted crown ethers are mainly used in the analysis of transition-metal ions [11], [43]; they are suitable for the masking and decontamination of environmentally and physiologically damaging heavy-metal ions [44]. A microassay of Na^+ and K^+ in blood and urine has also been developed [11].

Crown ethers can be used as aids in the detergent and surfactant industries, in electroorganic synthesis and galvanizing, and in the manufacture of organic conductors, liquid-crystalline phases, and anticorrosives [11]. Derivatives with special structures, e.g., rigid rings or functional groups, are suitable for highly selective complexation and ion extraction. Chiral crown ethers can be used for racemate splitting, and polymeric or polymer-bound crown ethers are chromatography materials for cations [5], [11].

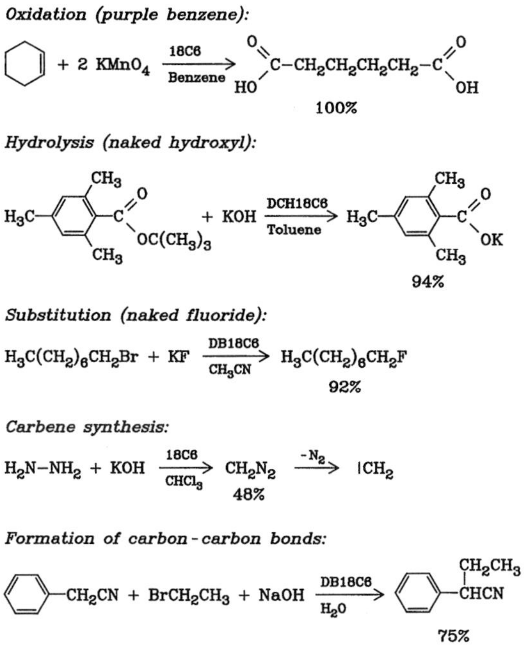

Oxidation (purple benzene):

$$\text{C}_6\text{H}_{10} + 2\,\text{KMnO}_4 \xrightarrow[\text{Benzene}]{18\text{C}6} \text{HO-CO-CH}_2\text{CH}_2\text{CH}_2\text{CH}_2\text{-CO-OH} \quad 100\%$$

Hydrolysis (naked hydroxyl):

2,4,6-trimethylbenzoic acid tert-butyl ester + KOH $\xrightarrow[\text{Toluene}]{\text{DCH18C6}}$ potassium 2,4,6-trimethylbenzoate 94%

Substitution (naked fluoride):

$$\text{H}_3\text{C(CH}_2)_6\text{CH}_2\text{Br} + \text{KF} \xrightarrow[\text{CH}_3\text{CN}]{\text{DB18C6}} \text{H}_3\text{C(CH}_2)_6\text{CH}_2\text{F} \quad 92\%$$

Carbene synthesis:

$$\text{H}_2\text{N-NH}_2 + \text{KOH} \xrightarrow[\text{CHCl}_3]{18\text{C}6} \text{CH}_2\text{N}_2 \xrightarrow{-\text{N}_2} \text{:CH}_2 \quad 48\%$$

Formation of carbon–carbon bonds:

$$\text{C}_6\text{H}_5\text{-CH}_2\text{CN} + \text{BrCH}_2\text{CH}_3 + \text{NaOH} \xrightarrow[\text{H}_2\text{O}]{\text{DB18C6}} \text{C}_6\text{H}_5\text{-CH(CH}_2\text{CH}_3)\text{CN} \quad 75\%$$

Figure 2. Applications of crown ethers in organic syntheses

7. Toxicology and Occupational Health

The oral and skin toxicity of crown ethers is much greater than that of ordinary ethers [10], [45]. As complexing agents they cause damage to the body's ion balance [44]. The main effects are therefore on the central nervous system [4]. Typical sublethal intoxication phenomena include rapid breathing, nervous tension, lack of coordination, tremor, and lassitude. Many crown ethers also damage the eyes and skin [21]. Lethal toxicity data for common crown ethers can be found in Table 6.

Protective clothing and goggles must be worn when crown ethers are handled [46]. Contact, especially with the eyes, should be avoided. The danger of intoxication by inhalation is greatest for 12-crown-4 [4], [47].

According to the RID and ADR transport regulations, crown ethers fall into **class 6.1**. Incineration is prescribed for the disposal of residues and waste.

Table 6. Toxicological data for crown ethers

Crown	Acute toxicity			
	oral		dermal	
12C4 (**1**)	LD_{50}	3.15 ± 0.06 g/kg^a		
15C5 (**2**)	LD_{50}	1.02 ± 0.1 g/kg^a		
18C6 (**3**)	LD_{50}	0.705 ± 0.08 g/kg^a	LD_{50}	200 mg/kg^b
DB18C6 (**6**)	LD_{50}	> 300 mg/kg^a	ALD^e	7.50 g/kg^c
	LD_{50}	11 g/kg^b		
DB24C8 (**7**)	LD_{50}	> 300 mg/kg^a		
DCH18C6 (**8**)	LD_{50}	> 300 mg/kg^a	ALD^e	130 mg/kg^d
	ALD^e	300 mg/kg^b		

^{a} Mouse.
^{b} Rat.
^{c} Guinea pig.
^{d} Rabbit.
^{e} ALD = approximate lethal dose.

8. References

General References

[1] R. M. Izatt, J. J. Christensen: *Synthetic Multidentate Macrocyclic Compounds*, Academic Press, New York, San Francisco, London 1978.

[2] R. M. Izatt, J. J. Christensen: *Progress in Macrocyclic Chemistry*, vols. **1–3**, Wiley, New York 1979, 1981, 1987.

[3] G. A. Melson: *Coordination Chemistry of Macrocyclic Compounds*, Plenum Press, New York, London 1979.

[4] M. Hiraoka: *Crown Compounds, Their Characteristics and Applications*, Studies in Organic Chemistry 12, Elsevier, Amsterdam, Oxford, New York 1982.

[5] F. Vögtle, E. Weber: *Host Guest Complex Chemistry–Macrocycles: Synthesis, Structures, Applications*, Springer Verlag, Berlin, Heidelberg, New York, Tokyo 1985.

[6] H. Klamberg, *Chem. Lab. Betrieb* **29** (1971) 97–101.

[7] C. J. Pedersen, H. K. Frensdorff, *Angew. Chem.* **84** (1972) 16–26; *Angew. Chem. Int. Ed. Engl.* **11** (1972) 16–26.

[8] J. J. Christensen, D. J. Eatough, R. M. Izatt, *Chem. Rev.* **74** (1974) 351–384.

[9] C. Kappenstein, *Bull. Soc. Chim. Fr.* 1974, 89–109.

[10] G. W. Gokel, H. D. Durst, *Synthesis* 1976, 168–184.

[11] E. Weber, *Kontakte (Merck)* 1984, no. 1, 26–43 and earlier articles of this series.

[12] E. Weber, F. Vögtle, *Top. Curr. Chem.* **98** (1981) 1–41.

[13] D. J. Cram, *Angew. Chem.* **98** (1986) 1041–1060; *Angew. Chem. Int. Ed. Engl.* **25** (1986) 1039–1057.

Specific References

[14] F. Vögtle, E. Weber, *Angew. Chem.* **91** (1979) 813–837; *Angew. Chem. Int. Ed. Engl.* **18** (1979) 753–776.

[15] J. M. Lehn, *Acc. Chem. Res.* **11** (1978) 49–57.

[16] F. L. Cook, T. C. Caruso, M. P. Byrne, C. W. Bowers et al., *Tetrahedron Lett.* 1974, 4029–4032.

[17] C. J. Pedersen, *J. Am. Chem. Soc.* **89** (1967) 7017–7036;US 3 562 295, 1971 (C. J. Pedersen).
[18] G. Johns, C. J. Ransom, C. B. Reese, *Synthesis* 1976, 515–516.
[19] G. W. Gokel, D. J. Cram, C. L. Liotta, H. P. Harris et al., *Org. Synth.* **57** (1977) 30–33.
[20] J. Dale, P. O. Kristiansen, *Acta Chem. Scand.* **26** (1972) 1471–1478.
[21] C. J. Pedersen, *Org. Synth.* **52** (1972) 66–74; US 3 678 978, 1972 (C. J. Pedersen).
[22] B. Dietrich, J. M. Lehn, J. P. Sauvage, J. Blanzat, *Tetrahedron* **29** (1973) 1629–1645.
[23] J. R. Dann, P. P. Chiesa, J. W. Gates, Jr., *J. Org. Chem.* **26** (1961) 1991–1995.
[24] B. Dietrich, J. M. Lehn, J. P. Sauvage, *J. Chem. Soc. Chem. Commun.* **1970**, 1055–1056; US 3 888 877, 1975 (J. M. Lehn).
[25] E. Weber, F. Vögtle, *Chem. Ber.* **109** (1976) 1803–1831.
[26] E. Weber, F. Vögtle, *Inorg. Chim. Acta* **45** (1980) L65–L67.
[27] F. Vögtle, P. Neumann, *Tetrahedron* **26** (1970) 5847–5873.
[28] B. Agai, I. Bitter, Z. Hell. A. Szöllosy et al., *Tetrahedron Lett.* **26** (1985) 2705–2708.
[29] G. S. Heo, R. A. Bartsch, *J. Org. Chem.* **47** (1982) 3557–3559.
[30] S. G. Bott, U. Kynast, J. L. Atwood, *J. Incl. Phenom.* **4** (1986) 241–246.
[31] E. Weber in *Progress in Macrocyclic Chemistry*, vol. **3** Wiley, New York 1987, pp. 337–419.F. Vögtle, W. M. Müller, W. H. Watson, *Top. Curr. Chem.* **125** (1984) 131–164.
[32] C. J. Pedersen, *J. Am. Chem. Soc.* **92** (1970) 386–391.
[33] D. E. Fenton, M. Mercer, N. S. Poonia, M. R. Truter, *J. Chem. Soc. Chem. Commun.* 1972, 66–67.
[34] N. S. Poonia, A. V. Bajaj, *Chem. Rev.* **79** (1979) 389–445.
[35] R. Hilgenfeld, W. Saenger, *Top. Curr. Chem.* **101** (1982) 1–82.
[36] F. Vögtle, E. Weber in S. Patai (ed.): *The Chemistry of the Ether Linkage*, Suppl. E, Part 1, Wiley, London 1981, pp. 59–156.
[37] R. M. Izatt, J. S. Bradshaw, S. A. Nielsen, J. D. Lamb et al., *Chem. Rev.* **85** (1985) 271–339.
[38] W. Burgermeister, R. Winkler-Oswatitsch, *Top. Curr. Chem.* **69** (1977) 91–196.
[39] G. W. Gokel, S. H. Korzeniowski: *Macrocyclic Polyether Synthesis*, Springer Verlag, Berlin, Heidelberg, New York 1982.
[40] CH 601 284, 1977 (J. Dale, K. Daasvatu).
[41] G. W. Gokel, G. W. Weber: *Phase Transfer Catalysis in Organic Synthesis*, Springer Verlag, Berlin, Heidelberg, New York 1977.
[42] L. J. Mathias, C. E. Carraher, Jr.: *Crown Ethers and Phase Transfer Catalysis in Polymer Science*, Polymer Science and Technology, vol. **24**, Plenum Press, New York, London 1984.
[43] I. M. Kolthoff, *Anal. Chem.* **51** (1979) 1R–22R.
[44] R. M. Izatt, J. D. Lamb, D. J. Eatough, J. J. Christensen et al. in E. J. Ariens (ed.): *Medicinal Chemistry*, vol. **11-VIII**, Drug Design, Academic Press, New York, San Francisco, London 1979, pp. 355–400.
[45] M. Hiraoka: *Oligomer Handbook*, Kagakukogyo Nippo, 1977, p. 240.
[46] E. I. du Pont de Nemours Co., *Elastomer Area-Chambers Works*, Chemical Hazard Sheet, Chemical No. 148, Sept. 1 (1972).
[47] B. K. J. Leong, T. O. T. Ts'o, M. B. Chenoweth, *Toxicol. Appl. Pharmacol.* **27** (1974) 342–354.

Cyanamides

PETER S. FORGIONE, American Cyanamid Company, Stamford, Connecticut 06905 United States

1.	Introduction	1735	3.4.	Chemical Analysis	1752
2.	Calcium Cyanamide	1736	3.5.	Storage and Transportation	1753
2.1.	Physical Properties	1736	3.6.	Uses	1753
2.2.	Chemical Properties	1737	4.	Dicyandiamide	1754
2.3.	Production	1737	4.1.	Physical Properties	1755
2.3.1.	Overall Process	1737	4.2.	Chemical Properties	1755
2.3.2.	Manufacture	1739	4.3.	Production	1758
2.3.3.	Processing of Technical Calcium Cyanamide	1743	4.4.	Quality Specifications and Chemical Analysis	1759
2.4.	Quality Specifications	1744	4.5.	Storage and Transportation	1759
2.5.	Storage and Transportation	1744	4.6.	Uses	1760
2.6.	Uses	1745	5.	Economic Aspects	1761
3.	Cyanamide	1745	6.	Toxicology and Occupational Health	1762
3.1.	Physical Properties	1746			
3.2.	Chemical Properties	1746			
3.3.	Production	1751	7.	References	1763

1. Introduction

Cyanamide and its alkaline-earth derivatives may have played a role in the formation of life on earth. It has been postulated that under primitive conditions these materials functioned as intermediates in phosphorylation and in peptide-bond formation [8]. Until early in this century, these materials were of little practical interest. However, with the commercial development of calcium cyanamide [156-62-7] in 1905, a whole family of derivatives evolved. The more important ones are cyanamide [420-04-2], its dimer dicyandiamide [461-58-5], and its trimer melamine [108-78-1]. Calcium cyanamide was developed at the same time as synthetic ammonia. Initially it was used principally as a fertilizer and, under some conditions, as a starting material for ammonia. With the introduction of the Haber ammonia process, its use as a fertilizer has diminished, particularly in North America. However, it is still used in certain applications in agriculture and industry.

2. Calcium Cyanamide

Calcium cyanamide was probably first obtained in the laboratory in 1877 by heating calcium carbamate to red heat [9]. In 1889 it was prepared in larger quantities by heating thoroughly mixed, finely pulverized urea and calcium oxide [10]. A commercial process for the nitrogenation of calcium carbide, the Frank–Caro process, was patented in Germany in 1895 [11]. A calcium cyanamide plant using a batch oven furnace was erected at Piano d'Orta, Italy, in 1905. At about the same time, the Polzeniusz–Krauss channel furnace was put into service. By 1910, plants were established in Germany (Bayerische Kalkstickstoffwerke; AG für Stickstoffdünger), France, Japan, Sweden, Switzerland, and the United States (Amer. Cyanamid).

Calcium cyanamide, [156-62-7] $CaNC \equiv N$, M_r 80.10, is the neutral salt of cyanamide, NH_2CN (see Chap. 3); it is also known as lime nitrogen, nitrolime, and kalkstickstoff.

Industrial-grade calcium cyanamide contains, in addition to $CaCN_2$, ca. 20% CaO and 10–12% free carbon, which gives the product its gray-black color. It also contains a small amount of nitrides formed from silica and alumina. The total nitrogen content varies from 22 to 25%, depending on the raw materials used. Of the total nitrogen, 92–95% is present as cyanamide and 0.1–0.4% as dicyandiamide; the remainder is present as nitrides.

2.1. Physical Properties [12a]

Pure calcium cyanamide is a colorless hygroscopic salt, which forms rhombohedral crystals. In a nitrogen atmosphere, it melts with decomposition at 1300 °C. Other physical properties are given below.

Density at 25 °C	2.36 g/cm^3
Heat of fusion	54 kJ/kg
Specific heat, 20–100 °C	909 J/kg
Enthalpy of formation ΔH_{298}	−348 kJ/mol
Free energy of formation ΔF_{298}	−303 kJ/mol
Normal entropy S	87.1 J mol^{-1} K^{-1}

2.2. Chemical Properties [12b]

Calcium cyanamide decomposes above 1000 °C in a way that depends on the temperature, the partial pressure of nitrogen, and impurities present. The principal decomposition products formed by heating under vacuum or in an inert gas are calcium carbide, calcium metal, and nitrogen:

$$2\ CaCN_2 \longrightarrow CaC_2 + Ca + 2\ N_2$$

Decomposition at higher nitrogen pressure gives primarily cyanide products. Calcium nitride and carbon are always formed.

Calcium cyanamide reacts with oxygen and carbon dioxide, starting at ca. 475 °C, with formation of nitrogen and calcium carbonate; calcium oxide forms above 850–900 °C. Carbon present as an impurity cannot be removed by oxidation, which attacks calcium cyanamide preferentially.

Carbon monoxide and calcium cyanamide react above 1000 °C to form CaO and calcium carbide.

The reactions of calcium cyanamide in aqueous solution are determined primarily by temperature and pH [13]. At room temperature, monocalcium cyanamide, $Ca(HCN_2)_2$, forms; when heated at pH 9–10, this is converted to calcium hydroxide and dicyandiamide, $(NH_2)_2CNCN$ (see Chap. 4). Below 40 °C, cyanamide is obtained at pH 6–8; lime is precipitated with carbon dioxide.

Urea forms in the presence of acid and catalysts, thiourea in the presence of sulfides.

At 200 °C under pressure, calcium cyanamide is hydrolyzed to ammonia and calcium carbonate in an alkaline medium. The reaction was employed to produce ammonia early in this century.

2.3. Production

2.3.1. Overall Process

Calcium cyanamide is manufactured in three steps. First, lime is made by heating high-grade limestone:

$$CaCO_3 \xrightarrow{\Delta} CaO + CO_2$$

Second, calcium carbide is synthesized from lime and coke or coal in an electric furnace:

$$3\ C + CaO \xrightarrow{\Delta} CaC_2 + CO$$

A smothered electric arc is used to melt the lime and effect the reaction with the coke.

Third, calcium cyanamide is synthesized from calcium carbide and nitrogen. This reaction is exothermic but requires heating of a portion of the reaction mixture to the initiation temperature of 900–1000 °C:

$$CaC_2 + N_2 \xrightarrow{\Delta} CaCN_2 + C$$

The heat source is then removed [13]. The reaction continues by controlled addition of nitrogen; it produces 286.6 kJ/mol at 1100 °C and 295 kJ/mol at 0 °C.

Mechanism. It is believed that calcium cyanamide is formed through a number of intermediates, such as $Ca(CN)_2$, CaC_2N_2, CaC, and Ca_2N_2 [14]–[16]. However, in the nitrogenation of calcium carbide, the main products are calcium cyanamide and carbon. Above 1000 °C, cyanide also forms and is in equilibrium with cyanamide and carbon:

$$CaCN_2 + C \rightleftharpoons Ca(CN)_2 \qquad \Delta H_{298} = 163 \text{ kJ/mol}$$

This reaction is endothermic, whereas the nitrogenation is exothermic.

Above 1160 °C, the system $CaCN_2/C/Ca(CN)_2$ melts, with over 60% cyanamide present at equilibrium.

Small amounts of cyanide present at the usual reaction temperatures of 1000–1100 °C are rapidly converted to calcium cyanamide by slow cooling; the cooled product is practically free of cyanide.

When the cyanide-containing melt is cooled in the presence of alkali-metal compounds, complete conversion to cyanamide does not take place. This is the basis of the fusion cyanide process, which is still used at the present time. Commercially, cyanide (ca. 50% NaCN equivalent) is prepared by melting and cooling a mixture of calcium cyanamide and sodium chloride [17]; the product is used mainly for precious metal extraction.

Nitrogenation Byproducts. Technical-grade calcium carbide contains impurities that, during nitrogenation, lower the yield of desired product. This is due to the fact that silica and alumina, which are only partially reduced in the carbide furnace, are present in the carbide raw materials. In the nitrogenation, Al_2O_3 and SiO_2 react with $CaCN_2$ to form nitrides.

$3\ CaCN_2 + Al_2O_3 \longrightarrow 3\ CaO + 2\ AlN + 3\ C + 2\ N_2$
$9\ CaCN_2 + 3\ SiO_2 \longrightarrow 6\ CaO + Ca_3N_2 + Si_3N_4 + 9\ C + 6\ N_2$

Industrial-grade calcium carbide also contains calcium hydroxide, $Ca(OH)_2$, and calcium carbonate, $CaCO_3$. On heating, water and carbon dioxide are liberated; reaction with carbide gives acetylene. As a result, hydrogen, formed by decomposition of acetylene, is always found in the exhaust gas of the nitrogenation furnaces.

These side reactions and the inadvertent introduction of some oxygen during the nitrogenation reduce the overall yield by ca. 10%. The purest possible raw materials give the most favorable results economically.

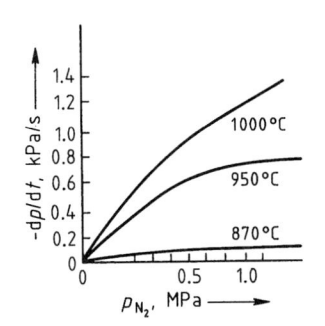

Figure 1. Rate of calcium carbide nitrogenation as a function of nitrogen pressure [19]
Powdered CaC_2 (67%) containing 1.2% CaF_2 used.

Catalysts. Various catalysts that act as fluxes are used to accelerate the reaction or lower the required temperature. The most commonly used are calcium chloride and calcium fluoride. Their function is not completely clear, but they may provide a liquid phase in which the reaction can occur. The product is in the form of a well-sintered pig, indicating the presence of a liquid phase at some stage of the reaction.

Rate studies have shown that calcium fluoride reduces the temperature of optimum reactivity and increases the reaction rate 4.5 times at 1000 °C [18].

Reaction Kinetics. The conversion rate depends on temperature, partial nitrogen pressure, carbide purity, and additives present. The crystallite size also plays a part. High-grade carbide with a coarse crystal structure reacts more slowly than fine material of a lower grade. As indicated above, metal halides accelerate the reaction.

In the nitrogenation of calcium carbide, the reaction proceeds inward from the grain surface. The rate of nitrogen transport through the porous layer is a determining factor at lower temperature, whereas the chemical reaction at the boundary layer governs at higher temperature and in the presence of additives. The dependence of the rate constants on temperature, additives, and nitrogen pressure has been reported [19]. Figure 1 shows the effect of nitrogen pressure.

2.3.2. Manufacture

Both batch and continuous processes have been used to prepare calcium cyanamide [28]. The more important processes include the Frank–Caro batch furnace process, the Polzeniusz–Krauss channel furnace process, the Trostberg rotary furnace process, and the Fujiyama shaft furnace process. Other processes have been explored, but have not achieved comparable importance.

In these processes, the exothermal reaction between carbide and nitrogen takes place between 1000 and 1150 °C. In the continuous rotary furnace process, the reactants, except at the beginning, are heated exclusively by the heat of reaction, whereas the batch furnace process requires ignition for each batch.

All processes use finely ground high-grade calcium carbide and control of the reaction temperature by the addition of lime nitrogen (crude calcium cyanamide). This

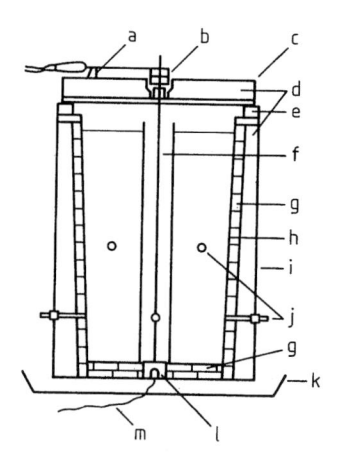

Figure 2. Self-heating nitrogenation oven
a) Insulator; b) Graphite contact; c) Steel cover; d) Diatomaceous earth; e) Sand seal; f) Graphite rod; g) Firebrick; h) Paper lining; i) Steel shell; j) Nitrogen inlets and outlets; k) Sand tray; l) Graphite ground contact; m) Ground lead

dilution of the carbide prevents a temperature rise that would decompose the cyanamide and promotes homogeneity and nitrogen diffusion. This type of temperature control is of particular importance in the Trostberg rotary furnace process to prevent caking on the furnace wall.

Frank–Caro Batch Oven Process [18]. The batch oven process is widely used in North America. It employs a batch reactor filled with ground calcium carbide. After an initial ignition, the reaction proceeds spontaneously. Large stationary furnaces with and without basket inserts have been used in several variations. A basket insert with a capacity up to 10 t is filled with ground carbide and inserted into a steel furnace equipped with nitrogen inlets and a current supply. Graphite heating rods (up to 3.5 m long) are inserted into the carbide, and contact is established with the cover and grounded furnace shell. Before application of the electrical charge, nitrogen is introduced through inlets in the lower part of the furnace shell. When the walls of the channels glow after being heated for 3–4 h, the current is turned off and the reaction is allowed to go to completion. After being initiated, the exothermic reaction is self-sustaining.

During the reaction, the contents sinter to a block or pig. In an 8–10-t furnace, the reaction is completed and the mixture partially cooled in ca. 70 h. Although calcium cyanamide is white, the pig is black because of graphite formed during the reaction. It contains ca. 70% calcium cyanamide, 10% free lime, 12% graphite, and 0.5% unreacted carbide. The pig is broken into pieces and milled.

Furnaces and Electrodes. The batch oven furnaces used in North America by Amer. Cyanamid are cylindrical, firebrick-lined steel shells (Fig. 2). The firebrick floor of the furnace (20 cm is adequate) is covered with carbon blocks cemented with pitch. This refractory material withstands the high temperature and alkalinity of the molten lime. The sides of the furnaces are not subjected to vigorous conditions because they are insulated by a mass of charge and product. The external dimensions of a typical large furnace are 12.5 × 13.8 m at the top, 8.8 × 3.3 m at the bottom, and a height of 5.5 m [18].

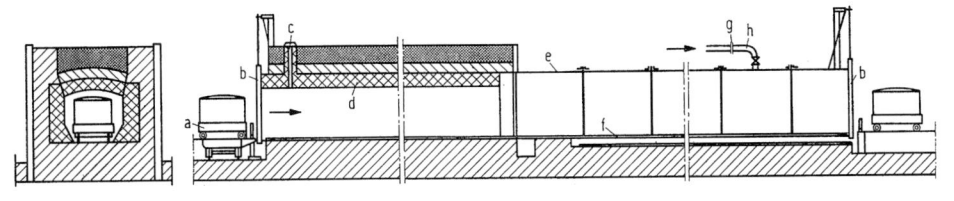

Figure 3. Channel furnace
a) Cast steel carriage; b) Doors; c) Temperature measuring connection; d) Refractory walls; e) Sheet metal muffler; f) Rails; g) Orifice gauge; h) Nitrogen inlet tube

A number of electrode systems have been used in batch furnaces. In one system the electrodes are equipped with one to three composite rectangular rods made of electrolytic-grade carbon held in a line to give a composite cross-section of 510×280 cm when all are used. Another system uses sliding contacts, permitting the addition of electrode components without removing the assembly from the furnace. Component electrodes have also been replaced with continuous electrodes of the Söderberg type, where a paste of carbonized anthracite, coke fines, and tar is packed into the top of a 12.2-m thin steel tube. The tube serves as an electrode and is fed into the furnace as it burns away from the bottom. As the paste moves closer to the surface of the charge, it is slowly baked, conferring great mechanical strength when it reaches the 2-m section below the sliding electrical contacts [18].

Polzeniusz–Krauss Channel Furnace Process. In the channel furnace process, the carbide mixture is diluted with lime nitrogen (ca. 67% carbide solids) and 2–3% calcium chloride or fluorspar. This mixture is loaded into iron boxes of ca. 1800 kg capacity and rolled into the channel furnace by a rail assembly. The furnace is 50–80 m long and can be closed by gastight doors made of brickwork in the reaction zone and of sheet metal muffles in the cooling zone (Fig. 3). A significant advantage of the channel furnace is its flexibility of operation, which can be modified, depending on the quality of the carbide, the duration of the reaction, and the temperature of the cooling zone. To accelerate ignition at 750 °C, a small amount of calcium nitrate is added.

After conversion, the carriages are pulled from the furnace and cooled. The product is easily crushed because the blocks have a cokelike structure and readily fall apart. Up to 30 t/d of calcium cyanamide can be produced in a single furnace.

Trostberg Rotary Furnace Process. In this continuous nitrogenation process, ground lime nitrogen is metered onto a material bed in a broadened part of the rotary furnace in such a way as to keep the carbide concentration in the bed as low as possible. The process is controlled by varying the carbide content.

The rotary furnace developed by Süddeutsche Kalkstickstoff-Werke is ca. 20 m long and includes the broadened furnace head, where the main reaction takes place [20]. The furnace is lined with fireclay (Fig. 4).

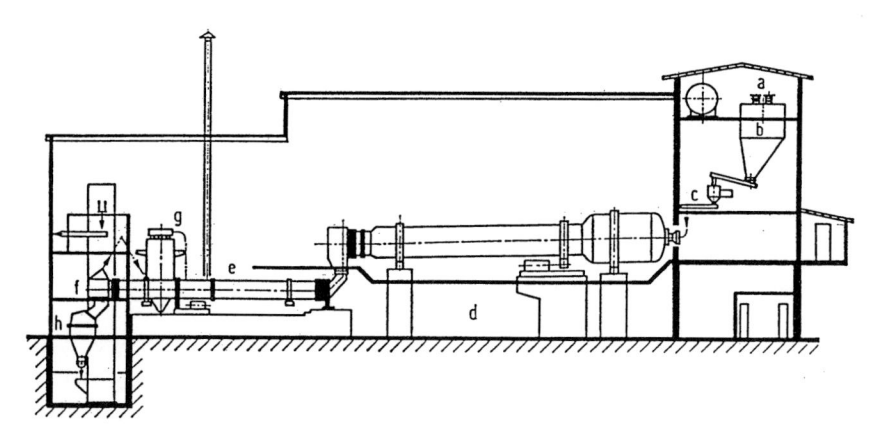

Figure 4. Trostberg rotary furnace
a) Feed; b) Silo for ground material; c) Metering; d) Reaction section; e) Cooling unit; f) Discharge; g) Filter; h) Hammer mill

Ground carbide with a CaC_2 content of 55–60% is continuously mixed with calcium fluoride and ground lime nitrogen and is blown into the furnace with nitrogen. The average residence time of the solid is between 5 and 6 h. The resultant reaction product is a granular or powdered lime nitrogen, which is transferred from the rotary furnace to a cooling drum. The heat released by the reaction is sufficient to bring the starting materials to the desired temperature of 1000–1100 °C. After the furnace has been started, it remains in operation without external heating for many months. The capacity of a unit is about 25 t/d of fixed nitrogen. For the manufacture of calcium cyanamide, crude calcium carbide (ca. 3.36 × 1.68 mm or 6 × 12 mesh) can be used; for cyanamide and dicyandiamide a 74-μm (200-mesh) anhydrous carbide is used.

Knapsack Rotary Furnace Process [21]. The Knapsack rotary kiln process was developed 20 years before the Trostberg process, but has not been in operation at Knapsack since 1971. It is the only industrial process that does not use finely ground calcium carbide and that operates without dilution with lime nitrogen. Granular carbide, up to 2 mm in diameter, is nitrogenated in the presence of 1–2% of calcium chloride in a cylindrical rotary drum, giving a product of the same grain size as the starting carbide. A typical kiln can produce 12–13 t/d of fixed nitrogen; the product can be sold without further processing.

Fujiyama Process. This Japanese process employs a shaft furnace equipped with a continuous charging unit for ground carbide and a scraping device at the lower end of the furnace where the calcium cyanamide is formed.

Carlson Process. This fluidized-bed process, operated with a furnace with stirrers, was used for a few years in Sweden. It failed because of the tendency of the reaction material to agglomerate [22]. A combination of this approach with a rotary furnace was explored, but was not adapted commercially [23].

Other Processes. Because large amounts of energy are consumed, particularly in the preparation of calcium carbide, less energy intensive methods for manufacturing calcium cyanamide have been examined. Between 600 and ca. 1000 °C many reactions of lime with nitrogen compounds containing hydrogen or carbon lead to calcium cyanamide. Compounds such as hydrocyanic acid, dicyanogen, urea, and dicyandiamide form calcium cyanamide with lime. Another example is the reaction of limestone with ammonia [24]:

$$CaCO_3 + 2\,NH_3 \xrightarrow{700-800\,°C} CaCN_2 + 3\,H_2O$$

Lime and urea form calcium cyanate [25], [26], which on heating is converted to calcium cyanurate and finally to calcium cyanamide:

$$CaO + 2\,NH_2CONH_2 \xrightarrow{200\,°C} Ca(OCN)_2 + H_2O + 2\,NH_3$$

$$3\,Ca(OCN)_2 \xrightarrow{400\,°C} 2\,[\text{cyanurate ring with } OCa_{1/2}, Ca_{1/2}O, OCa_{1/2}] \xrightarrow{700\,°C} 3\,CaCN_2 + 3\,CO_2$$

Lime reacts with hydrogen cyanide [27]:

$$CaO + 2\,HCN \xrightarrow{750-850\,°C} CaCN_2 + CO + H_2$$

An ammonia–carbon monoxide mixture produces 99 % calcium cyanamide:

$$CaO + 2\,NH_3 + 3\,CO \xrightarrow{700-900\,°C} CaCN_2 + 2\,CO_2 + 3\,H_2$$

The above processes give white calcium cyanamide, whereas the product obtained from limestone and coal always contains carbonaceous impurities. None of these processes have yet been commercialized because of cost or poor yield [28].

2.3.3. Processing of Technical Calcium Cyanamide

Crude calcium cyanamide is reduced in size and ground in tube mills to allow passage through a 0.2-mm screen. If calcium cyanamide is sold in granular form, grinding is omitted and the desired grain size is separated by screening.

When the carbide content is above 0.1 %, as in the Frank–Caro process, calcium cyanamide is degassed by treatment with water to convert the carbide to acetylene and calcium hydroxide. For safety reasons, the acetylene is dispersed with an inert gas [18].

Granulation. Because of the dust problems associated with finely ground calcium cyanamide in fertilizer applications, the product is oiled. Attempts have also been made to convert it to a more compact form by granulation or compression. A stable, abrasion-resistant granular material is produced by a two-stage process, in which ground calcium cyanamide is initially treated with a small amount of water or an aqueous solution to hydrate any free calcium oxide. In the second stage, the material is moistened, granulated, and dried or pressed [29].

Beads. Granulation with calcium nitrate solutions gives a fine, finished calcium cyanamide bead. This form affords a nitrate nitrogen system that can be used directly in the plant as starter nitrogen. In addition, it affords a slowly acting cyanamide nitrogen source [30]. This product has an almost unlimited storage life, because the free calcium oxide is completely hydrated and no expansion can occur. In Europe, the Trostberg plant of Süddeutsche Kalkstickstoff-Werke is the main producer of beaded calcium cyanamide.

Encapsulation of Granular Calcium Cyanamide. Various coating methods have been used to protect crushed granular calcium cyanamide against decomposition. Waxes and resins have been investigated, but only sulfur is used in practice [30]. It functions by limiting the diffusion of water into the material, and it permits fertilizing action in use.

2.4. Quality Specifications

Commercial-grade calcium cyanamide has approximately the following composition:

Total nitrogen	19–20%
Cyanamide nitrogen	ca. 15%
Nitrate nitrogen	ca. 2%
Dicyandiamide nitrogen	0.4–0.7%
Other nitrogen	2.0–2.5%
Total calcium oxide	53–55%
Water (chemically bound)	7–8%

Total nitrogen is determined by the Kjeldahl or Dumas method.

2.5. Storage and Transportation

Calcium cyanamide is stored in warehouses or silos, but storage life with free calcium oxide is limited because the latter is expanded by moisture.

Production-grade ground calcium cyanamide, with a nitrogen content of 22–24%, is adjusted to the usual commercial nitrogen content of 19–20% by adding a diluent such as ground limestone. Oiling may be used to prevent dusting.

Valve bags are used almost exclusively for packaging. Bags used for ground and granular calcium cyanamide must be moisture-tight to prevent grain breakdown and expansion. Polyethylene valve bags in combination with multilayer paper bags are excellent for granular calcium cyanamide. Paper and plastic bags are used for the less sensitive, beaded product. Loose beaded calcium cyanamide may be transported in special containers in the Federal Republic of Germany. For agricultural applications, 50-kg bags are used; smaller packages of finely beaded product are sold for horticultural applications.

2.6. Uses

In Europe, cyanamide and calcium cyanamide are used as fertilizers, weed killers, and defoliants. In North America these applications have been practically discontinued. In fertilizer applications, calcium cyanamide is broken down by soil moisture into highly reactive lime and free cyanamide; the latter is converted by soil microbes to urea and then ultimately to ammonia. Nitrifying bacteria convert ammonia to nitrate [31]. Calcium cyanamide is particularly valuable for acid soils in need of lime. It can also be used in mixed fertilizers, although if it is used in excess, the resultant high alkalinity reduces soluble phosphate. Calcium cyanamide is used on asparagus and onions as a weed killer. Heavy applications approximately 1 month before planting control soil-borne plant disease and weed seed.

Calcium cyanamide has been used to control animal pests. On grasslands it is used to kill the dwarf water snail, the intermediate host of the liver fluke. It also kills gastric and intestinal parasites in domestic animals and destroys salmonellae in liquid sewage.

Nitrogen oxides are removed from waste gases with over 99% efficiency by scrubbing with calcium cyanamide [32]. In portland cement, cyanamide and calcium cyanamide improve set characteristics, increase compression strength, and reduce freeze–thaw damage [33].

Pharmaceutical-grade calcium cyanamide is used to treat alcoholism. A small pill, taken once a day, subjects the alcohol user to an unpleasant cyanamide flush (see Chap. 6), which discourages drinking [34]. Calcium cyanamide is used for steel nitridation [35] and, to some extent, for desulfurization [36]. It has various uses in the production of cyanamide, dicyandiamide, melamine and other substituted triazines, thiourea, and guanidines.

3. Cyanamide

Cyanamide [420-04-2], $NH_2C \equiv N$, M_r 42.04, was obtained in 1838 by passing gaseous ammonia over cyanogen chloride [37]. Its physical properties are associated with the tautomeric forms:

$$H_2N-C\equiv N \quad \rightleftharpoons \quad HN=C=NH$$
Nitrile form   Carbodiimide form

Until the early 1960s, its instability prevented commercial use, but this situation changed when process improvements afforded a product with a high degree of purity and stability.

3.1. Physical Properties [12]

Cyanamide forms colorless, deliquescent orthorhombic crystals or long transparent needles (from water) [38], large platelike crystals when allowed to crystallize spontaneously from water [25], and crystals from dimethyl phthalate solution [39]. Physical properties are listed below.

mp	45 °C
bp at 1.6 kPa	132–138 °C
Density at 57 °C	1.069 g/cm^3
Refractive index n_D^{48}	1.4418
Heat of solution	−15.1 kJ/mol
Heat of neutralization	15.5 kJ/mol
Specific heat, 0–39 °C	2.291 J g^{-1} K^{-1}
Latent heat of fusion	208.5 J/g
Latent heat of vaporization	68 kJ/mol
Heat of combustion	718 kJ/mol
Dimerization energy	31.4 kJ/mol
Base K_B at 25 °C	2.5×10^{-15}
Acid K_A at 25 °C	5.42×10^{-11}
Conductivity of HNCN ion at 25 °C	54.4 Ω^{-1} cm^{-1}

Cyanamide is a weak acid that is very highly soluble in water. It is completely soluble at 43 °C, and has a minimum solubility at −15 °C (Fig. 5). It is highly soluble in polar solvents, such as the lower molecular mass alcohols, esters, and ketones, and less soluble in nonpolar solvents.

The IR spectrum of cyanamide exhibits an intense, poorly resolved doublet at 2225–2260 cm^{-1}. The UV spectrum shows only weak absorption below 230 nm, with no maximum observable above 208 nm. The infrared and Raman spectra support the N-cyanoamine structure, $NH_2-C\equiv N$ [33].

3.2. Chemical Properties [40]

Cyanamide is a highly reactive compound. The nitrile group is activated by the electrophilic nitrogen atom containing acidic hydrogens. Many addition reactions with nucleophiles are known. The reactivity in aqueous solution is highly dependent on pH.

Figure 5. Solubility of cyanamide in water

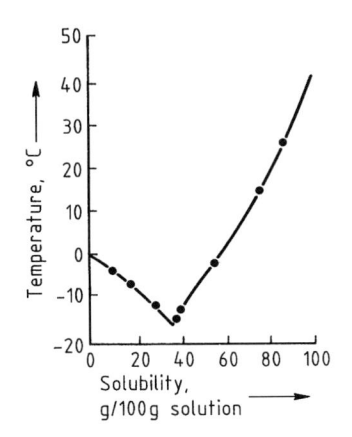

Reaction with Water. Reaction with water at pH 7 is sluggish; hydration is accelerated by acid, base, or heat and gives urea [41]:

$$H_2NCN + H_2O \longrightarrow H_2NCONH_2$$

At elevated temperature, cyanamide is hydrolyzed via urea to form ammonia and carbon dioxide.

Polymerization. The most important reactions of cyanamide are those leading to the dimer, cyanoguanidine [42] or dicyandiamide, and the trimer, melamine [43]. Unless stabilized by weak acids, crystalline cyanamide may polymerize violently. Polymerization in aqueous solution depends on pH and temperature.

Dimerization is best effected in solution at pH 8–10, up to ca. 80 °C.

$$2\ NH_2CN \xrightarrow{OH^-} NH_2\overset{NH_2}{\underset{|}{C}}=NCN$$

Melamine [108-78-1] is produced from cyanamide or dicyandiamide at elevated temperature and pressure. A mixture of methanol and liquid ammonia is a suitable reaction medium [44]. The mechanism of trimerization has not been established, but it is believed to involve either the reaction of a molecule of cyanamide with a molecule of dicyandiamide or the reaction of two molecules of dicyandiamide to form guanylmelamine, which breaks down into melamine and cyanamide (→ Melamine and Guanamines).

Reaction with Hydrogen Sulfide or Hydrogen Selenide. Thiourea [62-56-6] may be prepared by the action of hydrogen sulfide on an aqueous solution of cyanamide with ammonia as catalyst. The optimum conditions are pH 8–9 and a temperature above 50 °C [45].

$$H_2NCN + H_2S \longrightarrow H_2NCSNH_2$$

Selenourea [630-10-4] is obtained similarly from hydrogen selenide [46].

$$H_2NCN + H_2Se \longrightarrow H_2NCSeNH_2$$

Reaction with Thiols and Thiophenols. Cyanamide reacts with thiols and thiophenols more readily than with alcohols and phenols; the products are pseudothioureas. The reaction is conveniently conducted in diethyl ether.

$$H_2NCN + RSH \longrightarrow H_2N\overset{NH}{\underset{}{C}}-SR$$

Reaction with Alcohols and Phenols. Under suitable conditions, cyanamide reacts with alcohols and phenols to give O-alkyl- and O-arylpseudoureas (formerly called isoureas). Alkylpseudoureas are usually prepared as hydrochlorides by treating a solution of cyanamide in the appropriate anhydrous alcohol with dry hydrogen chloride; sulfuric acid has also been recommended [47].

$$NH_2CN + ROH + H^+ \longrightarrow NH_2C\begin{smallmatrix}NH_2^+\\OR\end{smallmatrix}$$

The alkylpseudourea bases, which have basic strengths (K_B) near 1×10^{-4} [48], can be liberated from their salts by treatment with moist potassium hydroxide under ether or with sodium alkoxides in anhydrous alcohols [25]. They are fairly stable under normal conditions, but are sensitive to moisture and dissociate when heated.

Reaction with Ammonia and Amines. Ammonia adds readily to the nitrile group of cyanamide to form guanidine. Weakly basic conditions help to avoid dimerization and, in aqueous systems, hydrolysis to urea.

$$NH_2CN + NH_4X \xrightarrow{NH_3} \left[NH_2-\overset{NH_2}{\underset{}{C}}-NH_2\right]^+ X^-$$

where X = halogen

Guanidine salts of common acids have been prepared by variations of this process starting from cyanamide or metal cyanamides [40].

Primary and secondary amines react with cyanamide to form substituted guanidines under the above conditions.

$$NH_2CN + R_1R_2NH \xrightarrow{H^+} \left[NH_2-\overset{NR_1R_2}{\underset{}{C}}=NH_2\right]^+$$

Reaction with Amino Acids. Cyanamide adds to amino acids to form carboxyalkylguanidines. The conversion of glycine to glycocyamine was the first such reaction reported [49].

$$NH_2CN + NH_2CH_2COOH \longrightarrow NH_2\overset{\overset{+}{N}H_2}{\underset{\|}{C}}NHCH_2COO^-$$

Cyanamide reacts with the ethyl ester of sarcosine to give creatinine [50].

Cyanamide and dialkylcyanamides induce formation of peptide bonds when heated with N-carbobenzoxyamino acids and amino acid esters [51]. This dehydrating effect is ascribed to the intermediate formation of an O-acylpseudourea or amino anhydride, which reacts further with the amino ester [52].

$$\underset{\underset{RCHCOOH}{|}}{C_6H_5CH_2OCONH} + NH_2CN \longrightarrow \underset{\underset{RCHCOO\overset{\|}{C}NH_2}{|}}{C_6H_5CH_2OCONH}\overset{NH}{\underset{\|}{}}$$

$$\downarrow \overset{NH_2}{\underset{|}{R^1CHCOOC_2H_5}}$$

$$\underset{\underset{R^1CHCOOC_2H_5}{|}}{\underset{\underset{RCHCONH}{|}}{C_6H_5CH_2OCONH}} + NH_2CONH_2$$

Reaction with Hydrazines and Hydrazones. Cyanamide reacts with hydrazines to form aminoguanidines, which are versatile intermediates for the synthesis of nitrogen-rich compounds, particularly heterocyclics. Aminoguanidine [79-17-4], itself unstable, is conveniently isolated as the relatively insoluble hydrogen carbonate after reaction of aqueous hydrazine with cyanamide at pH $\approx$ 6 [53], [54].

$$NH_2CN + NH_2NH_3^+ \longrightarrow \left[NH_2NH\overset{\overset{+}{N}H_2}{\underset{\|}{C}}NH_2 \right]^+$$

Aminoguanidine salts react further with hydrazine in aqueous or alcoholic solution to give salts of 1,2,3-triaminoguanidine [2203-24-9] [55].

$$\left[NH_2NH\overset{\overset{+}{N}HNH_2}{\underset{\|}{C}}NHNH_2 \right]^+$$

The reaction of cyanamide with phenylhydrazine hydrochloride in boiling alcohol produces two isomeric aminoguanidines [56].

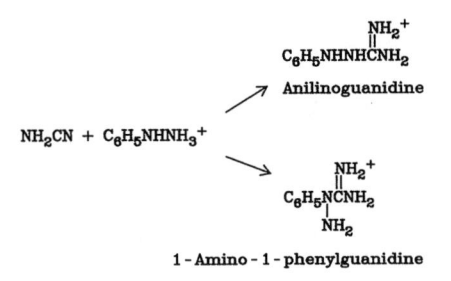

$$NH_2CN + C_6H_5NHNH_3^+ \nearrow \underset{\text{Anilinoguanidine}}{C_6H_5NHNH\overset{\overset{+}{N}H_2}{\underset{\|}{C}}NH_2}$$

$$\searrow \underset{\underset{NH_2}{|}}{C_6H_5\overset{\overset{+}{N}H_2}{\underset{\|}{N}}CNH_2}$$

1 - Amino - 1 - phenylguanidine

Reaction with Hydroxylamines. Hydroxylamine hydrochloride reacts slowly with aqueous cyanamide to give hydroxyguanidine [*13115-21-4*] [57].

$$NH_2OH + HCl + NH_2CN \xrightarrow[25\,°C]{H_2O} \left[\begin{array}{c} NH_2 \\ \| \\ NH_2CNHOH \end{array} \right]^+ Cl^-$$

Reaction with Amides and Sulfonamides. Amides do not react directly with cyanamide [25]. Cyanamide does, however, react with sulfanilamide to form both N^4-guanylsulfanilamide [58] and sulfaguanidine [*57-67-0*] [59].

Equimolar quantities of cyanamide and a primary thioamide, such as thiobenzamide or 2-phenylthioacetamide, when heated in an alcoholic solution, give good yields of thiourea and a nitrile [60].

$$\underset{\|}{\overset{S}{R\overset{\|}{C}NH_2}} + NH_2CN \longrightarrow RCN + H_2N\underset{\|}{\overset{}{C}}NH_2 \atop S$$

Alkylation with Alkyl Halides. Cyanamide salts can be dialkylated, especially with allyl and benzyl chlorides in water or alcohol [61].

$$2\,CH_2=CHCH_2Cl + NaHNCN + NaOH$$
$$\xrightarrow[60\,°C]{H_2O} (CH_2=CHCH_2)_2NCN + 2\,NaCl + H_2O$$

Reaction with Sulfonyl Chlorides. N-Acetylsulfanilyl chloride reacts readily with the cyanamide anion in water to give a good yield of N^4-acetylsulfanilylcyanamide salt, a convenient intermediate for the production of sulfaguanidine [62].

$$CH_3CONH-\!\!\!\!\left\langle\!\!\bigcirc\!\!\right\rangle\!\!\!-SO_2Cl + NHCN^- + OH^- \longrightarrow$$

$$CH_3CONH-\!\!\!\!\left\langle\!\!\bigcirc\!\!\right\rangle\!\!\!-SO_2N-CN^- + H_2O + Cl^-$$

Higher alkyl groups confer useful surfactant properties [63].

3.3. Production

Cyanamide is manufactured from calcium cyanamide by continuous carbonation in an aqueous medium [64].

$CaNCN + H_2O + CO_2 \longrightarrow NH_2CN + CaCO_3$

The following reactions take place:

$2\ CaCN_2 + 2\ H_2O \longrightarrow Ca(NHCN)_2 + Ca(OH)_2$
$Ca(OH)_2 + 2\ CO_2 \longrightarrow Ca(HCO_3)_2 \longrightarrow CaCO_3 + H_2O + CO_2$
$Ca(HNCN)_2 + CO_2 + H_2O \longrightarrow 2\ NH_2CN + CaCO_3$

In the continuous cyanamide process, the amount of calcium cyanamide and water added is equivalent to the aqueous cyanamide removed (Fig. 6) [65].

A carbonated slurry of cyanamide solution, solid calcium carbonate, and graphite is cooled to remove the heat of reaction (d). Part of the slurry is recycled to facilitate temperature control; the remainder is filtered, yielding a cyanamide solution and a cake of calcium carbonate and graphite. The final concentration of cyanamide is normally 25%.

The calcium cyanamide feed is well mixed with the recycled slurry and filtrate in a feed vessel (b) and added at a rate to maintain a pH of 6.0–6.5 in the cooling tank. After carbonation, the slurry is held at 30–40 °C to complete the formation of calcium carbonate. The equipment is preferably stainless steel. To produce a commercial 50% solution and recover crystalline cyanamide. This process is somewhat modified to produce a commercial 50% solution and recover crystalline cyanamide. Calcium and iron are removed by ion exchange, if desired.

Commercial Forms. The two principal producers of cyanamide are SKW Trostberg, Federal Republic of Germany, in Europe and Amer. Cyanamid, in North America. Both offer two grades:

1) Cyanamide in 50% aqueous solution (trade names SKW-Cyanamid L-500 and Aero-Cyanamide-50), a clear slightly yellowish solution that can change reversibly to blue under the action of light.
Specifications:

Cyanamide	50%
Dicyandiamide, urea, and related compounds	max. 4%
Iron	max. 300 mg/L
Calcium	max. 100 mg/L
Stabilizer	max. 2%

The maximum storage life is ca. 6 months; the product is packaged in polyethylene containers or polyethylene-lined steel drums.
2) Cyanamide, crystalline, ca. 97% (trade names: SKW-Cyanamid F-100 and Aero Cyanamide-100), colorless orthorhombic crystals with a tendency to deliquesce.

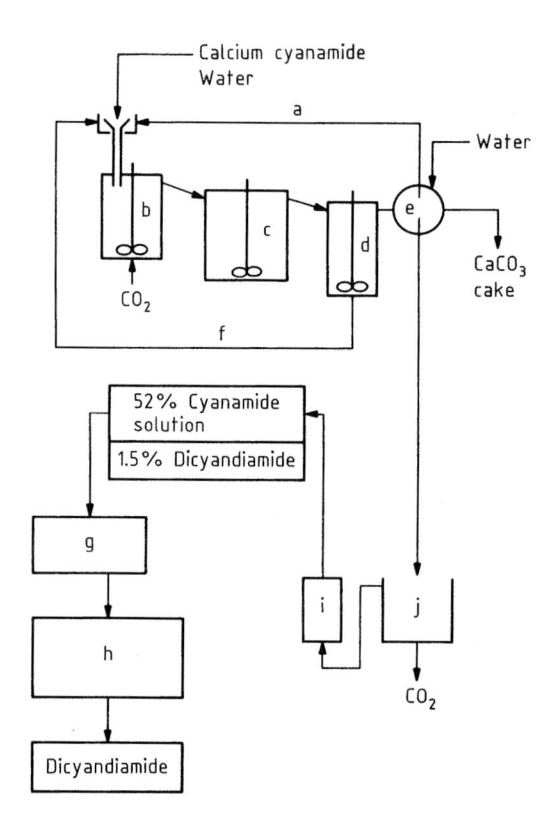

Figure 6. Continuous manufacture of cyanamide and dicyandiamide from calcium cyanamide
a) Liquid return line; b) Reactor; c) Reprecipitation tank; d) Cooling tank; e) Filtration equipment; f) Slurry return line; g) Cooling and crystallization of cyanamide; h) Solution with less than 38% cyanamide; i) Ion exchange; j) Storage tank

Specifications:

Cyanamide	95–97%
Dicyandiamide	max. 1%
Water	max. 0.5–1%
Iron	traces
Stabilizer	max. 0.05

Storage life at 10 °C is ca. 12 months. The product is packaged in an air-tight polyethylene bag inside a steel drum.

3.4. Chemical Analysis [66]

Cyanamide, especially in solution, always contains the dimer. Thus, the determination of cyanamide in the presence of dicyandiamide may be difficult. Cyanamide is determined by precipitation as the silver salt in an ammoniacal medium; excess silver nitrate is back-titrated. The dicyandiamide silver salt precipitated in an alkaline medium is determined by the Kjeldahl nitrogen.

Table 1. Decomposition of cyanamide*

t, °C	Decomposition in the first month, %
20	0.3
30	0.5
40	5.6
45**	100

* Containing 0.5% water.
** Molten.

3.5. Storage and Transportation

The commercial 50% solution produced by Amer. Cyanamid is stabilized at pH 4.5 – 5.0 with 2% monosodium phosphate and contains < 1.5% dicyandiamide and 0.2% urea. Such solutions show < 1% change in cyanamide content per month of storage below 10 °C. Periodic pH adjustment is recommended during extended storage. Attempts to concentrate this solution are hazardous and are not recommended.

Stainless steels 202 and 304, glass, polyethylene, and polyester-based materials are suitable for storing cyanamide solution. Iron, copper, brass, lead, Monel alloy, mild steel, Duriron alloy, and tin plate are attacked.

Crystalline cyanamide is stable below 25 °C. At higher temperature, particularly above the melting point, decomposition to dicyandiamide and urea becomes rapid (Table 1).

The decomposition of crystalline cyanamide to dicyandiamide and urea is exothermic. The molten form may decompose violently, particularly above 50 °C.

Moisture, acid, and alkali accelerate decomposition. Solid cyanamide is stored in tightly sealed containers and kept dry and cool (25 °C or lower). Temperature must not exceed 45 °C.

3.6. Uses

Cyanamide is used in the production of intermediates for drugs and agricultural chemicals, e.g., pesticides, via guanidine or pseudoourea intermediates (see Section 3.2).

In addition to calcium cyanamide, lead cyanamide ($PbCN_2$) and the mono- and disodium salts ($NaHNCN \cdot 2\,H_2O$; Na_2NCN) are of commercial importance. The yellow lead salt (decomposition at ca. 580 °C) is used as a rust-inhibiting pigment and in mirror-coating. More recently, it has found an important application in the synthesis of the antiulcer drug cimetidine (Tagamet, SKF) [67], [68]. The crystalline sodium salts are used in organic syntheses in the chemical and pharmaceutical industry, for chemical hop pruning (removal of shoots), and for defoliation and weed control in hop orchards [62]. Because of rapid degradation, cyanamide does not cause any residue problems. Fungitoxic and herbicidal action has been demonstrated [69]. A more recent application is the acceleration of the opening of dormant grapevine buds. For this purpose cyanamide is more potent than its calcium salt. This application is important in warm areas where many grapevine buds do not open because of inadequate winter chilling essential for the bud-break cycle [70].

Cyanamide–phosphorus pentachloride mixtures confer flame resistance to cotton textiles that are then treated with ammonia and heat-cured [71]–[73]. Cotton textiles are crease-proofed with cyanamide and phosphoric acid [74]. Cyanamide has been used to shrink-proof wool [75] and in stable bleach compositions [76].

Cyanamide-modified starch products are used in pulp sizing for paper and newsprint production [77]. A cyanamide–aldehyde combination gives thermally stable papers for transformers [78].

Cyanamide has also been used to cure epoxy resins [79].

4. Dicyandiamide

The dimeric form of cyanamide is known as dicyandiamide or cyanoguanidine. The occurrence of dimerization on evaporation of an aqueous solution was known as early as 1851; the product was later named param (from paracyanamide). The name dicyandiamide was coined in 1862 [80] and, together with the abreviation dicy, is now firmly established. The name dicyandiamide is not as descriptive as the name cyanoguanidine, which is preferred by Chemical Abstracts; however, the former is used in this article.

Raman spectra support the cyanoguanidine structure (**1**) [81]; the crystalline form apparently has the tautomeric form (**2**) [82].

$$\underset{\mathbf{1}}{H_2N-\overset{NH}{\underset{\|}{C}}-NHCN} \rightleftharpoons \underset{\mathbf{2}}{H_2N-\overset{NH_2}{\underset{|}{C}}=NCN}$$

Tautomerism undoubtedly exists between two or more forms of dicyandiamide. The preferred formulation appears to be tautomer **2**.

Dicyandiamide has achieved industrial importance as an intermediate for the production of guanidine salts (→ Guanidine) and melamine (→ Melamine and Guanamines); it is produced from calcium cyanamide. During World War II, dicyandiamide was used mainly for the production of nitroguanidine (picrite) from guanidine nitrate. Subsequently, production shifted primarily to melamine. In recent years, the production of melamine from dicyandiamide has decreased and is now based on urea.

4.1. Physical Properties [12]

Dicyandiamide [461-58-5], M_r 84.084, ϱ 1.404 g/cm³, mp 210–212 °C, is an odorless, colorless, nonvolatile powder with a monoclinic prismatic crystal structure. Solubilities are shown in Table 2. Solubility is poor in nonpolar solvents, such as benzene and kerosene. Other physical properties are given below.

Enthalpy of formation ΔH_{298}	24.9 kJ/mol
Enthalpy of combustion Q_v	−1387.0 kJ/mol
Q_p at 25 °C and Washburn corrected	−1384.0 kJ/mol
Average specific heat capacity c_v (0–204 K)	1.91 J g⁻¹ K⁻¹

Molecular heat capacity c_p in J mol⁻¹ K⁻¹:

T,K	14.40	54.82	100.62	200.49	246.37	294.63
c_p	2.26	30.0	52.3	87.0	102.1	117.8

4.2. Chemical Properties

Dicyandiamide reacts with a variety of reagents. It is the simplest organic compound containing the C–N, C=N, and C≡N groupings. Reaction may occur at one or more of these sites. The literature on the chemistry of dicyandiamide is extensive, and many more reactions than those indicated below have been reported.

Salt Formation. Dicyandiamide is amphoteric and forms salts with organic and inorganic acids and with alkali in an anhydrous medium [25]. In very concentrated aqueous acid, the corresponding salt precipitates. Such salts are almost completely hydrolyzed in dilute solution.

$$(H_2N)_2C=NCN + RCOOH \longrightarrow H_2NC(=NH)NHCN \cdot RCOOH$$

$$(H_2N)_2C=NCN + HCl \xrightarrow{HOAc} H_2NC(=NH)NHCN \cdot HCl$$

With alkali and alkaline-earth hydroxides, dicyandiamide forms salts [84], which are stable in concentrated aqueous solutions but hydrolyze in dilute solutions. The sodium salt is 94.5% hydrolyzed in 0.1 M solution.

$$(H_2N)_2C=NCN + KOH \longrightarrow [H_2NC(=NH)NCN]^- K^+ + H_2O$$

Hydrolysis. At elevated temperature in the presence of an equivalent amount of a mineral acid, dicyandiamide yields guanylurea salts. This reaction is quantitative and can be used as a basis for the determination of dicyandiamide [85]:

Table 2. Solubility of dicyandiamideSolubility, of dicyandiamide

Solvent	t, °C	Solubility, g/100 g of solvent
Acetone	31	1.73
	50	2.20
Ammonia	−33.3	71.9
	8.5	105.0
Ethylene glycol monobutyl ether (butyl cellosolve)	34	< 0.7
	134	6.0
Ethylene glycol monoethyl ether acetate	108	0.35
	151	1.25
Diethyl ether	0	0.0006
	25	0.0015
Dimethylformamide	28	35.1
	98	63.1
Ethylene glycol monoethyl ether (ethyl cellosolve)	34	< 3.0
	110	6.5
Ethylene glycol	30	8.0
	63	16.0
Methanol	1	3.46
	50	9.45
Ethylene glycol monomethyl ether (methyl cellosolve)	33	13.0
	50	17.0
Triethanolamine	85	3.5
	105	13.2
Water	25	4.13
	49.8	11.80
	100	96.8

$$(H_2N)_2C=NCN + HX + H_2O \xrightarrow{90-100\ °C} H_2NC(NH)NHC(O)NH_2 \cdot HX$$

where X = halogen

At atmospheric pressure and moderate temperature in the presence of aqueous ammonia, reaction of dicyandiamide with hydrogen sulfide gives guanylthiourea. With increased pressure, an additional atom of sulfur may be introduced to yield dithiobiuret [86]. Unlike dicyandiamide, guanylurea is basic and readily forms salts with organic and inorganic acids.

The reaction of dicyandiamide with concentrated sulfuric acid or glacial acetic acid containing a small amount of water proceeds to yield guanidine and ammonium salts [87] instead of guanylurea. Treatment of dicyandiamide with nitric acid gives nitroguanylurea [88].

$$(H_2N)_2C=NCN + HNO_3 \xrightarrow{H_2SO_4} H_2NC(NH)NHC(O)NHNO_2$$

The imino group of dicyandiamide is hydrolyzed by barium hydroxide to give cyanourea [2208-89-1].

$$(H_2N)_2C{=}NCN + H_2O \xrightarrow{Ba(OH)_2} H_2N\overset{\overset{O}{\|}}{C}NHCN + NH_3$$

Reaction with Amines. Heating dicyandiamide with aromatic amines in water in the presence of an equivalent amount of mineral acid gives high yields of aryl biguanide salts [89].

$$(H_2N)_2C{=}NCN + RNH_2 + HCl \xrightarrow{H_2O} H_2N\overset{\overset{NH}{\|}}{C}NH\overset{\overset{NH}{\|}}{C}NHR \cdot HCl$$

where R = aryl

Yields of biguanide salts from the reaction of dicyandiamide with ammonium salts are poor. The biguanide obtained reacts further with the ammonium salt, and the product forms the guanidine salt. This reaction is used for the commercial preparation of guanidine nitrate [506-93-4] [90]:

$$(H_2N)_2C{=}NCN + 2\,NH_4NO_3 \longrightarrow 2\,H_2N\overset{\overset{NH}{\|}}{C}NH_2 \cdot HNO_3$$

Reaction with Formaldehyde. Treatment with formaldehyde [91] produces resinous compositions of varying properties under acidic or alkaline conditions. The reaction can be controlled to give mainly the water-soluble hydroxymethyldicyandiamide, or carried further to yield amorphous, infusible solids.

$$(H_2N)_2C{=}NCN + HCHO \longrightarrow HOCH_2NH\overset{\overset{NH}{\|}}{C}NHCN$$

Heterocyclic Compounds. Under pressure and in the presence of ammonia, dicyandiamide cyclizes to melamine, of which large amounts have been made in this manner. At present, however, melamine is produced primarily by the urea process [92].

$$3\,(NH_2)_2C{=}NCN \xrightarrow{NH_3} 2\;\text{melamine}$$

Excellent yields of guanamines are obtained when dicyandiamide is heated with alkyl or aryl nitriles in the presence of a small amount of alkali [93].

$$(NH_2)_2C{=}NCN + RCN \xrightarrow[100-200\,°C]{Alkali} \text{guanamine}$$

Reaction with urea affords ammeline [645-92-1].

$$(NH_2)_2C{=}NCN + NH_2\overset{\overset{O}{\|}}{C}NH_2 \longrightarrow \text{ammeline} + NH_3$$

Dicyandiamide reacts with a wide variety of compounds to yield substituted pyrimidines. Acetoacetic ester reacts with dicyandiamide in the presence of sodium ethoxide and ethanol to give 2-cyanoamino-4-hydroxy-6-methylpyrimidine [94]. Derivatives of barbituric acid are synthesized by condensation of dicyandiamide with disubstituted malonic or cyanoacetic acid; their ester, chloride, or nitrile derivatives yield cyaniminopyrimidines [95]. These can be alkylated in the 3-position and hydrolyzed to N-monoalkyl-5,5-disubstituted barbituric acids [88].

4.3. Production

Dicyandiamide is formed in almost quantitative yield by the dimerization of cyanamide in alkaline solution. The reaction occurs most readily at pH 8–10 by the addition of the anionic cyanamide species to the nitrile group. The product is a weaker acid than cyanamide; it is protonated immediately with the generation of a new cyanamide anion, which continues the process.

$$H_2NC\equiv N \rightleftharpoons [HNC\equiv N]^- + H^+$$

$$H_2NC\equiv N + [HNC\equiv N]^- \longrightarrow \left[\begin{array}{c} NH \\ \parallel \\ H_2NC-NC\equiv N \end{array} \right]^-$$

$$\left[\begin{array}{c} NH \\ \parallel \\ H_2NC-NC\equiv N \end{array} \right]^- + H_2NC\equiv N \longrightarrow [HNC\equiv N]^- + \begin{array}{c} H_2N \\ \diagdown \\ C=N-C\equiv N \\ \diagup \\ H_2N \end{array}$$

In the industrial production of dicyandiamide, cyanamide is first prepared from calcium cyanamide by continuous carbonation in an aqueous medium (see Chap. 3) at pH 7–8 and is then dimerized at ph 8–10.

$$CaNCN + H_2O + CO_2 \longrightarrow H_2NCN + CaCO_3$$

$$2\,H_2NCN \xrightarrow{\text{pH 8–10}} (NH_2)_2C=NCN$$

To obtain a clean product, dimerization is carried out after separation of the precipitated calcium carbonate and other insoluble matter such as graphite [96]. The filtered solution containing ca. 25% cyanamide is adjusted to ca. pH 9 and held at ca. 80 °C for 2 h to effect the desired conversion. The resulting hot liquor is filtered and cooled in a vacuum crystallizer. The dicyandiamide crystals are centrifuged and transferred to a rotary drier. The pH of the reaction must be controlled. An increase in alkalinity accelerates decomposition by hydrolysis (Fig. 7). The dried product is stored in bulk or packaged in multiwall bags.

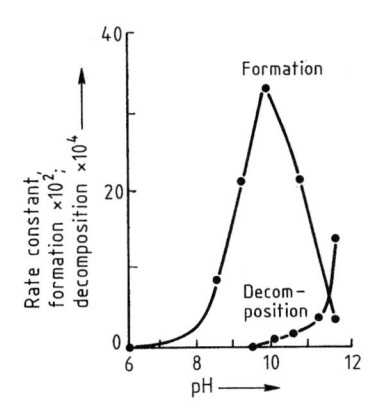

Figure 7. Effect of pH on rates of formation and decomposition of dicyandiamide

4.4. Quality Specifications and Chemical Analysis

Dicyandiamide is determined quantitatively by UV spectrometry and is identified qualitatively by paper chromatography [25]. It is also hydrolyzed to guanylurea and determined gravimetrically as the nickel salt [97]. Dicyandiamide has also been precipitated as the silver picrate [101]. It is often determined by titration with tetrabutylammonium hydroxide in pyridine [25]. More commonly, a total nitrogen analysis and melting point determination are used to ascertain purity (Table 3).

4.5. Storage and Transportation

Dry crystalline dicyandiamide is stable at ordinary temperature. Slow decomposition begins above 130 °C, becoming rapid above the melting point, with the evolution of ammonia and the formation of melamine, melam, and other nitrogen compounds. Dicyandiamide is stable in aqueous solution up to 80 °C; above 80 °C ammonia is liberated. Dicyandiamide is not hygroscopic, flammable, or corrosive.

On prolonged exposure to light, dicyandiamide gradually turns pink, but with no effect on its chemical properties.

In North America, technical-grade dicyandiamide is shipped in 45-kg net multiwall bags; the pulverized grade is supplied in 45-kg Leverpak bags. The filled bags occupy a volume of ca. 0.06 m^3, and the Leverpak bag occupies ca. 0.14 m^3.

Table 3. Typical analysis of commercial dicyandiamide*

Assay	Value
Total nitrogen, as wt% dicyandiamide, min.	99.0
Water, %	0.1–1.0
Melamine, %	0.1–0.7
mp, °C, min.	206
Thiourea, mg/kg	200
Heavy metals, mg/kg	10

* White, crystalline.

4.6. Uses

Dicyandiamide is an intermediate for a number of resins and organic nitrogen compounds, such as guanamines, biguanides, and guanidine salts. Guanidine phosphate is used as a fire-retardant in water-based systems; guanamines are employed as copolymers in resin compositions. Melamine is used primarily as a cross-linking agent in coatings and laminates. Substituted biguanides, prepared from dicyandiamide and arylamines, are used as stabilizers and antioxidants for soaps and other fatty materials [87].

The condensation of dicyandiamide with formaldehyde gives resins with broad industrial applications. The reaction can be controlled to give a water-soluble intermediate for impregnation of paper, wood, leather, and textiles [99]. These resins are effective as direct-dye fixing agents for textiles and leather products. Such agents allow dyeing and printing of fabrics with direct dyes, rather than with more expensive vat dyes. They also improve wash- and lightfastness [87]. Dicyandiamide itself prevents odor formation in resin-treated fabrics.

Dicyandiamide–formaldehyde condensates modified with cationic polyaminopolyamides have been used in paper manufacture [100].

Under certain conditions, formaldehyde and mixtures of dicyandiamide with casein or other proteins, urea, melamine, or aromatic amines give resinous compositions varying from white powders to creamy pastes or clear, glasslike products [87].

Dicyandiamide reduces the viscosity of certain colloidal solutions [101]. This property is commercially significant in the manufacture of glues and adhesives, in the coating and sizing of paper and textiles, and in the conditioning of phosphate drilling muds. This action could be useful in other applications requiring viscosity control.

Dicyandiamide fluidizes adhesives and glues that are based on converted starches, casein, α-protein (soya protein), and hide proteins or collagens, thus extending the useful life of the adhesive. Dicyandiamide is 2–5 times more effective than urea on the basis of weight [87].

Dicyandiamide, in combination with phosphoric acid, is used as a flame-retardant for cellulosic materials [102]. It is used commercially as a fire-retardant for wood, particularly shingles [103], [104].

Dicyandiamide is stable at room temperature and is used in coating formulations as a flame-retardant. When exposed to a flame, the coating composition bubbles up or intumesces to a foam that protects the substrate from ignition.

An intumescent formulation including dicyandiamide, pentaerythritol monoammonium phosphate, titanium dioxide, latex binder, and water gives good foam volume at low coating weights with reasonable stability of the foam structure under ignition [105], [106].

Dicyandiamide powder is used as a curing agent for epoxy laminating resins, especially when a long pot life is necessary [107], [108]. The dicyandiamide may be added as a powder (6–15 wt%) to liquid epoxy resins or introduced into solid resins in solution form. The concentration depends on the curing temperature and time cycle.

Dicyandiamide can be used as a component in copolymers and cocondensates [109]. It is used for the thermal stabilization of poly(vinyl chloride) (PVC), for which it is introduced in the solid form at 170 °C into the PVC melt or as an aqueous solution to a PVC emulsion [110]. It stabilizes cellulose and homopolymers and copolymers of formaldehyde.

Dicyandiamide has a preservative effect on protein solutions. A bactericidal, fungicidal, and insecticidal composition is based on the reaction of dicyandiamide with formaldehyde and an aromatic amine [111].

Barbiturates are produced from dicyandiamide and malonic esters. The copper complex of guanylurea has been used as a fungicide in viticulture, and the hydrochloride or hydrobromide of guanylurea as a fire-retardant [112]. The chromate of dicyandiamide has been used as a rust inhibitor, and the phosphate salt for cleaning metal surfaces. Cyanourea has been proposed as a stabilizer for polyacrylonitrile spinning solutions.

5. Economic Aspects

Calcium cyanamide production probably peaked in 1962, when world production for fertilizer was ca. 10^6 t/a; 3×10^5 t was directed to industrial uses. The largest producers are in the Federal Republic of Germany, Canada, and Japan. Approximately 9×10^4 t of nitrogen was produced in 1974, primarily by Süddeutsche Kalkstickstoff-Werke (Federal Republic of Germany). About 7×10^4 t of nitrogen was produced by Japan (1974) and smaller amounts ($< 10^4$ t) by Belgium, Spain, and Portugal.

A total of ca. 40 plants produce calcium cyanamide. In North America, the only producer is Cyanamid Canada, Inc., in Niagara Falls, Ontario, serving the Canadian and United States markets. This company, a subsidiary of Amer. Cyanamid, also produces dicyandiamide, cyanamide, and their derivatives. Cyanamide is occasionally marketed in North America in crystalline form, but primarily as a 50% solution. In Europe, cyanamide is available as a 50% solution and in crystalline form from the Süddeutsche Kalkstickstoff-Werke, Trostberg, Federal Republic of Germany.

In 1985, the total production of cyanamide products was less than half that of 1962.

6. Toxicology and Occupational Health

The manufacture of cyanamide and calcium cyanamide does not present any serious health hazard. However, ingestion of alcoholic beverages by workmen within several hours of leaving the workplace sometimes causes a vasomotor reaction known as *cyanamide flush,* i.e., reddening of the face.

Drinking of small amounts of alcoholic beverages after exposure to calcium cyanamide can cause severe hyperemia of the upper half of the body, tachycardia, accelerated respiration, headache, and nausea [113]. Large amounts of alcohol can lower blood pressure significantly with possible collapse. Cyanamide interferes with the oxidation of alcohol; the accumulation of acetaldehyde may account for these symptoms, which are unpleasant but transient.

Commercial grades of calcium cyanamide contain lime and are skin irritants. Contact with or ingestion of cyanamide and inhalation of dust must be avoided. Experiments with rats gave the following results [87]:

1) The toxicity of crystalline cyanamide ranges from an oral LD_{50} of 280 mg/kg to a dermal LD_{50} of 590 (420–820) mg/kg. The compound is, therefore, considered to be moderately toxic both by ingestion and by skin application. An aqueous paste is corrosive to rabbit skin. Small quantities of the dry product caused severe irritation of the conjunctival sac of the rabbit eye.
2) A 50% solution of cyanamide is considered to be slightly toxic by ingestion and moderately toxic by skin application. The irritation to the rabbit skin and eye is only slightly less than that observed with crystalline cyanamide.

Dicyandiamide is essentially nontoxic, but it may cause dermatitis. The acute oral LD_{50} of dicyandiamide for young, male albino rats is higher than 10 g/kg; therefore, the product is considered to be nontoxic. Albino rabbits tolerate single dosages of 10 g/kg of the product as an aqueous paste in contact with the closely clipped skin of the abdomen for 24 h with no evidence of systemic toxicity or skin irritation. Patch tests using the dry powdered material on 200 human subjects did not indicate sensitizing or primary irritant properties.

7. References

General References

[1] M. Wildhagen: *Handbook of Industrial Electrochemistry*, vol. **5,** Akademie-Verlag, Leipzig 1953, Part 1.
[2] A. Brauer, J. D'Ans: *Progress in Industrial Chemistry*, Springer, Berlin 1937.
[3] "SKW-Cyanamide Products," Südd. Kalkstickstoff-Werke, Trostberg, Federal Republic of Germany, 1973, 82 pp.
[4] "Cyanamide," Amer. Cyanamid, Wayne, N.J., 1966, 49 pp.
[5] *Beil.* **3,** 91; **3 (2),** 42; **3 (2),** 75; **3 (3),** 167.
[6] "Dicyandiamide," Südd. Kalkstickstoff-Werke, Federal Republic of Germany, 1973.
[7] "Aero Dicyandiamide," Amer. Cyanamid, Wayne, N.J., 1970, 20 pp.

Specific References

[8] B. E. Turner et al., *Astrophys. J.* **201** (1975)3.
[9] E. Drechsel, *J. Prakt. Chem.* **16** (1877) 180–200.
[10] F. Emich, *Monatsh. Chem.* **10** (1889) 321–352.
[11] DE 88 363, 1895(N. Caro, A. Frank).
[12] *Gmelin,* system no. **28**, Calcium, Parts B 1–2 (1956) pp. 179–195.
Gmelin, system no. **14**, Carbon, Part D 1 (1971) pp. 258–297.
[13] *Gmelin,* system no. **28**, Calcium, Part B 3 (1961) pp. 962–971.
[14] T. Aono, *Bull. Chem. Soc. Jpn.* **16** (1941) 92–98.
[15] V. Ehrlich, *Z. Elektrochem.* **32** (1926) 187–188.
[16] F. E. Polzeniusz, *Chem. Ztg.* **31** (1907) 958.
[17] H. H. Frank, H. Heimann, *Z. Elektrochem.* **33** (1927) 469–475.
[18] M. L. Kastens, W. G. McBurney, *Ind. Eng. Chem.* **43** (1951) 1020–1033.
[19] H. Rock, *Chem. Ztg. Chem. Appar.* **88** (1964) 191–271.
[20] Südd. Kalkstickstoff-Werke, DE 917 543, 1952 (F. Kaess et al.).
[21] Südd. Kalkstickstoff-Werke, US 2 838 379, 1958 (F. Kacss ct al.).
[22] BASF, DE 965 992, 1950 (G. Hamprecht, H. Gettert).
[23] Südd. Kalkstickstoff-Werke, DE 972 048, 1953 (T. Fischer et al.).
[24] A. A. Pimenova et al., *Tr. Tashk. Politekh. Inst.* **107** (1973) 49.
[25] Amer. Cyanamid, unpublished results.
[26] V. G. Golov et al., *Tr. N.-I i Proekt. In-Ta Azot. Prom.-Sti i Produktov Organ. Sinteza* **28** (1974) 49–52; *Chem. Abstr.* **84** (1975) 159 006 n.
[27] "Cyanamide by Cyanamid," Amer. Cyanamid, Wayne, N. J., 1961. O. I. Polyanshikov et al., *Vopr. Khim. Tekhnol.* **39** (1975) 136.
[28] *Gmelin,* system no. 28, Calcium, Part B 1–2, 1956–1957, pp. 179–196.
[29] Ferto Chemical Sales Co., DE 623 600, 1932.
[30] Südd. Kalkstickstoff-Werke, DE 1 097 457, 1961 (T. Fischer et al.).
[31] K. Rathsack, *Landwirtsch. Forsch.* **6** (Special ed.) (1954) 116–132.
[32] Nukem, DE 2 926 107, 1981 (H. Qui Umann).
[33] Amer. Cyanamid, US 3 503 766, 1970 (F. De Maria). R. E. Dwyer, US 4 049 465, 1977.
[34] K. Arikawa, K. Inanaga, *Folia Psychiat. Neurol. Jpn.* **27** (1973) no. 1, 9.

[35] Goerig Co., DE 1 771 827, 1973 (P. Birk, K. Wohlgemuth). K. Deutzmann, DE 2 136 450, 1973 (W. Blank, O. Vorbach).
[36] Südd. Kalkstickstoff-Werke, DE 2 252 795, DE 2 252 796, 1974 (W. Meichsner).
[37] A. Bineau, *Ann. Chim. Paris* **67** (1838) 225–272.
[38] M. Frund, A. Schander, *Ber. Dtsch. Chem. Ges.* **29** (1896) 2500–2505.
[39] D. Costa, C. Bolis-Cannella, *Ann. Chim. (Rome)* **43** (1954) 769–778.
[40] Cyanamide, Technical Bulletin, Amer. Cyanamid Co., Wayne, N.J., 1966.
[41] Amer. Cyanamid, US 1 758 641, 1930 (G. Barsky, P. W. Griffith).
[42] Aero Dicyandiamide Bulletin, Amer. Cyanamid, Wayne, N.J., 1970.
[43] The Chemistry of Melamine, Amer. Cyanamid, Wayne, N.J., 1954.
[44] M. M. Hooves, *Chem. Eng.* **57** (1950) 132–134.
[45] S. A. Miller, B. Bann, *J. Appl. Chem.* **6** (1956) 89–93.
[46] H. Hope, *Acta Chem. Scand.* **18** (1964) 1800.
[47] M. L. Kilpatrick, *J. Am. Chem. Soc.* **69** (1947) 40–46.
[48] S. Basterfield, J. Tomecko, *Can. J. Research* **8** (1933) 458–462.
[49] International Minerals & Chemicals Corp., US 2 620 354, 1952 (B. Vassel, W. D. Janssens).
[50] E. Abderhalden, H. Sickel, *Z. Physiol. Chem.* **175** (1928) 68–74.
[51] G. Losse, H. Weddige, *Angew. Chem.* **72** (1960) 323.
[52] G. Losse, H. Weddige, *Justus Liebigs Ann. Chem.* **636** (1960) 144–149.
[53] J. T. Thurston, L. P. Ferris: *Inorganic Synthesis*, vol. **3**, McGraw-Hill, New York 1950, pp. 45–47.
[54] Bayerische Stickstoff-Werke, DE 730 331, 1942 (K. Zieke, G. H. Hoffmann).
[55] W. Sauermilch, *Explosivstoffe* **12** (1965) no. 9, 197–199.
[56] P. Fanti, H. Silbermann, *Justus Liebigs Ann. Chem.* **467** (1928) 274–284.
[57] P. Adams et al., *J. Org. Chem.* **18** (1953) 934–940.
[58] C. H. Andrewes et al., *Proc. R. Soc. (London)* Ser. **B 133** (1946) 20–62.
[59] S. Birtwell et al., *J. Chem. Soc.* 1946, 491–494.
[60] P. Chabrier et al., *C. R. Acad. Sci.* **235** (1952) 64–66.
[61] E. B. Vliet: *Organic Synthesis*, Collective volume 1, J. Wiley & Sons, New York 1941, pp. 203–204.
[62] Amer. Cyanamid, US 2 357 181, 1944 (H. E. Faith, P. S. Winnek).
[63] Farbenfabriken Bayer, DE 1 002 330, 1957 (W. Hagge et al.).
[64] Amer. Cyanamid, US 3 300 281, 1967 (D. R. May).
[65] Amer. Cyanamid, DE 1 568 032, 1966 (D. R. May).
[66] Cyanamide Products Bulletin, Südd. Kalkstickstoff-Werke, Federal Republic of Germany, 1973, pp. 44–46.
[67] SKF, DE 2 344 779, 1974; US 3 876 647, 1975 (G. J. Durant et al.).
[68] *Chem. Week* **121** (1977) Aug. 24, no. 8, 23.
[69] Südd. Kalkstickstoff-Werke, FR 1 492 380, 1966.
[70] Y. Shulman, G. Nir, L. Fanberstein, S. Lavee, *Sci. Hortic. (Amsterdam)* **19** (1983) 97–104.
[71] Chem. Fabr. Pfersee, DE 1 906 389, 1969 (W. Bernheim, H. Deiner).
[72] S. J. O'Brien, *Test. Res. J.* **38** (1968) no. 3, 256–266.
[73] Monsanto, US 4 000 191, 1976; US 4 040 780, 1977 (A. V. Garner).
[74] Amer. Cyanamid assignee, JP-Kokai 47 974, 1982.
[75] H. Prietzel, *Melliand Textilber.* **52** (1971) 950–953.
[76] Shell Oil, US 4 086 175, 1978 (L. Kravetz et al.).Shell International, GB 1 573 144, 1980 (L. Kravetz et al.).

[77] Südd. Kalkstickstoff-Werke, DE 2 031 720, 1972 (H. Prietzel).
[78] Bayer AG, DE 1 619 047, 1967.
[79] Amer. Cyanamid, US 4 273 686, 1981 (J. Noland, M. Hajek).
[80] J. Haag, *Justus Liebigs Ann. Chem.* **122** (1862) 22.
[81] L. Kahovec, K. W. Kohlraush, *Z. Phys. Chem. (Leipzig)* **193** (1944) 188–195.
[82] E. W. Hughes, *J. Am. Chem. Soc.* **62** (1940) 1258–1267.
[83] H. C. Hetherington, J. M. Braham, *Ind. Eng. Chem.* **15** (1923) 1063.
[84] Amer. Cyanamid Co., US 2 357 261, 1944 (D. W. Kaiser).
[85] Du Pont, GB 434 961, 1935.
[86] Amer. Cyanamid, US 2 371 112, 1945.
[87] "Aero Dicyanamide," Technical Bulletin, Amer. Cyanamid, Wayne, N.J., 1970.
[88] J. Thiele, E. Uhfelder, *Justus Liebigs Ann. Chem.* **303** (1898) 108.
[89] ICI, GB 577 843, GB 581 346, 1946 (F. H. S. Curd, F. L. Rose).
[90] J. H. Paden, K. C. Martin, R. C. Swain, *Ind. Eng. Chem.* **39** (1947) 952.
[91] C. Hasegawa, *J. Soc. Chem. Ind. (Jpn.)* **45** (1942) 416.
[92] T. C. Ponder, *Hydrocarbon Process.* **48** (1969) no. 11, 200.
[93] P. Ostrogovich, *Atti Accad. Lincei* **20** (1911) no. 1, 249. W. Zerwick, W. Brunner (Alien Property Custodian), US 2 302 162, 1942.
[94] F. Pohl, *J. Prakt. Chem.* **77** (1908) no. 2, 533.
[95] I.G. Farbenind., DE 590 175, 1933 (L. Taub, W. Kropp).
[96] Südd. Kalkstickstoff-Werke, DE 1 758 250, 1968.
[97] C. D. Garby, *Ind. Eng. Chem.* **17** (1925) 226.
[98] R. N. Harger, *Ind. Eng. Chem.* **12** (1920) 1107. E. Johnson, *Ind. Eng. Chem.* **13** (1921) 533.
[99] F. Wolf, D. Spiethoff, *Melliand Textilber.* **48** (1967) no. 12, 1456–1460.
[100] Hercules, US 4 380 603, 1981; US 4 383 077, 1983 (R. A. Bankert).
[101] Le Page Glue Co., US 2 282 364, 1942 (W. G. Kunze, R. B. Evans).Amer. Cyanamid, US 2 581 111, 1951 (C. G. Landers, J. Studeny).
[102] J. W. Lyons: *The Chemistry and Uses of Fire Retardants,* J. Wiley & Sons, New York 1970, p. 136. Amer. Cyanamid, US 3 625 753, 1969 (S. J. O'Brien, R. G. Weyker).
[103] Koppers Co., US 3 159 503, 1964 (I. S. Goldstein, W. A. Dreher).
[104] Ciba-Geigy, DE 2 942 788, 1980 (P. Rohringer, H. Wegmueller).
[105] I. J. Cummings, *Ind. Eng. Chem.* **46** (1954) 1985.
[106] Dixon International Ltd., US 4 210 725, 1980 (C. A. Redfard).
[107] P. Eyerer, *J. Appl. Polym. Sci.* **15** (1971) no. 12, 3067–3088.
[108] M. Fedtke, K. Bieraegel, *Plaste Kautsch.* **28** (1981) 253–255.
[109] Ciba-Geigy, DE 2 017 114, 1970 (K. T. Shen, G. W. Jarvis).
[110] R. Schlimper, *Plaste Kautsch.* **10** (1963) 19–21.
[111] General Aniline & Film, US 2 536 983, 1951 (E. M. Owen).
[112] Soc. Mussy, DE 2 124 757, 1971.
[113] F. Koelsch, *Zentralbl. Gewerbehyg. Unfallverhüt.* **4** (1916) 113.

Cyanuric Acid and Cyanuric Chloride

NORBERT KRIEBITZSCH, Degussa-Antwerpen N.V., Antwerp, Belgium

HERBERT KLENK, Degussa AG, Wolfgang–Hanau, Federal Republic of Germany

1.	Introduction	1767	3.	Cyanuric Chloride	1773
2.	Cyanuric Acid	1768	3.1.	Physical Properties	1773
2.1.	Physical Properties	1768	3.2.	Chemical Properties	1774
2.2.	Chemical Properties	1768	3.3.	Production	1775
2.3.	Production	1770	3.4.	Environmental Protection	1776
2.4.	Environmental Protection	1771	3.5.	Quality Specifications and Chemical Analysis	1776
2.5.	Quality Specifications and Chemical Analysis	1772	3.6.	Storage and Transportation	1777
2.6.	Storage and Transportation	1772	3.7.	Uses	1777
2.7.	Uses	1772	3.8.	Economic Aspects	1778
2.8.	Economic Aspects	1773	3.9.	Toxicology and Occupational Health	1778
2.9.	Toxicology and Occupational-Health	1773	4.	References	1779

1. Introduction

Many acyclic compounds containing C–N double or triple bonds tend to condense to form six-membered rings with alternating C and N atoms:

$$3\ HO-C\equiv N \longrightarrow \mathbf{1a} \rightleftharpoons \mathbf{1b}$$

These derivatives are called s-triazines or 1,3,5-triazines. A large number of their reactions are described in the literature, mostly substitution reactions with ring retention. (For aromaticity of the triazine ring, see [6, p. 466].)

Compound **1** is commonly called cyanuric acid, even though the oxo (keto) structure **1b** is dominant over the hydroxy (enol) structure **1a**. Most, by far, of its substitution reactions give derivatives with structure **1b** (isocyanurates, e.g., **2–6**). Triazine compounds of structure **1a** (cyanurates, e.g., **9**) are made from cyanuric chloride (**8**). Compounds of industrial importance of both types are shown in Table 1.

Compounds that are mixtures of tautomers are known, although the keto structure is generally more stable than the hydroxy structure. Some alkyl cyanurates rearrange to give the corresponding isocyanurates [1], [7], [8]. Cyanuric acid and cyanuric chloride are easily interconvertible [9], [10]:

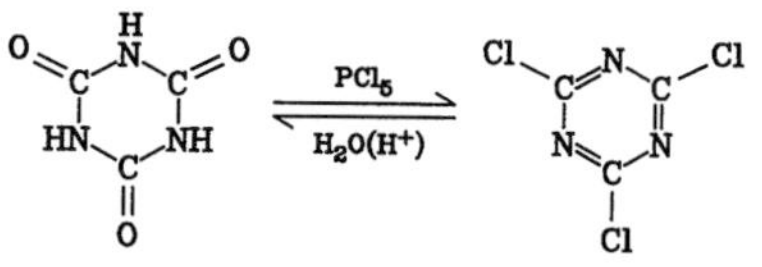

Although cyanuric acid and cyanuric chloride have been known since 1776 and 1827, respectively, not until the 1950s did they gain industrial importance, mainly in the production of disinfectants and herbicides.

2. Cyanuric Acid

2.1. Physical Properties

Cyanuric acid (isocyanuric acid), 1,3,5-triazine-2,4,6(1H,3H,5H)-trione [*108-80-5*] (**1**), $C_3H_3N_3O_3$, M_r 129.8, *mp* 320–330 °C (decomp.), is an odorless white solid, d_0^4 = 1.8. Solubility in water is 0.15 wt% at 10 °C and 2.6 wt% at 90 °C. The compound is only slightly soluble in common organic solvents except those with a high dipole moment (17 wt% in dimethyl sulfoxide, 7 wt% in dimethylformamide at 25 °C).

Cyanuric acid crystallizes from water as a colorless crystalline dihydrate, which loses water in dry air. Cyanuric acid is a weak tribasic acid (K_a's ca. 10^{-7}, 10^{-11}, and 10^{-14} [3, D1]). The solubilities of the mono-, di-, and trisodium salts in water are 0.9, 5.7, 14.1 g per 100 mL of solution, respectively [11]. For other physical data, see [1] and [3, D1].

2.2. Chemical Properties

Only a few reactions of cyanuric acid occur by replacement of hydrogen at oxygen: in alkaline solution the hydroxy structure is favored over the keto structure. Many metal and organic salts are known [1]. The reaction with ammonia produces melamine [*108-78-1*] [12]. In the presence of alkaline catalysts, ketene reacts with cyanuric acid to afford triacetyl cyanurate [*13483-16-4*] [13].

The typical reactions of cyanuric acid are those of a cyclic imide; they usually result in trisubstitution.

Table 1. Triazine derivatives

Name	Abbreviation	CAS Registry Number	Structure
Cyanuric acid		[108-80-5]	1
Trichloro isocyanurate	TCCA	[87-90-1]	2
Triallyl isocyanurate	TAIC	[1025-15-6]	3
Tris(2-hydroxyethyl) isocyanurate	THEIC	[839-90-7]	4
Triglycidyl isocyanurate	TGIC	[2451-62-9]	5
Tris(2-carboxyethyl) isocyanurate	TCEIC	[2904-41-8]	6
Trisodium cyanurate		[3047-33-4]	7
Cyanuric chloride	CC	[108-77-0]	8
Triallyl cyanurate	TAC	[101-37-1]	9
Chlorobis(ethylamino)triazine		[122-34-9]	10
Bis(triazinylamino)stilbenedisulfonic acids			11
Procion Brilliant Orange M-G		[6522-74-3]	12

Epoxides undergo addition to form hydroxy-alkyl isocyanurates when heated in dimethylformamide [14]. Epichlorohydrin forms tris-(chlorohydroxypropyl) isocyanurate [*7423-53-2*]in dioxane at 110 °C in the presence of basic catalysts [15]; its dehydrochlorination in aqueous alkali produces triglycidyl isocyanurate (**5**) [16].

A typical addition reaction of cyanuric acid across a double bond is the formation of bis(cyanoethyl) isocyanurate [*2904-27-0*] and tris(cyanoethyl) isocyanurate [*2904-28-1*] by reaction with acrylonitrile in dimethylformamide containing Triton B (a surfactant) at 130 °C [17]. Saponification of tris(cyanoethyl) isocyanurate gives tris(2-carboxyethyl) isocyanurate (**6**) [17].

Cyanuric acid dissolves readily in aqueous formaldehyde [11], [18] with formation of tris(hydroxymethyl) isocyanurate [*10471-40-6*].

Chlorine-containing compounds react with cyanuric acid at elevated temperature in the presence of a proton acceptor with elimination of HCl. Thus, allyl chloride forms triallyl isocyanurate (**3**) in dichlorobenzene–trimethylamine at 130 °C [19], [20]; 2-chloroethanol yields tris(2-hydroxyethyl) isocyanurate (**4**) in aqueous NaOH [21]. Chlorine reacts with cyanuric acid in water at controlled pH to afford *N*-dichloro and *N*-trichloro compounds (**2**).

Cyanuric acid decomposes slowly above 200 °C and rapidly above its melting point (320–330 °C). The main product is isocyanic acid [*75-13-8*] [22], [23]. At higher temperature, some reactions are likely to proceed via a cyanic acid intermediate; i.e., they involve ring cleavage. Another example of ring cleavage is the conversion of long-chain fatty acids into nitriles above 250 °C [24].

2.3. Production

Laboratory quantities of pure cyanuric acid are produced by hydrolysis of cyanuric chloride [10] or melamine [25]. The acid is further purified by recrystallization from dimethylformamide or via the sodium or ammonium salts.

In commercial production *urea* is decomposed between 200 and 300 °C, affording ammonia and cyanuric acid:

$$3 \text{ OC(NH}_2)_2 \longrightarrow \text{(HOCN)}_3 + 3 \text{ NH}_3$$

In some processes, urea is heated in a kiln for several hours. The process starts with molten raw material and ends with solid cyanuric acid. Numerous patents, therefore, deal with methods to improve heat and mass transfer with specially designed reactors or with a modified reaction mixture. The reaction can be carried out in heated screws, revolving drums [26], steel conveyor belts [27], rotary-tube furnaces [28], or ball mills [29]. Pyrolysis in a fluidized bed of AlF_3 particles [30] or cyanuric acid crystals [31] and heating of molten urea by high-frequency induction have also been reported [32]. The caking problem is solved by premixed powders with definite urea:cyanuric acid ratios [28], or preformed urea cyanurate [31], [33], or by recycling the crude product [28].

Other methods use molten metals [34], [35] or salts [36] as the heat-transfer medium. Urea is fed into the bottom of a crucible filled with agitated molten tin or tin alloy, and the cyanuric acid collects at the metal surface.

Other methods employ solvents with high boiling points, such as polyglycol ethers [37] and sulfolanes [38], that dissolve urea but not cyanuric acid. These methods avoid localized overheating, which reduces yields by formation of isocyanic acid. Other solvents are reported [3, D1, p. 369].

Byproduct formation is suppressed by rapidly removing the ammonia at reduced pressure, sparging with inert gas, or codistillation.

Crude cyanuric acid contains up to 30% impurities consisting mainly of melamine and its precursors:

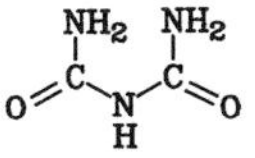

Biuret
[108-19-0]

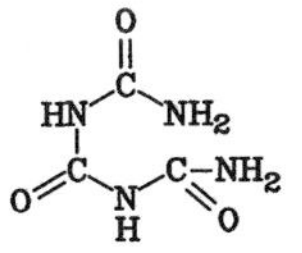

Triuret
[556-99-0]

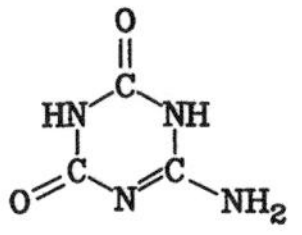

Ammelide
[645-93-2]

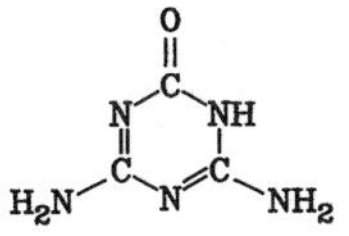

Ammeline
[645-92-1]

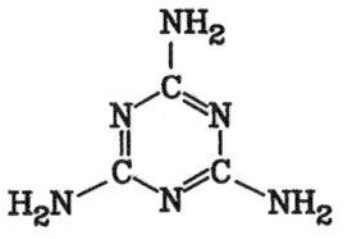

Melamine
[108-78-1]

Dilute mineral acids convert the impurities to ammonia and carbon dioxide or cyanuric acid [28], [39]. Another technique of purification is to dissolve the crude acid in NaOH or dilute aqueous ammonia, to filter the organic impurities off, and to precipitate the acid by adding mineral acids [40], [41].

2.4. Environmental Protection

Cyanuric acid biodegrades readily under anaerobic conditions [42], in the soil [43], and in aerated systems [44], [45]. The principal environmental impact comes from salt in the effluents resulting from the manufacture of cyanuric acid and its derivatives.

2.5. Quality Specifications and Chemical Analysis

Manufacturers' specifications are in the following range:

cyanuric acid	> 98.5 to > 99%
ammelide + ammeline	0 – 0.5%
moisture	0.1 – 1%
ash	50 – 1000 ppm
pH	4 – 5 (0.25% solution or 10% slurry)

Thin-layer chromatography on cellulose is suitable for quality control. Cyanuric acid is determined gravimetrically as its melamine salt [46] or by potentiometric titration with tetrabutylammonium hydroxide [47], [48]. The acid and the isocyanurates are determined by nitrogen analysis (Kjeldahl or Dumas).

2.6. Storage and Transportation

Cyanuric acid should be stored in a dry place. Silos and pneumatic conveyors should be grounded to prevent electrostatic discharges. The loose bulk density is 800 – 900 g/L. The product is packed in polyethylene-lined fiber drums or paper bags. The shipping classification in the United States is "Acid, NOIBN, Dry" (NOIBN, not dangerous according to DOT CFR Title 49).

2.7. Uses

Cyanuric acid is widely used for the manufacture of N-chlorinated isocyanurates, which are employed in swimming pool disinfectants, bleaches, cleansers, and sanitizers. Cyanuric acid serves as a stabilizer of available chlorine in swimming pools. In concentrations of 20 – 40 ppm, it reduces the deactivation rate of chlorine by a factor of 5 – 10 [49]. Cyanuric acid is approved in ruminant feed as a source of nitrogen [50], [51]. The use of cyanuric acid as a source of isocyanic acid by heating to 450 – 650 °C has been claimed [52].

Isocyanurates are used in the plastics industry [11], [53]. Triallyl isocyanurate (TAIC; **3**) can be homopolymerized and copolymerized. It serves as a crosslinking agent in polyethylene and poly(vinyl chloride) and in laminate formulations [54], [55]. Tris-(2-hydroxyethyl) isocyanurate (THEIC; **4**) is a cross-linking agent for polyurethanes, polyesters, and alkyd resins used for wire enamels and electrical varnishes [56]. Triglycidyl isocyanurate (TGIC; **5**) is a component in epoxy resins [15], [57]. Tris-

(2-carboxyethyl) isocyanurate (TCEIC; **6**) is used in making water-soluble alkyd resins; the esters serve as plasticizers for poly(vinyl chloride) and lubricants [58], [59].

2.8. Economic Aspects

World capacity for cyanuric acid is estimated to be 50 000 t/a (1981). More than 90 % is used to make N-chlorinated isocyanurates. In France, cyanuric acid is manufactured by CdF; in Japan by Nissan Chemical Industries and Shikoku Kasei Co.; in the United Kingdom by Chlor-chem. Ltd.; and in the United States by FMC, Monsanto, and Olin. It is also manufactured in the Soviet Union. All major producers also make the chlorinated derivatives.

2.9. Toxicology and OccupationalHealth

Cyanuric acid has been classified as essentially nontoxic (acute oral and dermal), nonirritating to the skin, and essentially nonirritating to the eye [11], [60]; LD_{50} (rat, oral), 5000 mg/kg [61]. Inhalation of the dust, contact with the eye, and ingestion should be avoided.

Above 200 °C, cyanuric acid forms toxic isocyanic acid.

3. Cyanuric Chloride

3.1. Physical Properties

Cyanuric chloride, 2,4,6-trichloro-1,3,5-triazine [*108-77-0*] (**8**), $C_3N_3Cl_3$, M_r 184.5, triple point 145.7 °C at 255 kPa; bp 194 °C, forms colorless monoclinic crystals of pungent odor, reminiscent of acetamide and acid chlorides; densities are $d_4^{20} = 1.92$ (solid) and $d_4^{150} = 1.48$ (melt). The compound is soluble in acetonitrile, ether, ketones, and chlorinated hydrocarbons, but insoluble in water. Heat capacity c_p is 0.99 kJ kg^{-1} K^{-1} at 150 °C, enthalpy of fusion is 123 kJ/kg, and enthalpy of vaporization is 256 kJ/kg. The vapor pressure (kPa) of the liquid as a function of temperature (°C) is $p = 0.3918 \cdot \exp 0.02866\, t$. Cyanuric chloride sublimes; the vapor pressure of the solid is $p = 0.00492 \cdot \exp 0.0587\, t$.

3.2. Chemical Properties

Under anhydrous conditions, cyanuric chloride can act as a chlorinating agent. Alcohols and tertiary amines are converted to alkyl chlorides [62], [63].

$$\text{(triazine-Cl)} + R_3N \longrightarrow \text{(triazine-NR}_2\text{)} + RCl$$

Carboxylic acids form acid chlorides in anhydrous acetone in the presence of triethylamine. The acid chloride can be isolated or treated in situ to give the ester, amide, etc. [64].

Cyanuric chloride usually reacts like an acid chloride with formation of hydrogen chloride.

$$\text{(triazine-Cl)} + 2\, R_2NH \longrightarrow \text{(triazine-NR}_2\text{)} + R_2NH \cdot HCl$$

A suspension of cyanuric chloride in ice water remains fairly stable for ca. 12 h. At 30 °C more than 40 % of the compound is hydrolyzed within 1 h to cyanuric acid. In an alkaline medium, hydrolysis stops at dihydroxychlorotriazine, even with excess NaOH. Trisodium cyanurate forms above 125 °C.

A number of processes take advantage of the temporary stability of cyanuric chloride in water by reacting a suspension of finely divided cyanuric chloride in ice water with a strong nucleophilic agent such as an alcohol, a mercaptan, or a primary or secondary amine [65]. These reactions often proceed stepwise to trisubstitution at a well-defined temperature. An empirical rule for amine substituents states that the first chlorine is replaced at 0–5 °C, the second at 30–50 °C, and the third at 70–100 °C [66]. Sodium hydroxide, sodium hydrogen carbonate, disodium hydrogen phosphate, and tertiary amines can be used as HCl scavengers (also see [6, p. 483]).

$$\text{(triazine-Cl)} + HNR_2 + NaOH \longrightarrow \text{(triazine-NR}_2\text{)} + NaCl + H_2O$$

The triazine product is usually isolated by filtration or centrifugation.

Solvents such as acetone, methyl ethyl ketone, or toluene are often used for these reactions. Cyanuric chloride, molten or powder, is dissolved or suspended in the solvent with the nucleophile and treated with aqueous NaOH. The product is filtered or is isolated from the organic layer by evaporation.

Both methods are used to introduce different substituents successively at different temperatures. Replacement of the third chlorine is sometimes carried out by employing high-boiling solvents or simply the molten reactant.

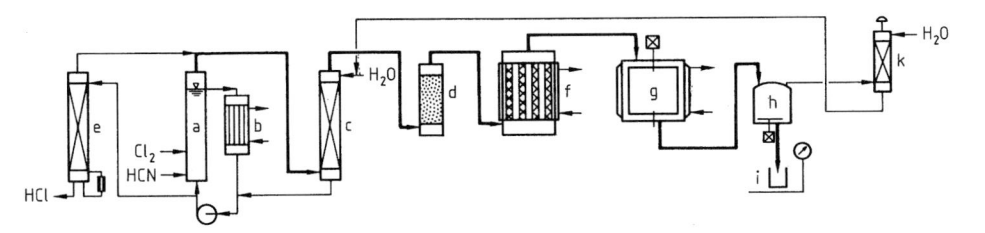

Figure 1. Production of cyanuric chloride from chlorine and hydrocyanic acid
a) ClCN reactor; b) Heat exchanger; c) Scrubber; d) Dryer; e) Effluent stripper; f) Trimerizer; g) Condenser; h) Storage; i) Drum; k) Tail gas scrubber

Care must be taken to neutralize the liberated hydrochloric acid and to dissipate the heat of reaction. Otherwise, the exothermal hydrolysis of cyanuric chloride (ΔH = -2164 kJ/kg) can become uncontrollable, especially in the manufacture of cyanurates from alcohols [67]. Hydrolysis is accelerated by water-miscible solvents. Storage of solutions of cyanuric chloride in aqueous acetone without heat removal can result in a run-away reaction, even starting at room temperature.

The reactions of cyanuric chloride are surveyed in [68].

3.3. Production

Most cyanuric chloride facilities are found at plants producing hydrocyanic acid, the key starting material. Sodium cyanide is a minor source.

The *hydrocyanic acid* is chlorinated to cyanogen chloride, which is trimerized directly to cyanuric chloride:

$$HCN + Cl_2 \longrightarrow ClCN + HCl \qquad \Delta H = -89 \text{ kJ/mol}$$
$$3 \text{ ClCN (g)} \longrightarrow (ClCN)_3 \text{ (g)} \qquad \Delta H = -233 \text{ kJ/mol}$$

In a typical process, hydrocyanic acid and chlorine are added to a reaction loop between 20 and 40 °C (see Fig. 1). Cyanogen chloride saturates the aqueous reaction medium and leaves the loop (a and b) [69]. It is washed with water (c) and dried (d) and is then ready for trimerization.

The wash water from c dissolves the hydrogen chloride and delivers it to a stripper (e), which recycles dissolved cyanogen chloride and releases hydrochloric acid. Sections e, a, and c can also be combined in a single apparatus [70].

The dry cyanogen chloride is trimerized $\geq$ 300 °C on activated charcoal (f) [71], [72]. The cyanuric chloride vapors are condensed to molten or solid product (g) [73], [74], which is dissolved in a solvent for captive use or filled from a hopper (h) into containers (i). Tail gases are scrubbed and recycled (k).

The ClCN yield in this process exceeds 95 %; the (ClCN)$_3$ yield exceeds 90 %.

A number of processes eliminate the formation of dilute hydrochloric acid during cyanogen chloride manufacture: a one-step gas-phase reaction of hydrocyanic acid and chlorine on charcoal [75], [76] or in tetrachloroethane [77], and a one-step reaction of cyanogen and chlorine on charcoal [78].

Catalysts for cyanogen chloride trimerization other than activated carbon include fused cyanuric chloride [79], [80], molten aluminum chloride [81], eutectic mixtures of tetrachloroaluminates [82], aluminum silicates [83], and zeolites [84] doped with metal oxides or sulfides [85].

In some of these processes, "tetrameric cyanogen chloride" (2,4-dichloro-6-isocyanodichloro-5-triazine) [877-83-8] is a byproduct; its conversion into cyanuric chloride is catalyzed by iron oxide [86], [87].

In the laboratory, ClCN in benzene or chloroform solution trimerizes in the presence of hydrogen chloride [88]. The use of the hydrogen chloride–dimethyl ether azeotrope on a commercial scale has been patented [89].

Equipment for cyanogen chloride production is lined with glass or fluoro polymers, or made of graphite or special resins. For cyanuric chloride, nickel, stainless steels, or aluminum is used. Moisture causes severe corrosion.

3.4. Environmental Protection

Normally the only effluent is dilute hydrochloric acid from cyanogen chloride production. This acid is usually kept below the concentration of the H_2O–HCl azeotrope, since the unwanted hydrolysis of ClCN depends on the concentration of the acid. Uses for this weak acid are very limited; appreciable amounts are neutralized with caustic soda or limestone, or in ion exchangers.

Residual gases from the trimerization are washed or condensed to recycle excess chlorine and unreacted ClCN. The tail gases consist of nitrogen and carbon dioxide; when they contain chlorine, they can be used to make bleach (NaOCl).

Spilled molten cyanuric chloride is cooled by covering it with foam to prevent sublimation. The solid is collected and hydrolyzed to cyanuric acid. Fumes from venting of molten cyanuric chloride tanks are scrubbed with the anhydrous solvent used in the consumer's process.

3.5. Quality Specifications and Chemical Analysis

The industrial product contains 99 % cyanuric chloride (minimum) and less than 1 % insolubles. In many applications, cyanuric chloride is in suspension, and thus, particle size is important. To facilitate reaction and avoid dust formation, a granule size between 10 and 160 μm (> 80 mesh Tyler) is optimal. Different sieving methods give different results because of abrasion or agglomeration.

To test reactivity, cyanuric chloride and an amine are reacted under standard conditions and the unchanged material is determined.

Hydrolysis caused by improper handling may result in contamination with cyanuric acid, which is found as insoluble matter in dried solvents such as toluene.

Cyanuric chloride can be assayed via the melamine complex by IR spectroscopy after hydrolysis of a chloroform extract [90], or by titration of excess reagent after reaction with morpholine [91].

3.6. Storage and Transportation

The solid is shipped in drums, with or without polyethylene bag liners, or in bulk bins. Free flowing cyanuric chloride is also transported in fluidized bulk trailers; molten product is shipped in heated tank trailers.

Cyanuric chloride should be stored in strictly dry conditions. The solid should be stored in a cool place to avoid an increase in particle size. Tank farms for molten product must have a well-designed heat-tracing system that includes the vent and relief lines.

The following transportation regulations apply: IMDG/ICAO-TI, class 8; UN no. 2670 PG III for solid, 2922 for liquid; ADR/RID, class 8 marg. 27 c; EEC no. 613-009-00-5; CFR 49, corrosive solid, corrosive liquid, poisonous, N.O.S.

3.7. Uses

The most important cyanuric chloride derivatives are the aminotriazines.

Alkylaminotriazines are used as pesticides, especially herbicides. The triazines are among the largest selling herbicides; most of them are 2-chloro- or 2-methylthio-4,6-dialkylaminotriazines. Simazine (**10**) is one of the oldest.

Reaction products of aminostilbenes with substituted triazines are used as brightening agents for fabrics and paper, especially bis(triazinylamino)stilbenedisulfonic acids (**11**).

Dye derivatives prepared from cyanuric chloride react chemically with fabrics. The Procion dyes (**12**) were among the first.

Cyanuric chloride and cyanurates are used for cross-linking. Cyanuric chloride is used in gelatin and glues [92]; triallyl cyanurate (**9**) is used in the rubber and plastics industry (→ Allyl Compounds) [93]. Other uses of triazines include modifiers, accelerators, and UV stabilizers.

Table 2. Manufacturers of cyanuric chloride

Country	Company
Belgium	Degussa
Federal Republic of Germany	Degussa
	SKW Trostberg
Japan	Kyowa Gas
	Musashino
Romania *	
Soviet Union *	
Switzerland	Lonza
United States	Ciba-Geigy *
	Degussa

* Captive use only.

3.8. Economic Aspects

Worldwide annual production exceeds 100 000 t/a; capacities corresponding to demand. Producers are listed in Table 2. The capacity of production units ranges from hundreds to several thousand tons per month.

Approximately 80% of production is used for pesticides, especially herbicides. Triazine herbicides are produced all over the world. More than 10% of the production is converted into optical brighteners. Approximately 2% is used for reactive dyes and certain anthraquinone dyes. Applications based on cross-linking chemicals like triallyl cyanurate (**9**) are increasing in interest.

3.9. Toxicology and Occupational Health

Cyanuric chloride strongly irritates the skin, mucous membranes, including those of the eye, and the respiratory and gastrointestinal tracts [61, pp. 2763–2765]. The 1-min threshold irritation effect on mucous membranes is produced by 0.3 mg/m^3. Allergic reactions are possible.

Direct contact with cyanuric chloride should be avoided; ventilation in the work place is necessary. A full-face gas mask with an active carbon canister or a self-contained breathing apparatus must be used during exposure to vapor and dust.

Funnels for adding cyanuric chloride powder to flammable solvents must be grounded.

4. References

General References

[1] E. M. Smolin, L. Rapoport in A. Weissberger (ed.): *The Chemistry of Heterocyclic Compounds*, vol. **13**, Interscience, New York 1959.
[2] E. J. Modest in R. C. Elderfield (ed.): *Heterocyclic Compounds*, vol. **7**, J. Wiley & Sons, New York 1961, pp. 627–719.
[3] *Gmelin*, system no. **14**, Carbon, D 1, 366–381; **D 3,** 272–287.
[4] *Kirk-Othmer*, **7,** 397–410.
[5] *Beilstein*, **E III/II, 26,** 632 ff.
[6] J. M. E. Quirke in A. J. Boulton, A. McKillop (eds.): *Comprehensive Heterocyclic Chemistry*, vol. **3**, Pergamon Press, Oxford 1984, pp. 457–530.

Specific References

[7] M. L. Tosato, *J. Chem. Soc. Perkin Trans. 2* 1979, 1371.
[8] L. Paoloni, M. L. Tosato, M. Cignitti, *J. Heterocycl. Chem.* **5** (1968) 533.
[9] F. Beilstein, *Justus Liebigs Ann. Chem.* **116** (1860) 357–358.
[10] F. Zobrist, H. Schinz, *Helv. Chim. Acta* **35** (1952) 2387.
[11] FMC Chemicals: *Cyanuric Acid*, Product Promotion Bulletin 10 B, 1965.
[12] American Cyanamid, GB 598 175, 1948.
[13] FMC, US 3 318 888, 1967 (J. H. Blumbergs, D. G. MacKellar).
[14] R. W. Cummins, *J. Org. Chem.* **28** (1963) 85–89.
[15] Devoe & Raynolds, US 2 809 942, 1957 (H. G. Cooke).
[16] Ciba-Geigy, US 3 547 918, 1967 (D. Porret, K. Metzger, A. Heer).
[17] Th. C. Frazier, E. D. Little, B. E. Lloyd, *J. Org. Chem.* **25** (1960) 1944–1946.
[18] Z. N. Pasenko, L. I. Chovnik, *Ukr. Khim. Zh.* **30** (1963) no. 2, 195–198; *Chem. Abstr.* **61** (1964) 1866.
[19] Allied: *Triallyl Isocyanurate*, Product Bulletin.
[20] FMC, US 3 075 979, 1963 (J. J. Tazuma, R. Miller).
[21] A. A. Sayigh, H. Ulrich, *J. Chem. Soc.* 1961, 3148–3149.
[22] F. W. Hoover, H. B. Stevenson, H. S. Rothrock, *J. Org. Chem.* **28** (1963) 1825–1830.
[23] M. Linhard, *Z. Anorg. Allg. Chem.* **236** (1938) 200–208.
[24] Sun Chem., US 2 444 828, 1948 (W. Kaplan).
[25] G. Ostrogovich, R. Bacaloglu, E. Fliegl, *Tetrahedron* **24** (1968) 2701–2705.
[26] Allied, US 3 093 641, 1963 (R. L. Formaini).
[27] BASF, GB 1 042 174, 1962 (O. B. Claren, L. Unterstenhoefer).
[28] FMC, US 2 943 088, 1959 (R. H. Westfall).
[29] Toa Gosei Chem. Ind. Co., JP 27 182, 1963; *Chem. Abstr.* **64** (1966) 9748.
[30] Lentia, DE-AS 1 179 215, 1961 (A. Schmidt, F. Weinrotter, W. Muller).
[31] Allied, US 3 394 136, 1966 (W. P. Moore, D. E. Elliott).
[32] Nippon Soda Co., EP 0 012 773, 1978 (H. Kizawa, R. Ichihara, T. Yao, I. Kikuchi).
[33] Monsanto, US 3 154 545, 1964 (W. F. Symes, S. Vazopolos).
[34] BASF, DE-AS 1 065 420, 1958 (O. B. Claren).
[35] Shikoku Kasei Kogyo Co., US 3 275 631, 1966 (H. Yanagizawa).
[36] Stamicarbon, US 4 112 232, 1978 (M. G. R. T. de Cooker).

[37] Dow Chemical, US 3 810 891, 1974 (J. M. Lee).
[38] FMC, US 3 563 987, 1971 (S. Berkowitz).
[39] Allied, US 3 296 262, 1967 (H. F. Scott).
[40] Monsanto, US 3 357 979, 1967 (E. C. Sobocinski, W. F. Symes).
[41] Allied, US 3 172 886, 1965 (I. Christoffel, D. P. Schulz).
[42] J. Saldick, *Appl. Microbiol.* **28** (1974) no. 6, 1004–1008.
[43] D. C. Wolf, J. P. Martin, *J. Environ. Qual.* **4** (1975) no. 1, 134–139.
[44] H. L. Jensen, A. S. Abdel-Ghaffar, *Arch. Microbiol.* **67** (1969) 1–5.
[45] Ciba-Geigy, US 4 274 955, 1978 (J. Zeyer, R. Hutter, P. Mayer).
[46] L. Nebbia, F. Guerrieri, B. Pagani, *Chim. Ind. (Milan)* **39** (1957) no. 2, 81–83; *Chem. Abstr.* **51** (1957) 7241.
[47] G. A. Harlow, C. M. Noble, G. E. A. Wyld, *Anal. Chem.* **28** (1956) 784–791.
[48] R. C. Cundiff, P. C. Markunas, *Anal. Chem.* **28** (1956) 792–797.
[49] FMC, US 2 988 471, 1961 (R. J. Fuchs, J. A. Lichtman).
[50] Feed Service Corp., US 2 808 332, 1957 (P. C. Anderson, J. L. C. Rapp).
[51] J. Kamlet, US 2 768 895, 1956 (J. Kamlet).
[52] Olin Mathieson, FR 1 328 696, 1962 (R. B. East).
[53] Monsanto: *Cyanuric Acid,* Bulletin no. IC/505/233.
[54] Boing Airplane Co., US 3 108 902, 1963 (J. R. Galli, R. B. Johnson).
[55] American Cyanamid, US 3 044 913, 1962 (L. A. Lundberg).
[56] Allied, US 3 088 948, 1963 (E. D. Little, B. T. Poon).
[57] Ciba-Geigy: *Merkblatt Araldit* PT 810, Publ. no. 24 384.
[58] Spencer Chemical Co., GB 912 563, 1961.
[59] H. Yanagizawa, *Plast. Ind. News* 1968, 16–18.
[60] E. Canelli, *Am. J. Public Health* **64** (1974) 155–162.
[61] C. F. Reinhardt, M. R. Brittelli in G. D. Clayton, F. E. Clayton (eds.): *Patty's Industrial Hygiene and Toxicology,* 3rd ed., vol. **2 A,** Wiley-Interscience, New York 1981, pp. 2765–2769.
[62] S. R. Sandler, *J. Org. Chem.* **35** (1970) 3967–3968.
[63] E. Kober, R. Raetz, *J. Org. Chem.* **27** (1962) 2509.
[64] K. Venkataraman, D. R. Wagle, *Tetrahedron Lett.* 1979, 3037.
[65] Degussa: *Cyanuric Chloride,* Product Bulletin, 1985.
[66] H. Fierz-David, M. Matter, *J. Soc. Dyers Colour.* **53** (1937) 424.
[67] Institution of Chemical Engineers: *Loss Prevention Bulletin,* 1979, no. 025.
[68] V. L. Mur, *Russ. Chem. Rev. (Engl. Transl.)* **33** (1964) 92–103.
[69] Degussa, US 2 672 398, 1950 (H. Huemer, H. Schulz, W. Pohl).
[70] Geigy Chem., US 3 197 273, 1961 (E. B. Trickey).
[71] American Cyanamid, US 2 491 459, 1945 (J. T. Thurston).
[72] Degussa, DE 842 067, 1950 (H. Huemer, H. Schulz).
[73] American Cyanamid, US 2 734 058, 1956 (H. Schulz, H. Huemer).
[74] Degussa, US 4 245 090, 1981 (R. Goedecke, M. Liebert, W. Nischk, W. Plötz, K. Puschner).
[75] Nitto, JP 2282-3, 1959.
[76] Monsanto, US 2 965 642, 1960 (E. E. Drott, G. D. Oliver).
[77] Saint Gobain, FR 1 251 359, 1959 (A. J. Courtier, H. Jean).
[78] Röhm & Haas, DE-AS 1 068 265, 1958 (H. Zima).
[79] Krupp, DE-AS 1 291 745, 1965 (W. Kirchhof, K. Schauerte); *Chem. Abstr.* **70** (1969) 106 573.
[80] Bayer, GB 718 806, 1954.
[81] ICI, GB 566 827, 1945 (T. P. Metcalfe).

[82] Toyama, GB 1 086 706, 1967.
[83] Nitto, JP 23 509, 1962 (N. Umehara, J. Ishikura).
[84] Nitto, JP 9 136, 1958.
[85] Bayer, DE 1 179 213, 1964 (A. v. Friedrich).
[86] Bayer, US 3 080 367, 1959 (A. v. Friedrich, P. Schmitz).
[87] Lonza, CH 396 020, 1966 (C. Zinsstag, R. Gentili).
[88] O. Diels, *Ber. Dtsch. Chem. Ges.* **32** (1899) 691.
[89] Lonza, US 2 838 512, 1958 (M. Teysseire, H. P. Sieber).
[90] R. Bacaloglu, E. Fliegl, G. Ostrogovich, *Fresenius' Z. Anal. Chem.* **257** (1971) 273–274; *Chem. Abstr.* **76** (1972) 67 940.
[91] Y. V. Lyande, V. V. Savenkova, A. A. Cherkasskii, *Zh. Anal. Khim.* **30** (1975) no. 9, 1817–1821; *Chem. Abstr.* **84** (1976) 53 629.
[92] E. Zerlotti, *Nature (London)* **214** (1967) 1304–1306.
[93] Degussa, *Triallylcyanurat,* Product Bulletin, 1980.

Cyclododecanol, Cyclododecanone, and Laurolactam

Hans Rademacher, Hüls AG, Marl, Federal Republic of Germany

1.	Cyclododecanol and Cyclododecanone 1783	1.3.	Uses	1786
		2.	Laurolactam	1786
1.1.	Properties 1784	3.	Toxicology	1787
1.2.	Production 1785	4.	References	1787

1. Cyclododecanol and Cyclododecanone

Commercial production of cyclododecanol [1724-39-6] (**3**) and cyclododecanone [830-13-7] (**4**) began when Wilke et al. [1], [2] found an efficient synthesis for 1,5,9-cyclododecatriene [4904-61-4] (**1**) by cyclotrimerization of butadiene [106-99-0]. Made by Du Pont in the United States and Hüls in Europe, world production capacity for cyclododecanol and cyclododecanone is ca. 10^4 t/a. Cyclododecanol and cyclododecanone are used primarily to synthesize dodecanedioic acid [693-23-2] (**5**) and azacyclotridecan-2-one [947-04-6] (laurolactam, **8**).

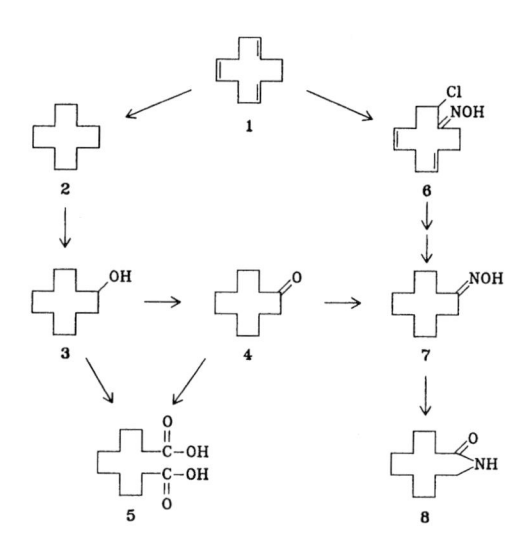

1.1. Properties

At room temperature, both cyclododecanol, $C_{12}H_{24}O$, and cyclododecanone, $C_{12}H_{22}O$, are colorless crystals that are soluble in organic solvents but only slightly soluble in water (0.004 wt% at 25 °C). Cyclododecanol is slightly hygroscopic and smells like camphor; cyclododecanone smells like peppermint. Table 1 lists physical properties for both.

Cyclododecanol dehydrates, esterifies, oxidizes, dehydrogenates, and hydrogenates like an aliphatic secondary alcohol. When dehydrated in dilute sulfuric acid or in the gas phase with catalysts, cyclododecanol yields cyclododecene [1501-82-2]. When treated with organic acids or anhydrides, cyclododecanol forms esters. Oxidation of cyclododecanol with nitric acid gives dodecanedioic acid (5). Liquid-phase dehydrogenation of cyclododecanol at ca. 230 °C yields cyclododecanone, and hydrogenation of cyclododecanol produces cyclododecane [294-62-2] (2).

In most cases, cyclododecanone reacts like an aliphatic ketone [3]. Hydroxylamine converts cyclododecanone into cyclododecanone oxime [946-89-4] (7), which rearranges in concentrated sulfuric acid to give laurolactam (8). Cyclododecanone reacts with elemental bromine to yield 2-bromocyclododecanone and 2,11-dibromocyclododecanone [4]. Addition of hydrocyanic acid to cyclododecanone produces cyanohydrin (95% yield) [5]. Although cyclododecanone gives pure cyclododecanol when hydrogenated under pressure in the presence of copper–chromium catalysts, more severe conditions produce cyclododecane. Upon oxidation with nitric acid, cyclododecanone, like cyclododecanol, yields dodecanedioic acid; however, the oxidation of cyclododecanol is more efficient.

Table 1. Physical properties

Property	Cyclododecanol	Cyclododecanone	Laurolactam
Molecular mass M_r	184.3	182.3	197.3
Melting point, °C	78.4	60.8	151.8
Boiling point, °C			
at 101.3 kPa	278	274.8	
at 8.0 kPa	190.8	178.4	
at 1.3 kPa	146.2	130.4	
Vapor pressure at 180 °C, kPa			0.35
Density ϱ, g/cm^3			
at 100 °C	0.909	0.888	
at 160 °C			0.90
at 200 °C	0.817	0.819	
Refractive index, D line, at 80 °C	1.4692	1.4581	
Heat of fusion, kJ/mol	15.1	16.6	16.1
Heat of vaporization, kJ/mol	65.1	56.2	91.4
Heat of formation, kJ/mol	−557	−482	
Specific heat, J g^{-1} K^{-1}			
at 100 °C	2.92	2.39	
at 200 °C	3.02	2.81	
Thermal conductivity, kJ m^{-1} h^{-1} K^{-1}	0.741	0.477	
Viscosity, mPa s			
at 100 °C	6.24	2.28	
at 150 °C	1.60	1.10	
at 200 °C	0.70	0.70	
Flash point, °C	137	119	192
Ignition point, °C	380	280	320–330

1.2. Production

Cyclododecane (**2**), which is obtained by catalytic hydrogenation of 1,5,9-cyclodo-decatriene (**1**) (→ Cyclododecatriene and Cyclooctadiene), is the starting material for producing cyclododecanol and cyclododecanone. Oxidizing cyclododecane with air or oxygen in the presence of boric acid gives cyclododecanol and cyclododecanone [6], [7]. The standard enthalpies of reaction are −205 and −395 kJ/mol, respectively.

Continuous or batch air oxidation (boric acid catalyst, atmospheric pressure, and 160–180 °C) converts 25–30% of the cyclododecane. Because boric acid esterifies the cyclododecanol produced, the reaction mixture (cyclododecane, cyclododecanyl borate, and cyclododecanone) is treated with water at 90–100 °C to hydrolyze the ester. After separation of the aqueous and organic phases, cyclododecane is distilled from the organic phase under reduced pressure and returned for further oxidation. Boric acid, which is in the aqueous phase, is concentrated under vacuum, crystallized as orthoboric acid, dried, and also returned to the oxidation section. The selectivity of the overall reaction producing a mixture of cyclododecanol and cyclododecanone from cyclodecane is 80–82 mol%.

Physical separation of the *cyclododecanol–cyclododecanone mixture* (mole ratio of 10:1) is tedious and costly. Pure cyclododecanol is made by hydrogenating cyclodecanone or the cyclododecanol–cyclododecanone mixture in the presence of copper–

chromium catalysts at 30 MPa and 160 °C. Pure cyclododecanone is produced by dehydrogenating the cyclododecanol–cyclododecanone mixture over copper or copper chromium catalysts on an active carrier at 230–245 °C and atmospheric pressure [8]. After dehydrogenation (endothermic, 75.4 kJ/mol), the reaction mixture is distilled to give pure cyclododecanone, cyclododecanol, and a slight amount of resin.

Storage and Transportation. Cyclododecanone and cyclododecanol are stored and transported in iron or, preferably, stainless steel containers in a dry inert atmosphere. To prevent crystallization, tanks and tank cars must be heated.

Analysis. Although the melting point generally indicates the purity of cyclododecanone, gas chromatography is the best method for evaluating the purity of cyclododecanone or cyclododecanol.

1.3. Uses

The cyclododecanol–cyclododecanone mixture from boric acid-catalyzed oxidation is the starting material for dodecanedioic acid (**5**), which is used to prepare polyamides, copolyamides, polyesters, and esters. Pure cyclododecanone is the starting material for production of laurolactam (**8**). Cyclododecanone and cyclododecanol are intermediates in organic syntheses. The esters and ethers of cyclododecanol are used in fragrances.

2. Laurolactam

At room temperature, laurolactam [*947-04-6*] (azacyclotridecan-2-one, **8**), $C_{12}H_{23}NO$, forms colorless crystals that are soluble in organic solvents but only slightly soluble in water (0.03 wt% at 20 °C). Table 1 lists the physical properties of laurolactam.

Laurolactam is produced in a manner similar to caprolactam by Beckmann rearrangement of cyclododecanone oxime (**7**) [9]–[12].

Because the oxime melts at 135 °C but begins to decompose at 110 °C, the cyclododecanone (purity 99%), which is dissolved in a hydrocarbon, is reacted at 90–100 °C with an aqueous solution of hydroxylamine sulfate. After removal of the aqueous phase, the oxime is extracted from the organic phase with cold concentrated sulfuric acid.

Heating the acid solution of the oxime to 100–120 °C produces laurolactam. The enthalpy of reaction (cyclododecanone oxime → laurolactam) is 166 kJ/mol. Water is added, and the crude lactam is extracted into an organic solvent, which is distilled to give the pure laurolactam (95% yield). Hüls produces laurolactam using this procedure.

Aquitaine Total, in France, and Toray, in Japan, produce laurolactam by a second procedure [13]. They add nitrosyl chloride, NOCl, to cyclododecatriene (**1**) to form 2-chlorocyclododeca-5,9-dien-1-one oxime (**6**); this is hydrogenated and dehydrochlorinated to give cyclododecanone oxime (**7**), which rearranges to laurolactam (**8**).

Laurolactam is used to make polyamide-12.

3. Toxicology

The oral LD_{50} of both cyclododecanol and cyclododecanone for rats is >2.5 g/kg. Contact of cyclododecanol with skin, eyes, or mucous membranes causes no irritation. Although cyclododecanone has no effect on the eyes or mucous membranes, it does irritate the skin slightly [14].

The oral LD_{50} of laurolactam for rats is 2.33 g/kg [14]; laurolactam does not irritate the skin or mucous membranes.

4. References

[1] G. Wilke, *Angew. Chem.* **69** (1957) 397, **75** (1963) 10; *Angew. Chem. Int. Ed. Engl.* **2** (1963) 105.
[2] H. Breil, P. Heimbach, M. Kröner, H. Müller, G. Wilke, *Makromol. Chem.* **69** (1963) 18.
[3] K. Kosswig, *Chem. Ztg.* **96** (1972) 373.
[4] W. Ziegenbein, *Chem. Ber.* **94** (1961) 2989.
[5] W. Kirchhoff, W. Stumpf, W. Franke, *Justus Liebigs Ann. Chem.* **681** (1965) 32.
[6] F. I. Novak, V. V. Kamsolkin, Y. A. Talysenkow, A. N. Bashkirow, *Neftekhimiya* **7** (1967) 248.
[7] F. Broich, H. Grasemann, *Erdöl Kohle Erdgas Petrochem.* **18** (1965) 360.
[8] Chemische Werke Hüls, DE 1 248 650, 1964 (M. zur Hausen, W. Knepper).
[9] G. Strauß, *Chem. Eng. (N.Y)* **76** (1969) no. 16, 106.
[10] Chemische Werke Hüls, DE 1 545 653, 1965 (G. Strauß, W. Thomas).
[11] Chemische Werke Hüls, DE 1 545 696, 1965 (K.-H. Simmrock, G. Strauß).
[12] Chemische Werke Hüls, DE 1 545 703, 1965 (G. Strauß).
[13] K. Weissermel, H.-J. Arpe: *Industrielle Organische Chemie*, Verlag Chemie, Weinheim 1976, p. 219.
[14] Hüls AG, unpublished results.

Cyclododecatriene and Cyclooctadiene

THEODORE A. KOCH, E. I. Du Pont de Nemours & Co., Inc., Wilmington, Delaware 19898, United States

1.	Introduction 1789	5.	Storage and Transportation ..	1791
2.	Physical Properties 1790	6.	Uses	1791
3.	Chemical Properties....... 1790			
4.	Production 1791	7.	References..............	1792

1. Introduction

Cyclododecatriene was first synthesized by H. W. B. REED [1] in 1954, but the discoveries of G. WILKE [2] led to the commercial catalytic process for producing cyclododecatriene and cyclooctadiene from butadiene (→ Butadiene). The first plant for cyclododecatriene production was brought on stream by Hüls in March of 1965 with a capacity of 7.3×10^3 t/a [3]. Additional plants were erected by Cities Service at Lake Charles, Louisiana, by Shell Chemical Corp. at Berre, France, by E. I. Du Pont de Nemours & Co. at Victoria, Texas, and by Mitsubishi at Tokkaichi. The Cities Service plant was later decommissioned, leaving the estimated total capacity of ca. 41×10^3 t/a of cyclododecatriene. The Shell plant can produce cyclooctadiene as well as cyclododecatriene.

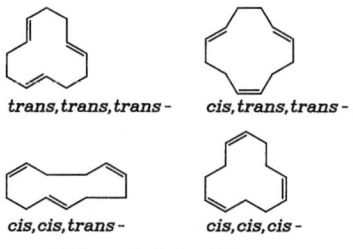

trans,trans,trans -   *cis,trans,trans -*

cis,cis,trans -   *cis,cis,cis -*

1,5,9 - Cyclododecatriene

Cyclooctadiene was first obtained by WILLSTÄTTER and VERAGUTH in 1905 as the unstable cis,trans isomer. The work of ZIEGLER and WILMS in 1950 [4] led to a thermal synthesis of the stable *cis, cis*-cyclooctadiene, which is the product obtained by the commercial catalytic dimerization [5].

Table 1. Physical properties of cyclododecatriene, cyclooctadiene, and derivatives *

Component	CAS registry no.	bp, °C 2.7 kPa	bp, °C 101.3 kPa	mp, °C	Relative density d_4^{20}	Refractive index n_D^{20}
trans,trans,trans -1,5,9-Cyclododecatriene	[676-22-1]	110	234	+34	0.864 (40 °C)	1.500
cis,trans,trans-1,5,9-Cyclododecatriene	[706-31-0]	116	240	−17	0.892	1.508
cis,cis,trans -1,5,9-Cyclododecatriene	[2765-29-9]	117	244	−8		1.513
cis,cis,cis-1,5,9-Cyclododecatriene	[4736-48-5]	110		−1		1.510 (25 °C)
trans-Cyclododecene	[1486-75-5]	118			0.872	1.485
cis-Cyclododecene	[1129-89-1]	121			0.875	1.486
Cyclododecane	[294-62-2]	122	243	+61	0.830 (65 °C)	1.455 (65 °C)
cis,trans-1,5-Cyclooctadiene	[5259-71-2]			−62	0.873	1.493
cis,cis-1,5-Cyclooctadiene	[1552-12-1]	50	151	−70	0.881	1.494
trans-Cyclooctene	[931-89-5]	42	145	−59	0.848	1.476
cis-Cyclooctene	[931-87-3]	43	145	−12	0.847	1.476
Cyclooctane	[292-64-8]	48	151	+14	0.835	1.459

* Adapted from [5].

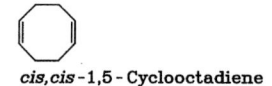

cis,cis-1,5-Cyclooctadiene

2. Physical Properties

The physical properties of the known isomers of 1,5,9-cyclododecatriene, $C_{12}H_{18}$, M_r 162.28, and 1,5-cyclooctadiene, C_8H_{12}, M_r 108.18, and their reduction derivatives are listed in Table 1 [6]. The materials are colorless with a typical terpene-like odor.

3. Chemical Properties

The cis,trans,trans and trans,trans,trans isomers of 1,5,9-cyclododecatriene and the cis-cis isomer of 1,5-cyclooctadiene have been studied extensively [7]. They are generally characterized by their tendency to form complexes with transition metals [8] and to undergo transannular reactions and isomerization. Selective hydrogenation to form cyclododecene and cyclooctene is a facile reaction [9], indicating a rapid decline in reactivity of the remaining unsaturation. The declining activity also leads to good yields of many classical double bond reactions, such as epoxidation, nitrosation [10], hydrocarboxylation, and carbonylation [11].

The saturated ring compounds are expectedly unreactive. Oxidation using boric acid leads predominantly to the alcohol, with some ketone and degradation product [12]. The expected products are obtained by nitrosation [13] and nitration [14].

4. Production

Cyclododecatriene is produced by the cyclotrimerization of butadiene, using catalysts based on titanium, chromium, or nickel [15]. The catalysts are prepared by reducing an appropriate salt in a hydrocarbon solvent with an aluminum alkyl. The titanium catalysts produce predominantly the cis,trans,trans isomer, whereas the nickel and chromium catalysts produce predominantly the trans,trans,trans isomer. The yield of cyclododecatriene is greater than 80%, with the byproducts being 1,5-cyclooctadiene, vinyl cyclohexene, and butadiene polymers and oligomers. Before distillation of the reactor product, the catalyst must be deactivated, generally by an aqueous caustic wash.

Cyclooctadiene is a coproduct of cyclododecatriene manufacture, but it is the primary product when nickel catalysts that are modified with appropriate ligands are used [2], [15].

5. Storage and Transportation

Cyclododecatriene, cyclooctadiene, and their hydrogenation products can be stored and transported in steel or aluminum containers. The materials are DOT classed as flammable liquids with a UN/NA no. UN 1993. International shipping classifies the C_{12} material in hazard class A III and the C_8 in A II. The UN numbers are 2518 and 2520, respectively. All of the materials are in Flammability Group G–3.

Because of the potential of autoxidation, the materials should always be blanketed with an inert gas. The unsaturated materials should contain an inhibitor unless they are to be processed immediately.

6. Uses

Cyclododecatriene and cyclooctadiene are precursors to many products; however, only 12-aminododecanoic acid [693-57-2] and dodecanedioic acid [693-23-2] are economically significant. Both of these materials are used in polymer production.

The noncatalytic manufacture of nylon 12 from 12-aminododecanoic acid, which is produced by the selective ozonization of cyclododecatriene, followed by catalytic rearrangement of the ozonide, reductive amination, and acidification, has been described [16]. Dodecanedioic acid is produced by the nitric acid oxidation of cyclododecanol – cyclododecanone, which is obtained by the boric acid catalyzed air oxidation of cyclododecane.

Many potential applications of the C_{12} and C_8 compounds have been disclosed in patent literature. They cover such diverse topics as polymers [17], rubbers [18], plasticizers [19], perfumes [20], lubrication oil additives [21], coal processing [22], sizing agents [23], corrosion inhibitors [24], and flame retardants [25]. The commercialization of these potential products or processes has not led to large volume consumption of cyclododecatriene or cyclooctadiene.

7. References

[1] H. W. B. Reed, *J. Chem. Soc.* 1954, 1931–1941.
[2] G. Wilke, *Angew. Chem.* **75** (1963) 10; *Angew. Chem. Int. Ed. Engl.* **2** (1963) 105–164.
[3] G. Bosmajian et al., *Ind. Eng. Chem. Prod. Res. Dev.* **3** (1964) no. 2, 117.
[4] K. Ziegler, H. Wilms, *Ann. Chem.* **1** (1950) 567.
[5] Cities Service Research Development Corp., US 2 964 575, 1959 (A. A. Sekul, H. G. Sellers, Jr.); 2 991 317, 1959 (H. G. Seller, Jr., A. A. Sekul); 2 972 640, 1959 (R. E. Burks, Jr., A. A. Sekul); 3 004 081, 1959 (G. Bosmajian). Studienges. Kohle, DE 1 140 569, 1959 (G. Wilke, E. W. Müller). BASF, DE 1 144 268, 1960 (N. von Kutepow, H. Seibt, F. Meier).
[6] *Ullmann*, 4th ed., **9**, 676.
[7] K. Kosswig, *Chem. Ztg.* **96** (1972) 373–383. T. Yamaguchi et al., *J. Chem. Soc. Jpn. Ind. Chem. Sec.* **4** (1970) 727. Columbia Carbon Co., US 3 360 578, 1965 (S. F. Chappel, III); US 3 398 205, 1965 (S. F. Chappel, III); 3 393 246, 1965 (F. L. George); US 3 418 386, 1966 (W. K. Hayes).
[8] E. O. Fischer, H. Werner: *Metal Pi Complexes with Di- and Oligomeric-Olefinic Ligands*, Verlag Chemie Weinheim 1963, pp. 76–82, 103–104. F. G. A. Stone, R. West:
Advances in Organometallic Chemistry, vol. **4**, Academic Press, New York–London 1966, pp. 369–387.
[9] Columbia Carbon Co., US 3 418 386, 1966 (W. K. Hayes). A. Misono, J. Ogata, *Faraday Soc.* **46** (1968) 72.
[10] Studienges. Kohle, DE 1 058 987, 1957 (G. Wilke).
[11] K. Bittler et al., *Angew. Chem.* **80** (1968) 352; *Angew. Chem. Int. Ed. Engl.* **7** (1968) no. 5, 329–335. J. Graefe et al., *Tetrahedron* **26** (1970) 2677.
[12] L. I. Zakhrein et al., *Pet. Chem. USSR* **2** (1963) 83.
[13] BASF, DE 1 079 036, 1958 (O. von Schickh, H. Metzger).
[14] Studienges. Kohle, DE 1 060 859, 1957 (H. Koch, K. E. Möller).
[15] C. W. Bird: *Oligomerization of Olefins in Transition Metal Intermediates in Organic Synthesis*, Logos Press – Academic Press, London 1966. P. Heimbach, R. Traummüller:
Ring Synthesis with Nickel Catalysts in Chemistry of the Metal-Olefin Complexes, Verlag Chemie, Weinheim 1970. P. Heimbach, P. W. Jolly, G. Wilke: "Pi-Allyl-Nickel Intermediates in Organic Synthesis," in E. G. A. Stone, R. West (eds.): Advances in Organometallic Chemistry, vol. **8**, Academic Press, New York–London 1970.
[16] P. Rossi et al., *Hydrocarbon Process.* **58** (1979) no. 8, 93–96.
[17] G. Natta, G. Dall'Asta: "Elastomers from Cyclic Olefins," in J. P. Kennedy, E. Tornquest (eds.): *High Polymers*, vol. **23**, Interscience Publ., New York 1969.
[18] Goodyear Tire & Rubber Co., DE 2 553 954, 1976 (C. Bancroft).
[19] B. R. Currell et al., *Adv. Chem. Ser.* **140** (1975) 1–17.
[20] International Flavors & Fragrances, Inc., US 4 397 789, 1982 (R. M. Poden).
[21] Cooper, Edwin & Co., DE 2 508 623, 1975 (G. Jayne, H. F. Askew).

[22] M. Dichter et al., *Org. Coat. Plast. Chem.* **43** (1980) 319–324.
[23] BASF, DE 2 234 908, 1973 (E. Scharf, K. Wendel, C. Taubitz).
[24] A. G. Khanlarova et al., *Korroz. Zashch. Nefteguzov. Promsti.* 1974, no. 7, 3–10.
[25] Cities Service Co., US 3 819 575, 1974 (J. Green).BASF, DE 2 247 654, 1974 (P. Dimroth, F. Dost).

Cyclohexane

M. Larry Campbell, Exxon Chemical Co., Florham Park, New Jersey 07932, United States

1.	Introduction	1795	6.	Uses, Economic Aspects	1803
2.	Physical Properties	1795	7.	Homologues	1803
3.	Chemical Properties	1796	7.1.	Cyclohexene	1803
4.	Production	1797	7.2.	Methylcyclohexane	1804
4.1.	Hydrogenation of Benzene	1798	7.3.	Methylcyclopentane	1804
4.2.	Separation from Naphtha	1802	8.	Toxicology and Occupational Health	1804
4.3.	Cost Considerations	1802			
5.	Quality Specifications and Tests	1802	9.	References	1805

1. Introduction

Cyclohexane [110-82-7], C_6H_{12}, is a cycloalkane, which was synthesized by Baeyer in 1893 and discovered in Caucasian crude oil by Markovnikov soon afterward. Its presence in U.S. crude oils was reported in 1931 [1]. Cyclohexane was first synthesized by hydrogenation of benzene in 1898 [2]. The predominant use of cyclohexane is in the manufacture of nylon.

2. Physical Properties

Like all cycloalkanes and related cycloolefins, cyclohexane is a clear liquid with a pungent petroleum-like odor. It is essentially insoluble in water but miscible in most organic liquids, and is noncorrosive, easily flammable, and considered to be much less toxic than benzene.

Cyclohexane can exist in two conformations, the chair and boat forms:

Chair Boat

Transformation from one isomer to the other involves rotation about carbon–carbon single bonds. The energy barriers associated with the transformation are low, but the boat form is estimated to have a potential energy 21–25 kJ/mol higher than the chair form [3]. As a result, cyclohexane is almost exclusively in the chair form at room temperature. Stereochemistry has no influence on its use as a raw material for nylon manufacture or in other applications.

Table 1 lists the important physical properties of cyclohexane; two closely related naphthenes, methylcyclopentane and methylcyclohexane; and the olefinic homologue, cyclohexene. Table 2 compares the key properties of some alkylcyclohexanes with cyclohexane. Tables 3, 4, and 5 give boiling points and compositions for azeotropes of cyclohexane with various compounds.

Although cyclohexane does not ignite spontaneously at ambient temperature, it must be kept away from an open flame. Information on the explosive limits and ignition properties for cyclohexane and related compounds is given in Table 6.

3. Chemical Properties

Cyclohexane is a relatively stable, nonpolar ring compound. Therefore, with thermal treatment, little isomerization to methylcyclopentane or ring-opening occurs, but benzene is formed by dehydrogenation. At high temperature (700–800 °C) cyclohexane decomposes to butadiene and other products. Aluminum chloride readily isomerizes cyclohexane to methylcyclopentane [7].

The most important commercial reaction of cyclohexane is its liquid-phase *oxidation* with air in the presence of soluble cobalt or boric acid catalysts to produce a mixture of cyclohexanol and cyclohexanone (KA oil) (→ Cyclohexanol and Cyclohexanone). Cyclohexanol is reduced to cyclohexanone, which is used to produce caprolactam, a monomer for nylon-6 production (→ Caprolactam). Alternatively, KA oil can be converted to adipic acid and hexamethylenediamine, the monomers for nylon-66 production (→ Adipic Acid; → Hexamethylenediamine) [8].

Table 1. Physical properties of cyclohexane, methylcyclohexane, cyclohexene, and methylcyclopentane

	Cyclohexane	Methylcyclohexane	Cyclohexene	Methylcyclopentane
CAS registry number	[110-82-7]	[108-87-2]	[110-83-8]	[96-37-7]
Formula	C_6H_{12}	C_7H_{14}	C_6H_{10}	C_6H_{12}
M_r	84.157	98.183	82.141	84.157
bp, °C	80.738	100.934	82.979	71.812
fp, °C	6.554	−126.593	−103.512	−142.455
ϱ_4^{20}, g/cm^3	0.778 55	0.769 39	0.810 96	0.74864
n_D^{20}	1.426 23	1.423 12	1.446 54	1.40970
Heat of fusion, kJ/mol	2.6787	6.7550	3.2954	6.9333
Entropy of fusion, J mol^{-1} K^{-1}	9.575	46.088		53.046
Heat of vaporization, kJ/mol				
at 25 °C	33.059	35.383		31.610
at bp	29.977	31.150		29.098
Entropy of vaporization, J mol^{-1} K^{-1}				
at 25 °C	110.879	118.675		106.022
at bp	84.70	83.275		84.351
Heat of combustion^a at 25 °C, kJ/mol	3 922.45	4 568.34	3 752.92^b	3 940.36
Heat of formation^c, kJ/mol	−123.22	−154.870	−5.35	−106.76
Entropy at 25 °C J mol^{-1} K^{-1}	298.435	343.569		340.136
$\Delta G°$, kJ/mol	31.778	27.298	106.86	35.80
Critical temperature, °C	280.3	298.97	287.4	259.58
Critical pressure, MPa	4.07	3.469	4.347	3.785
Critical density, g/cm^3	0.273	0.267	0.281	0.264
Dynamic viscosity at 20 °C, mPa · s	0.977	0.732	0.980	0.505
Kinematic viscosity at 20 °C, mm^2/s	1.259			
c_p at 300 K J mol^{-1} K^{-1}	107.098	136.11	105.84	110.66
at 800 K J mol^{-1} K^{-1}	279.51	329.67	249.07	267.12
Surface tension, mN/m	24.98	23.70	26.56	22.19
Constants for Antoine equation^d, kPa [4]				
A	13.7377	196.174	13.8093	13.7873
B	2766.3	146.6	2813.53	2731.00
C	−50.50	374.1	−49.98	−47.11
Dielectric constant at 20 °C and 10^5 Hz	2.021			
Cryoscopic constant^e, mole fraction/ °C	0.00411			

^{a} High heating value.
^{b} At 30 °C.
^{c} Gaseous.
^{d} $\ln p = A - B/(C + T)$.
^{e} For use in calculating mol% purity, p, by using the equation $\log p = 2.00000 - (A/2.302\ 59)(T' - T)$ where T' is the freezing point of a given sample [5], [6].

4. Production

Virtually all cyclohexane is produced commercially by hydrogenation of benzene (→ Benzene). A small amount is produced by superfractionation of the naphtha fraction from crude oil. Naturally occurring cyclohexane can be supplemented by fractionating methylcyclopentane from naphtha and isomerizing it to cyclohexane.

Table 2. Some physical properties of alkylcyclohexanes

Compound, CAS registry number	bp*, °C	fp, °C	d_4^{20}	n_D^{20}
Cyclohexane [110-82-7]	80.738	+ 6.554	0.77855	1.42623
Methylcyclohexane [108-87-2]	100.934	− 126.593	0.76939	1.42312
1,1-Dimethylcyclohexane [590-66-9]	119.543	− 33.495	0.78094	1.42900
cis-1,2-Dimethylcyclohexane [2207-01-4]	129.728	− 50.023	0.79627	1.43596
trans-1,2-Dimethylcyclohexane [6876-23-9]	123.419	− 88.194	0.77601	1.42695
cis-1,3-Dimethylcyclohexane [638-04-0]	120.088	− 75.573	0.76603	1.42294
trans-1,3-Dimethylcyclohexane [2207-03-6]	124.450	− 90.108	0.78472	1.43085
cis-1,4-Dimethylcyclohexane [624-29-3]	124.321	− 87.436	0.78285	1.42966
trans-1,4-Dimethylcyclohexane [2207-04-7]	119.351	− 36.962	0.76255	1.42090
1,1,3-Trimethylcyclohexane [3073-66-3]	136.626	− 65.750	0.77883	1.42955
1,1,2-Trimethylcyclohexane [7094-26-0]	145.2	− 29	0.8000	1.4382
1,1,4-Trimethylcyclohexane [7094-27-1]	135		0.7722	1.4251
Ethylcyclohexane [1678-91-7]	131.783	− 111.323	0.78792	1.43304
n-Propylcyclohexane [1678-92-8]	156.724	− 94.900	0.79360	1.43705
Isopropylcyclohexane [696-29-7]	154.762	− 89.390	0.80221	1.44087
n-Butylcyclohexane [1678-93-9]	180.947	− 74.725	0.79918	1.44075
Isobutylcyclohexane [1678-98-4]	171.221	− 94.785	0.79521	1.43861
sec-Butylcyclohexane [52993-54-1]	179.335		0.81314	1.44673
tert-Butylcyclohexane [3178-22-1]	171.591	− 41.158	0.81267	1.44694
n-Decylcyclohexane [1795-16-0]	297.589	− 1.726	0.81858	1.45338
cis-1-Methyl-4-isopropylcyclohexane [6069-98-3]	172.7	− 89.80		
trans-1-Methyl-4-isopropylcyclohexane [1678-82-6]	170.5	− 86.35		

* At 101.3 kPa.

4.1. Hydrogenation of Benzene

Benzene can be hydrogenated catalytically to cyclohexane in either the liquid or the vapor phase in the presence of hydrogen. Several cyclohexane processes, which use nickel, platinum, or palladium as the *catalyst*, have been developed. Usually, the catalyst is supported, e.g., on alumina, but at least one commercial process utilizes Raney nickel.

Hydrogenation proceeds readily and is highly exothermic ($\Delta H_{500\,K} = -216.37$ kJ/mol). From an equilibrium standpoint, the *reaction temperature* should not exceed 300 °C. Above this, the equilibrium begins to shift in favor of benzene so that high-purity cyclohexane cannot be produced. As a result of these thermodynamic considerations, temperature control of the reaction is critical to obtaining essentially complete conversion of benzene to cyclohexane.

Temperature control requires economic and efficient heat removal. This has been addressed in a number of ways by commercial processes. The earlier vapor-phase processes used multistage reactors with recycle of cyclohexane as a diluent to provide a heat sink, staged injection of benzene feed between reactors, and interstage steam generators to absorb the exothermic heat of hydrogenation [9], [10]. More recently, processes have been developed that use only one reactor [11], [12] or a combination of a liquid-and a vapor-phase reactor [13]. The objectives of the later processes were to

Table 3. Binary azeotropes with cyclohexane

Component	Component bp, °C	Cyclohexane, wt%	Azeotropic, bp, °C
Water	100.0	91.6	69.5
2,4-Dimethylpentane	80.5	48.6	80.2
2,2,3-Trimethylbutane	80.9	46.6	80.0
Benzene	80.1	48.8	77.6
		48.1	77.56
Ethanol	78.3	68.7	64.8
1-Propanol	97.2	81.5	74.7
2-Propanol	82.3	68.0	69.4
1-Butanol	117.7	90.5	79.8
Allyl alcohol	97.0	42.0	74.0
Acetone	56.3	32.5	53.0
2-Butanone	79.7	54.6	71.6
		56.0	71.5
Methyl acetate	57.0	17.0	54.9
Ethyl acetate	77.1	44.0	71.6
Vinyl acetate	72.2	37.7	67.4
2-Methoxyethanol	125.0	85.0	77.5
Thiophene	84.2	58.8 *	77.9
Isopropyl methyl sulfide	84.8	70.0 *	79.8
Nitromethane	100.8	73.5 *	69.5

* vol%.

Table 4. Ternary azeotropes with cyclohexane

Component		Component, wt%			Azeotropic bp, °C
A	B	A	B	C*	
Water	ethanol	4.7	19.7	75.5	62.6
Water	ethanol	7.0	17.0	76.0	62.1
Water	1-propanol	8.5	10.0	81.5	66.6
Water	2-butanone	5.0	35.0	60.0	63.6
Ethanol	benzene	30.4	10.8	58.8	63.1
1-Propanol	benzene	18.0	28.0	54.0	73.8
1-Butanol	benzene	4.0	48.0	48.0	77.4
2-Butanol	benzene	8.0	42.0	50.0	76.7
2-Methoxyethanol	benzene	9.0	39.1	51.9	73.0

* C = cyclohexane.

reduce capital cost and improve energy utilization. However, all of the commercial processes have comparably low capital cost and good energy efficiency.

A generalized flow scheme for a *vapor-phase process with multistage reactors* in series is shown in Figure 1. Benzene feed is divided and fed to each of the first two reactors. Recycled cyclohexane is introduced to the first reactor along with hydrogen. The recycled cyclohexane enables higher conversion in the reactors by absorbing part of the heat of hydrogenation. Steam generators between the reactors remove the heat of hydrogenation. The outlet temperature of the last reactor is controlled to achieve essentially 100% conversion of benzene to cyclohexane. The effluent from the last reactor is cooled, and the vapor and liquid are separated. Part of the hydrogen-rich vapor is recycled to the first reactor, and the rest is purged to fuel gas or hydrogen recovery facilities. The liquid from the separator

Table 5. Quaternary azeotrope

	Component, wt%			Azeotropic *bp*, °C
Water	Ethanol	Benzene	Cyclohexane	
7.1	17.4	21.5	54.0	62.19
6.1	19.2	20.4	54.3	62.14

Table 6. Explosive limits and ignition properties

	Explosive limits				Autoignition temp, °C	Flash point, °C
	Lower, vol%	Upper, vol%	Lower, g/m^3	Upper, g/m^3		
Ethylcyclohexane	0.9	6.6	42	310	260	− 21
Benzene	1.2	8.0	39	270	555	− 11
Cyclohexane	1.2	8.3	40	290	260	− 18
Cyclohexene					310	< − 20
Methylcyclohexane	1.1		45		260	− 4

goes to a stabilizer where the overhead gas is sent to fuel gas; the remaining material is cyclohexane product, part of which is recycled to the first reactor.

A simplified flow scheme for a *process with liquid- and vapor-phase reactors* is provided in Figure 2. Benzene and hydrogen are fed to the liquid-phase reactor, which contains a slurry of finely divided Raney nickel. Temperature is maintained at 180–190 °C by pumping the slurry through a steam generator and by vaporization in the reactor. Roughly 95% of the benzene is converted in this reactor. The vapor is fed to a fixed-bed reactor where the conversion of benzene is completed. The effluent from the fixed-bed reactor is processed as described previously for the vapor-phase process.

Benzene hydrogenation is done typically at 20–30 MPa. The maximum reactor temperature is limited to ca. 300 °C so that a typical specification of < 500 mg/kg benzene and < 200 mg/kg methylcyclopentane in the product can be achieved. This is necessary because of the thermodynamic equilibrium between cyclohexane–benzene and cyclohexane–methylcyclopentane. Actually, equilibrium strongly favors methylcyclopentane, but the isomerization reaction is slow enough with the catalysts employed to avoid a problem if the temperature is controlled.

The *hydrogen content* of the makeup hydrogen has no effect on product purity but it does determine the makeup, recycle, and purge gas rates. Streams with as low as 65 vol% hydrogen can be used.

Carbon monoxide and sulfur compounds are *catalyst deactivators*. Both can be present in the hydrogen from catalytic naphtha reformers or ethylene units, which are typical sources of makeup hydrogen. Therefore, the hydrogen-containing stream is usually passed through a methanator to convert carbon monoxide to methane and water. Prior to methanation, hydrogen-containing gas can be scrubbed with caustic to remove sulfur compounds. Commercial benzene contains less than 1 mg/kg sulfur. In some cases, the recycle gas is also scrubbed with caustic to prevent buildup of hydrogen sulfide from the small amount of sulfur in the benzene. With properly treated hydrogen and specification benzene, a catalyst life in excess of three years can be achieved easily

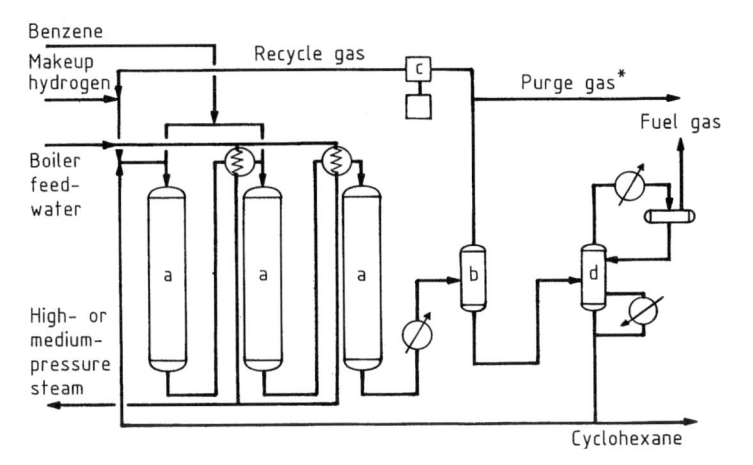

Figure 1. Multistage reactor, vapor-phase process
a) Reactors; b) Separator drum; c) Compressor; d) Stabilizer
* To hydrogen recovery (optional).

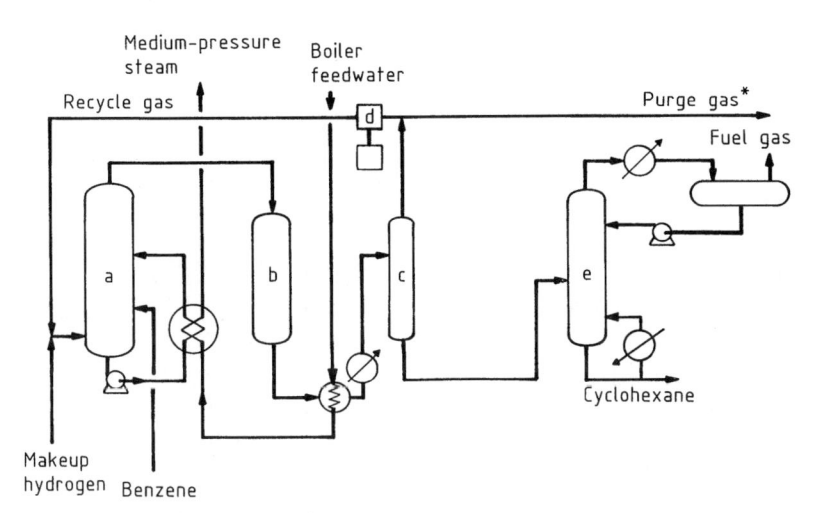

Figure 2. Liquid- and vapor-phase reactor process
a) Liquid-phase reactor; b) Vapor-phase reactor; c) Separator drum; d) Compressor; e) Stabilizer
* To hydrogen recovery (optional).

in fixed-bed reactors that use noble-metal catalysts supported on a base. The catalyst in the process that uses Raney nickel in suspension is reported to have a typical life of about six months before it must be replaced [14].

4.2. Separation from Naphtha

The cyclohexane content of the naphtha fraction of crude oil can vary from 0.5 to 5.0 vol%. However, *n*-hexane, isohexanes, methylcyclopentane, benzene, and dimethylpentanes have normal boiling points very close to cyclohexane; and the freezing points of cyclohexane and benzene are separated by < 2 °C. Consequently, recovery of cyclohexane from naphtha by distillation or crystallization is difficult and uneconomic.

Cyclohexane can be separated from benzene by solvent extraction or extractive distillation, but no practical solvent system has been developed for separating cyclohexane from hexanes or methylcyclopentane. Trace quantities of benzene and *n*-hexane can be removed with zeolite X and zeolite A sieves, respectively. The difficulty of separating cyclohexane from naphtha is underscored by the fact that only one company, Phillips Petroleum, produces cyclohexane by distillation (ca. 10% of U.S. capacity). This route is not used commercially elsewhere in the world.

4.3. Cost Considerations

The investment for a benzene hydrogenation unit that produces 100 000 t/a of cyclohexane was ca. $(5-7) \times 10^6$ \$ for a unit completed in 1985. This excludes investment for hydrogen recovery or manufacture and associated facilities, such as utility systems and tankage.

The net operating cost is low because of recovery of the heat of hydrogenation by steam generation and heat integration. The major costs are maintenance expenses, labor, capital recovery charges for the plant and working capital, benzene feedstock, and hydrogen. The cost of hydrogen can vary widely and can affect cyclohexane manufacturing cost by as much as \$ 40/t.

The selling price of cyclohexane is tied to the selling price of benzene. In the early 1980s, its price was ca. \$ 100/t above the benzene price.

5. Quality Specifications and Tests

The typical purity specification for cyclohexane is 99.9 wt% min. Impurities are unconverted benzene, aliphatic hydrocarbons in the benzene feed to the hydrogenation unit, methylcyclopentane formed by isomerization of cyclohexane, and methylcyclohexane formed by hydrogenation of the small amount of toluene in benzene. Freezing point, relative density, and a distillation range of 1 °C, with 80.7 °C in the range, are generally specified. A sulfur level of less than 1 mg/kg as measured by ASTM method D-3120 is also specified. In addition, specifications of phenol, chloride, and basic nitrogen content exist, but they are of little importance because the levels of these

Table 7. Estimated production capacity and consumption of cyclohexane, in 10^3 t/a

Year	Capacity*				Consumption			
	United States	Western Europe	Japan	World	United States	Western Europe	Japan	World
1979					1000	750	510	3150
1980					785	600	500	2850
1981					720	650	480	2800
1982	1480	940	580	4250	470	600	430	2465
1983	1510	940	580	4280	665	725	460	2890
1984	1330	940	530	4050	860	710	480	3140

* At end of year.

in product cyclohexane are generally less than 0.1 mg/kg. Although these specifications, with the appropriate ASTM test method, are used in contractual agreements between cyclohexane producers and their customers, gas chromatography is generally employed for quality control in manufacturing because of its precision and reproducibility.

6. Uses, Economic Aspects

Over 98% of the cyclohexane produced is used to make nylon intermediates: adipic acid, caprolactam, and hexamethylenediamine, with the first two consuming ca. 95% of the cyclohexane used in nylon manufacture. Minor miscellaneous uses, such as solvents and polymer reaction diluents, consume the remainder of the cyclohexane produced.

Estimated worldwide production capacity and consumption of cyclohexane are given in Table 7. Consumption dropped significantly in the early 1980s because of the generally depressed economy. The future growth in cyclohexane consumption is forecast to be 2–3%/a. Its growth is limited because nylon is a mature product and other routes to nylon intermediates, which use butadiene, acrylonitrile, or phenol, have been developed. The utilization of cyclohexane recovered to slightly over 75% of capacity in 1984 because consumption increased and several units were shut down.

7. Homologues

7.1. Cyclohexene

Cyclohexene can be used to synthesize many interesting products of potential commercial value. However, no viable process has been developed for producing it in large volume. The reason for this is the difficulty in hydrogenating benzene to cyclohexene because the $\Delta G°$ (298 K) for converting benzene to cyclohexene is −22.8 kJ/mol, which is considerably less than that for benzene to cyclohexane

[$\Delta G°$ (298 K) = – 97.9 kJ/mol]. Consequently, the hydrogenation reaction tends to go all the way to cyclohexane.

Despite the thermodynamic difficulty, considerable work on heterogeneous catalyst systems has been reported [15]–[18]. Ruthenium-based systems appear to be the best, but selectivities for cyclohexene are < 30%; the rest of the product is cyclohexane.

7.2. Methylcyclohexane

The primary use of methylcyclohexane is as a component in the feed to catalytic naphtha reformers where it is readily dehydrogenated to toluene, a high-octane motor gasoline component. Some methylcyclohexane is used for solvent applications in which it is one component of a narrow boiling range fraction of naphtha. Like cyclohexane, it can be produced easily by hydrogenation of toluene, but no current market exists for large-scale quantities.

7.3. Methylcyclopentane

The primary use of methylcyclopentane, like methylcyclohexane, is as a component in the feed to catalytic naphtha reformers where it is isomerized to cyclohexane and, in turn, dehydrogenated to benzene. Methylcyclopentane is a minor component in some light aliphatic solvents, but it is not produced as a pure component in significant volume.

8. Toxicology and Occupational Health

Cyclohexane is mildly irritating to the skin. Both the TLV–TWA and MAK are 300 ppm, 1050 mg/m^3; the TLV–STEL was listed in 1986–1987 as (375 ppm), (1300 mg/m^3), with a note of intention to delete this value [19], [20]. At levels above the TWA, cyclohexane can cause irritation of the eyes and mucous membranes. When high concentrations are inhaled, narcosis may occur. In contrast to benzene, little or no effect on the blood, even after long exposure, has been reported.

Methylcyclohexane is somewhat less narcotic than cyclohexane so a higher TLV–TWA of 400 ppm, 1600 mg/m^3 has been established; the MAK is 500 ppm, 2000 mg/m^3; the TLV–STEL was (500 ppm), (2000 mg/m^3) in 1986–1987, with the same notation as for cyclohexane. The toxic properties of cyclohexene (TLV–TWA and

MAK of 300 ppm, 1015 mg/m^3) and methylcyclopentane (TLV and MAK have not been established) are thought to be similar to those of cyclohexane.

9. References

[1] J. H. Bruun, M. M. Hicks-Bruun, *J. Res. Natl. Bur. Stand., Sect. A* **7** (1931) 607.
[2] P. Sabatier, *Ind. Eng. Chem.* **18** (1926) 1005.
[3] W. C. Dauber, K. S. Pitzer in M. S. Newman (ed.): *Steric Effects in Organic Chemistry*, J. Wiley & Sons, New York 1956, pp. 13–15.
[4] R. C. Reid, J. M. Prausnitz, J. K. Sherwood: *The Properties of Gases and Liquids*, 3rd ed., McGraw-Hill, New York 1977, App. A.
[5] *Selected Values of Hydrocarbons and Related Compounds*, A.P.I. Research Project 44, Carnegie Press, Pittsburgh, Pa., 1960, p. 1.
[6] P. R. Dreisbach, *Adv. Chem. Ser.* **15** (1955) 441.
[7] N. I. Shuikin et al., *Neftekhimiya* **1** (1961) 756; *Chem. Abstr.* **57** (1962) 8452.
[8] G. Nawata, *Chem. Econ. Eng. Rev.* **4** (1972) no. 5, 47–61.
[9] UOP, US 3 054 833, 1962 (G. R. Donaldson, V. Haensel).
[10] UOP, US 2 755 317, 1956 (L. S. Kasel).
[11] Texaco, US 3 767 719, 1973 (J. H. Colvert, E. F. Jones, H. C. Kaufman, R. Smith).
[12] S. Field, M. H. Dolson, *Hydrocarbon Process.* **46** (1967) no. 5, 169–174.
[13] Institut Français de Pétrole, US 3 597 489, 1971 (O. D. Vu, R. Odello).
[14] B. Cha, J. Cosyns, M. Derrien, A. Forge, *AIChE Symp. Ser.* **70** (1974) no. 142, 128–130.
[15] Stamicarbon, US 3 391 206, 1966 (F. Hartog).
[16] Du Pont, US 3 767 720, 1973 (W. C. Drinkard).
[17] M. M. Johnson, G. P. Nowak, *J. Catal.* **38** (1975) 518.
[18] C. U. I. Oldenbrand, S. T. Lundin, *J. Chem. Technol. Biotechnol.* **30** (1980) 677–687; **31** (1981) 660–669; **32** (1982) 365–375.
[19] ACGIH (ed.): *Threshold Limit Values (TLV)*, ACGIH, Cincinnati, Ohio, 1986–1987.
[20] DFG (ed.): *Maximum Concentration at the Workplace (MAK)*, VCH Verlagsgesellschaft, Weinheim 1986.

Cyclohexanol and Cyclohexanone

MICHAEL T. MUSSER, E. I. Du Pont de Nemours & Co., Orange, Texas 77631, United States

1.	Introduction 1807	6.	Storage and Transportation ..	1815
2.	Physical and Chemical Properties 1808	7.	Uses and Trade Names	1816
3.	Production 1808	8.	Derivatives	1816
3.1.	Phenol Hydrogenation 1809	8.1.	Esters of Cyclohexanol......	1816
3.2.	Liquid-Phase Oxidation of Cyclohexane............. 1810	8.2.	Cyclohexanone Oxime	1817
3.3.	Boric Acid Modified Oxidation of Cyclohexane 1812	8.3.	Methylcyclohexanols and Methylcyclohexanones	1817
3.4.	Cyclohexene Hydration....... 1813	8.4.	Trimethylcyclohexanols and Trimethylcyclohexanones	1818
3.5.	Dehydrogenation of Cyclohexanol 1814	9.	Economic Aspects	1820
4.	Plant Safety.............. 1814	10.	Toxicology and Occupational Health..................	1820
5.	Quality Specifications and Chemical Analysis 1815	11.	References...............	1821

1. Introduction

Cyclohexanol [108-93-0], $C_6H_{11}OH$, M_r 100.16, is a colorless hygroscopic crystalline substance having a camphor-like odor. In its pure state, it melts at room temperature, 25.15 °C. *Cyclohexanone* [108-94-1] $C_6H_{10}O$, M_r 98.15, is a colorless to pale yellow liquid having an odor similar to that of acetone.

Both cyclohexanol and cyclohexanone are produced on a large commercial scale; the vast majority of these compounds is consumed in the production of caprolactam and adipic acid, intermediates in the manufacture of nylon 6 and nylon 66 (→ Adipic Acid; → Caprolactam).

2. Physical and Chemical Properties

Physical Properties. The most important physical properties of cyclohexanol and cyclohexanone are shown in Table 1.

Chemical Properties. *Cyclohexanol* undergoes chemical reactions typical of secondary alcohols. It can esterify most organic acids or acyl halides, whereas halogen acids convert the alcohol to the corresponding cyclohexyl halide. Dehydration with sulfuric acid or vapor-phase treatment with alumina at 300–400 °C leads to cyclohexene. Mild oxidation or catalytic dehydrogenation gives cyclohexanone. Stronger oxidizing agents, such as nitric acid, give a good yield of adipic acid.

Cyclohexanone undergoes most of the reactions expected of aliphatic ketones. At room temperature 0.02% exists in the enol form [1]. Cyclohexanone reacts with bisulfite to give the addition product, and it reacts with hydrogen cyanide to give the cyanohydrin. It can be hydrogenated to cyclohexanol, which under more vigorous conditions is further reduced to cyclohexane. Reactions with chlorine or bromine lead to substitution at either or both the 2- and 6-positions. The most important commercial reactions of cyclohexanone are that with hydroxylamine to give cyclohexanone oxime (an intermediate to caprolactam) and its oxidation with nitric acid to give adipic acid.

3. Production

Cyclohexanol and cyclohexanone were first produced on an industrial scale by the hydrogenation of phenol. This was followed in the 1940s by a transition metal-catalyzed liquid-phase air oxidation of cyclohexane, which gave a mixture of cyclohexanol and cyclohexanone, the ratio of which could be partially controlled by the choice of metal catalyst. In the 1950s, Halcon developed a boric acid modified liquid-phase air oxidation of cyclohexane. This process, in which the intermediate cyclohexyl hydroperoxide was trapped as the peroxyborate ester, led to a higher yield and a higher ratio of cyclohexanol to cyclohexanone. Several recent patents discuss the preparation of cyclohexanol from cyclohexene, an intermediate obtained from the partial hydrogenation of benzene. Cyclohexanone can be prepared by dehydrogenation of cyclohexanol; the reverse reaction, hydrogenation of cyclohexanone to cyclohexanol, is also practical. Cyclohexane oxidation processes comprised 98% of the United States and 91% of the Western European cyclohexanol–cyclohexanone production in 1979 versus 95% and 60%, respectively, in 1970 [2]. European companies still using the phenol hydrogenation route include Hüls and Montedison. Only Allied and one small unit at Monsanto use this process in the United States. All large-scale facilities employ some form of cyclohexane oxidation technology.

Table 1. Physical properties of cyclohexanol and cyclohexanone

Property	Cyclohexanol	Cyclohexanone
mp, °C	25.15	−47
bp, °C	161.1	156.4
Vapor pressure, kPa	0.15 (20 °C)	0.52 (20 °C)
	0.48 (35 °C)	
d_4^{20}	0.9493	0.9455
n_D^{20}	1.4656	1.4552
Expansion coefficient	8.25×10^{-4}	9.14×10^{-4}
Dynamic viscosity, mPa · s	41.07 (30 °C)	1.803 (30 °C)
	17.20 (45 °C)	
Specific heat, J/g	1.747	1.811
Surface tension, mN/m	33.47 (30 °C)	33.51 (30 °C)
Heat of fusion, kJ/mol	1.791	1.501
Heat of combustion, MJ/mol	−3.722	−3.521
Heat of vaporization, kJ/mol	45.51	44.92
Heat of formation, kJ/mol	−352	−272
Flash point, °C	68	54
Autoignition temp., °C	300	420
Solubility at 20 °C,		
g of compound/100 g of water	3.6	9.0
g of water/100 g of compound	12.6	5.7
Azeotropes		
With water: bp, °C	97.8	96.6
composition, water : compound	80 : 20	56 : 44
With phenol: bp, °C	183	
composition, phenol : compound	87 : 13	

3.1. Phenol Hydrogenation

The product of phenol hydrogenation, i.e., cyclohexanol, cyclohexanone, or a mixture of the two, is determined by the metal catalyst. *Cyclohexanol* can be prepared in high yield by either vapor- or liquid-phase hydrogenation of phenol [*108-95-2*], C_6H_5OH. Vapor-phase hydrogenation can be carried out by using a supported nickel catalyst or a catalyst of nickel containing copper, cobalt, or manganese. The support is usually either alumina or silicic acid. Yields as high as 98 % have been reported [3]–[6]. A flow diagram of the vapor-phase hydrogenation is shown in Figure 1.

Hydrogenation of phenol to *cyclohexanone* can be carried out in the vapor phase with a large variety of noble-metal catalysts including palladium, platinum, iridium, ruthenium, rubidium, or osmium. The hydrogenation is normally carried out at 140 – 170 °C and atmospheric pressure. Yields as high as 95 % at 100 % conversion can be obtained [7], [8].

Liquid-phase hydrogenation of phenol under mild conditions, when a palladium on carbon catalyst is used, gives greater than 99 % yield of cyclohexanone at 80 % conversion [9]. By adjustment of catalyst and operating conditions, a mixture of cyclohexanol and cyclohexanone, which is preferred by adipic acid producers, can be obtained.

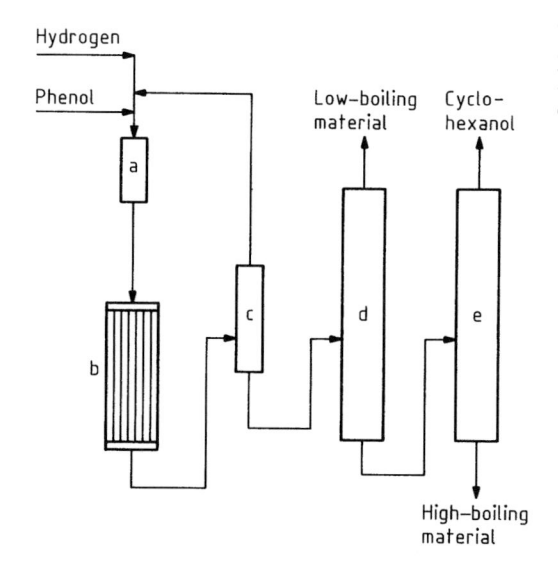

Figure 1. Vapor-phase phenol hydrogenation process
a) Phenol evaporator; b) Reactor; c) Condenser; d) Low-boiler column; e) Cyclohexanol column

3.2. Liquid-Phase Oxidation of Cyclohexane

The liquid-phase air oxidation of cyclohexane [110-82-7] to cyclohexanol and cyclohexanone was developed in the 1940s. The oxidation is usually carried out with a soluble cobalt catalyst in a series of stirred autoclaves at 140–180 °C and at 0.8–2 MPa, although a tower oxidizer can be used [10], [11]. The residence time in the oxidizers is 10–40 min. The initial product formed, cyclohexyl hydroperoxide [766-07-4] (**1**), is converted by the same homogeneous catalyst to cyclohexanol and cyclohexanone:

$$C_6H_{12} + O_2 \longrightarrow \underset{\mathbf{1}}{C_6H_{11}OOH}$$

$$3\,C_6H_{11}OOH \longrightarrow 2\,C_6H_{11}OH + C_6H_{10}O + H_2O + O_2$$

The cyclohexanol:cyclohexanone ratio can be substantially decreased under certain circumstances. The normal cobalt-catalyzed air oxidation and "deperoxidation" gives a cyclohexanol:cyclohexanone ratio of (2.5–4):1. If chromium (III) is added to the air oxidizer, it appears to promote dehydration of **1** to cyclohexanone, as compared to the free-radical decomposition mechanism of cobalt, leading to a lower cyclohexanol:cyclohexanone ratio [12]–[14].

Because the intermediate (**1**) and the products, cyclohexanol and cyclohexanone, are more readily oxidized than cyclohexane, the conversion must be kept low (usually under 10%) in order to maximize yield. Depending on the conversion, the yield of alcohol and ketone varies from 70 to 90%. Byproducts of the oxidation include a wide

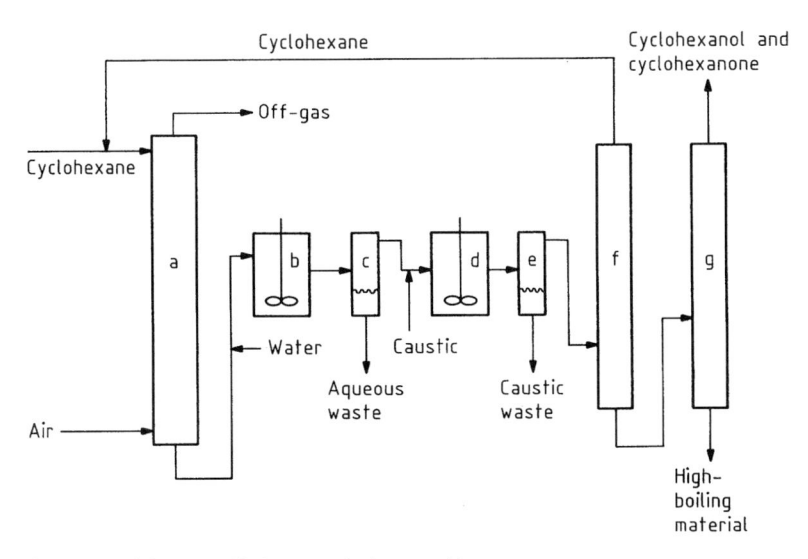

Figure 2. Cyclohexane oxidation – caustic decomposition process
a) Air oxidizer; b) Water wash; c) Decantation; d) Caustic contacting; e) Decantation; f) Cyclohexane recovery column; g) Cyclohexanol – cyclohexanone recovery column

range of mono- and dicarboxylic acids, esters, aldehydes, and other oxygenated materials.

Significant amounts of **1** usually remain in the stream exiting the air oxidizer. To maximize the amount of **1** leaving the air oxidizer, oxidation can be carried out with no catalyst or with a modified catalyst that minimizes hydroperoxide decomposition in the oxidizer, permitting the use of a higher yield peroxide decomposition process [15]–[18]. Some companies react this residual hydroperoxide to give higher product yields than would normally occur from a homogeneous cobalt catalyst. For example, the air oxidate may or may not be washed with water and then the hydroperoxide converted with either a homogeneous or heterogeneous metal catalyst [19]–[30], or the oxidate may be contacted with a caustic aqueous phase, which decomposes the hydroperoxide to give a higher yield of ketone and alcohol [31]–[35]. These modified hydroperoxide decomposition processes can result in a lower alcohol:ketone ratio in the final product. A flow diagram of a modified process using caustic is shown in Figure 2.

Several patents claim that the hydroperoxide in the oxidizer effluent can be hydrogenated in high yield to cyclohexanol [36]–[41]; this technology is not known to have been commercialized.

Table 2 shows the cyclohexanol:cyclohexanone ratios that can be obtained by using a range of conditions and a variety of hydroperoxide decomposition methods.

Table 2. Ratio of cyclohexanol to cyclohexanone produced from cyclohexyl hydroperoxide

Solution * – Modification	Catalyst	Temperature, °C	Ratio **
A	none	155	0.8
A	Cr	155	0.4
A	Ni	155	1.4
A	V	155	1.4
A	Co	155	3.0
A	Mn	155	3.3
B	Co	160	2.5
A – hydrogenation of **1**	Pd	150	5 – 10
A – separate caustic aqueous phase	Co	120	0.2

* A = cyclohexyl hydroperoxide (**1**) in cyclohexane; B = **1** in air oxidizer tails.
** Cyclohexanol:cyclohexanone.

3.3. Boric Acid Modified Oxidation of Cyclohexane

In the 1950s, Scientific Design (now Halcon International) developed a process in which anhydrous metaboric acid [*13460-50-9*] (**2**) was added as a slurry to the first of several staged air oxidation vessels (see Fig. 3) [42], [43]. The borate ester of cyclohexanol (**3**) was formed, presumably through the intermediate formation of the peroxyborate. This ester is relatively stable in the air oxidizer and is protected from over oxidation to ring-opened products. The ester is subsequently hydrolyzed to cyclohexanol and boric acid [*10043-35-3*] (**4**). The boric acid is dehydrated to **2** and is recycled to the air oxidizer.

$$18\ C_6H_{12} + 9\ O_2 + 2\ H_3B_3O_6 \longrightarrow 6\ B(OC_6H_{11})_3 + 12\ H_2O$$
$$\phantom{18\ C_6H_{12} + 9\ O_2 + 2\ H_3B_3O_6 \longrightarrow}\mathbf{2}\phantom{\longrightarrow 6\ B(OC_6H_{11})_3}\mathbf{3}$$

$$B(OC_6H_{11})_3 + 3\ H_2O \longrightarrow 3\ C_6H_{11}OH + H_3BO_3$$
$$\phantom{B(OC_6H_{11})_3 + 3\ H_2O \longrightarrow 3\ C_6H_{11}OH + H_3BO_3\ \ }\mathbf{4}$$

$$3\ H_3BO_3 \longrightarrow H_3B_3O_6 + 3\ H_2O$$

The chemistry of this process is very similar to metal-catalyzed oxidation except that the cyclohexyl rings are protected from further attack.

Compared to other commercial processes, the boric acid process is characterized not only by a higher investment and a higher operating cost (to recover and recycle the boric acid), but also by a higher yield of cyclohexanol and cyclohexanone (up to 87%) and an alcohol:ketone ratio as high as 10:1. The technology was further developed by Institut Français du Pétrole [44] – [45]. The Halcon process was licensed by several companies around the world, with Monsanto, ICI, Bayer, and Rhône-Poulenc having the major operating plants that use this technology.

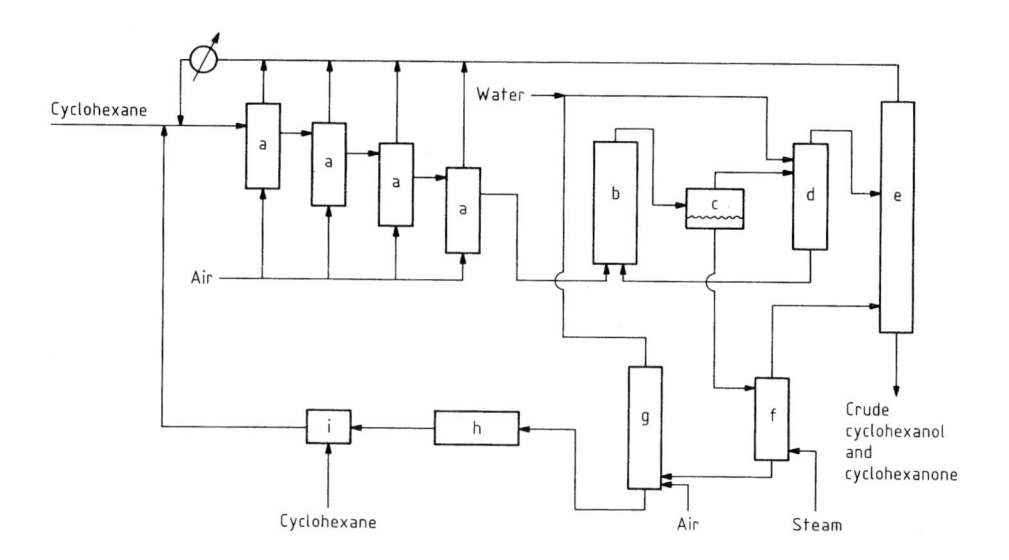

Figure 3. Boric acid modified oxidation of cyclohexane
a) Staged air oxidation vessels; b) Hydrolysis vessel; c) Decanter; d) Extraction column; e) Cyclohexane recovery column; f) Column to strip cyclohexanol–cyclohexanone from the aqueous phase; g) Boric acid oxidation vessel; h) Dehydration vessel; i) Boric acid slurry tank

3.4. Cyclohexene Hydration

In 1972 the first patent covering the selective hydrogenation of benzene [71-43-2] to cyclohexene [110-83-8] appeared [46]. Shortly thereafter, two patents were issued in which cyclohexene was hydrated to cyclohexanol by using various catalysts [47], [48]. The major problems with this technology are (1) that the hydrogenation of benzene results in significant excessive hydrogenation to cyclohexane (→ Cyclohexane), and (2) that separating the resultant mixture of cyclohexane, cyclohexene, and benzene is difficult. To minimize the first problem, the hydrogenation is not run to greater than ca. 60% conversion. To avoid the second problem, the cyclohexene is selectively reacted from the mixture of the three components.

There is no known commercial application of this technology, but a significant amount of research continues on methods for preparing and utilizing cyclohexene.

3.5. Dehydrogenation of Cyclohexanol

Cyclohexanol can be dehydrogenated to cyclohexanone (the desired starting material for caprolactam manufacture) without catalyst by passing the vapors through a tube furnace at 400–450 °C. The vapors are cooled, the hydrogen separated, and the cyclohexanone, which contains cyclohexene and water, is purified in a distillation column. The purity of the resulting cyclohexanone is 98–99% [49]–[52].

The catalytic dehydrogenation can be carried out at milder conditions over a variety of catalysts including chromium oxide–copper [53], copper chromate [54], nickel [55], [56], zinc sulfide or a zinc–iron catalyst [57]–[59], cobalt carbonate [60], or other heavy metals [61].

4. Plant Safety

The major industrial hazard associated with producing cyclohexanol and cyclohexanone is oxidation of cyclohexane at elevated temperature and pressure. Inside the air oxidation unit, the vapor is usually too rich in fuel to be explosive. However, in the air inlet piping, an explosive mixture can result if hot cyclohexane backs up into this system. Some of the solids formed in the oxidation, usually called "coffee grounds," can be pyrophoric. At least one incident in which the air sparger, entering the oxidizer, was observed to be "glowing red," indicated a fire inside the sparger.

A sudden release of the hot cyclohexane to the atmosphere, (e.g., from the rupture of a process line or a break in the air oxidation vessel itself) results in an extremely hazardous situation. An incident of this type occurred at a Monsanto plant at Pensacola, Florida in 1971. Fortunately, the resulting vapor cloud was dispersed by the prevailing winds before it could ignite.

In 1974, a more serious incident occurred at a Nipro plant in Flixborough, England. A section of 50-cm pipe connecting two air oxidation vessels ruptured, resulting in the release of a very large cloud of cyclohexane vapor. About 30 s after the break occurred, the vapor cloud ignited, and the resulting explosion destroyed the plant. A total of 28 lives were lost.

5. Quality Specifications and Chemical Analysis

High-grade commercially available *cyclohexanol*, usually obtained from hydrogenation of phenol, has the following specifications:

mp	19–23 °C
bp	159–162 °C
Water content	0.1%
Phenol content	0.0001%

The best analytical method for cyclohexanol is gas chromatography using Carbowax 20M on Chromosorb. Common impurities, such as cyclohexane, cyclohexanone, phenol, pentanol, and benzene, do not interfere. Cyclohexanol can be characterized as the 4-nitrobenzoyl derivative, *mp* 52 °C. It can be determined colorimetrically by using a solution of 4-hydroxybenzaldehyde in sulfuric acid [62]. The preferred acylating agent for microscale quantities is acetic anhydride [63].

Commerical *cyclohexanone* is available in several grades. A typical material is colorless to pale yellow, contains 98–99% cyclohexanone and 0.2% water, and has a boiling range of 151–157 °C.

Cyclohexanone, like cyclohexanol, is best analyzed by using a Carbowax 20M on Chromosorb gas chromatography column.

Cyclohexanone can be characterized as the oxime (*mp* 90.5 °C), the semicarbazone (*mp* 167 °C), or the nitrophenyl hydrazone (*mp* 146 °C).

It can be quantitatively determined by reaction with hydroxylamine hydrochloride, followed by basic titration of the hydrogen chloride that is liberated [64], [65].

6. Storage and Transportation

Cyclohexanol and cyclohexanone are not corrosive to iron or steel and can be stored or transported in drums, containers, tank trucks, or tank cars. An inert gas (nitrogen) blanket is not required. As of mid-1986, the U.S. Department of Transportation (DOT) classifies both cyclohexanol and cyclohexanone as combustible liquids. Drums containing < 416 L of either chemical do not require a hazardous material label, but larger quantities must be labeled "Combustible Liquid" [66].

7. Uses and Trade Names

Cyclohexanol has numerous industrial applications, the major two being an intermediate in the production of both adipic acid (for use in nylon 66) and cyclohexanone, which is then converted to caprolactam (the monomer for nylon 6). The United States market for refined cyclohexanol, other than as a nylon intermediate, was 4.3×10^3 t in 1971 [67]. The major uses are:

1) As a solvent for lacquers, varnishes, oils, alkyd resins, gums, shellacs, ethyl cellulose, acid dyes, and natural resins
2) In the preparation of esters for plasticizers, e.g., dicyclohexyl phthalate
3) In the dry cleaning, textile cleaning, and laundry industries and in soaps and synthetic detergents as a homogenizer and stabilizer
4) In paint and varnish removers
5) In the insecticide, fragrance, polish, and rubber cement industries

Trade names: Cyclohexyl Alcohol, Hexahydrophenol, Hydrophenol, Hexalin, and A.

Cyclohexanone. Over 96% of all cyclohexanone produced is either oxidized to adipic acid or converted to cyclohexanone oxime, which is then rearranged with sulfuric acid to caprolactam.

Other uses for cyclohexanone include that of thinner or solvent for synthetic resins, polymers, and lacquers. It is also used as a starting material in the synthesis of many insecticides, herbicides, and pharmaceuticals.

Trade names: Hexanon, Hytrol-O, Sextone, Pimelic Ketone, Ketohexamethylene, and K

8. Derivatives

8.1. Esters of Cyclohexanol

Numerous esters of cyclohexanol are of commercial significance, the more important being dicyclohexyl phthalate, dicyclohexyl adipate, and cyclohexyl acrylate.

Dicyclohexyl phthalate [*84-61-7*], $C_{20}H_{26}O_4$, M_r 330.45, *mp* 65 °C, *bp* 218 °C (0.6 kPa), d^{20} 1.148. This diester is normally prepared from the reaction of phthalic anhydride with cyclohexanol in an inert solvent like toluene at ca. 130 °C.

Dicyclohexyl phthalate is used as a plasticizer to modify the properties of synthetic resins, is used in alkyd resins and cellulose nitrate to increase their stability to light and weathering, and improves the chemical and physical properties of plastics by preventing creep. It is very stable to heat and light and imparts a glossy finish to extruded materials. It is also used in paper finishes and makes printers ink water resistant.

Dicyclohexyl adipate [*849-99-0*], $C_{18}H_{30}O_4$, M_r 294.50, *mp* 35.5 °C, *bp* 324 °C, d^{20} 1.037, n^{20} 1.4720. This diester is prepared by esterification of adipic acid, transesterification of dimethyl adipate, or direct reaction of adipic acid with cyclohexene [68]. It is used as a plasticizer in vinyl copolymers, polystyrene, and paints. A recent patent indicates that it can be oxidized in high yield to adipic acid [68].

8.2. Cyclohexanone Oxime

Cyclohexanone oxime [*100-64-1*], $C_6H_{10}NOH$, M_r 113.16, *mp* 90.5 °C, *bp* 204 °C, sublimes at room temperature (0.6 kPa); it is a white crystalline solid that can be crystallized as prisms from hydrocarbon solvents. This compound is the most important commercial derivative of cyclohexanone.

It can be prepared by warming cyclohexanone with an aqueous mixture of hydroxylamine hydrochloride and sodium bicarbonate.

Large commercial quantities of cyclohexanone oxime are used as the intermediate in the preparation of caprolactam. The oxime undergoes a Beckmann rearrangement in the presence of sulfuric acid to give caprolactam.

The cyclohexanone oxime can also be reduced to a mixture of cyclohexylamine and dicyclohexylamine or it can be hydrolyzed to cyclohexanone.

8.3. Methylcyclohexanols and Methylcyclohexanones

Methylcyclohexanol, $CH_3C_6H_{10}OH$, M_r 114.19, is usually available as a mixture of the cis and trans isomers of 2-, 3-, and 4-methylcyclohexanol. These derivatives are prepared by hydrogenation of individual or mixed cresols. Table 3 gives the physical properties of the individual isomers and of the commercial mixture.

The commercial mixture is a colorless liquid and is used as a solvent for resins, oils, and waxes, as a blending agent in soaps, and as an antioxidant in lubricants [69], [70]. It is low in toxicity, and the ACGIH has set TLVs of 75 ppm, 350 mg/m^3 for short-term exposure (STEL) and 50 ppm, 235 mg/m^3 for repeated 8-h exposures (TWA); the TLVs – STEL are intended to be deleted (for 1986 – 1987) [71]. The MAKs are the same as the TLVs – TWA [72]. Excessive exposure causes headaches and eye and nose irritation.

Table 3. Physical properties of methylcyclohexanols and methylcyclohexanones

Isomer	CAS registry number	mp, °C	bp, °C	d^{20}	n	Flash point, °C
Methylcyclohexanols						
DL-cis-2-Methylcyclohexanol	[7443-70-1]	7	165.2	0.934		
DL-trans-2-Methylcyclohexanol	[7443-52-9]	−4	166.5	0.924		
D-cis-3-Methylcyclohexanol	[5454-79-5]	−4.7	174.0	0.917		
L-trans-3-Methylcyclohexanol	[7443-55-2]	−1	175.0	0.915		
cis-4-Methylcyclohexanol	[7731-28-4]	−9.2	174.0	0.912		
trans-4-Methylcyclohexanol	[7731-29-5]		173.0	0.916		
Commercial mixture	[25639-42-3]	−50	173 – 175	0.913		154
Methylcyclohexanones						
DL-2-Methylcyclohexanone	[583-60-8]	−13.9	165.1	0.925	1.4440 (25 °C)	48
DL-3-Methylcyclohexanone	[591-24-2]	−73.5	170.0	0.920	1.4449 (20 °C)	48
4-Methylcyclohexanone	[589-92-4]	−40.6	171.3	0.916	1.4451 (20 °C)	48
Commercial mixture	[1331-22-2]		165 – 171	0.918 – 0.925		

Methylcyclohexanone is normally sold as a mixture of three isomers (2-methylcyclohexanone, 3-methylcyclohexanone, and 4-methylcyclohexanone). The individual isomers can be isolated; their physical properties, along with those of the commercial mixture, are shown in Table 3.

Methylcyclohexanones can be prepared by mild oxidation of the corresponding alcohols or by direct hydrogenation of cresol, in a process similar to the hydrogenation of phenol to cyclohexanone.

The technical-grade commercial mixture contains all three isomers. At 20 °C its solubility in water is 1.5%, and 3% water is soluble in the methylcyclohexanones.

Methylcyclohexanones are used as solvents in the dye and resin industries. These compounds are moderately toxic to laboratory animals and cause eye, nose, and throat irritation; they have produced corneal damage to the eyes of rabbits. The single dose oral LD_{50} for rats is ca. 2 g/kg [73].

2-Methylcyclohexanone may be absorbed through the skin. Its TLVs – TWA and MAKs are 50 ppm, 230 mg/m^3; the TLVs–STEL are 75 ppm, 345 mg/m^3.

8.4. Trimethylcyclohexanols and Trimethylcyclohexanones

3,3,5-Trimethylcyclohexanol (6), $C_9H_{18}O$, M_r 142.23, is a colorless, crystalline material, which is commercially produced as a mixture of the cis and trans isomers. The individual isomers can be separated by careful distillation or crystallization.

The physical properties of the two isomers are as follows: cis isomer [933-48-2], mp 37.3 °C, n^{60} 1.4390, d^{60} 0.860, vapor pressure 1.3 kPa (at 86 °C), flash point 88 °C, solubility in water 0.19% (at 60 °C), water solubility in alcohol 4.0% (at 60 °C); trans isomer [767-54-4], mp 57.3 °C, n^{60} 1.4390, d^{60} 0.862, vapor pressure 1.3 kPa (at 76 °C), flash point 76 °C, solubility in water 0.17% (at 60 °C), water solubility in alcohol 2.6% (at 60 °C).

Trimethylcyclohexanol is miscible in all proportions with most organic solvents. It is a cyclic alcohol with similar properties to cyclohexanol. Oxidation of trimethylcyclohexanol with nitric acid gives a mixture of 2,2,4-trimethyladipic acid and 2,4,4-trimethyladipic acid.

Trimethylcyclohexanol (**6**) is prepared by hydrogenating 3,5,5-trimethyl-2-cyclohexen-1-one [78-59-1] (**5**), isophorone (a condensation product of acetone):

$$\underset{5}{\underset{\substack{H_3C \\ H_3C}}{\bigodot}{\overset{O}{\diagup}}{\diagdown}{CH_3}} + 2\,H_2 \xrightarrow{Cat.} \underset{6}{\underset{\substack{H_3C \\ H_3C}}{\bigodot}{\overset{OH}{\diagup}}{\diagdown}{CH_3}}$$

The best catalyst for the hydrogenation is either nickel or a mixture of nickel and copper on either silica gel or alumina. The hydrogenation is carried out in high yield at 120–200 °C and 1–2 MPa. The byproducts of the reaction are small amounts of trimethylcyclohexanone, trimethylcyclohexane, and trimethylcyclohexene.

The most important use for the alcohol is as a feedstock for trimethyladipic acid [74]–[78]. This acid is used as a mixture of 2,2,4-trimethyladipic acid [3586-39-8] and 2,4,4-trimethyladipic acid [3937-59-5] in the manufacture of plasticizers and lubricants, and can be converted to dinitriles, diamines, or glycols. Trimethylcyclohexanol can be used instead of cyclohexanol in polishes and waxes. Because of its lower vapor pressure and unique structure, trimethylcyclohexanol is better than cyclohexanol in varnishes and shellacs.

The trimethylcyclohexanol that is sold commercially is a mixture of the cis and trans isomers. Analysis of a typical commercial product indicates 98% trimethylcyclohexanols, 1% trimethylcyclohexanone, 0.4% trimethylcyclohexane, 0.3% trimethylcyclohexene, 0.1% isophorone, and 0.2% water.

Trimethylcyclohexanol can be stored and transported in iron or steel drums, containers, tank cars, or tank trucks. It is stable indefinitely and does not need to be blanketed with an inert gas.

Trimethylcyclohexanol has the same toxicological properties as methylcyclohexanol, but because of its lower vapor pressure, it presents less hazards. Neither TLVs nor MAKs have been adopted.

3,3,5-Trimethylcyclohexanone [873-94-9], $C_9H_{16}O$, M_r 140.24, is a colorless liquid having properties similar to those of cyclohexanone. It has the following physical properties: mp −10 °C, bp 188.8 °C, d^{20} 0.888, n^{20} 1.4455, flash point 72 °C, solubility (at 20 °C) of ketone in water 0.3%, water in ketone 1.4%. It is miscible in all proportions with most organic solvents.

Trimethylcyclohexanone is also prepared by hydrogenating trimethylcyclohexenone (**5**). However, the olefin hydrogenation can be carried out without hydrogenating the ketone by using a palladium-supported catalyst at 110–170 °C and 0.2–6 MPa [79]. The combination of catalyst, reaction conditions, and amount of hydrogen determines whether the ketone or alcohol is produced.

Uses. Trimethylcyclohexanone is a good solvent for low molecular mass poly(vinyl chloride)-type alkyd resins and for nitrocellulose. It can be used in place of methylcyclohexanone, where its higher boiling point results in superior performance in some applications.

The toxicity of trimethylcyclohexanone is similar to that of methylcyclohexanone, but its lower vapor pressure makes it less dangerous to handle.

9. Economic Aspects

The mid-1982 price for *cyclohexanol* was $ 1.42/kg in the United States as opposed to $ 0.62/kg in 1974 [80]. The quoted price for technical-grade cyclohexanol in November 1986 was $ 1.32/kg [81]. The United States sales of cyclohexanol in 1971 were 4.3×10^3 t [67]. The remaining Western world sales were believed to be about half the United States sales. The sales volume is not believed to have changed substantially between 1971 and 1986.

In 1981, the United States sales of *cyclohexanone* were 16×10^3 t, not substantially different from 15×10^3 t in 1970 [67]. The mid-1982 cost of cyclohexanone was $ 1.32/kg [80], whereas the quote for technical-grade cyclohexanone in November 1986 was $ 1.25/kg [81].

This sales volume for either chemical does not represent capacity, as the vast majority of both are consumed within the manufacturing companies to make adipic acid or caprolactam. In 1983, the worldwide capacity of cyclohexanol–cyclohexanone for conversion to caprolactam was ca. 1.6×10^6 t [82], and that available for adipic acid manufacture was ca. 1.2×10^6 t. Of the total 2.8×10^6 t consumed, ca. 60% was cyclohexanone with the remainder being cyclohexanol.

10. Toxicology and Occupational Health

Cyclohexanol and cyclohexanone are moderately toxic [83] and have no apparent carcinogenicity. For cyclohexanol the OSHA TLV for exposure to vapor for 8 h is 50 ppm [84]. The ACGIH has established TLVs–TWA for exposure to cyclohexanol vapor of 50 ppm, 200 mg/m^3 [71]. The TLVs–TWA for cyclohexanone are 25 ppm,

100 mg/m^3; the TLVs–STEL of 100 ppm and 400 mg/m^3 are intended to be deleted (in 1986–1987). The MAKs for both compounds are 50 ppm, 200 mg/m^3.

Excessive exposure to vapor of either chemical results in irritation to the eyes, nose, and throat. Exposure of animals to a high concentration of cyclohexanol vapor resulted in irritation, lacrimation, salivation, lethargy, incoordination, narcosis, mild convulsions, and, in some cases, death. Degenerative changes were noted in the brain, kidney, and liver of rabbits exposed repeatedly to high vapor concentrations of cyclohexanol [83]. Exposure of animals to high concentrations of cyclohexanone vapor also resulted in irritation, lacrimation, and salivation. Decreased heart rate and depression of the central nervous system were also noted [85]. Degenerative changes were noted in the kidney and liver of monkeys and rabbits exposed repeatedly to high vapor concentrations of cyclohexanone.

For single oral dosages of cyclohexanol, the LD$_{50}$ for a rat is 2.06 g/kg [86] and the minimum lethal dose for a rabbit is 2.4 g/kg [73]. With cyclohexanone, the LD$_{50}$ for a rat is 1.62 g/kg [87] and the minimum lethal dose for a rabbit is 1.6–1.9 g/kg [73].

Cyclohexanol and cyclohexanone are moderately irritating to the eye and can cause corneal injury [87], [88]. Cyclohexanol can be absorbed through the skin in toxic amounts, and extensive exposure can cause tremors, narcosis, hypothermia, and death [73]. Cyclohexanone is only moderately irritating to the skin of animals when exposed for prolonged periods of time [89].

11. References

[1] G. Schwarzenbach, E. Felder, *Helv. Chim. Acta* **27** (1944) 1044. G. Schwarzenbach, C. Witwer, *Helv. Chim. Acta* **30** (1947) 659, 669.
[2] C. S. Hughes (compiler): *Chemical Economics Handbook,* SRI International, Menlo Park, Calif., Sept., 1983, 638-5062-0.
[3] BASF, DE 352 439, 1913.
[4] Tetralin Ges., DE 299 012, 1916.
[5] *Chem. Tech. (Leipzig)* **18** (1966) no. 10, 608–613.
[6] Stamicarbon, US 3 305 586, 1967 (P. Bernard).
[7] Stamicarbon, GB 890 095, 1960.
[8] G. D. Lyubarskii, G. K. Kervalishvili, *Khim. Promst. (Moscow)* **7** (1972) 491.
[9] Allied, US 2 829 166, 1958 (G. G. Joris, J. Vitrone).
[10] Du Pont, US 3 530 185, 1970 (K. Pugi).
[11] Du Pont, US 3 957 876, 1976 (M. Rapoport, J. O. White).
[12] Union Carbide, US 3 404 185, 1968 (T. Adams).
[13] Celanese, US 3 598 869, 1971 (P. J. Volpe, W. J. Humphrey).
[14] Du Pont, US 3 987 100, 1976 (W. J. Barnett, D. L. Schmitt, J. O. White).
[15] Rhône-Poulenc, US 3 510 526, 1970 (J. P. M. Bonnart, Y. Bonnet, P. P. M. Rey).
[16] Arco, US 3 949 004, 1976 (H. A. Sorgenti, S. N. Rudnick).
[17] Rhône-Poulenc, BE 731 125, 1969.
[18] Arco, GB 1 335 296, 1973 (H. A. Sorgenti, S. N. Rudnick).

[19] Du Pont, US 2 851 496, 1958 (H. L. Cates, R. W. Wheatcroft, A. B. Stiles, J. O. Punderson).
[20] Rhône-Poulenc, US 3 925 316, 1975 (J-C. Brunie, N. Crenne, F. Maurel).
[21] Rhône-Poulenc, US 3 923 895, 1975 (M. Costantini, N. Crenne, M. Jouffret, J. Nouvel).
[22] Rhône-Poulenc, US 3 928 452, 1975 (J-C. Nouvel, M. Costantini, N. Crenne, M. Jouffret).
[23] Rhône-Poulenc, US 3 927 105, 1975 (J-C. Brunie, N. Crenne).
[24] Stamicarbon, US 3 941 845, 1976 (W. Vvoskuil, J. J. M. van der Donck).
[25] Stamicarbon, US 3 987 101, 1976 (J. Wolters, J. L. J. P. Hennekens).
[26] Stamicarbon, US 4 042 630, 1977 (J. Wolters, J. L. J. P. Hennekens).
[27] Du Pont, GB 777 087, 1957.
[28] Rhône-Poulenc, BE 715 662, 1968.
[29] Rhône-Poulenc, GB 1 229 734, 1971 (J-C. Brunie, N. Crenne).
[30] Rhône-Poulenc, US 3 923 895, 1975 (M. Costantini, N. Crenne, M. Jouffret, J. Nouvel).
[31] Phillips, US 2 931 834, 1960 (W. W. Crouch, J. C. Hillyer).
[32] BASF, GB 1 382 849, 1975.
[33] BASF, GB 1 382 849, 1975 (O. A. Grosskinsky, G. Herrmann, R. Kaiser, A. Kuessner).
[34] Stamicarbon, US 4 238 415, 1980 (W. O. Bryan).
[35] Stamicarbon, EP 0 004 105, 1979 (W. O. Bryan).
[36] Rhône-Poulenc, US 3 557 215, 1971 (J. P. M. Bonnart, Y. Bonnet, P. P. M. Rey).
[37] Rhône-Poulenc, US 3 694 511, 1972 (J. Nouvel).
[38] Stamicarbon, US 3 927 108, 1975 (C. G. M. van der Moesdijk, A. M. J. Thomas).
[39] Stamicarbon, US 3 937 735, 1976 (P. L. M. Dois).
[40] Rhône-Poulenc, GB 1 112 837, 1968.
[41] Rhône-Poulenc, GB 1 221 629, 1971 (J. Nouvel).
[42] Halcon, US 3 317 614, 1967 (W. C. Long).
[43] Halcon, US 3 665 028, 1972 (J. L. Russell).
[44] Institut Français du Pétrole, FR 1 556 979, 1969 (J. Alagy, A. Zuech, B. Cha).
[45] Institut Français du Pétrole, FR 1 573 834, 1969 (J. Alagy, L. Asselineau, C. Busson, B. Cha).
[46] Du Pont, BE 782 721, 1972 (W. C. Drinkard).
[47] BASF, DE 2 124 590, 1972 (R. Platz, W. Fuchs, C. Dudeck).
[48] Ube Industries, JP 7 809 746, 1978 (T. Kawahito, M. Tamure).
[49] F. Andreas, *Chem. Tech. (Leipzig)* **18** (1966) no. 10, 608.
[50] J. W. Bruce, *Ind. Chem.* 1963, no. 3, 121.
[51] Scientific Design, BE 598 094, 1960.
[52] Monsanto, US 2 970 172, 1959 (D. R. Cova).
[53] Wingfoot Corp., US 2 218 457, 1937 (C. F. Winans).
[54] Du Pont, US 2 163 284, 1937 (W. A. Lazier).
[55] Monsanto, US 2 970 172, 1959 (D. R. Cova).
[56] L. K. Filippenko, K. N. Belonogov, V. P. Gostikin, *Izv. Vyssh. Uchebn. Zaved. Khim. Tekhnol.* **13** (1970) 3.
[57] F. Laucht, US 2 338 445, 1939.
[58] I. G. Farben, DE 743 004, 1940 (F. Laucht, O. Klopfer).
[59] K. Smeykal, H.-J. Naumann, *Chem. Tech. (Leipzig)* **13** (1961) no. 3, 132.
[60] Allied, US 3 974 221, 1976 (J. Duggan).
[61] BASF, US 3 149 166, 1964 (G. Poehler, A. Wegerich, H. Giehne, O. Goerhre).
[62] S. D. Nogare, J. Mitchell, Jr., *Anal. Chem.* **25** (1953) 1376.
[63] N. A. Cheronics, T. S. Ma: *Organic Functional Group Analysis by Micro and Semimicro Methods*, Interscience, New York 1964, pp. 493–496.

[64] J. Mitchell, Jr., D. M. Smith, W. M. D. Bryant, *J. Am. Chem. Soc.* **63** (1941) 573.
[65] W. M. D. Bryant, D. M. Smith, *J. Am. Chem. Soc.* **57** (1935) 57.
[66] R. M. Graziano, Tariff 31, Hazardous Material Regulations of the Department of Transportation, Association of American Railroads, Bureau of Explosives, Mar. 31, 1977.
[67] C. S. Hughes (compiler): *Chemical Economics Handbook*, SRI International, Menlo Park, Calif., Apr., 1983, 638.7020-A.
[68] Toray, GB 1 402 480, 1975.
[69] A. K. Doolittle: *The Technology of Solvents and Plasticizers*, J. Wiley & Sons, New York 1954.
[70] I. Mellon: *Industrial Solvents*, Reinhold Publ., New York, 1950, pp. 521–522.
[71] ACGIH (ed.): *Threshold Values (TLV) and Biological Exposure Indices*, ACGIH, Cincinnati, Ohio, 1986–1987.
[72] DFG (ed.): *Maximum Concentrations at the Workplace (MAK)*, VCH Verlagsgesellschaft, Weinheim 1986.
[73] J. F. Treon, W. E. Crutchfield, Jr., K. V. Kitzmiller, *J. Ind. Hyg. Toxicol.* **25** (1943) 199.
[74] VEBA, DE 1 111 163, 1959.
[75] VEBA, DE 1 229 510, 1963 (K. Schmitt, H. Heumann, W. Pollack).
[76] VEBA, DE 1 418 067, 1959.
[77] VEBA, DE 1 418 074, 1959 (E. Rindtorff, K. Schmitt, H. Heumann).
[78] VEBA, DE 1 468 693, 1963 (K. Schmitt, H. Neumann).
[79] VEBA, US 3 361 822, 1968 (K. Schmitt, J. Disteldorf, H. Schnurbusch, W. Hilt).
[80] C. S. Hughes (compiler): *Chemical Economics Handbook*, SRI International, Menlo Park, Calif., Apr., 1983, 638.7020-F.
[81] *Chem. Mark. Rep.* 1986 (Nov. 10), 38.
[82] C. S. Hughes (compiler): *Chemical Economics Handbook*, SRI International, Menlo Park, Calif., Aug., 1983, 625.2032-A.
[83] G. D. Clayton, F. E. Clayton (eds.): *Patty's Industrial Hygiene and Toxicology*, 3rd ed., Wiley-Interscience, New York 1982, pp. 4643–4649.
[84] U.S. Government, CFR 29, Washington, D.C., Jul. 1, 1978, Section 1910.1000.
[85] H. Specht, J. W. Miller, P. J. Valaer, R. R. Sayers: "Acute Response of Guinea Pigs to the Inhalation of Ketone Vapors," U.S. Public Health Service, NIH Bulletin, no. 176, Division of Industrial Hygiene, 1940, pp. 1–66.
[86] F. Bar, F. Griepentrog, *Med. Ernaehr.* **8** (1967) 244.
[87] H. F. Smyth, Jr., C. P. Carpenter, C. S. Weil, U. C. Pozzani et al.: "Range Finding Toxicity Data VII," *Am. Ind. Hyg. Assoc. J.* **30** (1969) 470–476.
[88] J. Pohl, *Zentralbl. Gewerbehyg. Unfallverhuet.* **12** (1925) 91.
[89] P. K. Gupta, W. H. Lawrence, J. E. Turner, J. Autian: "Toxicological Aspects of Cyclohexanone," *Toxicol. Appl. Pharmacol.* **49** (1979) 525–533.

Cyclopentadiene and Cyclopentene

KARL GRIESBAUM, Universität Karlsruhe (TH), Karlsruhe, Federal Republic of Germany (Chaps. 1 and 2)
DIETER HÖNICKE, Universität Karlsruhe (TH), Karlsruhe, Federal Republic of Germany (Chaps. 1 and 2)
MICHAEL OLSON, General Motors Research Laboratories, Warren, Michigan 48090, United States (Chap. 3)

1.	Cyclopentadiene and Dicyclopentadiene.............. 1825	2.1.	Physical Properties 1832
1.1.	Physical Properties 1825	2.2.	Chemical Properties........ 1833
1.2.	Chemical Properties........ 1826	2.3.	Production 1835
1.3.	Production 1829	2.4.	Uses 1837
1.4.	Uses 1831	3.	Toxicology............... 1837
2.	Cyclopentene 1832	4.	References............... 1839

1. Cyclopentadiene and Dicyclopentadiene

The physical and chemical properties of cyclopentadiene and dicyclopentadiene have been reviewed elsewhere [1], [3]–[4]

1.1. Physical Properties

Cyclopentadiene [542-92-7], C_5H_6, M_r 66.10, is a colorless liquid having a sweet terpenic odor. It is also known as pentole and pyropentylene. It is soluble in most organic solvents, including ethanol, diethyl ether, acetone, petroleum ether, benzene, and toluene, but is relatively insoluble in water.

The boiling points and compositions of three hydrocarbon azeotropes with cyclopentadiene are listed below:

Second component	bp (0.1 MPa), °C	Cyclopentadiene, mole fraction
n-Pentane	35.3	0.31
2-Methyl-2-butene	38.0	0.37
cis-2-Pentene	36.9	0.03

Dicyclopentadiene [77-73-6], $C_{10}H_{12}$, M_r 132.20, colorless crystals, has a camphoraceous odor and is soluble in ethanol, diethyl ether, and acetic acid,

The physical properties of cyclopentadiene and dicyclopentadiene are listed in Table 1; thermal and thermodynamic properties are summarized in Table 2. The vapor pressure of cyclopentadiene (MPa) over the range 220–320 K is represented by

$$\log p = 3.8892 - 1530.49/T$$

The vapor pressure of dicyclopentadiene (MPa) over the range 305–450 K is represented by

$$\log p = 3.6172 - 2056.49/T$$

Additional properties including enthalpy, entropy, and free enthalpy of formation are available [5]. The IR [6], ^{1}H NMR [7], and mass spectra [8] of cyclopentadiene and the ^{1}H NMR spectrum [9] of dicyclopentadiene have been reported.

1.2. Chemical Properties

Cyclopentadiene (**1**) is an extremely reactive compound. Neat it can only be preserved at ≤ -80 °C, and then only for a limited amount of time. At or above room temperature, it dimerizes spontaneously to give **2**.

The dimerization rate r_D is a function of temperature [1, p. 423]:

T, °C	−20	0	10	15	20	25	30	40
r_D, mol%/h	0.05	0.5	1	1.5	2.5	3.5	6	15

Above 160 °C, this dicyclopentadiene dissociates into cyclopentadiene again. Therefore, cyclopentadiene is stored and handled as its stable dimer. The dimerization of cyclopentadiene is exothermic and may get out of control; leading to polymerization and even coking in sealed vessels that lack effective heat removal.

Reactions of cyclopentadiene with aggressive reagents, such as concentrated sulfuric acid, fuming nitric acid, or ozone, can lead to decomposition or even explosion, depending on the conditions, particularly the reagent:substrate ratio. If proper care is taken, however, selective reaction can be achieved with most reagents. Major reaction

Table 1. Physical properties of cyclopentadiene and dicyclopentadiene

Physical property	Cyclopentadiene	Dicyclopentadiene
Boiling point at 101.3 kPa, °C	40.2	172.8
Freezing point, °C	− 97.2	32.5
Density, g/cm^3	0.8021 (20 °C)	0.9770 (35 °C)
Refractive index n_D	1.440 (20 °C)	1.5050 (35 °C)
Dielectric constant at 40 °C	2.43	
Dipole moment, vapor, D	0.416	
Surface tension at 40 °C, N/m	3.15 × 10^{-2}	
Critical pressure, MPa	5.17	3.04
Critical temperature, °C	232	383

Table 2. Thermal and thermodynamic properties of cyclopentadiene and dicyclopentadiene

Property	Cyclopentadiene	Dicyclopentadiene
Ignition temperature. °C	640	510
Flash point, °C	< − 50	41
Heat of combustion, kJ/kg	53 649	41 870
Heat of vaporization, kJ/kg	418.05 (40.2 °C)	321.8 (172.8 °C)
Heat of fusion at 32.5 °C, kJ/kg		95 385
Heat of dimerization at 25 °C, kJ/kg	584 ± 20	
Specific heat capacity, kJ kg^{-1}K^{-1}		
liquid	1.780	1.311
gas, 0.1 MPa	1.074 (25 °C)	0.935 (33 °C)
Vapor pressure at 20 °C, kPa	45.1	0.04

types are cycloadditions and additions to the double bonds, oligo- and polymerizations, substitutions at the reactive methylene group, and oxidations.

A variety of *cycloaddition reactions* have been realized (Fig. 1). The formation of **3** [10] and **4** [11] are examples of [2 + 2] and [2 + 3] cycloadditions, respectively. By far the most important and most versatile types are [4 + 2] cycloadditions, the Diels – Alder reactions. They afford the ring systems **5, 6, 7, 8,** and **9** by reaction of cyclopentadiene with monoolefins, acetylenes [12], [13], conjugated diolefins [14], [15], [16], cumulated diolefins [17], and activated ketones [18]. Diels – Alder reactions with monoolefins and acetylenes are particularly facile with substrates bearing one (R^1) or two (R^1 and R^2) electron-withdrawing groups, such as carbonyl-, cyano-, or halomethane moieties. Reactions with a large variety of such substrates have been reported. Diels – Alder additions of cyclopentadiene with cyclic substrates afford tricyclic ring systems that exhibit predominantly endo geometry, as in the case of dicyclopentadiene (**12**) and the Diels – Alder adduct **13** of cyclopentadiene with maleic anhydride.

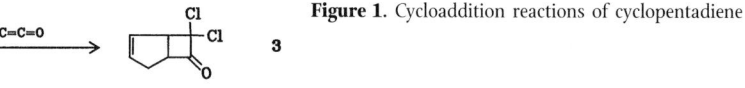

Figure 1. Cycloaddition reactions of cyclopentadiene

The formation of compounds **10** [19] and **11** [20] are examples of [4 + 3] and [4 + 4] cycloadditions of cyclopentadiene, respectively.

Addition reactions have been realized with a variety of reagents. Examples are the selective hydrogenation to give cyclopentene, hydrohalogenations to give 3-halocyclopentenes, catalyzed hydrocyanation to give cyanocyclopentene and dicyanocyclopentane [21], and halogenations to give 3,5-dihalocyclopentenes and 1,2,3,4-tetrahalocyclopentanes. Tetrachlorocyclopentane can be readily converted into hexachlorocyclopentadiene: catalytic chlorination of the tetrachlorocyclopentane gives octachlorocyclopentane, which is dehydrochlorinated.

Oligo- and polymerizations can be carried out thermally or catalytically. Thermal reactions have been reported to occur by a series of consecutive cycloadditions of cyclopentadiene with Diels–Alder adducts like **14** to give products of structure **15** [22], [23].

A special case is the thermally induced homopolymerization of cyclopentadiene via dicyclopentadiene (**12**) to give crystalline oligomers or amorphous polymers, depending on the reaction temperature and time. Polymerization of cyclopentadiene with cationic catalysts gives cross-linked elastomeric materials [24]; with Ziegler-type catalysts, these polymerizations give elastomers. In addition, numerous copolymers of cyclopentadiene, e.g., with aromatics and/or olefins, drying oils, and indene – cumaron resins, have been reported.

Substitution reactions at the reactive methylene group occur predominantly by initial deprotonation, giving the resonance-stabilized cyclopentadienyl anion, which exhibits a quasi-aromatic character. Products derived from it are organometallics, such as alkali-metal and Grignard compounds and especially transition-metal complexes. A variety of the latter have been synthesized following the discovery of ferrocene. Other notable substitution reactions at the methylene group are mono- and dialkylations with alkyl halides, base-catalyzed carboxylation with carbon dioxide, and base-catalyzed condensations with aldehydes and ketones to give fulvenes. The last has been used to prepare a wide variety of fulvenes by starting from acyclic, carbocyclic, and heterocyclic [25] carbonyl compounds in efficient, high-yield syntheses [26].

Oxidation reactions with molecular oxygen occur spontaneously and provide intractable peroxidic materials. Controlled photochemical oxidation with oxygen gives the cyclic peroxide **16** [27], which can be converted into **17, 18,** or **19**

by reduction. Reaction with oxygen in the gas phase catalyzed heterogeneously by various mixed metal oxides has been reported to produce maleic anhydride [28], [29] and in one case a mixture of maleic and phthalic anhydride [30]. Heterogeneously catalyzed ammoxidation of cyclopentadiene affords pyridine [31].

1.3. Production

Cyclopentadiene is available from coal tar as dicyclopentadiene in quantities of ca. 10 – 20 g per tonne of coal and can be readily isolated [32]. Considerably larger quantities are available from steam cracking operations, e.g., of naphtha, which affords ca. 14 kg of dicyclopentadiene per tonne of feed material [32] – [34]. In this case, the C_5

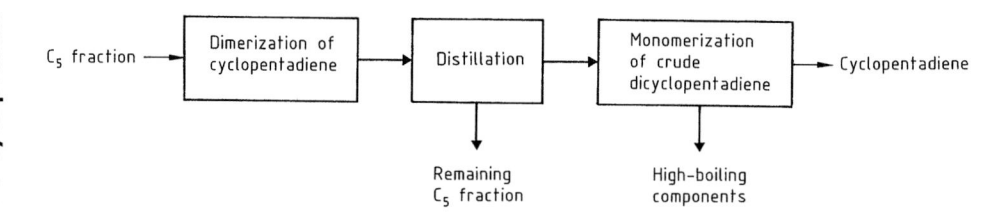

Figure 2. Block diagram for the separation of cyclopentadiene from a C_5 fraction

fraction of pyrolysis gasoline contains ca. 25 wt% cyclopentadiene with approximately equal proportions of isoprene and of *n*-pentane, along with other C_5 hydrocarbons.

From the C_5 fraction, cyclopentadiene can be isolated by the sequence of steps shown in Fig. 2.

In the first step, cyclopentadiene is dimerized to give dicyclopentadiene by heat soaking the entire C_5 fraction, either at normal pressure and 30–100 °C within 5–24 h or at elevated pressure and 140–150 °C [32], [35]. In the second step, the remaining components of the original C_5 fraction, which have boiling points of 28–50 °C, are distilled overhead, and crude dicyclopentadiene (*bp* 172.8 °C) is obtained in 85–90% purity [32] at the bottom of the column. In the third step, the crude dicyclopentadiene is monomerized, either in the liquid phase at 170–200 °C [33] or in the gas phase at 300–400 °C in the presence of diluents such as steam, hydrogen, nitrogen, or methane [1, p. 424], to give cyclopentadiene in ca. 95% purity [32].

Cyclopentadiene can be further purified by sequential dimerizations under carefully controlled slow heat soaking [32] and subsequent monomerizations. This procedure largely avoids the codimerization of cyclopentadiene with residual unsaturated C_5 impurities, allowing the latter to be removed from dicyclopentadiene by stripping.

Alternate modes of formation of cyclopentadiene are heterogeneously catalyzed monomerizations of dicyclopentadiene [36], [37]; catalyzed dehydrocyclization of piperylene [38], [39]; catalyzed dehydrogenations of cyclopentene [40], [41] and cyclopentane [42]; catalyzed hydrodenitrification of pyridine, quinoline [43], and other aromatic amines [44], and the pyrolysis of phenol [45], [46]. An alternate possibility for the isolation of cyclopentadiene from the C_5 fraction of pyrolysis gasoline is the selective formation of a Diels–Alder adduct by introducing the C_5 fraction into aqueous maleic acid. The pure Diels–Alder adduct precipitates, whereas the remaining dienes of the C_5 fraction do not react [47]–[49].

1.4. Uses

Cyclopentadiene is the single major precursor for the production of cyclopentene by selective hydrogenation. Ethylidenenorbornene (**8**; R = CH$_3$), which can be produced by Diels–Alder addition of cyclopentadiene with 1,2-butadiene or with 1,3-butadiene followed by isomerization of the ensuing adduct **7**, serves as a termonomer in the production of ethylene–propylene terpolymer elastomers. Polymerization of **8** occurs through the more reactive endocyclic double bond, whereas the exocyclic double bond imparts the active site for vulcanization. The facile reversibility of the Diels–Alder reaction has been used for the transient protection of α,β-unsaturated carbonyl compounds from polymerization while the compounds were chemically modified at or near the carbonyl functions [50]. By this method, isoprenic alcohols such as linalool and nerolidol have been synthesized in high yields starting from methyl vinyl ketone [51]. Substituted pentacycles such as **21**, which are readily obtained from Diels–Alder adducts **20** of cyclopentadiene with substituted paraquinones, have been shown to undergo facile acid-catalyzed ring opening to the tricyclic system **22**, which in turn is completely reconverted to **21** in the presence of light. Thus, the system bears prospects for light–energy conversion by storing energy in the strained pentacyclic system **21** [52]. Diels–Alder adducts of fulvenes with activated olefins have been converted into bicyclic ketones such as **23**, which are useful in perfume formulations.

Homopolymers of cyclopentadiene have been reported to be useful as modifiers for high molecular mass polymers as base materials for coatings, adhesives, paper-sizing agents, and printing inks. The Diels–Alder adduct **13** of cyclopentadiene and maleic anhydride is useful for the production of modified polyester resins [32].

Hexachlorocyclopentadiene, which was a substantial commercial product derived from cyclopentadiene, has decreased in importance considerably. Major products derived from it were the chlorinated pesticides aldrin, dieldrin, chlordane, and heptachlor, but their application has been drastically restricted due to their toxicity. Hexachlorocyclopentadiene has also been used for the preparation of flame- and fire-retardant chemicals. In particular, the Diels–Alder adduct with maleic anhydride has been used as a comonomer in the synthesis of fire-resistant polyesters.

Dicyclopentadiene has attracted attention as a building block for the production of modified hydrocarbon resins, which show increased reactivity in copolymerizations with drying oils and produce paint resins with improved drying rate, gloss, and hardness. Modifications of unsaturated polyester resins with dicyclopentadiene have also been reported [32]. Hydrogenation of dicyclopentadiene and subsequent acid-catalyzed isomerization of the ensuing saturated product **24** has opened a facile synthetic route to adamantane **25**.

2. Cyclopentene

The physical and chemical properties of cyclopentene have been reviewed elsewhere [2, pp. 593–605], [4].

2.1. Physical Properties

Cyclopentene [*142-29-0*], C_5H_8, M_r 68.12, is a colorless, pungent liquid. It is soluble in most organic solvents, including ethanol, diethyl ether, petroleum ether, and benzene. However, at 23 °C only 160 ppm of cyclopentene is soluble in water and 400 ppm of water is soluble in cyclopentene. Two typical organic azeotropes with cyclopentene are listed below:

Second component	bp (0.1 MPa), °C	Cyclopentene, mole fraction
Methanol	37	ca. 0.68
Methyl acetate	41.7	0.74

The physical properties of cyclopentene are listed in Table 3; thermal and thermodynamic properties are summarized in Table 4. The vapor pressure of cyclopentene (MPa) over the range 215–350 K is represented by

$$\log p = 3.6594 - 1482.13/T$$

Additional properties, including enthalpy, entropy, and free enthalpy of formation, can be found in [5]. The IR [6], ^1H NMR [7], and mass spectra [8] of cyclopentene have been reported.

Table 3. Physical properties of cyclopentene

Boiling point at 101.3 kPa, °C	44.2
Freezing point, °C	−135.1
Density at 20 °C, g/cm^3	0.7720
Dynamic viscosity at 13.5 °C, Pa s	4.29
Refractive index n_D^{20}	1.4225
Effect of temperature on n_D over the range 10–25 °C, K^{-1}	−0.0006
Dielectric constant at 20 °C	2.095
Dipole moment in CCl$_4$, D	0.93
Surface tension at 13.5 °C, N/m	2.356×10^{-2}
Critical pressure, MPa	4.75
Critical temperature, °C	232.94
Critical volume, cm^3/g	3.613

Table 4. Thermal and thermodynamic properties of cyclopentene

Ignition temperature, °C	385
Flash point, °C	< −50
Explosive limits in air at 0.1 MPa and 20 °C,	
vol%	3.4–8.5
g/m^3	103–258
Heat of combustion at 0.1 MPa and 30 °C, kJ/kg	45 756
Heat of vaporization at 44.2 °C, kJ/kg	394.77
Heat of fusion at −135.1 °C, kJ/kg	49.37
Specific heat capacity, kJ kg^{-1} K^{-1}	
liquid	1.799
gas (25 °C, 0.1 MPa)	1.103
Vapor pressure at 25 °C, kPa	48.9

2.2. Chemical Properties

Cyclopentene undergoes the usual reactions of monoolefins, i.e. cycloadditions and additions to the double bond, substitutions in the allylic positions, and oxidative as well as metathetic cleavage of the ring system.

Cycloaddition reactions occur with allene [53] and butynone [54] in a [2 + 2] fashion and with conjugated dienes such as cyclopentadiene and 1,3-cyclohexadiene. Photochemically induced cycloadditions with aromatic substrates such as benzene [55] or toluene [56] occur by meta addition. Palladium complex-catalyzed reaction of cyclopentene with dimethyl acetylenedicarboxylate results in a cocyclization reaction [57].

Addition reactions are numerous, such as hydrogenation to give cyclopentane, halogenations to give *trans*-1,2-dihalocyclopentanes, formylation to give formylcyclopentane, and hydrocarboxylation (Koch reaction) to give cyclopentanecarboxylic acid. Acid-catalyzed reactions with substituted benzenes lead to alkylation products, whereas the joint reaction of cyclopentene, benzene, and carbon monoxide leads to simultaneous alkylation and formylation to give 4-cyclopentylbenzaldehyde [58]. Epoxidation of cyclopentene has been carried out with peracids, hydrogen peroxide [59], [60], and oxygen by cooxidation with acetaldehyde [61]. Direct hydroxylation of cyclopentene to give 1,2-dihydroxycyclopentane has been achieved with modified permanganate [62] and with hydrogen peroxide–formic acid [63]. Cyclopentanone is obtained either by direct catalytic oxidation [64] or by catalytic hydration of cyclopentene and subsequent oxidative dehydrogenation of the intermediate cyclopentanol. Cyclopentanol can also be obtained by hydroboration of cyclopentene and subsequent oxidation [65]. Cyclopentene can be dimerized by homogeneous [66] and oligomerized by heterogeneous catalysis [67].

Substitution reactions occur predominantly in the allylic positions. Reactions with molecular oxygen lead to **26** as the primary product, both under autoxidation [68] and photooxidation [69]. Secondary reaction products are epoxides and alcohols [70]. Catalyzed photooxidations of cyclopentene lead to **27** [71] or **28** [72], depending on the catalyst used. Catalyzed cooxidation of cyclopentene and acetic acid affords **29** [73].

Ring cleavage and ring enlargement reactions occur under drastic oxidation or by metathesis. Ozonolysis in methanol can be guided by selective acid catalysis such that **30a** [74], **30b**, or **30c** [75] is formed. Catalyzed oxidations of cyclopentene with hydrogen peroxide and noncatalyzed oxidation with oxygen yield **30d** [76], and oxidation with nitric acid at elevated temperature and pressure gives **30e**. Oxidation of cyclopentene with oxygen on heterogeneous V_2O_5-based catalysts yields maleic anhydride [77], and with special catalysts, mixtures of maleic and phthalic anhydride result [78]. Heterogeneously catalyzed ammoxidation yields pyridine and acyclic nitriles [79]. Metathesis of cyclopentene catalyzed homogeneously produces a polymer of structure **31**, called polypentenamer [34].

$R–(CH_2)_3–R'$
30

	a	b	c	d	e
R	CH_3OOC	CH_3OOC	$O=CH$	$O=CH$	$HOOC$
R'	CH_3OOC	$(CH_3O)_2CH$	$(CH_3O)_2CH$	$O=CH$	$HOOC$

$$\left[\begin{array}{c} H \\ | \\ C=C-CH_2CH_2CH_2 \\ | \\ H \end{array}\right]_n$$
31

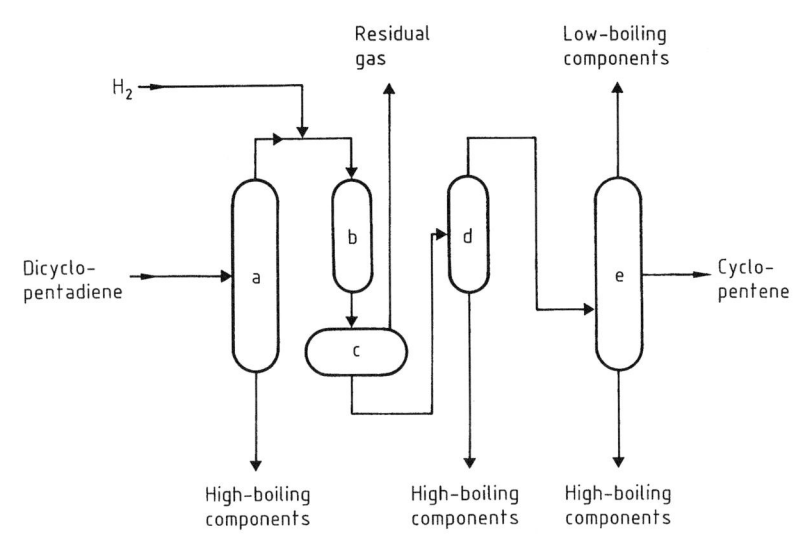

Figure 3. Bayer process for the production of cyclopentene from crude dicyclopentadiene
a) Cracking column; b) Hydrogenation reactor; c) Gas–liquid separator; d) Flash column; e) Distillation column

2.3. Production

Cyclopentene is formed in proportions of ca. 2.2 kg per tonne of feed in the steam cracking of naphtha and is present in ca. 4 wt % in the C_5 fraction of pyrolysis gasoline [32], [33]. Such a low proportion is insufficient for industrial production, and hence, cyclopentene is effectively produced solely by selective hydrogenation of cyclopentadiene, generally by heterogeneous catalysis of gas-phase reactions with a wide variety of supported metal catalysts [80]–[85] or by heterogeneous [86]–[89] or homogeneous [90]–[92] catalysis of liquid-phase reactions. Feed materials are either crude dicyclopentadiene or the entire C_5 fraction.

Figure 3 shows the flow scheme of the Bayer process for the production of cyclopentene from crude dicyclopentadiene derived either from coal tar or steam-cracking operations (Section 1.3).

Crude dicyclopentadiene is monomerized in cracking column a at 170–190 °C, whereby codimers function as a diluent for the feed and are continuously removed from the bottom of column a. Cyclopentadiene leaves the top of column a, is mixed with hydrogen, and hydrogenated on a palladium catalyst [93] at 70–80 °C under moderate pressure in reactor b. The effluent from b passes into c, where it is separated into residual gas and liquid products. The liquid fraction passes flash column d to remove the major part of high-boiling components and is subsequently fractionated in column e to give cyclopentene. The yield of cyclopentene is 85–90 %, based on monomeric cyclopentadiene; the selectivity is 90–95 %.

Figure 4 shows the flow scheme of the NMP process for the production of cyclopentene from a C_5 fraction.

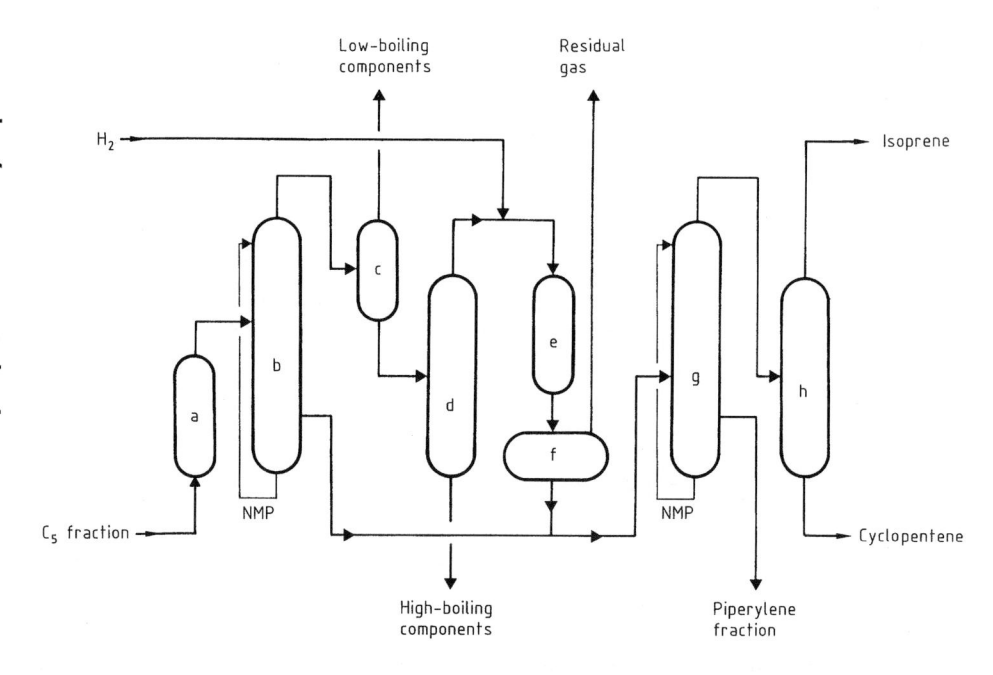

Figure 4. NMP process for the production of cyclopentene and isoprene from a C_5 fraction
a) Dimerization reactor; b) Liquid–liquid extractor; c) Distillation column; d) Cracking column; e) Hydrogenation reactor; f) Gas–liquid separator; g) Column for extractive distillation; h) Column for fractionated distillation

The C_5 fraction is introduced to reactor a, where cyclopentadiene is thermally dimerized. In column b the remaining diolefins of the C_5 fraction (mostly isoprene and piperylene) and cyclopentene are selectively extracted with *N*-methylpyrrolidone (NMP). The raffinate is distilled in column c to give the lower boiling olefinic and paraffinic C_5 hydrocarbons as distillates and crude dicyclopentadiene as bottoms. Crude dicyclopentadiene is monomerized in cracking column d to give cyclopentadiene at the top and high boilers at the bottom. Subsequently, cyclopentadiene is mixed with hydrogen and selectively hydrogenated in reactor e. In drum f, the effluent from e is separated into residual gas and liquid crude cyclopentene. The latter is combined with the mixture of C_5 diolefins and cyclopentene obtained after extraction in column b. The combined stream is processed by extractive distillation with NMP in column g to give piperylene at the bottom and a mixture of cyclopentene and isoprene at the top. This mixture is subsequently separated by fractional distillation in column h.

This process allows the simultaneous production of isoprene and cyclopentene as well as the isolation of the cyclopentene present in the C_5 fraction.

Alternate modes of formation of cyclopentene are catalytic cyclization of piperylene [94], metathesis of 1,6-cyclodecadiene and polypentenamer, and thermal isomerization of vinylcyclopropane.

2.4. Uses

At present, cyclopentene does not find substantial use in chemical syntheses. The broadest potential is offered by the production of polypentenamer, which has been tested extensively as an elastomer. Of potential use are also the methods for the selective oxidative cleavage of cyclopentene to give the various bifunctional products of structures **30a–30e**. Alkylcyclopentylbenzenes have low pour points and high flash points and are useful as working fluids, and cyclopentylphenols have been tested as disinfectants, antioxidants, and light stabilizers for polymers. Epoxycyclopentanol (**28**) is of potential use for the synthesis of prostaglandins.

3. Toxicology

Among the three chemicals, acute toxicity increases in the order cyclopentene, cyclopentadiene, dicyclopentadiene. However, cyclopentene is most readily absorbed through the skin.

Cyclopentadiene forms a dimer spontaneously, and the reported toxicity probably is that for the dimeric form. The oral LD_{50} in rats is 0.82 g/kg [95]. The dermal LD_{50} in rabbits is 6.72 mL/kg [95]. Subcutaneous injection of 0.5–1 mL of cyclopentadiene monomer had no detectable effects on rabbits; 3 mL produced narcosis and fatal convulsions [96]. Mild kidney and liver damage occurred after exposure of rats to 500 ppm for 35 daily exposures each of 7 h [97]. No toxicological effects were observed in rats, guinea pigs, rabbits, or dogs exposed to 250 ppm for 135 d, 7 h/d [97]. Dogs exposed to 400 ppm cyclopentadiene for 39 d, 6 h/d, then to 800 ppm for 16 d, 6 h/d, showed no abnormalities in hematological parameters, liver function, serum enzyme levels, electrocardiogram, or microscopic appearance of tissues [97]. There is no evidence to suggest mutagenic, carcinogenic, or teratogenic potential of cyclopentadiene. The odor of cyclopentadiene at 250 ppm is unpleasant; thus, human workplace exposure limits have been placed below this level. The ACGIH has adopted a TLV value for cyclopentadiene of 75 ppm and an STEL limit of 150 ppm [98]. The MAK value is 75 ppm [99].

Dicyclopentadiene, the tricyclic form of cyclopentadiene, is the most toxic of the cycloolefin compounds described under this keyword. Oral and intraperitoneal LD_{50} values in rats are 0.35 mL/kg and 0.31 mL/kg, respectively. The dermal LD_{50} value for rabbits is 5.08 mL/kg [100]. Exposure of rats to saturated dicyclopentadiene vapor for 1 h produced 100% mortality [100]. LC_{50} values for 4 h exposure to dicyclopentadiene vapor range from 146 ppm in mice to ca. 375 ppm in rats and 770 ppm in guinea pigs and rabbits [100]. Rats exposed to 0, 72, 146, and 332 ppm (7 h/d, 5 d/week, 10 d) exhibited treatment-related mortality only in the highest concentration group and no significant toxicological signs in the other groups [100]. Exposure to dicyclopentadiene at concentrations up to 74 ppm, 7 h/d, 5 d/week for 89 d, caused increased liver and kidney weights and renal tubular degeneration in male rats [100]. Dogs exposed to 0, 9, 23, and 32 ppm for 89 d showed no dose-related pathological changes and only minimal changes in serum enzyme levels [100]. Dicyclopentadiene is irritating to the eyes and respiratory tract of humans at concentrations as low as 1 ppm for 7 min. [100]. The olfactory threshold is ca. 0.003 ppm [97]. ACGIH has adopted a TLV of 5 ppm for dicyclopentadiene, but no STEL value has been proposed [98]. No MAK value has been adopted, but Soviet and Bulgarian workplace limits are 1 mg/m^3 or 0.185 ppm [99].

Cyclopentene is moderately toxic. In rats, the oral LD$_{50}$ is 2.14 mL/kg [101], and the 4-h LC$_{50}$ for cyclopentene vapor exceeds 8110 ppm [102]. However, concentrated vapors are lethal and exposure to 16 000 ppm for 4 h caused death in 4/6 rats [101]. Transdermal absorption occurs readily; the dermal LD$_{50}$ in rabbits is 1.59 mL/kg [101]. Subacute exposure (6 h/d, 5 d/week, 3 weeks) of rats to 0, 870, and 3100 ppm cyclopentene vapor produced no toxicologically significant effects [102]. Chronic inhalation (6 h/d, 5 d/week, 12 weeks) of up to 1139 ppm did not result in hematological alterations, changes in serum chemistry, or organ pathology in rats [102]. Cyclopentene has not been suggested to be mutagenic or carcinogenic and reproductive hazards are unknown. No specific human toxicity attributed to cyclopentene has been described; however, headache and unpleasant sensations were reported by humans exposed to 32 ppm for 10 min. [102]. The threshold for detection of the odor of cyclopentene is ca. 13 ppm [102]. No MAK value for cyclopentene has been officially adopted [99]; however, values in the range 10 – 15 ppm are suggested based on subjective effects [102]. No TLV or STEL limits for cyclopentene are in effect currently in the United States [98].

These chemicals are listed in the U.S. Environmental Protection Agency's inventory for the TSCA [103]. Little is known about the hazards of chronic exposure; however, these chemicals are not presently listed as candidates for genetic toxicity or carcinogenicity testing by the U.S. Public Health Service [104].

Signs and symptoms of human intoxication with these cycloolefins include irritation of the eyes, nose, throat, and skin, excitation, loss of coordination and equilibrium, hypothermia, dyspnea, central nervous system depression, stupor, and coma [105]. Toxic and irritant properties and the possibility of dermal absorption suggest strongly that prudent handling practices, including adequate ventilation, protective clothing, neoprene gloves, chemical goggles, and respirators (self-contained, full face-mask, organic vapor cartridge) be used when working with these chemicals.

4. References

[1] *Kirk-Othmer* **7,** 417–429.
[2] *Kirk-Othmer* **8,** 593–605.
[3] F. Asinger: *Die Petrolchemische Industrie,* Akademie-Verlag, Berlin 1971, pp. 447–451.
[4] C. T. Lin, F. K. Young, M. R. Brulé, L. L. Lee, K. E. Starling, J. Chao, *Hydrocarbon Process.* **59** (1980) 117–123.
[5] D. R. Stull, E. F. Westrum, G. C. Sinke: *The Chemical Thermodynamics of Organic Compounds,* J. Wiley and Sons, New York 1969, p. 346.
[6] API 41 Selected Infrared Spectral Data vol. II, Thermodynamics Research Center, Chem. Eng. Division of Texas Engineering Experiment Station, College Station, Texas 1983.
[7] W. Brügel: *Handbook of NMR Spectral Parameters,* vol. **1,** Heyden and Son Ltd., London 1979.
[8] A. Cornu, R. Massot: *Compilation of Mass Spectral Data,* vol. **2,** Heyden, London-New York-Rheine 1975.
[9] K. C. Ramey, D.C. Lini, *J. Magn. Reson.* **3** (1970) 94.
[10] T. Teiter, R. L. N. Harris, *Aust. J. Chem.* **32** (1979) 1329-1337.
[11] G. Bailo, P. Caramella, G. Cellerino, A. G. Invernizza, P. Gruenanger, *Gazz. Chim. Ital.* **103** (1973) 47–59.
[12] Esso Research and Engineering Co., DE-OS 2 234 087, 1972 (F. Baldwin, G. Sartori, J. Lefebvre).
[13] G. Maier, W. A. Jung, *Chem. Ber.* **115** (1982) 804–807.
[14] Asahi Chem. Industry Co., JP-Kokai 92 353, 1973 (S. Miyata).
[15] Mitsui Petrochem. Ind., JP 25 262, 1974 (M. Ogawa).
[16] ICI, DE-OS 2 441 433, 1975 (P. Jhawar).
[17] Japan. Geon Co., JP-Kokai 26 748, 1973 (R. Kita).
[18] B. F. Martynov, L. T. Lantseva, B. L. Dyatkin, *Zh. Org. Khim.* **11** (1975) 2282–2285.
[19] H. Mayr, I. Halberstadt, *Angew. Chem.* **92** (1980) 840–841; *Angew. Chem. Int. Ed. Engl.* **19** (1980) 814.
[20] R. L. Danheiser, S. K. Gee, H. Sard, *J. Am. Chem. Soc.* **104** (1982) 7670–7672.
[21] C. Y. Wu, H. E. Swift, *Prepr. Am. Chem. Soc. Div. Petr. Chem.* **25** (1980) no. 2, 372.
[22] Hitachi Chem, Co., JP-Kokai 31 970, 1972 (H. Kouchi, T. Akima).
[23] Daichi Kogyo Seiyaku Co., JP-Kokai 39 689, 1974 (I. Morita, H. Shimoda, I. Nishizawa).
[24] G. Heublein, B. Adelt, *Plaste Kautsch.* **19** (1972) 728–730.
[25] D. B. Knight, R. W. Hall, D. G. Cleary, *J. Heterocycl. Chem.* **18** (1981) 1649–1650.
[26] K. J. Stone, R. D. Little, *J. Org. Chem.* **49** (1984) 1849–1853.
[27] W. Adam, H. J. Eggelte, *J. Org. Chem.* **42** (1977) 3987–3988.
[28] Hitachi Chem. Co., JP 21 093, 1973 (T. Fujiki, H. Tsukumi, Y. Yamazaki).
[29] Ube Ind., JP-Kokai 13 113, 1974 (S. Umemura, F. Sakai).
[30] D. Hönicke, R. Bührer, A. Newrzella, *Erdöl Kohle Erdgas Petrochem.* **37** (1984) 569.
[31] Hoechst, DE-OS 3 244 032, 1984 (R. Bicker).
[32] W. Meyer, *Hydrocarbon Process.* **55** (1976) no. 9, 235–238.
[33] *Winnacker-Küchler,* **5,** 191, 247.
[34] W. Graulich, W. Swodenk, D. Theisen, *Hydrocarbon Process.* **51** (1972) no. 12, 71–75.
[35] K. Weissermel, H. J. Arpe: *Industrielle Organische Chemie,* Verlag Chemie, Weinheim-New York 1976, p. 98.

[36] Institute of Petrochemical Processes, SU 1 109 371, 1983 (M. R. Musaev, S. M. Sharifova, S. M. Mirzoev, T. A. Samedova, M. A. Aliev, Y. G. Mamedaliev).

[37] M. M. Ermilova, N. L. Basov, V. S. Smirnov, A. N. Rumyantsev, V. M. Gryaznov, *Izv. Akad. Nauk SSSR, Ser. Kim.* **8** (1979) 1773–1775; *Chem. Abstr.* **92** (1980) 22 104 p.

[38] G. V. Jsagulyants, K. M. Gitis, M. I. Rozengart, V. N. Kornyshev, G. L. Markaryan, *Geterog. Katal. 5th. 1983,* 511–516; Chem. Abstr. 100 (1984) 191 200 f.

[39] Acad. Sc. Belorussian SSR, SU 1 081 153, 1982 (Y. G. Egiazarov, M. G. Savon'kina, N. S. Bobchenok).

[40] R. W. Carr, J. O. Shoemaker, *Chem. Eng. Commun.* **19** (1982) no. 1–3, 91–98.

[41] M. Tokuda, H. Suginome, L. L. Miller, *Tetrahedron Lett.* **23** (1982) 4573–4576.

[42] P. Ausloos, S. G. Lias, R. E. Rebbert, *J. Phys. Chem.* **85** (1981) no. 2, 2322–2328.

[43] Z. Kafka, M. Kuras, L. Vodicka, *Sb. Vys. Sk. Chem. Technol. Praze Technol. Paliv* **D 47** (1983) 199–209; *Chem. Abstr.* **100** (1984) 141 776 c.

[44] M. J. Ledoux, *Appl. Catal.* **9** (1984) no. 1, 31–40.

[45] E. Weckman, K. Brezinsky, J. DeMay, I. Glassmann, *Chem. Phys. Processes Combust,* 1983, no. 35; *Chem. Abstr.* **101** (1984) 93 818 q.

[46] R. Cyprés, *Compend. Dtsch. Ges. Mineraloelwiss. Kohlechem. 1976/77* (1976) no. 1, 435–455.

[47] R. Kubicka, M. Goppoldova, CS 178 509, 1973.

[48] M. Goppoldova, M. Goppold, *Chem. Prum.* **30** (1980) no. 7, 362–366; *Chem. Abstr.* **94** (1981) 102 899 h.

[49] M. Goppoldova, *Chem. Prum.* **30** (1980) no. 8, 408–413; *Chem. Abstr.* **94** (1981) 102 872 u.

[50] Kohijin Co., DE-OS 2 217 623, 1972 (A. Oshima, K. Tsuboshima, N. Takahashi). JP-Kokai 45 011, 1974 (N. Takahashi, M. Nonaka).

[51] R. Bloch, *Tetrahedron* **39** (1983) 639–643.

[52] G. Mehta, D. S. Reddy, A. V. Reddy, *Tetrahedron Lett.* **25** (1984), 2275–2278.

[53] J. H. Lukas, A. P. Kouwenhoven, F. Baardman, *Angew. Chem.* **87** (1975) 740–741; *Angew. Chem. Int. Ed. Engl.* 14 (1975) 709.

[54] B. B. Snider, L. A. Brown, R. S. Eichen Conn, T. Λ. Killinger, *Tetrahedron Lett.* 1977, 2831–2832.

[55] D. Bryce-Smith, B. Foulger, J. Forrester, A. Gilbert, B. H. Orger, H. M, Tyrrell, *J. Chem. Soc. Perkin Trans I* 1980, 55–71.

[56] D. Bryce-Smith, W. M. Dadson, A. Gilbert, B. H. Orger, H. M. Tyrrell, *Tetrahedron Lett.* 1978, 1093–1096.

[57] L. D. Brown, K. Itoh, H. Suzuki, K. Hirai, J. Ibers, *J. Am. Chem. Soc.* **100** (1978) 8232–8238.

[58] B. L. Booth, T. A. El Fekky, G. F. M. Noori, *J. Chem. Soc. Perkin Trans I* 1980, 181–186.

[59] J. Rebek, R. Mc. Cready, S. Wolf, A. Mossman, *J. Org. Chem.* **44** (1979) 1485–1493.

[60] J. Rebek, R. Mc. Cready, *Tetrahedron Lett.* 1979, 1001–1002.

[61] H. Kropf, M. R. Yazdanbakhch, *Synthesis* 1977, 711–713.

[62] W. Reischl, E. Zbiral, *Tetrahedron Lett.* 1979, 1109–1110.

[63] Degussa, DE-OS 2 937 810, 1981 (G. Kaebisch, H. Malitius, S. Raupach, R. Truebe, H. Wittmann).

[64] H. Ogawa, H. Fujinami, K. Toya, *J. Chem. Soc. Chem. Commun.* 1981, 1274–1275.

[65] H. C. Brown, A. K. Mandal, *Synthesis* 1980, 153–155.

[66] B. F. Goodrich, US 3 808 283, 1974.

[67] Mobil Oil Corp., US 4 255 600, 1981 (L. Young).

[68] F. R. Mayo, P. S. Fredricks, T. Mill, J. K. Castleman, T. Delaney, *J. Org. Chem.* **39** (1974) 885–889.

[69] A. J. Bloodworth, H. J. Eggelte, *J. Chem. Soc. Perkin Trans I*, 1981, 1375–1382.
[70] K. Blau, U. Müller, W. Pritzkow, W. Schmitt-Renner, Z. Sedshaw, *J. Prakt. Chem.* **322** (1980) 915–932.
[71] E. O. Mihelich, D. J. Eickhoff, *J. Org. Chem.* **48** (1983) 4135–4137.
[72] Procter and Gamble Co., US 4 345 984, 1982 (E. Mihelich).
[73] A. Heumann, B. Akermark, *Angew. Chem.* **96** (1984) 443–444; *Angew. Chem. Int. Ed. Engl.* **23** (1984) 453.
[74] K. Griesbaum, J. Neumeister, M. P. Saxena, *Erdöl Kohle Erdgas Petrochem.* **36** (1983) 252–257.
[75] S. L. Schreiber, R. E. Claus, J. Reagan, *Tetrahedron Lett.* **23** (1982), 3867–3870.
[76] Bayer, DE-OS 2 201 456, 1973 (H. Waldmann, W. Schwerdtel, W. Swodenk). DE-OS 2 252 674, 1974 (H. Waldmann, W. Schwerdtel, W. Swodenk). DE-OS 2 261 657, 1974 (K. Wedemeyer, A. Klein).
[77] Nippon Zeon Co., JP 29 165, 1974.
[78] D. Hönicke, K. Griesbaum, *Chem. Ing. Tech.* **54** (1982) 497.
[79] Nissan Chem. Ind., JP-Kokai 64 020, 1973 (A. Murata, H. Suzuki).
[80] Gulf Research and Dev. Co., US 4 062 902, 1976 (A. Montagna).
[81] Anic S.p.A., DE-OS 2 622 917, 1975 (D. Sanfilippo, M. Morelli).
[82] A. V. Topchiev, DE-OS 3 100 631, 1981.
[83] M. B. Taghavi, G. Pajonk, S. J. Teichner, *Bull. Soc. Chim. Fr.* 1978, nos. 7–8, part 1, 285–293.
[84] Yu. S. Rozov, B. A. Grigorovich, V. Sh. Fel'dblyum, A. M. Kutin, I. M. Galperin, K. S. Solovev, V. S. Elizarov, SU 535 099, 1975.
[85] A. V. Topchiev, NL-A 7 408 819, 1974.
[86] L. Cerveny, J. Vopatova, V. Ruzicka, *React. Kinet. Catal. Lett.* **19** (1982) no. 1–2, 223–226; *Chem. Abstr.* **97** (1982) 40 091 q.
[87] Goodyear Tire and Rubber Co., US 4 131 627, 1977 (L. G. Wideman).
[88] Goodyear Tire and Rubber Co., US 4 131 629, 1977 (L. G. Wideman, E. A. Ofstead).
[89] Y. Nakamura, H. Hirai, *Chem. Lett.* 1976, no. 11, 1197–1202.
[90] L. I. Gvinter, V. Sh. Fel'dblyum, L. Kh. Freidlin, L. N. Suvorova, *Neftekhimiya* **23** (1981) no. 1, 41–44; *Chem. Abstr.* **98** (1983) 179 944 j.
[91] Goodyear Tire and Rubber Co., US 4 188 348, 1978 (H. R. Menapace).
[92] SABA, EP-A 9 035, 1978 (H. R. Menapace).
[93] Bayer, DE-OS 2 025 411, 1970 (W. Schwerdtel, W. Swodenk, P. Woernerle).
[94] I. M. Kolesnikov, N. N. Belov, *Zh. Fiz. Khim.* **52** (1978) no. 10, 2712; *Chem. Abstr.* **90** (1979) 86 800 d.
[95] H. F. Smyth, Jr., C. P. Carpenter, C. S. Weil, U. C. Pozzani, *AMA Arch. Ind. Hyg. Occup. Med.* **10** (1954) 61–68.
[96] W. F. von Oettingen, *U.S. Public Health Bulletin* no. 255 (1940) 40–41.
[97] Dow Chemical Company, Biochemical Laboratory, unpublished results (1964), cited in [98]
[98] American Conference of Governmental Industrial Hygenists: *Documentation of the Threshold Limit Values*, 4th ed., ACGIH, Cincinnati 1980.
[99] International Labor Office: *Occupational Safety and Health Series no. 37*, "Occupational Exposure Limits for Airborne Toxic Substances," 2nd ed., International Labor Office, Geneva 1980.
[100] E. R. Kinkead, U. C. Pozzani, D. L. Geary, C. P. Carpenter, *Toxicol. Appl. Pharmacol.* **20** (1971) 552–561.
[101] H. F. Smyth, Jr., C. P. Carpenter, C. S. Weil, U. C. Pozzani et al., *Am. Ind. Hyg. Assoc. J.* **30** (1969) 470–476.

[102] G. Kimmerle, J. Thyssen, *Int. Arch. Arbeitsmed.* **34** (1975) 177–184.
[103] U.S. Environmental Protection Agency, Office of Toxic Substances: *Candidate List of Chemical Substances,* Toxic Substances Control Act, U.S. EPA, Washington 1983.
[104] U.S. Department of Health and Human Services, Public Health Service: *National Toxicology Program,* Fiscal Year 1985 Annual Plan, NTP, Research Triangle Park 1985.
[105] E. E. Sandmeyer in *Patty's Industrial Hygiene and Toxicology* vol. **2 B,** Wiley Interscience, New York 1981 pp. 3221–3251.

Dextran

Anthony N. de Belder, Pharmacia AB, Uppsala, Sweden

1.	Introduction 1843	3.3.	Future Trends 1848
2.	Structure, Chemical, and	4.	Quality Specifications 1848
	Physicochemical Properties . . 1844	5.	Storage 1849
2.1.	Structure 1844	6.	Uses 1849
2.2.	Physicochemical Properties . . 1845	6.1.	Clinical Products 1849
2.3.	Reactivity 1845	6.2.	General Uses 1850
3.	Production 1846	7.	Economic Aspects 1850
3.1.	Biosynthesis 1846	8.	Toxicology 1851
3.2.	Production of Clinical Dextran 1847	9.	References 1851

1. Introduction

The term dextran [9004-54-0] refers to those polysaccharides which are composed primarily of 1 → 6 linked α-D-glucopyranose units. Many dextrans contain branches attached to C-2, C-3, or C-4. Others may also contain 1 → 3 linkages in the main chain.

The formation of slimes and jellies during the processing of wines and in sugar refining has long been an undesirable complication. In the middle of the 19th century, reports by Pasteur [1] and Van Tieghem [2] implicated a bacterial mechanism. The name dextran appears to have been assigned ca. 1870 by Scheibler [3], who also established that dextran was a polymer of glucose. Further references to the early history of dextran are available [4].

Dextrans are synthesized from sucrose by a large number of organisms. Jeanes and coworkers have characterized over 96 strains of dextran-producing bacteria [5]. These bacteria are confined to the family Lactobacillaceae and in particular to the genera *Lactobacillus*, *Leuconostoc*, and *Streptococcus*. Several members of the Streptococcus group are implicated in the development of cariogenic plaque on tooth surfaces [6], [7].

Leuconostoc and *Lactobacillus* dextrans are also undesirable contaminants in the sugar refining industry, in which they adversely affect the filterability and crystallization of sucrose [8].

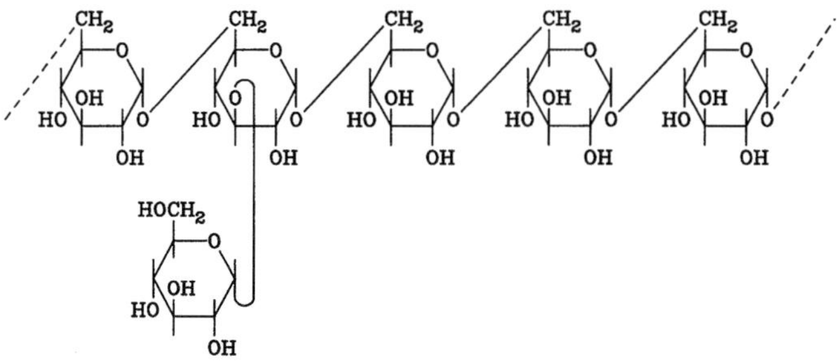

Figure 1. Structure of a fragment of the dextran B-512 (F) chain

Only the slightly branched dextran from *Leuconostoc mesenteroides B-512 (F)* is of commercial interest. Most of the chemical, physicochemical, and biological reports have therefore focused on this dextran. Unless otherwise stated, the discussions in this article are concerned only with B-512 (F). Figure 1 shows its structure.

2. Structure, Chemical, and Physicochemical Properties

2.1. Structure

The dextran elaborated by *Leuconostoc mesenteroides NRRL B-512 (F)* consists of an α (1 → 6)-linked glucan with side chains attached to the 3-positions of the backbone glucose units. From periodate and methylation analyses, the degree of branching is estimated to be 5%. Upon hydrolysis, the degree of branching is found to decrease slightly with decreasing molecular mass.

The ^{1}H- and ^{13}C-NMR spectra afford compelling evidence for the main structural features of dextran [9]. The degree of branching of clinical dextrans as determined by ^{13}C-NMR is 4.0–6.0%.

Many details of the fine structure are unresolved, in particular the length and distribution of the side chains. LARM and colleagues concluded that 40% of the side chains are one unit long, 45% are two units long, and the remaining 15% are still longer [10]. The preponderance of single unit branches in several other dextrans, *Leuconostoc mesenteroides B-1375, B-1415,* and *B-1416,* has been established [11]. The availability of dextranases with well characterized properties has permitted WALKER and colleagues to corroborate these results [12]. The presence of a small proportion of longer branches (up to 50 units) is implicit in the studies of the viscosity of dextran [13]. COVACEVICH and RICHARDS [14] concluded that the branches were distributed in a relatively regular manner.

2.2. Physicochemical Properties

Dextran B-512 (F) is freely soluble in water, methyl sulfoxide, formamide, ethylene glycol, glycerol, N-methylmorpholine-N-oxide, and hexamethylphosphoramide. Some dextran fractions may have adopted a certain degree of crystallinity and can only be brought into solution by strong heating.

The molecular mass of hydrolyzed natural dextran NRRL B-512 (F) is 9×10^6 to 500×10^6 [13], [15]. Measurements in a variety of solvents (e.g., 4 M aqueous sodium chloride and 6 M aqueous urea) failed to reveal any evidence of association.

The relationship between the mass-average molecular mass $\overline{M}_w$ and the intrinsic viscosity is shown in Figure 2 [13], [16]. GRANATH obtained the viscosity dependence of dextran in the clinical range [16]:

$$[\eta] = 2.23 \times 10^{-3} \, \overline{M}_v^{0.43} \quad (\text{for } M_r < 5 \times 10^5)$$

where $[\eta]$ = intrinsic viscosity and $\overline{M}_v$ = viscosity-average molecular mass.

The colloid osmotic pressure of dextran solutions significantly affects their plasma volume expansion [17].

The α (1 → 6) linked polysaccharides represent a class of very flexible and extended polymers. Table 1 shows the relationship between mass-average molecular mass and the radius of gyration. At high molecular masses, the molecules display increased symmetry [13], [16].

At M_r 2000, the solution properties are best explained by a transition from a coil to a rodlike conformation. Several attempts have been made to crystallize dextran, and recently CHANZY et al. grew single crystals from a dextran fraction having M_r 19 900 and proposed a ribbon-like conformation [18].

2.3. Reactivity

As with other glucans, the reactivity of the hydroxyl group at position 2 toward alkylating agents is higher than at position 3 or position 4. Studies on the partial methylation of dextran revealed the following relative reactivities: $k_2 : k_3 : k_4 = 16 : 2 : 7$. Migration of the substituents may subject acylations to thermodynamic control. Thus, although partial acetylation of dextran with acetic anhydride–pyridine in formamide as solvent gave $k_2 \gg k_3 = k_4$, the reactivities were virtually identical when the reaction was carried out with acetic anhydride in aqueous alkali [19].

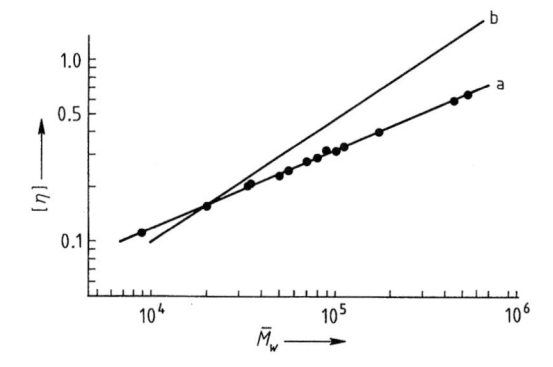

Figure 2. Log–log plot of intrinsic viscosity $[\eta]$ against mass-average molecular mass $\overline{M}_w$ for dextran B-512 (a) and for a hypothetical linear dextran (b)

Table 1. Molecular dimensions of dextran B-512(F)

Molecular mass $\overline{M}_w$	Radius of gyration, nm
2×10^6	38
5×10^5	20
1×10^5	9.5
5×10^4	6.8
(Albumin)	3.5

3. Production

3.1. Biosynthesis

The early work of BEIJERINCK [20] and HEHRE [21] showed that dextran was elaborated by an extracellular enzyme designated dextransucrase (sucrose-1,6-α-D-glucan 6-α-glucosyltransferase, E.C. 2.4.1.5).

The dextransucrases from many *Leuconostoc* and *Streptococcal* strains have recently been isolated and their properties studied. The purified B-512 (F) enzyme, which binds dextran strongly, rapidly loses activity at 4 °C and even at – 15 °C. Its activity decreases by 60% over 20 d. The enzyme can be stabilized by adding dextran (> 4 mg/mL).

Dextransucrase appears to be a glycoprotein (M_r 64 000) with mannose being the primary sugar. Calcium ions appear to be essential for the activity of the enzyme.

A mechanism for the biosynthesis has been proposed whereby the enzyme serves two functions [22]. It first hydrolyzes the sucrose and binds the glucosyl moiety, and thereafter it builds up the dextran chain by an insertion mechanism (Fig. 3).

The biosynthesis can be terminated by any one of a large number of acceptors, and if so, the dextran chain is released. Examples of acceptors are maltose, isomaltose, and methyl-α-D-glucoside. Dextran itself can also function as an acceptor [23].

The formation of branches is also an acceptor reaction in which a ring hydroxyl group on a dextran molecule is inserted in the growing dextran chain.

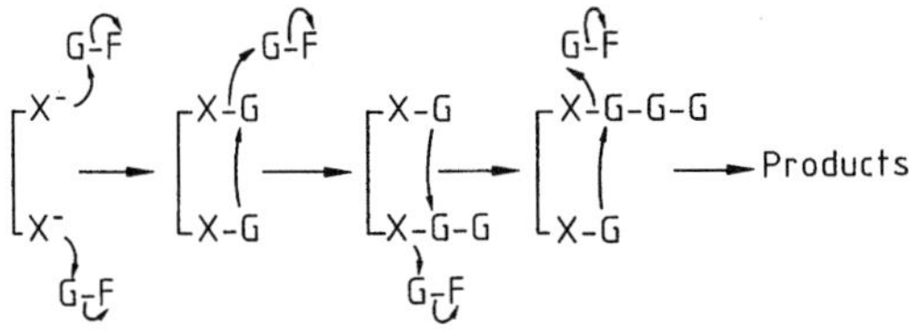

Figure 3. Two-site mechanism for the biosynthesis of dextran chains by dextransucrase
G–F represents a sucrose molecule composed of glucose (G) and fructose (F). G–G represents α (1 → 6) linked glucose residues. The binding sites X may not be identical.

3.2. Production of Clinical Dextran

In the West, most major manufacturers of dextran employ a process based on the batchwise culture of *Leuconostoc mesenteroides B-512 (F)* in the presence of sucrose. The viscous culture fluid is precipitated in an alcohol, and the native dextran then is hydrolyzed and fractionated to the desired molecular mass range. In Japan and East-bloc countries, different strains of *Leuconostoc* are used, but the dextrans produced are of similar structure. The preparation of dextran on a laboratory scale has been described in detail [24], and the industrial production of dextran has been reviewed [25]–[27].

To obtain vigorous growth of the organism, the culture medium must contain various nutrients in addition to sucrose. In practice, these are supplied by addition of either yeast extracts, corn steep liquors, or malt extracts with peptone or tryptone broth. The pH of the medium affects both the production of dextransucrase by the organism and the activity and stability of the enzyme; Figure 4 shows the pH changes. The influence of sucrose concentration on production of dextran has been reviewed [25]. A sucrose concentration of ca. 10% is satisfactory; at higher concentrations the yield of dextran with $\overline{M}_w > 5000$ decreases. The cultures are maintained at a temperature of ca. 25 °C, at which the fermentation is complete after 24–48 h. Prolongation of the cultures for more than 48 h may lead to a decrease in molecular mass of the dextran.

The native dextran that is obtained is then subjected to partial hydrolysis to form products of appropriate molecular size, and fractionation by ethanol or methanol to give clinical fractions.

Extensive studies on the acid hydrolysis and fractionation of native B-512 (F) dextran have been carried out and conditions for maximal yields have been established [28], [29]. Following the partial hydrolysis, the clinical fraction can be precipitated with 39–46% aqueous ethanol under careful temperature control (25 °C). The product is then redissolved and spray-dried.

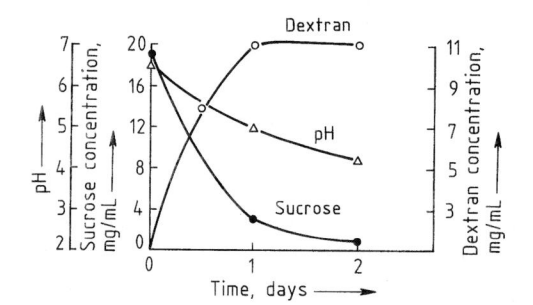

Figure 4. Changes in sucrose concentration, pH, and dextran concentration during the fermentation of *Leuconostoc mesenteroides*

3.3. Future Trends

Several alternative processes have been devised over the years:

1) Fermentation with cell-free extracts
2) Fermentation in the presence of acceptors
3) Continuous processes
4) Use of immobilized dextransucrase

About 30 years ago, a supplement (1 mg/mL) of low molecular mass dextran was found to yield a product dextran of lower molecular mass [30]. Few producers have adopted this type of process for clinical dextrans. One drawback is that it requires a production facility for low molecular mass dextran.

Promising results have been obtained with immobilized dextransucrase but only on a laboratory scale [31]. A process in which ethanol precipitation has been replaced by ultrafiltration has been reported [32].

4. Quality Specifications

Two dextran products are available in most countries for medical purposes, one with a M_r of ca. 70 000 and the other with a M_r of ca. 40 000.

In most official monographs, the specifications for the molecular size are stipulated in terms of limits of the average values for the total distribution, for the highest 10% and for the lowest 10% [33]. Size exclusion chromatography will eventually be adopted universally for quality control.

The specific optical rotation $[\alpha]_D^{25}$ of dextran provides a useful test of purity and is given as +196° to +201°. The acceptance of clinical dextrans is also based on a series of such impurity tests as loss on drying; pH of a 10% aqueous solution; determination of heavy metals, sulfate, and nitrogen; assay for *Leuconostoc mesenteroides* antigens [34]; assay for pyrogens; and microbial limit tests.

5. Storage

Dextran is stable indefinitely when stored in the absence of light, excessive heat, and moisture. Sterile solutions of clinical dextrans with pH 4–7 have excellent chemical stability (at least 10 years) when stored at 4–40 °C.

Clinical dextran solutions tend to form small amounts of flakes if kept in glass bottles and subjected to considerable temperature fluctuations. These flakes have been found to consist of dextran with the same M_r as the substance in solution. They redissolve upon heating.

6. Uses

6.1. Clinical Products

Dextran 70 (M_r ca. 70 000) is marketed in most countries as a 6% solution in normal saline and, as such, represents the plasma volume expander of choice. It is the only plasma volume expander included in the WHO list of essential drugs. Clinical experience supports its use in the treatment of shock or impending shock as a result of hemorrhage, burns, or surgery.

Dextran 40 ($\overline{M}_w$ ca. 40 000) was introduced in 1961 following studies of the erythrocyte disaggregating and blood flow improvement properties of lower molecular mass dextrans. It is marketed as a 10% solution in normal saline or 5% glucose which provides rapid plasma volume expansion and promotes microcirculatory blood flow. Both products are used extensively to prevent fatal pulmonary complications of surgery, trauma, and shock.

In 1982, the monovalent hapten Dextran 1 (a fraction of $\overline{M}_w$ ca. 1000) was launched as prophylaxis against dextran-induced anaphylactoid reactions, which, although rare, may cause undesirable complications in the clinic. Dextrans are commonly added to infusion and perfusion solutions, for example, in connection with transplants. A dextran sulfate of $\overline{M}_w$ 7000 and sulfur content of 16% was tested clinically as an anticoagulant, but at the doses used, it caused toxic symptoms [35]. A diethylaminoethyl derivative of dextran ($\overline{M}_w$ ca. 500 000) has been found to be effective in reducing serum cholesterol.

An iron–dextran complex (Imferon) prepared from dextran of M_r ca. 5000 is used for the treatment of iron deficiency. It is particularly valuable for treating anemic newborn piglets.

In 1977, a cross-linked dextran in bead form (Debrisan) was introduced as a woundcleansing agent for secreting and infected wounds in particular.

6.2. General Uses

Purified dextran fractions are currently used in the cosmetic and photographic industries. A comprehensive bibliography covering dextran literature and patents is available [36]. Dextran fractions find application for partitioning subcellular particles and as cryoprotective agents in two-phase polymer systems.

Sephadex gels, prepared by emulsion polymerization of dextran, have occupied a dominant place among gel filtration media for many years. Sephadex G-25 is used on an industrial scale for desalting operations during the purification of biopolymers of medical importance, e.g., insulin. Many of the ion exchangers derived from Sephadex are also used commercially for separations [37].

Dextran fractions have been used extensively for preparing conjugates with biologically active substances, e.g., drugs, enzymes, and hormones [38]. Conjugation prolongs the lifetime of the active component in vivo, increases its stability, and facilitates the targeting of the active component. Many derivatives of dextran have been described and numerous patents have appeared. However, few have attained significant commercial interest. Dextran sulfate ($\overline{M}_w$ ca. 500 000; sulfur content 18%) and diethylaminoethyl dextran ($\overline{M}_w$ ca. 500 000) have revealed an astonishing range of biological activities [39].

Several microcarriers based on cross-linked dextran substituted with cationic substituents (e.g., quaternary ammonium) are used commercially for culturing anchorage-dependent animal cells. As such, they have proved invaluable for the production of vaccines, viruses, and interferon.

7. Economic Aspects

Exact figures for the annual world production are not available, but an estimate based on the steadily increasing demand for clinical dextrans would lead to a figure somewhat in excess of 500 t.

The major producer of dextran and dextran solutions is Pharmacia (Uppsala, Sweden); other producers of dextran substance are Fisons (United Kingdom), Meito (Japan), Pfeiffer & Langen (Federal Republic of Germany), VEB Serumwerke (German Democratic Republic), and Polfa (Poland). A selected list of producers of clinical dextran solutions is as follows: Pharmacia (United States and Sweden), Knoll/Schiwa (Federal Republic of Germany), Abbott (United States), Travenol (United States), Fisons (United Kingdom), Pfrimmer (Federal Republic of Germany), Polfa (Poland), Leceiva (Czechoslovakia), Soviet Union, Green Cross Corp. (Japan), Il Sung (South Korea), and Pisa (Mexico).

8. Toxicology

The infusion of clinical dextran solutions in humans is now approved by the regulatory authorities in most countries. No adverse effects are to be expected, provided the recommended doses are observed. However, a low incidence of dextran-induced anaphylactoid reactions has been reported. The risk of this type of reaction has been virtually eliminated by the preinjection of 20 mL of a 15% Dextran 1 (M_r ca. 1000) solution [40].

Dextran B-512 (F) is degraded by bacteria in the gut and the products produce a rapid increase in blood sugar and liver glycogen.

Dextrans are not permitted as food additives in the United States and Western Europe [41], [42], but they are considered safe as components of food packing materials. Dextrans may be used as additives in pharmaceutical and cosmetic formulations provided that the necessary safety documentation is presented.

9. References

[1] L. Pasteur, *Bull. Soc. Chim. Fr.* **1861**, 30–31.
[2] P. van Tieghem, *Ann. Sci. Nat. Botan. Biol. Veg. s 2*, **7** (1878) 180–203.
[3] G. Scheibler, *Z. Ver. Dtsch. Zucker-Ind.* **19** (1869) 472; **24** (1874) 309.
[4] T. H. Evans, H. Hibbert, *Adv. Carbohydr. Chem.* **2** (1946) 209–219.
[5] A. Jeanes, W. C. Haynes, C. A. Wilham, J. C. Rankin et al., *J. Am. Chem. Soc.* **76** (1964) 5041–5052.
[6] R. L. Sidebotham, *Adv. Carbohydr. Chem. Biochem.* **30** (1974) 435–444.
[7] G. J. Walker: "Int. Rev. of Biochem.," in D. J. Manners (ed.): *Biochem. of Carbohydrates II*, vol. **16**, Univ. Park Press, Baltimore 1978, pp. 75–110.
[8] P. A. Inkerman in M. B. Lopez, C. M. Madrazo (eds.): *Proceedings XVII Congress, International Society of Sugar Cane Technologists*, vol. **3**, Executive Committee of ISSCT, Metro Manila 1980, pp. 2411–2427.
[9] D. Gagnaire, M. Vignon, *Makromol. Chem.* **178** (1977) 2321–2333.
[10] O. Larm, B. Lindberg, S. Svensson, *Carbohydr. Res.* **20** (1971) 39–48.
[11] E. J. Bourne, D. H. Hutson, H. Weigel, *Biochem. J.* **86** (1963) 555–562.
[12] C. Taylor, N. W. H. Cheetham, G. J. Walker, *Carbohydr. Res.* **137** (1985) 1–3.
[13] F. R. Senti, N. N. Hellman, N. H. Ludwig, G. E. Babcock et al., *J. Polym. Sci.* **17** (1955) 527–546.
[14] M. T. Covacevich, G. N. Richards, *Carbohydr. Res.* **54** (1977) 311–315.
[15] L. H. Arond, H. P. Frank, *J. Phys. Chem.* **58** (1954) 953–958.
[16] K. A. Granath, *J. Colloid Sci.* **13** (1958) 308–328.
[17] U. F. Gruber: *Blutersatz*, Springer Verlag, Berlin 1968, p. 68.
[18] C. Guizard, H. Chanzy, A. Sarko, *Macromolecules* **17** (1984) 100–107.
[19] A. N. de Belder, B. Norrman, *Carbohydr. Res.* **8** (1968) 1–6.
[20] M. W. K. Beijerinck, *Akad. Wetensch. Amsterdam Proc. Sect. Sc.* **12** (1910) 635.
[21] E. J. Hehre, *Science (Washington, D.C., 1883)* **93** (1941) 237–238.

[22] J. F. Robyt, B. K. Kimble, T. F. Walseth, *Arch. Biochem. Biophys.* **165** (1974) 634–640.
[23] J. F. Robyt, S. H. Eklund, *Bioorg. Chem.* **11** (1982) 115–132.
[24] A. Jeanes in R. L. Whistler (ed.): *Methods in Carbohydrate Chemistry*, vol. **5**, Academic Press, New York 1965, p. 118.
[25] R. M. Alsop, *Prog. Ind. Microbiol.* **18** (1983) 1–44.
[26] G. H. Bixler, G. H. Hines, R. M. McGhee, R. A. Shurter, *Ind. Eng. Chem.* **45** (1953) 692–705.
[27] F. H. Foster, *Process Biochem.* **3** (1968) 15–19.
[28] I. A. Wolff, C. L. Mehltretter, R. L. Mellies, P. R. Watson et al., *Ind. Eng. Chem.* **46** (1954) 370–377.
[29] M. Zief, G. Brunner, J. Metzendorf, *Ind. Eng. Chem.* **48** (1956) 119–121.
[30] H. M. Tsuchiya, N. N. Hellman, H. J. Koepsell, J. Corman et al., *J. Am. Chem. Soc.* **77** (1955) 2412–2419.
[31] A. Lopez, P. Monsan, *Biochimie* **62** (1980) 323–329.
[32] R. M. Alsop, P. E. Barker, G. J. Vlachogiannis, *Chem. Eng. (Rugby, Engl.)* **399** (1984) 24–25.
[33] K. Nilsson, G. Soederlund, *Acta Pharm. Suec.* **15** (1978) 439–454.
[34] W. Richter, *Int. Arch. Allergy Appl. Immunol.* **39** (1970) 469–478.
[35] P. Hjort, H. Stormorken, O. Gilje, *J. Clin. Lab. Invest.* **9**, Suppl. 29 (1957).
[36] A. Jeanes: *Dextran Bibliography*, Misc. Publication no. 1355, U. S. Department of Agriculture, Washington 1978.
[37] J. M. Curling (ed.): *Separation of Plasma Proteins*, Pharmacia Fine Chemicals AB, Uppsala 1983.
[38] L. Molteni in H. Bundgaard, A. B. Hansen, H. Kofod (eds.): *Optimization of Drug Delivery*, Munksgaard, Copenhagen 1982, pp. 285–301. Z. A. Rogovin, A. D. Virnik, K. P. Khomiakov, O. P. Laletina et al., *J. Macromol. Sci. Chem.* **A 6** (1972) 569–593.
[39] A. N. de Belder in R. L. Whistler (ed.): *Industrial Gums*, 3rd ed., in press.
[40] A. W. Richter, H. I. Hedin, *Immunol. Today* **3** (1982) 132–138.
[41] M. Glicksman in M. Glicksman (ed.): *Food Hydrocolloids*, vol. **1**, CRC Press, Boca Raton, Florida, 1982, p. 157.
[42] *Fed. Regist.* **42** (1977) Nov. 18, FR 59 518-21.

Dialkyl Sulfates and Alkylsulfuric Acids

KARL WEISENBERGER, Hoechst AG, Frankfurt, Federal Republic of Germany, (Chaps. 2–9)

DIETER MAYER, Hoechst AG, Frankfurt, Federal Republic of Germany, (Chap. 10)

1.	Introduction 1853	5.	Quality Specifications and Analysis	1860
2.	Physical Properties 1854	6.	Storage, Transportation, and	
3.	Chemical Properties. 1855		Handling.	1862
4.	Production 1857	7.	Environmental Protection . . .	1864
4.1.	From Ethers and Sulfur	8.	Uses	1864
	Trioxide. 1857	8.1.	Alkylation	1864
4.2.	From Olefins and Sulfuric Acid 1857	8.2.	Other Uses	1869
4.3.	From Alcohols and	9.	Economic Aspects	1870
	Chlorosulfuric Acid 1858	10.	Toxicology and Occupational	
4.4.	From Alcohols and Sulfuryl		Health.	1870
	Chloride or Thionyl Chloride . 1859	10.1.	Dimethyl Sulfate	1870
4.5.	From Alcohols and Sulfuric	10.2.	Diethyl Sulfate.	1871
	Acid 1860	11.	References.	1871

1. Introduction

Dialkyl sulfates (**1**) are neutral esters of sulfuric acid and aliphatic alcohols. The corresponding monoesters (**2**) (alkyl hydrogen sulfates, monoalkyl sulfates) are known as alkylsulfuric acids. They are available only as the stable alkali-metal salts. Cyclic sulfates (**3**) (alkylene sulfates) represent a special group of neutral sulfuric acid esters. They result from the reaction of one molecule of a diol with one molecule of sulfuric acid.

$$\underset{\mathbf{1}}{\text{RO}-\overset{\overset{\text{O}}{\|}}{\underset{\underset{\text{O}}{\|}}{\text{S}}}-\text{OR}} \qquad \underset{\mathbf{2}}{\text{RO}-\overset{\overset{\text{O}}{\|}}{\underset{\underset{\text{O}}{\|}}{\text{S}}}-\text{OH}} \qquad \underset{\mathbf{3}}{(\text{CH}_2)_n\overset{\text{O}}{\underset{\text{O}}{\diagdown}}\text{SO}_2}$$

where R = alkyl

The most commercially important dialkyl sulfates, dimethyl sulfate [77-78-1] and diethyl sulfate [64-67-

5], were first prepared in 1835 by DUMAS and PELIGOT, who distilled a mixture of sulfuric acid and the corresponding alcohol. The general applicability of dimethyl sulfate as an alkylating agent was recognized by ULLMANN and WENNER in 1900. Around this time, commercial production of dimethyl sulfate from sulfur trioxide and dimethyl ether began in Lyon, France.

The alkali salts of higher alkylsulfuric acids (fatty alcohol sulfates) are important surfactants and are not treated in this article.

2. Physical Properties

Up to dihexyl sulfate, the dialkyl sulfates are liquids at room temperature. They are completely miscible with polar organic solvents and aromatic hydrocarbons, but their solubility in water is poor. The first members of the homologous series of dialkyl sulfates are only partially miscible with saturated hydrocarbons. Boiling points are relatively high. Some physical properties of symmetrical dialkyl sulfates are listed in Table 1.

Dimethyl sulfate forms an azeotrope with water (73 wt% water, bp 98.6 °C).

Viscosity of dimethyl sulfate [1]:

t, °C	20	30	40	50	60	70	80	90	100
η, mPa · s	1.181	1.54	1.33	1.16	1.02	0.90	0.81	0.73	0.66

Solubility of dimethyl ether in dimethyl sulfate at 0.1 MPa [1]:

t, °C	0	10	20	30	40	50
Solubility, wt%	6.3	5.2	4.3	3.6	3.0	2.6

Vapor pressure of dimethyl sulfate (figures in brackets are extrapolated):

t, °C	0	20	40	60	80	100	120	140	160
p, kPa	(0.01)	(0.06)	(0.25)	0.8	2.2	5.5	12.5	25.0	47.0

The lower cyclic sulfates have comparatively high melting points, which decrease with increasing molecular mass (see Table 2).

Alkylsulfuric acids, which are usually described as viscous oils, are unstable compounds. They are strong acids. Methyl hydrogen sulfate [75-93-4], for example, is easily soluble in water; is is strongly dissociated in aqueous solution [2]. The alkali and alkaline-earth metal salts of alkylsulfuric acids are stable, crystalline compounds, which often contain water of crystallization. The ammonium salts have well-defined melting points (ammonium methyl sulfate [19803-43-1], mp 135 °C; ammonium ethyl sulfate [513-13-3], mp 97 °C; ammonium n-propyl sulfate, mp 132 °C) and can be used to identify the alcohols.

Table 1. Physical properties of dialkyl sulfates, $(RO)_2SO_2$

R	CAS registry number	Formula	M_r	mp, °C	ϱ (20 °C), g/cm^3
Methyl	[77-78-1]	$C_2H_6O_4S$	126.11	−31	1.329
Ethyl	[64-67-5]	$C_4H_{10}O_4S$	154.18	−25	1.180
sec-Propyl	[2973-10-6]	$C_6H_{14}O_4S$	182.24	ca.−19	1.096
n-Propyl	[598-05-0]	$C_6H_{14}O_4S$	182.24		1.109
n-Butyl	[625-22-9]	$C_8H_{18}O_4S$	210.29		1.062
n-Pentyl	[5867-98-1]	$C_{10}H_{22}O_4S$	238.34	14	1.029
n-Hexyl	[7722-57-8]	$C_{12}H_{26}O_4S$	266.40	ca. 12	1.006
n-Heptyl	[50632-94-5]	$C_{14}H_{30}O_4S$	294.45		0.986

3. Chemical Properties

Pure dialkyl compounds are not acidic. Traces of acid can be removed with cold hydrogen carbonate solution. Dimethyl sulfate may be stored for several days over dry anhydrous sodium carbonate, dry calcium oxide, or dry calcium carbonate, without undergoing any changes.

An equilibrium exists between monoalkyl sulfates and dialkyl sulfates:

$$2\ RHSO_4 \rightleftharpoons R_2SO_4 + H_2SO_4$$

This equilibrium can be shifted completely to the right by chemical binding of the sulfuric acid or by distilling the dialkyl sulfate.

Even at temperatures as low as 100 °C, unsymmetrical dialkyl sulfates are converted partially into the symmetrical esters:

$$2\ RR'SO_4 \rightleftharpoons R_2SO_4 + R'_2SO_4$$

Above 100 °C, acidic decomposition products such as $RHSO_4$ and SO_3 also form.

Dialkyl sulfates undergo acid- or base-catalyzed hydrolysis. In the acid-catalyzed reaction, free sulfuric acid forms rapidly. In alkaline solution, only the alkali salt of the alkylsulfuric acid is produced at room temperature. Higher temperatures are necessary for further hydrolysis. The alkyl groups are split off as alcohols, but small amounts of the corresponding ethers may also form.

The most important property of dialkyl sulfates is their ability to transfer alkyl groups to other molecules (for examples, see Chap. 8). Reactivity decreases with increasing chain length. In unsymmetrical dialkyl sulfates, the shorter alkyl group reacts before the longer one. Above 160 °C, the lower alkyl sulfates react with metal oxides, halides, carbonates, and other inorganic compounds to form metal sulfates.

Under ultraviolet light and at high temperature, one methyl group of dimethyl sulfate can be chlorinated to yield the monochloro derivative, which is more reactive than dimethyl sulfate. In alkylation, only the chloromethyl group is transferred.

Table 2. Physical properties of cyclic sulfates

Name	Abbreviation	CAS Registry Number	Structure
Cyanuric acid		[108-80-5]	**1**
Trichloro isocyanurate	TCCA	[87-90-1]	**2**
Triallyl isocyanurate	TAIC	[1025-15-6]	**3**
Tris(2-hydroxyethyl) isocyanurate	THEIC	[839-90-7]	**4**
Triglycidyl isocyanurate	TGIC	[2451-62-9]	**5**
Tris(2-carboxyethyl) isocyanurate	TCEIC	[2904-41-8]	**6**
Trisodium cyanurate		[3047-33-4]	**7**
Cyanuric chloride	CC	[108-77-0]	**8**
Triallyl cyanurate	TAC	[101-37-1]	**9**
Chlorobis(ethylamino)triazine		[122-34-9]	**10**
Bis(triazinylamino)stilbenedisulfonic acids			**11**
Procion Brilliant Orange M-G		[6522-74-3]	**12**

Alkylsulfuric acids are hydrolyzed in water or aqueous mineral acid to sulfuric acid and the corresponding alcohol. The alkali-metal salts serve as mild alkylating agents. Alkyl fluorosulfates, $ROSO_2F$, can be prepared readily from dialkyl sulfates and fluorosulfuric acid; they have greater alkylating potential than dialkyl sulfates [3].

4. Production

4.1. From Ethers and Sulfur Trioxide

The addition of sulfur trioxide to dimethyl ether is used industrially for the production of dimethyl sulfate.

$$(CH_3)_2O\ (g) + SO_3\ (l) \longrightarrow (CH_3)_2SO_4 \qquad \Delta H_{76\ °C} = -92 \pm 7\ \text{kJ/mol}$$

The reaction is carried out in water-cooled, vertical aluminum or stainless steel tubes (diameter, 0.2 m; height, 4 m). In a continuous process at ca. 70 °C, stoichiometric quantities of liquid sulfur trioxide and gaseous dimethyl ether are fed from the top and bottom, respectively, into the dimethyl sulfate that is already present. The two reactants quickly form an acidic adduct, which slowly rearranges to the neutral ester. The product drawn off from the bottom of the tubes still contains a considerable amount of the acidic adduct. To achieve a quantitative rearrangement, the product is held at 70 °C for several hours. A tower and its associated holding vessel have an output of up to 120 kg of dimethyl sulfate per hour [4].

Technical-grade dimethyl sulfate contains small amounts of acidic compounds: water leads to formation of an equivalent amount of sulfuric acid; methanol introduced with the dimethyl ether results in methylsulfuric acid; and saturated hydrocarbons present in the dimethyl ether can be converted to sulfonic acids. Technical-grade dimethyl sulfate may also contain small amounts of dimethyl ether. Most of the impurities can be removed by vacuum distillation over anhydrous sodium sulfate.

This process is unsuitable for higher dialkyl sulfates because ethers with carbon atoms that are not bound to either oxygen or chlorine will undergo oxidation.

4.2. From Olefins and Sulfuric Acid

Sulfuric acid can add to the double bond of olefins:

$$2\ (CH_3)_2C=CH_2 + H_2SO_4 \longrightarrow [(CH_3)_3CO]_2SO_2$$

An excess of olefin leads almost exclusively to formation of the dialkyl sulfate. Further addition of sulfuric acid results in a decrease in the concentration of dialkyl sulfate and an increase in the concentration of alkylsulfuric acid [5].

Generally, tertiary olefins react much more readily than secondary or primary ones. According to Markovnikov's rule, the hydrogen of sulfuric acid always adds to the carbon atom with the greater number of hydrogen atoms. Therefore, even olefins with a terminal double bond do not form *n*-alkyl sulfates.

The reaction of olefins in aqueous sulfuric acid leads not to a single product, but to a mixture of dialkyl sulfate, alkylsulfuric acid, sulfuric acid, water, alcohol, ether, and olefin. The mixture is characterized customarily by the concentration of acid used and the molar ratio of absorbed olefins. A higher olefin pressure and a higher sulfuric acid concentration favor formation of the corresponding dialkyl sulfate. Thus, the reaction between ethylene and 96% sulfuric acid at standard pressure leads almost exclusively to ethyl hydrogen sulfate, whereas anhydrous sulfuric acid at 0.35 MPa (3.5 bar) yields a mixture of 35% diethyl sulfate and 51% ethyl hydrogen sulfate. At 1 MPa (10 bar), 62% of diethyl sulfate and 30% of ethyl hydrogen sulfate are formed.

Sulfuric acid always causes some polymerization of olefins. Therefore, the sulfuric acid concentration and the reaction temperature should be kept as low as possible. Indeed, the reaction is most often carried out at or below ambient temperature.

The reaction of olefins with sulfuric acid is used industrially to extract isobutene [*115-11-7*] (2-methylpropene) from mixtures of C_4 hydrocarbons (65% sulfuric acid) (→ Butenes); 80% sulfuric acid is required to remove isoamylene [*563-45-1*] (3-methyl-1-butene) from mixtures of C_5 hydrocarbons [6].

4.3. From Alcohols and Chlorosulfuric Acid

The reaction between chlorosulfuric acid [*7790-94-5*] and an alcohol at low temperature yields the corresponding alkylsulfuric acid [7].

$$ROH + ClSO_3H \longrightarrow ROSO_3H + HCl \quad \text{where} \quad R = \text{alkyl}$$

The alkylsulfuric acid cannot be isolated by distillation because it is less volatile than other constituents and rearranges rapidly in the acidic environment. Crystalline methylsulfuric acid (*mp* –30 to –35 °C) is obtained if the reaction is carried out at –40 °C and the hydrogen chloride generated is driven off with nitrogen [2].

Reaction between an alcohol and amidosulfuric acid leads directly to the ammonium salt of the corresponding alkylsulfuric acid.

Conversion of alkylsulfuric acids into dialkyl sulfates by vacuum distillation produces high yields only with methylsulfuric acid [8].

$$2\ RHSO_4 \rightleftharpoons R_2SO_4 + H_2SO_4$$

The isolation of dialkyl sulfates from the equilibrium mixture progresses more smoothly if a neutral sodium salt is added, which binds the liberated sulfuric acid as sodium hydrogen sulfate [9]. The following procedure is used to prepare diethyl sulfate:

Ethylsulfuric acid (1100 kg, d 1.41 at 15 °C) is distilled under vacuum (0.65 kPa) in an enameled, stirred tank of 3-m^3 capacity over 1950 kg of anhydrous sodium sulfate. The condenser and recipient vessel are made of aluminum. The yield is 580 kg of diethyl sulfate (99.8 % purity), which is equivalent to 86% of the theoretical yield.

The drawback of such processes is the production of relatively large amounts of sodium hydrogen sulfate contaminated with residual dialkyl sulfate.

4.4. From Alcohols and Sulfuryl Chloride or Thionyl Chloride

If alcohols and sulfuryl chloride react in a molar ratio of 1:1, alkyl chlorosulfates are produced.

$$ROH + Cl-SO_2-Cl \longrightarrow RO-SO_2-Cl + HCl \tag{1}$$

Alcohols and thionyl chloride react readily to produce dialkyl sulfites [10].

$$2\ ROH + Cl-SO-Cl \longrightarrow RO-SO-OR + 2\ HCl \tag{2}$$

Dialkyl sulfites prepared according to Equation (2) react directly with the alkyl chlorosulfates to form dialkyl sulfates. Prior purification is unnecessary.

$$RO-SO_2-Cl + RO-SO-OR \longrightarrow RO-SO_2-OR + RO-SO-Cl \tag{3}$$

Reaction (3) can be carried out at 130 °C or, if zinc chloride is used as catalyst, at 90 °C; the reaction time is ca. 6 h [11]. Under these conditions, an exchange of alkyl groups takes place (see Chap. 3); therefore, only symmetrical dialkyl sulfates can be produced.

A more recent process operates at 45 °C, with a reaction time of 2 h [12]. This enables a reaction between sulfites and chlorosulfates with different alkyl groups; no exchange of alkyl groups occurs. However, the alkyl chlorosulfates, especially those with long-chain alkyl groups, must be stabilized, e.g., by an alkyl chlorosulfite. Chlorosulfites are always present when the dialkyl sulfite rather than the alcohol is used in reaction (1) [13]:

$$RO-SO-OR + Cl-SO_2-Cl \longrightarrow RO-SO-Cl + RO-SO_2-Cl \tag{4}$$

Despite an increased amount of valueless byproduct (RO–SO–Cl), this variant of Equation (1) is often preferred in the process operating at higher temperatures because of its smoother progress.

To produce a symmetrical dialkyl sulfate, only one sulfite is needed. If an unsymmetrical dialkyl sulfate is desired, however, the reaction of two different alcohols according to Equation (2) is necessary.

At the relatively high temperatures required in the older process, alkyl chlorosulfites decompose to sulfur dioxide and alkyl chloride. In the low-temperature process subsequent decomposition in aqueous alkali is necessary.

In the production of unsymmetrical dialkyl sulfates, yields are considerably higher if the longer-chain sulfite is used in Equation (4) and the sulfite with the smaller alkyl group in Equation (3). The overall reaction is as follows:

$$4\ ROH + 2\ SOCl_2 + SO_2Cl_2 \longrightarrow R_2SO_4 + 4\ HCl + 2\ RCl + 2\ SO_2$$

This process is especially suitable for the production of higher, unsymmetrical dialkyl sulfates.

For further production methods, see [14], [15].

4.5. From Alcohols and Sulfuric Acid

Reaction of alcohols with sulfuric acid can be used to produce alkylsulfuric acids.

$$ROH + H_2SO_4 \rightleftharpoons ROSO_3H + H_2O$$

With excess sulfuric acid, the equilibrium can be shifted to the right. The alkylsulfuric acid may be isolated in the form of its alkali-metal salt, after neutralization in the cold and precipitation of the sulfate anion as barium sulfate. At higher temperature (120–140 °C), dialkyl ethers are produced. Use of oleum or sulfur trioxide leads to sulfonation. Thus ethionic acid [461-42-7], $HO_3S–CH_2CH_2–O–SO_3H$, is produced readily from ethanol and sulfur trioxide [7].

5. Quality Specifications and Analysis

Commercial dimethyl sulfate contains 0.01–1.5% acid, calculated as methylsulfuric acid, and dissolved dimethyl ether in concentrations from a few parts per million to about 2 wt%. Other impurities are sulfuric acid and alkanesulfonic acids (see Section 4.1). Although the following specifications and analytical procedures pertain to dimethyl sulfate they are, in principle, also applicable to its nearest homologues.

Assay. (1) A sample is boiled for at least 30 min under reflux with dilute hydrochloric acid. The hydrochloric acid is then removed completely by evaporation with repeated addition of water. The sulfuric acid left behind is taken up in water and titrated [16].

(2) A sample is stirred for 12 h at 20 °C in water to which an approximately equivalent amount of sodium hydroxide has been added. The excess hydroxide is then back-titrated. Because hydrolysis stops at the monomethyl sulfate, free sulfate ions can be determined with barium chloride in the cold solution after a small amount of hydrochloric acid has been added.

In both methods, the free acid must be determined separately and subtracted from the result.

Another method is based on the formation of methyl iodide from dimethyl sulfate and potassium iodide.

$(CH_3)_2SO_4 + KI \longrightarrow CH_3SO_4^- + K^+ + CH_3I$

An aqueous methanol solution of a measured amount of potassium iodide is added to the sample with mild heating. Excess potassium iodide is titrated according to the Volhard method or potentiometrically with silver nitrate [16].

Determination of the Free Acid. The sample is suspended in ice–water or extracted by shaking with ice–water. The dissolved acid is then titrated with sodium hydroxide, by using methyl orange as indicator. Rapid assay at low temperature is essential to prevent significant formation of acids by hydrolysis.

The free acid can also be titrated with sodium acetate in glacial acetic acid, by using crystal violet as indicator. This accounts only for the first dissociation step of any sulfuric acid present [16].

The alkylsulfuric acid and sulfuric acid contents of dimethyl sulfate can be determined separately by potentiometric titration (glass electrode) with tributylethylammonium hydroxide [*16208-32-5*] in a water-free medium. This procedure is also suitable for diethyl sulfate [17].

Determination of Dimethyl Ether. The liquid sample is dripped by means of a syringe onto diatomaceous earth; a given amount of air is then passed over the absorbed liquid and collected in an evacuated bottle. The ether content of the air is determined in the usual manner by gas chromatography [16].

Determination of Dimethyl Sulfate Vapor in Air. Small amounts of dimethyl sulfate in air can be determined by using special test tubes in which concentrations can be read directly from color changes (Manufacturer: Drägerwerk AG, Lübeck, Federal Republic of Germany). Reliable detection of levels as low as 0.005 ppm can be achieved.

A process that satisfies current requirements for sensitivity involves the following steps (sample volume 500 L): (1) absorption from air onto silica gel; (2) elution with acetone; (3) reaction with nitrophenol; (4) separation of the methylation product by thin-layer chromatography; (5) staining of the elution zone by reduction, diazotization, and coupling; and (6) comparison of the color intensity with a standard [18]. In one variant of this process, nitroanisole is determined by gas chromatography after step 3.

Continuous monitoring of rooms in which dimethyl sulfate is processed can be achieved with flame photometry using a sulfur detector (Messer Griesheim AG, Düsseldorf). If no other sulfur compounds are present (particularly SO_2 and CS_2), dimethyl sulfate levels from 0 to 0.5 ppm can be indicated. The last two methods are recognized by the chemical industry trade association in the Federal Republic of Germany [19].

Table 3. Safety data for dimethyl sulfate and diethyl sulfate

	Dimethyl sulfate	Diethyl sulfate
Solubility in water, g/100 g	2.8 (18 °C)	0.7
Boiling point, °C	188 (decomp.)	208 (decomp.)
Density of vapor (air = 1)	4.35	5.31
Flash point, °C	83	104
Ignition temperature (DIN 51 794), °C	470	395
Temperature class (VDE 0165)	T 1	T 2
Explosive limits, vol%		
Lower	3.6	1.8
Upper	23.2	12.2
Heat of combustion at constant pressure [25 °C; CO_2, H_2O, H_2SO_4 (aqueous)], J/g	11 970	18 110
Heat of vaporization at 25 °C, J/g [2]	385	368
Heat of formation at 25 °C (liquid), kJ/mol	736	812
Shipping regulations		
RID, ADR	class 6.1/13 b	class 6.1/13 b
IMDG code	class 6, p. 6182	class 6, p. 6182
	UN no. 1595	UN no. 1594
TLV–TWA	A2, 0.1 ppm	
TRK (FRG), mg/m^3	0.2* (use)	0.2*
	0.1* (manufacture)	

* Carcinogenic in animals, III A2.

6. Storage, Transportation, and Handling

Storage. Dimethyl sulfate does not attack iron and can be stored in ordinary steel drums or tanks. However, the rust that covers iron surfaces may produce a brownish discoloration of the product. In addition, traces of moisture will cause further rusting. Therefore, dimethyl sulfate is sometimes kept in aluminum or stainless steel containers or in resin-coated tanks. The organic coating can absorb dimethyl sulfate, and when such tanks are cleaned, traces of dimethyl sulfate can escape even after decontamination with ammonia water.

Rubber is rapidly destroyed by dimethyl sulfate. Gaskets made from asbestos–rubber materials, with metal ring inserts and coated with polytetrafluoroethylene (PTFE), have proved most satisfactory [20]. Polyethylene and polypropylene are also resistant to attack; ethylene–propylene copolymers can be used in place of soft rubber. They do not deform, but are difficult to decontaminate.

Shipping regulations and safety data for dimethyl sulfate and diethyl sulfate are given in Table 3.

Handling. Contact with liquid or gaseous dimethyl sulfate must be prevented. Suitable safety equipment includes impermeable gloves and close-fitting goggles. If any risk of dimethyl sulfate spilling into the work area exists, e.g., during a production breakdown, a gas mask with a filter suitable for acid gases or even full protective clothing and compressed-air respirators are required.

The presence of dimethyl sulfate can be detected by using the color reaction with 4-(4-nitrobenzyl)pyridine [1083-48-3] [21]. The sensitivity is about 20 mg/L. This method is also suitable for spot analysis.

First aid treatment after accidents with liquid or gaseous dimethyl sulfate involves thorough flushing of affected skin areas with 5 % ammonia solution. All clothing soiled with dimethyl sulfate should be removed immediately and decontaminated. A doctor must be consulted; symptoms may develop after a latency period of several hours (see Chap. 10).

Absorption by kieselguhr is the best way to clean up spilled dimethyl sulfate; 1 kg of kieselguhr binds 5 – 6 kg of dimethyl sulfate to form a doughlike mass.

A 5 – 10% aqueous ammonia solution is commonly used to break down dimethyl sulfate. The resulting compounds are methylamine, dimethylamine, and methyl sulfate anion. Without mixing and at ambient temperature, the dimethyl sulfate tends to settle (ϱ 1.329 g/cm^3) and several days may be necessary for the reaction to be complete. However, with effective stirring, breakdown is complete in ca. 3 min. Methylamine and dimethylamine have very unpleasant, persistent odors. The odor nuisance can be avoided by using an approximately 2 M aqueous solution of hexamethylenetetramine [100-97-0] instead of ammonia. In this case, a nonvolatile methyl sulfate salt is formed [22]. If hexamethylenetetramine is dissolved in a mixture of two parts water and one part diethylene glycol monobutyl ether, a homogeneous phase is obtained with dimethyl sulfate, and stirring is unnecessary. Because the breakdown of dimethyl sulfate with ammonia or hexamethylenetetramine is always an exothermic reaction, the volume of breakdown agent should be five times that of dimethyl sulfate. Solid hexamethylenetetramine reacts explosively with dimethyl sulfate. Dry alkali such as soda or lime is unsuitable as a breakdown agent.

The higher homologues of dimethyl sulfate behave similarly, but longer reaction times are required.

If dimethyl sulfate gets into insulation, complete dismantling and destruction of the porous materials are necessary.

Large quantities of water and dimethyl sulfate must not be allowed to come into contact in an uncontrolled manner: the acid and heat produced during hydrolysis may result in a continuously increasing speed of reaction. In addition, the acid attacks iron containers, and a substantial buildup of hydrogen pressure may result.

Technical-grade dimethyl sulfate contains small amounts of dimethyl ether, a flammable gas, which can collect in the air space in containers during prolonged storage. Some dimethyl ether may also be generated during reaction in alkaline solution. In this case, it may be found in either the waste gas or the final product; a considerable amount of dimethyl ether can accumulate in the solvent.

Further information on safety precautions can be found in [23]–[25].

7. Environmental Protection

Buildings in which large amounts of dimethyl sulfate are stored or handled must not have drainage that leads directly into the public sewer system; traps must be installed to test and, if necessary, detoxify effluents.

In most alkylations with dialkyl sulfates in aqueous alkaline media, the effluent contains half of the organic component of the dialkyl sulfate in the form of salts of alkylsulfuric acid. These salts are biodegradable. Alternatively, they can be hydrolyzed completely under pressure and at higher temperature; the alcohol produced is recovered by distillation. Acid hydrolysis at low temperature is also possible but subsequent neutralization is required.

Activated carbon may be used to absorb dimethyl sulfate vapors in waste gas. The carbon can be regenerated by treatment with dilute ammonia solution [26] or with water for several hours. Drying is not necessary.

8. Uses

8.1. Alkylation

The most important use of dialkyl sulfates is in alkylation. In this application, dialkyl sulfates must compete with alkyl halides, which are also important alkylating agents. Halides, especially chlorides, are considerably cheaper; in addition, chlorides result in a lower salt load in the effluent. However, dimethyl and diethyl sulfate can be processed at atmospheric pressure because of their higher boiling points, and their high reactivity enables conversion to occur at lower temperatures. Dimethyl sulfate is by far the most important dialkyl sulfate because it is cheaper and more reactive than its homologues. Diethyl sulfate also has commercial significance, whereas higher homologues or unsymmetrical and cyclic dialkyl sulfates are used only in special applications.

Alkylation is usually carried out in stirred vessels. The alkylating agent is added slowly to an aqueous alkaline solution or suspension of the compound to be alkylated. Where large volumes are concerned, gradual addition of the alkali during the reaction is preferred, to maintain a certain pH. In the alkaline range, steel equipment may be used. Stainless steel or enamel is recommended for weakly acidic media or for sensitive products (e.g., quaternized dyes).

If the reaction does not proceed in an aqueous alkaline medium, treatment of the dry alkali-metal salt with dialkyl sulfate, in the presence of an inert solvent if necessary, is often successful. Occasionally, the pure substance is fed into dialkyl sulfate above 100 °C. However, increased side reactions (e.g., sulfonations) must be expected under these conditions.

Etherification. The classic use of dimethyl sulfate is to etherify an aromatic hydroxyl group in aqueous alkaline solution:

$$C_6H_5OH + (CH_3)_2SO_4 + NaOH \longrightarrow C_6H_5OCH_3 + CH_3OSO_3^- Na^+ + H_2O$$

Dimethyl sulfate is added at 10 °C over 1 h, and the mixture is heated at 80 °C for 30 min; the yield is 90 % [27]. This reaction requires the presence of phenolate ions and, therefore, proceeds only at sufficiently high pH. The second methyl group of dimethyl sulfate may also be utilized if the mixture is refluxed for 15 h, but this is accompanied by reduced yield.

In strongly alkaline solution, the hydroxyl group reacts more rapidly than the carboxyl group:

$$\text{3,4,5-trihydroxybenzoic acid} + 3\,(CH_3)_2SO_4 + 3\,NaOH \longrightarrow \text{3,4,5-trimethoxybenzoic acid} + 3\,CH_3OSO_3^-\,Na^+$$

Small amounts of trimethoxybenzoic acid ester formed as byproduct are hydrolyzed by heating at 100 °C for 2 h.

Near neutral pH, however, the carboxyl group is alkylated more rapidly than the phenolic hydroxyl group. An example is the production of methyl salicylate [29]:

$$\text{salicylic acid} + (CH_3)_2SO_4 + NaHCO_3 \longrightarrow \text{methyl salicylate} + CH_3OSO_3^-\,Na^+$$

Because of steric hindrance, the two ortho hydroxyl groups of ethyl dihydroxyterephthalate [610-92-4] can be etherified only under special conditions:

$$\text{ROOC-C}_6H_2(OH)_2\text{-COOR} \xrightarrow{DMF} \text{ROOC-C}_6H_2(OCH_3)_2\text{-COOR}$$

where R = methyl

Sodium methoxide is used as the base, and the alkylation is carried out in dimethylformamide (DMF) at 65 °C. The reaction is completed by heating rapidly to 100 °C. The product shows a strong bluish fluorescence and can be used as a brightener for polyester fibers [30].

Steric hindrance also affects alkylation of 2,4,6-trihydroxybenzoic acid [83-30-7]. Both hydroxyl groups in the ortho position remain free although esterification occurs [31]:

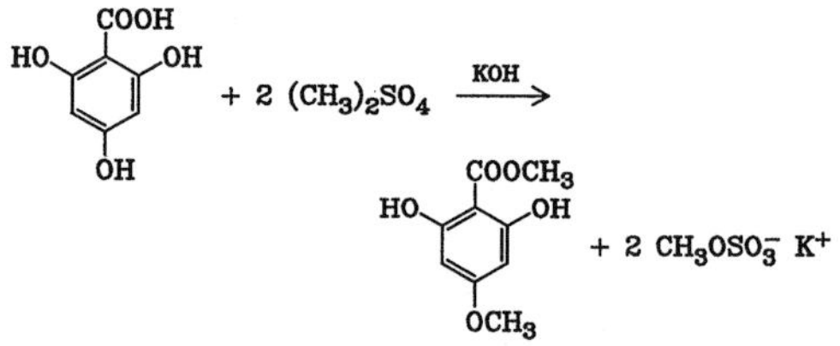

2,4,6-(HO)$_3$C$_6$H$_2$COOH + 2 (CH$_3$)$_2$SO$_4$ $\xrightarrow{\text{KOH}}$ 2,6-(HO)$_2$-4-(CH$_3$O)C$_6$H$_2$COOCH$_3$ + 2 CH$_3$OSO$_3^-$ K$^+$

Selective methylation of one hydroxyl group is also possible. For example, hydroquinone monomethyl ether is obtained in 84% yield if alkylation is carried out in a relatively large amount of an inert solvent (benzene), which extracts the monomethyl ether from the reaction mixture [32]:

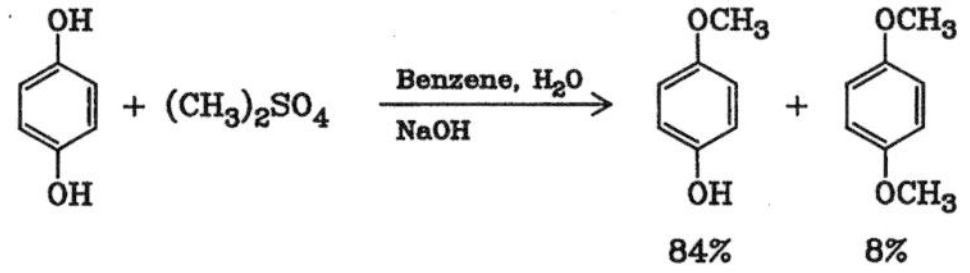

1,4-(HO)$_2$C$_6$H$_4$ + (CH$_3$)$_2$SO$_4$ $\xrightarrow[\text{NaOH}]{\text{Benzene, H}_2\text{O}}$ 4-CH$_3$O-C$_6$H$_4$-OH (84%) + 1,4-(CH$_3$O)$_2$C$_6$H$_4$ (8%)

The aliphatic hydroxyl group can also be alkylated with dialkyl sulfates in the usual way if it is sufficiently activated by neighboring groups:

HO−CH$_2$−C≡N $\xrightarrow{8-13\,°C}$ CH$_3$O−CH$_2$−C≡N

Dimethyl sulfate is partially recovered, and the low temperature reduces losses from hydrolysis [34]. In a strongly acidic medium, the nonactivated aliphatic hydroxyl group is selectively methylated before the amino group, although the yield is low (30%) [35]:

HO−CH$_2$CH$_2$CH$_2$−NH$_2$ $\xrightarrow{\text{CH}_3\text{COOH, HCl}}$ CH$_3$O−CH$_2$CH$_2$CH$_2$−NH$_2$

An unexpected byproduct of the methylation of phenol and other hydroxyaromatic compounds by dimethyl sulfate in the presence of tetrahydrofuran (THF) and potassium carbonate is the 4-methoxybutyl ether. In some cases, this is obtained in considerable quantities in addition to the normal methyl ether [36]:

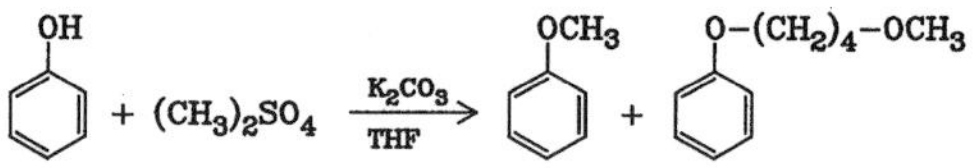

C$_6$H$_5$OH + (CH$_3$)$_2$SO$_4$ $\xrightarrow[\text{THF}]{\text{K}_2\text{CO}_3}$ C$_6$H$_5$OCH$_3$ + C$_6$H$_5$O−(CH$_2$)$_4$−OCH$_3$

Esterification. Fatty acid ethyl esters, which can be used as plasticizers, are produced according to the following reaction [33]:

2 R−COONa + (C$_2$H$_5$)$_2$SO$_4$ ⟶ 2 R−COOC$_2$H$_5$ + Na$_2$SO$_4$

Both ethyl groups of the sulfate are utilized and, therefore, prolonged heating at 150 °C is necessary. The yield drops sharply outside the pH range of 7.5 – 8.5. Diethyl sulfate should be added at temperatures between 70 and 90 °C, never above 95 °C. If calcium salts are used instead of sodium salts, the second ethyl group does not react, because calcium bis(ethyl sulfate) [926-03-4] is unstable above 100 °C. Monomers of acrylic resins (raw materials for varnishes) and acrylic glass can also be produced by esterification with diethyl sulfate. The reaction is of commercial interest because diethyl sulfate is continually regenerated by feeding in ethylene. The process is also applicable to lower saturated fatty acids [33].

N-Alkylation. Ammonia and aliphatic or aromatic amines are alkylated readily by lower dialkyl sulfates. The reaction may progress to the tetraalkylammonium ion (quaternization). Quaternization leads to the formation of an organic cation; i.e., a positively charged, strongly hydrophilic center is created within the molecule. The corresponding anion is usually the alkyl sulfate ion (RSO_4^-) or the sulfate ion (SO_4^{2-}) produced during the reaction, although in some cases the cation can be associated with other suitable anions.

Quaternization can increase the water solubility of a drug or the affinity of a dye molecule for a negatively charged fiber. Numerous organic ammonium derivatives are used as raw materials for detergents, textile auxiliaries, and antistatic agents.

Most quaternizations can also be achieved with methyl chloride. However, methylsulfuric acid salts (methosulfates) often yield preparations with better consistency and greater solubility in water than the chlorides. The sodium methyl sulfate produced during methylation can often be tolerated in the product and does not have to be removed. For plastics additives that are put into the melt, the greater heat stability of methyl sulfates is a desirable property. In addition, many products must be free of chloride ions to prevent corrosion during subsequent processing.

Quaternization is feasible with aromatic, aliphatic, and heterocyclic compounds [37]. For example, pyridine reacts with dimethyl sulfate in the absence of a solvent to form methylpyridinium methyl sulfate [34]. Treatment of nitro-2-hydroxyquinolines with dimethyl sulfate also yields the quaternary base (**4**). If the nitro group is in the 8-position, i.e., adjacent to the nitrogen, steric hindrance inhibits N-alkylation, and 2-methoxyquinoline is the main product (**5**) [38].

Alkylation of the dry potassium salt of carbazole in chlorobenzene at 100 °C gives *N*-ethylcarbazole [86-28-2] in 99 % yield; both alkyl groups react [39]:

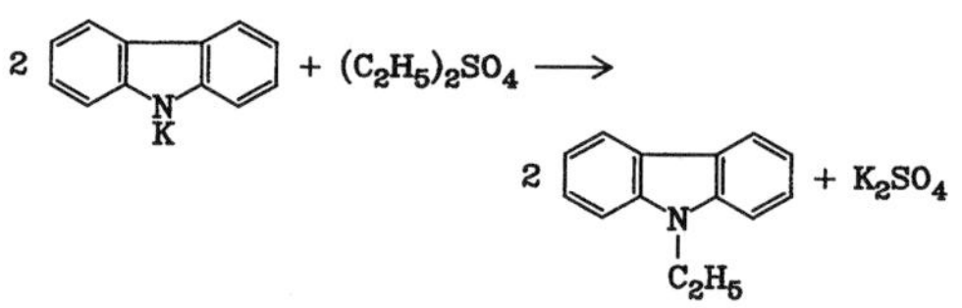

$$2\; \underset{K}{\text{(carbazole)}} + (C_2H_5)_2SO_4 \longrightarrow$$

$$2\; \underset{C_2H_5}{\text{(N-ethylcarbazole)}} + K_2SO_4$$

Indirect methods must often be used to obtain the desired product. Thus, *N*-methylbutylamine [*110-68-9*] is produced from butylamine [*109-73-9*] in 49% yield by the following reaction sequence [40]:

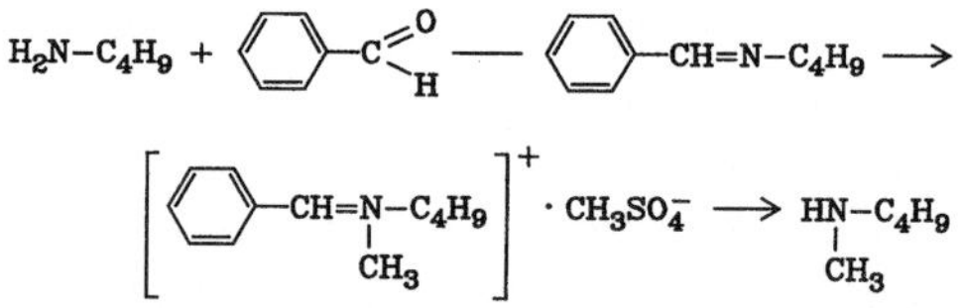

$$H_2N-C_4H_9 + Ph-C(=O)H \longrightarrow Ph-CH=N-C_4H_9 \longrightarrow$$

$$\left[\underset{CH_3}{Ph-CH=N-C_4H_9} \right]^+ \cdot CH_3SO_4^- \longrightarrow \underset{CH_3}{HN-C_4H_9}$$

To prevent more than one methyl group from attaching to the nitrogen atom, the first step is formation of the Schiff base with benzaldehyde. After methylation, the Schiff base is then hydrolyzed. Reaction of urea with dialkyl sulfates usually gives the alkyl compound of the enol form, –C(OR)=NH. The cyclic amide caprolactam behaves similarly [41]:

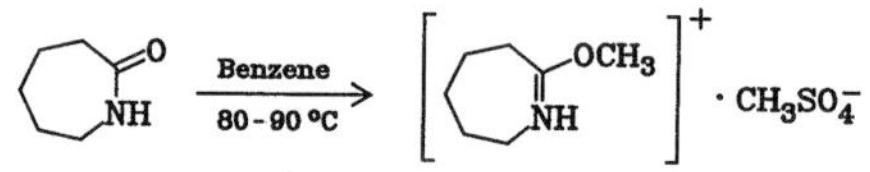

$$\text{(caprolactam)} \xrightarrow[80-90\,°C]{\text{Benzene}} \left[\text{(methoxy-iminium)} \right]^+ \cdot CH_3SO_4^-$$

However, some urea derivatives react in the keto form, –CO–NH$_2$, which yields N-alkylated compounds. An example is the final step in the synthesis of a chlorinated *N*-phenylurea [42]:

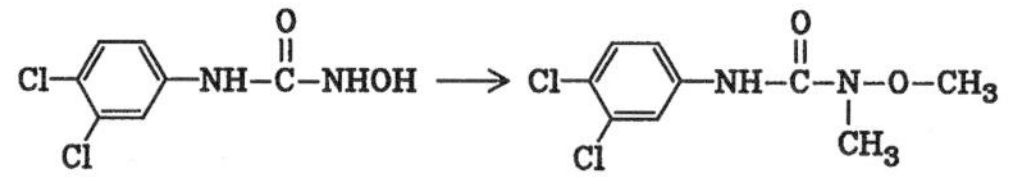

$$\text{Cl}_2C_6H_3-NH-\overset{O}{\underset{}{C}}-NHOH \longrightarrow \text{Cl}_2C_6H_3-NH-\overset{O}{\underset{}{C}}-\underset{CH_3}{N}-O-CH_3$$

The hydroxyl group and the amino group are methylated in a single operation. The resulting *N*-dichlorophenyl-*N'*-methyl-*N'*-methoxyurea and compounds of similar structure are used in large quantities as herbicides.

Hydrazine is also methylated readily. An indirect pathway via a diacyl compound is preferred for the synthesis of symmetrical dimethylhydrazine [*540-73-8*] [43]:

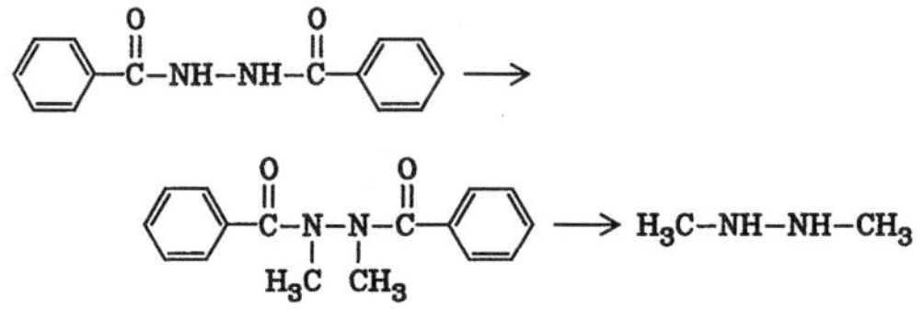

$$Ph-\overset{O}{\underset{}{C}}-NH-NH-\overset{O}{\underset{}{C}}-Ph \longrightarrow$$

$$Ph-\overset{O}{\underset{}{C}}-\underset{H_3C}{N}-\underset{CH_3}{N}-\overset{O}{\underset{}{C}}-Ph \longrightarrow H_3C-NH-NH-CH_3$$

C-Alkylation. Compounds that can react as either ketones or enols (ambident nucleophiles, e.g., acetoacetic ester) may undergo C-alkylation (–CO–CHR–) in addition to

O-alkylation (–C(OR)=CH–). Aromatic compounds with several hydroxyl groups in the meta position, e.g., resorcinol [108-46-3] and phloroglucinol [6099-90-7], react accordingly. In addition to ethers, small amounts of compounds in which the alkyl group is located at the aromatic nucleus are formed [44].

S-Alkylation. Phenols carrying a thiocyano group react with dimethyl sulfate in aqueous alkali to form thioethers.

$$HO-C_6H_4-S-C\equiv N \longrightarrow HO-C_6H_4-S-CH_3$$

The hydroxyl group remains unaltered and the cyano group is split off [45].

Alkylation of Inorganic Salts. Water-soluble metal halides can be converted to alkyl halides by reaction with dialkyl sulfates. A well-known example is the production of methyl iodide from potassium iodide [46]. Other salts can be alkylated similarly, e.g., sodium sulfide to diethyl sulfide or sodium sulfite to methanesulfonic acid [47]. Di-*tert*-butyl peroxide, which is used as a polymerization catalyst, is produced continuously in a cooled, vertical reaction tube from hydroperoxide and a mixture of *tert*-butyl hydrogen sulfate and di-*tert*-butyl sulfate [48].

Sulfonation. At high temperature, reaction with dialkyl sulfates may lead to sulfonation. This results from the equilibrium between dialkyl sulfates, on the one hand, and sulfur trioxide and the corresponding ether, on the other (see Section 4.1).

$$R_2SO_4 \rightleftharpoons SO_3 + R_2O$$

For example, if diphenyl ether or diphenylmethylamine is heated with diethyl sulfate for several hours at 150 °C, phenylsulfonic acids are produced [49].

8.2. Other Uses

The lower dialkyl sulfates are used occasionally as a source of acidifiers in thermosetting resins, as stabilizers for various chemicals, and as selective solvents for aromatic compounds in petroleum fractions.

An interesting application in preparative chemistry is the protection of the hydroxyl group against oxidation by esterification with sulfuric acid. For example, *tert*-butyl sulfate is oxidized with nitrogen dioxide at about 15 °C, leading to the conversion of a methyl group into a carboxyl group. After hydrolysis, a sulfuric acid solution of α-hydroxyisobutanoic acid [594-61-6] is obtained [50].

9. Economic Aspects

The production of dimethyl sulfate in Western Europe currently totals ca. 35 000 t/a. Diethyl sulfate production is estimated to be much lower. The remaining dialkyl sulfates are not commercially available, although they may be produced as intermediates in the manufacture of drugs or textile auxiliaries. The following companies produce dimethyl sulfate: Hoechst (Federal Republic of Germany); Rhône-Poulenc (France); Säurefabrik Schweizerhall (Switzerland); Du Pont (United States); Lee Chang Yung (Taiwan); and some producers in India, in the Soviet Union, and in Czechoslovakia. Diethyl sulfate is produced by British Celanese (United Kingdom) and Union Carbide (United States).

10. Toxicology and Occupational Health

10.1. Dimethyl Sulfate

Acute Toxicity. Inhalatory LC_{50} values in laboratory rodents (30–60 min) are between 30 and 75 ppm in the inhaled air [51], [52]. When humans are exposed to low concentrations, conjunctivitis with photophobia and irritation of the mucosal membranes of the upper respiratory tract result. Severe intoxication is characterized first by minor symptoms of local irritation of eyes, nose, throat, trachea, and bronchi; after a latency period of several hours lung edema may develop. If the lung edema is not lethal, bronchopneumonia and lung emphysema are typical secondary symptoms [53]–[64].

Chronic Toxicity, Carcinogenicity, and Mutagenicity. Inhalation by mice, rats, and hamsters of 0.5 ppm over a period of 15 months caused severe inflammation in the mucosa of the nose and in the lungs as well as increased mortality [65]. Long-term inhalation of 0.5 ppm of dimethyl sulfate induced malignant tumors in the nose and the sinusoidal membranes in rats, mice, and hamsters [66], [67]. Compared to other compounds with alkylating properties, dimethyl sulfate has a rather weak methylating activity, probably because it is quickly hydrolyzed after contact with wet mucosal membranes. Therefore, it induces local tumors but no tumors remote of the site of the primary contact. Dimethyl sulfate is a directly acting mutagen [68]–[72]. It methylates DNA in rat liver, lung, and intestine [73].

10.2. Diethyl Sulfate

In animal experiments, diethyl sulfate is a more potent alkylating agent than dimethyl sulfate.

Acute Toxicity. The acute toxicity of diethyl sulfate is lower than that of dimethyl sulfate: LC_{50} (rat, 4 h) > 250 ppm [74], LD_{50} (rat, oral) 880 mg/kg [74], LD_{50} (rat, subcutaneous) 350 mg/kg [75], LD_{50} (mouse, i.p.) 150 mg/kg [76]. The corrosive effect is weaker than that of dimethyl sulfate because hydrolysis occurs slower, but solutions containing 40% burn the eyes of rabbits [77]. Lethal intoxications are characterized by lung edema occurring after a latency period of 10–20 h.

Chronic Toxicity and Carcinogenicity. After chronic oral administration of diethyl sulfate to rats (25 and 50 mg/kg in weekly intervals) benign tumors of the forestomach and one squamous cell carcinoma, but no systemic tumors, occurred. This probably results from rapid hydrolysis of diethyl sulfate in the stomach [75]. Subcutaneous administration also produced only local tumors at the site of injection [53], [78], [79]. In offsprings of dams who had been treated with diethyl sulfate during pregnancy, tumors of the central nervous system and multiple mammary cancer occurred [75].

Mutagenicity. Diethyl sulfate is mutagenic in bacteriophages [80], in *E. coli* [81], [82], in *Drosophila* [83], [84], and in the dominant lethal assay in mice [85]. Cytogenetic studies in mice induced increased frequencies of aneuploidy in bone marrow but no structural chromosomal aberrations [86]. Diethyl sulfate is also positive in the point-mutation assay in Chinese hamster ovary cells [79].

11. References

[1] G. Seydel, *Chem.-Ing.-Tech.* **43** (1971) 140.
[2] F. Klages, H. A. Jung, P. Hegenberg, *Chem. Ber.* **99** (1966) 1704.
[3] M. G. Ahmed, G. W. Alder, G. H. James, M. C. Sinnot, M. G. Whiting, *Chem. Commun.* 1968, 1533.
[4] *Bios* **986**, 175.
[5] E. Clippinger, *Ind. Eng. Chem. Prod. Res. Develop.* **3** (1964) 3.
[6] F. Asinger: *Die petrochemische Industrie,* Akademie-Verlag, Berlin 1971, Part 1, pp. 279 ff, 379 ff, 396 ff.
[7] Union Chimique–Chemische Bedrijven S. A., GB 1 101 238, 1967 (A. v. Gysel, J. Colle).
[8] *Houben-Weyl,* **6**/2, p. 481.
[9] K. Venkataraman, *J. Sci. Ind. Res., Sect. A,* **8** (1949) 141.
[10] *Houben-Weyl,* **6**/2, p. 437.
[11] R. Levaillant, *C. R. Acad. Sci.* **189** (1929) 465; **200** (1935) 140.

[12] VEB Arzneimittelwerk Dresden, DD 32 781; DE-AS 1 200 290, 1962 (W. Lugenheim, E. Carstens, H. Fürst).
[13] *Houben-Weyl,* **6**/2, pp. 470.
[14] G. A. Sokolskij, N. N. Bogdanov, V. M. Pavlov, *Zh. Obshch. Khim.* 1971, 2530; *Chem. Abstr.* **76** (1972) 112 598.
[15] *Houben-Weyl,* **6**/2, pp. 475 ff.
[16] Hoechst, Analytical Laboratory, personal communication.
[17] W. M. Banick, E. C. Francis, *Talanta* **13** (1966) 979. D. K. Banerjee, M. J. Fuller, H. Y. Chen, *Anal. Chem.* **36** (1964) 2046.
[18] J. Keller, *Fresenius Z. Anal. Chem.* **269** (1974) 206.
[19] *Krebserzeugende Arbeitsstoffe, Anerkannte Analysenverfahren,* no. ZH 1 120.7, Carl Heymanns Verlag, Köln 1983.
[20] H. Wendt, *Chem.-Tech.* **9** (1980) no. 1, 31–34.
[21] J. Epstein, R. W. Rosenthal, R. J. Ess, *Anal. Chem.* **27** (1955) no. 9, 1435–1439.
[22] F. L. Hahn, H. Walter, *Ber. Dtsch. Chem. Ges.* **54** (1921) 1532.
[23] *Merkblatt für Arbeiten mit Dimethylsulfat,* Berufsgenossenschaft der Chemischen Industrie, no. M 013, Verlag Chemie, Weinheim 1982.
[24] *Unfallverhütungsvorschrift der Berufsgenossenschaft der Chemischen Industrie,* Appendix 4, no. 34, Jedermann-Verlag Dr. Otto Pfeffer, Heidelberg 1982.
[25] *Jahrbuch „Betriebswacht",* Chap. MAK-Werte ff, Universum-Verlagsanstalt, Wiesbaden.
[26] Hoechst DE-OS 2 063 071, 1970 (R. Grimm, W. Herzog, R. Lademann).
[27] *Org. Synth. Coll. Vol.* **I**, p. 58.
[28] *Org. Synth. Coll. Vol.* **I**, p. 537.
[29] M. Pailler, P. Bergthaller, *Monatsh. Chem.* **99** (1968) 103. U. N. Hirwe et al., *J. Univ. Bombay, Sect. A,* **22**, part 5; *Chem. Abstr.* **49** (1953) 11594 h.
[30] Alpine Chem. AG, FR 1 533 584 1967 (J. Hrach, W. Zeschmar).
[31] Kabushi Kaisha Ricoh, DE-OS 1 917 055, 1969.
[32] Universal Oil Products, US 3 274 260, 1964 (J. Levy, A. Friedmann).
[33] B. K. Zeinalov, J. A. Panteeva, *Vses. Soveshch. Sin. Zhirosdmen, Poverkhùoslnoaktiv. Vesheheslvam Moyushch. Sredslvam, 3rd Shebekino* 1965, pp. 225–229; *Chem. Abstr.* **66** (1967) 30 233 h; **67** (1967) 11 140.
[34] *Org. Synth. Coll. Vol.* **II**, p. 387.
[35] C. Pfizer Co. Inc., GB 1 070 307, 1966.
[36] Hori Zenichi et al., *Chem. Pharm. Bull.* **20** (1972) 827.
[37] H. Quast, E. Schmitt, *Chem. Ber.* **101** (1968) 4012.
[38] H. Decker, *Ber. Dtsch. Chem. Ges.* **38** (1905) 1148.
[39] Hoechst, DE-OS 2 132 961, 1971 (Th. Papenfuhs).
[40] *Org. Synth. Coll. Vol.* **V**, p. 736.
[41] *Org. Synth. Coll. Vol.* **IV**, p. 588.
[42] O. Scherer, G. Hörlein, K. Härtel, *Angew. Chem.* **75** (1963) 852.
[43] *Org. Synth. Coll. Vol* **II**, p. 208.
[44] E. Ott, E. Nauen, *Ber. Dtsch. Chem. Ges.* **55** (1922) 928.
[45] Dow Chemical, US 3 655 773, 1970 (W. Reifenschneider).
[46] *Org. Synth. Coll. Vol.* **II**, p. 404.
[47] O. I. Kajurin, SU 335 240, 1969; *Chem. Abstr.* **77** (1972) 33 931.
[48] G. Nettesheim, *Erdöl Kohle Erdgas Petrochem.* **19** (1966) 265.
[49] V. N. Belov, M. Z. Finkelstein, *Khim. (USSR)* **16** (1946) 1248; *Chem Abstr.* **41** (1947) 3065 i.

[50] Nissan Chem. Ind. Co., JP 6 726 611, 1962; 6 726 612, 1962.
[51] N. Hein, *Med. Inaug.-Diss.*, Würzburg 1969.
[52] L. Ghiringhelli, U. Colombo, A. Monteverde, *Med. Lav.* **48** (1957) 634.
[53] S. Weber, *Naunyn-Schmiedeberg's Arch. Exp. Path. Pharmakol.* **47** (1902) 113.
[54] F. D. Mohlau, *J. Ind. Hyg.* **2** (1920) 238.
[55] H. Strothmann, *Klin. Wochenschr.* **8** (1929) 493.
[56] J. Balazs in H. Fühner, *Sammlung von Vergiftungsfällen* **5** (1934) 47.
[57] S. v. Nida, *Klin. Wochenschr.* **24/25** (1947) 633.
[58] S. Tara, A. Cavigneaux, Y. Delplace, *Arch. Mal. Prof.* **15** (1954) 291
[59] T. R. Littler, R. B. McConnel, *Br. J. Ind. Med.***12** (1955) 54.
[60] M. G. Duverneul, *Arch. Mal. Prof.* **18** (1957) 306.
[61] W. Nebelung, *Arch. Gewerbepath. Gewerbehyg.* **15** (1957) 581.
[62] W. Tilling, B. Knick, *Med. Klin.* **35** (1960) 1534.
[63] H. Leithoff, J. Weinreich, *Beitr. gerichtl. Med.* **22** (1963) 196.
[64] L. Roche, J. M. Robert, P. Paliard, *Can. J. Chem.* **44** (1966) 1728.
[65] F. A. Schlögel, *Med. Inaug.-Diss.*, Würzburg 1972.
[66] H. Druckrey, H. Kruse, R. Preussmann, S. Ivankovic, Ch. Landschütz, *Z. Krebsforsch.* **74** (1970) 241.
[67] F. A. Schlögel, P. Bannasch, *Naunyn-Schmiedeberg's Arch. Exp. Path. Pharmakol.* **266** (1970) 441.
[68] J. R. Connell, A. S. C. Medcalf, *Carcinogenesis* **3** (1982) 385.
[69] M. Fox, J. Brennand, *Carcinogenesis* **1** (1980) 795.
[70] K. Hemminki, K. Falck, H. Vainio, *Arch. Toxicol.* **46** (1980) 277.
[71] H. Kubinski, G. E. Gutzke, Z. O. Kubinski, *Mutat. Res.* **89** (1981) 95.
[72] R. B. Painter, R. Howard, *Mutat. Res.* **92** (1982) 427.
[73] P. F. Swann, P. N. Magee, *Biochem. J.* **110** (1968) 39.
[74] H. F. Smyth, Jr., C. P. Carpenter, C. S. Weil, *J. Ind. Hyg.* **31** (1949) 60.
[75] H. Druckrey, H. Kruse, R. Preussmann, S. Ivankovic, Ch. Landschütz, *Z. Krebsforsch.* **74** (1970) 241.
[76] A. M. Malashenko, *Sov. Genet.* **7** (1973) no. 1, 59.
[77] C. P. Carpenter, H. F. Smyth, Jr., *Amer. J. Ophthal.* **29** (1946) 1363.
[78] W. B. Deichmann, H. W. Gerarde: *Toxicology of Drugs and Chemicals,* Academic Press, New York 1969, p. 217.
[79] D. B. Couch, N. L. Forbes, A. W. Hsie, *Mutat. Res.* **57** (1978) 217.
[80] A. Loveless, *Proc. R. Soc. B* **150** (1959) 497.
[81] Yu. A. Ratiner, S. K. Kanareikina, V. M. Bondarenko, I. V. Golubeva, *Zh. Mikrobiol. (Moscow)* **5** (1976) 117.
[82] C. H. Clarke, D. M. Shankel, *Mutat. Res.* **53** (1978) 168.
[83] M. Pelecanos, *Nature (London)* **210** (1966) 1294.
[84] M. Pelecanos, *Experientia (Basel)* **27** (1971) 473.
[85] A. M. Malashenko, I. K. Egorov, *Sov. Genet.* **4** (1972) no. 1, 14.
[86] N. I. Surkova, A. M. Malashenko, *Sov. Genet.* **8** (1974) no. 11, 1386.

Diazo Compounds and Diazo Reactions

HASSO HERTEL, Hoechst Aktiengesellschaft, Werk Offenbach, Federal Republic of Germany

1.	Introduction	1875	3.5.	Other Aromatic Diazo Compounds ... 1890
2.	Aliphatic Diazo Compounds	1876	3.6.	Analysis ... 1891
3.	Aromatic Diazo Compounds	1876	4.	Diazo Reactions ... 1892
3.1.	Diazonium Salts	1877	4.1.	Reactions in Which the Diazo Group is Retained ... 1892
3.1.1.	Properties	1877		
3.1.2.	Diazotization	1878	4.2.	Reactions of the Diazo Group 1893
3.1.3.	Other Methods of Preparation	1883	4.3.	Replacement of the Diazo Group (Dediazotization) ... 1895
3.1.4.	Uses	1883		
3.2.	Diazotates	1884	5.	Environmental Protection ... 1899
3.3.	Diazosulfonates	1886	6.	Safety ... 1900
3.4.	Diazoamino Compounds (Triazenes)	1887	7.	Toxicology ... 1900
			8.	References ... 1901

1. Introduction

A diazo compound is an aliphatic, aromatic, or heterocyclic compound in which an $-N_2$ group is attached to a carbon atom. This group can be present as such (diazo compounds in the narrowest sense), it can carry a positive charge (diazonium salts), or the nitrogen atom in the β-position can carry a substituent (diazo compounds in the general sense).

Diazo compounds readily take part in a variety of reactions, which gives them their value to the chemical industry. Diazo reactions, which normally take place rapidly and smoothly, lead to a number of important intermediates and end products.

2. Aliphatic Diazo Compounds

Aliphatic diazo compounds have an open chain structure, which may be represented by the following mesomeric formulas:

$$\underset{R'}{\overset{R}{>}}\overset{-}{C}-\overset{+}{N}=N \longleftrightarrow \underset{R'}{\overset{R}{>}}\overset{-}{C}-\overset{+}{N}\equiv N \longleftrightarrow$$

$$\underset{R'}{\overset{R}{>}}C=\overset{+}{N}=\overset{-}{N} \longleftrightarrow \underset{R'}{\overset{R}{>}}\overset{+}{C}-N=\overset{-}{N}$$

With few exceptions, aliphatic diazo compounds lack the ability for azo coupling and as a result they are of little industrial importance.

The simplest diazo compound is *diazomethane* [334-88-3], CH_2N_2, bp −23 °C, fp −145 °C, a yellow gas, which is explosive both as a gas and in a concentrated solution. It is prepared by reacting (1) an alkali-metal hydroxide with *N*-nitroso-*N*-methylurea [684-93-5] or *N*-nitroso-*N*-methyl-4-toluenesulfonamide [80-11-5], e.g., in diethyl ether,

$$CH_3-\underset{\underset{R}{|}}{N}-NO + NaOH \longrightarrow CH_2N_2 + RONa + H_2O$$

$$R = NH_2CO \quad \text{or} \quad CH_3C_6H_4SO_2$$

(2) methylamine with nitrosyl chloride in diethyl ether at −80 °C, or (3) chloroform and hydrazine with potassium hydroxide in ethanol. The solutions decompose slowly, even at room temperature, liberating nitrogen.

Diazomethane has been used in the esterification of unstable carboxylic acids and generally as a methylating agent. Because of its toxicity (see Chap. 7) and explosive properties, however, its use has been largely discontinued.

The Arndt–Eistert synthesis, in which diazomethane is used, converts an acid chloride into the next higher acid (RCOCl → RCH$_2$COOH) in a multistage process. An extra CH$_2$ group can also be introduced into cyclohexanone by reaction with diazomethane, converting it into cycloheptanone.

Other aliphatic diazo reagents used in organic syntheses are diphenyldiazomethane [883-40-9], which is obtained by the oxidation of benzophenone hydrazone with yellow mercuric oxide, and diazoacetic ester [623-73-4], a yellow oil produced by the room-temperature reaction of glycolic ester with nitrous acid.

3. Aromatic Diazo Compounds

The diazonium salts, Ar–N≡N$^+$Z$^-$, are the most important aromatic diazo compounds. (Ar represents an aryl group throughout this article.) Apart from these, the only industrially important compounds are the aryldiazotates, which contain the anion Ar–N=N–O$^-$, the aryldiazosulfonates, which contain the anion Ar–N=N–SO$_3^-$, and the aryldiazoamino compounds, Ar–N=N–NRR′.

3.1. Diazonium Salts

The diazonium cation is a mesomeric system that may be represented by the following limiting structures:

$$\text{Ph}-\overset{+}{N}=N \longleftrightarrow \text{Ph}-\overset{+}{N}\equiv N \longleftrightarrow$$

$$\overset{+}{=}\overset{+}{N}=\overset{-}{N} \longleftrightarrow +\overset{+}{=}\overset{+}{N}=\overset{-}{N}$$

The inclusion of the π-electrons of the aromatic system increases the possibilities for resonance, and because the aryldiazonium ion is thereby rendered much more stable than the alkyldiazonium ion, the aromatic diazonium salts can be isolated as moderately stable compounds.

3.1.1. Properties

Diazonium cations form salts with many inorganic and organic anions. The salts of strong acids are completely dissociated in aqueous solution; therefore, such solutions are neutral. A large number of salts with complex anions are known [1], [2], [9]. Salts of industrial importance are formed with the following anions: chloride, hydrogen sulfate, tetrafluoroborate, tetrachlorozincate, and naphthalene-1,5-disulfonate, their solubility decreasing in that order. The diazonium compounds formed from aromatic sulfonic or carboxylic acids, being inner salts, or zwitterions, are sparingly soluble if no solubilizing groups are present.

With naphthalene and its derivatives, including the industrially important sulfonic acids, diazonium ions form donor-acceptor complexes (charge-transfer or π-complexes), which additionally stabilizes the diazonium ion [4].

Many diazonium salts are *explosive* when dry, some even when moist, an extreme example being diazotized aminotetrazole, which can explode as an aqueous suspension. Explosion can be initiated by heat, friction, or impact. Extremely unstable salts occasionally explode without any apparent external cause. Nitrates, perchlorates, and iodides are very explosive. The risk of explosion decreases in the order iodide, bromide, and chloride and continues to decrease in the order tetrachlorozincate, hydrogen sulfate, tetrafluoroborate, and naphthalene-1,5-disulfonate. Nitro groups on the aromatic nucleus increase explosiveness while alkylamino and especially arylamino groups reduce it. The pressure increase in the exothermic decomposition was studied [10], and the explosion properties were studied calorimetrically [11].

Apart from explosion, which completely destroys the molecule, diazonium salts at or just above room temperature undergo a *gradual decomposition* in which nitrogen is split off from the rest of the molecule [12]. For many industrial diazonium salts, decomposition is <1 % after 1 month at 40 °C.

Diazonium salt solutions are also not completely stable, especially at elevated temperatures. Decompositon in a dilute mineral acid is a first-order reaction. The rate of decomposition, which is largely independent of the anion and almost entirely determined by the cation, increases about 4-fold for a 10 °C rise in temperature [13]. Solutions of diazonium naphthalenesulfonates are an exception because of π-complex formation [4]. Figure 1 shows the rate of decomposition of some important diazonium chloride solutions at various temperatures.

For mineral acid concentrations between 0.1 and 10 wt%, the rate of decomposition is practically constant, but higher concentrations can have a stabilizing effect: the stability of diazotized 2,4-dinitroaniline in 85 wt% sulfuric acid is ca. 60 times that in 63 wt% sulfuric acid. Many diazonium salts of heterocyclic amines, e.g., of 2-aminobenzothiazole, are stable only in concentrated sulfuric acid; dilution with water leads to rapid decomposition.

Above the pH characteristic of mineral acids, decomposition takes place much faster. Although such a higher pH is suitable for azo coupling, the rate of decomposition can exceed that in mineral acids by several orders of magnitude. The maximum rate of decomposition occurs at the pH where the concentrations of diazonium ion and diazotate ion are equal (see Section 3.2) [14].

3.1.2. Diazotization

By far the most important method of industrial production of diazonium salts is by treating amines with nitrous acid in an acid medium.

Although continuous diazotization processes are known [15]–[20], crystalline diazonium salts are still made by the batch method without exception. Apart from the well-known problems associated with continuous operation of a multiphase process, the small production amounts could well be the reason. For the details of specific industrial diazotization procedures, see [21]–[23].

The most important process is *direct diazotization* in dilute hydrochloric acid with sodium nitrite. The amine must either be completely dissolved in the liquid phase or be in the form of a fine, lump-free suspension of either the free amine or the amine hydrochloride. Amines that are liquid at room temperature, or that melt below ca. 70 °C, are slowly added as liquids with vigorous agitation to dilute hydrochloric acid at a temperature about equal to or higher than the melting point of the amine. For each mole of amine ca. 2.3–3.0 mol of hydrochloric acid and 180–500 mL of water are required, depending on the molecular mass of the amine. To prevent lump formation, only the amount of hydrochloric acid that is necessary to form the hydrochloride ($\approx$ 1.2 mol) is used; the remainder required for diazotization (1.1–1.8 mol) is added at a later stage.

Amines that melt $\geq$ 80 °C are first suspended in finely crystalline or powder form in water at room temperature. Agglomerates that are not broken down by stirring alone

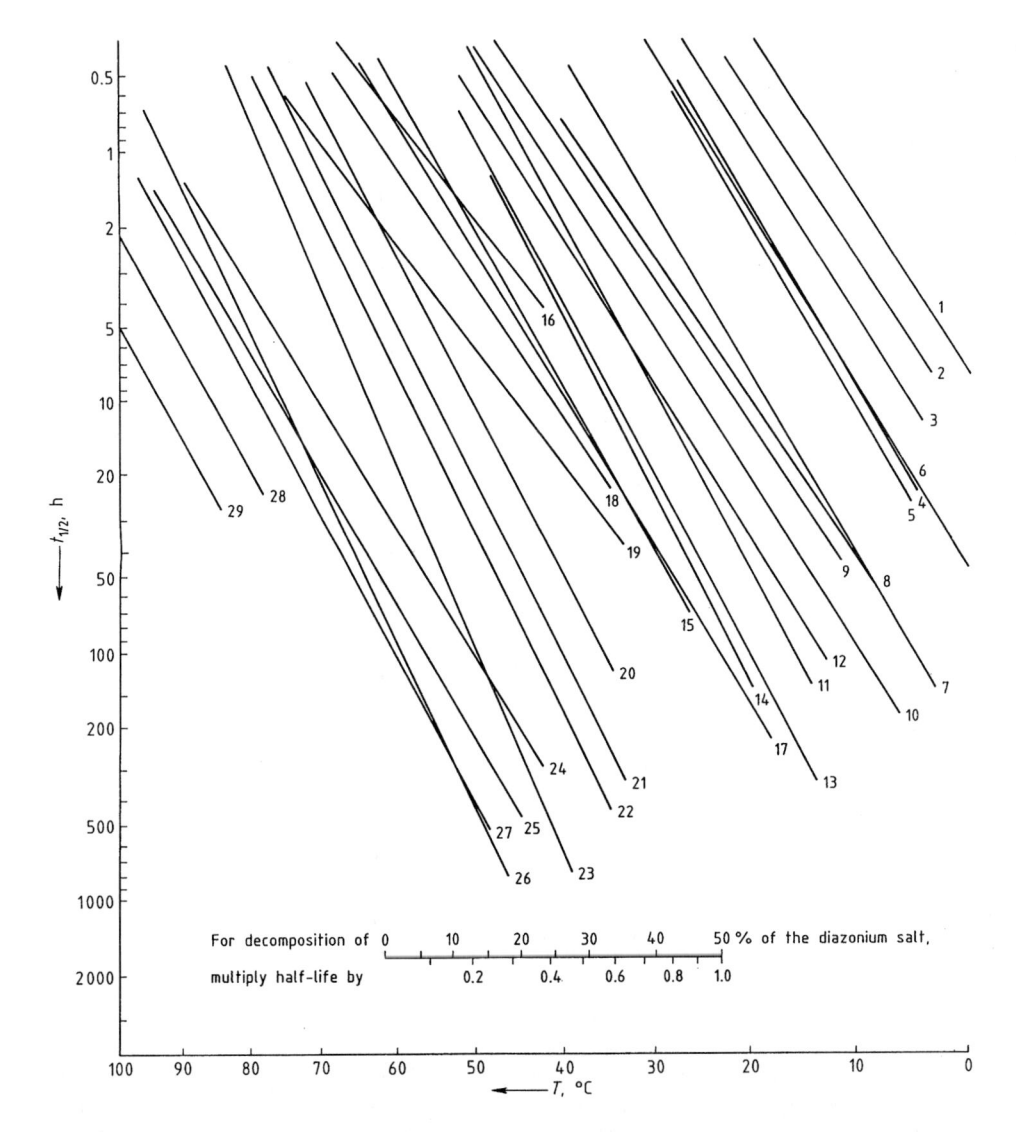

Figure 1. Decomposition half-life of mineral acid solutions of diazonium salts as functions of temperature (partly extrapolated)
Diazonium chloride solutions (ca. 0.1 mol/L, pH ca. 1.5) prepared from: 1) 5-Amino-1,4-dimethylbenzene; 2) 3-Aminophenol; 3) 3-Ethylaniline; 4) 2-Aminotoluene; 5) 3-Aminotoluene; 6) 3-Aminoanisole; 7) Aniline; 8) 4-Amino-1,3-dimethylbenzene; 9) 4-Chloro-2-aminotoluene; 10) 2-Aminonaphthalene; 11) 3-Aminobenzenesulfonic acid; 12) 6-Chloro-2-aminotoluene; 13) 4-Aminotoluene; 14) 2-Aminobenzenesulfonic acid; 15) 4-Aminobenzenesulfonic acid; 16) 4-Aminoazobenzene; 17) 3-Chloroaniline; 18) 5-Nitro-2-aminotoluene; 19) 4-Nitro-2-aminotoluene; 20) 2-Aminophenol; 21) 4-Nitroaniline; 22) 4-Chloroaniline; 23) 3-Nitroaniline; 24) 4-Aminophenol; 25) 2-Nitroaniline; 26) 4-Aminoanisole; 27) 2-Chloroaniline; 28) 1,4-Diaminobenzene (only one amine group diazotized); 29) 2,4-Dichloroaniline

are homogenized with a high-speed rotor – stator system or with an auxiliary disk mill. The hydrochloric acid is then added slowly.

The diazotization is usually carried out at 0–15 °C with external cooling to avoid side reactions and decomposition. When this temperature has been reached, a concentrated solution of sodium nitrite is added. The rate of addition is usually adjusted to match the rate of reaction, which is tested by sulfone reagent (3,6-diamino-9*H*-thioxanthene-10,10-dioxide [*10215-25-5*]) or redox measurement. A large excess of nitrite leads to the formation of nitrogen oxides, which should be avoided. The heat of reaction, ca. 100 kJ/mol, must be removed continuously, which frequently imposes a limit on the speed of the process.

Certain amines require special reaction conditions. For example, many diazotizations take place satisfactorily only if all the nitrite is added quickly, in practice within 1–2 min, 4-nitroaniline being a well-known example. In such cases, the heat of reaction cannot be removed quickly enough by external cooling; thus, either the starting temperature must be very low (–10 to –15 °C) or ice must be added, or both. In the diazotization of aminoazo compounds or aminonaphtholsulfonic acids and in some continuous diazotization processes, a higher reaction temperature, 30–50 °C, is sometimes used.

Amines that form sparingly soluble diazonium chlorides must be diazotized in a more diluted state; otherwise, the solution does not clarify. For many diazoazobenzenes, concentrations of only 0.03 mol/L are used, compared with ca. 1 mol/L for diazonium compounds of chlorobenzenes or chlorotoluenes.

Amines that tend to form triazenes require more acid, (e.g., 2,5-dichloroaniline requires 4–5 mol HCl), whereas the easily oxidized aminonaphthols are diazotized satisfactorily only with the stoichiometric amount of acid.

Amines that contain sulfonic acid groups are usually only slightly soluble in water, as are their diazonium salts. Sulfanilic acid may be diazotized like a normal amine. For larger molecules, *indirect diazotization* must be used: the aminosulfonic acid is dispersed in water and then dissolved as its sodium salt by the addition of an equivalent amount of sodium hydroxide solution. Sodium nitrite solution is then added, and the mixture poured into dilute ice-cooled hydrochloric acid.

Very weakly basic amines (dinitroaniline and many trifluoromethylanilines) cannot be diazotized in dilute acid. They are dissolved in 70–95 % sulfuric acid and diazotized with a solution of nitrosylsulfuric acid in sulfuric acid. The reaction often requires several hours, with the temperature reaching 40 °C. Pouring the product onto ice produces a diazonium salt solution that can be further processed in the same way as a product made by the hydrochloric acid–nitrite route.

Clarification of diazonium salt solutions produced by one of the above routes removes undiazotized material and byproducts. The addition of a filtration aid such as diatomite is recommended, and a bright solution can be obtained by using activated carbon. The solution then usually goes directly to the diazo reaction stage. The diazonium salt is isolated only if special products, such as pure salt-free materials for reprography or the textile industry, are required.

Crystalline diazonium salts may be precipitated from solution by the addition of a solution of zinc chloride, arylsulfonate, sodium chloride, etc. Solid precipitants should be introduced gradually. In a few cases, the diazonium salt solution is added to the precipitant. Temperature, time, and stirring intensity are chosen to give crystals as large and compact as possible, so that they filter well on a press or vacuum filter. The adhering mother liquor can be removed largely by a basket centrifuge, giving residual moistures of 5–30%, depending on the product. Some materials that do not separate readily in a centrifuge, e.g., those with platelet or needle-like crystals, may be dewatered in a hydraulic press. Continuous centrifuges enable filtration and dewatering to be done in a single operation, but care must be taken to avoid heating the diazonium salt excessively, with the consequent danger of fire when the diazonium salt is removed by the scraper blade.

The *yield* depends on the choice of amine and the method of isolation. Tetrachlorozincates give the best results, with yields as high as 96%.

Reaction vessels can have a volume up to and sometimes exceeding 20 m^3. Preferably they should be enameled and have a cooling jacket for water or brine. With rubber- or brick-lined vessels, the only cooling method possible is by ice addition. Pipework can consist of rubber-lined steel, polypropylene, or glass-fiber-reinforced polyester resin. The filtration equipment (leaf filters, pressure filters, and filter presses) usually consists of rubber-lined steel.

The commercially important diazonium salts include the chlorides; the hydrogen sulfates; the tetrachlorozincates, $ZnCl_4^{2-}$; the tetrafluoroborates, BF_4^-; and the arylsulfonates, $ArSO_3^-$.

Of the *diazonium chlorides*, benzenediazonium chloride and the simple substituted derivatives have solubilities so high that they effectively cannot be crystallized from solution. Only higher molecular mass derivatives give satisfactory yields, e.g., those with two or more benzene rings per diazonium group. Because the crystalline diazonium chlorides tend to decompose relatively easily, only the most stable diazonium cations are produced industrially as chlorides. Diphenylamine-4-diazonium chloride [*101-56-4*] and 4′-methoxydiphenylamine-4-diazonium chloride [*101-69-9*] are especially important because of their use as diazo components of blue dyes. The amino group in the para position has a stabilizing effect, giving the diazonium chlorides unusually good storage properties, which surpass those of most other industrial diazonium salts.

Diazonium hydrogen sulfates are more sparingly soluble than the corresponding chlorides and can be precipitated from the solution obtained in a diazotization reaction by adding sulfuric acid or sodium hydrogen sulfate (in practice, 78% sulfuric acid plus sodium sulfate). The second method of isolation has proved to be very successful in the production of azobenzenediazonium salts.

Diazonium tetrachlorozincates display extremely good crystallization properties. They have high solubility in water or dilute mineral acids but not in concentrated sodium chloride solution, which allows them to be isolated in good yield. Their large, hard, compact crystals are suitable for further processing and are usually very stable, with the result that these are the most common solid diazonium salts.

The first step is the clarification of the diazonium salt solution by filtration. Over a period of ca. 1 h zinc chloride (0.6–0.7 mol per mole of diazonium salt) is added in the form of a ca. 65% technical grade solution. For the less-soluble tetrachlorozincates, a dilute solution is used to avoid formation of an amorphous precipitate. Addition of 10–20 vol% of sodium chloride completes the precipitation.

Tetrachlorozincates may also be produced from the solutions in concentrated sulfuric acid that are obtained by adding nitrosyl diazotization products to ice. Zinc chloride solution and sodium chloride are added, precipitating the tetrachlorozincate. However, usually the diazonium hydrogen sulfate is precipitated, filtered off, dissolved in water, and then precipitated as the tetrachlorozincate.

Diazonium tetrafluoroborates may be obtained from diazonium salt solutions in hydrochloric acid by addition of tetrafluoroboric acid or its salts, usually the sodium salts. For good yields, an excess of 30–50% is required. Solubilities are similar to those of the tetrachlorozincates, but crystallization properties and yields are inferior. They are used in industry because they are less explosive than other salts.

Diazonium arylsulfonates are obtained by addition of arylsulfonic acids or their alkali-metal salts to diazonium salt solutions. If necessary, precipitation can be completed by addition of sodium chloride. Their solubility, for a given cation, depends strongly on the type of arylsulfonic acid.

Important compounds include the salts of benzenesulfonic acid and its substitution products, e.g., the *p*-chloro and *p*-acetylamino derivatives, and especially the acid naphthalene-1,5-disulfonates. These last compounds are produced by addition of free naphthalene-1,5-disulfonic acid to diazonium salt solutions in mineral acids.

Drying of diazonium salts must be carried out gently because of their instability. Usually drying cabinets with trays are ventilated with air at ca. 45–70 °C and provided with emergency water sprays that come into operation at a predetermined temperature. Vacuum drying is avoided because the gas that is evolved when the material decomposes can destroy equipment. For safety reasons, no drying processes are used that subject the product to mechanical stress. Production rates are usually too small to justify the use of continuous equipment such as belt dryers.

Only a few diazonium salts are stable enough to be dried in the form in which they are produced. Many are first blended with inert salts, such as sodium sulfate or sodium chloride, to reduce the danger of explosion. Any adhering acid is neutralized by mildly alkaline materials such as borax or magnesium oxide.

A drying process of great industrial importance is the addition of an anhydrous salt that picks up water to form a hydrate. Suitable salts are zinc sulfate, magnesium sulfate, and especially partially dehydrated aluminum sulfate [24]. The last is a hexahydrate, and quickly absorbs the remaining water, forming a · 9 H_2O (sensitive diazonium salts) or · 12 H_2O hydrate. The partially dehydrated salt is placed in a mixer with a lining resistant to hydrochloric acid, and the moist product is slowly added. The released heat of hydration must be removed by jacket cooling.

The final processing of a diazonium salt consists of grinding, plus some other processes if the diazonium salt is to be used as a dye intermediate, i.e., incorporation of additives, adjustment of product content, and removal of fines. The main object of the grinding is to ensure product uniformity, mainly by breaking down agglomerates formed during drying; fine grinding is not necessary. The process must be gentle because of the sensitivity of the products. Fairly low-speed air-swept mills without screens are often used.

Additives suppress undesirable properties in the product and widen the range of uses. The prime concern is to reduce the risk of explosion, e.g., by addition of inert materials such as sodium chloride or sodium sulfate. Another approach is to add compounds with a large heat of fusion or vaporization. In practice, only hydrates, usually aluminum sulfate hydrates, are used. These methods desensitize the commercial products, rendering them completely safe even under severe stress. Some metal salts, such as chromium acetate or aluminum sulfate, or boric acid [25], [26] can be added as buffering agents. Metaboric acid or boric anhydride improves the stability and avoids discolorization or caking [27].

The diazonium salt content is usually adjusted with one of the above additives. If these substances are unsuitable, owing to their "salting-out" effect, dextrin can be used as an extender. Dextrin behaves somewhat as a protective colloid, holding insoluble byproducts and decomposition products in solution, but its adhesive properties can cause trouble.

Dust suppressants must not be allowed to react with the diazonium ion. Silicone oils and diethyl carbonate have been suggested, but a satisfactory solution has not yet been found.

3.1.3. Other Methods of Preparation

Other methods of producing diazonium salts have some importance, e.g., the introduction, replacement, and modification of substituent groups in diazonium salts that are already formed (see Section 4.1), and the reaction of nitrous acid with arylamine-N-sulfonic acids.

$$Ar-NH-SO_3H + HNO_2 \longrightarrow Ar-\overset{+}{N} \equiv N + HSO_4^- + H_2O$$

This last reaction is carried out in mineral acid at low temperature just as for the diazotization of an amine. Weakly basic amines may be diazotized more easily in this way than in the unsubstituted form, probably because of the solubility imparted by the N-sulfonic acid group.

3.1.4. Uses

Solid diazonium salts with blended additives are used mainly for the manufacture of developing dyes for textile dyeing and printing.

Another application is in diazo-type photocopying processes. In contrast to developing dyes, which usually use compounds with electron-withdrawing substituents, this process uses compounds with electron-releasing substituents because they give darker shades and usually better light sensitivity. The main products are 4′-methoxydiphenylamine-4-diazonium chloride [*101-69-9*], 4-dimethylaminobenzene-1-diazonium tetrachlorozincate [*6023-44-5*], and 5-benzoylamino-1,4-diethoxybenzene-2-diazonium tetrachlorozincate [*5486-84-0*]. These substances give blue, violet, or black shades, while the equally common 1,2-diazonaphthol-4-sulfonic acid [*4857-47-0*] gives a red-brown shade.

3.2. Diazotates

The diazonium ion is a Lewis acid and, therefore, is capable of accepting a hydroxyl ion, thereby becoming a diazohydroxide, which dissociates immediately to form a diazotate [4]:

$$Ar-\overset{+}{N}\equiv N \underset{H^+}{\overset{OH^-}{\rightleftharpoons}} Ar-N=N-OH \underset{H^+}{\overset{OH^-}{\rightleftharpoons}} Ar-N=N-O^-$$

Diazohydroxide   Diazotate

The change from diazonium ion to diazotate takes place in weakly alkaline media if strong electron-withdrawing substituents are present, but otherwise moderately or strongly alkaline media is required. The equilibrium pH values for 50% conversion of various diazonium ions [28] are

$p-NO_2-C_6H_4-N_2^+$	9.4
$p-Cl-C_6H_4-N_2^+$	11.2
$C_6H_5-N_2^+$	11.9
$p-CH_3-C_6H_4-N_2^+$	12.6

The normal or *syn*-diazotates in aqueous solution are changed to the isomeric iso- or *anti*-diazotates by the action of caustic alkalis. According to the generally accepted Hantzsch theory, this is steric isomerism, *syn*-diazotates having the cis form and *anti*-diazotates the trans:

Ar＼N=N＼O⁻   Ar＼N=N＼O⁻ (trans)

syn / cis / Z   *anti / trans / E*

The rate of the rearrangement to the *anti*-diazotate depends strongly on the substituents present in the aromatic nucleus, electron-withdrawing substituents having a strongly accelerating effect. For example, *p*-nitrobenzenediazotate rearranges at least 1000 times faster than benzenediazotate. Only the *anti*-diazotates are used in industry.

Properties. The sodium and potassium *anti*-diazotates are mostly light yellow crystalline substances that store well. They decompose on strong heating, though not violently.

In alkaline solution, sodium and potassium *anti*-diazotates do not undergo the coupling reaction. With excess acid, they change slowly and irreversibly to the diazonium ion, which couples in the usual way. This coupling forms the basis for their use in the textile printing industry.

Preparation. The *anti*-diazotates of the *o*-and *p*-nitranilines are prepared by mixing the diazonium salt solution with a room-temperature 5–15 wt% solution of caustic alkali at 0–10 °C. The *anti*-diazotates precipitate as golden yellow crystals, this separation being completed by the addition of an alkali-metal chloride. *anti*-Diazotates of strongly basic amines, such as those of the chlorotoluidines or chloroanisidines, require a higher alkali concentration, up to 75 wt%, as well as a high temperature, up to 135 °C. The diazonium salt solution can sometimes be run directly into the hot alkaline solution, but usually the *syn*-diazotate is first formed at 0–10 °C and is then heated or added to hot, concentrated caustic alkali solution [7, 10/3, 555–557], [23, C1]. The completion of the chemical rearrangement is indicated by a negative spot test with *β*-naphthol solution. Slight dilution with water is often necessary to precipitate well-crystallized diazotate.

Sodium 4-Nitrobenzene-anti-diazotate [61889-43-8]. The crystalline amine is added to water with vigorous agitation, after which 32% hydrochloric acid is added; stirring is then continued for several hours until hydrochloride formation is complete. Ice is added, followed by rapid addition of a concentrated solution of sodium nitrite. After further agitation, diatomite is added and the solution filtered. The filtrate is given bulk by addition of ice water and then is run into 16 wt% sodium hydroxide solution at 6 °C. The temperature is kept constant by external cooling. After several minutes, sodium chloride is added. The *anti*-diazotate that crystallizes is filtered off.

Cast iron has proved to be a suitable construction material for vessels, pipes, and filtration equipment. The amounts of alkali-metal hydroxide required are considerable, sometimes exceeding 10 000% of the stoichiometric amount. Naturally, the less expensive sodium hydroxide is used whenever possible, but in many cases potassium hydroxide must be used.

The isolated *anti*-diazotates can be dried by gentle heating. Preparations for the cotton printing industry are first centrifuged or hydraulically pressed to remove adhering mother liquor and are then mixed with anhydrous sodium acetate, which can absorb up to 3 mol of water as water of crystallization. Yields can be as high as 90%.

Uses. The main use of the diazotates is textile printing. For this purpose they are marketed in commercial preparations that also contain the coupling component, usually 2-hydroxy-3-naphthoic acid anilide.

3.3. Diazosulfonates

Diazosulfonates exist in a labile cis form and a stable trans form.

$$\underset{syn\,/\,cis\,/\,Z}{\overset{Ar}{\underset{}{\diagdown}}N=N\overset{SO_3^-}{\diagup}} \qquad \underset{anti\,/\,trans\,/\,E}{\overset{Ar}{\underset{}{\diagdown}}N=N\underset{SO_3^-}{\diagdown}}$$

Although they are usually termed *syn* and *anti* these compounds are sometimes called normal and iso diazosulfonates. The syn forms exist in equilibrium with the starting materials:

$$Ar-\overset{+}{N}\equiv N + SO_3^{2-} \rightleftharpoons Ar-N=N-SO_3^-$$

As the diazonium ion becomes more electrophilic, the equilibrium increasingly favors the formation of diazosulfonate. On the other hand, the rearrangement of the syn form to the anti form is irreversible. Only the sodium *anti*-diazotates have industrial application.

Properties. The sodium *anti*-diazosulfonates are yellow to orange and crystallize well. Being quite stable, they may be stored for years without decomposition. Only *anti*-diazosulfonates with electron-releasing substituents, such as the derivatives of 4-aminodiphenylamine, dissociate when warmed to give azo dyes with coupling components. Unsubstituted *anti*-diazosulfonates and *anti*-diazosulfonates with electron-withdrawing substituents, on the other hand, react only slightly or not at all. Reducing agents, including an excess of sulfite, convert the diazosulfonates more or less smoothly into phenylhydrazine-*N*-sulfonic acid.

Preparation. Diazosulfonates are formed by the reaction of diazonium ions with sulfite anions, SO_3^{2-}, but not with hydrogen sulfite anions, HSO_3^-. Therefore, a weakly alkaline medium, pH 8–10, is required. More alkaline conditions favor production of decomposition products; in addition, diazotates are formed in a side reaction. In order to avoid the formation of arylhydrazine-*N,N'*-disulfonic acid, only a small excess (ca. 10%) of alkali-metal sulfite is used. In industrial production, because of the solubility of the resulting diazosulfonates, diazonium solutions of the highest possible concentration are run at 0–10 °C into the prepared sodium sulfite solution. To neutralize excess acid, an alkali (usually sodium carbonate or sodium hydrogen carbonate) may be added, either at the beginning or in small amounts during the addition of the diazonium solution.

The two reactants first form the *syn*-diazosulfonates, which can crystallize if the reaction is carried out so rapidly that their concentration is high enough. They are not isolated, but after some time, change in situ to the stable *anti*-diazosulfonates. Often the diazonium solution is added over several hours so that the *syn* form is never present in more than a low concentration. When the rearrangement is complete, again indicated by the spot test with β-naphthol solution, the *anti*-diazosulfonate is precipitated by the addition of sodium chloride, filtered off, and gently dried.

Uses. Sodium 4-dimethylaminobenzene-*anti*-diazosulfonate [*140-56-7*] has fungicidal properties and is used as a soil disinfectant to combat damage to seedlings and roots. Dexon is a tradename. Diphenylamine-4-*anti*-diazosulfonate and its 4-methoxy derivative are used in textile printing for producing blue and black shades. They are marketed as mixtures with the coupling components, usually 2-hydroxynaphthalene-3-carboxylic acid anilide. The azo dyes are formed by steaming the prints.

3.4. Diazoamino Compounds (Triazenes)

Diazoamino compounds, called triazenes by their systematic name, are produced by the reaction of a diazonium ion with ammonia, a primary amine, or a secondary amine:

$$Ar-\overset{+}{N}\equiv N + HN{<}^{R^1}_{R^2} \rightleftharpoons Ar-N=N-N{<}^{R^1}_{R^2} + H^+$$

With few exceptions, they exist in a pH-dependent equilibrium with their starting components, the position of the equilibrium depending on the reactivities of the two starting components.

Properties. The aromatic triazenes are mostly solid, colorless-to-deep-yellow compounds. In the absence of solubilizing groups, triazenes are practically insoluble in water. Carboxylic acid groups and especially sulfonic acid groups increase the solubility, often drastically.

Diazoamino compounds are fairly stable at room temperature. At higher temperatures they decompose, in some cases violently. When 1,3-diaryltriazenes are gently heated, 40–100 °C, in a weakly acidic medium, they rearrange to form aminoazo compounds.

$$\text{Ph}-N=N-\underset{H}{N}-\text{Ph} \longrightarrow \text{Ph}-N=N-\text{C}_6\text{H}_4-NH_2$$

The arylazo group migrates to the para position or, if this is occupied, to the ortho position.

Acids cleave the triazene chain. 1,3-Diaryltriazenes revert back to amine and diazonium salt, but because of the labile hydrogen atom, some products have the amine and diazonium groups interchanged (see [4]). The rate of the cleavage reaction is proportional to the hydrogen ion concentration, but no data are available on the pK values, particularly for important industrial products. Qualitative observations indicate that the pK values are lower for more basic amines and more electrophilic diazonium ions.

The triazene chain is stable in alkaline media. Compounds that still have a hydrogen atom on the triazene group can form salts, often intensely colored. Thiazol Yellow G [1829-00-1] turns red in caustic alkali solutions, but not in alkali-metal carbonate solutions; therefore, it is used as an indicator for caustic alkalis.

Thiazol Yellow G

Preparation. The only method of industrial importance is the reaction of diazonium salts with amines, usually accomplished by dissolving or suspending the amine in water and adding the diazonium salt solution. The hydrogen ions that are produced may be taken up by an excess of amine or by acid binders such as sodium acetate, sodium hydrogen carbonate, or sodium carbonate. The total required quantity of acid binder is often premixed with the amine. If a narrow pH range is necessary, only a small amount of acid binder is added as a buffer, and the pH is kept constant by continuous addition of sodium hydroxide solution.

Decomposition products are avoided by operating below room temperature, 0–10 °C. The heat of reaction can be absorbed by addition of ice if the products are only sparingly soluble. Otherwise, to avoid dilution, jacket cooling is used.

If the triazenes do not precipitate at the end of the reaction, they can usually be forced to do so by addition of common salt. Very soluble triazenes are isolated by spray drying their concentrated solutions, [29, p. 28]. Drum dryers may also be used. The use of the very gentle freeze drying process is limited due to cost.

At this stage the triazenes are moist products. They are dried like diazonium salts (see Section 3.1.2). The equipment must be resistant to alkali; stainless steel is suitable. Alternatively, drying can be carried out by addition of a salt that forms a hydrate, provided the moisture level is low. Sodium acetate is suitable, and its capacity to combine with water, 54 g per mole (82 g), is high.

The reaction of diazonium salts with secondary aliphatic amines takes place smoothly and usually with an excellent yield; thus, industrially important triazenes are produced from amines with solubilizing groups such as sarcosine and methyltaurine.

CH₃–NH–CH₂–COOH   CH₃–NH–CH₂–CH₂–SO₃H
Sarcosine   Methyltaurine
[107-97-1]   [107-68-6]

A good example is the preparation of sodium 3-methyl-1-(5-chloro-2-methoxyphenyl)-3-(carboxylmethyl)triazene. 4-Chloro-2-amino-1-methoxybenzene hydrochloride is mixed with water, and 32 % hydrochloric acid is added. Ice is added to bring the temperature to 0 °C, and nitrite is then added as a 30 % solution, maintaining the temperature at 0 °C by further ice addition. After the diazotization, a small amount of diatomite is added, and the liquid is filtered with a clarifying press. A mixture is prepared of sarcosine, sodium carbonate, sodium chloride, water, and ice. The diazotized solution is added over ca. 30 min, keeping the temperature at 0 °C by ice addition. The precipitated triazene is filtered and dried at 80 °C.

When diazonium salts are reacted with primary or secondary aromatic amines, triazene formation competes with azo coupling. The dominant form depends on the nucleophilicity of the amine. If the nucleophilicity is not increased by electron-releasing substituents or fused rings, triazene tends to be formed. Triazene formation is also favored by electron-withdrawing substituents and by having the position para to the amine group occupied. With amines that react to form triazenes as well as azoamino compounds, triazene formation is favored by high pH because the splitting of the triazene chain is catalyzed by acid. Secondary aromatic amines have a greater tendency to form triazenes than the corresponding primary amines. Water-soluble triazenes are formed without difficulty from **1** and especially from **2** and **3**:

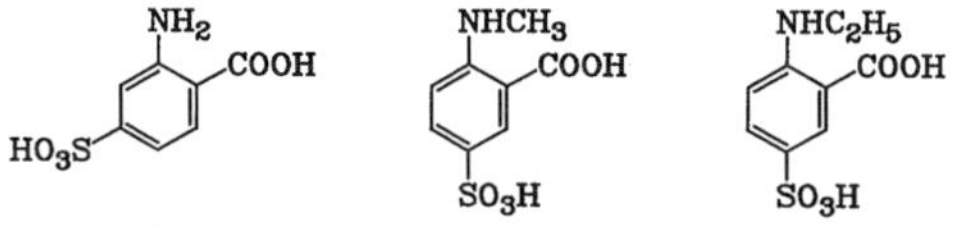

The sodium derivative of 3-methyl-1-(4-nitro-2-methoxyphenyl)-3-(4-sulfo-2-carboxyphenyl)triazene is prepared by first diazotizing 5-nitro-2-amino-1-methoxybenzene at 10–15 °C in dilute sulfuric acid and adding this to a solution of 2-methylamino-5-sulfobenzoic acid that has been neutralized with sodium hydroxide solution and mixed with calcium carbonate. The pH is adjusted to ca. 12 with sodium hydroxide solution, thus precipitating calcium sulfate, which can be filtered off. The filtrate is then spray dried [29, p. 25].

Symmetrical 1,3-diaryltriazenes, which are to be rearranged to *p*-aminoazo compounds, can be prepared by reacting the amine with half an equivalent mass of nitrous acid in a moderately acidic medium. The reaction is usually carried out with an excess of amine, which acts as a reaction medium and speeds up the rearrangement [6].

Uses. Some triazenes are active against bacterial infections and especially so against protozoal infections. Berenil [536-71-0] has chemotherapeutic properties and is useful for treating diseases caused by trypanosomes and babesias.

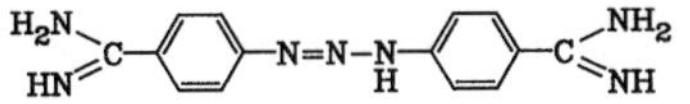

Triazenes are used in the preparation of aminoazo compounds, such as *p*-aminoazobenzene [60-09-3] and 4'-amino-2,3'dimethylazobenzene [97-56-3].

The ability of triazenes to break down to yield diazonium compounds with coupling properties is widely used to produce insoluble azo dyes on fibers. Water-soluble types are used for cotton dyeing, while water-insoluble types are used for hydrophobic fibers. Alkali-soluble mixtures are marketed for cotton printing, consisting of a naphthol AS type mixed with a triazene derivative.

3.5. Other Aromatic Diazo Compounds

Diazotized *o*- or *p*-aminophenols (**4**) exist only in mineral acid solution. The diazonium group makes the hydrogen ion much more acidic, increasing the ionization constant by many orders of magnitude, e.g., from 10^{-10} for phenol to 10^{-3} for compound **4**. The mesomeric system formed, **5,** is known as *quinone diazide*.

$$HO-\underset{4}{\bigcirc}-\overset{+}{N}\equiv N$$

$$-H^+ \updownarrow +H^+$$

$$^-O-\underset{5a}{\bigcirc}-\overset{+}{N}\equiv N \longleftrightarrow O=\underset{5b}{\bigcirc}=\overset{+}{N}=\overset{-}{N}$$

These compounds are often termed diazophenols in the literature, but are distinctly different from normal diazonium salts. They have a yellow color, are rather sparingly soluble in water, and are quite explosive, especially when nitro groups are present in the molecule. Picramic acid [*96-91-3*] may be diazotized to give dinitrobenzoquinone diazide [*7008-81-3*], which is used as a detonator.

$$O_2N-\underset{NO_2}{\overset{O}{\bigcirc}}=\overset{+}{N}=\overset{-}{N}$$

Quinone diazides do not form tetrachlorozincates, tetrafluoroborates, or diazotates, but they do undergo the Sandmeyer reaction and they do form azo dyes. However, their energy of coupling is considerably less than that of the corresponding substituted diazonium salts, and they react only with quite reactive components. When quinone diazides couple with substituents that themselves couple in the ortho position to a hydroxyl group, *o-o'*-dihydroxyazo dyes are formed. These are capable of forming complexes, either normally or on the fiber surface.

By the action of light, *o*-quinone diazides are converted to indenecarboxylic acids, with liberation of nitrogen and ring contraction [6]; thus, they can be used in reprography.

Quinone diazides in the benzene series are formed by the normal diazotization of aminophenols. In some cases, buffering with sodium acetate is necessary. The products are not isolated, but are used in the form of a suspension to avoid the danger of explosion.

If *o*-aminonaphthols are treated with nitrite in a mineral acid medium, they are oxidized to naphthoquinones. Therefore, they must be diazotized in weak acid solution in the presence of a small amount of metal salts, preferably copper(II) [7, 10/3, 51], [30]; under these conditions the reaction proceeds smoothly. The naphthoquinone diazides formed are stable compounds; for example, they can be crystallized from boiling water, be nitrated, or be chlorinated (see p. 1893). If moist, these compounds can be safely stored and marketed. For details of the production of 4-diazo-3,4-dihydro-3-oxonaphthalene-1-sulfonic acid [4857-47-0], see [30].

Naphthoquinone diazides are important starting materials in the manufacture of metal-complex dyes, and are used in reprography (especially films for offset litho printing).

Aryldiazo ethers, esters, sulfides, thioethers, sulfones, cyanides, and phosphoric acid esters have very few practical applications. One or two of these have been suggested as polymerization catalysts and modifiers. The acaricide Milbex [2274-74-0] is a tetrachloroaryldiazo thioether.

3.6. Analysis

The best way to detect aromatic diazo compounds is by the azo coupling reaction. The substance under investigation is treated with (1) a phenolic azo component, such as resorcinol or β-naphthol dissolved in sodium hydroxide or carbonate solution, or (2) an aromatic amine capable of azo coupling dissolved in dilute acetic acid. The most convenient method is to use a spot test [31].

anti-Diazotates and triazenes must first be converted to diazonium salts, if necessary by heating. Noncoupling *anti*-diazosulfonates may be converted to diazonium salts by treatment with oxidizing agents.

To identify the aromatic part of the molecule, the diazonium salts are first converted to chlorinated aromatic compounds by the Sandmeyer reaction, which are then identified by standard methods.

The universal method for quantitative analysis of diazonium salts is to decompose them by treatment with acid and copper(I) ions and measure the nitrogen evolved [32], [33].

In industry, diazonium salts and their solutions are often determined by volumetric analysis, by which the diazonium salt solution is gradually added to a known amount of coupling component, e.g., R salt (disodium 3-hydroxy-2,7-naphthalene-sulfonate [135-51-3]), β-naphthol, or 2,4-diaminotoluene. The end point is detected by the disappearance of the coupling reaction or the appearance of the diazonium salt, as detected by a spot test. Unreactive or easily decomposed diazonium compounds may be treated with an excess of the coupling component, which may then be back-titrated.

Quantitative determination may also be accomplished by spectrophotometry: the extinction coefficient of the absorption peak is measured. The products must be fairly pure so that secondary or decomposition products do not interfere with the absorption maximum of the diazo compound.

The quantity of a diazo compound in a commercial product is determined by one of the above methods, the solubility is checked by solution–filtration tests, and the coloring properties are checked by dyeing and printing tests.

4. Diazo Reactions

From a practical point of view, the most important reactions of the diazo compounds can be placed in three groups: (1) reactions in which the diazo group is retained but substituents are introduced or replaced, (2) reactions in which the diazo group is modified, and (3) reactions in which the diazo group is replaced.

4.1. Reactions in Which the Diazo Group is Retained

Substituents can be introduced or replaced without changing the diazo group.

Introduction of Substituents. The diazonium group is an electron-withdrawing substituent, surpassing all other groups in this respect. Therefore, electrophilic substitutions in the same nucleus are very difficult. Such substitutions are successful only at the meta position, which is least affected by the diazonium group, provided that this position is activated by an effective ortho–para-orienting substituent, as in the case of *o*- and *p*-diazophenols. Under these circumstances, halogen atoms or nitro or sulfonic acid groups may be introduced into the benzene ring. The reaction can also be carried out in the naphthalene series. An example is the production of 6-nitro-1-diazo-2-naphthol-4-sulfonic acid [5366-84-7] [34].

Replacement of Substituents. Owing to the strongly electron-withdrawing character of the diazonium group, nucleophilic replacement of ortho and para substituents is possible. Ease of replacement increases in the order $-SO_3H$, $-OCH_3$, $-Cl$, $-NO_2$, and $-N \equiv N^+$. In the benzene-1,4-bis(diazonium) ion, one of the diazonium groups can be easily replaced by a halogen or hydroxyl group. Replacement of a nitro group can take place only in the presence of a further electron-withdrawing substituent; therefore, diazotized solutions of *o*- and *p*-nitroanilines, which are heavily used in industry, are stable.

On the other hand, the *o*-nitro group in diazotized 2,4-dinitroaniline can be replaced by a hydroxyl in a weakly acidic medium. This reaction muddies the color of the pigment dye from 2,4-dinitroaniline and β-naphthol during production. Whenever diazonium salts with strong electron-withdrawing substituents are prepared or reacted, a nucleophilic substitution reaction is always possible.

The replacement of chlorine by a hydroxyl group can also be of value, e.g., in the manufacture of 4-hydroxy-3,5-bis(diazo)benzenesulfonic acid, a component for complex-forming azo dyes.

4.2. Reactions of the Diazo Group

Reactions of the diazo group include azo coupling, formation of formazan, cyclization, and reduction.

Azo Coupling. Diazonium salts combine with many aromatic phenols and amines to form azo compounds.

Formazan Formation. Diazonium ions react with active hydrogen on a carbon atom. With an aldehyde hydrazone, the product is a formazan:

$$Ar-\overset{+}{N}\equiv N + R-CH=N-NH-Ar' \xrightarrow{OH^-} \begin{array}{c} Ar\diagdown N\diagup H\diagdown N\diagup Ar' \\ \| \quad \| \\ N\diagdown C \diagup N \\ | \\ R \end{array}$$

Compounds with a doubly activated methylene group, e.g., acetoacetyl compounds, first form an arylhydrazone, which can react in an alkaline medium with another diazonium ion to give the formazan. In this reaction one of the two activating radicals is split off.

$$Ar-\overset{+}{N}\equiv N + H_2C\diagdown^R_{R'} \longrightarrow Ar-\underset{H}{N}-N=C\diagdown^R_{R'} \xrightarrow{Ar'N_2^+}$$

$$\begin{array}{c} Ar\diagdown N-N \\ H\diagdown \quad \diagdown C-R \\ N=N\diagup \\ Ar' \end{array}$$

R, R' = CHO, COCH$_3$, CN, COOCH$_3$, CONH$_2$, COOH

If the second coupling reaction goes more slowly than the first, unsymmetrical formazans can be produced by using two diazonium salts.

Formazans are intensely colored compounds. They have both weakly acidic and weakly basic properties and, hence, change color in the presence of acid or alkali. For this reason and because of their poor lightfastness, formazans are rarely used for textile dyeing. However, they are used in photography, e.g., for yellow filters and color developers.

Formazans form heavy-metal complexes, the industrially important ones being those of Co, Cr, Cu, and Ni. These complexes have considerably improved stability, owing to complex-forming groups (OH, COOH) in the ortho position of the aryl radicals attached to the nitrogen. They have excellent lightfastness and low sensitivity to acids and bases. Some of these complexes are valuable textile dyes (acid dyes and reactive dyes).

Cyclization Reactions. If the diazo group reacts with an ortho substituent, heterocyclic products are formed [35], [36]. In this way diazotized o-toluidines can be converted to *indazoles*. Although the parent substance o-toluidine and derivatives with an electron-releasing substituent undergo this reaction only to a minor extent, the mononitro derivatives in acetic acid solution or the dinitro derivatives in sulfuric acid solution give high yields of indazole. The optimum reaction conditions vary greatly. Thus, diazotized 4-nitro-2-aminotoluene requires warming, while diazotized 5-nitro-2-aminotoluene reacts better at room temperature. A frequent side product is a 3-arylazoindazole.

Indazole   Azinphos-methyl   Benzotriazole

o-Aminobenzamides give benzotriazines under diazotizing conditions. Azinphos-methyl [*86-50-0*], a component of the insecticide *Gusathion*, belongs to this class of compounds.

o-Diamines cannot normally be diazotized, but rather are converted to cyclic triazoles when treated with nitrous acid. This reaction also takes place when one of the two amino groups is alkylated, arylated, or acylated. The preparation of *benzotriazole* and its simple derivatives is best carried out in dilute acetic acid [37]. The whole of the nitrite is added at once, and the liberated heat of reaction is not removed by cooling. Unlike the open-chain triazenes, triazoles are stable toward mineral acid. Triazoles can form salts with acids or bases. Benzotriazole and tolutriazoles are corrosion inhibitors.

Reduction. Diazonium salts and covalent diazo compounds such as diazotates, diazosulfonates, diazosulfones, and triazenes may be reduced to arylhydrazines. Almost all common reducing agents are effective, but in industry sodium sulfite and sodium hydrogen sulfite are used almost exclusively. When the reaction proceeds sluggishly, it can be brought to completion by the addition of zinc dust.

The sulfite reduction of aryldiazonium salts takes place in stages: aryldiazosulfonate (**6**), arylhydrazine-α,β-disulfonic acid (**7**), arylhydrazine-β-sulfonic acid (**8**), and finally arylhydrazine.

$$Ar-N{=}N^{+} \xrightarrow{SO_3^{2-}} \underset{\mathbf{6}}{Ar-N{=}N-SO_3^-} \quad (1)$$

$$Ar-N{=}N-SO_3^- \xrightarrow{SO_3^{2-}} \underset{\mathbf{7}}{Ar-\underset{SO_3^-}{\overset{H}{N}}-\overset{}{N}-SO_3^-} \quad (2)$$

$$Ar-\underset{SO_3^-}{\overset{H}{N}}-\overset{}{N}-SO_3^- \xrightarrow{H^+} \underset{\mathbf{8}}{Ar-\overset{H}{N}-\overset{H}{N}-SO_3^-} \quad (3)$$

$$Ar-\overset{H}{\underset{H}{N}}-\overset{H}{N}-SO_3^- \xrightarrow{H^+} Ar-\overset{H}{\underset{H}{N}}-\overset{H}{N} \quad (4)$$

Arylhydrazine

Reactions (1) and (2) take place in a weakly acidic to weakly alkaline medium, reaction (3) takes place in weak to moderately strong acid, and reaction (4) takes place in the mineral acid range. Altogether, each diazonium cation requires two sulfite ions. To give optimum reaction, excesses of 10–20% are usually required, but in the preparation of phenylhydrazine, 102% of the theoretical amount just suffices [15].

Commercial 40% sodium hydrogen sulfite solution is converted completely or almost completely to sulfite by addition of a solution of sodium carbonate or hydroxide. The diazonium solution or suspension is then added. If no electron-withdrawing substituents are present, the mixture is then heated at 60–80 °C, giving compound **8**. The pH must be held between 5 and 7 throughout the reaction period, and any colored impurities can be reduced with a little zinc dust. If necessary, a clarifying filtration may be carried out at this point.

The second sulfonic acid group is much more firmly bonded than the first, and the conversion to arylhydrazine requires warming with mineral acids, usually accomplished by adding ca. 10 wt% of hydrochloric or sulfuric acid and heating at 60–100 °C.

Arylhydrazines with sulfonic acid groups in the aromatic nucleus precipitate as sparingly soluble zwitterions. If no sulfonic acid groups are present, the hydrazines are isolated as sulfates or hydrochlorides and are often further processed in this form. Free hydrazines can be precipitated by adding the solution to alkali. If solid, they are filtered off; if liquid, they are dissolved in xylene and isolated by fractional vacuum distillation.

Phenylhydrazine hydrochloride can also be manufactured by a continuous process [22], but this is not appropriate for the other arylhydrazines, for which the demand is too low. Arylhydrazines are starting materials for pyrazolones, which are dye and pharmaceutical intermediates. They react with dioxo compounds to give azo dyes.

4.3. Replacement of the Diazo Group (Dediazotization)

The diazo group can be replaced by a hydrogen, hydroxyl, halogen, cyanide, etc.

Replacement by Hydrogen. Although replacement by hydrogen is important in the laboratory, it is used in industry only for preparing compounds that are difficult or impossible to make in any other way. For example, the amino group causes another substituent to be directed to a location that would be otherwise avoided, the amino group later being removed.

Aqueous solutions are preferred in industry, with hypophosphorous acid as the reducing agent. For some special purposes salts, esters, or amides of formic acid are used [38]. For example, 3-hydroxynaphthalene-1-sulfonic acid [6357-85-3], a coupling component for dyes, can be manufactured simply and in excellent yield.

Replacement by Hydroxyl. Decomposition of diazonium ions in aqueous acid gives transient aryl ions, which react with water to give phenols:

$$\text{Ar–N}^+ \equiv \text{N} \longrightarrow \text{Ar}^+ + \text{N}_2$$
$$\text{Ar}^+ + \text{HOH} \longrightarrow \text{ArOH} + \text{H}^+$$

If other nucleophilic reagents are present, side reactions can occur:

$$\text{Ar}^+ + \text{Z}^- \longrightarrow \text{Ar–Z}$$

Therefore, diazotization for this process is carried out in sulfuric acid solution.

In its simplest form, the method is to heat the diazonium salt solution to 80–100 °C, but the products are usually extremely impure. Much higher purity and yield are obtained if the diazonium salt solution is added to hot, dilute sulfuric acid. If the latter has a concentration of 45–60 wt%, the reaction temperature can be as high as 118–142 °C. A part of the sulfuric acid can be replaced by sodium sulfate.

Secondary products, sometimes dark in color and insoluble, are often formed in considerable amounts, e.g., diphenyl sulfates, diphenyl ethers, hydrocarbons, nitrosophenols, indazenes, dioxazines, hydrazobenzenes, and arylsulfonic acids. This *diazo resin formation*, which visibly builds up during the course of the reaction, can be minimized by careful diazotization and avoidance of excess nitrite. Of course, continuous removal of the main product, e.g., by steam distillation, is very beneficial. Continuous reaction can also improve yield and quality.

Yields seldom exceed 75%. Improved yield and quality can sometimes be achieved by dissolving the reaction product in an organic phase such as xylene. Some diazonium salts that do not withstand treatment with sulfuric acid at elevated temperature are reacted in phosphoric acid at ca. 200 °C. The preparation of hydroxyanthraquinones from aminoanthraquinones can be accomplished in concentrated or fuming sulfuric acid.

The special industrial importance of this method lies in the introduction of chlorine atoms or nitro groups into phenols and cresols at the meta position, e.g., the preparation of 2,3-dichlorophenol [*576-24-9*] [39], or 4-methyl-3-nitrophenol [*2042-14-0*].

Uncatalyzed Replacement by Halogen. The halide ions F⁻, Cl⁻, and Br⁻ react as nucleophilic agents in an aqueous acid medium, combining with the aryl cations initially formed on decomposition of diazonium salts, to form halogenated hydrocarbons. However, this reaction is in competition with phenol formation.

In contrast, iodide ions are first converted to the intermediate periodide ions, which are then reacted directly with diazonium ions:

$$\underset{\text{I–I}_2}{\overset{\delta^+}{\text{C–N}}\equiv\overset{+}{\text{N}}} \longrightarrow \text{C–I} + \text{N}_2 + \text{I}_2$$

This reaction is carried out in mineral acid, usually hydrochloric or sulfuric acid, and is much more rapid than the uncatalyzed reaction of other halide ions. A cooled diazonium salt solution is prepared, and to each mole of this is added 1.1–2.0 mol of alkali iodide as a concentrated aqueous solution. In addition, the diazonium solution may be added to the iodide solution. Room temperature is usually adequate. The more volatile aromatic iodides may be recovered by steam distillation when the reaction is complete. Aromatic iodides of lower volatility must be isolated directly or extracted by solvents such as toluene. One example is the manufacture of the plant growth regulator 2-iodo-3-nitrobenzoic acid [*5398-69-6*] [40].

If solutions of diazonium fluorides in anhydrous hydrofluoric acid are warmed, aromatic fluorides are obtained. The amines are dissolved in 2–3 times the quantity of anhydrous hydrofluoric acid and diazotized by adding dry sodium nitrite at ca. 0 °C. When the reaction mixture is heated to boiling (ca. 30 °C), nitrogen is evolved. The reaction takes place smoothly over several hours. It is used to make fluorobenzene and gives excellent results with many other amines. However, diazonium fluorides that have a substituent in the ortho position with a free electron pair cannot be decomposed in this way [41].

Catalyzed Replacement by Halogen. Copper(I) chloride attaches to diazonium ions, forming labile complexes, which break down homolytically in two steps to form aromatic halides, nitrogen, and copper(I) ions.

$$[Ar-N \equiv N \rightarrow Cu^I Cl]^+ \longrightarrow [Ar \cdot + Cu^{II}Cl^+] + N_2$$
$$[Ar \cdot + Cu^{II}Cl^+] \longrightarrow Ar-Cl + Cu^+$$

In competition with this Sandmeyer reaction, other reactions form azo compounds and diphenyls [42], e.g.:

$$2\ Ar-N_2^+ + 2\ CuCl \longrightarrow Ar-N=N-Ar + N_2 + 2\ Cu^{2+} + 2\ Cl^-$$
$$2\ Ar-N_2^+ + 2\ CuCl \longrightarrow Ar-Ar + 2\ N_2 + 2\ Cu^{2+} + 2\ Cl^-$$

These can be limited by increasing the halide concentration. Enough hydrochloric acid is added so that its concentration does not fall below 5–10 wt% during addition of the diazonium salt solution. In addition, the concentration of the diazonium salt is kept as high as possible.

The two side reactions consume catalyst. In industry, a molar ratio of copper(I) halide to diazonium salt from 1:10 to 1:5 is used, although in exceptional cases considerably more can be used. For example, one mole equivalent is used in the manufacture of 2-chloro-5-nitrophenol [*619-10-3*] from 5-nitro-2-aminophenol.

Copper(I) chloride is insoluble in water, and is usually used in the form of the soluble chloro complex. To keep the cost low the chlorocopper complex is made from copper turnings or precipitated waste copper.

The reaction vessels are enameled, rubber coated, or lined with acid-resistant bricks. The filtration equipment and pipework are of rubber-lined steel, the pipework often also being plastic. The reaction temperature is usually 10–30 °C, although in a few instances, it is higher, e.g., 75 °C for 4-chlorotoluene-2-sulfonic acid.

The addition of the diazonium salt solution takes several hours. So far a continuous process has not been possible. The product is isolated by steam distillation or filtration. The copper is precipitated and recycled.

An example of the industrial production of an aryl halide by the Sandmeyer reaction is that of 3-chlorotoluene [*108-41-8*] [43].

The Sandmeyer reaction may also be carried out with copper(I) bromide and hydrobromic acid [44].

Replacement by Cyanide. The diazonium group can also be replaced by cyanide with the help of a copper catalyst [45]. The reaction is no more difficult than replacement by chlorine, but is carried out in a nearly neutral medium.

Because copper(I) cyanide is insoluble, sodium copper(I) cyanide, $NaCu(CN)_2$, is the usual catalyst. The amounts of catalyst are larger than those required for replacement by chloride, usually 25–30 mol%. The reaction temperature is usually near room temperature. The large excesses of cyanide (180–300% of theory) may give unwanted diazocyanides. To minimize this effect, only a part of the alkali cyanide is added with the copper cyanide; the rest is added gradually along with the diazonium salt solution, e.g., in the production of 5-chloro-4-nitro-2-methylbenzonitrile [46].

Replacement by a Nitro Group. In industry if a nitro group cannot be easily placed in the required location by direct nitration, one solution is to replace a diazo group with a nitro group. The reaction is performed with sodium nitrite in neutral or weakly alkaline medium in the presence of a copper catalyst. Generous proportions of nitrite and catalyst are necessary: 5 mol of sodium nitrite and 1 mol of catalyst per mole of diazo compound. Catalysts used include copper powder, CuO, $CuSO_4$, $CuCO_3$, or $Cu_3(SO_3)_2 \cdot 2 H_2O$, the last being especially effective.

Replacement by Sulfur-containing Groups. The addition of diazonium salt solutions to alkali-metal disulfide solutions first gives the unstable diazosulfides, which convert to aryl sulfides, liberating nitrogen. The reaction is carried out at an elevated temperature to decompose the explosive primary products as quickly as possible. However, these reactions are best avoided, even if their use on an industrial scale is discussed in the literature.

A reaction of some importance is the replacement of the diazonium group by the sulfonyl chloride group in a sulfur dioxide medium. The catalyst is a copper(II) compound (Meerwein reaction).

$$Ar-N \equiv N^+Cl^- + SO_2 \xrightarrow{Cu^{II}} ArSO_2Cl + N_2$$

The reaction is carried out in strong hydrochloric acid or in the absence of water to prevent the formation of diazosulfonates. Sulfur dioxide is present in large excess. Electron-withdrawing substituents in the aromatic nucleus favor the reaction and improve the yield.

Replacement by the Arsonic Acid Group. Diazonium salts react in weakly alkaline medium with arsenite to form arylarsonic acids [47].

$$Ar-N^+ \equiv N + HAsO_3^{2-} \longrightarrow Ar-As(OH)O_2^- + N_2$$

This is known as the Barth reaction and does not normally require a catalyst, although often it is accelerated by copper salts and copper powder, with improved yield and product quality. The copper–ammonia complex is especially suitable as it is soluble in weakly alkaline medium.

Replacement by Organic Radicals. Because of their reactivity, diazonium salts react with a broad range of organic compounds, forming carbon–carbon bonds and liberating nitrogen. Although this is seldom used in industry, two exceptions deserve mention: (1) diazotized 2-chloroaniline is reacted with benzoquinone to form 2-(2′-chlorophenyl)-1,4-benzoquinone, which is converted to 2-hydroxydibenzofuran [86-77-1], a dye intermediate, and (2) the reaction of chlorobenzenediazonium fluoroborate with benzonitrile to form 6-chloro-2,4-diphenylquinazoline [30169-34-7], which is converted to diazepam [439-14-5], an important psychopharmacologic drug.

5. Environmental Protection

In general, no problems arise with *waste gases* from the manufacture of diazo compounds or from industrial diazo reactions. The only operations producing appreciable amounts of toxic gases are the manufacture of certain azobenzene diazonium salts and the Sandmeyer nitrile synthesis. In the former, nitrogen oxides are given off because of the excess nitrite, and in the latter the liberated nitrogen carries some hydrogen cyanide. These substances may be easily removed by alkali washing; oxidizing agents are added in the second case.

Wastewater contains dissolved reaction and decomposition products, usually with quite a large amount of mineral acid or caustic alkali. If the diazonium salts are isolated as tetrachlorozincates, the wastewater also contains considerable quantities of zinc.

If all wastewaters are led into one collecting tank, partial neutralization takes place, which can be completed by the addition of a suspension of calcium hydroxide in water or powdered limestone. Many diazonium salts quickly decompose under these conditions, usually giving insoluble products that precipitate along with zinc hydroxide or carbonate. Remaining products may be converted to insoluble azo dyes by adding wastewater containing naphtholate or other coupling components. The solids are concentrated by sedimentation, separated by filtration, and then incinerated or dumped. The wastewater is discharged, sometimes via filtration through a sand filter, or it may be treated biologically to reduce the organic content still further.

Wastewater from the Sandmeyer reaction must be treated to remove all traces of copper, preferably by adding sodium sulfide solution. The precipitated copper sulfide is isolated and recycled. This treatment also precipitates arsenic, which is disposed of by deep burial to avoid groundwater contamination.

The mother liquor from the cyanide replacement reaction contains cyanide. It is usually oxidized with chlorine, hypochlorite, or hydrogen peroxide to form cyanate. It may be hydrolyzed to sodium formate and ammonia at a temperature of 200 °C [48].

Other solid wastes include filtration aids from diazotization processes, which mainly consist of activated carbon and diatomite. If the diazotization reactions are well controlled, only a small amount of undiazotized bases and insoluble decomposition products is present. These solid wastes can be incinerated like other organic waste products.

6. Safety

Many diazo compounds are explosive, some even having the brisance of detonators. Therefore, a diazo product whose explosive properties are unknown must first be prepared in the laboratory and subjected to ignition tests, impact tests, and differential thermal analysis [49]. Before full-scale or pilot-plant production is initiated, an industrial safety investigation should be performed, e.g., following the guidelines of the Bundesamt für Materialprüfung (Federal German Office for Materials Testing) [50]. Commercial products must always fall into the nonexplosive classification, because otherwise, they could be marketed only under restrictive conditions. The same restrictions should be applied to intermediates.

In addition, flammability must not be overlooked. Although commercial products are usually self-extinguishing because of the presence of extenders, many intermediates, especially those containing nitro groups, burn even with- out an air supply.

Aqueous solutions of diazonium salts can decompose, but do not explode. They are protected against dangerous increases in temperature by the high specific heat and heat of vaporization of water. However, diazonium salt solutions in concentrated sulfuric acid, as formed in nitrosyldiazotization, are a special case. Because of the low specific heat and the high boiling point of sulfuric acid, the heat of decomposition can elevate the temperature until explosion occurs [51].

7. Toxicology

Diazomethane is the most thoroughly researched diazo compound with regard to toxicology. *Acute poisoning* follows a course similar to that of phosgene poisoning: a latent period is followed by severe difficulty in breathing and, in severe cases, by pulmonary edema. Animal trials have revealed *carcinogenic properties;* therefore, diazomethane is included in Section III A2 of the *Technische Regeln für gefährliche Arbeitsstoffe*, and a MAC (maximal admissible concentration) value is not quoted. For manufacture and use of diazomethane special rules must be observed [52]. For another aliphatic diazo compound, ethyl diazoacetate [*623-73-4*], the LD_{50} value for acute oral toxicity (rat) is 400 mg/kg.

Table 1 provides a survey of the acute toxicity of some *aromatic diazonium salts* used as azoic diazo components

The acute oral toxicity of *aromatic diazo compounds* is somewhat greater: the LD_{50} values are 430 mg/kg for 3,3-dimethyl-1-phenyltriazene [*7227-91-0*], 224 mg/kg for 1,3-diphenyltriazene [*136-35-6*], and 63 mg/kg for4-dimethylaminobenzene-*anti*-diazosulfonate [*140-56-7*] [54].

Table 1. Acute toxicity of some diazonium salts

Diazonium salt [*]	Toxicity	
	LD_{50} (rat, oral), mg/kg	LC_0 (fish [**]), mg/L
2-Chlorobenzenediazonium tetrachlorozincate [14263-92-4]	1900	10–100
3-Chlorobenzenediazonium tetrafluoroborate [456-39-3]	1000	<10
5-Chloro-2-methoxybenzenediazonium tetrachlorozincate [68025-25-2]	1500	100–500
4-Chloro-2-nitrobenzenediazonium tetrachlorozincate [14263-89-9]	1600	10–100
4-Chloro-2-trifluormethylbenzenediazonium 1,5-naphthalenedisulfonate [85222-98-6]	3400	<10
2,5-Dichlorobenzenediazonium tetrachlorozincate [14239-23-7]	2200	<10
2,5-Dimethoxy-4-[(4-nitrophenylazo)]benzenediazonium tetrachlorozincate [64071-86-9]	3700	100–500
2-Methoxy-4-nitrobenzenediazonium 1,5-naphthalenedisulfonate [49735-71-9]	2500	
2-Methoxy-5-nitrobenzenediazonium tetrachlorozincate [61919-18-4]	1900	100–500
4-Methoxy-2-nitrobenzenediazonium tetrachlorozincate [14239-24-8]	2100	10–100
4-[(4-Methoxyphenyl)amino]benzenediazonium chloride [101-69-9]	200	
2-Methyl-4-[(2-methylphenyl)azo]benzenediazonium hydrogen sulfate [101-89-3]	2000	
2-Methyl-4-nitrobenzenediazonium tetrafluoborate [455-90-3]	1200	
2-Methyl-5-nitrobenzenediazonium 1,5-naphthalenedisulfonate [49735-69-5]	1800	10–100
2-Nitrobenzenediazonium tetrafluoroborate [365-33-3]	2000	
4-(Phenylamino)benzenediazonium chloride [101-56-4]	400	<10
3,3′-Dimethoxy-[1,1′-biphenyl]-4,4′-bis-(diazonium) tetrachlorozincate [14263-94-6]	1600	<10

[*] As a commercial azoic diazo component: the amount of diazonium salt in a commercial diazo component is ca. 25–50 wt%, the rest being inorganic salts such as sodium sulfate and aluminium sulfate;
[**] *Lanciscus idus.*

8. References

General References

[1] S. Patai: *The Chemistry of Diazonium and Diazo Groups,* J. Wiley & Sons, Chichester, England, 1978.
[2] K. Holzach: *Die aromatischen Diazoverbindungen,* Enke Verlag, Stuttgart 1947.
[3] K. H. Saunders: *The Aromatic Diazo Compounds,* 3rd ed., E. Arnold, London 1985.
[4] H. Zollinger: *Chemie der Azofarbstoffe,* Birkhäuser, Basel-Stuttgart 1958.
[5] H. Zollinger: *Azo and Diazo Chemistry,* Interscience Publ., London-New York 1961.
[6] H. R. Schweizer: *Künstliche organische Farbstoffe und ihre Zwischenprodukte,* Springer Verlag, Berlin-Göttingen-Heidelberg 1964, pp. 177–208.
[7] *Houben-Weyl,* **10/2,** 169–692; **10/3,** 1–212, 545–626, 631–743; **10/4,** 473–893.
[8] I. G. Laing: "Arenediazonium Salts and Diazo Compounds," in *Rodd's Chemistry of Carbon Compounds,* vol. **III,** Section C, Elsevier, Amsterdam-London-New York 1973, pp. 11–88.

Specific References

[9] A. N. Nesmejanow, K. A. Konzeschkow, W. A. Klimowa, *Ber. Dtsch. Chem. Ges.* **68** (1935) 1877–1883.
[10] O. Klais, T. Grewer, *Inst. Chem. Eng. Symp. Ser.* **82** (1983) C24–C34.

[11] P. D. Storey, *Inst. Chem. Eng. Symp. Ser.* **68** (1981) 3/P/1 – 3/P/9.
[12] J. Ribka, *Angew. Chem.* **70** (1958) 241 – 244.
[13] D. F. DeTar, A. Ballantine, *J. Am. Chem. Soc.* **78** (1956) 3916 – 3920.
[14] H. Zollinger, *Angew. Chem.* **90** (1978) 157; *Angew. Chem. Int. Ed. Engl.* **17** (1978) 147.
[15] H. Hupfer, *Angew. Chem.* **70** (1958) 244 – 246.
[16] H. Kindler, D. Schuler, *Chem. Ing. Tech.* **37** (1965) 402 – 405.
[17] Bayer, DE-OS 2 741 925, 1977 (K. Breig, G. Demel, N. Hamm).
[18] Hoechst, DE-OS 2 825 655, 1978 (H. Behringer, K. Karrenbauer).
[19] ICI, EP 3 656, 1979 (H. Alterton, I. Hodgkinson).
[20] Ciba, US 4 246 171, 1981 (A. Hamilton, C. Nelson).
[21] Fiat Final Report 1023, p. 6.
[22] Hoechst, DE 1 543 623, 1966 (H. Hertel).
[23] BIOS DOCS 1156/1121.
[24] Hoechst, DE 2 521 650, 1975 (H. Hertel, R. Kostka, H. Milz, R. Ullrich).
[25] Hoechst, DE 2 433 232, 1974 (H. Hertel).
[26] Hoechst, DE 2 527 264, 1975 (H. Hertel).
[27] Showa Chemical Co., JP-Kokai 82 139 146, 1981.
[28] E. S. Lewis, H. Suhr, *Chem. Ber.* **91** (1958) 2350 – 2358.
[29] BIOS Final Report 988.
[30] BIOS Final Report 986, p. 146.
[31] R. F. Muraca, *Treatise Anal. Chem. 1959 Part II* **15** (1976) 251 – 384.
[32] *Houben-Weyl,* **2,** 700.
[33] E. Bitterlin, *Z. Anal. Chem.* **253** (1971) 120 – 122.
[34] BIOS Final Report 1152, p. 91.
[35] R. C. Elderfield (ed.): *Heterocyclic Compounds,* vols. **5 – 8,** J. Wiley & Sons, New York-London 1957 – 1967.
[36] A. Weissberger (ed.): *The Chemistry of Heterocyclic Compounds,* vol. **4, 22,** and **27,** Interscience, New York-London-Sidney 1950 et seq.
[37] Hoechst, DE 2 351 595, 1973 (K. Gengnagel, T. Papenfuhs).
[38] Cassella, DE 901 175, 1943 (W. Zerweck, M. Schubart, F. Fleischhauer).
[39] BIOS Final Report 1153, p. 20.
[40] *Houben-Weyl,* **5/4,** pp. 639 – 647.
[41] *Houben-Weyl,* **5/3,** pp. 213 – 245.
[42] *Houben-Weyl,* **5/3,** 846 – 852.
[43] BIOS Final Report 1145, p. 19.
[44] *Houben-Weyl,* **5/4,** 438 – 447,
[45] *Houben-Weyl,* **7,** 311 – 313.
[46] BIOS Final Report 1149, p. 32.
[47] C. F. Hamilton, J. P. Morgan: „Die Darstellung von aromatischen Arsonsäuren," *Org. React. (N. Y.)* **2** (1944) 415 – 454.
[48] J. Hoerth, U. Schindewolf, W. Zbinden, *Chem. Ing. Tech.* **45** (1973) 641 – 646.
[49] J. Luetolf, *Staub Reinhalt. Luft* **31** (1971) 93 – 97.
[50] Bundesgesetzblatt, part 1, no. 85, pp. 1357 – 1404 (1969).
[51] P. Bersier, L. Valpiana, H. Zubler, *Chem. Ing. Tech.* **43** (1971) 1311.
[52] *Gefahrstoffverordnung,* Deutscher Bundes-Verlag, Bonn 1986.
[53] Hoechst, *Sicherheitsdatenblätter der Echtfärbesalze* (1978 – 1986).
[54] E. J. Fairchild (ed.): *Registry of Toxic Effects of Chemical Substances,* U.S. Dept. of Health (1977).

Dicarboxylic Acids, Aliphatic

Individual keywords: →*Adipic Acid;* →*Citric Acid;* →*Ethylenediaminetetraacetic Acid and Related Chelating Agents;* →*Maleic and Fumaric Acids;* →*Malonic Acid and Derivatives;* →*Nitrilotriacetic Acid;* →*Oxalic Acid;* →*Tartaric Acid.*

Boy Cornils, Hoechst AG, Frankfurt, Federal Republic of Germany

Peter Lappe, Ruhrchemie AG, Oberhausen, Federal Republic of Germany

1.	**Introduction**	1903	2.4.9.	1,12-Dodecanedioic Acid ... 1917
2.	**Saturated Dicarboxylic Acids**	1904	2.4.10.	1,13-Tridecanedioic Acid (Brassylic Acid) ... 1918
2.1.	**Physical Properties**	1904	2.4.11.	C_{19} Dicarboxylic Acids ... 1918
2.2.	**Chemical Properties**	1906	3.	**Unsaturated Dicarboxylic Acids** ... 1919
2.3.	**Production**	1908		
2.3.1.	Degradative Methods	1908	3.1.	**Physical Properties** ... 1919
2.3.2.	Processes Maintaining the Carbon Structure	1910	3.2.	**Chemical Properties** ... 1919
2.3.3.	Syntheses from Smaller Units	1912	3.3.	**Production** ... 1921
2.4.	**Individual Saturated Dicarboxylic Acids**	1913	3.4.	**Individual Unsaturated Dicarboxylic Acids** ... 1921
2.4.1.	Succinic Acid	1914	3.4.1.	Itaconic Acid ... 1922
2.4.2.	Glutaric Acid	1914	3.4.2.	Dimer Acids ... 1922
2.4.3.	Dimethylglutaric Acids	1915	4.	**Quality Specifications and Analysis** ... 1923
2.4.4.	Trimethyladipic Acid	1915		
2.4.5.	Pimelic Acid	1915	5.	**Storage, Transportation, and Handling** ... 1924
2.4.6.	Suberic Acid	1916		
2.4.7.	Azelaic Acid	1916		
2.4.8.	Sebacic Acid	1917	6.	**References** ... 1924

1. Introduction

Aliphatic ω,ω'-dicarboxylic acids (or diacids) can be described by the following general formula:

$HOOC-(CH_2)_n-COOH$

According to IUPAC nomenclature, dicarboxylic acids are named by adding the suffix dioic acid to the name of the hydrocarbon with the same number of carbon atoms, e.g., nonanedioic acid for $n = 7$. The older literature often uses another system based on the hydrocarbon for the $(CH_2)_n$ carbon segment and the suffix dicarboxylic acid, e.g., heptanedicarboxylic acid for $n = 7$. However, trivial names are commonly used for the saturated linear aliphatic dicarboxylic acids from $n = 0$ (oxalic acid) to $n = 8$ (sebacic acid) and for the simple unsaturated aliphatic dicarboxylic acids; these names are generally derived from the natural substance in which the acid occurs or from which it was first isolated.

Aliphatic dicarboxylic acids are found in nature both as free acids and as salts. For example, malonic acid is present in small amounts in sugar beet and in the green parts of the wheat plant; oxalic acid occurs in many plants and in some minerals as the calcium salt. However, natural sources are no longer used to recover these acids.

The main industrial process employed for manufacturing dicarboxylic acids is the ring-opening oxidation of cyclic compounds.

Adipic acid is the most important dicarboxylic acid. Oxalic, malonic, suberic, azelaic, sebacic, and 1,12-dodecanedioic acids, as well as maleic and fumaric acids, are also manufactured on an industrial scale.

Dicarboxylic acids are important feedstocks in the manufacture of polyamides or of di- and polyesters. Esters produced by the reaction of dicarboxylic acids with monofunctional alcohols serve as plasticizers or lubricants. In addition, dicarboxylic acids are used as intermediates in many organic syntheses.

2. Saturated Dicarboxylic Acids

The most important saturated aliphatic dicarboxylic acids are treated under separate keywords (see → Adipic Acid, → Malonic Acid and Derivatives, → Oxalic Acid).

2.1. Physical Properties

Dicarboxylic acids are colorless, odorless crystalline substances at room temperature. Table 1 lists the major physical properties of some saturated aliphatic dicarboxylic acids.

The lower dicarboxylic acids are stronger acids than the corresponding monocarboxylic ones. The first dissociation constant is considerably greater than the second. Density and dissociation constants decrease steadily with increasing chain length. By contrast, melting point and water solubility alternate: Dicarboxylic acids with an even number of carbon atoms have higher melting points than the next higher odd-num-

Table 1. Physical properties of saturated dicarboxylic acids

IUPAC name	Common name	CAS registry number	Formula	M_r	mp, °C	bp at 13.3 kPa, °C	ϱ at 25 °C, g/cm³	Solubility in H$_2$O at 20 °C, wt %	Decarboxylation temperature, °C	Ionization constants K_1	Ionization constants K_2
Ethanedioic acid	oxalic acid	[144-62-7]	HOOC–COOH	90.03	189.5	(sublimes)	1.653	8.0	166–180	5.36×10^{-2}	5.42×10^{-5}
Propanedioic acid	malonic acid	[141-82-2]	HOOC–CH$_2$–COOH	104.06	135		1.619 (16 °C)	73.5	140–160	1.42×10^{-3}	2.01×10^{-6}
Butanedioic acid	succinic acid	[110-15-6]	HOOC–(CH$_2$)$_2$–COOH	118.08	188	235^a	1.572	5.8	290–310	6.21×10^{-5}	2.31×10^{-6}
Pentanedioic acid	glutaric acid	[110-94-1]	HOOC–(CH$_2$)$_3$–COOH	132.11	99	200 (2.7 kPa)	1.424	63.9	280–290	4.58×10^{-5}	3.89×10^{-6}
2,2-Dimethylpentanedioic acid	2,2-dimethylglutaric acid	[681-57-2]	**1**^{d}	160.17	85			^{b}		5.25×10^{-5}	3.8×10^{-6}
Hexanedioic acid	adipic acid	[124-04-9]	HOOC–(CH$_2$)$_4$–COOH	146.14	153	265	1.360	1.6	300–320	3.85×10^{-5}	3.89×10^{-6}
2,4,4-Trimethylhexanedioic acid	2,4,4-trimethyladipic acid	[3937-59-5]	**2**	188.22	68			1.075^c			
Heptanedioic acid	pimelic acid	[111-16-0]	HOOC–(CH$_2$)$_5$–COOH	160.17	106	272	1.329 (15 °C)	5.0	290–310	3.19×10^{-5}	3.74×10^{-6}
Octanedioic acid	suberic acid	[505-48-6]	HOOC–(CH$_2$)$_6$–COOH	174.19	144	279	1.266	0.16	340–360	3.05×10^{-5}	3.85×10^{-6}
Nonanedioic acid	azelaic acid	[123-99-9]	HOOC–(CH$_2$)$_7$–COOH	188.22	108	287	1.225	0.24	320–340	2.88×10^{-5}	3.86×10^{-6}
Decanedioic acid	sebacic acid	[111-20-6]	HOOC–(CH$_2$)$_8$–COOH	202.25	134.5	295	1.207	0.10	350–370	3.1×10^{-5}	3.6×10^{-6}
Undecanedioic acid		[1852-04-6]	HOOC–(CH$_2$)$_9$–COOH	216.27	110			0.014			
Dodecanedioic acid		[693-23-2]	HOOC–(CH$_2$)$_{10}$–COOH	230.30	131	254 (2.0 kPa)		0.004			
Tridecanedioic acid	brassylic acid	[505-52-2]	HOOC–(CH$_2$)$_{11}$–COOH	244.33	114		1.150 (18 °C)	0.0025			
Tetradecanedioic acid		[821-38-5]	HOOC–(CH$_2$)$_{12}$–COOH	258.35	129						

^{a} Forms the anhydride (101.3 kPa).
^{b} Readily soluble in water.
^{c} Mixture of 40 % 2,2,4- and 60 % 2,4,4-trimethyladipic acid.
^{d} **1** = HOOC–C(CH$_3$)$_2$–(CH$_2$)$_2$–COOH   **2** = HOOC–CH(CH$_3$)–CH$_2$–C(CH$_3$)$_2$–CH$_2$–COOH

bered dicarboxylic acid. In the $n = 0-8$ range, dicarboxylic acids with an even number of carbon atoms are slightly soluble in water, while the next higher homologues with an odd number of carbon atoms are more readily soluble. As chain length increases, the influence of the hydrophilic carboxyl groups diminishes; from $n = 5$ (pimelic acid) onward, solubility in water decreases rapidly. The alternating solubility of dicarboxylic acids can be exploited to separate acid mixtures [1], [2]. Most dicarboxylic acids dissolve easily in lower alcohols; at room temperature, the lower dicarboxylic acids are practically insoluble in benzene and other aromatic solvents.

2.2. Chemical Properties

The chemical behavior of dicarboxylic acids is determined principally by the two carboxyl groups. The neighboring methylene groups are activated generally to only a minor degree; malonic acid derivatives (esters and nitriles) are an exception (→ Malonic Acid and Derivatives).

Thermal decomposition of dicarboxylic acids gives different products depending on the chain length. Acids with an even number of carbon atoms require higher decarboxylation temperatures than the next higher odd-numbered homologues; lower dicarboxylic acids decompose more easily than higher ones. To avoid undesired decomposition reactions, aliphatic dicarboxylic acids should only be distilled in vacuum. When heated above 190 °C, oxalic acid decomposes to carbon monoxide, carbon dioxide, and water. Malonic acid is decarboxylated to acetic acid at temperatures above 150 °C:

$$HOOC-CH_2-COOH \longrightarrow CH_3COOH + CO_2$$

When malonic acid is heated in the presence of P_2O_5 at ca. 150 °C, small amounts of carbon suboxide (C_3O_2) are also formed. Succinic and glutaric acids are converted into cyclic anhydrides on heating:

$$HOOC-(CH_2)_n-COOH \xrightarrow{-H_2O} \underset{\text{where } n = 2 \text{ or } 3}{\text{cyclic anhydride}}$$

When the ammonium salt of succinic acid is distilled rapidly, succinimide is formed, with the release of water and ammonia.

Higher dicarboxylic acids from $n = 4$ (adipic acid) to $n = 6$ (suberic acid) split off carbon dioxide and water to form cyclic ketones:

$$HOOC-(CH_2)_n-COOH \longrightarrow (CH_2)_n C=O + CO_2 + H_2O$$

where $n = 4-6$

The decomposition of still higher dicarboxylic acids leads to complex mixtures. With the exception of oxalic acid, dicarboxylic acids are resistant to oxidation. Oxalic acid is used as a reducing agent for both commercial and analytical purposes. Dicarboxylic acids react with dialcohols to form polyesters and with diamines to form polyamides. They also serve as starting materials for the production of the corresponding diamines. Reaction with monoalcohols yields esters. All of these reactions are commercially important. Several reactions with malonic and glutaric acids are of interest in organic syntheses: the Knoevenagel condensation, Michael addition, and malonic ester synthesis ($\rightarrow$ Malonic Acid and Derivatives) [3], [4].

Succinic acid ester reacts with aldehydes or ketones in the presence of sodium ethoxide or potassium *tert*-butoxide to form alkylidenesuccinic acid monoesters (Stobbe condensation). These can subsequently be converted into monocarboxylic acids by hydrolysis, decarboxylation, and hydrogenation [5]:

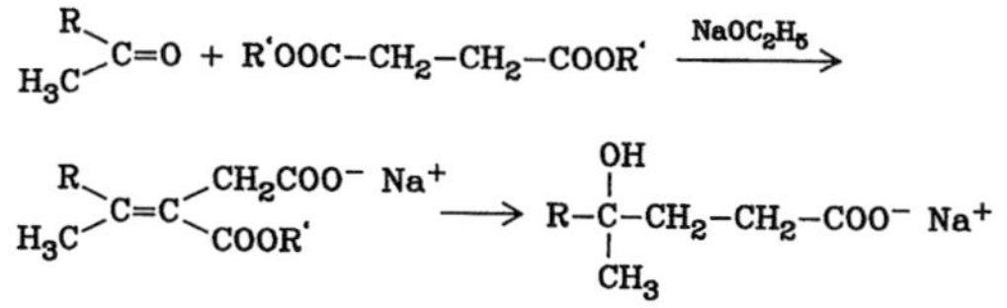

$$\underset{H_3C}{\overset{R}{>}}C=O + R'OOC-CH_2-CH_2-COOR' \xrightarrow{NaOC_2H_5}$$

$$\underset{H_3C}{\overset{R}{>}}C=C\underset{COOR'}{\overset{CH_2COO^- Na^+}{<}} \longrightarrow R-\underset{CH_3}{\overset{OH}{\underset{|}{C}}}-CH_2-CH_2-COO^- Na^+$$

where R = H, alkyl
R' = C_2H_5

Cyclic ketones are obtained from C_6–C_8 dicarboxylic acid esters and sodium methoxide (Dieckmann reaction) [6]. Esters of adipic, pimelic, and suberic acids can be converted in good yields; esters of higher dicarboxylic acids cannot be cyclized by this method.

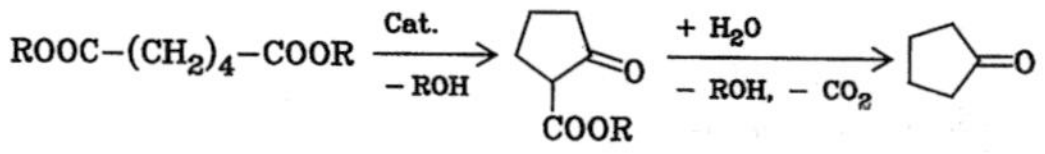

$$ROOC-(CH_2)_4-COOR \xrightarrow[-ROH]{Cat.} \underset{COOR}{\bigcirc}=O \xrightarrow[-ROH, -CO_2]{+H_2O} \bigcirc=O$$

where R = CH_3, C_2H_5

Acyloin condensation with metallic sodium gives cyclic acyloins; this method is particularly suitable for synthesis of large rings [7]:

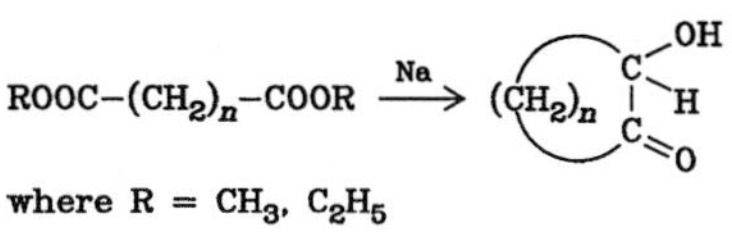

$$ROOC-(CH_2)_n-COOR \xrightarrow{Na} (CH_2)_n\underset{C=O}{\overset{C-OH}{<\!\!<}}{H}$$

where R = CH_3, C_2H_5

Detailed summaries of reactions with dicarboxylic acids can be found in [8].

2.3. Production

A number of straight-chain aliphatic dicarboxylic acids and their derivatives occur in nature. However, isolation from natural substances has no commercial significance. Although many syntheses for the production of aliphatic dicarboxylic acids are known, only a few have found industrial application. This is due partly to the shortage of raw materials.

The most important processes for the manufacture of saturated aliphatic dicarboxylic acids are the following:

1) Oxidative cleavage of cyclic compounds (e.g., adipic acid from cyclohexane, 1,12-dodecanedioic acid from 1,5,9-cyclododecatriene)
2) Oxidative cleavage of unsaturated monocarboxylic acids (e.g., azelaic acid from oleic acid)
3) Alkaline cleavage of substituted monocarboxylic acids (e.g., sebacic acid from ricinoleic acid)
4) Hydrogenation of unsaturated dicarboxylic acids (e.g., succinic acid from maleic acid)
5) Oxidation of ω,ω'-diols (e.g., pimelic acid from 1,7-heptanediol)
6) Carbonylation reactions (e.g., suberic acid from 1,6-hexanediol)

Some special syntheses are also of interest. The following sections treat the most important manufacturing processes, which can be subdivided into degradative methods, processes in which the carbon structure is maintained, and synthetic methods starting from smaller units.

2.3.1. Degradative Methods

Ozonolysis of Oleic Acid. Ozonolysis of oleic acid [112-80-1] followed by oxidative cleavage gives pelargonic acid [112-05-0] and azelaic acid [9]:

$$CH_3-(CH_2)_7-CH=CH-(CH_2)_7-COOH \xrightarrow[2)\ O_2,\ 70-110\,°C]{1)\ O_3/O_2,\ 20-40\,°C}$$
$$CH_3-(CH_2)_7-COOH + HOOC-(CH_2)_7-COOH$$

Figure 1 shows a commercial process for the production of azelaic acid from oleic acid.

Oleic acid is cleaved by ozonolysis (O_3 concentration in the air: 1.0 vol%) at 20–40 °C in pelargonic acid and water. The alkene residence time is about 10 min. The ozonide is then cleaved with oxygen at 70–110 °C. Pelargonic and azelaic acids are separated from higher boiling compounds by subsequent distillation. Azelaic acid is subjected to extraction to remove monocarboxylic acids; distillation of the extractant finally yields pure acid.

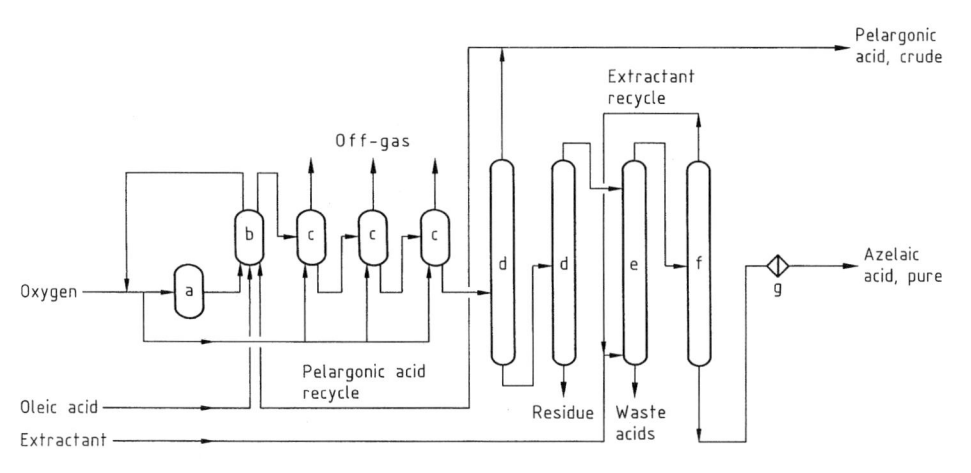

Figure 1. Manufacture of azelaic acid by ozonolysis of oleic acid
a) Ozone generator; b) Ozone absorber; c) Reactor; d) Distillation column; e) Extraction column; f) Distillation of the extractant; g) Flaking

Cleavage of Ricinoleic Acid. Alkaline cleavage of ricinoleic acid [*141-22-0*] (12-hydroxy-9-octadecenoic acid) under pressure and at high temperature leads to the formation of sodium sebacate and 2-octanol [10]:

$$CH_3-(CH_2)_5-CH(OH)-CH_2-CH=CH-(CH_2)_7-COOH$$
$$\downarrow NaOH$$
$$NaOOC-(CH_2)_8-COONa + CH_3-(CH_2)_5-CH(OH)-CH_3$$

In industry, castor oil [*8001-79-4*], which contains about 87 % ricinoleic acid, is normally used instead of the pure acid.

Oxidation with N_2O_4. Oxidative degradation of monocarboxylic acids generally produces dicarboxylic acid mixtures; the composition of the reaction products shifts toward the higher dicarboxylic acids as the chain length of the monocarboxylic acids increases. Oxidation of stearic acid [*57-11-4*] with N_2O_4 yields a mixture consisting mainly of sebacic and caprylic [*124-07-2*] acids. In the same way, palmitic acid [*57-10-3*] can be oxidized with nitric acid – N_2O_4 to form suberic acid [11].

Commercial production of adipic acid from cyclohexanol – cyclohexanone yields two major byproducts, succinic and glutaric acids, which can be separated easily (→ Adipic Acid).

Oxidation of Hydrocarbons. Oxidative degradation of hydrocarbons is also a common manufacturing process. The best-known example is the oxidation of benzene to maleic acid by means of vanadium pentoxide catalysts (→ Maleic and Fumaric Acids).

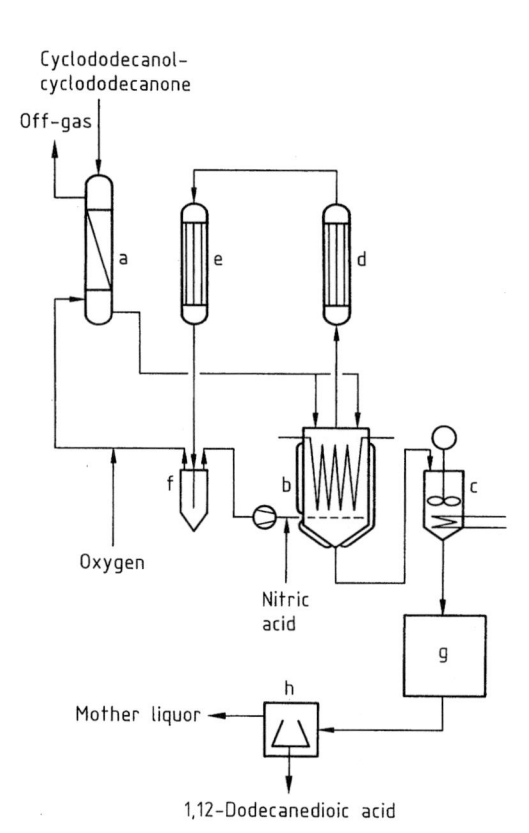

Figure 2. Manufacture of 1,12-dodecanedioic acid from cyclododecanol–cyclododecanone
a) Scrubber; b) Reactor; c) Downstream reactor; d) Condenser; e) Cooler; f) Separator; g) Crystallizer; h) Filter

2.3.2. Processes Maintaining the Carbon Structure

Cleavage of Cyclic Compounds. Many processes for the manufacture of dicarboxylic acids by oxidative cleavage of cyclic compounds are commercially significant; however, the oxidation of cyclohexane via cyclohexanol–cyclohexanone is the most important (→ Adipic Acid). Similar processes are employed to convert cyclopentanol–cyclopentanone to glutaric acid, cycloheptanone to pimelic acid, and cyclododecanol–cyclododecanone to 1,12-dodecanedioic acid [12]. Figure 2 shows a process for the manufacture of 1,12-dodecanedioic acid from cyclododecanol–cyclododecanone [13].

The oxidation is carried out in a stirred reactor (b) fed continuously with nitric acid and cyclododecanol–cyclododecanone; ammonium vanadate is used as a catalyst. The nitric oxides formed during oxidation are recycled via the condenser (d) and cooler (e). The reaction slurry passes to the postreaction stage and then to the crystallizer (g), in which most of the acid crystallizes. The solids are filtered off and the mother liquor is recycled to the reactor. In industrial-scale processes, selectivities toward the acid of about 90 % are achieved.

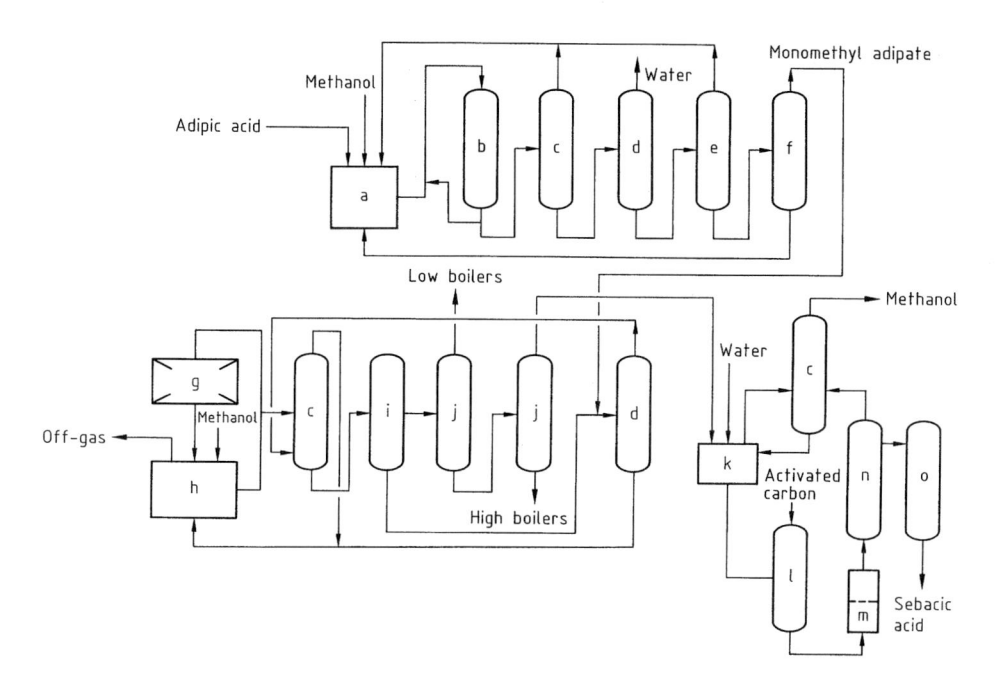

Figure 3. Manufacture of sebacic acid by electrochemical dimerization of monomethyl adipate
a) Mixing tank; b) Reaction column; c) Methanol stripper; d) Water stripper; e) Dimethyl adipate stripper;
f) Adipic acid cutting column; g) Electrolyzer; h) Electrolyte tank; i) Decanter; J) Distillation column; k) Reactor;
l) Mixing tank; m) Filter; n) Dehydrator; o) Prill tower

Cyclic ethers can also be used as starting materials. Thus, pimelic acid is obtained from potassium tetrahydrofurylpropionate [14] or from hydroxycyclohexanoic acid [15]. Hydrolysis of dihydropyran produces 5-hydroxypentanal, which is converted to glutaric acid by subsequent oxidation with nitric acid [16].

Ozonolysis of Cyclic Olefins. Cyclic olefins can be converted to dicarboxylic acids by ozonolysis and subsequent oxidative cleavage. For example, 1,12-dodecanedioic acid can be obtained by ozonolysis of cyclododecene (see Section 2.4.9) [17].

Oxidation of Bifunctional Compounds. Dicarboxylic acids can be produced by oxidation of bifunctional compounds with HNO_3 in the presence of ammonium vanadate, with N_2O_4, or with oxygen in the presence of palladium on carbon. Diols are preferred as bifunctional starting materials. Well-known examples of this process are the syntheses of pimelic acid from 1,7-heptanediol and of succinic acid from 1,4-butanediol [18].

Nitrile Hydrolysis. Saponification of dinitriles also yields dicarboxylic acids. Thus, glutaric acid can be produced from glutarodinitrile, which is obtained by the reaction of

1,3-dihalopropane with sodium cyanide. Saponification of the nitrile group can take place concurrently with oxidation of a carbonyl group; e.g., 4-cyano-2,2-dimethylbutanal, obtained by the addition of isobutanal to acrylonitrile, gives 2,2-dimethylglutaric acid [19].

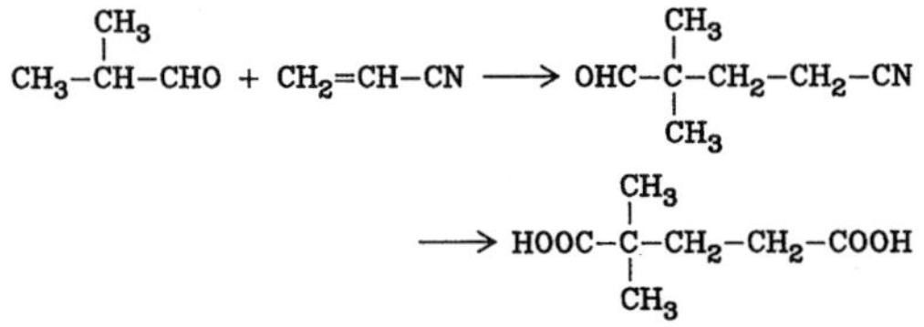

$$CH_3-CH(CH_3)-CHO + CH_2=CH-CN \longrightarrow OHC-C(CH_3)_2-CH_2-CH_2-CN$$

$$\longrightarrow HOOC-C(CH_3)_2-CH_2-CH_2-COOH$$

Hydrogenation. Hydrogenation of unsaturated dicarboxylic acids or their anhydrides produces good yields of the corresponding saturated compounds. Succinic acid is obtained by this method from maleic acid or maleic anhydride [20].

Fermentation. Numerous alkane-based fermentation processes have recently been described for the manufacture of dicarboxylic acids [21]. However, these biotechnical processes have not yet become standard commercial practice.

2.3.3. Syntheses from Smaller Units

The main addition reactions leading to dicarboxylic acids are variants of carbonylation. Diolefins, dialcohols, and unsaturated monocarboxylic acids are used as starting materials [22]. Reppe carbonylation of 1,6-hexanediol produces suberic acid; C_{19} dicarboxylic acids are obtained from oleic acid [23].

The dimerization of monomethyl adipate to sebacic acid is an electrochemical process which has achieved commercial significance. Figure 3 illustrates this process [24].

The reaction takes place in three stages:
Esterification:

$$HOOC-(CH_2)_4-COOH + CH_3OH \longrightarrow HOOC-(CH_2)_4-COOCH_3 + H_2O$$

Electrolysis:

$$^-OOC-(CH_2)_4-COOCH_3 - e^- \longrightarrow 1/2\ CH_3OOC-(CH_2)_8-COOCH_3 + CO_2$$
$$H^+ + e^- \longrightarrow 1/2\ H_2$$

Hydrolysis:

$$CH_3OOC-(CH_2)_8-COOCH_3 + 2\ H_2O \longrightarrow HOOC-(CH_2)_8-COOH + 2\ CH_3OH$$

In the first stage, adipic acid reacts with methanol at 80 °C to form monomethyl adipate. Ion exchangers containing sulfonic acid groups are used as catalysts and also prevent the formation of byproducts such as cyclopentanone. The monomethyl adipate is separated by distillation.

In the second stage, the potassium salt of monomethyl adipate is dimerized electrolytically either continuously or batchwise. The electrolyzer is equipped with bipolar electrodes. Electrolysis takes place at 50–60 °C. Aqueous methanol is used as solvent, the H_2O concentration being between 0.15 and 0.30 %. The resulting dimethyl sebacate solution is distilled, and unreacted potassium methyl adipate is returned to the electrolysis process. Dimethyl sebacate is purified by distillation.

In the third stage, dimethyl sebacate is hydrolyzed at 160–180 °C and a pressure of about 0.9 MPa (9 bar). Methanol is removed, and the crude sebacic acid is treated with activated carbon and then dried.

The Wurtz synthesis can also be used [25]:

$$2\,Cl-CH_2-COOR + 2\,Na \longrightarrow$$
$$ROOC-(CH_2)_2-COOR + 2\,NaCl \xrightarrow{H_2O}$$
$$HOOC-(CH_2)_2-COOH$$

The reactivity of the methylene group in malonic ester is exploited in many dicarboxylic acid syntheses (malonic ester synthesis; → Malonic Acid and Derivatives).

The Stetter dicarboxylic acid synthesis is another important process [26]:

$$\longrightarrow HOOC-(CH_2)_8-COOH$$

Long-chain dicarboxylic acids can be prepared in the following manner [27]:

$$\longrightarrow HOOC-(CH_2)_{n+12}-COOH$$

Many processes used to manufacture monocarboxylic acids are also suitable for the synthesis of dicarboxylic acids. This field has been reviewed extensively [28] (→ Carboxylic Acids, Aliphatic).

2.4. Individual Saturated Dicarboxylic Acids

Dicarboxylic acids are used mainly as intermediates in the manufacture of esters and polyamides. Esters derived from monofunctional alcohols serve as plasticizers or lubricants. Polyesters are obtained by reaction with dialcohols. In addition, dicarboxylic acids are employed in the manufacture of hydraulic fluids, agricultural chemicals, pharmaceuticals, dyes, complexing agents for heavy-metal salts, and lubricant additives (as metal salts).

2.4.1. Succinic Acid

Succinic acid is found in amber, in numerous plants (e.g., algae, lichens, rhubarb, and tomatoes), and in many lignites.

Production. A large number of syntheses are used to manufacture succinic acid. Hydrogenation of maleic acid, maleic anhydride, or fumaric acid produces good yields of succinic acid; the standard catalysts are Raney nickel [20], Cu, NiO, or CuZnCr [29], Pd–Al_2O_3 [30], Pd–$CaCO_3$ [31], or Ni–diatomite [32]. 1,4-Butanediol can be oxidized to succinic acid in several ways: (1) with O_2 in an aqueous solution of an alkaline-earth hydroxide at 90–110 °C in the presence of Pd–C; (2) by ozonolysis in aqueous acetic acid; or (3) by reaction with N_2O_4 at low temperature [18]. Succinic acid or its esters are also obtained by Reppe carbonylation of ethylene glycol, catalyzed with $RhCl_3$–pentachlorothiophenol [33]; Pd-catalyzed methoxycarbonylation of ethylene [34]; and carbonylation of acetylene, acrylic acid, dioxane, or β-propiolactone [35], [36].

Acid mixtures containing succinic acid are obtained in various oxidation processes. Examples include the manufacture of adipic acid [2], [37]–[39] (→ Adipic Acid); the oxidation of enanthic acid [40]; and the ozonolysis of palmitic acid [41].

Succinic acid can also be obtained by phase-transfer-catalyzed reaction of 2-haloacetates [42], electrolytic dimerization of bromoacetic acid or ester [43], oxidation of 3-cyanopropanal [44], and fermentation of *n*-alkanes [45].

Uses. Succinic acid is used as a starting material in the manufacture of alkyd resins, dyes, pharmaceuticals, and pesticides. Reaction with glycols gives polyesters; esters formed by reaction with monoalcohols are important plasticizers and lubricants.

2.4.2. Glutaric Acid

Glutaric acid occurs in washings from fleece and, together with malonic acid, in the juice of unripened sugar beet.

Production. Glutaric acid is obtained from cyclopentane by oxidation with oxygen and cobalt (III) catalysts [46], [47] or by ozonolysis [48]; and from cyclopentanol–cyclopentanone by oxidation with oxygen and $Co(CH_3CO_2)_2$, with potassium peroxide in benzene, or with N_2O_4 or nitric acid [12], [49]–[51]. Like succinic acid, glutaric acid is formed as a byproduct during oxidation of cyclohexanol–cyclohexanone (→ Adipic Acid).

Other production methods include reaction of malonic ester with acrylic acid ester [52], [53], carbonylation of γ-butyrolactone [22], oxidation of 1,5-pentanediol with N_2O_4 [18], and oxidative cleavage of γ-caprolactone [54].

Uses. The applications of glutaric acid, e.g., as an intermediate, are limited. Its use as a starting material in the manufacture of maleic acid has no commercial importance.

2.4.3. Dimethylglutaric Acids

2,2-Dimethylglutaric acid is manufactured from dimethyl-γ-butyrolactone by carbonylation using $HF-SbF_5$ as a catalyst or by reaction with formic acid in stronger acids such as $H_2SO_4-SO_3$ [55], [56]. 4-Cyano-2,2-dimethylbutanal, which is obtained by addition of isobutanal to acrylonitrile, can be converted to the acid by oxidation of the formyl group and subsequent hydrolysis of the nitrile group [19], [57].

2,2-Dimethylglutaric acid is used in the manufacture of diglycidyl esters (for coating materials) [58], pyrethroids (for insecticides and acaricides) [59], and antibiotics [60].

3,3-Dimethylglutaric acid [4839-46-7] is manufactured from isophorone by oxidation with H_2O_2 in the presence of concentrated sulfuric acid or by ozonolysis in methanolic solution and subsequent oxidation with H_2O_2 [61]. This acid is used in the manufacture of pesticides and lubricating oil additives.

2.4.4. Trimethyladipic Acid

Commercial trimethyladipic acid is a mixture of ca. 40% 2,2,4-trimethyladipic acid and 60% 2,4,4-trimethyladipic acid.

Production. Trimethyladipic acid is manufactured by oxidative cleavage of 3,3,5-trimethylcyclohexanol [116-02-9] (produced from acetone) with 65% nitric acid at 50 °C [62]. To separate the short-chain dicarboxylic acids, the mixture is heated to 180–250 °C and the cyclic anhydrides formed are distilled off [63].

Uses. Trimethyladipic acid is used in the production of synthetic lubricating oils [64], polyesters [65], and polyamides [65], and in the modification of terephthalic acid esters [66].

2.4.5. Pimelic Acid

Pimelic acid is an oxidation product of fats.

Production. Pimelic acid can be manufactured with good selectivity by oxidation of cycloheptanone [502-42-1] (suberone) with N_2O_4 [12]. It is also obtained in a mixture with other dicarboxylic acids by oxidative cleavage of palmitic acid [41]. Other manufacturing processes include oxidation of 1,7-heptanediol [67], carbonylation of ε-caprolactone [22], and acid cleavage of tetrahydrosalicylic acid with potassium hydroxide at 300 °C under pressure [15].

Uses. Pimelic acid has slight significance as a starting material in the manufacture of polyesters and polyamides.

2.4.6. Suberic Acid

Suberic acid is formed from the action of nitric acid on cork.

Production. Suberic acid is manufactured by oxidation of cyclooctene with ozone–oxygen [48], [68] or with ozone–H_2O_2 [69]. The acid is formed together with other dicarboxylic acids during ozonolysis of palmitic acid [41] as well as during cleavage of ricinoleic acid with nitric acid [70]. Other manufacturing processes include oxidation of cyclooctanol–cyclooctanone with N_2O_4 or HNO_3 [12], [71], carbonylation of 1,6-hexanediol [72], and oxidative cleavage of 2-(cyclohexanon-2-yl)acetic acid ethyl ester [26].

Uses. Suberic acid has been used in the manufacture of mono- and diesters as well as polyamides. Nylon 6,8 is obtained by reaction of suberic acid with hexamethylenediamine, and nylon 8,8 by reaction with octamethylenediamine. Polyamides of suberic acid with diamines such as 1,3-bis(aminomethyl)benzene, 1,4(bisaminomethyl)cyclohexane, and bis(4-aminocyclohexyl)methane are also of commercial interest. Esters of suberic acid with mono- and bifunctional alcohols are used as lubricants.

2.4.7. Azelaic Acid

Azelaic acid occurs in many natural substances containing long-chain fatty acids.

Production. Azelaic acid is obtained by oxidative cleavage of oleic acid with oxidants such as RuO_4 [73]; Cl_2–RuO_2 or Cl_2–$RuCl_2$ [74], [75]; $KMnO_4$ [76]; $NaOCl$–RuO_4–OsO_4 [77]; and HNO_3 [78]. The industrially most important process is the ozonolysis of oleic acid (see Section 2.3.1) [9], [79], [80].

Other means of synthesizing azelaic acid include carbonylation of 1,5-cyclooctadiene [81], oxidation of 1,9-nonanedial with oxygen [82], oxidative cleavage of 2-cyanoethylcyclohexanone [83], and fermentation of pelargonic acid [84]. A mixture of azelaic and other dicarboxylic acids is obtained during ozonolysis of palmitic acid [41].

Uses. Monoesters of azelaic acid with 2-ethylhexanol are used as plasticizers. Mono- and diesters with other alcohols act as hydraulic fluids and lubricating oils; their metal salts are recommended as lubricating oil additives. Reaction with hexamethylenediamine leads to nylon 6,9, which is used as extruded film for food packaging, as a coating for wire, and in the electronics and automobile industries. Unsaturated polyesters are employed as resins, laminates, and adhesives.

2.4.8. Sebacic Acid

Production. The most important processes for manufacturing sebacic acid are alkaline cleavage of ricinoleic acid (see Section 2.3.1) [85]–[88] and electrolytic dimerization of monomethyl adipate (see Section 2.3.3) [89]–[91]. 2-Octanol is formed as a byproduct during ricinoleic acid cleavage. Other methods used to manufacture sebacic acid are oxidation of stearic acid by N_2O_4 [11], oxidation of 1,10-decanediol [18], and various fermentation processes [21], [92]. A mixture of sebacic acid and other dicarboxylic acids is formed during ozonolysis of palmitic acid [41]. Processes for purifying sebacic acid are described in [93]–[95].

The C_{10} dicarboxylic acid mixture obtained by dimerization of butadiene and subsequent reaction with CO_2 is called isosebacic acid; it consists of ca. 75% 2-ethylsuberic acid, 15% diethyladipic acid, and 10% sebacic acid. Because of the varying composition of different production batches, this mixture has not been able to gain a foothold in the market.

Uses. The polyamide nylon 6,10 obtained by reaction of sebacic acid with hexamethylenediamine no longer has great industrial significance. The sebacates of various oxo and straight-chain alcohols are important plasticizers. Their main characteristics are high migration resistance and good low-temperature resistance. The esters are also used as components of lubricating oils and as diluents; because of their low toxicity they are important components of packaging films. Sebacic acid-based alkyd resins are characterized by marked flexibility.

2.4.9. 1,12-Dodecanedioic Acid

Over the past few years, 1,12-dodecanedioic acid has achieved industrial importance.

Production. The starting compound for industrial-scale production of 1,12-dodecanedioic acid is 1,5,9-cyclododecatriene (CDT), which is obtained by trimerization of butadiene (→ Cyclododecatriene and Cyclooctadiene). Cyclododecatriene can react to form the acid by two different processes. In a three-stage reaction sequence, 1,5,9-cyclododecatriene is first hydrogenated to cyclododecane on nickel catalysts; cyclododecane is then oxidized with oxygen or air to a cyclododecanol–cyclododecanone mixture; and this mixture is finally oxidized with nitric acid to 1,12-dodecanedioic acid (→ Cyclododecanol, Cyclododecanone, and Laurolactam) [96]–[100].

The second route consists of partial hydrogenation of 1,5,9-cyclododecatriene to cyclododecene and subsequent oxidative ozonolytic cleavage to the acid [101]–[103]; ozonolysis of cyclododecanol has been described [17].

$$\text{cyclododecatriene} \xrightarrow{H_2, \text{Cat.}} \text{cyclododecane} \xrightarrow{O_3/O_2} \text{HOOC-(CH}_2)_{10}\text{-COOH}$$

Other manufacturing processes such as the fermentation of *n*-dodecane [21], [104], [105] and the oxidation of analogous mono- and diformyl compounds [106] have no industrial importance. Processes for purifying the acid have been described [107]–[110].

Uses. 1,12-Dodecanedioic acid is used mainly in manufacturing polyamides and polyesters. Reaction with hexamethylenediamine gives nylon 6,12; reaction with *trans,trans*-bis-(4-aminocyclohexyl)methane yields the polyamide known as Qiana. 1,12-Dodecanedioic acid is also used for the manufacture of lubricating oils and plasticizers.

2.4.10. 1,13-Tridecanedioic Acid (Brassylic Acid)

Production. The most important raw material for the production of brassylic acid is erucic acid [112-86-7] (*cis*-13-docosenoic acid), which occurs in large quantities in the seed oil of rape, mustard, wallflowers, and cress. Yields of 82–92 % are obtained by ozonolysis of erucic acid in acetic acid and subsequent oxidation with oxygen [111], [112]. The byproduct, pelargonic acid, can be separated easily. Oxidative cleavage of erucic acid by reaction with nitric acid is also possible [78].

$CH_3\text{-}(CH_2)_7\text{-}CH=CH\text{-}(CH_2)_{11}\text{-}COOH \longrightarrow$
$CH_3\text{-}(CH_2)_7\text{-}COOH + HOOC\text{-}(CH_2)_{11}\text{-}COOH$

Fermentation of *n*-tridecane has been investigated over the past few years, especially in Japan [113]–[117]; purification of the resulting acid is described in [118]–[121].

Uses. Brassylic acid is used in the manufacture of polyamides (nylon 13,13) and esters which are employed as low-temperature plasticizers for poly(vinyl chloride) (PVC) and as lubricant components. It is also a starting material for synthetic musk.

2.4.11. C_{19} Dicarboxylic Acids

Production. The composition of C_{19} dicarboxylic acid mixtures depends on the manufacturing process. Three processes, based on oleic acid or oleic acid esters, are used industrially: (1) Reppe carbonylation catalyzed by $Ni(CO)_4$ or metal complexes such as $PdCl_2$–triphenylphosphine; (2) Koch reaction in concentrated sulfuric acid at 10–20 °C [23], [122], [123] or with HF catalysis at 30 °C [124]; and (3) hydroformylation. Hydroformylation gives a mixture of isomeric formylstearates, which are subsequently oxidized with air or oxygen. Oxidation takes place in an aqueous emulsion at 20 – 25 °C in the presence of calcium acetate or manganese naphthenate catalyst [125], [126]. Potassium permanganate [127] – [129] or potassium dichromate [128], [129] can also be used as the oxidizing agent.

Uses. Esters of C_{19} dicarboxylic acids are used as plasticizers for PVC. The esterification rate of the terminal carboxyl group is considerably higher than that of the central group. This allows selective synthesis of mixed esters.

The C_{19} dicarboxylic acids are also used as starting materials in the production of polyamides, epoxy resins, unsaturated polyester resins, lubricants, and adhesives.

3. Unsaturated Dicarboxylic Acids

The most important derivative of unsaturated dicarboxylic acids is maleic anhydride (→ Maleic and Fumaric Acids). Unsaturated C_{36} dicarboxylic acids containing cyclic structures, which are known as dimer acids, also have some industrial significance.

3.1. Physical Properties

Table 2 lists the most important physical properties of some unsaturated aliphatic dicarboxylic acids. The lower members of the series are colorless, crystalline substances at room temperature; the dimer acids, which are commercially available as isomeric mixtures, are viscous liquids at 25 °C. Melting point, solubility in water, and dissociation constants of the lower homologues are influenced by configuration. For example, the melting points of fumaric and mesaconic acid (trans) are considerably higher than those of the cis isomers maleic and citraconic acid, respectively; the trans isomers are also much less water soluble and less acidic. An indication of the higher stability of the trans form is the higher heat of combustion of maleic acid compared with fumaric acid.

3.2. Chemical Properties

The chemical behavior of unsaturated aliphatic dicarboxylic acids is determined primarily by the two carboxyl groups (see Section 2.2) and the olefinic double bond. Reactions of the carboxyl groups can also be influenced by the olefinic bond. For example, when maleic or citraconic acid is heated above 100 °C, water is split off, and maleic or citraconic anhydride is obtained (→ Maleic and Fumaric Acids). Fumaric acid, however, forms no anhydride; above 230 °C, decomposition occurs and maleic anhydride, water, and an appreciable amount of residue are formed.

Dicarboxylic Acids, Aliphatic

Table 2. Physical properties of unsaturated dicarboxylic acids

IUPAC name	Common name	CAS registry number	Formula	M_r	mp, °C	ϱ at 20 °C, g/cm³	Solubility in H₂O, wt %	Ionization constants $K_1 K_2$
cis-Butenedioic acid	maleic acid	[110-16-7]	H\C=C/H / HOOC COOH	116.07	143.5	1.590	78.9 (25 °C)	1.14×10^{-2} 5.95×10^{-7}
trans-Butenedioic acid	fumaric acid	[110-17-8]	H\C=C/COOH / HOOC H	116.07	296–298	1.635 (25 °C)	0.428 (15.5 °C)	9.57×10^{-4} 4.13×10^{-5}
2-Methyl-cis-butenedioic acid	citraconic acid	[498-23-7]	H₃C\C=C/H / HOOC COOH	130.10	93	1.617		1.10×10^{-3} 1.0×10^{-6}
2-Methyl-trans-butenedioic acid	mesaconic acid	[498-24-8]	HOOC\C=C/H / H₃C COOH	130.10	204.5	1.466	2.7 (18 °C)	4.78×10^{-4} 1.86×10^{-5}
2-Methylenebutanedioic acid	itaconic acid	[97-65-4]	CH₂=C(COOH)–CH₂–COOH	130.10	175	1.632	8.3	1.5×10^{-4} 2.2×10^{-6}
2-Methylenepentanedioic acid	2-methyleneglutaric acid	[3621-79-2]	CH₂=C(COOH)–(CH₂)₂–COOH	144.12	132–133			
2,4-Hexadienedioic acid	muconic acid (cis–cis)	[3588-17-8]	H\C=C/COOH, H\C=C/H / HOOC H	142.11	194–195			

Addition of halogen to the carbon–carbon double bond yields dihalodicarboxylic acids; reaction with ozone gives formylcarboxylic acids; and hydroxydicarboxylic acids are formed by addition of water. Catalytic hydrogenation leads to saturated dicarboxylic acids; the cis isomers generally react much more quickly than the trans isomers.

The Diels–Alder reaction of maleic anhydride with conjugated dienes is used both industrially and in preparative organic chemistry [130]; for example, tetrahydrophthalic anhydride is formed by reaction with butadiene:

Comprehensive information on reactions with unsaturated dicarboxylic acids can be found in [131].

3.3. Production

Only a few processes are used industrially for the production of unsaturated dicarboxylic acids:

1) Oxidation of hydrocarbons (maleic anhydride from benzene and C_4 hydrocarbons such as *n*-butane and *n*-butene; see → Maleic and Fumaric Acids)
2) Diels–Alder reaction of unsaturated acids (dimer acids from oleic or linoleic acid; see Section 3.4.2)
3) Fermentation (itaconic acid)

3.4. Individual Unsaturated Dicarboxylic Acids

Unsaturated dicarboxylic acids are used mainly to manufacture unsaturated polyester resins, copolymers, or polyamides, and as intermediates in the synthesis of herbicides, insecticides, fungicides, surfactants, lubricants, and plasticizers.

Maleic anhydride is also employed as a starting material for the manufacture of DL-tartaric acid, DL-malic acid, glyoxylic acid, and tetrahydrophthalic anhydride, as well as γ-butyrolactone, 1,4-butanediol, and tetrahydrofuran.

3.4.1. Itaconic Acid

Itaconic acid (2-methylenebutanedioic acid) is soluble in water; moderately soluble in chloroform, benzene, and ligroin; and slightly soluble in ether. When distilled at normal pressure, itaconic acid or itaconic anhydride yields citraconic anhydride.

Production. Itaconic acid is produced by fermentation [132]–[138]. A mixture of itaconic acid, citraconic acid, and citraconic anhydride is obtained by reaction of succinic anhydride with formaldehyde at 200–500 °C in the presence of alkali or alkaline-earth hydroxides [139]; SiO_2–Al_2O_3 or SiO_2–MgO can also be used as catalysts [140]. Other methods involve carbonylation of propargyl chloride with metal carbonyl catalysts [141] and thermal decomposition of citric acid.

Uses. Itaconic acid can be used as a comonomer and in the separation of triorganophosphine mixtures [142].

3.4.2. Dimer Acids

The only higher dicarboxylic acids of commercial importance are unsaturated cyclic C_{36} dicarboxylic acids known as dimer acids. The main difference among various standard proprietary products is the content of trimer compounds; the ratio of dimer to trimer acids can vary from 36:1 to about 0.7:1 [143]. The composition of dimer acids depends to a large degree on the feed materials and the manufacturing process.

Production. Dimer acids are produced by intermolecular condensation of unsaturated C_{18} carboxylic acids or their esters. Tall oil fatty acids are the main feed materials, but oleic and linoleic acids can also be used. The reaction is conducted preferably in the presence of special aluminum silicates (montmorillonites) at 190–240 °C; thermal dimerization at 270–290 °C is also possible [144]–[147]. Numerous catalyst modifications such as addition of alkali, amines, or sulfonic acid halides are described in the literature. A summary of the manufacturing processes for dimer acids can be found in [151]–[153]. Diels–Alder and free-radical reactions have been suggested as mechanisms for the thermal dimerization [148]; when aluminum silicates are used, ionic intermediates may be formed. Commercial processes yield mixtures of dimer acids, higher polycarboxylic acids, and various isomeric monomer acids, which are separated by distillation using film evaporators. After removal of the monomer acid fraction and the polycarboxylic acids, the remaining acids can be separated by further distillation into dimer and trimer acids. Figure 4 shows some dimer acids found in commercial mixtures. Investigations of the structures of dimer acids can be found in [149], [150].

Uses. The dimer acids produced on an industrial scale are used in the manufacture of polyamides, polyesters, epoxy resins, lubricants, plasticizers, and pesticides [154]–[157].

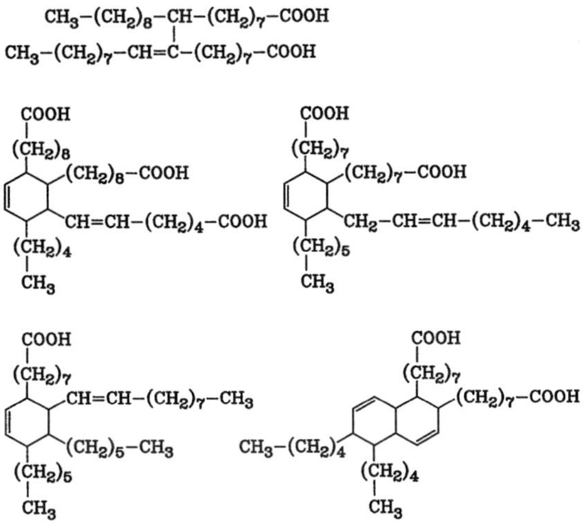

CH₃–(CH₂)₈–CH–(CH₂)₇–COOH
CH₃–(CH₂)₇–CH=C–(CH₂)₇–COOH

Figure 4. Dimer acids found in commercial mixtures

4. Quality Specifications and Analysis

Quality control of dicarboxylic acids covers the determination of content, melting point, color, traces of heavy metals, and solubility in water or other solvents. High purity is generally demanded of dicarboxylic acids.

The content of dicarboxylic acids is usually determined by acidimetric titration. Specific regulations exist for properties; for regulations concerning dimer acids, see [158]–[159]. Specifications may include condition, color, content, and ash. Melting point, density, refractive index, water content, steam pressure, specific heat, dissociation constants, and solubility in water and other solvents are also often determined.

The most important qualitative or quantitative analytical method used industrially is gas chromatography. Dicarboxylic acids are first converted into their esters (preferably methyl esters) because free acids generally undergo undesirable secondary reactions during chromatography. Both packed columns and capillary columns with stationary liquids of different polarities are used. Calcined kieselguhr is the most common carrier material. Other chromatographic methods such as HPLC, paper and thin-layer chromatography, and gel chromatography are also widely employed.

Alkalimetric titration is commonly used to monitor the different production steps and to identify pure dicarboxylic acids or their mixtures. In the absence of other reducing agents, oxalic acid is determined by titration with potassium permanganate. Crystalline derivatives such as phenacyl esters and amides are particularly suitable for chemical determination of dicarboxylic acids.

With infrared spectroscopy, dicarboxylic acids can be detected by the intense carbonyl stretching frequency in the range of 1650–1740 cm^{-1}. In the ^{1}H-NMR spectrum, the hydroxyl proton signals can be found at $\delta = 10-13$ ppm; the signals for the methylene groups of malonic and succinic acids are around 3.4 and 2.6 ppm, respectively. In the ^{13}C-NMR spectrum, the absorption range of the carboxyl carbon atom is around $\delta = 160-180$ ppm.

For further details on analysis, see [158], [160].

5. Storage, Transportation, and Handling

At room temperature, straight-chain dicarboxylic acids are solid compounds that are delivered and stored as crystals or — particularly if a melt or distillation is used to recover the pure substance — as flakes.

Dicarboxylic acids are stored and transported in drums made of plastic-coated steel, stainless steel, or aluminum. Polyethylene-lined paper sacks are also used. These acids are hygroscopic and should be stored in cool, dry rooms to avoid clumping. Shipping regulations and hazard classification for dicarboxylic acids depend on the specific properties of the compounds such as flash point, decomposition temperature, water solubility, toxicity, and ignition temperature.

Aliphatic dicarboxylic acids are local irritants, especially to the mucous membranes; this effect decreases with increasing chain length. Oxalic acid is absorbed readily by the outer layers of the skin and can upset the body's calcium balance through the formation of calcium oxalate. Therefore, gloves and safety glasses must be worn when dicarboxylic acids are handled. To avoid dust that can damage health, dicarboxylic acids are normally supplied in the form of flakes or laminated moldings.

Waste gases from dicarboxylic acid production facilities are generally drawn off at a central point and fed into a combustion chamber. The wastewater is subjected to chemical and biological treatment.

6. References

[1] BP Chemicals Ltd., JP-Kokai 77 19 618, 1977.
[2] Monsanto Co., US 4 254 283, 1981 (G. H. Mock).
[3] G. Jones, *Org. React. (N.Y.)* **15** (1967) 204.
[4] H. R. Kaenel, M. Brossi, *SLZ Schweiz. Lab. Z.* **41** (1984) 197–9.H. O. House, *Modern Synthetic Reactions*, W. A. Benjamin Inc., New York–Amsterdam 1965.
[5] R. N. Hurd, D. H. Shah, *J. Org. Chem.* **38** (1967) 607. W. S. Johnson, G. H. Daub, *Org. React. (N.Y.)* **6** (1951) 1.
[6] C. W. Schimelpfenig, V. T. Lin, J. F. Waller, *J. Org. Chem.* **28** (1963) 805.
[7] J. J. Bloomfield, D. C. Owsley, J. M. Nelke, *Org. React. (N.Y.)* **23** (1976) Chap. 2, 259.

[8] V. Matthews: "Aliphatic Dicarboxylic Acids and related Compounds," Chap. 17, in: Rodd's Chemistry of Carbon Compounds, Elsevier, Amsterdam 1973. A. Cox: "Dicarboxylic and Polycarboxylic Acids," in: Comprehensive Organic Chemistry, Pergamon Press, Oxford 1979. J. Falbe: "Carbonsäuren," in: Methodicum Chimicum, vol. **5**, Thieme Verlag, Stuttgart 1975. R. W. Johnson, *J. Am. Oil Chem. Soc.* **61** (1984) no. 2, 241.

[9] Emery Industries, US 2 813 113, 1957 (C. G. Goebel). N. A. Bogdanova, M. L. Kolesov, *Zh. Khim. Promsti* **4** (1978) no. 49; *Chem. Abstr.* **88** (1978) 191 525. A. Heins, M. Witthaus, *Henkel-Referate* **20** (1984) 42.

[10] D. D. Nanavati, *J. Sci. Ind. Res.* **35** (1976) 163–8. A. P. Kudchadker et al., *Encycl. Chem. Process. Des.* **6** (1978) 401–20. Intreprinderea de Sapun, RO 60 094, 1976 (C. Razvan, D. Lungu, T. Popescu, M. Petrescu).

[11] General Anilin and Film Corp., US 2 821 534, 1954 (W. N. Alexander).

[12] Chem. Werke Hüls, EP 48 476, 1982 (H. Heumann, W. Hilt, H. Liebing, M. Schweppe). J. G. D. Schulz, A. Onopchenko, *J. Org. Chem.* **45** (1980) 3716–19. Kokai Boshi Chosa Kenkyusho K. K., DE-OS 2 638 046, 1977 (W. Ando, I. Nakaoka).

[13] Chem. Werke Hüls, DE-OS 2 001 182, 1970 (H. Röhl, W. Eversmann, P. Hegenberg, E. Hellemanns). Chem. Werke Hüls, DE-OS 1 919 228, 1969 (H. Röhl, W. Eversmann, P. Hegenberg, E. Hellemanns). Chem. Werke Hüls, DE-OS 2 217 003, 1972 (E. Hellemanns, H. Röhl, P. Hegenberg, W. Eversmann). Du Pont, DE-OS 1 912 569, 1969 (J. O. White, D. D. Davis).

[14] F. Runge, R. Hueter, H. D. Wulf, *Chem. Ber.* **87** (1954) 1430.

[15] J. H. Pistor, H. Plieninger, *Justus Liebigs Ann. Chem.* **562** (1939) 239.

[16] J. English, J. E. Dayan, *Org. Synth., Coll. Vol.* **IV** (1963) 499.

[17] Agency of Ind. Sciences and Technology, JP-Kokai 82 64 637, 1982. Chem. Werke Hüls, DE-OS 2 942 279, 1981 (K. D. Dohm, P. Hofmann). Dainippon Ink and Chemicals Inc., JP-Kokai 82 32 245, 1982.

[18] Mitsui Toatsu Chemicals Inc., JP-Kokai 77 151 117, 1977 (M. Kawamata, S. Fujikake, K. Tanabe). S. Miyazaki, Y. Suhara, *J. Am. Oil Chem. Soc.* **55** (1978) 536. Kogai Boshi Chosa Kenkyusho K. K., DE-OS 2 826 065, 1979 (W. Ando, I. Nakaoka).

[19] BASF, DE-OS 1 618 177, 1975 (F. Merger). Dynamit Nobel, FR 2 008 138, 1970 (H. Aus der Fuenten, H. Richtzenhain). Dynamit Nobel, BE 742 378, 1970 (H. Nestler, H. Richtzenhain).

[20] Intreprinderea Chimica Dudesti, RO 79 020, 1982 (S. D. Paucescu, L. Toth, G. Munteanu, M. Badea, S. Kurti). H. Mueller, U. Resch, DD 146 454, 1981.

[21] Asahi Denka Kogyo K. K., JP 793 950, 1979 (K. Yamada, T. Hattori, Y. Shirakawa). Bio Research Center Co. Ltd., DE-OS 2 853 847, 1979 (A. Taoka, S. Uchida). Dainippon Ink and Chemical Inc., JP-Kokai 82 129 694, 1982. Mitsui Petrochemical Industries Inc., JP-Kokai 82 206 394, 1982.

[22] J. Falbe (ed.): *Carbon Monoxide in Organic Synthesis,* Springer Verlag, Berlin–Heidelberg–New York 1970.
J. Falbe (ed.): *New Syntheses with Carbon Monoxide,* Springer Verlag, Berlin–Heidelberg–New York 1980. Noguchi Research Foundation, JP-Kokai 79 92 913, 1979 (Y. Sado, K. Tajima).- Shell Intern. Res. Maats. B. V., EP 42 633, 1981 (E. Drent).

[23] N. E. Lawson, T. T. Cheng, F. B. Slezak, *J. Am. Oil Chem. Soc.* **54** (1977) 215. E. N. Frankel, E. H. Pryde, *J. Am. Oil Chem. Soc.* **54** (1977) 873A.

[24] M. Seko, A. Yomiyama, T. Isoya, *Hydrocarbon Process.* 1979, Dec., 117.

[25] J. W. Conolly, G. Urry, *J. Org. Chem.* **29** (1964) 619. *Houben-Weyl*, **V/1a**, Part 1, 480.

[26] H. Stetter, W. Dierichs, *Chem. Ber.* **86** (1953) 693. H. Stetter, R. Engl, H. Rauhut, *Chem. Ber.* **91** (1958) 2882. H. Stetter, R. Engl. H. Rauhut, *Chem. Ber.* **92** (1959) 1184.

[27] S. Hünig, H. Hoch, *Chem. Ber.* **105** (1972) 2197, 2216. S. Hünig, *Fortschr. Chem. Forsch.* **14** (1970) 235.

[28] *Winnacker-Küchler*, 4th ed., vol. **6**, Chap. 8.6. E. C. Leonard in: E. H. Pryde (ed.): *Fatty Acids*, Chap. 25, The American Oil Chemists' Society, Champaign, Ill., 1979. V. Matthews in: *Rodd's Chemistry of Carbon Compounds*, vol. **I**, Chap. 17, Elsevier, Amsterdam 1973. A. Cox in: *Comprehensive Organic Chemistry*, vol. **2**, Chap. 9.2, Pergamon Press, Oxford 1979. J. Falbe (ed.): "C–O-Verbindungen" in: *Methodicum Chimicum*, vol. **5**, Chap. 8, Thieme Verlag, Stuttgart–New York 1975.

[29] M. Polievka, V. Macho, L. Uhlar, J. Kordik, CS 195 860, 1982.

[30] P. Ruiz et al., *ACS Symp. Ser.* **237** (1984) 15–36; *Chem. Abstr.* **100** (1984) 23 918.

[31] H. Müller, U. Resch, DD 146 454, 1981.

[32] S. Loktev et al., *Neftepererab. Neftekhim. (Moscow)* **2** (1978) 34; *Chem. Abstr.* **88** (1978) 169 557.

[33] Shell Int. Res. Maats. B. V., EP 42 633, 1981 (E. Drent).

[34] J. Stille, R. Divakatumi, *J. Org. Chem.* **44** (1979) 3474–82.

[35] Lonza, DE 1 133 359, 1956 (G. Natta, P. Pino). Lonza, US 2 851 456, 1957 (E. S. Rothman, M. E. Wall).

[36] BASF, US 2 604 490, 1951 (W. Reppe).

[37] Gulf Research and Development Co., US 4 263 453, 1981 (J. G. D. Schulz, A. Onopchenko).

[38] BP Chemicals Ltd., GB 7 533 012, 1975.

[39] ICI Ltd., EP 2598, 1979 (B. Baker).

[40] B. L. Moldavskii, R. J. Rudakova, *Zh. Prikl. Khim. (Leningrad)* **33** (1960) 417.

[41] Agency of Ind. Sciences and Technology, JP-Kokai 78 46 914, 1978 (S. Miyazaki).

[42] V. N. Gogte et al., *Tetrahedron Lett.* **24** (1983) 4131–4.

[43] M. T. Ismail et al., *Bull. Fac. Sci. Assiut. Univ.* **11** (1982) 121–6.

[44] I. B. Blanshtein, V. Bunin, E. I. Leenson; SU 70 19 94, 1979.

[45] J. Tian et al., *Wei Sheng Wu Hsueh Pao* **21** (1981) 229–33; *Chem. Abstr.* **95** (1981) 113 347.

[46] All-Union Scientific Res. and Dev. Inst., SU 937 444, 1982 (R. Nesterova et al.).

[47] Gulf Research and Development Co., US 4 158 739, 1979 (J. G. D. Schulz, A. Onopchenko).

[48] R. M. Habib et al., *J. Org. Chem.* **49** (1984) 2780–4.

[49] Z. G. Petrova et al., *Deposited Doc.* 1981, 1336–81; *Chem. Abstr.* **96** (1982) 217 201.

[50] M. Lissell, E. V. Dehmlov, *Tetrahedron Lett.* **39** (1978) 3689.

[51] K. Isogai, K. Awano, *Yuki Gosei Kagaku Kyokaishi* **35** (1977) 280–4; *Chem. Abstr.* **87** (1977) 52 699.

[52] All-Union Scientific-Research Inst., SU 891 630, 1981 (L. V. Aleksanyan et al.).

[53] Erevan Plant of Chem. Reagents, SU 535 283, 1976 (K. G. Akopyan et al.).

[54] Mitsubishi Chemical Ind. Co. Ltd., JP-Kokai 78 12 807, 1978 (T. Maki, T. Ochiai, T. Yamaura).

[55] Y. Takahashi et al., *Prepr. Am. Chem. Soc. Div. Pet. Chem.* **28** (1983) 392–6.

[56] N. Yomeda et al., *Chem. Lett.* **6** (1981) 767–8.

[57] Dynamit Nobel, FR 2 017 352, 1970 (H. Richtzenhain, H. Aus der Fuenten).

[58] Ciba S. A., BE 739 526, 1968.

[59] Bayer, DE 2 923 775, 1979 (F. Maurer, U. Priesnitz, H. J. Riebel).

[60] Merck & Co. Inc., US 4 262 009, 1979 (B. G. Christensen, D. H. Shin).

[61] Lonza, EP 64 633, 1981 (P. Lehky, V. Franzen). Lonza, EP 64 634, 1981 (P. Lehky, P. Hardt).Lonza, EP 65 706, 1981 (P. Lehky).

[62] Veba-Chemie AG, DE 1 418 074, 1976 (E. Rindtorff, K. Schmitt, H. Heumann).

[63] Chem. Werke Hüls, DE-OS 3 200 065, 1983 (W. Huebel, J. Reiffer).

[64] F. Debuan, P. Häussle, *Erdöl Kohle Erdgas Petrochem.* **37** (1984) 511.

[65] Vianova Kunstharz AG, DE 2 638 464, 1975 (H. Lackner, J. Manger, P. Thaller). Chem. Werke Hüls, EP 25 828, 1979 (S. Maman).
[66] Ciba-Geigy, DE 2 651 650, 1975 (L. Buxbaum, J. Habermeier).
[67] W. Langenbeck, M. Richter, *Chem. Ber.* **89** (1956) 202.
[68] V. Odinokov et al., *Zh. Org. Khim.* **14** (1978) 54–9; *Chem. Abstr.* **88** (1978) 151 987.
[69] Dainippon Ink and Chemicals Inc., JP-Kokai 82 142 940, 1982.
[70] P. E. Verkade, *Rec. Trav. Chim. Pays Bas* **46** (1927) 137.
[71] T. Antonova et al., *Ref. Zh. Khim.* 1981, Abstr. No. 10N44; *Chem. Abstr.* **95** (1981) 61 415.
[72] W. Reppe, *Justus Liebigs Ann. Chem.* **582** (1953) 1–161.
[73] Y. Nakauo, T. A. Foglia, *J. Am. Oil Chem. Soc.* **59** (1982) 163–6.
[74] Nippon Oils and Fats Co., JP-Kokai 81 169 640, 1981.
[75] Nippon Oils and Fats Co., JP-Kokai 82 4940, 1982.
[76] N. Garti, E. Auni, *J. Am. Oil Chem. Soc.* **58** (1981) 840-1.
[77] T. A. Foglia et al., *J. Am. Oil Chem. Soc.* **54** (1977) 870A–872A.
[78] V. W. Advani et al., *J. Oil Technol. Assoc. India* **8** (1976) 27–30.
[79] Welsbach Corp., US 2 865 937, 1958 (A. Maggiolo).
[80] Welsbach Corp., US 2 897 231, 1959 (S. J. Niegowski, A. Maggiolo).
[81] Inst. of Heteroorganic Compounds, Acad. of Sc., SU 1 092 150, 1984 (L. I. Zakharin, V. V. Guseva).
[82] Kuraray Co., JP-Kokai 83 140 038, 1983.
[83] Brichima, DE-OS 3 027 111, 1981 (F. Minisci, P. Maggioni, A. Citterio).
[84] Mitsubishi Petrochemical Co., JP-Kokai 82 79 889, 1982.
[85] A. P. Kudehandker et al., *Encycl. Chem. Process. Des.* **6** (1978) 401–20.
[86] Intreprinderea de Sapun, RO 60 094, 1976 (C. Razvan, D. Lungu, T. Popescu, M. Petrescu).
[87] D. D. Nanavati, *J. Sci. Ind. Res.* **35** (1976) 163–8.
[88] M. J. Diamond et al., *J. Am. Oil Chem. Soc.* **44** (1967) 656.
[89] Asahi Chemical Ind. Co., US 4 237 317, 1980 (K. Yamataka, Y. Matsuoka, T. Isoya).
[90] Asahi Chemical Ind. Co., DE-OS 3 019 537, 1980 (K. Yamataka, T. Isoya, C. Kawamura).
[91] Asahi Chemical Ind. Co., DE-OS 2 830 144, 1979 (K. Yamataka, Y. Matsuoka, T. Isoya).
[92] Daicel Ltd., JP 78 25 032, 1978 (K. Yamada, A. Nishihara, Y. Shirakawa, T. Nakazawa).
[93] S. I. Popovick et al., *Plast. Massy* **9** (1983) 62; *Chem. Abstr.* **99** (1983) 176 712.
[94] Asahi Chemical Ind. Co. Ltd., JP-Kokai 79 46 710, 1979 (K. Yamataka, T. Isoya, Y. Matsuoka).
[95] Asahi Chemical Ind. Co., JP-Kokai 78 82 717, 1978 (J. Nishikido, N. Tamura).
[96] I. S. Chevilenko et al., *Zh. Prikl. Khim. (Leningrad)* **55** (1982) 2742–6.
[97] I. S. Chevilenko et al., *Khim. Promst. (Moscow)* **2** (1980) 123.
[98] Asahi Denka Kogyo K. K., JP 78 15 051, 1978 (H. Yamamoto, I. Hisano, T. Okamota).
[99] I. S. Chevilenko et al., *Khim. Promst. (Moscow)* **1** (1977) 14–16.
[100] BASF, GB 1 092 603, 1967 (G. Riegelbauer, A. Wegerich, A. Kuerzinger, E. Haarer).Chem. Werke Hüls, DE 1 643 854, 1968 (E. Vangermain).
[101] Dainippon Ink and Chemicals Inc., JP-Kokai 82 32 245, 1982.
[102] Chem. Werke Hüls, DE-OS 2 942 279, 1981 (K. D. Dohm, P. Hofmann).
[103] Y. N. Yurev et al., SU 322 984, 1981.
[104] Bio Research Center Co., JP-Kokai 81 154 993, 1981.
[105] Bio Research Center Co., DE-OS 2 909 420, 1979 (A. Watanabe, A. Taoka, S. Uchida).
[106] SNIA Viscosa, DE-OS 2 945 004, 1980 (F. Siclari, L. Canavesi, P. P. Rossi).
[107] Ube Industries, JP-Kokai 80 104 226, 1980.
[108] Du Pont, US 4 149 013, 1979 (D. A. Klein).

[109] Asahi Chem. Ind. Co., JP-Kokai 78 82 718, 1978 (J. Nishikido, N. Tamura).
[110] Toa Gosei Chemical Industry Co., JP-Kokai 76 108 015, 1976 (Y. Yasuda, T. Matsubara).
[111] J. J. Jaskierski et al., *Zesz. Probl. Postepow. Nauk Roln.* **211** (1981) 159–165, 311–319.
[112] K. D. Carlson et al., *Ind. Eng. Chem. Prod. Res. Dev.* **16** (1977) 95–101.
[113] Mitsui Petrochemical Ind. Ltd., JP-Kokai 83 121 797, 1983.
[114] Mitsui Petrochemical Ind. Ltd., JP-Kokai 83 60 994, 1983.
[115] Mitsubishi Petrochemical Ind. Ltd., JP-Kokai 82 102 191, 1982.
[116] Mitsubishi Petrochemical Co. Ltd., JP-Kokai 81 11 796, 1981.
[117] Nippon Mining Co. Ltd., JP-Kokai 81 11 796, 1981.
[118] Mitsui Petrochemical Ind., JP-Kokai 81 15 693, 1981.
[119] Mitsui Petrochemical Ind., JP-Kokai 81 15 694, 1981.
[120] Mitsui Petrochemical Ind., JP-Kokai 81 15 695, 1981.
[121] Bio Research Center Co., DE-OS 2 951 177, 1980 (A. Watanabe, S. Uchida, A. Toaka).
[122] United States of America, Secretary of Agriculture, US 3 270 035,1966 (E. T. Roe, D. Swern).
[123] Matsubara et al., *Kogyo Kagaku Zasshi* **71** (1968) 1179.
[124] Armour and Company, US 3 481 977, 1969 (E. J. Miller, A. Mais, D. Say).
[125] J. P. Friedrich, *J. Am. Oil Chem. Soc.* **53** (1976) 125.
[126] E. H. Pryde, *J. Am. Oil Chem. Soc.* **61** (1984) 419.
[127] E. N. Frankel, *J. Am. Oil Chem. Soc.* **48** (1971) 248.
[128] A. W. Schwab et al., *J. Am. Oil Chem. Soc.* **49** (1972) 75.
[129] E. J. Dufek et al., *J. Am. Oil Chem. Soc.* **49** (1972) 302.
[130] H. M. R. Hoffmann, *Angew. Chem.* **81** (1969) 597; *Angew. Chem. Int. Ed. Engl.* **8** (1969) 556.
[131] A. Cox in: *Comprehensive Organic Chemistry,* vol. **2**, Chap. 9.2, Pergamon Press, Oxford 1979. V. Matthews in:
Rodd's Chemistry of Carbon Compounds, vol. **1**, Chap. 17, Elsevier, Amsterdam 1973. E. C. Leonard in:
Fatty Acids, Chap. 25, The American Oil Chemists Society, Champaign, Ill., 1979.
[132] Iwata Kagaku Kogyo K. K., JP-Kokai 84 63 190, 1984.
[133] Shizuoka Prefecture, Banda Kagaku Kogyo K. K., JP-Kokai 81 137 893, 1981.
[134] N. Nakagawa et al., *J. Ferment. Technol.* **62** (1984) 201–203.
[135] Mitsubishi Chemical Industries Co. Ltd., JP-Kokai 80 34 017, 1980.
[136] S. Ikeda, *Hakko Kogaku Kaishi* **60** (1982) 208–210.
[137] L. B. Lockwood, *Microb. Technol.,* 2nd ed., vol. **1**, Academic Press, New York 1979, pp. 355–387.
[138] L. M. Miall, *Econ. Microbiol.* **2** (1978) 47–119.
[139] Nissan Chemical Industries Ltd., JP-Kokai 78 15 316, 1978 (A. Murata, T. Ishii).
[140] Denki Kagaku Kogyo K. K., JP-Kokai 74 101 326, 74 101 327, 1973 (T. Shimizui, C. Fujii).
[141] J. Tsuji, T. Nogi, *Tetrahedron Lett.* 1966, 1801–04.
[142] Union Carbide Corp., EP 19 296, 1980 (D. R. Bryant, R. A. Galley).
[143] E. C. Leonard in: *Fatty Acids,* Chap. 25, The American Oil Chemists Society, Champaign, Ill., 1979, p. 521.
[144] Yaroslavl Polyt. Inst., SU 1 057 515, 1983 (V. V. Solovev, B. N. Bychkov, G. N. Koshel, L. A. Rodivilova, R. D. Zhilina, E. A. Bondareva).
[145] Toei Chemical K. K., JP-Kokai 78 23 306, 1978 (N. Fujihana, M. Terauchi).
[146] Emery Industries Inc., CA 1 019 343, 1977 (R. J. Sturwold, H. J. Sharkey).
[147] Agency of Industrial Sciences and Technology, JP-Kokai 76 88 910, 1976 (O. Suzuki, K. Tanabe, T. Hashimoto).

[148] R. W. Johnson in: *Fatty Acids,* Chap. 17,The American Oil Chemists Society, Champaign, Ill., 1979,
[149] D. H. Wheeler, A. Milum, F. Linn, *J. Am. Oil Chem. Soc.* **47** (1970) 242–244.
[150] D. H. Wheeler, J. M. White, *J. Am. Oil Chem. Soc.* **44** (1967) 298–302.
[151] E. H. Pryde (ed.): *Fatty Acids,* Chaps. 17, 25, The American Oil Chemists Society, Champaign, Ill., 1979.
[152] E. C. Leonhard (ed.): *The Dimer Acids,* Humbko Sheffield Chemicals, 1975.
[153] E. C. Leonhard, *J. Am. Oil Chem. Soc.* **56** (1979) 782A–785A.
[154] Sankyo Co. Ltd., JP-Kokai 76 41 441, 1976 (H. Takehara).
[155] W. E. A. De Mear, R. J. Bellamy, *Polym. Paint Colour J.* **164** (1974) 233.
[156] *Aust. OCCA Proc. News* **9** (1972) 5–14.
[157] G. C. Guainazzi, *Boll. Assoc. Ital. Tec. Ind. Vernici Affini* **61** (1978) 3–12.
[158] J. P. Nelson, A. J. Milun, *J. Am. Oil Chem. Soc.* **52** (1975) 81–83.
[159] A. Zeman, H. Sharmann, *Fette, Seifen, Anstrichm.* **71** (1969) 957–960.
[160] E. Pretsch, T. Clerc, J. Seibl, W. Simon: *Tabellen zur Strukturaufklärung organischer Verbindungen,* Springer Verlag, Heidelberg 1976. D. H. Williams, J. Fleming: *Spektroskopische Methoden in der organischen Chemie,* Thieme Verlag, Stuttgart 1971.H. Roth: "Analytik der Carboxylgruppen," in: Houben-Weyl, 2, 487.F. Korte (ed.): "Analytik," in: Methodicum Chimicum, vol. **1,** Thieme Verlag, Stuttgart 1973.